Probability and Its Applications

Published in association with the Applied Probability Trust

Probability and Its Applications

Azencott et al.: Series of Irregular Observations. Forecasting and Model Building. 1986
Bass: Diffusions and Elliptic Operators. 1997
Bass: Probabilistic Techniques in Analysis. 1995
Berglund/Gentz: Noise-Induced Phenomena in Slow-Fast Dynamical Systems: A Sample-Paths Approach. 2006
Biagini/Hu/Øksendal/Zhang: Stochastic Calculus for Fractional Brownian Motion and Applications. 2008
Chen: Eigenvalues, Inequalities and Ergodic Theory. 2005
Costa/Fragoso/Marques: Discrete-Time Markov Jump Linear Systems. 2005
Daley/Vere-Jones: An Introduction to the Theory of Point Processes I: Elementary Theory and Methods. 2nd ed. 2003, corr. 2nd printing 2005
Daley/Vere-Jones: An Introduction to the Theory of Point Processes II: General Theory and Structure. 2nd ed. 2008
de la Peña/Gine: Decoupling: From Dependence to Independence, Randomly Stopped Processes, U-Statistics and Processes, Martingales and Beyond. 1999
Del Moral: Feynman-Kac Formulae. Genealogical and Interacting Particle Systems with Applications. 2004
Durrett: Probability Models for DNA Sequence Evolution. 2002, 2nd ed. 2008
Galambos/Simonelli: Bonferroni-Type Inequalities with Equations. 1996
Gani (ed.): The Craft of Probabilistic Modelling. A Collection of Personal Accounts. 1986
Gut: Stopped Random Walks. Limit Theorems and Applications. 1987
Guyon: Random Fields on a Network. Modeling, Statistics and Applications. 1995
Kallenberg: Foundations of Modern Probability. 1997, 2nd ed. 2002
Kallenberg: Probabilistic Symmetries and Invariance Principles. 2005
Last/Brandt: Marked Point Processes on the Real Line. 1995
Molchanov: Theory of Random Sets. 2005
Nualart: The Malliavin Calculus and Related Topics, 1995, 2nd ed. 2006
Rachev/Rueschendorf: Mass Transportation Problems. Volume I: Theory and Volume II: Applications. 1998
Resnick: Extreme Values, Regular Variation and Point Processes. 1987
Schmidli: Stochastic Control in Insurance. 2008
Schneider/Weil: Stochastic and Integral Geometry. 2008
Shedler: Regeneration and Networks of Queues. 1986
Silvestrov: Limit Theorems for Randomly Stopped Stochastic Processes. 2004
Thorisson: Coupling, Stationarity and Regeneration. 2000

Rolf Schneider · Wolfgang Weil

Stochastic and Integral Geometry

Springer

Rolf Schneider
Mathematisches Institut
Albert-Ludwigs-Universität
Eckerstr. 1
79104 Freiburg
Germany
rolf.schneider@math.uni-freiburg.de

Wolfgang Weil
Institut für Algebra und Geometrie
Universität Karlsruhe
Englerstraße 2
76128 Karlsruhe
Germany
weil@math.uka.de

Series Editors:

Joe Gani
Chris Heyde
Centre for Mathematics and its Applications
Mathematical Sciences Institute
Australian National University
Canberra, ACT 0200
Australia
gani@maths.anu.edu.au

Thomas G. Kurtz
Department of Mathematics
University of Wisconsin - Madison
480 Lincoln Drive
Madison, WI 53706-1388
USA
kurtz@math.wisc.edu

Peter Jagers
Mathematical Statistics
Chalmers University of Technology
and Göteborg (Gothenburg) University
412 96 Göteborg
Sweden
jagers@chalmers.se

ISBN: 978-3-540-78858-4 e-ISBN: 978-3-540-78859-1
DOI: 10.1007/978-3-540-78859-1

Probability and Its Applications ISSN print edition: 1431-7028

Library of Congress Control Number: 2008933565

Mathematics Subject Classification (2000): 60D05, 52A22, 60G55, 62M30

Cover design: *Erich Kirchner*, Heidelberg

Printed on acid-free paper

9 8 7 6 5 4 3 2 1

springer.com

Preface

Stochastic Geometry deals with mathematical models for random geometric structures and spatial data, as they frequently arise in modern applications. As a mathematical discipline, stochastic geometry came into life in the last third of the twentieth century, but its roots and the close connections between geometric probability and integration techniques using invariant measures (though not under this name) date back much farther. The famous Buffon needle problem of 1777 was solved by what seems to be the first application of integral calculus to a probability question. A variety of problems in *Geometric Probability* was treated in the late nineteenth and early twentieth century. After the role of invariant measures had become clear, the discipline of *Integral Geometry* was initiated in the 1930s, mostly by Wilhelm Blaschke and his school. The book *Integral Geometry and Geometric Probability* by Luis Santaló (1976) summarizes the concepts and results of the preceding development. Interpretations of integral geometric results in terms of geometric probability abound in that work. At that time, David Kendall and Georges Matheron had already developed, independently, a theory of *Random Sets*, and Roger Miles had written his pioneering thesis on Poisson processes of certain geometric objects. The book *Random Sets and Integral Geometry* by Matheron (1975) presented the new field of Stochastic Geometry in its intimate relation with Integral Geometry. Applications in *Spatial Statistics* and *Stereology*, later also in *Image Analysis*, contributed to a rapid development. The classical integral geometry of Euclidean spaces is well suited to the treatment of random sets and point processes with invariance properties, like stationarity and isotropy. The necessity of studying structures which exhibit anisotropy, or even without spatial homogeneity, grew hand in hand with new developments in integral geometry, coming from *Geometric Measure Theory*. In particular, Federer's local formulas for curvature measures proved useful, and *Translative Integral Geometry* was promoted, meeting the needs of stationary structures.

Over many years, we both gave courses on Integral Geometry or Stochastic Geometry in Freiburg and Karlsruhe. This led to the joint publication of lecture notes in German, under the titles of *Integralgeometrie* (1992) and

Stochastische Geometrie (2000). It was always our plan later to amalgamate both topics in one extended monograph in English. During the time we worked on this project, the field of stochastic geometry has expanded considerably in various directions, too many to include them all in one volume. We decided to concentrate on our original idea, namely to present the basic models of stochastic geometry and their properties, the fundamental concepts and formulas of integral geometry, and the interrelations between these two fields.

In this book, therefore, we have three main aims: to give a sound mathematical foundation for the most basic and general models of stochastic geometry, namely random closed sets, particle processes, and random mosaics, to introduce the reader to the parts of integral geometry that are relevant for the applications in stochastic geometry, and, naturally, to demonstrate such applications. Since the strength of integral geometry lies in the computation of mean values and in integral transformations, this means that we develop mainly a 'first order theory' of stochastic geometry, centering around expectations. This restricted concept, with its foundational character, implies that essential and interesting parts of stochastic geometry are missing: we do not treat special point process models other than Poisson processes, nor higher order moment measures, limit theorems, spatial statistics, practical procedures, simulations; however, we comment on some of these developments in the section notes. The integral geometry here is taylored to its use in stochastic geometry; this influences the selection of topics as well as the approach, which is measure theoretic rather than differential geometric. Another restriction may be seen in the predominance of *invariance* and *independence*. The first means that we study (except in one chapter providing an outlook) only random sets and geometric point processes that are stationary (spatially homogeneous) or even stationary and isotropic, in distribution. Invariance of measures and distributions is the leitmotiv of this volume; it underlies both the stochastic geometry parts and the integral geometric parts. On the stochastic side, there is a preference for independence assumptions, as for example in the prominent role of Poisson processes, with their strong independence properties. Very often, only invariance and independence assumptions allow simple approaches and lead to beautiful results. The confinement to the fundamentals of stochastic geometry leaves us room for emphasizing the geometry; in fact, in integral as well as in stochastic geometry, we draw a richer picture than sketched above, and we include various topics of geometric appeal. For example, there is a chapter on *Geometric Probability*, since this area has seen a recent revival with many interesting problems and results.

Naturally, this book employs notions and results from other fields. We make use of some basic facts from general topology, from the theory of topological groups and homogeneous spaces of Euclidean geometry and their invariant measures, and from the geometry of convex sets; further, some more specialized results concerning geometric inequalities and additive functionals on convex bodies are needed. Anticipating that the familiarity of the readers with these topics will not be uniform, we have collected the required material in an Appendix; this should be consulted whenever necessary. This also allows us to start directly with the fundamental notion in this book, the concept of a random closed set.

We are grateful to many colleagues for their helpful comments on early drafts of our book. Special thanks go to Paul Goodey, Günter Last and Werner Nagel, for providing useful hints after reading parts of the final manuscript, and in particular to Daniel Hug, who has carefully read all of it. He prevented us from including a number of flaws and made many suggestions for improvements. We also thank the Mathematisches Forschungsinstitut Oberwolfach for giving us the opportunity to spend some time, working on our mansucript, in their wonderful 'Research in Pairs' programme.

Freiburg i. Br., Karlsruhe
Spring 2008

Rolf Schneider
Wolfgang Weil

Contents

Part II Integral Geometry

Part III Selected Topics from Stochastic Geometry

1

Prolog

1.1 Introduction

Since this book is about relations between stochastic geometry and integral geometry, we begin with an imaginary experiment that demonstrates the need for and use of integral geometry for certain geometric probability questions and at the same time leads in a natural way to a basic model of stochastic geometry.

We assume that K and W are given convex bodies (nonempty compact convex sets) in d-dimensional Euclidean space $\mathbb{R}^d$. The body K serves to generate a random field of congruent copies of K, and the body W plays the role of an 'observation window'. The random field consists of countably many congruent copies of K which are laid out in space randomly and independently, overlappings being allowed. The number of bodies in the random field that hit (that is, have nonempty intersection with) the observation window W is a random variable. We ask for its distribution. This is, of course, not a meaningful question, as long as no stochastic model for the random field of convex bodies is specified. In a few steps, we shall introduce some natural assumptions, which motivate a precise model and lead to an explicit formula for the desired distribution.

In the first step, we consider a much simpler situation. We take a ball B_r of radius r and origin 0 that contains the observation window W, and we consider only one randomly moving copy of K, under the condition that it hits B_r. We ask for the probability that it also hits W. There is a geometrically very natural way of specifying a probability distribution of a randomly moving convex body that satisfies the side condition. A random congruent copy of K can be represented in the form $\widetilde{g}K$, where $\widetilde{g}$ is a random element of the group G_d of rigid motions. The locally compact group G_d carries an essentially unique Haar measure, that is, a locally finite Borel measure that is similarly under left and right multiplications and is not identically zero. We denote this measure, with a suitable normalization, by μ. A natural probability distribution of a random congruent copy of K hitting B_r is then obtained by restricting μ as required by the side condition, normalizing, and taking an image measure. Thus, in our situation we define a probability measure $\mathbb{Q}$ on the space $\mathcal{K}$ of convex bodies (with its usual topology) in $\mathbb{R}^d$ by

$$\mathbb{Q}(A) := \frac{\mu(\{g \in G_d : gK \cap B_r \neq \emptyset,\ gK \in A\})}{\mu(\{g \in G_d : gK \cap B_r \neq \emptyset\})}$$

for Borel sets $A \subset \mathcal{K}$. A random congruent copy of K hitting B_r is then, by definition, a random convex body with distribution $\mathbb{Q}$.

Now the probability, denoted by p, that a random congruent copy of K hitting B_r also hits W, is well defined. If we put

$$\mu(K, M) := \mu(\{g \in G_d : gK \cap M \neq \emptyset\})$$

for convex bodies K and M, this probability is given by

$$p = \frac{\mu(K, W)}{\mu(K, B_r)}. \tag{1.1}$$

The computation of $\mu(K, M)$ is a typical task of integral geometry. First, we assume that K is a ball of radius ρ. If the Haar measure μ is suitably normalized, the measure of all motions g that bring K into a hitting position with M is just the measure of all translations that bring the center of K into the parallel body

$$M + B_\rho := \{m + b : m \in M,\ b \in B_\rho\},$$

and hence is the volume of this body. By the **Steiner formula** of convex geometry, this volume is a polynomial of degree at most d in the parameter ρ. It is convenient to write it in thc form

$$\lambda_d(M + B_\rho) = \sum_{i=0}^{d} \rho^{d-i} \kappa_{d-i} V_i(M), \tag{1.2}$$

where λ_d is the Lebesgue measure on $\mathbb{R}^d$ and κ_j is the volume of the j-dimensional unit ball. This defines the **intrinsic volumes** $V_0, \dots, V_d$, which are important functionals on the space of convex bodies.

The intrinsic volumes, which appear naturally in the computation of the measure $\mu(K, M)$ for the special case $K = B_\rho$, are also sufficient to handle the general case. The **principal kinematic formula** of integral geometry, specialized to convex bodies, states that

$$\mu(K, M) = \sum_{i=0}^{d} \alpha_{di} V_i(K) V_{d-i}(M), \tag{1.3}$$

with certain explicit constants α_{di}. From (1.1) and (1.3) we obtain

$$p = \frac{\sum_{i=0}^{d} \alpha_{di} V_i(K) V_{d-i}(W)}{\sum_{i=0}^{d} \alpha_{di} V_i(K) V_{d-i}(B_r)}, \tag{1.4}$$

which depends only on the intrinsic volumes of K and W (and on r).

In the second step, we consider $m \geq 2$ independent, identically distributed random convex bodies, each with distribution $\mathbb{Q}$, thus each one is a random congruent copy of K hitting B_r. For $k \in \{0, 1, \dots, m\}$, we denote by p_k the probability that the fixed body W is hit by exactly k of the random congruent copies of K. By the independence, we obtain a binomial distribution, thus

$$p_k = \binom{m}{k} p^k (1-p)^{m-k},$$

with p given by (1.4).

In the third step, we choose m depending on the radius r and let r tend to ∞, in such a way that

$$\lim_{r\to\infty} \frac{m}{\lambda_d(B_r)} = \gamma > 0$$

with a constant γ. Since

$$\lim_{r\to\infty} \frac{\mu(K, B_r)}{\lambda_d(B_r)} = 1,$$

we obtain $\lim_{r\to\infty} mp = \gamma\mu(K, W) =: \theta$, and hence

$$\lim_{r\to\infty} p_k = \frac{\theta^k}{k!} \mathrm{e}^{-\theta} \tag{1.5}$$

with

$$\theta = \gamma \sum_{i=0}^{d} \alpha_{di} V_i(K) V_{d-i}(W). \tag{1.6}$$

We have found, not surprisingly, a Poisson distribution. Its parameter is expressed explicitly in terms of the constant γ, which can be interpreted as the number density of our random system of convex bodies, and the intrinsic volumes of K and W.

The original question and the answer given by (1.5) and (1.6) are found in a paper by Giger and Hadwiger [260]. The answer, though nice and explicit, is still not entirely satisfactory. We have computed a limit of probabilities and found a Poisson law. However, this Poisson distribution is not yet interpreted as the distribution of a well-defined random variable. What we would prefer, and what is needed for applications, is a model that allows us to consider from the beginning countably infinite systems of randomly placed convex bodies, with suitable independence properties.

This goal is readily achieved by employing suitable point processes. For the purpose of this introduction, a point process in $\mathbb{R}^d$ is a measurable map from the underlying probability space into the measurable space of locally finite subsets of $\mathbb{R}^d$. In particular, let Ξ be a Poisson point process of intensity γ in $\mathbb{R}^d$, with a translation invariant distribution. We choose a Poisson process since its built-in independence properties reflect the independence assumptions made above in the second step. With each point of Ξ, we associate a congruent copy of K, in the following way. For easier visualization, we suppose that $0 \in K$. We may assume that $\Xi = \{\xi_1, \xi_2, \ldots\}$, with a measurable numeration. Let $(\vartheta_1, \vartheta_2, \ldots)$ be an independent sequence of random rotations of $\mathbb{R}^d$, each with distribution given by the invariant probability measure on the rotation group SO_d; let this sequence be independent of Ξ. Then $\{\xi_i + \vartheta_i K,\ i = 1, 2, \ldots\}$ defines a random field X of convex bodies which are congruent copies of K. For this model one can compute that the probability, say q_k, of the event that the fixed observation window W is hit by precisely k bodies of the field X, is given by

$$q_k = \frac{\theta^k}{k!} \mathrm{e}^{-\theta}, \tag{1.7}$$

with θ according to (1.6).

This very special model can immediately be generalized. There is no particular reason for attaching to the points ξ_i of the Poisson process Ξ only rotation images $\vartheta_i K$ of a fixed convex body K. One may as well attach to $\xi_1, \xi_2, \ldots$ random convex bodies $K_1, K_2, \ldots$, chosen independently and independent from Ξ, according to some given rotation invariant probability distribution on the space $\mathcal{K}$ of convex bodies. Essentially equivalent is the assumption that X is a Poisson process in the locally compact space $\mathcal{K}$, which is **stationary** and **isotropic**, that is, whose distribution is invariant under translations and rotations. Again, let q_k denote the probability that the observation window W is hit by k bodies of the particle process X. The intrinsic volumes $V_i(K)$ appearing in (1.6), or rather $\gamma V_i(K)$, must now be replaced by suitable densities. Under a mild integrability condition on X (which is assumed in the following), it can be shown that the limit

$$\overline{V}_i(X) = \lim_{r \to \infty} \frac{1}{\lambda_d(rW)} \mathbb{E} \sum_{K \in X,\, K \subset rW} V_i(K) \tag{1.8}$$

exists for every convex body W with $\lambda_d(W) > 0$ and is finite and independent of W; here $\mathbb{E}$ denotes mathematical expectation. The number $\overline{V}_i(X)$ is called the **density of the** ith **intrinsic volume**, or the ith **specific intrinsic volume**, of the particle process X. If we now replace (1.6) by

$$\theta := \sum_{i=0}^{d} \alpha_{di} \overline{V}_i(X) V_{d-i}(W),$$

then (1.7) still holds.

Together with the Poisson particle process X, we consider its union set,

$$Z := \bigcup_{K \in X} K.$$

Under the mentioned integrability assumption, this is almost surely a closed set. Thus, we obtain an example of a random closed set. Generally, a **random closed set** in $\mathbb{R}^d$ is a measurable map from the underlying probability space into the space of closed subsets of $\mathbb{R}^d$, endowed with a suitable topology and the induced Borel σ-algebra. Random closed sets are, besides particle processes, the second basic model of stochastic geometry. The random closed set obtained here is of a special type: besides being stationary and isotropic, it is the union set of a Poisson particle process. Random closed sets generated in this way are known as **Boolean models**. Due to the strong independence properties of Poisson processes, Boolean models are mathematically more tractable than general random closed sets. We give one example, after introducing specific intrinsic volumes of the random set Z.

In a certain analogy to (1.8), we want to define the ith **specific intrinsic volume** of the random closed set Z by

$$\overline{V}_i(Z) = \lim_{r \to \infty} \frac{1}{\lambda_d(rW)} \mathbb{E}\, V_i(Z \cap rW). \tag{1.9}$$

This is indeed possible. By the properties of the generating particle process X, the set $Z \cap rW$ is, for a convex body W, almost surely the union of finitely many convex bodies. The intrinsic volume V_i has a unique additive and measurable extension from $\mathcal{K}$ to the lattice of finite unions of convex bodies. With this extension, also denoted by V_i, the random variable $V_i(Z \cap rW)$ is well defined, and the limit (1.9) exists for every convex body W with positive volume, it is finite and independent of W. The numbers $\overline{V}_0(Z), \dots, \overline{V}_d(Z)$ are, in several respects, the simplest and most basic parameters for a quantitative description of a stationary random set. They include the specific volume $\overline{V}_d(Z)$, the specific surface area $2\overline{V}_{d-1}(Z)$, and the specific Euler characteristic $\overline{V}_0(Z)$.

A special and remarkable property of the stationary and isotropic Boolean model Z is now the fact that the specific intrinsic volumes of Z can be expressed explicitly in terms of the specific intrinsic volumes of the generating particle process X, and conversely! The latter fact is rather surprising at first sight: it says that, in principle, the specific intrinsic volumes of the particle process can be determined by observing its union set. This is astonishing, since observation of the union set does not allow us to observe individual particles. The explanation for this seeming paradox lies in the strong independence properties of Poisson processes. The first two of the mentioned relations, connecting the specific volumes and the specific surface areas of the Poisson particle process X and of its union set Z, are given by

$$\overline{V}_d(Z) = 1 - \mathrm{e}^{-\overline{V}_d(X)},$$

$$\overline{V}_{d-1}(Z) = \overline{V}_{d-1}(X)\mathrm{e}^{-\overline{V}_d(X)}.$$

The remaining relations are more complicated. Their proof is a typical application of iterated kinematic formulas of integral geometry.

Poisson processes of convex bodies and their union sets, as described, are interesting and tractable models of stochastic geometry, but are, of course, too special for many applications. Part I of our book, on foundations of stochastic geometry, begins with an introduction to general random closed sets in a topological space. The basic space, as in the treatment of point processes, is assumed to be locally compact and to have a countable base. This generality is sufficient, but it is also required for the geometric models to be introduced. Some prerequisites from general topology are collected in the Appendix. Point processes and marked point processes are the subject of Chapter 3.

Since the point processes we introduce live in quite general spaces, the 'points' can themselves be geometric objects, such as compact or convex subsets of $\mathbb{R}^d$, submanifolds or planes of a fixed dimension. This leads to the geometric models which are the subject of Chapter 4. We study particle processes and their union sets, and the geometry of processes of flats. Geometric results are treated to an extent that does not yet require special knowledge from integral geometry, but considerable use is made of results from convex geometry. The latter are made available in the Appendix.

The quantitative description of random closed sets and particle processes in $\mathbb{R}^d$ requires the definition of suitable parameters. In the spatially homogeneous case one may hope that real-valued parameters already carry useful information. Let X be a stationary particle process, Z a stationary random closed set, and φ a suitable function. In analogy to (1.8) and (1.9) above, it is a plausible attempt to define **φ-densities** by a double averaging process, stochastically and spatially, in the form

$$\overline{\varphi}(X) = \lim_{r\to\infty} \frac{1}{\lambda_d(rW)} \mathbb{E} \sum_{K\in X,\, K\subset rW} \varphi(K) \tag{1.10}$$

and

$$\overline{\varphi}(Z) = \lim_{r\to\infty} \frac{1}{\lambda_d(rW)} \mathbb{E}\, \varphi(Z \cap rW), \tag{1.11}$$

where W is, say, a convex body with positive volume. Clearly, such a procedure requires appropriate assumptions. In general, $Z \cap rW$ will have a well-defined Lebesgue measure, but not, for example, a well-defined surface area or Euler characteristic, and other appropriate functions φ are even harder to think of. In most of the quantitative investigations we shall therefore restrict ourselves to particle processes X and random closed sets Z with the properties that $K \in X$ and $Z \cap W$, for a convex body W, are almost surely **polyconvex**, that is, can be represented as finite unions of convex bodies. From the viewpoint of modeling real materials and structures, this is not a severe restriction, since such objects can be approximately represented by unions of large numbers of small convex bodies. The advantage of this restriction is that a series of geometrically meaningful functions φ becomes available. Since we want to generate sets as unions of convex bodies, the functions φ to be considered must have a simple behavior under taking unions; therefore, we demand finite additivity. More precisely, a real function φ on the space $\mathcal{K}$ of convex bodies is called **additive** or a **valuation** if

$$\varphi(K \cup M) = \varphi(K) + \varphi(M) - \varphi(K \cap M)$$

whenever $K, M, K \cup M \in \mathcal{K}$. Every continuous valuation on $\mathcal{K}$ has a unique extension to an additive function on the system of polyconvex sets. For translation invariant, additive functions φ on polyconvex sets, suitable measurability and integrability conditions are sufficient to ensure the existence of the densities $\overline{\varphi}(Z)$ according to (1.11). The densities (1.10) already exist under weaker assumptions. In isotropic situations, the relevant functions φ are well known. By a remarkable theorem of Hadwiger, every continuous, rigid motion invariant valuation on $\mathcal{K}$ is a linear combination of the intrinsic volumes. This explains the predominant role of the intrinsic volumes in large parts of this book. The required facts about additive functionals on convex bodies and their proofs can be found in the Appendix.

Our emphasis on polyconvex sets and intrinsic volumes and their generalizations also affects our introduction to integral geometry, in Part II of the book. A main task of integral geometry is to compute mean values of geometric functions with respect to invariant measures. Some fundamentals about invariant measures are collected in the Appendix. Specifically, we need the invariant measures on the groups and homogeneous spaces of Euclidean geometry, namely the translation, rotation and rigid motion group, and spaces such as spheres and linear or affine Grassmannians. Typical formulas of integral geometry will evaluate the integral, with respect to an invariant measure, of a function taken at the intersection of a fixed and a moving polyconvex set. First we consider fairly general additive functions and the motion group; then we concentrate on intrinsic volumes and their local versions, the curvature measures, and also study the case of the translation group. The picture is enriched by also treating some related topics.

Another subject of integral geometry is integral transforms involving invariant measures. As an example, consider an integral, with respect to d-fold Lebesgue measure in $\mathbb{R}^d$, of a function of d points where the function does, in fact, depend only on the hyperplane that is spanned (up to a set of measure zero) by the d points. Then it may be of advantage to transform the integral into one with respect to the invariant measure on the space of hyperplanes. Integral geometry provides geometric techniques for obtaining a variety of such transformation results, which are known as **Blaschke–Petkantschin formulas**. They are extremely useful, often allowing explicit calculations in geometric probabilities and stochastic geometry.

Part III of the book, on selected topics from stochastic geometry, combines the first two parts, but also aims at giving a broader picture. With this goal in mind, in Chapter 8 we present some geometric probability problems. This topic is not only the origin of stochastic geometry, but remains to be an attractive subject of many investigations. Our presentation touches convex hulls of random points, random projections of polytopes, questions about randomly moving convex bodies and flats, touching probabilities for convex bodies, and extremal problems for probabilities and expectations coming from intuitive geometric settings. As this chapter intends to paint a colorful picture, the presentation is not very systematic, and much information is to be found in the section notes.

Chapter 9 returns to the mainstream of the book and proceeds with a quantitative treatment of stationary random closed sets and particle processes. We begin with a study of the Boolean model. For more general random closed sets and for particle processes, we then introduce, as basic descriptive parameters, densities of additive functionals, in particular the specific intrinsic volumes. In their further investigation, stochastic geometry and integral geometry come close together. Intersection formulas lead to unbiased estimators for such parameters, and some selected estimation procedures are described.

Chapter 10 gives a detailed treatment of stationary random mosaics, another basic model of stochastic geometry. After a careful introduction, particular attention is paid to tessellations induced by stationary Poisson processes, either as Voronoi or Delaunay tessellations corresponding to Poisson point processes, or as hyperplane tessellations generated by a Poisson process in the space of hyperplanes. Zero cells and typical cells of stationary random mosaics provide interesting examples of random polytopes and are studied in some detail.

Chapter 11 is an outlook to non-stationary models. While, as emphasized in the preface, invariance of measures and distributions, at least under translations, is an essential feature in this book, we want to conclude with extending some of the results in previous chapters to non-stationary situations. Naturally, the statements become more involved, but it is perhaps surprising to see how the structure of the translative results is still recognizable and how the tools developed in the stationary case remain indispensable.

Part IV, the Appendix, collects basic material from other fields that is needed in the different chapters of the book. In Chapters 12 to 14, the reader will find, when necessary, the employed notions and results from general topology, the theory of invariant measures, and the geometry of convex bodies.

1.2 General Hints to the Literature

As explained in the preface, our presentation of stochastic geometry in this book has restricted aims only: to lay sound foundations for the standard models of stochastic geometry, and to prepare and describe the use of integral geometry. Although several further topics of geometric interest are touched, we are necessarily far from giving a complete picture of stochastic geometry. Therefore, in the following we list monographs and collections where the reader may find what is missing here. We shall, with a few exceptions, mention only literature of the last forty years, the period in which stochastic geometry, as it is understood today, has developed. We order the references in thematic groups and then chronologically.

Stochastic geometry:
1974 Harding, Kendall (eds.) [321] (collection of articles)
1975 Matheron [462]
1987 Stoyan, Kendall, Mecke [743] (second ed. 1995)
1988 Hall [317] (coverage processes)
1990 Ambartzumian [35]
1990 Mecke, Schneider, Stoyan, Weil [500] (DMV seminar, in German)
1993 Ambartzumian, Mecke, Stoyan [36] (in German)
1999 Barndorff–Nielsen, Kendall, van Lieshout (eds.) [80] (collection)
2004 Beneš, Rataj [90]
2007 Baddeley, Bárány, Schneider, Weil [50] (C.I.M.E. course)

Integral geometry:
1957 Hadwiger [307] (chapter 6, in German)
1968 Stoka [738] (in French)
1972 Sulanke, Wintgen [749] (chapter 5, in German)
1976 Santaló [662]
1982 Ambartzumian [34] (combinatorial integral geometry)
1994 K. Mecke [505] (applications to statistical physics, in German)
1994 Ren [635]
1997 Klain, Rota [416] (combinatorial aspects)
2007 Voss [772] (applied to stereology and image processing, in German)

Geometric probability:
1963 Kendall, Moran [397]
1978 Solomon [731]
1999 Mathai [456]

Random sets:
1993 Molchanov [543] (limit theorems)
1997 Goutsias, Mahler, Nguyen (eds.) [284] (collection of articles)
1997 Jeulin (ed.) [384] (collection of articles)
2005 Molchanov [548]
2006 Nguyen [583]

Point processes with geometric applications:
1986 Kallenberg [385]
1986 Matérn [454]
1988 Daley, Vere–Jones [194]
1992 König, Schmidt [423] (in German)
1993 Kingman [413]
2005 Daley, Vere–Jones [195]

2008 Daley, Vere–Jones [196]
2008 Illian, Penttinen, H. Stoyan, D. Stoyan [376]

Stereology:
1980 Weibel [778]
1998 Jensen [379]
2005 Baddeley, Jensen [53]

Spatial and geometric statistics:
1981 Ripley [644]
1988 Ripley [645]
1983 Diggle [204]
1991 Karr [389]
1992 D. Stoyan, H. Stoyan [746] (in German 1992, in English 1994)
1993 Cressie [185]
1997 Molchanov [546] (statistics of the Boolean model)
1999 Kendall, Barden, Carne, Le [396] (shape theory and shape statistics)
2000 van Lieshout [439]
2002 Ohser, Mücklich [587] (materials science)
2002 Torquato [759] (materials science)
2004 Møller, Waagepetersen [556]
2006 Baddeley, Gregori, Mateu, Stoica, D. Stoyan [51] (collection)

Random tessellations:
1994 Møller [553]
2000 Okabe, Boots, Sugihara, Chiu [591]

Several areas involving random geometric structures overlap more or less with stochastic geometry, or can be subsumed under it (the more so as stochastic geometry is not clearly defined), or they apply stochastic geometry. The following list is certainly not exhaustive.

1981 Adler [1] (random fields)
1982 Serra [729] (image analysis and mathematical morphology)
1996 Meesters, Roy [509] (continuum percolation)
2003 Penrose [598] (random geometric graphs)
2007 Adler, Taylor [2] (random fields)

Introductory surveys, emphasizing different aspects of stochastic geometry, were written by Baddeley [44, 45, 49], Cruz–Orive [189], Stoyan [741, 742], Weil [785], Weil and Wieacker [806].

1.3 Notation and Conventions

We collect here some basic notation, which will be used throughout the book. More detailed explanations of fundamental notions are found in the Appendix. The reader is advised to consult Chapters 12 to 14 whenever the notions and results from general topology, the theory of invariant measures, or convex geometry that we use do not appear sufficiently familiar.

Let E be a set. We denote by $\mathbf{P}(E)$ the power set, that is, the system of all subsets of E. For a subset $A \subset E$, the complement of A is denoted by A^c and the indicator function by $\mathbf{1}_A$. When one of the latter two notions is used, it will be clear from the context to which basic set E it refers. We also write $\mathbf{1}\{x \in A\}$ instead of $\mathbf{1}_A(x)$, if convenient.

Let E be a topological space. Most of the considered spaces will be locally compact or compact; by definition, this includes the Hausdorff property. Let A a subset of E. Then $\operatorname{cl} A$, $\operatorname{int} A$, $\operatorname{bd} A$ are, respectively, the closure, the interior and the boundary of A. The system of closed, open, and compact subsets of E is denoted, in this order, by $\mathcal{F}$, $\mathcal{G}$, $\mathcal{C}$. If necessary to avoid ambiguities, we also write $\mathcal{F}(E)$, $\mathcal{G}(E)$, $\mathcal{C}(E)$. A prime always indicates the corresponding system of nonempty sets, thus $\mathcal{F}'$, $\mathcal{G}'$, $\mathcal{C}'$ are the systems of nonempty closed, open, compact subsets of E, respectively. The vector space of continuous real functions on E is denoted by $\mathbf{C}(E)$, and $\mathbf{C}_c(E)$ is the subspace of functions with compact support.

A measure or signed measure on a topological space E will always be defined on the σ-algebra $\mathcal{B}(E)$ of Borel sets of the space, unless a different domain is indicated. $\mathcal{B}(E)$ is the smallest σ-algebra in E containing the open sets. Also measurability, of sets or mappings, refers to Borel σ-algebras, if no other σ-algebras are mentioned explicitly. We write

$$\mu^r := \mu \otimes \ldots \otimes \mu \qquad (r \text{ factors})$$

for the r-fold product of a measure μ. The restriction of a measure μ to a measurable set A is denoted by $\mu \llcorner A$, thus $(\mu \llcorner A)(B) := \mu(B \cap A)$ for all B in the domain of μ. If X, Y are topological spaces, ρ is a measure on X and $f : X \to Y$ is a measurable map, we denote the image measure of ρ under f by $f(\rho)$.

In probabilistic considerations, the underlying probability space will generally be denoted by $(\Omega, \mathbf{A}, \mathbb{P})$. If ξ is a random variable, then $\mathbb{P}_\xi$ denotes its distribution. We employ the usual abbreviations, such as $\mathbb{P}(\xi \in A) := \mathbb{P}(\{\omega \in \Omega : \xi(\omega) \in A\})$. The expected value of a real random variable ξ is denoted by $\mathbb{E}\,\xi$.

Most of our investigations take place in Euclidean space. $\mathbb{R}^d$ is the d-dimensional real Euclidean vector space, with scalar product $\langle \cdot, \cdot \rangle$ and induced norm $\|\cdot\|$. The distance of two points $x, y \in \mathbb{R}^d$ is denoted by $d(x, y) := \|x - y\|$, the distance of two nonempty sets $K, L \subset \mathbb{R}^d$ by $d(K, L) := \inf\{d(x, y) : x \in K,\ y \in L\}$, and we write $d(K, x) = d(x, K) := d(\{x\}, K)$ for the distance of the point x from the set K.

For subsets $A, B \subset \mathbb{R}^d$, the set $A + B := \{a + b : a \in A,\ b \in B\}$ is the vector sum or Minkowski sum, $\lambda A := \{\lambda a : a \in A\}$ is the dilate of A by the number $\lambda \geq 0$, and $-A := \{-a : a \in A\}$ is the image of A under reflection in the origin. $A - B$ means $A + (-B)$. This has to be distinguished from the Minkowski difference of A and B, which is defined by

$$A \ominus B := \bigcap_{b \in B} (A - b) = \{x \in \mathbb{R}^d : B + x \subset A\}$$

(note that in some of the literature this is $A \ominus -B$). We denote by $\operatorname{conv} A$ the convex hull of the set A, and by $\operatorname{pos} A$ its positive hull.

If $A \subset \mathbb{R}^d$ and if $E \subset \mathbb{R}^d$ is an affine subspace, then $A|E$ denotes the image of A under orthogonal projection to E.

The following systems of subsets will play a prominent role. $\mathcal{K}$ is the family of compact convex subsets of $\mathbb{R}^d$. The convex ring $\mathcal{R}$ consists of all finite unions of compact convex sets; its elements are sometimes called polyconvex sets. A locally polyconvex set in $\mathbb{R}^d$ is defined by the property that its intersection with any compact convex set is polyconvex. The system of these sets is denoted by $\mathcal{S}$ and is called the extended convex ring. $\mathcal{P}$ is the family of (compact, convex) polytopes. Again, $\mathcal{K}'$, $\mathcal{R}'$, $\mathcal{S}'$, $\mathcal{P}'$ denote the corresponding systems of nonempty sets. On $\mathcal{C}'$ (and thus also on $\mathcal{K}'$) the Hausdorff metric δ is defined by

$$\delta(K, L) := \max \left\{ \max_{x \in K} \min_{y \in L} d(x, y),\ \max_{x \in L} \min_{y \in K} d(x, y) \right\}.$$

Some particular subsets of $\mathbb{R}^d$ will occur frequently. These are the unit ball $B^d := \{x \in \mathbb{R}^d : \|x\| \leq 1\}$, the unit sphere $S^{d-1} := \{x \in \mathbb{R}^d : \|x\| = 1\}$, and the unit cube $C^d := [0,1]^d$. The 'half-open' cube $C_0^d := [0,1)^d$ is useful since its translates by the vectors of $\mathbb{Z}^d$ form a decomposition of $\mathbb{R}^d$; moreover, the 'upper right' boundary $\partial^+ C^d := C^d \setminus C_0^d$ is an element of the convex ring.

Hyperplanes of $\mathbb{R}^d$ are written in the form

$$H(u, \tau) := \{x \in \mathbb{R}^d : \langle x, u \rangle = \tau\}$$

with $u \in S^{d-1}$ and $\tau \in \mathbb{R}$; this representation is unique if $\tau > 0$. For $H(u, 0)$ we often write $u^\perp$.

The following measures are used. Lebesgue measure on $\mathbb{R}^d$ is denoted by λ or, if there is danger of ambiguity, by λ_d. For $k \in \{0, \dots, d-1\}$, λ_k is the k-dimensional Lebesgue measure on a k-dimensional affine subspace of $\mathbb{R}^d$. If E is this subspace, the Lebesgue measure on E is also denoted by λ_E. If F is a compact convex set with affine hull E, then

$$\lambda_F := \lambda_E \llcorner F.$$

The spherical Lebesgue measure on a k-dimensional great subsphere of S^{d-1} is denoted by σ_k, and we write σ instead of σ_{d-1} if this does not cause ambiguities. Occasionally, the k-dimensional Hausdorff measure is used, which is denoted by $\mathcal{H}^k$. For the Lebesgue measure of a compact set C, we often use the notation $V_d(C)$ and call it the volume of C. The intrinsic volumes $V_0(M), \dots, V_{d-1}(M)$ of a compact convex set M are defined by the Steiner formula (1.2); they are discussed in more detail in Section 14.3.

A frequently occurring constant is the volume of the unit ball,

$$\kappa_d := \lambda_d(B^d) = \frac{\pi^{\frac{d}{2}}}{\Gamma\left(1+\frac{d}{2}\right)}.$$

The surface area of the unit sphere S^{d-1} is given by

$$\omega_d := \sigma_{d-1}(S^{d-1}) = d\kappa_d = \frac{2\pi^{\frac{d}{2}}}{\Gamma\left(\frac{d}{2}\right)}.$$

The standard groups operating on $\mathbb{R}^d$ are the translation group, which is the additive group of $\mathbb{R}^d$ and is denoted by T_d if a distinction is appropriate, the group SO_d of proper (orientation-preserving) rotations, and the group G_d of rigid motions, or orientation-preserving isometries. These groups carry their standard topologies.

The translation by the vector $x \in \mathbb{R}^d$ is denoted by t_x, thus $t_x y := y + x$ for $y \in \mathbb{R}^d$. For a set $A \subset \mathbb{R}^d$, we have $A + x := t_x A = \{a + x : a \in A\}$. If μ is a measure on $\mathbb{R}^d$, then the image measure $t_x(\mu)$ is also denoted by $t_x\mu = \mu + x$, thus $(\mu + x)(A) = \mu(t_x^{-1}A) = \mu(A - x)$ for $A \in \mathcal{B}(\mathbb{R}^d)$. Similarly, $(\vartheta\mu)(A) := \mu(\vartheta^{-1}A)$ for $\vartheta \in SO_d$.

For $k \in \{0, \dots, d\}$, the Grassmannian of k-dimensional linear subspaces of $\mathbb{R}^d$ is denoted by $G(d, k)$, and the affine Grassmannian of k-dimensional affine subspaces by $A(d, k)$; both are equipped with their standard topologies.

We denote by $\overline{\mathbb{R}} = \mathbb{R} \cup \{-\infty, \infty\}$ the extended system of real numbers, and by $\mathbb{R}^+$ the set of positive real numbers.

Part I

Foundations of Stochastic Geometry

2

Random Closed Sets

A random set in a space E is defined, in agreement with the usual approach of axiomatic probability, as a set-valued random variable, that is, as a measurable map from some abstract probability space into a system of subsets of E, endowed with a suitable σ-algebra. It has turned out to be particularly tractable to assume that E is a locally compact space with a countable base and to consider the system $\mathcal{F}$ of its closed subsets, equipped with the topology of closed convergence and the induced σ-algebra of Borel sets. This approach is described in Section 2.1.

The distribution of a random closed set is completely determined by certain hitting probabilities, in particular, by its capacity functional. This gives, for every compact set $C \subset E$, the probability that the random set has nonempty intersection with C. The capacity functional can be seen in a certain analogy to the distribution function of a real random variable. Like distribution functions, the possible capacity functionals can be completely characterized. This characterization is provided by the Theorem of Choquet, for which we give a proof in Section 2.2. Some applications of this theorem are treated in Section 2.3. Special features of random closed sets in Euclidean spaces are the subject of Section 2.4.

2.1 Random Closed Sets in Locally Compact Spaces

The basic space in this chapter is a locally compact topological space E with a countable base. We denote by $\mathcal{F}$, $\mathcal{G}$, $\mathcal{C}$ the system of the closed, open, and compact subsets of E, respectively. The empty set is always included; we write $\mathcal{F}' := \mathcal{F} \setminus \{\emptyset\}$, and similarly $\mathcal{G}'$ and $\mathcal{C}'$ are defined. If necessary to avoid ambiguities, we write $\mathcal{F}(E)$, $\mathcal{G}(E)$, $\mathcal{C}(E)$ for $\mathcal{F}$, $\mathcal{G}$, $\mathcal{C}$.

Since random sets will be investigated in terms of their hitting probabilities with given sets, the following notation is fundamental. For $A \subset E$ we write

$$\mathcal{F}^A := \{F \in \mathcal{F} : F \cap A = \emptyset\},$$

$$\mathcal{F}_A := \{F \in \mathcal{F} : F \cap A \neq \emptyset\},$$

and we set

$$\mathcal{F}^A_{A_1,\dots,A_k} := \mathcal{F}^A \cap \mathcal{F}_{A_1} \cap \dots \cap \mathcal{F}_{A_k}$$

($:= \mathcal{F}^A$ for $k = 0$), if $k \in \mathbb{N}_0$ and $A_1, \dots, A_k \subset E$.

Definition 2.1.1. *The* **topology of closed convergence** *on $\mathcal{F}$ is the topology generated by the set system*

$$\{\mathcal{F}^C : C \in \mathcal{C}\} \cup \{\mathcal{F}_G : G \in \mathcal{G}\}.$$

The topology of closed convergence is also known as the 'Fell topology'. It is an example of a 'hit-and-miss topology'.

In the following, $\mathcal{F}$ will always be equipped with the topology of closed convergence. Basic properties of this topology are proved in Chapter 12, which the reader is advised to consult when necessary. The space $\mathcal{F}$ is compact and has a countable base (Theorem 12.2.1), and the subspace $\mathcal{F}'$ is locally compact.

Lemma 2.1.1. *The σ-algebra $\mathcal{B}(\mathcal{F})$ of Borel sets of $\mathcal{F}$ is generated by either of the systems*

$$\{\mathcal{F}^C : C \in \mathcal{C}\} \qquad \text{and} \qquad \{\mathcal{F}_G : G \in \mathcal{G}\}.$$

Proof. As shown in the proof of Theorem 12.2.1, the topology of $\mathcal{F}$ is generated by a countable subsystem of $\mathcal{A} := \{\mathcal{F}^C : C \in \mathcal{C}\} \cup \{\mathcal{F}_G : G \in \mathcal{G}\}$. Therefore, $\mathcal{A}$ generates $\mathcal{B}(\mathcal{F})$.

Let $G \in \mathcal{G}$. According to Theorem 12.1.1, there is a sequence $(C_i)_{i\in\mathbb{N}}$ of compact sets with $\bigcup_{i\in\mathbb{N}} C_i = G$, hence

$$\mathcal{F}_G = \bigcup_{i\in\mathbb{N}} \mathcal{F}_{C_i} = \bigcup_{i\in\mathbb{N}} (\mathcal{F}^{C_i})^c.$$

This shows that the system $\{\mathcal{F}^C : C \in \mathcal{C}\}$ is sufficient to generate the σ-algebra $\mathcal{B}(\mathcal{F})$.

Let $C \in \mathcal{C}$. According to Theorem 12.1.1, there is a sequence $(G_i)_{i\in\mathbb{N}}$ of open neighborhoods of C such that every open set G with $C \subset G$ contains a suitable set G_i. This yields

$$\mathcal{F}^C = \bigcup_{i\in\mathbb{N}} \mathcal{F}^{G_i} = \bigcup_{i\in\mathbb{N}} (\mathcal{F}_{G_i})^c,$$

hence also the system $\{\mathcal{F}_G : G \in \mathcal{G}\}$ is sufficient to generate $\mathcal{B}(\mathcal{F})$. □

Remark. Similarly, also each of the systems $\{\mathcal{F}_C : C \in \mathcal{C}\}$ and $\{\mathcal{F}^G : G \in \mathcal{G}\}$ generates $\mathcal{B}(\mathcal{F})$.

The following consequence is important. If the map $\varphi : T \to \mathcal{F}$ from some topological space T to $\mathcal{F}$ is upper or lower semicontinuous (see Section 12.2), then it is Borel measurable. In fact, if φ is upper semicontinuous, then $\varphi^{-1}(\mathcal{F}^C)$ is open, and hence a Borel set, for every compact set $C \in \mathcal{C}$. Since $\{\mathcal{F}^C : C \in \mathcal{C}\}$ is a generating system of $\mathcal{B}(\mathcal{F})$, the measurability of φ follows. For lower semicontinuous maps, the proof is analogous.

Lemma 2.1.2. *$\mathcal{C}$ is a Borel set in $\mathcal{F}$.*

Proof. By Theorem 12.1.1, there is a sequence $(C_i)_{i\in\mathbb{N}}$ of compact sets with $C_i \subset \operatorname{int} C_{i+1}$ for $i \in \mathbb{N}$ and $\bigcup_{i\in\mathbb{N}} C_i = E$. This yields

$$\mathcal{C} = \bigcup_{i\in\mathbb{N}} \mathcal{F}^{C_i^c},$$

where each $\mathcal{F}^{C_i^c}$ is closed, hence $\mathcal{C}$ is a Borel set in $\mathcal{F}$. □

Now we introduce random closed sets.

Definition 2.1.2. *A* **random closed set** *in E is an $\mathcal{F}$-valued random variable, that is, an $(\mathbf{A}, \mathcal{B}(\mathcal{F}))$-measurable map $Z : \Omega \to \mathcal{F}$ from some probability space $(\Omega, \mathbf{A}, \mathbb{P})$ into $\mathcal{F}$. The* **distribution** *of Z is the image measure $\mathbb{P}_Z := Z(\mathbb{P})$ of $\mathbb{P}$ under Z.*

In the following, 'random closed set' always means 'random closed set in E'.

As usual in probability theory, the essential feature of a random variable is its distribution and what can be derived from it. Two random closed sets Z and Z', which may be defined on different probability spaces, are called **stochastically equivalent** if they have the same distribution. This is also written as $Z \stackrel{\mathcal{D}}{=} Z'$ (**equality in distribution**). Even though every random closed set Z has a canonical representation Z' with $Z' \stackrel{\mathcal{D}}{=} Z$, via the identical map on $(\mathcal{F}, \mathcal{B}(\mathcal{F}), \mathbb{P}_Z)$, it is still more convenient to use the general representation of Definition 2.1.2, with an abstract probability space.

For $\mathbb{P}_Z(A)$, where $A \in \mathcal{B}(\mathcal{F})$, we also use the notation $\mathbb{P}(Z \in A)$, as an abbreviation for $\mathbb{P}(\{\omega \in \Omega : Z(\omega) \in A\})$, etc. If $\mathbb{P}(Z \in A) = 1$, we say that '$Z \in A$ almost surely' (a.s.).

If, in the following, several (finitely or countably many) random closed sets are treated simultaneously, we always assume that they are defined on the same probability space $(\Omega, \mathbf{A}, \mathbb{P})$. If $Z_1, \dots, Z_k$ are random closed sets, their **joint distribution** is the probability measure $\mathbb{P}_{Z_1,\dots,Z_k}$ on $\mathcal{F}^k$ defined by

$$\mathbb{P}_{Z_1,\dots,Z_k}(A_1 \times \dots \times A_k) = \mathbb{P}(Z_1 \in A_1, \dots, Z_k \in A_k)$$

for $A_1, \dots, A_k \in \mathcal{B}(\mathcal{F})$. Analogously, the joint distribution $\mathbb{P}_{Z_1,Z_2,\dots}$ of a sequence $Z_1, Z_2, \dots$ of random closed sets is defined. It is a probability measure on $\mathcal{F}^{\mathbb{N}}$. As usual, the random closed sets $Z_1, \dots, Z_k$, respectively $Z_1, Z_2, \dots$, are called (**stochastically**) **independent** if their joint distribution is the product of their individual distributions, that is, if

$$\mathbb{P}_{Z_1,\ldots,Z_k} = \mathbb{P}_{Z_1} \otimes \ldots \otimes \mathbb{P}_{Z_k},$$

respectively

$$\mathbb{P}_{Z_1,Z_2,\ldots} = \bigotimes_{i\in\mathbb{N}} \mathbb{P}_{Z_i}.$$

From given random closed sets Z and Z', one can obtain new ones by means of set-theoretic or topological operations. If $\varphi : \mathcal{F} \to \mathcal{F}$ and $\psi : \mathcal{F}\times\mathcal{F} \to \mathcal{F}$ are measurable maps, then also the compositions $\varphi \circ Z$ and $\psi \circ (Z, Z')$ are measurable. Therefore, the continuity or semicontinuity results of Theorems 12.2.3, 12.2.6, 13.1.1 yield the following.

Theorem 2.1.1. *If Z and Z' are random closed sets, then also $Z\cup Z'$, $Z\cap Z'$, $\operatorname{bd} Z$ and $\operatorname{cl} Z^c$ are random closed sets. If the topological group G operates continuously on E, then for $g \in G$ also gZ is a random closed set.*

We mention some simple examples of random closed sets. Trivially, if $F \in \mathcal{F}$, the constant map $\omega \mapsto F$ from Ω into $\mathcal{F}$ is a random closed set. Therefore, Theorem 2.1.1 implies that for a random closed set Z also the intersection $Z \cap F$ with a fixed set $F \in \mathcal{F}$ is a random closed set (and similarly $Z \cup F$). If $\xi_1, \xi_2, \ldots$ is a sequence of random variables with values in E, then the countable set $Z = \{\xi_1, \xi_2, \ldots\}$ is a random closed set if the set $\{\xi_1(\omega), \xi_2(\omega), \ldots\}$ has no accumulation points, for almost all ω. If, as in this case, $Z \cap C$ is almost surely finite for every compact set $C \in \mathcal{C}$, we say that the random closed set Z is **locally finite**.

Now we introduce, for random closed sets, a functional which can be considered as an analog to the distribution function of a real random variable. We first recall this latter notion.

For a random variable ξ with values in $(-\infty, \infty]$, the distribution function $\varphi = \varphi_\xi$ is defined by

$$\varphi_\xi(t) := \mathbb{P}(\xi \le t) = \mathbb{P}(\{\xi\} \cap (-\infty, t] \ne \emptyset), \quad t \in [-\infty, \infty).$$

It has the following properties:

(a) $0 \le \varphi \le 1$, $\varphi(-\infty) = 0$,
(b) φ is continuous from the right, that is, $t_i \downarrow t$ implies $\varphi(t_i) \to \varphi(t)$,
(c) φ is increasing, that is, $\varphi(t_0 + t_1) - \varphi(t_0) \ge 0$ for all $t_1 \ge 0$ and all $t_0 \in [-\infty, \infty)$.

The distribution function φ_ξ determines the distribution $\mathbb{P}_\xi$ uniquely. For any function φ satisfying (a), (b), (c), there exists a random variable with distribution function φ.

A tool with analogous properties exists in the theory of random closed sets.

Definition 2.1.3. *The* **capacity functional** T_Z *of the random closed set* Z *is defined by*

$$T_Z(C) := \mathbb{P}_Z(\mathcal{F}_C) = \mathbb{P}(Z \cap C \neq \emptyset) \qquad \text{for } C \in \mathcal{C}.$$

The following theorem shows that the capacity functional has properties corresponding to the properties (a), (b), (c) of a distribution function. We denote by $A_i \downarrow A$ the monotone convergence of sets A_i to A; this means that $A_{i+1} \subset A_i$ for $i \in \mathbb{N}$ and $\bigcap_{i\in\mathbb{N}} A_i = A$. Similarly, $A_i \uparrow A$ means that $A_{i+1} \supset A_i$ for $i \in \mathbb{N}$ and $\bigcup_{i\in\mathbb{N}} A_i = A$. If a function $T : \mathcal{C} \to \mathbb{R}$ is given, we define

$$S_0(C) := 1 - T(C) \qquad \text{for } C \in \mathcal{C}$$

and then, by recurrence,

$$S_k(C_0; C_1, \dots, C_k) := S_{k-1}(C_0; C_1, \dots, C_{k-1}) - S_{k-1}(C_0 \cup C_k; C_1, \dots, C_{k-1})$$

for $C_0, C_1, \dots, C_k \in \mathcal{C}$ and $k \in \mathbb{N}$. It should be kept in mind that S_k depends on T, although the notation does not reveal this.

Theorem 2.1.2. *The capacity functional* $T = T_Z$ *of a random closed set* Z *has the following properties:*

(a) $0 \leq T \leq 1$, $T(\emptyset) = 0$,
(b) *if* $C_i, C \in \mathcal{C}$ *and* $C_i \downarrow C$, *then* $T(C_i) \to T(C)$,
(c) $S_k(C_0; C_1, \dots, C_k) \geq 0$ *for* $C_0, C_1, \dots, C_k \in \mathcal{C}$ *and* $k \in \mathbb{N}_0$.

Proof. Assertion (a) follows immediately from the definition.

(b) If $C_i \downarrow C$, then the sequence $(\mathcal{F}_{C_i})_{i\in\mathbb{N}}$ is decreasing, and $\mathcal{F}_C \subset \bigcap_{i\in\mathbb{N}} \mathcal{F}_{C_i}$. We show that $\mathcal{F}_{C_i} \downarrow \mathcal{F}_C$. Let $F \in \bigcap_{i\in\mathbb{N}} \mathcal{F}_{C_i}$, then $F \cap C_i \neq \emptyset$ for all $i \in \mathbb{N}$. From $\bigcap_{i\in\mathbb{N}} C_i = C$ and the intersection property of compact sets it follows that $F \cap C = \bigcap_{i\in\mathbb{N}}(F \cap C_i) \neq \emptyset$. Hence, $F \in \mathcal{F}_C$ and thus $\bigcap_{i\in\mathbb{N}} \mathcal{F}_{C_i} = \mathcal{F}_C$. Assertion (b) now follows from the fact that the probability measure $\mathbb{P}_Z$ is continuous from above.

(c) Clearly, $S_0 \geq 0$. Using the relation

$$\mathcal{F}^{C_0}_{C_1,\dots,C_k} = \mathcal{F}^{C_0}_{C_1,\dots,C_{k-1}} \setminus \mathcal{F}^{C_0\cup C_k}_{C_1,\dots,C_{k-1}}, \tag{2.1}$$

one shows by induction with respect to k that

$$S_k(C_0; C_1, \dots, C_k) = \mathbb{P}_Z(\mathcal{F}^{C_0}_{C_1,\dots,C_k}), \qquad k \in \mathbb{N}. \tag{2.2}$$

The assertion follows. □

A real function T on $\mathcal{C}$ satisfying (a) and (b) of Theorem 2.1.2 is called a **Choquet capacity**. (The reason for this terminology comes from the fact that T can be extended to a set function on the power set $\mathbf{P}(E)$ of E which has the properties of a capacity; see Choquet [174].) A Choquet capacity satisfying (c) is called **alternating of infinite order**. The distribution of a random closed set is uniquely determined by its capacity functional.

Theorem 2.1.3. *If Z, Z' are random closed sets with $T_Z = T_{Z'}$, then $Z \stackrel{\mathcal{D}}{=} Z'$.*

Proof. The equality $T_Z = T_{Z'}$ means that $\mathbb{P}_Z(\mathcal{F}^C) = 1 - \mathbb{P}_Z(\mathcal{F}_C) = 1 - \mathbb{P}_{Z'}(\mathcal{F}_C) = \mathbb{P}_{Z'}(\mathcal{F}^C)$. Since the system $\{\mathcal{F}^C : C \in \mathcal{C}\}$ is $\cap$-stable and by Lemma 2.1.1 generates the σ-algebra $\mathcal{B}(\mathcal{F})$, the assertion follows from a well-known uniqueness theorem of measure theory. □

Notes for Section 2.1

1. Random sets were systematically developed by Matheron [459, 460] and D.G. Kendall [395]. Important fundamental ideas can already be found in Choquet's [173] theory of capacities. The introduction given in this chapter is essentially based on Matheron's seminal book [462].

2. General introductions to the theory of random closed sets are found in the monographs by Molchanov [548] and by Nguyen [583]. As the reader is advised to consult these volumes, the section notes in this chapter will be very brief.

3. Several different aspects of the theory of random sets are described in the surveys [542, 547] of Molchanov. The volumes edited by Jeulin [384] and by Goutsias, Mahler and Nguyen [284] contain various contributions to theory and applications of random sets.

2.2 Characterization of Capacity Functionals

The capacity functional $T = T_Z$ of a random closed set Z has the properties listed in Theorem 2.1.2. These properties of a function T on the system $\mathcal{C}$ of compact sets are also sufficient for T to be the capacity functional of a random closed set. This result is known as Choquet's Theorem.

Theorem 2.2.1 (Theorem of Choquet). *Let $T : \mathcal{C} \to \mathbb{R}$ be a function with the following properties:*

(a) $0 \le T \le 1$, $T(\emptyset) = 0$,
(b) *if* $C_i, C \in \mathcal{C}$ *and* $C_i \downarrow C$, *then* $T(C_i) \to T(C)$,
(c) $S_k(C_0; C_1, \dots, C_k) \ge 0$ *for* $C_0, C_1, \dots, C_k \in \mathcal{C}$ *and* $k \in \mathbb{N}_0$.

Then there exists a uniquely determined probability measure $\mathbb{P}$ on $\mathcal{F}$ with

$$T(C) = \mathbb{P}(\mathcal{F}_C)$$

for all $C \in \mathcal{C}$.

Consequently, the function T is the capacity functional of a random closed set Z. For example, one can take for Z the identical map on the probability space $(\mathcal{F}, \mathcal{B}(\mathcal{F}), \mathbb{P})$.

The stated uniqueness is clear from Theorem 2.1.3. For the existence, we shall present a proof due to Matheron [462], with a simplification taken from

Salinetti and Wets [655]. The proof is not short, but we want to give a complete and comprehensible presentation, not presupposing more than first principles from probability theory.

The basic idea of the proof is to define a function $\mathbb{P}$ on the set system

$$\mathsf{A} := \{\mathcal{F}^{C_0}_{C_1,\ldots,C_k} : C_i \in \mathcal{C},\, k \in \mathbb{N}_0\}$$

by means of

$$\mathbb{P}(\mathcal{F}^{C_0}_{C_1,\ldots,C_k}) := S_k(C_0; C_1, \ldots, C_k), \qquad C_i \in \mathcal{C}.$$

Since the sets $C_0, C_1, \ldots, C_k$ are not uniquely determined by $\mathcal{F}^{C_0}_{C_1,\ldots,C_k}$, it must be shown that this definition is unambiguous. It is then possible to prove that A is a semialgebra generating $\mathcal{B}(\mathcal{F})$ and that $\mathbb{P}$ is σ-additive on A. By the measure extension theorem, $\mathbb{P}$ can be extended to a probability measure on $\mathcal{B}(\mathcal{F})$. This probability measure satisfies $\mathbb{P}(\mathcal{F}_C) = T(C)$ for all $C \in \mathcal{C}$.

The first part of the proof (consisting of three lemmas) is combinatorial in nature and does not use topological properties; therefore, we formulate it for a general set system with appropriate properties. Let $\mathcal{V} \subset \mathbf{P}(E)$ be a nonempty, $\cup$-stable system of subsets of E, and let $T : \mathcal{V} \to \mathbb{R}$ be an arbitrary function. Then we define

$$S_0(V) := 1 - T(V) \qquad \text{for } V \in \mathcal{V}$$

and, recursively,

$$S_k(V_0; V_1, \ldots, V_k) := S_{k-1}(V_0; V_1, \ldots, V_{k-1}) - S_{k-1}(V_0 \cup V_k; V_1, \ldots, V_{k-1})$$

for $k \in \mathbb{N}$ and $V_i \in \mathcal{V}$. It will be clear from the context from which function T the functions S_k are derived. Explicitly, for $k \geq 1$, S_k is given by

$$S_k(V_0; V_1, \ldots, V_k) = \sum_{r=0}^{k} (-1)^{r-1} \sum_{1 \leq i_1 < \ldots < i_r \leq k} T(V_0 \cup V_{i_1} \cup \ldots \cup V_{i_r}), \quad (2.3)$$

as follows by induction. For $r = 0$, the inner sum has to be read as $T(V_0)$. In the following, we assume that a system $\mathcal{V}$ as above is given. Sets V, V_i, W, W_i appearing below are always elements of $\mathcal{V}$.

Lemma 2.2.1. *If* $\emptyset \neq \mathcal{F}^V_{V_1,\ldots,V_m} \subset \mathcal{F}^W_{W_1,\ldots,W_k}$, *then*

(a) $W \subset V$,
(b) *to each* $i \in \{1, \ldots, k\}$ *there exists* $j(i) \in \{1, \ldots, m\}$ *with* $V_{j(i)} \subset W_i \cup V$.

Proof. Suppose (a) were false. Then there is $x \in W \cap V^c$. Since $\mathcal{F}^V_{V_1,\ldots,V_m} \neq \emptyset$, to each $i \in \{1, \ldots, m\}$ there is $x_i \in V_i \cap V^c$. Then $\{x, x_1, \ldots, x_m\} \in \mathcal{F}^V_{V_1,\ldots,V_m} \subset \mathcal{F}^W_{W_1,\ldots,W_k}$ and thus $\{x, x_1, \ldots, x_m\} \cap W = \emptyset$, which contradicts $x \in W$.

Suppose (b) were false. Then there exists $i \in \{1, \ldots, k\}$ with

$$V_j \not\subset W_i \cup V \qquad \text{for } j = 1, \ldots, m.$$

Hence, to each $j \in \{1, \ldots, m\}$ there is $x_j \in V_j \cap W_i^c \cap V^c$, and it follows that $\{x_1, \ldots, x_m\} \in \mathcal{F}^V_{V_1,\ldots,V_m} \subset \mathcal{F}^W_{W_1,\ldots,W_k}$, hence $\{x_1, \ldots, x_m\} \cap W_i \neq \emptyset$, a contradiction. □

We recall that a **semialgebra** in $\mathcal{F}$ is a set system $\mathsf{A} \subset \mathbf{P}(\mathcal{F})$ with the following properties:

(a) $\emptyset \in \mathsf{A}$, $\mathcal{F} \in \mathsf{A}$,
(b) A is $\cap$-stable,
(c) for every $A \in \mathsf{A}$, the complement A^c is the union of a finite family of pairwise disjoint sets from A.

With the system $\mathcal{V} \subset \mathbf{P}(E)$ as given above, we define

$$\mathsf{A} := \{\mathcal{F}^V_{V_1,\ldots,V_k} : V, V_1, \ldots, V_k \in \mathcal{V},\, k \in \mathbb{N}_0\}. \tag{2.4}$$

The representation

$$A = \mathcal{F}^V_{V_1,\ldots,V_k}$$

of an element $A \in \mathsf{A}$ with $k \in \mathbb{N}_0$ and $V, V_1, \ldots, V_k \in \mathcal{V}$ is called **reduced** if

$$V_i \not\subset V_j \cup V \qquad \text{for } i, j \in \{1, \ldots, k\} \text{ with } i \neq j.$$

The following lemma provides the main structural information about the set system A.

Lemma 2.2.2. *Let $\emptyset \in \mathcal{V} \subset \mathbf{P}(E)$ be $\cup$-stable and let A be defined by* (2.4).

(a) *A is a semialgebra in $\mathcal{F}$.*
(b) *Every $A \in \mathsf{A}$ has a reduced representation.*
(c) *If*

$$A = \mathcal{F}^V_{V_1,\ldots,V_m} = \mathcal{F}^W_{W_1,\ldots,W_k}$$

are two reduced representations of an element $A \in \mathsf{A} \setminus \{\emptyset\}$, then $V = W$, $m = k$ and

$$V_i \cup V = W_{\pi(i)} \cup V \qquad \text{for } i = 1, \ldots, m$$

with a permutation π of $\{1, \ldots, m\}$.
(d) *If $A, B \in \mathsf{A}$ and $A \subset B$, there are elements $D_0, D_1, \ldots, D_r \in \mathsf{A}$ with*

$$A = D_0 \subset D_1 \subset \ldots \subset D_r = B$$

and $D_i \setminus D_{i-1} \in \mathsf{A}$ for $i = 1, \ldots, r$.

Proof. (a) From $\emptyset \in \mathcal{V}$ it follows that $\emptyset = \mathcal{F}^\emptyset_\emptyset \in \mathsf{A}$ and $\mathcal{F} = \mathcal{F}^\emptyset \in \mathsf{A}$. Since

$$\mathcal{F}^V_{V_1,\ldots,V_m} \cap \mathcal{F}^W_{W_1,\ldots,W_k} = \mathcal{F}^{V \cup W}_{V_1,\ldots,V_m,W_1,\ldots,W_k}$$

and $V \cup W \in \mathcal{V}$ for $V, W \in \mathcal{V}$, the system A is closed under finite intersections. The complement of $A = \mathcal{F}^V_{V_1,\dots,V_m} \in \mathsf{A}$ can be written in the form

$$A^c = \mathcal{F}^{\emptyset}_V \cup \mathcal{F}^{V\cup V_1} \cup \mathcal{F}^{V\cup V_2}_{V_1} \cup \mathcal{F}^{V\cup V_3}_{V_1,V_2} \cup \ldots \cup \mathcal{F}^{V\cup V_m}_{V_1,\dots,V_{m-1}},$$

where the right side is a disjoint union of elements of A. Thus, A is a semialgebra.

(b) Let $A = \mathcal{F}^V_{V_1,\dots,V_m}$. If there are indices $i \neq j$ with $V_i \subset V_j \cup V$, then

$$\mathcal{F}^V_{V_1,\dots,V_m} = \mathcal{F}^V_{V_1,\dots,V_{j-1},V_{j+1},\dots,V_m}.$$

In fact, if $F \in \mathcal{F}^V_{V_1,\dots,V_{j-1},V_{j+1},\dots,V_m}$, then $F \cap V_i \neq \emptyset$ and $F \cap V = \emptyset$, hence $F \cap V_j \neq \emptyset$ and, therefore, $F \in \mathcal{F}^V_{V_1,\dots,V_m}$. The converse inclusion is trivial. Hence, V_j can be deleted in the representation of A. Repeating this procedure, we finally obtain a reduced representation.

(c) Suppose that

$$A = \mathcal{F}^V_{V_1,\dots,V_m} = \mathcal{F}^W_{W_1,\dots,W_k}$$

are two reduced representations of some $A \in \mathsf{A} \setminus \{\emptyset\}$. Lemma 2.2.1 yields $V = W$ and, if $k \geq 1$, the existence of an index $j(1) \in \{1,\dots,m\}$ with $V_{j(1)} \subset W_1 \cup V$. By the same argument (interchanging the two representations and replacing the index 1 by $j(1)$), there is an index $i(1) \in \{1,\dots,k\}$ with

$$W_{i(1)} \subset V_{j(1)} \cup W = V_{j(1)} \cup V.$$

Since $W_{i(1)} \subset V_{j(1)} \cup V \subset W_1 \cup V = W_1 \cup W$ and the representations are reduced, we conclude that $i(1) = 1$. Thus, $V_{j(1)} \cup V = W_1 \cup V$. Repeating the procedure with each of the sets $W_2,\dots,W_k$ (and observing, for example, that $W_2 \cup V \neq W_1 \cup V$ since the representations are reduced and $V = W$), we deduce the truth of (c).

(d) Let $A, B \in \mathsf{A}$ be elements with $A \subset B$, say

$$\emptyset \neq A = \mathcal{F}^V_{V_1,\dots,V_m} \subset \mathcal{F}^W_{W_1,\dots,W_k} = B$$

(the case $A = \emptyset$ is trivial). By Lemma 2.2.1, $W \subset V$, and to each $i \in \{1,\dots,k\}$ there is $j(i) \in \{1,\dots,m\}$ with

$$V_{j(i)} \subset W_i \cup V. \tag{2.5}$$

Renumbering $V_1,\dots,V_m$, we may assume that

$$\{j(1),\dots,j(k)\} = \{1,\dots,p\}$$

with $p \in \{1,\dots,m\}$. For $q \in \{1,\dots,p\}$, let $W(q)$ be the tuple of the W_i for which $j(i) = q$. Then (with $\mathcal{F}^V_{W(1),V_2,\dots,V_p} := \mathcal{F}^V_{W_1,\dots,W_s,V_2,\dots,V_p}$ if $W(1) = (W_1,\dots,W_s)$, etc.),

$$\begin{aligned}&\mathcal{F}^V_{V_1,\dots,V_m} \subset \mathcal{F}^V_{V_1,\dots,V_p,V_{p+2},\dots,V_m} \subset \mathcal{F}^V_{V_1,\dots,V_p,V_{p+3},\dots,V_m} \subset \dots \subset \mathcal{F}^V_{V_1,\dots,V_p}\\ &\subset \mathcal{F}^V_{W(1),V_2,\dots,V_p} \subset \mathcal{F}^V_{W(1),W(2),V_3,\dots,V_p} \subset \dots \subset \mathcal{F}^V_{W(1),\dots,W(p)}\\ &= \mathcal{F}^V_{W_1,\dots,W_k} \subset \mathcal{F}^W_{W_1,\dots,W_k}.\end{aligned}$$

The inclusions of the first line are trivial; generally

$$\mathcal{F}^V_{U_1,\dots,U_r} \subset \mathcal{F}^V_{U_1,\dots,U_{r-1}}$$

and

$$\mathcal{F}^V_{U_1,\dots,U_{r-1}} \setminus \mathcal{F}^V_{U_1,\dots,U_r} = \mathcal{F}^{V\cup U_r}_{U_1,\dots,U_{r-1}} \in \mathsf{A}$$

for $V, U_1, \dots, U_r \in \mathcal{V}$. To prove the inclusion

$$\mathcal{F}^V_{V_1,\dots,V_p} \subset \mathcal{F}^V_{W(1),V_2,\dots,V_p}$$

connecting the first and the second line, we may assume, after renumbering, that $W(1) = (W_1, \dots, W_s)$. For $F \in \mathcal{F}^V_{V_1,\dots,V_p}$ we have $F \cap V = \emptyset$, $F \cap V_j \neq \emptyset$ for $j = 1, \dots, p$ and $V_1 \subset W_i \cup V$ for $i = 1, \dots, s$, by (2.5), hence $F \cap W_i \neq \emptyset$ for $i = 1, \dots, s$ and thus $F \in \mathcal{F}^V_{W(1),V_2,\dots,V_p}$. Moreover,

$$\mathcal{F}^V_{W(1),V_2,\dots,V_p} \setminus \mathcal{F}^V_{V_1,\dots,V_p} = \mathcal{F}^{V\cup V_1}_{W(1),V_2,\dots,V_p} \in \mathsf{A}.$$

In the same way, the corresponding relations for the second line are obtained. The inclusion of the third line follows from $W \subset V$, and we have

$$\mathcal{F}^W_{W_1,\dots,W_k} \setminus \mathcal{F}^V_{W_1,\dots,W_k} = \mathcal{F}^W_{W_1,\dots,W_k,V} \in \mathsf{A}.$$

This completes the proof of Lemma 2.2.2. □

As a first step towards the existence proof for the probability measure $\mathbb{P}$, we construct a finitely additive function on the semialgebra A.

Lemma 2.2.3. *Let $T : \mathcal{V} \to \mathbb{R}$ be a function satisfying $T(\emptyset) = 0$ and*

$$S_k(V; V_1, \dots, V_k) \geq 0$$

for all $k \in \mathbb{N}_0$ and all $V, V_1, \dots, V_k \in \mathcal{V}$. Then there is a finitely additive function $\mathbb{P} : \mathsf{A} \to [0, 1]$ with

$$\mathbb{P}\left(\mathcal{F}^V_{V_1,\dots,V_k}\right) = S_k(V; V_1, \dots, V_k)$$

for $V, V_1, \dots, V_k \in \mathcal{V}$ and $k \in \mathbb{N}_0$. In particular, $\mathbb{P}(\emptyset) = 0$, $\mathbb{P}(\mathcal{F}) = 1$, and $\mathbb{P}(\mathcal{F}_V) = T(V)$.

This lemma is a consequence of Propositions 2.2.1 and 2.2.2 below. The first of them removes the ambiguity in the definition of $\mathbb{P}$.

Proposition 2.2.1. *If* $\mathcal{F}^V_{V_1,\ldots,V_m} = \mathcal{F}^W_{W_1,\ldots,W_k} =: A \neq \emptyset$, *then*

$$S_m(V;V_1,\ldots,V_m) = S_k(W;W_1,\ldots,W_k).$$

Proof. We show that the value of $S_m(V;V_1,\ldots,V_m)$ does not change under a stepwise reduction of the representation $\mathcal{F}^V_{V_1,\ldots,V_m}$ as performed in the proof of Lemma 2.2.2. For example, suppose that $V_1 \subset V_m \cup V$. Then $A = \mathcal{F}^V_{V_1,\ldots,V_{m-1}}$. Generally, an inclusion $V_i \subset V_0$ for some $i \in \{1,\ldots,n\}$ implies $S_n(V_0;V_1,\ldots,V_n) = 0$. In fact, since S_n is symmetric in its last n arguments, we can assume $i = n$ and then obtain

$$S_n(V_0;V_1,\ldots,V_n) = S_{n-1}(V_0;V_1,\ldots,V_{n-1}) - S_{n-1}(V_0 \cup V_n;V_1,\ldots,V_{n-1}) = 0.$$

Hence, in our case we have

$$S_{m-1}(V \cup V_m;V_1,\ldots,V_{m-1}) = 0$$

and, therefore,

$$S_m(V;V_1,\ldots,V_m) = S_{m-1}(V;V_1,\ldots,V_{m-1}).$$

This shows that the value of $S_m(V;V_1,\ldots,V_m)$ does not change under stepwise reduction. We may, therefore, assume that the two given representations of A are already reduced. Then, by Lemma 2.2.2, $V = W$, $m = k$, and without loss of generality (that is, after a permutation of indices, by which the value of $S_m(V;V_1,\ldots,V_m)$ is not changed) $V_i \cup V = W_i \cup V$ for $i = 1,\ldots,m$. Now the relation

$$S_m(V;V_1,\ldots,V_m) = S_m(V;V \cup V_1,\ldots,V \cup V_m),$$

which follows from (2.3), yields the truth of Proposition 2.2.1. □

Under the assumptions of Lemma 2.2.3, it is now possible to define

$$\mathbb{P}\left(\mathcal{F}^V_{V_1,\ldots,V_m}\right) := S_m(V;V_1,\ldots,V_m) \qquad \text{for } m \in \mathbb{N}_0, V, V_i \in \mathcal{V}.$$

Then $\mathbb{P} \geq 0$. The recursion formula for S_m and the assumption $S_m \geq 0$ yield

$$\begin{aligned}\mathbb{P}\left(\mathcal{F}^V_{V_1,\ldots,V_m}\right) &= S_m(V;V_1,\ldots,V_m) \leq S_{m-1}(V;V_1,\ldots,V_{m-1}) \\ &\leq \ldots \leq S_0(V) = 1 - T(V) = 1 - S_1(\emptyset;V) \leq 1.\end{aligned}$$

Thus $\mathbb{P}$ maps A into $[0,1]$. Moreover, $\mathbb{P}(\emptyset) = \mathbb{P}(\mathcal{F}^\emptyset_\emptyset) = S_1(\emptyset;\emptyset) = 0$, $\mathbb{P}(\mathcal{F}) = \mathbb{P}(\mathcal{F}^\emptyset) = S_0(\emptyset) = 1 - T(\emptyset) = 1$, and $\mathbb{P}(\mathcal{F}_V) = \mathbb{P}(\mathcal{F}^\emptyset_V) = S_1(\emptyset;V) = T(V)$.

Proposition 2.2.2. $\mathbb{P}$ *is finitely additive on* A.

Proof. We have to show that

$$\mathbb{P}(A_1 \cup \ldots \cup A_r) = \sum_{i=1}^{r} \mathbb{P}(A_i) \tag{2.6}$$

whenever $r \geq 2$ and $A_1, \ldots, A_r \in \mathsf{A}$ are pairwise disjoint elements with $A_1 \cup \ldots \cup A_r \in \mathsf{A}$. First we consider the case $r = 2$.

Let $A, B \in \mathsf{A} \setminus \{\emptyset\}$ be elements with representations

$$A = \mathcal{F}^V_{V_1,\ldots,V_m}, \qquad B = \mathcal{F}^W_{W_1,\ldots,W_k}$$

and satisfying $A \cap B = \emptyset$ and $A \cup B \in \mathsf{A}$.

Suppose first that $m = 0$, say, so that $A = \mathcal{F}^V$. Then $\emptyset \in A \subset A \cup B$, hence $A \cup B = \mathcal{F}^U$ with some $U \in \mathcal{V}$. It follows that

$$A = (A \cup B) \cap \mathcal{F}^V = \mathcal{F}^{U \cup V},$$
$$B = (A \cup B) \cap \mathcal{F}_V = \mathcal{F}^U_V$$

and, therefore,

$$\begin{aligned}\mathbb{P}(A) + \mathbb{P}(B) &= \mathbb{P}(\mathcal{F}^{U \cup V}) + \mathbb{P}(\mathcal{F}^U_V) \\ &= S_0(U \cup V) + S_0(U) - S_0(U \cup V) \\ &= \mathbb{P}(\mathcal{F}^U) \\ &= \mathbb{P}(A \cup B).\end{aligned}$$

Hence, we can now assume that $m, k \geq 1$ and, therefore, $\emptyset \notin A \cup B$.

Because of $A \cup B \in \mathsf{A}$, there is a representation

$$A \cup B = \mathcal{F}^U_{U_1,\ldots,U_p}$$

with $U, U_1, \ldots, U_p \in \mathcal{V}$, and $\emptyset \notin A \cup B$ implies that $p \geq 1$. By Lemma 2.2.1, $U \subset V$ and $U \subset W$, thus

$$U \subset V \cap W. \tag{2.7}$$

We assert that

$$V \subset U \quad \text{or} \quad W \subset U. \tag{2.8}$$

Suppose this were false. Then there are points $x \in V \cap U^c$ and $y \in W \cap U^c$. Since $A \cup B \neq \emptyset$, we can choose points $z_i \in U_i \cap U^c$ for $i = 1, \ldots, p$. Then $\{x, y, z_1, \ldots, z_p\} \in \mathcal{F}^U_{U_1,\ldots,U_p} = A \cup B = \mathcal{F}^V_{V_1,\ldots,V_m} \cup \mathcal{F}^W_{W_1,\ldots,W_k}$. Since $\{x, y\} \cap V \neq \emptyset$ and $\{x, y\} \cap W \neq \emptyset$, this is a contradiction. Thus (2.8) holds, say $V \subset U$. Then (2.7) implies $V = U$ and

$$V \subset W. \tag{2.9}$$

By assumption,

$$\mathcal{F}^{V \cup W}_{V_1,\dots,V_m,W_1,\dots,W_k} = A \cap B = \emptyset.$$

This implies $V_i \subset V \cup W$ for some $i \in \{1,\dots,m\}$ or $W_j \subset V \cup W$ for some $j \in \{1,\dots,k\}$. The second case would imply $W_j \subset W$ by (2.9) and thus $B = \emptyset$, a contradiction. Hence, $V_i \subset V \cup W$ for some i and thus

$$V_i \subset W. \tag{2.10}$$

Now let $F \in A \cup B$. If $F \in \mathcal{F}_{V_i}$, then $F \cap W \neq \emptyset$ by (2.10), hence $F \notin B$ and thus $F \in A$. If, on the other hand, $F \in \mathcal{F}^{V_i}$, then $F \notin A$, hence $F \in B$. This shows that

$$A = (A \cup B) \cap \mathcal{F}_{V_i} = \mathcal{F}^{V}_{U_1,\dots,U_p,V_i},$$

$$B = (A \cup B) \cap \mathcal{F}^{V_i} = \mathcal{F}^{V \cup V_i}_{U_1,\dots,U_p}.$$

Observing that $U = V$, we conclude that

$$\begin{aligned}
\mathbb{P}(A \cup B) &= \mathbb{P}\left(\mathcal{F}^{V}_{U_1,\dots,U_p}\right) \\
&= S_p(V; U_1,\dots,U_p) \\
&= S_{p+1}(V; U_1,\dots,U_p, V_i) + S_p(V \cup V_i; U_1,\dots,U_p) \\
&= \mathbb{P}\left(\mathcal{F}^{V}_{U_1,\dots,U_p,V_i}\right) + \mathbb{P}\left(\mathcal{F}^{V \cup V_i}_{U_1,\dots,U_p}\right) \\
&= \mathbb{P}(A) + \mathbb{P}(B),
\end{aligned}$$

which establishes (2.6) for $r = 2$.

The case $r = 2$ does not trivially imply the general case of (2.6), since A is merely a semialgebra. To obtain the general case and thus establish finite additivity, we observe that assertion (d) of Lemma 2.2.2 shows that the semialgebra A is also a semiring (in the sense of Halmos [318, p. 22]). That $\mathbb{P}$ satisfies (2.6) for all $r \geq 2$, now follows from [318, pp. 31–32]. □

Proof of Theorem 2.2.1. We apply Lemma 2.2.3 to the function T of Theorem 2.2.1 and the system $\mathcal{V} = \mathcal{C}$. The subsequent steps serve the purpose of extending the finitely additive function $\mathbb{P}$ on the semialgebra

$$\mathsf{A} := \{\mathcal{F}^{C_0}_{C_1,\dots,C_k} : C_i \in \mathcal{C},\ k \in \mathbb{N}_0\}$$

to a probability measure on $\mathcal{B}(\mathcal{F})$.

Let $\mathsf{A}^\bullet$ be the algebra in $\mathcal{F}$ generated by A. Then $\mathsf{A}^\bullet$ consists of all finite unions of pairwise disjoint elements of A, and the unique additive extension of $\mathbb{P}$ to $\mathsf{A}^\bullet$ is given by

$$\mathbb{P}(A) = \sum_{i=1}^{m} \mathbb{P}(A_i)$$

if $A = A_1 \cup \ldots \cup A_m$ and $A_1, \ldots, A_m$ are pairwise disjoint elements of A (see Neveu [581, Proposition I.6.1]).

We state the following approximation property: for every $A \in \mathsf{A}^\bullet$ and every $\epsilon > 0$ there exist a set $A^0 \in \mathsf{A}^\bullet$ and a compact set $B \subset \mathcal{F}$ such that

$$A^0 \subset B \subset A \tag{2.11}$$

and

$$\mathbb{P}(A \setminus A^0) < \epsilon. \tag{2.12}$$

Since every element of $\mathsf{A}^\bullet$ is a finite union of pairwise disjoint elements of A, it suffices to prove this for $A \in \mathsf{A}$, say $A = \mathcal{F}^{C_0}_{C_1,\ldots,C_k}$ with $C_i \in \mathcal{C}$. We can choose a sequence $(G_i)_{i\in\mathbb{N}}$ of open, relatively compact sets in E with $\operatorname{cl} G_{i+1} \subset G_i$ and $G_i \downarrow C_0$ (by Theorem 12.1.1). Then

$$\mathcal{F}^{\operatorname{cl} G_i}_{C_1,\ldots,C_k} \subset \mathcal{F}^{G_i}_{C_1,\ldots,C_k} \subset \mathcal{F}^{C_0}_{C_1,\ldots,C_k}.$$

The sets $\mathcal{F}^{G_i}_{C_1,\ldots,C_k}$ are closed and hence compact, since $\mathcal{F}$ is compact. From $\operatorname{cl} G_i \downarrow C_0$ and condition (b) of Theorem 2.2.1 we have $T(\operatorname{cl} G_i) \to T(C_0)$. The function T is isotone, since $C \subset C'$ implies $0 \le S_1(C; C') = -T(C) + T(C \cup C')$, hence $T(C) \le T(C')$. It follows that $T(\operatorname{cl} G_i) \downarrow T(C_0)$ and hence that $\mathbb{P}(\mathcal{F}^{\operatorname{cl} G_i}) \uparrow \mathbb{P}(\mathcal{F}^{C_0})$, thus

$$\lim_{i\to\infty} \mathbb{P}(\mathcal{F}^{\operatorname{cl} G_i}_{C_1,\ldots,C_k}) = \mathbb{P}(\mathcal{F}^{C_0}_{C_1,\ldots,C_k}).$$

Therefore, for given $\epsilon > 0$, there exists a number $j \in \mathbb{N}$ with

$$\mathbb{P}(\mathcal{F}^{C_0}_{C_1,\ldots,C_k}) < \mathbb{P}(\mathcal{F}^{\operatorname{cl} G_j}_{C_1,\ldots,C_k}) + \epsilon.$$

The sets $A^0 := \mathcal{F}^{\operatorname{cl} G_j}_{C_1,\ldots,C_k}$ and $B := \mathcal{F}^{G_j}_{C_1,\ldots,C_k}$ satisfy (2.11) and (2.12).

Now let $(A_i)_{i\in\mathbb{N}}$ be a sequence in $\mathsf{A}^\bullet$ with $A_i \downarrow \emptyset$, and let $\epsilon > 0$. To each $i \in \mathbb{N}$, we can choose a set $A_i^0 \in \mathsf{A}^\bullet$ and a compact set $B_i \subset \mathcal{F}$ such that

$$A_i^0 \subset B_i \subset A_i \tag{2.13}$$

and

$$\mathbb{P}(A_i \setminus A_i^0) < \epsilon 2^{-i}. \tag{2.14}$$

From $\bigcap_{i\in\mathbb{N}} A_i = \emptyset$ and (2.13) we have $\bigcap_{i\in\mathbb{N}} B_i = \emptyset$. The finite intersection property of compact sets yields the existence of a number $m \in \mathbb{N}$ with $\bigcap_{i=1}^m B_i = \emptyset$, and (2.13) gives $\bigcap_{i=1}^m A_i^0 = \emptyset$. Since the A_i are decreasing, for $j \ge m$ we have

$$A_j = \bigcap_{i\le j} A_i \subset \bigcup_{i\le j} (A_i \setminus A_i^0).$$

This yields

$$\mathbb{P}(A_j) \le \sum_{i\le j} \mathbb{P}(A_i \setminus A_i^0) < \epsilon$$

by (2.14). Since $\epsilon > 0$ was arbitrary, we get

$$\lim_{i\to\infty} \mathbb{P}(A_i) = 0.$$

Thus, $\mathbb{P}$ is $\emptyset$-continuous, which implies that $\mathbb{P}$ is σ-additive on $\mathsf{A}^\bullet$.

By the measure extension theorem, $\mathbb{P}$ has an extension to a measure on the σ-algebra generated by $\mathsf{A}^\bullet$, namely $B(\mathcal{F})$. □

Notes for Section 2.2

1. We have presented here an elaborate version of Matheron's proof for Choquet's Theorem (Theorem 2.2.1, with an alternate proof for the σ-additivity taken from Salinetti and Wets [655]), in a form we found suitable for class room use. One reason for the choice of this proof was the fact that it requires only basic measure theory and does not need tools from functional analysis. Several proofs in the literature are seemingly much shorter, but we found them either less elementary or less detailed.

2. Choquet's [173] original proof of Theorem 2.2.1 used Choquet's functional-analytic generalization of the theorem of Krein–Milman. For further proofs, we refer to Berg, Christensen and Ressel [91, Th. 6.19], Norberg [586], Kallenberg [386, Th. 24.22], Molchanov [548, ch. 1, sect. 1]. An extension of Choquet's theorem to spaces without a countable base is due to Ross [648].

Kallenberg's proof uses the following basic idea (adjusted to our notation). Let $C_1, C_2, \dots$ be a sequence of compact sets such that, for each pair $C \in \mathcal{C}$, $G \in \mathcal{G}$ with $C \subset G$ there is an $i \in \mathbb{N}$ with $C \subset C_i \subset G$ (such a sequence exists by Theorem 12.1.1). Let $\mathcal{C}_n$ be the collection of finite unions of the sets $C_1, \dots, C_n$, $n \in \mathbb{N}$. By induction on n, one can show that, for each functional T_n on $\mathcal{C}_n$ satisfying conditions (a), (b), (c) of Theorem 2.2.1 (on $\mathcal{C}_n$), there exists a locally finite random set Z_n (a simple point process in the terminology of the next chapter) such that the distribution $\mathbb{P}_n$ of Z_n satisfies

$$\mathbb{P}_n(\mathcal{F}_C) = T_n(C)$$

for all $C \in \mathcal{C}_n$. Choose T_n to be the restriction of T to $\mathcal{C}_n$. Since $\mathcal{F}$ is compact, the space $\mathcal{M}_1(\mathcal{F})$ of probability measures on $\mathcal{F}$ is weakly compact. Hence there is a subsequence $\mathbb{P}_{i_1}, \mathbb{P}_{i_2}, \dots$ that converges weakly to a probability measure $\mathbb{P}$ on $\mathcal{F}$. This measure satisfies

$$\mathbb{P}(\mathcal{F}_C) = T(C)$$

for all $C \in \bigcup_{n=1}^\infty \mathcal{C}_n$, but then also for all $C \in \mathcal{C}$.

3. For simple point processes, a characterization of the capacity functional was also obtained by Kurtz [430].

2.3 Some Consequences of Choquet's Theorem

In order to construct a random closed set in E, it may in some cases be possible first to define it locally, that is, in compact subsets of E, and then to extend the definition through a sequence of compact sets covering E. This works if suitable compatibility assumptions are satisfied.

Theorem 2.3.1. *Let $(Z_i)_{i\in\mathbb{N}}$ be a sequence of random closed sets in E with the following property. There is a sequence $(G_i)_{i\in\mathbb{N}}$ of open, relatively compact sets in E with $\mathrm{cl}\, G_i \subset G_{i+1}$ for $i \in \mathbb{N}$ and $G_i \uparrow E$ such that*

$$Z_m \cap \mathrm{cl}\, G_i \stackrel{\mathcal{D}}{=} Z_i \qquad \text{for } m > i$$

(that is, $Z_m \cap \mathrm{cl}\, G_i$ and Z_i are stochastically equivalent). Then there exists a random closed set Z in E with

$$Z \cap \mathrm{cl}\, G_i \stackrel{\mathcal{D}}{=} Z_i \qquad \text{for } i \in \mathbb{N}.$$

Proof. Let T_i be the capacity functional of Z_i. For given $C \in \mathcal{C}$ there exists $i \in \mathbb{N}$ with $C \subset \mathrm{cl}\, G_i$. For $m > i$ we then have

$$\begin{aligned} T_i(C) &= \mathbb{P}(Z_i \cap C \neq \emptyset) \\ &= \mathbb{P}(Z_m \cap \mathrm{cl}\, G_i \cap C \neq \emptyset) \\ &= T_m(\mathrm{cl}\, G_i \cap C) \\ &= T_m(C). \end{aligned}$$

Therefore, it is possible to define $T(C) := T_i(C)$. The obtained function T on $\mathcal{C}$ satisfies $0 \leq T \leq 1$ and $T(\emptyset) = 0$. If $C_j \downarrow C$ in $\mathcal{C}$, then there exists $m \in \mathbb{N}$ with $C_j, C \subset G_m$ for all $j \in \mathbb{N}$, hence $T(C_j) = T_m(C_j)$ and $T(C) = T_m(C)$, which yields $T(C_j) \to T(C)$. Similarly one shows that $S_k(C; C_1, \dots, C_k) \geq 0$ for $k \in \mathbb{N}_0$ and $C, C_1, \dots, C_k \in \mathcal{C}$. By Theorem 2.2.1, there exists a random closed set Z in E with capacity functional T.

Let $i \in \mathbb{N}$. For $C \in \mathcal{C}$ and $m > i$, we have

$$\begin{aligned} T_{Z\cap \mathrm{cl}\, G_i}(C) &= \mathbb{P}(Z \cap \mathrm{cl}\, G_i \cap C \neq \emptyset) \\ &= T(\mathrm{cl}\, G_i \cap C) \\ &= T_m(\mathrm{cl}\, G_i \cap C) \\ &= T_i(C). \end{aligned}$$

Thus, $Z \cap \mathrm{cl}\, G_i$ and Z_i have the same capacity functional and, therefore, by Theorem 2.1.3, also the same distribution. □

In the following, we consider (Borel) measures on $\mathcal{F}' = \mathcal{F} \setminus \{\emptyset\}$, the space of nonempty closed subsets of E. The space $\mathcal{F}'$ (with the trace topology) is locally compact. A measure μ on $\mathcal{F}'$ is called **locally finite** if it is finite on compact sets.

Lemma 2.3.1. *The measure μ on $\mathcal{F}'$ is locally finite if and only if*

$$\mu(\mathcal{F}_C) < \infty \qquad \text{for all } C \in \mathcal{C}.$$

A locally finite measure μ on $\mathcal{F}'$ is uniquely determined by its values on the system $\{\mathcal{F}_C : C \in \mathcal{C}\}$.

Proof. Since $\{\mathcal{F}^C : C \in \mathcal{C}\}$ is a neighborhood base of $\emptyset$ in $\mathcal{F}$, every compact subset of $\mathcal{F}'$ is contained in some $\mathcal{F}_C$, $C \in \mathcal{C}$, and every such set $\mathcal{F}_C$ is compact. This yields the first assertion.

If μ is a locally finite measure on $\mathcal{F}'$ and if $C_0, C_1, \dots, C_k \in \mathcal{C}$, $k \in \mathbb{N}$, then

$$\mu\left(\mathcal{F}^{C_0}_{C_1,\dots,C_k}\right) = \sum_{r=0}^{k} (-1)^{r-1} \sum_{1 \le i_1 < \dots < i_r \le k} \mu(\mathcal{F}_{C_0 \cup C_{i_1} \cup \dots \cup C_{i_r}}).$$

For $k = 1$, this follows from $\mathcal{F}^{C_0}_{C_1} = \mathcal{F}_{C_0 \cup C_1} \setminus \mathcal{F}_{C_0}$, and the general case is obtained by induction, using (2.1). Hence, the values of μ on $\{\mathcal{F}_C : C \in \mathcal{C}\}$ determine its values on the system $\{\mathcal{F}_{C_1,\dots,C_k} : C_1, \dots, C_k \in \mathcal{C}, k \in \mathbb{N}\}$. The latter is a $\cap$-stable generating system of the σ-algebra $\mathcal{B}(\mathcal{F}')$, hence μ is uniquely determined. □

By Theorem 12.1.1, there exists a sequence $(C_i)_{i \in \mathbb{N}}$ of compact sets with $C_i \uparrow E$, hence with $\mathcal{F}_{C_i} \uparrow \mathcal{F}'$. Therefore, every locally finite measure on $\mathcal{F}'$ is σ-finite.

For later application, we extend Theorem 2.2.1 to functions T not necessarily satisfying $T \le 1$. Instead of a probability measure, we then obtain a locally finite measure.

Theorem 2.3.2. *Let $T : \mathcal{C} \to \mathbb{R}$ be a function with the following properties:*

(a) $T \ge 0$, $T(\emptyset) = 0$,
(b) *if $C_i \downarrow C$ for $C_i, C \in \mathcal{C}$, then $T(C_i) \to T(C)$,*
(c) *$S_k(C_0; C_1, \dots, C_k) \ge 0$ for all $C_i \in \mathcal{C}$ and all $k \in \mathbb{N}$.*

Then there exists a uniquely determined locally finite measure Θ on $\mathcal{F}'$ with

$$T(C) = \Theta(\mathcal{F}_C) \qquad \text{for all } C \in \mathcal{C}. \tag{2.15}$$

Proof. The case $k = 1$ in (c) shows that T is isotone. By Theorem 12.1.1, there is a sequence $(K_m)_{m \in \mathbb{N}}$ of compact sets in E such that $K_m \subset \operatorname{int} K_{m+1}$ for $m \in \mathbb{N}$ and $K_m \uparrow E$. If $T(C) = 0$ for all $C \in \mathcal{C}$, then $\Theta = 0$ is the required measure. Therefore, we can assume $T(K_m) > 0$ for all m. We define

$$T^{(m)}(C) := T(K_m)^{-1}[T(C) + T(K_m) - T(C \cup K_m)] \qquad \text{for } C \in \mathcal{C}.$$

From $S_2(\emptyset; C, K_m) \ge 0$ and the isotony of T it follows that $0 \le T^{(m)} \le 1$; moreover, $T^{(m)}(\emptyset) = 0$. If $C_i \downarrow C$ in $\mathcal{C}$, then $T^{(m)}(C_i) \to T^{(m)}(C)$. Let $S_k^{(m)}$ denote the function derived from $T^{(m)}$ in the same way as S_k is derived from T. Then

$$S_k^{(m)}(C_0; C_1, \dots, C_k) = T(K_m)^{-1} S_{k+1}(C_0; K_m, C_1, \dots, C_k) \tag{2.16}$$

for $C_i \in \mathcal{C}$, as can be seen by induction. Thus $T^{(m)}$ satisfies the assumptions of Theorem 2.2.1. By that theorem, there exists a uniquely determined probability measure $\mathbb{P}^{(m)}$ on $\mathcal{F}$ with

$$\mathbb{P}^{(m)}(\mathcal{F}_C) = T^{(m)}(C) \qquad \text{for } C \in \mathcal{C}.$$

The finite measure $\Theta^{(m)} := T(K_m)\mathbb{P}^{(m)}$ satisfies

$$\Theta^{(m)}(\mathcal{F}_C) = T(C) + T(K_m) - T(C \cup K_m) \qquad \text{for } C \in \mathcal{C}.$$

For $C \in \mathcal{C}$ we get, from (2.2) and (2.16),

$$\begin{aligned}\left(\Theta^{(m+1)} \llcorner \mathcal{F}_{K_m}\right)(\mathcal{F}_C) &= \Theta^{(m+1)}(\mathcal{F}_C \cap \mathcal{F}_{K_m}) \\ &= T(K_{m+1})\mathbb{P}^{(m+1)}\left(\mathcal{F}^{\emptyset}_{C,K_m}\right) \\ &= T(K_{m+1})S_2^{(m+1)}(\emptyset; C, K_m) \\ &= S_3(\emptyset; K_{m+1}, C, K_m) \\ &= T(C) + T(K_m) - T(C \cup K_m) \\ &= \Theta^{(m)}(\mathcal{F}_C).\end{aligned}$$

By the second assertion of Lemma 2.3.1, $\Theta^{(m+1)} \llcorner \mathcal{F}_{K_m} = \Theta^{(m)}$. Since $K_m \uparrow E$ implies $\mathcal{F}_{K_m} \uparrow \mathcal{F}'$, we can define

$$\Theta(A) := \lim_{m\to\infty} \Theta^{(m)}(A) \qquad \text{for } A \in \mathcal{B}(\mathcal{F}').$$

Then $\Theta \geq 0$ and $\Theta(\emptyset) = 0$. Being a monotone limit of measures, Θ is σ-additive: for a disjoint sequence $(A_i)_{i\in\mathbb{N}}$ in $\mathcal{B}(\mathcal{F}')$ we have

$$\begin{aligned}\Theta\left(\bigcup_{i\in\mathbb{N}} A_i\right) &= \lim_{m\to\infty} \Theta^{(m)}\left(\bigcup_{i\in\mathbb{N}} A_i\right) = \lim_{m\to\infty} \sum_{i\in\mathbb{N}} \Theta^{(m)}(A_i) \\ &= \sum_{i\in\mathbb{N}} \lim_{m\to\infty} \Theta^{(m)}(A_i) = \sum_{i\in\mathbb{N}} \Theta(A_i)\end{aligned}$$

by the theorem of monotone convergence (applied to the counting measure on $\mathbb{N}$).

For given $C \in \mathcal{C}$, there is $m \in \mathbb{N}$ with $C \subset K_m$ and thus $\mathcal{F}_C \subset \mathcal{F}_{K_m}$. It follows that

$$\Theta(\mathcal{F}_C) = \Theta^{(m)}(\mathcal{F}_C) = T(C) + T(K_m) - T(C \cup K_m) = T(C).$$

Thus, (2.15) is satisfied and Θ is locally finite. The uniqueness is clear by Lemma 2.3.1. □

Infinitely Divisible Random Sets

We turn our attention to an interesting special class of random closed sets, which we shall later come across again in a different context. In probability theory, when studying limit distributions of sums of independent random variables, one is led to infinitely divisible distributions. The prominent role of Poisson and normal distributions can then be explained by the representation of infinitely divisible distributions as compositions of normal and (generalized) Poisson distributions (see, for example, Araujo and Giné [37, p. 68]). A comparable phenomenon exists for random closed sets, with sums replaced by unions.

Definition 2.3.1. *A random closed set Z in E is called* **infinitely divisible** *if to each $m \in \mathbb{N}$ there are independent, identically distributed random closed sets $Z_1, \ldots, Z_m$ such that*

$$Z \stackrel{\mathcal{D}}{=} Z_1 \cup \ldots \cup Z_m.$$

The point $x \in E$ is called a **fixed point** *of the random closed set Z if $\mathbb{P}(x \in Z) = 1$.*

Let $(x_i)_{i\in\mathbb{N}}$ be a sequence of fixed points of Z with $x_i \to x$. For $m \in \mathbb{N}$, the set $C_m := \operatorname{cl}\{x_i : i \geq m\}$ is compact and satisfies $T_Z(C_m) = 1$. Since $C_m \downarrow \{x\}$, we conclude that $T_Z(\{x\}) = 1$, hence x is a fixed point. Thus the set of fixed points of Z is closed. For that reason we can, if Z is a random closed set with fixed point set $F \neq E$, replace E by the locally compact space $E \cap F^c$ and replace Z by $Z \cap F^c$, which is a random closed set in $E \cap F^c$. Therefore, it is no restriction to assume in the following that Z has no fixed points.

Lemma 2.3.2. *Let Z be a random closed set in E.*

(a) *Z is infinitely divisible if and only if for each $m \in \mathbb{N}$ the function*

$$T^{(m)} := 1 - (1 - T_Z)^{1/m}$$

is an alternating Choquet capacity of infinite order.

(b) *If Z is infinitely divisible and has no fixed points, then*

$$\mathbb{P}(Z \cap C \neq \emptyset) = T_Z(C) < 1 \qquad \text{for } C \in \mathcal{C}.$$

Proof. (a) If Z is infinitely divisible, then to each $m \in \mathbb{N}$ there exist independent, identically distributed random closed sets $Z_1, \ldots, Z_m$ with $Z \stackrel{\mathcal{D}}{=} Z_1 \cup \ldots \cup Z_m$. For $C \in \mathcal{C}$ it follows that

$$\begin{aligned}
1 - T_Z(C) &= 1 - T_{Z_1 \cup \ldots \cup Z_m}(C) \\
&= \mathbb{P}(Z_1 \cap C = \emptyset, \ldots, Z_m \cap C = \emptyset) \\
&= \mathbb{P}(Z_1 \cap C = \emptyset)^m \\
&= (1 - T_{Z_1}(C))^m,
\end{aligned}$$

thus $T_{Z_1} = 1 - (1 - T_Z)^{1/m} = T^{(m)}$. Therefore, $T^{(m)}$ has the properties (a), (b), (c) of Theorems 2.1.2 and 2.2.1, that is, it is an alternating Choquet capacity of infinite order. Conversely, if this is satisfied, then Theorem 2.2.1 ensures the existence of a random closed set $Z^{(m)}$ with capacity functional $T^{(m)}$. Let $Z_1, \dots, Z_m$ be independent copies of $Z^{(m)}$. As just shown, $T_{Z_1 \cup \dots \cup Z_m} = 1 - (1 - T^{(m)})^m$, hence $T_{Z_1 \cup \dots \cup Z_m} = T_Z$, which implies $Z_1 \cup \dots \cup Z_m \stackrel{\mathcal{D}}{=} Z$, by Theorem 2.1.3.

(b) Let Z be infinitely divisible and without fixed points. Assume to the contrary that there exists a set $C \in \mathcal{C}$ with $T_Z(C) = 1$. The system $\mathcal{T} := \{C' \in \mathcal{C} : C' \subset C, T_Z(C') = 1\}$ is ordered by inclusion. Let $\mathcal{S} \subset \mathcal{T}$ be a linearly ordered subset, and put $C_{\mathcal{S}} := \bigcap_{C' \in \mathcal{S}} C'$. In the space $\mathcal{F}(E)$, $C_{\mathcal{S}}$ is an accumulation point of $\mathcal{S}$ and thus the limit of a sequence in $\mathcal{S}$. Hence, there exists a sequence $(C_i)_{i \in \mathbb{N}}$ in $\mathcal{S}$ with $C_i \downarrow C$. This yields $T_Z(C_i) \to T_Z(C_{\mathcal{S}})$. Thus $\mathcal{S}$ has a lower bound in $\mathcal{T}$. By Zorn's lemma, there exists a minimal element $C_0 \in \mathcal{T}$, and $C_0 \neq \emptyset$ because of $T_Z(C_0) = 1$. Since Z has no fixed points, C_0 must contain more than one point. By Theorem 12.1.1, there exist sets $C_1, C_2 \in \mathcal{C}$ with $C_1, C_2 \notin \{\emptyset, C_0\}$ and $C_0 = C_1 \cup C_2$. Since C_0 is minimal, we must have $T_Z(C_1) < 1$ and $T_Z(C_2) < 1$.

Since Z is infinitely divisible, by (a) there exists for each $m \in \mathbb{N}$ a random closed set $Z^{(m)}$ with capacity functional $T^{(m)} = 1 - (1 - T_Z)^{1/m}$. Also $T^{(m)}$ satisfies $T^{(m)}(C_0) = 1$ and $T^{(m)}(C_1) < 1$, $T^{(m)}(C_2) < 1$. From

$$\mathbb{P}_{Z^{(m)}}(\mathcal{F}^{C_1} \cap \mathcal{F}^{C_2}) = \mathbb{P}_{Z^{(m)}}(\mathcal{F}^{C_1 \cup C_2}) = \mathbb{P}_{Z^{(m)}}(\mathcal{F}^{C_0}) = 1 - T_{Z^{(m)}}(C_0) = 0$$

we get

$$\begin{aligned}
\mathbb{P}_{Z^{(m)}}(\mathcal{F}^{C_1} \cup \mathcal{F}^{C_2}) &= \mathbb{P}_{Z^{(m)}}(\mathcal{F}^{C_1}) + \mathbb{P}_{Z^{(m)}}(\mathcal{F}^{C_2}) \\
&= (1 - T_{Z^{(m)}}(C_1)) + (1 - T_{Z^{(m)}}(C_2)) \\
&= (1 - T_Z(C_1))^{1/m} + (1 - T_Z(C_2))^{1/m},
\end{aligned}$$

where both summands are positive. For sufficiently large m, this exceeds one, a contradiction. □

Now we can give an explicit description of the capacity functional of an infinitely divisible random closed set without fixed points.

Theorem 2.3.3. *If Z is an infinitely divisible random closed set without fixed points in E, then there exists a locally finite measure Θ on $\mathcal{F}'$ with*

$$T_Z(C) = 1 - e^{-\Theta(\mathcal{F}_C)}$$

for $C \in \mathcal{C}$.

Proof. Let Z be infinitely divisible and without fixed points. By Lemma 2.3.2, $T^{(m)} := 1 - (1 - T_Z)^{1/m}$ is an alternating Choquet capacity of infinite order $(m \in \mathbb{N})$, and we have $T_Z(C) < 1$ for all $C \in \mathcal{C}$. We deduce that

$$\begin{aligned}
S(C) &:= \lim_{m\to\infty} mT^{(m)}(C) \\
&= \lim_{m\to\infty} m\left[1-(1-T_Z(C))^{1/m}\right] \\
&= -\log(1-T_Z(C)) \\
&< \infty
\end{aligned}$$

for all $C \in \mathcal{C}$, hence

$$T_Z(C) = 1 - e^{-S(C)}. \tag{2.17}$$

We have defined a function S on $\mathcal{C}$ which satisfies the conditions for T in Theorem 2.3.2. In fact, (a) holds by the definition of S, (b) follows from (2.17), and (c) is obtained by taking the limit $m \to \infty$ in the corresponding property of $T^{(m)}$. By Theorem 2.3.2 there exists a locally finite measure Θ on $\mathcal{F}'$ with $\Theta(\mathcal{F}_C) = S(C)$ for all $C \in \mathcal{C}$. This yields

$$T_Z(C) = 1 - e^{-S(C)} = 1 - e^{-\Theta(\mathcal{F}_C)}$$

for $C \in \mathcal{C}$. □

The converse of Theorem 2.3.3 is also true. This will be proved later in Section 3.6, together with a further characterization of infinitely divisible random closed sets.

Note for Section 2.3

Infinitely divisible random closed sets referring to the operation of union are further investigated by Molchanov [543, 548]; he also studies union stable random closed sets. Random closed sets with values in the space of convex bodies of $\mathbb{R}^d$ which are infinitely divisible with respect to Minkowski addition, are characterized by Giné and Hahn [262]; a more special result is due to Mase [451].

2.4 Random Closed Sets in Euclidean Space

In this section, the basic space is $E = \mathbb{R}^d$, the d-dimensional Euclidean space. Random closed sets in $\mathbb{R}^d$ show some additional features, due to the character of $\mathbb{R}^d$ as a linear space and a homogeneous space with respect to the group of isometries.

We begin with a remark on compact sets. The set $\mathcal{C}$ of compact subsets of $\mathbb{R}^d$ is often equipped with the Hausdorff metric δ (see Section 12.3), which is convenient for geometric considerations. By Lemma 2.1.2, $\mathcal{C}$ is a Borel subset of $\mathcal{F}$, with respect to the topology of closed convergence. The following theorem shows that for the Borel σ-algebras on $\mathcal{C}$ it does not matter whether one uses the topology induced from $\mathcal{F}$ or from δ.

Theorem 2.4.1. *The trace σ-algebra $\mathcal{B}(\mathcal{F})_{\mathcal{C}}$ of $\mathcal{B}(\mathcal{F})$ on $\mathcal{C}$ coincides with the Borel σ-algebra $\mathcal{B}(\mathcal{C})$ of $\mathcal{C}$ when $\mathcal{C}$ is equipped with the Hausdorff metric.*

Proof. By Theorem 12.3.2, the topology of $(\mathcal{C}, \delta)$ is finer than the trace topology induced from $\mathcal{F}$. This implies $\mathcal{B}(\mathcal{F})_{\mathcal{C}} \subset \mathcal{B}(\mathcal{C})$. To show the reverse inclusion, let $U_\epsilon(C)$ be a closed ϵ-neighborhood of C in $\mathcal{C}'$. By Theorem 12.3.2, $U_\epsilon(C)$ is closed in $\mathcal{F}$ and hence a Borel set in $\mathcal{F}$, thus $U_\epsilon(C) \in \mathcal{B}(\mathcal{F})_{\mathcal{C}}$. For $C = \emptyset$, we have $U_\epsilon(C) = \{\emptyset\} = \mathcal{F}^{\mathbb{R}^d} \in \mathcal{B}(\mathcal{F})_{\mathcal{C}}$. There is a countable system of such neighborhoods generating the topology of $\mathcal{C}$, for example all $U_\epsilon(C)$ where ϵ is rational and C is a set of finitely many points with rational coordinates. It follows that every open set in $\mathcal{C}$ belongs to $\mathcal{B}(\mathcal{F})_{\mathcal{C}}$. Hence, the Borel sets in $\mathcal{C}$ are also Borel sets in $\mathcal{F}$. □

For our later investigations, we need further special subsystems of $\mathcal{F}$, whose elements have simple geometric properties. One of these systems is the set $\mathcal{K}$ of convex bodies (compact convex subsets) of $\mathbb{R}^d$. We emphasize that $\emptyset \in \mathcal{K}$ in this book, which is convenient, but differs from common usage. We denote by $\mathcal{R}$ the **convex ring**, whose elements are the finite unions of convex bodies, also called **polyconvex sets.** The **extended convex ring** is the system

$$\mathcal{S} := \{F \in \mathcal{F} : F \cap K \in \mathcal{R} \text{ for all } K \in \mathcal{K}\}.$$

The elements of $\mathcal{S}$ are the countable unions of convex bodies with the property that every compact set hits only finitely many of the bodies. Clearly, $\mathcal{K} \subset \mathcal{R} \subset \mathcal{S} \subset \mathcal{F}$ and $\mathcal{R} \subset \mathcal{C} \subset \mathcal{F}$, where each inclusion is strict. However, $\operatorname{cl} \mathcal{R} = \mathcal{F}$; in fact, every element of $\mathcal{F}$ is the limit of a sequence of finite sets.

Theorem 2.4.2. *$\mathcal{K}$, $\mathcal{R}$ and $\mathcal{S}$ are Borel sets in $\mathcal{F}$.*

Proof. The set $\mathcal{K} = \{C \in \mathcal{C} : C = \operatorname{conv} C\}$ is closed, by the continuity of the map $C \mapsto \operatorname{conv} C$ (Theorem 12.3.5); hence, it is a Borel set in $\mathcal{C}$ and thus, by Theorem 2.4.1, also in $\mathcal{F}$.

For $k, m \in \mathbb{N}$ let

$$\mathcal{R}_k^m := \{K_1 \cup \ldots \cup K_m : \mathcal{K} \ni K_i \subset kB^d \text{ for } i = 1, \ldots, m\}.$$

The set $\mathcal{R}_k^1 = \mathcal{K} \cap \mathcal{F}^{(kB^d)^c}$ is closed in $\mathcal{C}$ and hence, by Theorem 12.3.2, also in $\mathcal{F}$, thus it is compact. Together with Theorem 12.2.3 and induction, this yields the compactness of $\mathcal{R}_k^m$. Since

$$\mathcal{R} = \bigcup_{k \in \mathbb{N}} \bigcup_{m \in \mathbb{N}} \mathcal{R}_k^m,$$

$\mathcal{R}$ is a Borel set.

The set

$$\mathcal{S}_k := \{F \in \mathcal{F} : F \cap kB^d \in \mathcal{R}\}$$

is the pre-image of the Borel set $\mathcal{R}$ under the map $F \mapsto F \cap kB^d$. The latter is upper semicontinuous by Theorem 12.2.6 and hence measurable. Therefore, $\mathcal{S}_k$ is a Borel set, and since $\mathcal{S} = \bigcap_{k \in \mathbb{N}} \mathcal{S}_k$, the same holds for $\mathcal{S}$. □

Other geometrically defined subsets of $\mathcal{F}$ are the Grassmannian $G(d,k)$ and the affine Grassmannian $A(d,k)$. Both are Borel sets in $\mathcal{F}$, as follows from Section 13.2.

If a random closed set Z in $\mathbb{R}^d$ satisfies $Z \in \mathcal{C}$ almost surely, we call it a **random compact set**. Similarly, **random convex bodies**, **random $\mathcal{R}$-sets** (or **random polyconvex sets**) and **random $\mathcal{S}$-sets** are defined. The random closed set Z is a **random k-subspace**, respectively a **random k-flat**, if its distribution $\mathbb{P}_Z$ is concentrated on $G(d,k)$, respectively $A(d,k)$.

From the continuity and lower semicontinuity results of Theorem 12.3.1, the following is deduced.

Theorem 2.4.3. *Let Z and Z' be random closed sets in $\mathbb{R}^d$. Then also αZ for $\alpha \geq 0$, $-Z$, $\mathrm{cl}\,(Z+Z')$ and $\mathrm{cl\,conv}\,Z$ are random closed sets. If Z is compact, then $\mathrm{conv}\,Z$ and $Z+Z'$ are random closed sets.*

If Z is a random closed set in $\mathbb{R}^d$ and E is a fixed affine subspace of $\mathbb{R}^d$, then the intersection $Z \cap E$ is a random closed set (in $\mathbb{R}^d$ and in E). Also every parallel set $Z + \epsilon B^d$, $\epsilon > 0$, is a random closed set.

Now we consider invariance properties for random closed sets in $\mathbb{R}^d$; they will play an essential role later on.

Definition 2.4.1. *The random closed set Z in $\mathbb{R}^d$ is called* **stationary** *if $Z + t \overset{\mathcal{D}}{=} Z$ for all $t \in \mathbb{R}^d$. It is called* **isotropic** *if $\vartheta Z \overset{\mathcal{D}}{=} Z$ for all rotations $\vartheta \in SO_d$.*

Theorem 2.4.4. *A nonempty stationary random closed set Z is a.s. unbounded. A stationary convex random closed set attains a.s. only the values $\emptyset$ and $\mathbb{R}^d$.*

Proof. If Z is a stationary random closed set in $\mathbb{R}^d$, then

$$\mathrm{cl\,conv}\,(Z+t) = (\mathrm{cl\,conv}\,Z) + t \qquad \text{for } t \in \mathbb{R}^d$$

shows that also $\mathrm{cl\,conv}\,Z$ is stationary. Therefore, it suffices to prove the second assertion.

Let Z be a stationary convex random closed set, and assume that $\mathbb{P}(Z \notin \{\emptyset, \mathbb{R}^d\}) > 0$. Let $0 < \alpha < \pi/2$. For $x, y \in \mathbb{R}^d$ with $y \neq 0$, let

$$K(x,y) := \{z \in \mathbb{R}^d : \langle z, y\rangle \geq \|z\| \cos\alpha\} + x;$$

this is a convex cone with apex x. We assert that there are rational vectors $x, y \in \mathbb{Q}^d$, $y \neq 0$, with

$$\mathbb{P}(\emptyset \neq Z \cap K(x,y) \subset x + \|y\| B^d) =: p > 0. \tag{2.18}$$

Suppose this were false. Then

$$\mathbb{P}\left(\bigcup_{x,y\in\mathbb{Q}^d,\,y\neq 0}\{\emptyset\neq Z\cap K(x,y)\subset x+\|y\|B^d\}\right)=0. \tag{2.19}$$

For every $\omega\in\Omega$ with $Z(\omega)\notin\{\emptyset,\mathbb{R}^d\}$ we have $\operatorname{bd} Z(\omega)\neq\emptyset$, hence there are a point $x\in\operatorname{bd} Z(\omega)$ and a supporting hyperplane H of $Z(\omega)$ at x. Let y be an outer normal vector of H. Then $Z(\omega)\cap K(x,y)=\{x\}$ and hence

$$\emptyset\neq Z(\omega)\cap K(x,y)\subset x+\|y\|B^d.$$

This inclusion can also be satisfied with suitable vectors $x,y\in\mathbb{Q}^d$, $y\neq 0$. Now (2.19) implies $\mathbb{P}(Z\notin\{\emptyset,\mathbb{R}^d\})=0$, a contradiction. Hence, there are $x,y\in\mathbb{Q}^d$, $y\neq 0$, satisfying (2.18).

Now we consider, for $k\in\mathbb{N}_0$, the events

$$\begin{aligned}A_k&:=\{\emptyset\neq Z\cap K(x+2ky,y)\subset x+2ky+\|y\|B^d\}\\&=\{\emptyset\neq(Z-2ky)\cap K(x,y)\subset x+\|y\|B^d\}.\end{aligned}$$

Since Z is stationary, we have $\mathbb{P}(A_k)=p$, hence

$$\sum_{k\in\mathbb{N}_0}\mathbb{P}(A_k)=\infty.$$

But the events A_k are pairwise disjoint, which yields

$$\sum_{k\in\mathbb{N}_0}\mathbb{P}(A_k)=\mathbb{P}\left(\bigcup_{k\in\mathbb{N}_0}A_k\right)\leq 1,$$

a contradiction. This completes the proof. □

To obtain nontrivial examples of stationary and isotropic random closed sets, we can proceed as follows. The set

$$C^d:=\{x=(x^1,\dots,x^d)\in\mathbb{R}^d:0\leq x^i\leq 1,\,i=1,\dots,d\}$$

is the unit cube. Let Z' be a compact set, deterministic or random, which is a.s. contained in C^d, and let

$$\widetilde{Z}:=\bigcup_{z\in\mathbb{Z}^d}(Z'+z).$$

Let ξ be a random vector, independent of Z', which is uniformly distributed in C^d (that is, its distribution is $\lambda\llcorner C^d$, the restriction of the Lebesgue measure to C^d). Then

$$Z:=\widetilde{Z}+\xi$$

is a stationary random closed set. More generally, one could replace Z' by an independent sequence $Z'_1,Z'_2,\dots$ with $\mathbb{P}_{Z'_i}=\mathbb{P}_{Z'}$ and define

$$\widetilde{Z} := \bigcup_{i \in \mathbb{N}} (Z_i' + z_i),$$

where $\{z_1, z_2, \dots\}$ is an enumeration of $\mathbb{Z}^d$.

Starting from a stationary random closed set Z, we obtain a stationary and isotropic random closed set ϑZ by taking for ϑ a random rotation, independent of Z and with uniform distribution (that is, a random variable with values in SO_d, whose distribution is the unique rotation invariant probability measure ν on SO_d; see Section 13.2).

The stationary random closed sets obtained in this way have too special a structure to be of interest and use for applications. More flexible constructions will be developed later. Nevertheless, the above considerations show that for the theoretical investigation of a structure observed in some 'observation window' it is often legitimate to assume that the observed specimen comes from a realization of a stationary and isotropic random set, if only position and orientation of the observation window have been chosen in a suitable random fashion.

The invariance properties of a random closed set are reflected in corresponding properties of its capacity functional.

Theorem 2.4.5. *The random closed set Z in $\mathbb{R}^d$ is stationary if and only if its capacity functional T_Z is translation invariant, and it is isotropic if and only if T_Z is rotation invariant.*

Proof. The first assertion follows from

$$T_{Z+t}(C) = \mathbb{P}((Z+t) \cap C \neq \emptyset) = \mathbb{P}(Z \cap (C-t) \neq \emptyset) = T_Z(C-t)$$

for $t \in \mathbb{R}^d$, the second from

$$T_{\vartheta Z}(C) = \mathbb{P}(\vartheta Z \cap C \neq \emptyset) = \mathbb{P}(Z \cap \vartheta^{-1} C \neq \emptyset) = T_Z(\vartheta^{-1} C)$$

for $\vartheta \in SO_d$, together with Theorem 2.1.3. □

We give a further example showing how properties of a random closed set can possibly be read off from its capacity functional. The following theorem characterizes the capacity functionals of almost surely convex random closed sets.

For sets $K, K', C \subset \mathbb{R}^d$ we say that C **lies between** K and K' if every segment $[x, x']$ joining two points $x \in K$, $x' \in K'$ satisfies $[x, x'] \cap C \neq \emptyset$.

Theorem 2.4.6. *For a random closed set Z in $\mathbb{R}^d$ and its capacity functional T the following assertions are equivalent:*

(a) *Z is almost surely convex.*
(b) *If $K, K', C \in \mathcal{C}$ and C lies between K and K', then*

$$T(K \cup K' \cup C) + T(C) = T(K \cup C) + T(K' \cup C).$$

(c) *The functional T is additive on $\mathcal{K}$, that is,*

$$T(K \cup K') + T(K \cap K') = T(K) + T(K')$$

holds for all $K, K' \in \mathcal{K}$ with $K \cup K' \in \mathcal{K}$.

Proof. First we prove the implication (a) $\Rightarrow$ (b). Let $K, K', C \in \mathcal{C}$ be compact sets such that C lies between K and K'. If $Z(\omega)$ is a convex realization of Z and if

$$Z(\omega) \cap K \neq \emptyset, \quad Z(\omega) \cap K' \neq \emptyset,$$

then also $Z(\omega) \cap C \neq \emptyset$. It follows that

$$\mathbb{P}_Z(\mathcal{F}^C_{K,K'}) = 0,$$

thus $S_2(C; K, K') = 0$ by (2.2) and hence

$$-T(C) + T(C \cup K) + T(C \cup K') - T(C \cup K \cup K') = 0.$$

This shows that (a) implies (b).

Suppose that (b) holds. Let $K, K' \in \mathcal{K}$ be convex bodies such that $K \cup K' \in \mathcal{K}$. Then $K \cap K'$ lies between K and K'. In fact, suppose that $x \in K$ and $x' \in K'$, then $[x, x'] \cap K$ and $[x, x'] \cap K'$ are closed and not empty, and their union is $[x, x']$. It follows that $[x, x'] \cap K \cap K' \neq \emptyset$. Now (b), applied to $C := K \cap K'$, shows that (c) holds.

To prove the implication (c) $\Rightarrow$ (a), let $F \in \mathcal{F}$ be a set which is not convex. Then there are points $x, x' \in F$ with $[x, x'] \cap F^c \neq \emptyset$, and we can choose a ball $B(y_0, \epsilon)$ with rational center y_0 and rational positive radius ϵ such that

$$B(y_0, \epsilon) \subset F^c \qquad \text{and} \qquad [x, x'] \cap \operatorname{int} B(y_0, \epsilon) \neq \emptyset.$$

Because of the second relation, there are rational points $x_0, x_0' \in \mathbb{R}^d$ with $y_0 \in [x_0, x_0']$, $x \in B(x_0, \epsilon)$, $x' \in B(x_0', \epsilon)$. Putting

$$\begin{aligned} C &:= \operatorname{conv}\left(B(x_0, \epsilon) \cup B(y_0, \epsilon)\right), \\ C' &:= \operatorname{conv}\left(B(x_0', \epsilon) \cup B(y_0, \epsilon)\right), \end{aligned}$$

we have $C, C', C \cup C' \in \mathcal{K}$ and $F \in \mathcal{F}^{C \cap C'}_{C,C'}$. From (c) we get

$$\mathbb{P}_Z\left(\mathcal{F}^{C \cap C'}_{C,C'}\right) = -T(C \cap C') + T(C) + T(C') - T(C \cup C') = 0$$

and thus $\mathbb{P}_Z\left(\bigcup \mathcal{F}^{C \cap C'}_{C,C'}\right) = 0$, where the union extends over all the countably many possible pairs C, C'. Hence, with probability 1 we have $Z \notin \bigcup \mathcal{F}^{C \cap C'}_{C,C'}$, which means that Z is almost surely convex. $\square$

Finally in this chapter, we consider first examples of simple parameters for a quantitative description of a random closed set in $\mathbb{R}^d$.

In the following, $\mathbb{E}$ denotes mathematical expectation. Since one can replace the random closed set Z by the stochastic process (with parameter domain $\mathbb{R}^d$) given by its indicator function, it is natural to define a mean value function of Z by

$$m(x) := \mathbb{E}\mathbf{1}_Z(x) \qquad \text{for } x \in \mathbb{R}^d.$$

Explicitly,

$$m(x) = \int_\Omega \mathbf{1}_{Z(\omega)}(x)\,\mathbb{P}(\mathrm{d}\omega) = \int_{\mathcal{F}} \mathbf{1}_F(x)\,\mathbb{P}_Z(\mathrm{d}F).$$

The measurability of the integrand follows from Theorem 12.2.7. We may also write

$$m(x) = \mathbb{P}(x \in Z).$$

Further, the **covariance function** k of Z is defined by

$$k(x,y) := \mathbb{E}(\mathbf{1}_Z(x) - m(x))(\mathbf{1}_Z(y) - m(y)) \qquad \text{for } x, y \in \mathbb{R}^d.$$

For stationary Z, the function m is constant, thus

$$m(x) = m(0) =: p,$$

and k satisfies

$$k(x,y) = k(x-y,0).$$

Theorem 2.4.7. *If Z is a stationary random closed set in $\mathbb{R}^d$, then*

$$p = \mathbb{P}(0 \in Z) = T_Z(\{0\}) = \mathbb{E}\lambda(Z \cap C^d)$$

and

$$k(x,0) = \mathbb{P}(0 \in Z, x \in Z) - p^2 = \mathbb{E}\lambda(Z \cap (Z-x) \cap C^d) - p^2.$$

Proof. By definition,

$$p = \mathbb{E}\mathbf{1}_Z(0) = \mathbb{P}(0 \in Z) = T_Z(\{0\}).$$

By Fubini's theorem,

$$p = \int_{C^d} \mathbb{E}\mathbf{1}_Z(x)\,\lambda(\mathrm{d}x) = \mathbb{E}\int_{C^d} \mathbf{1}_Z(x)\,\lambda(\mathrm{d}x) = \mathbb{E}\lambda(Z \cap C^d).$$

(The measurability of the map $(F,x) \mapsto \mathbf{1}_F(x)$ from $\mathcal{F} \times \mathbb{R}^d$ to $\mathbb{R}$ is proved in Theorem 12.2.7.) The second assertion is obtained similarly. □

For the quantity p, several names are in use: it has been called the 'volume fraction', the 'volume density', the 'intensity of the volume', or the 'specific volume' of the stationary random closed set. We shall talk here of the **volume density** and also of the **specific volume**. We also write $\overline{V}_d(Z) := p$ for the specific volume, and we remark that the above proof more generally shows that

$$\overline{V}_d(Z) = \frac{\mathbb{E}\lambda(Z \cap B)}{\lambda(B)} \tag{2.20}$$

for every Borel set $B \subset \mathbb{R}^d$ with $0 < \lambda(B) < \infty$.

For a stationary random closed set Z, the function defined by

$$C(x) := \mathbb{P}(0 \in Z,\, x \in Z), \qquad x \in \mathbb{R}^d,$$

is called the **covariance** of Z. For $x \in \mathbb{R}^d$, $C(x)$ is the specific volume of the random closed set $Z \cap (Z - x)$. If Z is isotropic, then $C(x)$ depends only on the norm $\|x\|$.

A further means of quantifying size and shape aspects of a stationary random closed set are the contact distributions. Let $K \in \mathcal{K}'$ be a convex body containing the origin, and for $F \in \mathcal{F}$ and $x \in \mathbb{R}^d$ let

$$d_K(x, F) := \min\{r \geq 0 : (x + rK) \cap F \neq \emptyset\}; \tag{2.21}$$

this is the **K-distance** of x from F (with $\min \emptyset := \infty$). For given K and x, the function $F \mapsto d_K(x, F)$ is lower semicontinuous and hence measurable. Now let Z be a stationary random closed set in $\mathbb{R}^d$ with specific volume $p < 1$ and define

$$H_K(r) := \mathbb{P}(d_K(0, Z) \leq r \mid 0 \notin Z) = \mathbb{P}(0 \in Z - rK \mid 0 \notin Z)$$

for $r \geq 0$. Thus, $H_K(r)$ is the distribution function of the K-distance of 0 from Z, conditional to $0 \notin Z$ (due to the stationarity, the point 0 could be replaced by any other point x). The function H_K is called the **contact distribution function** of Z (with respect to the **structuring element** or **gauge body** K). By definition,

$$\begin{aligned} H_K(r) &= 1 - \mathbb{P}(0 \notin Z - rK \mid 0 \notin Z) \\ &= 1 - \frac{\mathbb{P}(0 \notin Z - rK)}{\mathbb{P}(0 \notin Z)} = 1 - \frac{\mathbb{P}(0 \notin Z - rK)}{1 - p} \end{aligned}$$

and thus

$$H_K(r) = 1 - \frac{1 - \overline{V}_d(Z - rK)}{1 - \overline{V}_d(Z)} = \frac{\overline{V}_d(Z - rK) - \overline{V}_d(Z)}{1 - \overline{V}_d(Z)}.$$

Thus, the quantity $H_K(r)$ can be estimated by estimating specific volumes.

Two special cases of K are particularly relevant for applications. For $K = B^d$ we get the **spherical contact distribution function**, denoted by H; this is the distribution function of the Euclidean distance of 0 from the set Z under the condition that $0 \notin Z$. The case $K = [0, u]$ with a unit vector u (and $[x, y]$ denoting the closed line segment with endpoints x and y) gives the **linear contact distribution function** $H_{[0,u]}$ of Z in the direction u.

Notes for Section 2.4

1. Limit theorems. For random compact sets Z in $\mathbb{R}^d$, the operation of Minkowski addition motivates the search for notions and results which are in analogy to the classical addition of real- or vector-valued random variables. The characterization of random convex bodies which are infinitely divisible with respect to Minkowski addition, mentioned in the Note for Section 2.3, is an example of that kind. The first result in this direction was the **strong law of large numbers** for random compact sets, established by Artstein and Vitale [40] (here we exclude the trivial case of constant random sets):

Let $Z, Z_1, Z_2, \ldots$ be a sequence of independent, identically distributed (i.i.d.) random compact sets in $\mathbb{R}^d$ with $\mathbb{E}\|Z\| < \infty$. Then

$$\frac{1}{n}(Z_1 + \ldots + Z_n) \to \mathbb{E}Z \quad a.s.,$$

as $n \to \infty$.

Here, convergence is in the Hausdorff metric, $\|Z\| := \delta(\{0\}, Z)$, and the expectation $\mathbb{E}Z$ of a random compact set Z is defined as

$$\mathbb{E}Z := \{\mathbb{E}\xi \;:\; \xi : \Omega \to \mathbb{R}^d \text{ measurable with } \xi \in Z \text{ a.s.}\}.$$

Measurable mappings $\xi : \Omega \to \mathbb{R}^d$ satisfying $\xi \in Z$ a.s. are called **measurable selections** of Z. Since $\mathbb{E}Z$ is the set of expectation vectors of all measurable selections of Z, it is also called the **selection expectation** (or **Aumann expectation**) (see also Subsection 8.2.4). The condition $\mathbb{E}\|Z\| < \infty$ implies that $\mathbb{E}Z$ is compact. Moreover, if the underlying probability space $(\Omega, \mathbf{A}, \mathbb{P})$ does not have atoms (which in the above situation is guaranteed by the existence of an i.i.d sequence of non-trivial sets $Z, Z_1, Z_2, \ldots$), then $\mathbb{E}Z$ is convex and its support function satisfies

$$h(\mathbb{E}Z, \cdot) = \mathbb{E}h(Z, \cdot)$$

(for a compact set C, the support functions of C and $\operatorname{conv} C$ are the same).

For random convex bodies Z_i, the strong law of large numbers follows from a corresponding result for Banach-space-valued random variables (see, e.g., the book by Araujo and Giné [37]) since $h(Z_i, \cdot)$ is a random element of $\mathbf{C}(S^{d-1})$ and the mapping $K \mapsto h(K, \cdot)$ is linear and injective on $\mathcal{K}'$. The extension from random convex bodies to random compact sets uses the fact that the Minkowski addition is a convexifying operation, as expressed in the Shapley–Folkman–Starr theorem (see, e.g., Schneider [695, Theorem 3.1.6]). Namely, for $C_1, \ldots, C_n \in \mathcal{C}$,

$$\delta(C_1 + \ldots + C_n, \operatorname{conv}(C_1 + \ldots + C_n)) \le \sqrt{d} \max_{i=1,\ldots,n} \|C_i\|.$$

In a similar vein, a **central limit theorem** for random compact sets was proved independently by Weil [784] and Giné, Hahn and Zinn [263]. It requires that Z is square integrable, that is $\mathbb{E}\|Z\|^2 < \infty$, and involves the covariance function

$$\Gamma_\zeta : (u,v) \mapsto \mathbb{E}[\zeta(u)\zeta(v)] - \mathbb{E}\zeta(u)\mathbb{E}\zeta(v)$$

of a random element $\zeta \in \mathbf{C}(S^{d-1})$.

Let $Z, Z_1, Z_2, \ldots$ be a sequence of square integrable and i.i.d. random compact sets in $\mathbb{R}^d$. Then

$$\sqrt{n}\delta\left(\frac{1}{n}(Z_1 + \ldots + Z_n), \mathbb{E}Z\right) \to \max_{x \in S^{d-1}} \|\zeta(x)\|,$$

in distribution, as $n \to \infty$. Here, ζ is a centered Gaussian variable in $\mathbf{C}(S^{d-1})$ with covariance function $\Gamma_\zeta = \Gamma_{h(Z,\cdot)}$.

Subsequently, various further results have been obtained (ergodic theorems, laws of the iterated logarithm, characterization of stable random sets), and also extensions to random compact or compact and convex sets in Banach spaces have been considered. We refer to Molchanov [548, ch. 3] for details and references.

It should be mentioned that the Gaussian law appearing in such limit results does not allow a simple geometric interpretation, in general. This is due to the fact that a Gaussian measure on the convex cone $\mathcal{K}'$ (which corresponds to the closed convex cone of support functions in $\mathbf{C}(S^{d-1})$) is degenerate (i.e., concentrated on points). This was shown by Lyashenko [444] and Vitale [768].

For a random compact set Z in $\mathbb{R}^d$, Vitale [770] proved a **Brunn–Minkowski theorem** showing that

$$\lambda^{1/d}(\mathbb{E}Z) \geq \mathbb{E}\lambda^{1/d}(Z).$$

2. Set-valued expectations. For random compact sets Z, the Aumann expectation mentioned in Note 1 is especially adapted to Minkowski addition and the representation by support functions. Various other notions of set-valued expectations have been studied in the literature (e.g. the Fréchet, Vorob'ev and Herer expectations), some of which make sense in more general spaces E (even without a vector space structure). We refer to Molchanov [548, ch. 2], for a detailed discussion. Such set-valued expectations are of particular interest for structures like Boolean models, which will be treated later, since the estimation of a mean particle is a first step to adapt a random set model to a given spatial structure. Unfortunately, for random compact sets Z which are isotropic, most of the mentioned expectations (including the Aumann expectation) yield balls, hence they do not reflect the shapes attained by Z. This problem was discussed by Stoyan and Molchanov [744], who proposed to transform the random sets into a standard form before performing an average.

3

Point Processes

The notion of a random closed set, as developed so far, is still very general. To obtain tractable models for applications, one has to restrict the admissible set classes suitably. One possibility consists in considering sets which are generated as the union set of a countable family of simpler sets, such as compact sets, convex bodies, curves, lines, or flats. The appropriate notion for randomizing such families is that of a point process in a space of geometric objects. Point processes are, besides random sets, the second basic object of stochastic geometry. In many applications, the 'points' of the process are ordinary points of $\mathbb{R}^d$, but in others, like those employing random closed sets, the 'points' may themselves be sets. For that reason, we study point processes in a general locally compact space E.

The basic idea of a point process in E is that of a random collection of isolated points in E. Therefore, a point process in E could be defined as a random closed set in E which is almost surely locally finite. This leads to a 'simple point process' in E. For some purposes, however, this model is too narrow, since constructions leading to point processes may produce 'points with multiplicity'. One can deal with this problem by introducing 'marked point processes', where the multiplicity of a point is attached to it as its 'mark'. Another possibility consists in replacing a locally finite set X by the measure which attains the value one at each point of X and is zero elsewhere, a 'counting measure'. Then multiplicities can easily be treated, by allowing nonnegative integer values at single points for the measure. In this way, point processes appear as special random measures.

As this measure-theoretic approach has some advantages, we shall introduce point processes as random counting measures and will therefore treat general random measures as well. This is done in Section 3.1. As for point processes, a particular role is played by the Poisson processes; they are the subject of Section 3.2. Palm distributions of random measures and point processes are treated, with two different approaches, in Sections 3.3 and 3.4. Section 3.5, on marked point processes, lays some foundations for the treatment of geometric models. Marked point processes also allow us to discuss some special types of point processes in $\mathbb{R}^d$. Finally in this chapter, we consider point processes of closed sets.

3.1 Random Measures and Point Processes

As in Chapter 2, we assume that a locally compact space E with a countable base is given. Its Borel σ-algebra is denoted by $\mathcal{B} = \mathcal{B}(E)$. The argument E will often be deleted, also in similar terms.

Let $\mathsf{M} = \mathsf{M}(E)$ be the set of all Borel measures η on E which are **locally finite**, that is, satisfy $\eta(C) < \infty$ for all $C \in \mathcal{C}$. Note that this implies that the measures $\eta \in \mathsf{M}$ are σ-finite. We supply M with the σ-algebra $\mathcal{M}$ generated by the evaluation maps

$$\begin{aligned} \Phi_A : \mathsf{M} &\to \mathbb{R} \cup \{\infty\} \\ \eta &\mapsto \quad \eta(A) \end{aligned}$$

with $A \in \mathcal{B}$. Thus, $\mathcal{M}$ is the smallest σ-algebra for which all maps Φ_A, $A \in \mathcal{B}$, are measurable. To obtain a convenient generating system, we denote by $\mathcal{G}_c$ the system of open, relatively compact subsets of E, and for $A \in \mathcal{B}$ and $r \geq 0$ we define

$$\mathsf{M}_{A,r} := \{\eta \in \mathsf{M} : \eta(A) \leq r\}.$$

Lemma 3.1.1. *The σ-algebra $\mathcal{M}$ is generated by the system*

$$\mathcal{E} := \{\mathsf{M}_{G,r} : G \in \mathcal{G}_c,\ r \geq 0\}.$$

Proof. Let $\mathcal{M}'$ be the σ-algebra that $\mathcal{E}$ generates in M. It is easy to check that

$$\mathcal{A} := \{A \in \mathcal{B} : \Phi_{A \cap G} \text{ is } \mathcal{M}'\text{-measurable for all } G \in \mathcal{G}_c\}$$

is a Dynkin system. For $G \in \mathcal{G}_c$ and $r \geq 0$, we have $\Phi_G^{-1}([0,r]) = \mathsf{M}_{G,r} \in \mathcal{M}'$, hence $\mathcal{A}$ contains the Dynkin system generated by $\mathcal{G}_c$. Since $\mathcal{G}_c$ is $\cap$-stable, this Dynkin system is equal to the σ-algebra generated by $\mathcal{G}_c$, which is $\mathcal{B}$. Thus, for every Borel set $A \in \mathcal{B}$, all functions $\Phi_{A \cap G}$, $G \in \mathcal{G}_c$, are $\mathcal{M}'$-measurable. Since in $\mathcal{G}_c$ there exists a sequence increasing to E, also Φ_A is $\mathcal{M}'$-measurable. As $\mathcal{M}$ is the smallest σ-algebra in M with this property, we conclude that $\mathcal{M} \subset \mathcal{M}'$. From $\mathsf{M}_{G,r} \in \mathcal{M}$ for $G \in \mathcal{G}_c$ and $r \geq 0$ we now get $\mathcal{M} = \mathcal{M}'$, as asserted. □

A special class of measures on E is given by the counting measures. A **counting measure** on E is a measure $\eta \in \mathsf{M}$ with $\eta(A) \in \mathbb{N}_0 \cup \{\infty\}$ for all $A \in \mathcal{B}$ (observe that a counting measure is locally finite by definition). Let N be the set of all counting measures on E. The following lemma shows that N is a measurable subset of M. By $\mathcal{N}$ we then denote the trace σ-algebra of $\mathcal{M}$ on N.

Lemma 3.1.2. *We have* $\mathsf{N} \in \mathcal{M}$. *The trace σ-algebra* $\mathcal{N}$ *is generated by the system*

$$\mathcal{E}' := \{\mathsf{N}_{G,k} : G \in \mathcal{G}_c,\ k \in \mathbb{N}_0\}$$

with

$$\mathsf{N}_{G,k} := \{\eta \in \mathsf{N} : \eta(G) = k\}.$$

Proof. By Theorem 12.1.1, there exists a countable base $\{D_1, D_2, \ldots\}$ of the topology, consisting of open relatively compact sets $D_i \subset E$. Then,

$$\mathsf{N} = \bigcap_{i=1}^{\infty} \{\eta \in \mathsf{M} : \eta(D_i) \in \mathbb{N}_0\},$$

as is easy to see. Namely, for a measure η in the right-hand set, we have $\eta(D) \in \mathbb{N}_0 \cup \{\infty\}$, for every open set D, and thus for every Borel set, by the usual extension argument.

The assertion about the generating system follows as in the proof of Lemma 3.1.1. □

Examples of counting measures are the locally finite sums of Dirac measures,

$$\eta = \sum_{i=1}^{k} \delta_{x_i}, \qquad k \in \mathbb{N}_0 \cup \{\infty\},$$

where the x_i are points in E. Here, for $x \in E$, the Dirac measure δ_x is defined by

$$\delta_x(A) := \begin{cases} 1, & \text{if } x \in A, \\ 0, & \text{if } x \notin A, \end{cases}$$

for $A \in \mathcal{B}$. It is a probability measure on E. We do not assume here that the x_i are pairwise distinct; thus $\eta(\{x\}) > 1$ is possible. If $k = 0$, then $\eta = 0$, the zero measure.

The following lemma shows that every counting measure is such a finite or countable sum of Dirac measures. It even allows us to enumerate the corresponding points $x_1, x_2, \ldots$ in a measurable way.

Lemma 3.1.3. *There exist measurable mappings* $\zeta_i : \mathsf{N} \to E$ *such that*

$$\eta = \sum_{i=1}^{\eta(E)} \delta_{\zeta_i(\eta)}$$

for $\eta \in \mathsf{N}$.

Proof. We use a metric generating the topology of E. From the proof of Theorem 12.1.1, we obtain for each $k \in \mathbb{N}$ the existence of a sequence $A_1^k, A_2^k, \ldots$ of pairwise disjoint, relatively compact Borel sets in E of diameters less than $1/k$ and such that $E = \bigcup_{i \in \mathbb{N}} A_i^k$.

Let $\eta \in \mathsf{N}$. An **atomic pair** (x, m) of η consists of a point $x \in E$ with $\eta(\{x\}) > 0$ (an **atom**) and the multiplicity $m = \eta(\{x\}) \in \mathbb{N}$. Since η is finite on compact sets, it has at most countably many atoms. Let $B \in \mathcal{B}(E)$ be a Borel set with $\eta(B) > 0$ and hence $\eta(B) \in \mathbb{N} \cup \{\infty\}$. We can inductively define a sequence $i_1, i_2, \dots$ such that $\eta(B \cap A^1_{i_1} \cap \dots \cap A^r_{i_r}) \in \mathbb{N}$ for $r \in \mathbb{N}$. It follows that $\eta\left(B \cap \bigcap_{k\in\mathbb{N}} A^k_{i_k}\right) \in \mathbb{N}$. Since the intersection has diameter zero, B contains an atom. Now it is clear that

$$\eta = \sum_{(x,m)} m\,\delta_x$$

where the sum runs over the atomic pairs (x, m) of η.

For $x \in E$, the relations

$$x \in A^k_{j_k(x)}, \qquad k \in \mathbb{N},$$

define uniquely a sequence $(j_1(x), j_2(x), \dots)$ in $\mathbb{N}$. By

$$x \prec y :\Leftrightarrow \quad (j_1(x), j_2(x), \dots) \le (j_1(y), j_2(y), \dots),$$

where $\le$ on the right side denotes the lexicographical order, we define a linear order $\prec$ on E. We construct, for each $p \in \mathbb{N}$, a measurable map $\zeta_p : \mathsf{N} \to E$. It will associate with every counting measure its pth atom (counted with multiplicity). Let $\eta \in \mathsf{N}$, and let (x, m) be an atomic pair of η. All atoms y of η with $y \prec x$, $y \neq x$, lie in the relatively compact set $\bigcup_{i=1}^{j_1(x)} A^1_i$, hence their number is finite and so is the sum of their multiplicities, say n. We define $\zeta_{n+j}(\eta) := x$ for $j = 1, \dots, m$. If this is done for all atomic pairs (x, m) of η, then $\zeta_p(\eta)$ is defined, for all $p \in \mathbb{N}$ if $\eta(E) = \infty$, respectively for $p = 1, \dots, q$, if $\eta(E) = q < \infty$. In the latter case, we put $\zeta_p(\eta) := a$ for $p > q$, where $a \in E$ is an arbitrary given point. Now for $p \in \mathbb{N}$ and $B \in \mathcal{B}(E)$, the set $\{\eta \in \mathsf{N} : \eta(E) < p,\ \zeta_p(\eta) \in B\}$ is either empty or equal to $\{\eta \in \mathsf{N} : \eta(E) < p\}$, and we have

$$\begin{aligned}
&\{\eta \in \mathsf{N} : \eta(E) \ge p,\ \zeta_p(\eta) \in B\} \\
&= \bigcup_{j=1}^{\infty} \bigcup_{i_1,\dots,i_j=1}^{\infty} \Bigg\{ \eta \in \mathsf{N} : \eta(B \cap A^1_{i_1} \cap \dots \cap A^j_{i_j}) = \eta(A^1_{i_1} \cap \dots \cap A^j_{i_j}) \in \mathbb{N}, \\
&\quad \eta\left(\bigcup_{(r_1,\dots,r_j)<(i_1,\dots,i_j)} A^1_{r_1} \cap \dots \cap A^j_{r_j} \right) \le p-1, \\
&\quad \eta\left(\bigcup_{(r_1,\dots,r_j)\le(i_1,\dots,i_j)} A^1_{r_1} \cap \dots \cap A^j_{r_j} \right) \ge p \Bigg\}.
\end{aligned}$$

This shows that ζ_p is measurable. □

The counting measure η is **simple** if $\eta(\{x\}) \leq 1$ for all $x \in E$. Since $\eta \in \mathsf{N}$ is simple if and only if the mappings ζ_i in the preceding lemma satisfy $\zeta_i(\eta) \neq \zeta_j(\eta)$, for all pairs $j \neq i$, we see that the set N_s of simple counting measures is a measurable subset of N. Let $\mathcal{N}_s$ denote the induced σ-algebra.

For a measure $\eta \in \mathcal{M}$, the **support** $\operatorname{supp} \eta$ is the smallest closed set A in E such that $\eta(E \setminus A) = 0$. If η is a counting measure, then $\operatorname{supp} \eta$ is locally finite and satisfies

$$\operatorname{supp} \eta := \{x \in E : \eta(\{x\}) \geq 1\}.$$

The mapping $i : \eta \mapsto \operatorname{supp} \eta$ maps the set N to the set $\mathcal{F}_{\ell f}$ of locally finite sets in $\mathcal{F}$. The restriction $i_s : \mathsf{N}_s \to \mathcal{F}_{\ell f}$ of i is bijective.

Lemma 3.1.4. *The set $\mathcal{F}_{\ell f}$ is measurable, that is, $\mathcal{F}_{\ell f} \in \mathcal{B}(\mathcal{F})$. The mapping $i : \mathsf{M} \to \mathcal{F}$ is measurable. The trace σ-algebra $\mathcal{B}(\mathcal{F})_{\ell f}$ of $\mathcal{B}(\mathcal{F})$ on $\mathcal{F}_{\ell f}$ satisfies $\mathcal{N}_s = i_s^{-1}(\mathcal{B}(\mathcal{F})_{\ell f})$ and $\mathcal{B}(\mathcal{F})_{\ell f} = i_s(\mathcal{N}_s)$.*

Proof. To prove the measurability of $\mathcal{F}_{\ell f}$, let $(G_k)_{k \in \mathbb{N}}$ be a sequence in $\mathcal{G}_c$ increasing to E. The set

$$\{F \in \mathcal{F} : |F \cap G_k| \leq m\}$$

is closed, as follows from Theorem 12.2.2, hence

$$\mathcal{F}_{\ell f} = \bigcap_{k=1}^{\infty} \bigcup_{m=1}^{\infty} \{F \in \mathcal{F} : |F \cap G_k| \leq m\}$$

is a Borel set.

For the measurability of the map $i : \mathsf{M} \to \mathcal{F}$ it suffices by Lemma 2.1.1 to consider only the sets $\mathcal{F}_G$, $G \in \mathcal{G}$. From

$$i^{-1}(\mathcal{F}_G) = \{\eta \in \mathsf{M} : \operatorname{supp} \eta \cap G \neq \emptyset\} = \{\eta \in \mathsf{M} : \eta(G) > 0\}$$

it follows that $i^{-1}(\mathcal{F}_G) \in \mathcal{M}$, hence i is measurable.

This also implies that $i_s^{-1}(A) \in \mathcal{N}_s$ for $A \in \mathcal{B}(\mathcal{F})_{\ell f}$. For the converse direction, we note that the σ-algebra $\mathcal{N}_s$ is generated by the system

$$\{\mathsf{N}_{G,k} \cap \mathsf{N}_s : G \in \mathcal{G}_c,\ k \in \mathbb{N}_0\}.$$

For $G \in \mathcal{G}_c$ and $k \in \mathbb{N}_0$, the set

$$i_s(\mathsf{N}_{G,k} \cap \mathsf{N}_s) = \{F \in \mathcal{F} : |F \cap G| = k\} \cap \mathcal{F}_{\ell f}$$

is the intersection of a closed set, an open set and a Borel set and hence is a Borel set. Thus, for a generating system of $\mathcal{N}_s$, the images under i_s are Borel sets. This shows that also $i_s(\mathcal{N}_s) \subset \mathcal{B}(\mathcal{F}_{\ell f})$. □

This lemma implies, as a counterpart to Lemma 3.1.2, that

$$\mathcal{E}_0' := \{\mathsf{N}_{G,0} : G \in \mathcal{G}_c\}$$

is a ($\cap$-stable) generating system of $\mathcal{N}_s$.

Now we define random measures and point processes.

Definition 3.1.1. *A* **random measure** X *on* E *is a measurable map from some probability space* $(\Omega, \mathbf{A}, \mathbb{P})$ *into the measurable space* $(\mathsf{M}, \mathcal{M})$ *of locally finite measures on* E*. If* X *is a random measure, the image measure* $\mathbb{P}_X := X(\mathbb{P})$ *is the* **distribution** *of* X.

A random measure X *which is almost surely concentrated on* N *is called a* **point process** *(in* E*). The point process* X *is* **simple** *if* $X \in \mathsf{N}_s$ *almost surely.*

A point process X in $\mathbb{R}^d$ is also called an **ordinary point process**.

For a random measure X, for $\omega \in \Omega$, and $A \in \mathcal{B}(E)$, we shall often write $X(\omega, A)$ instead of $X(\omega)(A)$.

Lemma 3.1.5. *The mapping* $X : (\Omega, \mathbf{A}, \mathbb{P}) \to (\mathsf{M}, \mathcal{M})$ *is a random measure if and only if* $\{X(G) \leq r\}$ *is measurable for all* $G \in \mathcal{G}_c$ *and all* $r \geq 0$.

The mapping $X : (\Omega, \mathbf{A}, \mathbb{P}) \to (\mathsf{N}, \mathcal{N})$ *is a point process if and only if* $\{X(G) = k\}$ *is measurable for all* $G \in \mathcal{G}_c$ *and all* $k \in \mathbb{N}_0$.

Proof. Since both assertions have a similar proof, we concentrate on the second.

The condition is clearly necessary. If the condition is satisfied, then

$$X^{-1}(\mathsf{N}_{G,k}) = \{\omega \in \Omega : X(\omega, G) = k\} \in \mathbf{A}$$

for $G \in \mathcal{G}_c$ and $k \in \mathbb{N}_0$. By Lemma 3.1.2, X is measurable. □

By Lemma 3.1.4, a simple point process X is isomorphic to the locally finite random closed set $\operatorname{supp} X$. For that reason, we shall often identify a simple point process X with its support $\operatorname{supp} X$, so that $X(\omega)$ is understood as a counting measure and also as a locally finite set. For example, the notations $X(\{x\}) = 1$ and $x \in X$ will be used synonymously.

Since simple point processes can be identified with locally finite random closed sets, results from Section 2.1 carry over to simple point processes. In this way, we obtain the following measurability and uniqueness result (using analogous notation).

Theorem 3.1.1. *The mapping* $X : (\Omega, \mathbf{A}, \mathbb{P}) \to (\mathsf{N}_s, \mathcal{N}_s)$ *is a point process if and only if* $\{X(C) = 0\}$ *is measurable for all* $C \in \mathcal{C}$.

Let X, X' *be simple point processes in* E*. If*

$$\mathbb{P}(X(C) = 0) = \mathbb{P}(X'(C) = 0)$$

for all $C \in \mathcal{C}$*, then* $X \stackrel{\mathcal{D}}{=} X'$.

Proof. The first assertion follows from the fact that the mapping $Z := \operatorname{supp} X : (\Omega, \mathbf{A}, \mathbb{P}) \to \mathcal{F}$ is measurable if and only if $Z^{-1}(\mathcal{F}^C)$ is measurable for all $C \in \mathcal{C}$.

By Theorem 2.1.3, the distribution of X is uniquely determined by the probabilities $\mathbb{P}(X \cap C \neq \emptyset)$, $C \in \mathcal{C}$. Therefore, also the probabilities $\mathbb{P}(X \cap C = \emptyset) = \mathbb{P}(X(C) = 0)$, $C \in \mathcal{C}$, determine the distribution of X. □

If X and X' are random measures (point processes) on E, then also $X+X'$ is a random measure (a point process) on E (the measurability is easy to check). If $X+X'$ is a simple point process, the sum corresponds to taking the union of the corresponding random closed sets. The random measure $X+X'$ is called the **superposition** of X and X'.

If $A \in \mathcal{B}$, the mapping $\eta \mapsto \eta \llcorner A$ is measurable on $(\mathsf{M}, \mathcal{M})$. Therefore, the **restriction** $X \llcorner A$ of a random measure (a point process) X to A is again a random measure (a point process). For a simple point process X, this corresponds to taking the intersection of X with A.

If a topological group G operates measurably on the space E, then G operates in a canonical way on M, by letting $g\eta$ for $g \in G$ and $\eta \in \mathsf{M}$ be the image of η under g,

$$g\eta(B) := \eta(g^{-1}B) \qquad \text{for } B \in \mathcal{B}.$$

It follows from Lemma 3.1.1 that the mapping $\eta \mapsto g\eta$ from $(\mathsf{M}, \mathcal{M})$ into itself is measurable. Hence, for a random measure (a point process) X on E and for $g \in G$, also gX is a random measure (a point process) on E. This will be used later, mainly for $E = \mathbb{R}^d$ or $E = \mathcal{F}'(\mathbb{R}^d)$, where G is the group G_d of rigid motions of $\mathbb{R}^d$. In both cases, the operation is continuous. In the second case, the continuity follows from Theorem 13.1.1. If $E = \mathbb{R}^d$ or $E = \mathcal{F}'(\mathbb{R}^d)$ and t_x is the translation by the vector x, we denote the image measures $t_x\eta$ and t_xX also by $\eta + x$ and $X + x$, respectively. This notation is extended to sets A of measures in the obvious way, that is, $A+x$ is the set of all measures $\mu + x$ with $\mu \in A$.

Definition 3.1.2. *The random measure X on $E = \mathbb{R}^d$ or $E = \mathcal{F}'(\mathbb{R}^d)$ is* **stationary** *if $X \stackrel{\mathcal{D}}{=} X + x$ for all $x \in \mathbb{R}^d$. It is* **isotropic** *if $X \stackrel{\mathcal{D}}{=} \vartheta X$ for all rotations $\vartheta \in SO_d$.*

We return to the general situation and introduce, for a random measure X on E, a quantity corresponding to the expectation of a real random variable.

Definition 3.1.3. *The* **intensity measure** *of the random measure X is the measure on E defined by*

$$\Theta(A) := \mathbb{E}X(A) \qquad \textit{for } A \in \mathcal{B}.$$

Since $X(A) \geq 0$, $\Theta(A)$ is always defined, but may be infinite. It follows from the theorem of monotone convergence that Θ is indeed a measure on E. If X is a simple point process, then $\Theta(A)$ is the mean number of points of X lying in A. Although the random measure X is locally finite a.s., it may well happen that $\Theta(C) = \infty$ for some compact sets C. Later, we shall concentrate on random measures and point processes with locally finite intensity measure.

If X is a stationary random measure on $\mathbb{R}^d$, its intensity measure Θ, which is now a measure on $\mathbb{R}^d$, is invariant under translations. This follows immediately from the definition of Θ. The only translation invariant, locally finite measure on $\mathbb{R}^d$ is, up to a constant factor, the Lebesgue measure λ. Hence, if Θ is locally finite, then

$$\Theta = \gamma\lambda$$

with a constant $\gamma \in [0, \infty)$. The number γ is called the **intensity** of the (stationary) random measure X. The case $\gamma = 0$ means that $\Theta = 0$, thus $X = 0$ almost surely. As the zero measure is not very interesting, we may assume $\gamma > 0$ where necessary.

For a stationary random measure X on $\mathcal{F}'(\mathbb{R}^d)$, the intensity measure Θ is a translation invariant measure on $\mathcal{F}'(\mathbb{R}^d)$. If X (and hence Θ) is supported by certain subclasses of $\mathcal{F}'(\mathbb{R}^d)$, the class $\mathcal{C}'(\mathbb{R}^d)$ of compact sets (particles) or the class $A(d, k)$ of k-dimensional affine flats, the translation invariance of Θ will lead to important decomposition results, as we shall see in Chapter 4.

The following simple observation will be used frequently.

Theorem 3.1.2 (Campbell). *Let X be a random measure on E with intensity measure Θ, and let $f : E \to \mathbb{R}$ be a nonnegative, measurable function. Then $\int_E f \, \mathrm{d}X$ is measurable, and*

$$\mathbb{E} \int_E f \, \mathrm{d}X = \int_E f \, \mathrm{d}\Theta.$$

Proof. For $A \in \mathcal{B}$,

$$X(A) = \int_E \mathbf{1}_A \, \mathrm{d}X;$$

this is a nonnegative measurable function, and

$$\mathbb{E} \int_E \mathbf{1}_A \, \mathrm{d}X = \mathbb{E}X(A) = \Theta(A) = \int_E \mathbf{1}_A \, \mathrm{d}\Theta.$$

Thus, the assertion holds for indicator functions of Borel sets and therefore also for linear combinations of such functions. By a standard argument of integration theory, it holds for nonnegative measurable functions. □

Remark. We have formulated Campbell's theorem only for nonnegative measurable functions, but it is clear that it holds also for Θ-integrable functions. The same remark refers to the subsequent relatives of Campbell's theorem; they will later tacitly be applied to integrable functions.

For a simple point process X, Campbell's theorem can be written in the form

$$\mathbb{E} \sum_{x \in X} f(x) = \int_E f \, \mathrm{d}\Theta.$$

The intensity measure Θ is also known as the **first moment measure**. Higher moment measures can be introduced in a similar way. The **second moment measure** of the random measure X is defined by

$$\Gamma^{(2)}(A) := \mathbb{E}(X \otimes X)(A)$$

for $A \in \mathcal{B}(E) \otimes \mathcal{B}(E) = \mathcal{B}(E \times E)$ (for the latter equality see, for example, Cohn [177, Proposition 7.6.2]). In particular,

$$\Gamma^{(2)}(A \times A') = \mathbb{E}(X \otimes X)(A \times A') = \mathbb{E}\, X(A)X(A') \quad \text{for } A, A' \in \mathcal{B}.$$

Generally, the **mth moment measure** $\Gamma^{(m)}$ of X is the Borel measure on E^m for which

$$\Gamma^{(m)}(A_1 \times \ldots \times A_m) = \mathbb{E}\, X^m(A_1 \times \ldots \times A_m) = \mathbb{E}\, X(A_1) \cdots X(A_m)$$

for $A_1, \ldots, A_m \in \mathcal{B}$. Here, the product measure X^m is a random measure on the locally compact space E^m, therefore $\Gamma^{(m)}$ is just the intensity measure of X^m.

For $m \in \mathbb{N}$, let

$$E^m_{\neq} := \{(x_1, \ldots, x_m) \in E^m : x_i \text{ pairwise distinct}\};$$

then $E^m_{\neq}$ is an open set in E^m. For a random measure X on E, one defines the **mth factorial moment measure** as the Borel measure $\Lambda^{(m)}$ on E^m for which

$$\Lambda^{(m)}(A_1 \times \ldots \times A_m) := \mathbb{E}\, X^m(A_1 \times \ldots \times A_m \cap E^m_{\neq})$$

for $A_1, \ldots, A_m \in \mathcal{B}$. In particular, for a simple point process X and for $A \in \mathcal{B}$,

$$\begin{aligned}\Lambda^{(m)}(A^m) &= \mathbb{E} \sum_{x_1 \in X \cap A} \;\sum_{x_2 \in X \cap A,\, x_2 \neq x_1} \cdots \sum_{x_m \in X \cap A,\, x_m \neq x_1, \ldots, x_{m-1}} 1 \\ &= \mathbb{E}[X(A)(X(A)-1) \cdots (X(A)-m+1)]\end{aligned}$$

is the mth factorial moment of the random variable $X(A)$. We observe that $\Lambda^{(m)}$ is the intensity measure of the random measure

$$X^m_{\neq} := X^m \llcorner E^m_{\neq}$$

(which is different from X^m, in general). It is clear that also the mth moment measure $\Gamma^{(m)}$ and the mth factorial moment measure $\Lambda^{(m)}$ satisfy the Campbell theorem. We only formulate the corresponding result for simple point processes.

Theorem 3.1.3. *Let X be a simple point process in E, let $f : E^m \to \mathbb{R}$ be a nonnegative measurable function ($m \in \mathbb{N}$). Then $\sum_{(x_1,\ldots,x_m) \in X^m} f(x_1, \ldots, x_m)$ and $\sum_{(x_1,\ldots,x_m) \in X^m_{\neq}} f(x_1, \ldots, x_m)$ are measurable, and*

$$\mathbb{E} \sum_{(x_1,\ldots,x_m) \in X^m} f(x_1, \ldots, x_m) = \mathbb{E} \int_{E^m} f \,\mathrm{d}X^m = \int_{E^m} f \,\mathrm{d}\Gamma^{(m)}$$

and

$$\mathbb{E} \sum_{(x_1,\ldots,x_m) \in X^m_{\neq}} f(x_1, \ldots, x_m) = \mathbb{E} \int_{E^m} f \,\mathrm{d}X^m_{\neq} = \int_{E^m} f \,\mathrm{d}\Lambda^{(m)}.$$

From this one obtains, for example, the following connection between $\Gamma^{(2)}$ and $\Lambda^{(2)}$. For a simple point process X in E and for $A_1, A_2 \in \mathcal{B}$,

$$\begin{aligned}\Gamma^{(2)}(A_1 \times A_2) &= \mathbb{E} \sum_{(x_1,x_2)\in X^2} \mathbf{1}_{A_1\times A_2}(x_1,x_2) \\ &= \mathbb{E}\left(\sum_{(x_1,x_2)\in X^2_{\neq}} \mathbf{1}_{A_1\times A_2}(x_1,x_2) + \sum_{x\in X} \mathbf{1}_{A_1}(x)\mathbf{1}_{A_2}(x) \right),\end{aligned}$$

hence

$$\Gamma^{(2)}(A_1 \times A_2) = \Lambda^{(2)}(A_1 \times A_2) + \Theta(A_1 \cap A_2). \tag{3.1}$$

Examples of random measures will occur later in connection with particle processes and random mosaics. Concerning point processes, simple examples are easily constructed. A starting point can be random variables or sequences of random variables with values in E. For example, if $\xi_1, \dots, \xi_m$ are random points in E, that is, E-valued random variables, then

$$X := \sum_{i=1}^{m} \delta_{\xi_i}$$

is a point process in E (not necessarily a simple one). In fact, for $A \in \mathcal{B}$ and $k \in \mathbb{N}_0$ we have

$$\begin{aligned}&\{X(A) = k\} = \{\text{precisely } k \text{ of the } \xi_i \text{ are in } A\} \\ &= \bigcup_{1\le i_1<\dots<i_k\le m} \{\xi_i \in A,\, i \in \{i_1,\dots,i_k\},\, \xi_j \notin A,\, j \notin \{i_1,\dots,i_k\}\},\end{aligned}$$

hence $\{X(A) = k\}$ is measurable, and Lemma 3.1.5 implies the measurability of X. Similarly, one can start with a sequence $\xi_1, \xi_2, \dots$ of random points in E; then

$$X := \sum_{i\in\mathbb{N}} \delta_{\xi_i}$$

defines a point process in E, provided that additional conditions ensure the local finiteness of the measure $\sum_{i\in\mathbb{N}} \delta_{\xi_i(\omega)}$ for almost all ω. Using Lemma 3.1.3 and putting $\xi_i = \zeta_i \circ X$, $i \in \mathbb{N}$, we see that every point process X can be represented in this way.

We conclude this section with a motivation for the next one. For that, we return to a point process $X = \sum_{i=1}^{m} \delta_{\xi_i}$ with finitely many random points $\xi_1, \dots, \xi_m$. If these points are independent and identically distributed, then, for $k = 0, 1, \dots, m$,

$$\begin{aligned}&\mathbb{P}(X(A) = k) \\ &= \sum_{1\le i_1<\dots<i_k\le m} \mathbb{P}(\xi_i \in A,\, i \in \{i_1,\dots,i_k\},\, \xi_j \notin A,\, j \notin \{i_1,\dots,i_k\})\end{aligned}$$

$$= \binom{m}{k} p_A^k (1 - p_A)^{m-k}$$

with $p_A := \mathbb{P}(\xi_1 \in A)$. Thus, $X(A)$ has a binomial distribution.

Letting $m \to \infty$ and $p_A \to 0$ in such a way that mp_A converges, one is led to the class of Poisson processes. For example, consider a locally finite measure Θ on E and a sequence $(C_i)_{i \in \mathbb{N}}$ in $\mathcal{C}$ with $C_i \subset \operatorname{int} C_{i+1}$, $\Theta(C_i) > 0$ for $i \in \mathbb{N}$, $C_i \uparrow E$ and $\Theta(C_i) \to \infty$ for $i \to \infty$. For $i \in \mathbb{N}$, let $\xi_1^i, \dots, \xi_{m(i)}^i$ be independent random points in E, each with distribution

$$\frac{\Theta \llcorner C_i}{\Theta(C_i)}.$$

As before, we define the point process

$$X_i := \sum_{j=1}^{m(i)} \delta_{\xi_j^i}.$$

For $k \in \mathbb{N}_0$ and $A \in \mathcal{B}$, we have

$$\mathbb{P}(X_i(A) = k) = \binom{m(i)}{k} p_{i,A}^k (1 - p_{i,A})^{m(i)-k}$$

with $p_{i,A} := \Theta(A \cap C_i)/\Theta(C_i)$. If we choose the numbers $m(i)$ in such a way that

$$\frac{m(i)}{\Theta(C_i)} \to 1 \qquad \text{for } i \to \infty,$$

then we get, for every relatively compact Borel set A, the relation

$$\lim_{i \to \infty} \mathbb{P}(X_i(A) = k) = \mathrm{e}^{-\Theta(A)} \frac{\Theta(A)^k}{k!}.$$

It is plausible to conjecture the existence of a 'limit process', a point process X satisfying

$$\mathbb{P}(X(A) = k) = \mathrm{e}^{-\Theta(A)} \frac{\Theta(A)^k}{k!}$$

for all $k \in \mathbb{N}_0$ and all $A \in \mathcal{B}$ with $\Theta(A) < \infty$. Such a process X, if it is simple, is called a Poisson process; the measure Θ is then its intensity measure. In the next section, we prove the existence of these Poisson processes, though not by a limit procedure, but by a direct construction, which permits us to derive further properties of Poisson processes in a straightforward way.

General assumption. From now on, the intensity measures Θ of all random measures and point processes, which are considered, are assumed to be locally finite, that is, to satisfy

$$\Theta(C) < \infty \qquad \text{for all } C \in \mathcal{C}.$$

Notes for Section 3.1

1. From the general theory of point processes, we treat in this chapter only some basic notions and results, mainly to lay the foundations for our later considerations. For comprehensive presentations of the theory, we refer to Daley and Vere–Jones [195, 196], Karr [389], Kerstan, Matthes and Mecke [400] (the extended English version is Matthes, Kerstan and Mecke [465]), König and Schmidt [423], Neveu [582], Reiss [626]. A short introduction to point processes in measurable spaces is found in Appendix A of Mecke, Schneider, Stoyan and Weil [500].

2. The theory of random measures is developed in much more detail in some of the mentioned books and particularly in Kallenberg [385].

3. If the set M of locally finite measures on E is equipped with the vague topology, then M is a Polish space and the σ-algebra $\mathcal{M}$, generated by the evaluation mappings $\Phi_A : \eta \mapsto \eta(A)$, $A \in \mathcal{B}$, turns out to be the Borel σ-algebra. Thus, also $\mathcal{N}$ is the Borel σ-algebra with respect to the vague topology on N. See Kallenberg [386, Th. A2.3].

3.2 Poisson Processes

Poisson processes can be introduced in quite general measurable spaces with suitable additional structures. Here we restrict ourselves, as in the previous sections, to a locally compact space E with a countable base, and we also concentrate on simple processes. We first introduce the two characteristic properties of Poisson processes.

A point process X in E with intensity measure Θ has **Poisson counting variables** if for each $A \in \mathcal{B}$ with $\Theta(A) < \infty$ the (a.s. real) random variable $X(A)$ has a Poisson distribution. In that case, $\mathbb{E}X(A) = \Theta(A)$, so that $\Theta(A)$ is the parameter of the Poisson distribution. Thus, X has Poisson counting variables if and only if

$$\mathbb{P}(X(A) = k) = \mathrm{e}^{-\Theta(A)} \frac{\Theta(A)^k}{k!} \tag{3.2}$$

for $k \in \mathbb{N}_0$ and $A \in \mathcal{B}$. The equation (3.2) is also correct for $\Theta(A) = \infty$, if the right side is read as 0. In fact, for such $A \in \mathcal{B}$ there is a sequence $A_1, A_2, \dots$ of Borel sets increasing to A and satisfying $\Theta(A_i) < \infty$. This gives

$$\mathbb{P}(X(A) = k) \le \mathbb{P}(X(A_i) \le k) = \mathrm{e}^{-\Theta(A_i)} \sum_{j=0}^{k} \frac{\Theta(A_i)^j}{j!} \to 0 \quad (i \to \infty)$$

for $k \in \mathbb{N}_0$, hence $X(A) = \infty$ a.s. Thus (3.2) holds in this case, too.

We shall see later that the probabilities (3.2) do not determine the distribution of X uniquely if the intensity measure Θ has **atoms**, hence if $\Theta(\{x\}) > 0$ for some $x \in E$. (Observe that, under our assumptions on E and Θ, the existence of atoms is equivalent to the existence of point masses.) For that reason and in view of the applications which we have in mind, we shall restrict ourselves to the case of intensity measures without atoms. For a point process with Poisson counting variables, this condition is equivalent to being simple.

Lemma 3.2.1. *The point process X with Poisson counting variables is simple if and only if its intensity measure Θ has no atoms.*

Proof. From $\Theta(\{x\}) > 0$ and (3.2) it follows that $\mathbb{P}(X(\{x\}) = k) > 0$ for $k \in \mathbb{N}$, hence the point process X is not simple.

Conversely, suppose that $\Theta(\{x\}) = 0$ for all $x \in E$. Assume that X were not simple, that is, $\mathbb{P}_X(\mathsf{N}_s) < 1$. Then there is a compact set $C \in \mathcal{C}$ with

$$\alpha := \mathbb{P}(X \llcorner C \text{ not simple}) > 0;$$

in particular, $\epsilon := \Theta(C) > 0$. Let $k \in \mathbb{N}$. Since the range of a finite measure without atoms is a closed interval (see, for example, Neveu [581, exercise 1.4.3] or, without Zorn's lemma, Gardner and Pfeffer [246, Lemma 9.1]), there exist pairwise disjoint Borel sets $C_1^{(k)}, \dots, C_k^{(k)} \in \mathcal{B}$ with

$$\Theta(C_i^{(k)}) = \frac{\epsilon}{k}, \qquad i = 1, \dots, k,$$

and

$$\bigcup_{i=1}^{k} C_i^{(k)} = C.$$

There must be a number $i \in \{1, \dots, k\}$ with

$$\mathbb{P}\left(X(C_i^{(k)}) > 1\right) \geq \frac{\alpha}{k}.$$

This yields

$$\frac{\alpha}{k} \leq 1 - \mathrm{e}^{\Theta(C_i^{(k)})}\left(1 + \Theta(C_i^{(k)})\right)$$

and thus

$$\alpha \leq k - \mathrm{e}^{-\epsilon/k}(k + \epsilon).$$

For $k \to \infty$, the right side converges to 0, a contradiction. □

The second characteristic property of Poisson processes on the real line is that they have independent increments. For a point process X in E, we say that X has **independent increments**, if for pairwise disjoint Borel sets $A_1, \dots, A_m$ in E, $m \in \mathbb{N}$, the random variables $X(A_1), \dots, X(A_m)$ are independent.

Definition 3.2.1. *A* **Poisson process** *in E is a simple point process in E with Poisson counting variables and independent increments.*

The following theorem shows that the class of Poisson processes is quite large.

Theorem 3.2.1. *Let Θ be a locally finite measure without atoms on E. Then there exists a Poisson process in E with intensity measure Θ; it is uniquely determined (up to equivalence).*

Proof. We first prove the existence by construction.

Since Θ is locally finite, there are (by Theorem 12.1.1) pairwise disjoint Borel sets $A_1, A_2, \ldots$ in E with $E = \bigcup_{i \in \mathbb{N}} A_i$, $\Theta(A_i) < \infty$, and such that to each $C \in \mathcal{C}$ there exists $k \in \mathbb{N}$ with $C \subset \bigcup_{i=1}^{k} A_i$. In each A_i, we define a point process $X^{(i)}$ in the following way. For $r \in \mathbb{N}$, write $A_i^r = A_i \times \ldots \times A_i$ (r factors) and let

$$\Gamma_r : A_i^r \to \mathsf{N},$$

be the map defined by

$$\Gamma_r(x_1, \ldots, x_r) := \sum_{j=1}^{r} \delta_{x_j}.$$

Then Γ_r is $(\mathcal{B}(A_i^r), \mathcal{N})$-measurable. Let Δ_0 denote the Dirac measure on N concentrated at the zero measure. Then

$$\mathbb{P}_i := \mathrm{e}^{-\Theta(A_i)} \left(\Delta_0 + \sum_{r \in \mathbb{N}} \frac{1}{r!} \Gamma_r \left((\Theta \llcorner A_i)^r \right) \right)$$

is a probability measure on N which is concentrated on the counting measures η with $\operatorname{supp} \eta \subset A_i$. (For the normalization, observe that $\Gamma_r \left((\Theta \llcorner A_i)^r \right)(\mathsf{N}) = \Theta(A_i)^r$.) Let $X_1, X_2, \ldots$ be an independent sequence of point processes in E such that X_i has distribution $\mathbb{P}_i$, for $i \in \mathbb{N}$ (for example, we may take as underlying probability space Ω the space $\mathsf{N}^{\mathbb{N}}$ with the product σ-algebra and the probability measure $\mathbb{P} = \bigotimes_{i \in \mathbb{N}} \mathbb{P}_i$, and for X_i the ith coordinate mapping). Finally, we put

$$X := \sum_{i \in \mathbb{N}} X_i.$$

Then $X \in \mathsf{N}$ almost surely. In fact, if $C \in \mathcal{C}$ and $k \in \mathbb{N}$ is chosen such that $C \subset \bigcup_{i=1}^{k} A_i$, then

$$X(\omega, C) \le X\left(\omega, \bigcup_{i=1}^{k} A_i\right) = \sum_{i=1}^{k} X^{(i)}(\omega, A_i),$$

and the right side is finite for $\mathbb{P}$-almost all ω.

Since X is locally a finite sum of point processes, it is measurable and thus a point process in E. We show that X has Poisson counting variables.

For this, let $A \in \mathcal{B}$ be a set with $\Theta(A) < \infty$ and put $A_i' := A \cap A_i$ for $i \in \mathbb{N}$, then

$$X(A) = \sum_{i \in \mathbb{N}} X_i(A) = \sum_{i \in \mathbb{N}} X_i(A_i').$$

By construction, the random variables $X_1(A_1'), X_2(A_2'), \ldots$ are independent. Again from construction, for each $k \in \mathbb{N}$ we have

$$\begin{aligned}
\mathbb{P}(X_i(A_i') = k) &= \mathbb{P}_i(\{\eta \in \mathsf{N} : \eta(A_i') = k\}) \\
&= \sum_{r=k}^{\infty} \mathbb{P}_i(\{\eta \in \mathsf{N} : \eta(A_i') = k,\ \eta(A_i \setminus A_i') = r - k\}) \\
&= \mathrm{e}^{-\Theta(A_i)} \sum_{r=k}^{\infty} \binom{r}{k} \frac{1}{r!} \Theta(A_i')^k \Theta(A_i \setminus A_i')^{r-k} \\
&= \mathrm{e}^{-\Theta(A_i)} \frac{\Theta(A_i')^k}{k!} \sum_{r=k}^{\infty} \frac{\Theta(A_i \setminus A_i')^{r-k}}{(r-k)!} \\
&= \mathrm{e}^{-\Theta(A_i)} \frac{\Theta(A_i')^k}{k!} \mathrm{e}^{\Theta(A_i \setminus A_i')} \\
&= \mathrm{e}^{-\Theta(A_i')} \frac{\Theta(A_i')^k}{k!}.
\end{aligned}$$

The corresponding result for $k = 0$ is obtained similarly. Thus, the random variable $X_i(A_i') =: \xi_i$ has a Poisson distribution with parameter $\Theta(A_i') =: \alpha_i$, and the sequence $(\xi_i)_{i \in \mathbb{N}}$ is independent. For $k \in \mathbb{N}_0$,

$$\begin{aligned}
\mathbb{P}(\xi_1 + \xi_2 = k) &= \sum_{j=0}^{k} \mathbb{P}(\xi_1 = j,\ \xi_2 = k - j) \\
&= \sum_{j=0}^{k} \mathrm{e}^{-\alpha_1} \frac{\alpha_1^j}{j!} \mathrm{e}^{-\alpha_2} \frac{\alpha_2^{k-j}}{(k-j)!} \\
&= \mathrm{e}^{-(\alpha_1+\alpha_2)} \frac{(\alpha_1 + \alpha_2)^k}{k!}.
\end{aligned}$$

Hence, $\xi_1 + \xi_2$ is Poisson distributed with parameter $\alpha_1 + \alpha_2$. By induction, it follows that $S_m := \xi_1 + \ldots + \xi_m$ is Poisson distributed with parameter $\sigma_m := \alpha_1 + \ldots + \alpha_m$ ($m \in \mathbb{N}$). For the sum $S := \sum_{j \in \mathbb{N}} \xi_j$ we have $\{S_m \le k\} \downarrow \{S \le k\}$ for $m \to \infty$ and hence

$$\begin{aligned}
\mathbb{P}(S \le k) &= \lim_{m \to \infty} \mathbb{P}(S_m \le k) \\
&= \lim_{m \to \infty} \sum_{j=0}^{k} \mathrm{e}^{-\sigma_m} \frac{\sigma_m^j}{j!} \\
&= \sum_{j=0}^{k} \mathrm{e}^{-\Theta(A)} \frac{\Theta(A)^j}{j!},
\end{aligned}$$

where $\sum_{j \in \mathbb{N}} \Theta(A_j') = \Theta(A) < \infty$ was used. Thus, for the sets $A \in \mathcal{B}$ with $\Theta(A) < \infty$, the random variable $S = X(A)$ has a Poisson distribution with parameter $\Theta(A)$, which means that X is a point process with Poisson counting variables and with intensity measure Θ. Since Θ has no atoms, X is simple, by Lemma 3.2.1.

Because of (3.2) and Theorem 3.1.1, the simple point process X with Poisson counting variables which we just constructed is uniquely determined in distribution.

In order to see that X is a Poisson process, it remains to show that X has independent increments. For $m \in \mathbb{N}$, let $A_1, \dots, A_m$ be pairwise disjoint Borel sets in E. Each A_i can be divided into (countably many) pairwise disjoint Borel sets with finite Θ-measure. Hence, for the independence of $X(A_1), \dots, X(A_m)$, we may assume that $\Theta(A_i) < \infty$, $i = 1, \dots, m$. Repeating the construction for the existence of X, given above, we may put $A'_i := A_i$ for $i = 1, \dots, m$ and then choose A'_i for $i > m$ so that we obtain a sequence $A'_1, A'_2, \dots$ of pairwise disjoint Borel sets with $\Theta(A'_i) < \infty$, $E = \bigcup_{i \in \mathbb{N}} A'_i$, and such that to each $C \in \mathcal{C}$ there exists $k \in \mathbb{N}$ with $C \subset \bigcup_{i=1}^k A'_i$. With this sequence, and the intensity measure Θ of X, we obtain a point process X' with intensity measure Θ and Poisson counting variables. By the uniqueness assertion, we have $X \overset{\mathcal{D}}{=} X'$. By construction, the point processes $X' \llcorner A'_1, X' \llcorner A'_2, \dots$ are independent, thus also $X \llcorner A_1, \dots, X \llcorner A_m$ are independent; this implies the independence of $X(A_1), \dots, X(A_m)$. □

Corollary 3.2.1. *Let $\gamma \in [0, \infty)$. Then there is (up to equivalence) precisely one stationary Poisson process X in $\mathbb{R}^d$ with intensity γ. The process X is also isotropic.*

Proof. Since $\Theta = \gamma\lambda$ has no atoms, the first part follows from Theorem 3.2.1. If $\vartheta \in SO_d$ is a rotation, then ϑX has the intensity measure $\vartheta\Theta = \Theta$. By the uniqueness of Poisson processes with given intensity measure, $\vartheta X \overset{\mathcal{D}}{=} X$, which means that X is isotropic. □

Whereas a stationary Poisson process in $\mathbb{R}^d$ is automatically isotropic, there are other stationary point processes in $\mathbb{R}^d$ which are not isotropic. It is clear from the general existence theorem 3.2.1 that there exist non-stationary Poisson processes in $\mathbb{R}^d$, and also non-stationary Poisson processes that are isotropic.

As we have seen in the proof of Theorem 3.2.1, for a simple point process X in E, the condition that X has Poisson counting variables already implies that X has independent increments. We state this important fact explicitly.

Corollary 3.2.2. *A simple point process in E with Poisson counting variables is a Poisson process.*

The constructive proof even permits us to establish some further properties of Poisson processes, which we collect in the following theorem.

Theorem 3.2.2. *Let X be a Poisson process in E with intensity measure Θ.*

(a) *Let $A_1, A_2, \dots$ be pairwise disjoint Borel sets in E. Then the point processes $X \llcorner A_1, X \llcorner A_2, \dots$ are independent. If $m, k \in \mathbb{N}$ and $\bigcup_{i=1}^m A_i =: A$ with $0 < \Theta(A) < \infty$, then under the condition $X(A) = k$ the random vector $(X(A_1), \dots, X(A_m))$ has a multinomial distribution.*

(b) *Let $A \subset E$ be a Borel set with $0 < \Theta(A) < \infty$, and let $k \in \mathbb{N}$. Then*

$$\mathbb{P}(X \llcorner A \in \cdot \mid X(A) = k) = \mathbb{P}\left(\sum_{i=1}^k \delta_{\xi_i} \in \cdot\right),$$

where $\xi_1, \dots, \xi_k$ are independent, identically distributed random points in E with distribution

$$\mathbb{P}_{\xi_i} := \frac{\Theta \llcorner A}{\Theta(A)}, \qquad i = 1, \dots, k.$$

Proof. (a) We need to show the independence of $X \llcorner A_1, \dots, X \llcorner A_m$, for each $m \in \mathbb{N}$. As was already explained at the end of the proof of Theorem 3.2.1, we may assume $\Theta(A_i) < \infty$, $i = 1, \dots, m$, and then the independence follows from the construction in the proof together with the uniqueness.

For $j_1 + \dots + j_m = k$, the established independence yields

$$\begin{aligned} &\mathbb{P}(X(A_1) = j_1, \dots, X(A_m) = j_m \mid X(A) = k) \\ &= \frac{\prod_{i=1}^m \mathrm{e}^{-\Theta(A_i)} \Theta(A_i)^{j_i}/j_i!}{\mathrm{e}^{-\Theta(A)} \Theta(A)^k/k!} \\ &= \frac{k!}{j_1! \cdots j_m!} \left(\frac{\Theta(A_1)}{\Theta(A)}\right)^{j_1} \cdots \left(\frac{\Theta(A_m)}{\Theta(A)}\right)^{j_m}. \end{aligned}$$

(b) We can take A as the first member A_1 of a sequence $A_1, A_2, \dots$ which we use to construct the Poisson process X, up to distribution. By construction, $X \llcorner A$ has the distribution

$$\mathbb{P}_1 := \mathrm{e}^{-\Theta(A)} \left(\Delta_0 + \sum_{r \in \mathbb{N}} \frac{1}{r!} \Gamma_r((\Theta \llcorner A)^r)\right).$$

For $k \in \mathbb{N}$, this gives

$$\mathbb{P}(X \llcorner A \in \cdot \mid X(A) = k) = \frac{\mathrm{e}^{-\Theta(A)} \Gamma_k((\Theta \llcorner A)^k)/k!}{\mathrm{e}^{-\Theta(A)} \Theta(A)^k/k!} = \frac{\Gamma_k((\Theta \llcorner A)^k)}{\Theta(A)^k}.$$

Let $\xi_1, \dots, \xi_k$ be independent random points in E with distribution

$$\mathbb{P}_{\xi_i} := \frac{\Theta \llcorner A}{\Theta(A)}, \qquad i = 1, \dots, k.$$

By the independence, we have

$$\mathbb{P}_{\xi_1,\dots,\xi_k} = \frac{(\Theta \llcorner A)^k}{\Theta(A)^k},$$

hence

$$\mathbb{P}_{\sum_{i=1}^k \delta_{\xi_i}} = \frac{\Gamma_k((\Theta \llcorner A)^k)}{\Theta(A)^k},$$

which proves (b). □

The proof of Theorem 3.2.1 and assertion (b) of Theorem 3.2.2 suggest a way for simulating a Poisson process X in $\mathbb{R}^d$, stationary or not, within a given observation window. In the stationary case, for example, we prescribe an intensity γ and an observation window W, say convex and with $\lambda(W) = 1$. First one generates a random number ν which has a Poisson distribution with parameter γ. If the outcome is $\nu(\omega) = k$, one generates k independent random points $\xi_1, \dots, \xi_k$ uniformly in W, that is, with distribution $\lambda \llcorner W$. The point process $\widetilde{X}$ constructed in this way (which is concentrated in W), has the same distribution as X restricted to W, that is, $\widetilde{X} \overset{\mathcal{D}}{=} X \llcorner W$. Therefore, the realization $\{\xi_1(\omega), \dots, \xi_k(\omega)\}$ can be considered as a realization of X in W.

The construction in the proof of Theorem 3.2.1 works also for measures Θ with atoms; it then produces a point process with Poisson counting variables. However, the uniqueness assertion no longer holds in this case, as shown by the subsequent example. For that reason, point processes satisfying (3.2) in general do not have the strong independence properties of Theorem 3.2.2.

Example. The basic space is $E = \{0, 1\}$ with the discrete topology. Every $\eta \in \mathsf{N}$ can be identified with the pair $(\eta(0), \eta(1)) \in \mathbb{N}_0^2$. With a real number $c \in [-\mathrm{e}^{-2}/2, \mathrm{e}^{-2}/2]$, we define a probability distribution $\mathbb{P}_c$ on $\mathbb{N}_0^2$ by the counting density

$$p(0,1) := \mathrm{e}^{-2} + c, \qquad p(1,0) := \mathrm{e}^{-2} - c,$$

$$p(0,2) := \frac{\mathrm{e}^{-2}}{2} - c, \qquad p(2,0) := \frac{\mathrm{e}^{-2}}{2} + c,$$

$$p(1,2) := \frac{\mathrm{e}^{-2}}{2} + c, \qquad p(2,1) := \frac{\mathrm{e}^{-2}}{2} - c,$$

and

$$p(i,j) := \mathrm{e}^{-2}\frac{1}{i!j!}$$

for all other pairs $(i,j) \in \mathbb{N}_0^2$. For the corresponding point process X_c we have

$$\mathbb{P}(X_c(\{0\}) = k) = \mathbb{P}(X_c(\{1\}) = k) = \frac{\mathrm{e}^{-1}}{k!}$$

and

$$\mathbb{P}(X_c(\{0,1\}) = k) = e^{-2}\frac{2^k}{k!}$$

for $k \in \mathbb{N}_0$, independently of the parameter c. Hence, all point processes X_c, $c \in [-e^{-2}/2, e^{-2}/2]$, have Poisson counting variables and have different distributions, but they yield the same distributions for the counting variables $X_c(A), A \subset \{0,1\}$.

We note two particular formulas for Poisson processes, which will be applied later. The second one even characterizes Poisson processes.

Theorem 3.2.3. *Let X be a Poisson process in E with intensity measure Θ. Let $A \in \mathcal{B}$ be a Borel set with $\Theta(A) < \infty$, and let $f : \mathsf{N} \to \mathbb{R}$ be a nonnegative measurable function. Then*

$$\begin{aligned}&\mathbb{E}f(X \llcorner A)\\ &= e^{-\Theta(A)}\left(f(0) + \sum_{k\in\mathbb{N}} \frac{1}{k!}\int_A \cdots \int_A f\left(\sum_{i=1}^k \delta_{x_i}\right)\Theta(dx_1)\cdots\Theta(dx_k)\right).\end{aligned}$$

Proof. The result trivially holds if $\Theta(A) = 0$, hence we now assume $\Theta(A) > 0$. Using Theorem 3.2.2(b), we get

$$\begin{aligned}&\mathbb{E}f(X \llcorner A)\\ &= \sum_{k\in\mathbb{N}_0} \mathbb{P}(X(A) = k)\,\mathbb{E}(f(X \llcorner A) \mid X(A) = k)\\ &= e^{-\Theta(A)}\\ &\quad\times\left(f(0) + \sum_{k\in\mathbb{N}} \frac{\Theta(A)^k}{k!}\Theta(A)^{-k}\int_A \cdots \int_A f\left(\sum_{i=1}^k \delta_{x_i}\right)\Theta(dx_1)\cdots\Theta(dx_k)\right)\end{aligned}$$

and thus the assertion. □

Theorem 3.2.4. *Let X be a point process in E, the intensity measure Θ of which has no atoms. Then X is a Poisson process if and only if*

$$\mathbb{E}\prod_{x\in X} f(x) = \exp\left(\int_E (f-1)\,d\Theta\right) \tag{3.3}$$

holds for all measurable functions $f : E \to [0,1]$.

Proof. Let X be a Poisson process. For $\Theta = 0$, the assertion is trivial (the product is empty a.s.), hence we assume $\Theta \neq 0$. First we suppose that $A \subset E$ is a compact set with $\Theta(A) > 0$ and such that $f(x) = 1$ for $x \in E \setminus A$. Using Theorem 3.2.2(b), we obtain

$$\begin{aligned}\mathbb{E}\prod_{x\in X} f(x) &= \mathbb{E}\prod_{x\in X\llcorner A} f(x)\\ &= \sum_{k\in\mathbb{N}_0}\mathbb{P}(X(A)=k)\,\mathbb{E}\left(\prod_{x\in X\llcorner A} f(x)\,\middle|\, X(A)=k\right)\\ &= \sum_{k\in\mathbb{N}_0}\mathrm{e}^{-\Theta(A)}\frac{\Theta(A)^k}{k!}\left(\int_A f\,\mathrm{d}\Theta\right)^k\Theta(A)^{-k}\\ &= \exp\left(-\Theta(A)+\int_A f\,\mathrm{d}\Theta\right) = \exp\left(\int_A (f-1)\,\mathrm{d}\Theta\right)\\ &= \exp\left(\int_E (f-1)\,\mathrm{d}\Theta\right).\end{aligned}$$

The general assertion (3.3) is now obtained if we choose a sequence of compact sets increasing to E and apply the monotone convergence theorem.

Conversely, suppose that (3.3) holds for all measurable functions $f : E \to [0,1]$. Again, the assertion is trivial for $\Theta = 0$ (the empty process is Poisson), hence we assume $\Theta \neq 0$. We choose

$$f(x) := t^{\mathbf{1}_B(x)},$$

where $t \in (0,1)$ and $B \in \mathcal{B}$ with $\Theta(B) < \infty$. Then,

$$\begin{aligned}\mathbb{E}\,t^{X(B)} &= \mathbb{E}\prod_{x\in X} t^{\mathbf{1}_B(x)}\\ &= \exp\left(\int_E (t^{\mathbf{1}_B(x)}-1)\,\Theta(\mathrm{d}x)\right)\\ &= \exp\left((t-1)\Theta(B)\right)\\ &= \mathrm{e}^{-\Theta(B)}\mathrm{e}^{t\Theta(B)}.\end{aligned}$$

We expand both sides and get

$$\sum_{k\in\mathbb{N}_0} t^k\mathbb{P}(X(B)=k) = \mathrm{e}^{-\Theta(B)}\sum_{k\in\mathbb{N}_0} t^k\frac{1}{k!}\Theta(B)^k$$

for all $t \in (0,1)$. Comparing coefficients yields

$$\mathbb{P}(X(B)=k) = \mathrm{e}^{-\Theta(B)}\frac{\Theta(B)^k}{k!}$$

for all $k \in \mathbb{N}_0$ and all $B \in \mathcal{B}$ with $\Theta(B) < \infty$. This shows that X has Poisson counting variables. Since Θ has no atoms, it follows that X is simple and therefore a Poisson process, by Corollary 3.2.2. □

We next deduce a characterization of Poisson processes which will be useful in Section 3.3 when we study Palm distributions.

Theorem 3.2.5 (Mecke). *Let X be a point process in E, the intensity measure Θ of which has no atoms. Then X is a Poisson process if and only if*

$$\mathbb{E}\sum_{x\in X} f(X,x) = \int_E \mathbb{E}\, f(X+\delta_x, x)\,\Theta(\mathrm{d}x), \tag{3.4}$$

for all nonnegative measurable functions f on $\mathsf{N}\times E$.

Proof. First let X be a Poisson process. Since X is simple and since the sets $\mathsf{N}_{C,0}$, $C\in\mathcal{B}$, generate the σ-algebra $\mathcal{N}_s$ (see the remark after Lemma 3.1.4), it is sufficient to consider the case $f=\mathbf{1}_{\mathsf{N}_{C,0}\times B}$, where $C,B\in\mathcal{B}$ are Borel sets with $\Theta(B)<\infty$.

The left side of (3.4) then equals

$$\mathbb{E}[\mathbf{1}\{X(C)=0\}X(B)].$$

For the right side, we obtain

$$\begin{aligned}
&\int_B \mathbb{E}\,\mathbf{1}\{(X+\delta_x)(C)=0\}\,\Theta(\mathrm{d}x)\\
&= \int_{B\cap C} \mathbb{P}(X(C)+1=0)\,\Theta(\mathrm{d}x) + \int_{B\setminus C} \mathbb{P}(X(C)=0)\,\Theta(\mathrm{d}x)\\
&= \mathbb{P}(X(C)=0)\Theta(B\setminus C)\\
&= \mathbb{E}\,\mathbf{1}\{X(C)=0\}\mathbb{E}\,X(B\setminus C)\\
&= \mathbb{E}\,[\mathbf{1}\{X(C)=0\}X(B)],
\end{aligned}$$

due to the independence properties of X (Theorem 3.2.2(a)).

For the converse direction, we assume that (3.4) holds for all measurable functions $f:\mathsf{N}\times E\to[0,\infty)$. We choose

$$f(\eta,x)=t^{\eta(B)}\mathbf{1}_B(x)$$

for $t\in(0,1]$ and $B\in\mathcal{B}$ with $\Theta(B)<\infty$. Then we obtain

$$\mathbb{E}\,t^{X(B)}X(B) = \int_E \mathbb{E}\,t^{X(B)+1}\mathbf{1}_B(x)\,\Theta(\mathrm{d}x) = \mathbb{E}\,t^{X(B)+1}\Theta(B).$$

It follows that $h:t\mapsto \mathbb{E}\,t^{X(B)}$ is differentiable in $(0,1]$ and satisfies the differential equation

$$h'(t)=\Theta(B)h(t),\quad h(1)=1.$$

Therefore

$$h(t)=\exp((t-1)\Theta(B)),$$

which gives

$$\mathbb{E}\,t^{X(B)}=\mathrm{e}^{-\Theta(B)}\mathrm{e}^{t\Theta(B)}.$$

As we have seen in the proof of Theorem 3.2.4, this implies that X is a Poisson process. □

Formula (3.4) can be iterated and then yields a result which will be useful, for instance, in Section 10.2.

Corollary 3.2.3 (Slivnyak–Mecke formula). *Let X be a Poisson process in E with intensity measure Θ, let $m \in \mathbb{N}$, and let $f : \mathsf{N} \times E^m \to \mathbb{R}$ be a nonnegative measurable function. Then*

$$\mathbb{E} \sum_{(x_1,\dots,x_m)\in X^m_{\neq}} f(X, x_1, \dots, x_m)$$

$$= \int_E \cdots \int_E \mathbb{E} f\left(X + \sum_{i=1}^m \delta_{x_i}, x_1, \dots, x_m\right) \Theta(\mathrm{d}x_1)\cdots\Theta(\mathrm{d}x_m).$$

Proof. One uses induction, writing

$$\sum_{(x_1,\dots,x_m)\in X^m_{\neq}} f(X, x_1, \dots, x_m) = \sum_{x\in X} g(X, x)$$

with

$$g(X, x) := \sum_{\substack{(x_2,\dots,x_m)\in X^{m-1}_{\neq} \\ x_2,\dots,x_m \neq x}} f(X, x, x_2, \dots, x_m),$$

and Fubini's theorem. □

If we choose

$$f(\eta, x_1, \dots, x_m) := \mathbf{1}_{A_1\times\ldots\times A_m}(x_1, \dots, x_m)$$

with $A_1, \dots, A_m \in \mathcal{B}$, then

$$\mathbb{E} \sum_{(x_1,\dots,x_m)\in X^m_{\neq}} f(X, x_1, \dots, x_m) = \mathbb{E}\, X^m(A_1 \times \ldots \times A_m \cap E^m_{\neq})$$

$$= \Lambda^{(m)}(A_1 \times \ldots \times A_m),$$

hence Corollary 3.2.3 immediately gives the following result.

Corollary 3.2.4. *For a Poisson process X in E with intensity measure Θ and for $m \in \mathbb{N}$, the mth factorial moment measure $\Lambda^{(m)}$ of X satisfies*

$$\Lambda^{(m)} = \Theta^m.$$

We finally mention a generalization of Poisson processes, the Cox processes. Let Y be a random measure on E; we assume that Y is not identically 0 and has a.s. no atoms. The **Cox process** (or **doubly stochastic Poisson process**) X **directed by** Y is a Poisson process in E with random intensity measure Y. More precisely, the distribution of X is specified by

$$\mathbb{P}(X(A) = k) = \int_{\mathsf{M}} \mathrm{e}^{-\eta(A)} \frac{\eta(A)^k}{k!} \, \mathbb{P}_Y(\mathrm{d}\eta)$$

for $k \in \mathbb{N}_0$ and $A \in \mathcal{G}_c$. The right side is well defined since

$$\varphi_A : \mathsf{M} \to [0, \infty), \quad \eta \mapsto \mathrm{e}^{-\eta(A)} \frac{\eta(A)^k}{k!},$$

is measurable.

Of course, it still remains to show the existence of such a point process X. For $\eta \in \mathsf{M}$, let X_η be the Poisson process with intensity measure η. The mapping $\eta \mapsto \mathbb{P}_{X_\eta}$ is a (probability) kernel; for fixed η it is a probability measure on N_s, and for fixed $A \in \mathcal{N}_s$, the mapping $\eta \mapsto \mathbb{P}_{X_\eta}(A)$ is measurable. In order to see this, we remark that

$$\eta \mapsto \mathbb{P}_{X_\eta}\left(\bigcap_{i=1}^{m} \mathsf{N}_{G_i,k_i}\right) = \prod_{i=1}^{m} \mathrm{e}^{-\eta(G_i)} \frac{\eta(G_i)^{k_i}}{k_i!}$$

is measurable, as we have mentioned above, for pairwise disjoint $G_1, \dots, G_m \in \mathcal{G}_c$. The measurability extends to intersections $\bigcap_{i=1}^m \mathsf{N}_{G_i,k_i}$ with arbitrary sets $G_1, \dots, G_m \in \mathcal{G}_c$, since the probability on the left is then a finite sum of such products. It therefore holds for all $A \in \mathcal{N}_s$, by the usual extension argument. Since $\mathbb{P}_{X_\eta}$ is a kernel,

$$\mathbb{P}_X := \int_{\mathsf{M}} \mathbb{P}_{X_\eta} \, \mathbb{P}_Y(\mathrm{d}\eta)$$

is the distribution of a point process X on E.

We thus have obtained the following result.

Theorem 3.2.6. *Let Y be a random measure on E which is not identically* 0 *and has a.s. no atoms. Then there exists a Cox process X in E directed by Y.*

As follows from the definition of the Cox process X and Fubini's theorem for kernels, the intensity measure Θ of X is given by the intensity measure of the random measure Y.

Notes for Section 3.2

1. Poisson processes are treated in the books on point processes listed in the Notes for Section 3.1; they are also the subject of Kingman [413]. In our treatment of Poisson processes, we emphasized and made heavy use of the fact that the distribution property (3.2) together with the simplicity implies the strong independence properties of Theorem 3.2.2. This observation goes back to Rényi [638]. Conversely, for a point process X with an intensity measure without atoms one can show that the independence of the random variables $X(A_1), X(A_2), \dots$, for all sequences $A_1, A_2, \dots$ of pairwise disjoint Borel sets, implies the Poisson distribution property (3.2) (see, for example, Daley and Vere–Jones [194, Lemma 2.VI]). However, the usual definition of a Poisson process is based on both properties, thus requiring that for pairwise disjoint Borel sets $A_1, \dots, A_k$, $k \in \mathbb{N}$, the random variables $X(A_1), \dots, X(A_k)$

are independent and Poisson distributed. Equivalent to this is the assumption that the random vectors $(X(A_1), \dots, X(A_k))$ have multivariate Poisson distributions. In Mecke *et al.* [500, Appendix A] a corresponding definition is given which uses the generating functional of X. For general (non-simple) Poisson processes, uniqueness can still be proved, based on the finite-dimensional distributions.

2. The example showing that for intensity measures with atoms the distribution property (3.2) is in general not sufficient for uniqueness, is taken from Kerstan, Matthes and Mecke [400, p. 17].

Remarkably, there are simple point processes in $\mathbb{R}^2$ satisfying (3.2) (with $\Theta = \lambda$) for all convex Borel sets, without being Poisson processes (an example due to Moran [560] shows even more).

3. Theorem 3.2.4 presents the 'generating functional' of a Poisson process. How the generating functional is used as a tool in the theory of point processes can be seen, for instance, in Daley and Vere–Jones [194], König and Schmidt [423], Mecke *et al.* [500, Appendix].

4. For Theorem 3.2.5 we refer to Mecke [476], where this property is used in the construction of Poisson processes.

5. The 'complementary theorem'. The complementary theorem of Miles [522, 523] allows one to find Gamma distributions for the 'contents' of certain random sets constructed from the points of a Poisson process. Whereas Miles used an ergodic approach, Møller and Zuyev [555] employed Palm distributions in their more general investigation. We use their words to describe their aims. They consider a family of Poisson processes Φ_ρ defined on an arbitrary space S with intensity measure $\rho\vartheta$ where $\rho > 0$ is a scale parameter and ϑ is a σ-finite measure. For functionals $f(\Phi_\rho)$ which only depend on Φ_ρ through a finite subprocess Ψ_ρ they study derivatives of the mean $\mathbb{E}f(\Phi_\rho)$ with respect to ρ and establish various distributional results for random subsets $\Delta_\rho \subset S$ determined by Ψ_ρ. In particular, under certain equivariance conditions with respect to the scale ρ, the conditional distribution of the 'content' $\vartheta(\Delta_\rho)$ given the 'cardinality' of Ψ_ρ is shown to be a Gamma distribution.

Another more comprehensive version of the complementary theorem is discussed by Cowan [182].

3.3 Palm Distributions

In this section, we consider stationary random measures and point processes on $\mathbb{R}^d$ and introduce the Palm calculus, a powerful tool to describe conditional distributions of random measures and point processes.

In order to motivate the idea of Palm distributions, we start with an intuitive example.

If X is a simple stationary point process in $\mathbb{R}^d$, then X (identified with supp X, by convention) is at the same time a stationary, locally finite closed random set. Hence, we can consider its contact distributions as introduced in Section 2.4. The spherical contact distribution function of X is defined by

$$H(r) = \mathbb{P}(0 \in X + rB^d) \mid 0 \notin X).$$

Since $\mathbb{P}(0 \in X) = 0$ (by Theorem 2.4.7), the condition $0 \notin X$ is satisfied a.s. If δ denotes the distance of the origin 0 to the nearest point in X, then

$$H(r) = 1 - \mathbb{P}(0 \notin X + rB^d) = 1 - \mathbb{P}(rB^d \cap X = \emptyset)$$
$$= \mathbb{P}(X(rB^d) > 0) = \mathbb{P}(\delta \leq r).$$

Thus, H is the distribution function of δ. For a stationary Poisson process with intensity γ,

$$H(r) = 1 - \mathrm{e}^{-\gamma\kappa_d r^d}$$

(where $\kappa_d := \lambda(B^d)$).

The spherical contact distribution function is used in statistical investigations of stationary point processes (for example, in testing the Poisson hypothesis). Further statistical data involve mutual distances between points. For that reason, one wants to consider the distribution of the distance Δ of a 'typical' point of X from its nearest neighbor in X. This requires a suitable notion of distributions with respect to a typical point of X. One might think of conditional distributions, under the condition that a prescribed point, say 0, is a point of the process X. But since $\mathbb{P}(X(\{0\}) = 1) = 0$ for a stationary process, such conditional distributions do not exist in the elementary sense. For a stationary Poisson process X, an alternative procedure is possible. To define the distribution of the random variable Δ which we have in mind, we replace the point 0 by the ball ϵB^d and define $\mathbb{P}(\Delta \leq r)$ as the limit, for $\epsilon \to 0$, of the well-defined conditional probabilities $\mathbb{P}(X(rB^d \setminus \epsilon B^d) > 0 \mid X(\epsilon B^d) = 1)$. Using the independence assertion of Theorem 3.2.2, we get

$$\begin{aligned}
\mathbb{P}(\Delta \leq r) &= \lim_{\epsilon \to 0} \mathbb{P}(X(rB^d \setminus \epsilon B^d) > 0 \mid X(\epsilon B^d) = 1) \\
&= \lim_{\epsilon \to 0} \mathbb{P}(X(rB^d \setminus \epsilon B^d) > 0) \\
&= \lim_{\epsilon \to 0} \left(1 - \mathrm{e}^{-\gamma\lambda(rB^d \setminus \epsilon B^d)}\right) \\
&= 1 - \mathrm{e}^{-\gamma\kappa_d r^d} \\
&= \mathbb{P}(\delta \leq r).
\end{aligned}$$

Thus, Δ and δ have the same distribution, in the case of a stationary Poisson process.

In probability theory, it is possible under suitable assumptions to introduce conditional probabilities, which are not defined in an elementary way, by means of disintegration procedures. A somewhat similar procedure is possible for random measures and point processes and leads to the notion of the Palm distribution $\mathbb{P}^x$ with respect to a given point $x \in \mathbb{R}^d$. Since we shall consider only stationary random measures in this section, it suffices to consider $\mathbb{P}^0$, with respect to the origin 0; the distribution $\mathbb{P}^x$ is then obtained as the image $t_x\mathbb{P}^0$ of $\mathbb{P}^0$ under the translation by the vector x.

We begin with a measurability assertion, for which stationarity does not yet play a role.

Lemma 3.3.1. *Let X be a random measure on $\mathbb{R}^d$, and let $f : \mathbb{R}^d \times \mathsf{M} \to \mathbb{R}$ be a nonnegative $\mathcal{B} \otimes \mathcal{M}$-measurable function. Then the mappings*

$$\varphi_1 : \omega \mapsto \int_{\mathbb{R}^d} f(x, X(\omega))\, X(\omega, \mathrm{d}x)$$

and

$$\varphi_2 : \omega \mapsto \int_{\mathbb{R}^d} f(x, X(\omega) - x)\, X(\omega, \mathrm{d}x)$$

are measurable.

Proof. Since $(x, \eta) \mapsto (x, \eta - x)$ is measurable, together with f also the function $g : (x, \eta) \to f(x, \eta - x)$ is measurable (and nonnegative). Therefore, observing that

$$\int_{\mathbb{R}^d} g(x, X(\omega))\, X(\omega, \mathrm{d}x) = \int_{\mathbb{R}^d} f(x, X(\omega) - x)\, X(\omega, \mathrm{d}x),$$

it is sufficient to prove the measurability of φ_1.

For that, it is sufficient to consider indicator functions $f = \mathbf{1}_{B \times A}$ with $B \in \mathcal{B}$ and $A \in \mathcal{M}$. In this case,

$$\varphi_1(\omega) = \int_{\mathbb{R}^d} \mathbf{1}_B(x) \mathbf{1}_A(X(\omega))\, X(\omega, \mathrm{d}x) = X(\omega)(B) \mathbf{1}_A(X(\omega)),$$

thus φ_1 is a product of measurable functions. □

Due to the lemma, for a random measure X on $\mathbb{R}^d$, we can define two measures μ and $\mathbf{C}$ on $\mathcal{B} \otimes \mathcal{M}$ in the following way. For $\widetilde{A} \in \mathcal{B} \otimes \mathcal{M}$, let

$$\mu(\widetilde{A}) := \mathbb{E} \int_{\mathbb{R}^d} \mathbf{1}_{\widetilde{A}}(x, X - x)\, X(\mathrm{d}x)$$

and

$$\mathbf{C}(\widetilde{A}) := \mathbb{E} \int_{\mathbb{R}^d} \mathbf{1}_{\widetilde{A}}(x, X)\, X(\mathrm{d}x).$$

The σ-additivity is a consequence of the monotone convergence theorem. We observe that $\mathbf{C}$ is the image measure of μ under $(x, \eta) \mapsto (x, \eta + x)$, and μ is the image measure of $\mathbf{C}$ under $(x, \eta) \mapsto (x, \eta - x)$. The measure $\mathbf{C}$ is called the **Campbell measure** of X. It has a simple meaning on product sets:

$$\mathbf{C}(B \times A) = \mathbb{E}[X(B) \mathbf{1}_A(X)]$$

for $B \in \mathcal{B}$ and $A \in \mathcal{M}$. The definition of $\mathbf{C}$, Lemma 3.3.1, and monotone convergence immediately yield an extension of the Campbell theorem,

$$\mathbb{E}\int_{\mathbb{R}^d} f(x,X)\,X(\mathrm{d}x) = \int_{\mathbb{R}^d\times\mathsf{M}} f(x,\eta)\,\mathbf{C}(\mathrm{d}(x,\eta)) \tag{3.5}$$

for nonnegative measurable functions $f : \mathbb{R}^d \times \mathsf{M} \to \mathbb{R}$.

Now we show that in the stationary case the measures μ and $\mathbf{C}$ can be decomposed.

Theorem 3.3.1. *Let X be a stationary random measure on $\mathbb{R}^d$ with intensity $\gamma > 0$. Then*

$$\mu = \gamma\lambda \otimes \mathbb{P}^0$$

with a (uniquely determined) probability measure $\mathbb{P}^0$ on $\mathcal{M}$. For $B \in \mathcal{B}$ and $A \in \mathcal{M}$,

$$\mathbf{C}(B \times A) = \gamma \int_B \mathbb{P}^0(A - x)\,\lambda(\mathrm{d}x).$$

Proof. Let $B \in \mathcal{B}$, $A \in \mathcal{M}$ and $y \in \mathbb{R}^d$ be given. Using the stationarity of X, we get

$$\begin{aligned}
\mu((B+y)\times A) &= \mathbb{E}\int_{\mathbb{R}^d} \mathbf{1}_{B+y}(x)\mathbf{1}_A(X-x)\,X(\mathrm{d}x)\\
&= \mathbb{E}\int_{\mathbb{R}^d} \mathbf{1}_{B}(x)\mathbf{1}_A(X-y-x)\,(X-y)(\mathrm{d}x)\\
&= \mathbb{E}\int_{\mathbb{R}^d} \mathbf{1}_{B}(x)\mathbf{1}_A(X-x)\,X(\mathrm{d}x)\\
&= \mu(B\times A).
\end{aligned}$$

Thus, for fixed A the measure $\mu(\cdot \times A)$ is translation invariant (and locally finite). The uniqueness of Lebesgue measure implies that $\mu(B\times A) = \alpha(A)\lambda(B)$ for $B \in \mathcal{B}$, with a factor $\alpha(A)$. Since

$$\mu(B \times \mathsf{M}) = \mathbb{E}\int_{\mathbb{R}^d} \mathbf{1}_B(x)\,X(\mathrm{d}x) = \mathbb{E}X(B) = \gamma\lambda(B)$$

for $B \in \mathcal{B}$, the definition $\mathbb{P}^0 := \gamma^{-1}\mu(C^d \times \cdot)$ (where $C^d = [0,1]^d$ is the unit cube) yields the product representation of μ.

The representation of $\mathbf{C}$ is obtained by applying the map $(x,\eta) \mapsto (x, \eta + x)$. □

The probability measure $\mathbb{P}^0$ on $\mathcal{M}$ is called the **Palm distribution** of X. Theorem 3.3.1 together with the definition of μ gives the following explicit representation of $\mathbb{P}^0$, which reveals its intuitive meaning.

Theorem 3.3.2. *Let X be a stationary random measure on $\mathbb{R}^d$ with intensity $\gamma > 0$. For an arbitrary Borel set $B \in \mathcal{B}$ with $\lambda(B) = 1$ and for $A \in \mathcal{M}$,*

$$\gamma\mathbb{P}^0(A) = \mathbb{E}\int_{\mathbb{R}^d} \mathbf{1}_B(x)\mathbf{1}_A(X-x)\,X(\mathrm{d}x).$$

More generally, we can choose an arbitrary Borel set $B \in \mathcal{B}$ with $0 < \lambda(B) < \infty$ and represent the Palm distribution at A in the form

$$\mathbb{P}^0(A) = \frac{\mathbb{E} \int_B \mathbf{1}_A(X - x)\, X(\mathrm{d}x)}{\mathbb{E} X(B)} \tag{3.6}$$

and thus as a quotient of intensities.

Using the Palm distribution, the extension (3.5) of the Campbell theorem can be simplified in the stationary case.

Theorem 3.3.3 (Refined Campbell theorem). *Let X be a stationary random measure on $\mathbb{R}^d$ with intensity $\gamma > 0$, and let $f : \mathbb{R}^d \times \mathsf{M} \to \mathbb{R}$ be a nonnegative measurable function. Then $\int_{\mathbb{R}^d} f(x, X)\, X(\mathrm{d}x)$ is measurable, and*

$$\mathbb{E} \int_{\mathbb{R}^d} f(x, X)\, X(\mathrm{d}x) = \gamma \int_{\mathbb{R}^d} \int_{\mathsf{M}} f(x, \eta + x)\, \mathbb{P}^0(\mathrm{d}\eta)\, \lambda(\mathrm{d}x).$$

Proof. The measurability was already proved in Lemma 3.3.1. The further assertion then follows from (3.5), using the decomposition of **C**. □

We use the refined Campbell theorem to prove an inversion formula by which the distribution $\mathbb{P}_X$ of a random measure X is expressed in terms of its Palm distribution $\mathbb{P}^0$. For this purpose, we make use of a measurable function $h : \mathbb{R}^d \times \mathsf{M} \to [0, \infty)$ which satisfies

$$\int_{\mathbb{R}^d} h(x, \eta)\, \eta(\mathrm{d}x) = 1 \tag{3.7}$$

for all $\eta \in \mathsf{M} \setminus \{0\}$. Such a function h can be easily constructed as follows. There exist pairwise disjoint Borel sets $A_1, A_2, \dots$ in $\mathbb{R}^d$ which are relatively compact and cover $\mathbb{R}^d$. For $\eta \in \mathsf{M} \setminus \{0\}$ and $x \in \mathbb{R}^d$, let

$$\tilde{h}(x, \eta) := \sum_{n \in \mathbb{N},\, \eta(A_n) > 0} \frac{1}{2^n} \eta(A_n)^{-1} \mathbf{1}_{A_n}(x),$$

then

$$0 < \alpha(\eta) := \int_{\mathbb{R}^d} \tilde{h}(x, \eta)\, \eta(\mathrm{d}x) = \sum_{n \in \mathbb{N},\, \eta(A_n) > 0} \frac{1}{2^n} \le 1.$$

Hence

$$h(x, \eta) := \begin{cases} \alpha(\eta)^{-1} \tilde{h}(x, \eta), & \text{if } \eta \neq 0, \\ 0, & \text{if } \eta = 0, \end{cases}$$

satisfies (3.7).

Theorem 3.3.4. *Let X be a stationary random measure on $\mathbb{R}^d$ with intensity $\gamma > 0$, and let $h : \mathbb{R}^d \times \mathsf{M} \to \mathbb{R}$ be a nonnegative measurable function satisfying (3.7). Then*

$$\mathbb{P}_X(A) = \gamma \int_{\mathbb{R}^d} \int_{\mathsf{M}} \mathbf{1}_A(\eta + x) h(x, \eta + x) \, \mathbb{P}^0(\mathrm{d}\eta) \, \lambda(\mathrm{d}x)$$

for all $A \in \mathcal{M}$ *with* $0 \notin A$.

Proof. For $A \in \mathcal{M}$ with $0 \notin A$, we apply Theorem 3.3.3 to the measurable function $f : \mathbb{R}^d \times \mathsf{M} \to \mathbb{R}$ defined as

$$f(x, \eta) := h(x, \eta + x) \mathbf{1}_A(\eta + x)$$

and get, using (3.7),

$$\begin{aligned} \mathbb{P}_X(A) &= \int_A \int_{\mathbb{R}^d} h(x, \eta) \, \eta(\mathrm{d}x) \, \mathbb{P}_X(\mathrm{d}\eta) \\ &= \int_{\mathsf{M}} \int_{\mathbb{R}^d} f(x, \eta - x) \, \eta(\mathrm{d}x) \, \mathbb{P}_X(\mathrm{d}\eta) \\ &= \mathbb{E} \int_{\mathbb{R}^d} f(x, X - x) \, X(\mathrm{d}x) \\ &= \gamma \int_{\mathbb{R}^d} \int_{\mathsf{M}} f(x, \eta) \, \mathbb{P}^0(\mathrm{d}\eta) \, \lambda(\mathrm{d}x) \\ &= \gamma \int_{\mathbb{R}^d} \int_{\mathsf{M}} \mathbf{1}_A(\eta + x) h(x, \eta + x) \, \mathbb{P}^0(\mathrm{d}\eta) \, \lambda(\mathrm{d}x), \end{aligned}$$

as stated. □

For $x \in \mathbb{R}^d$ we now put $\mathbb{P}^x := t_x \mathbb{P}^0$. Then Theorem 3.3.1 yields the representation

$$\mathbf{C}(B \times A) = \gamma \int_B \mathbb{P}^x(A) \, \lambda(\mathrm{d}x) = \int_B \mathbb{P}^x(A) \, \Theta(\mathrm{d}x). \tag{3.8}$$

Thus, the family $\{\mathbb{P}^x : x \in \mathbb{R}^d\}$ is a disintegration of the Campbell measure $\mathbf{C}$ with respect to the intensity measure. Since in the classical case of real (or vector-valued) random variables, (regular) conditional probabilities are defined by a disintegration of the distribution (cf., for example, Kallenberg [386, Th. 6.3]), it is plausible to interpret $\mathbb{P}^x$ as the distribution of X, given that x is a point in the support of X. For simple point processes X this intuitive meaning becomes clearer since we can interpret $\mathbb{P}^x$ as the conditional distribution $\mathbb{P}(X \in \cdot \mid x \in X)$.

If the random measure X is ergodic (some ergodicity considerations are the topic of Section 9.3), the interpretation of $\mathbb{P}^0$ can be made even more explicit. As a counterpart to (3.6), without expectations, we then have

$$\mathbb{P}^0(A) = \lim_{r \to \infty} \frac{\int_{rB^d} \mathbf{1}_A(X - x) \, X(\mathrm{d}x)}{X(rB^d)} \qquad \text{a.s.} \tag{3.9}$$

If X is, in particular, a simple point process, this can be interpreted as follows: $\mathbb{P}^0$ is the asymptotic distribution (for $r \to \infty$) of $X - \xi$, where ξ is a point randomly chosen from $X \cap rB^d$. For that reason, $\mathbb{P}^0$ is often considered as the distribution of X with respect to a 'typical point' of the process. Also the disintegration (3.8) can be made more explicit in the ergodic case; we have

$$\mathbb{P}_X(A) = \lim_{r\to\infty} \frac{1}{\lambda(rB^d)} \int_{rB^d} \mathbb{P}^x(A)\,\lambda(\mathrm{d}x) \qquad \text{a.s.,} \tag{3.10}$$

which makes the mentioned analogy of Palm distributions to (regular) conditional probabilities even clearer. Relation (3.10) follows from (3.8) and the definition of the Campbell measure, since in the ergodic case

$$\lim_{r\to\infty} \frac{1}{\lambda(rB^d)} X(rB^d) = \gamma \qquad \text{a.s.}$$

For further details, and a proof of (3.9), we refer to König and Schmidt [423, ch. 12].

Now we show that the Palm distribution can be used to obtain a simplified representation of the second moment measure of a simple stationary point process X in $\mathbb{R}^d$. Using its Palm distribution $\mathbb{P}^0$ and intensity $\gamma > 0$, we define the **reduced second moment measure** $\mathbb{K}$ by

$$\gamma\mathbb{K}(A) := \int_{\mathsf{N}} \eta(A \setminus \{0\})\,\mathbb{P}^0(\mathrm{d}\eta) \qquad \text{for } A \in \mathcal{B}.$$

We assume that it is locally finite. For the second factorial moment measure of X we then obtain from Theorem 3.3.3, for $A_1, A_2 \in \mathcal{B}$,

$$\begin{aligned}
\Lambda^{(2)}(A_1 \times A_2) &= \mathbb{E} \sum_{(x,y)\in X^2_{\neq}} \mathbf{1}_{A_1}(x)\mathbf{1}_{A_2}(y) \\
&= \mathbb{E} \sum_{x\in X} \mathbf{1}_{A_1}(x) X(A_2 \setminus \{x\}) \\
&= \gamma \int_{\mathbb{R}^d} \int_{\mathsf{N}} \mathbf{1}_{A_1}(x)\eta((A_2 - x) \setminus \{0\})\,\mathbb{P}^0(\mathrm{d}\eta)\,\lambda(\mathrm{d}x) \\
&= \gamma^2 \int_{A_1} \mathbb{K}(A_2 - x)\,\lambda(\mathrm{d}x) \\
&= \gamma^2 \int_{A_1} \int_{\mathbb{R}^d} \mathbf{1}_{A_2}(x+t)\,\mathbb{K}(\mathrm{d}t)\,\lambda(\mathrm{d}x).
\end{aligned}$$

Together with (3.1), for the second moment measure of X this yields the representation

$$\Gamma^{(2)}(A_1 \times A_2) = \Theta(A_1 \cap A_2) + \gamma^2 \int_{\mathbb{R}^d} \int_{\mathbb{R}^d} \mathbf{1}_{A_1}(x)\mathbf{1}_{A_2}(x+t)\,\lambda(\mathrm{d}x)\,\mathbb{K}(\mathrm{d}t),$$

which involves only measures on $\mathbb{R}^d$.

For a stationary Poisson process X, the Palm distribution $\mathbb{P}^0$ is closely related to the distribution of X. To formulate this very useful connection, it is convenient to identify again a simple point process X with its support $\operatorname{supp} X$. Correspondingly, the Palm distribution $\mathbb{P}^0$ of X is interpreted as a measure on $\mathcal{F}$. In other words, we identify $\mathbb{P}^0$ with its image measure under the bijective measurable mapping $i_s : \mathsf{N}_s \to \mathcal{F}_{\ell f}$ (cf. Lemma 3.1.4), but without introducing a new notation, and view it as a measure on $\mathcal{F}$.

Theorem 3.3.5 (Theorem of Slivnyak). *Let $\mathbb{P}^0$ be the Palm distribution of a stationary simple point process X in $\mathbb{R}^d$ with intensity $\gamma > 0$. Then X is a Poisson process if and only if*

$$\mathbb{P}^0(A) = \mathbb{P}(X \cup \{0\} \in A) \tag{3.11}$$

holds for all $A \in \mathcal{B}(\mathcal{F})$.

Proof. Theorem 3.2.5 (applied to this special situation and formulated in set-theoretic language) and the stationarity of X imply that X is a Poisson process if and only if

$$\begin{aligned}\mathbb{E} \sum_{x \in X} g(x, X) &= \gamma \int_{\mathbb{R}^d} \mathbb{E}\, g(x, X \cup \{x\})\, \lambda(\mathrm{d}x) \\ &= \gamma \int_{\mathbb{R}^d} \int_{\mathsf{N}} g(x, (\eta \cup \{0\}) + x)\, \mathbb{P}_X(\mathrm{d}\eta)\, \lambda(\mathrm{d}x)\end{aligned}$$

holds for all measurable functions $g : \mathbb{R}^d \times \mathsf{N} \to [0, \infty)$. Applying the refined Campbell theorem (Theorem 3.3.3) to the left side, we obtain that X is a Poisson process if and only if

$$\int_{\mathbb{R}^d} \int_{\mathsf{N}} g(x, \eta + x)\, \mathbb{P}^0(\mathrm{d}\eta)\, \lambda(\mathrm{d}x) = \int_{\mathbb{R}^d} \int_{\mathsf{N}} g(x, (\eta \cup \{0\}) + x)\, \mathbb{P}_X(\mathrm{d}\eta)\, \lambda(\mathrm{d}x)$$

holds for all g. Choosing $g(x, \eta) := \mathbf{1}_A(\eta - x)\mathbf{1}_B(x)$ with $A \in \mathcal{B}(\mathcal{F})$ and $B \in \mathcal{B}(\mathbb{R}^d)$, we see that the latter equality is equivalent to

$$\mathbb{P}^0(A) = \mathbb{P}(X \cup \{0\} \in A)$$

for all $A \in \mathcal{B}(\mathcal{F})$. □

With the aid of Slivnyak's theorem, for the spherical contact distribution function of the stationary Poisson process X we obtain

$$\begin{aligned}\mathbb{P}(\delta \le r) &= \mathbb{P}(d\,(0, (X \cup \{0\}) \setminus \{0\}) \le r) \\ &= \mathbb{P}^0(\{F \in \mathcal{F}_{\ell f} : d\,(0, F \setminus \{0\}) \le r\}) \\ &= \lim_{s \to \infty} \frac{\operatorname{card}\{x \in X \cap sB^d : d\,(x, X \setminus \{x\}) \le r\}}{\operatorname{card}\,(X \cap sB^d)} \\ &=: F(r),\end{aligned}$$

where the limit relation holds a.s., because of (3.9). If we interpret the distribution function F as the distribution function of the distance Δ of a typical point of X to its nearest neighbor, we see that δ and Δ have the same distribution, as we already found heuristically.

Notes for Section 3.3

1. An essential precondition for the existence of the Palm distribution $\mathbb{P}^0$ was our general assumption that the intensity measure Θ of a random measure X is locally finite. Thus, we only considered in this section stationary random measures X on $\mathbb{R}^d$ with finite intensity γ. The measure $\rho^0 := \gamma\mathbb{P}^0$ is usually called the **Palm measure** of X. If one skips the assumption that $\gamma < \infty$, the decompositions of μ and the Campbell measure $\mathbf{C}$ in Theorem 3.3.1 still hold true and define the Palm measure ρ^0, which in this case need not be locally finite. Some of the results following Theorem 3.3.1 are valid in this more general situation.

2. If the interpretation of point processes as random counting measures is maintained, then there is another concise formulation of equation (3.11) in Slivnyak's theorem, namely

$$\mathbb{P}^0 = \mathbb{P}_X * \delta_{\delta_0},$$

where $*$ denotes the convolution of measures on N.

3. Besides the Palm distribution $\mathbb{P}^0$, one often considers the **reduced Palm distribution** $\mathbb{P}_!^0$. For a simple point process X in $\mathbb{R}^d$ it is defined by

$$\mathbb{P}_!^0(A) := \mathbb{P}^0(\{F \cup \{0\} : F \in A \cap \mathcal{F}^{\{0\}}\})$$

for $A \in \mathcal{B}(\mathcal{F})$; it describes the distribution of $X \setminus \{0\}$ under the condition $0 \in X$. Equation (3.11) in Slivnyak's theorem then takes the simple form

$$\mathbb{P}_!^0 = \mathbb{P}_X.$$

4. Palm distributions were introduced for stationary point processes on the real line $\mathbb{R}$ by Palm [593] and developed further by many authors. For a brief description of the historical development, including the names Khinchin (1955), Kaplan (1955), Ryll–Nardzewski (1961) and Slivnyak (1962, 1966), we refer to Daley and Vere–Jones [194]. Of special importance were the paper by Mecke [476] and the book by Kerstan, Matthes and Mecke [400]. Most modern books on random measures or point processes contain chapters on Palm measures.

5. The interpretation of Palm distributions given after (3.9), where X is a simple point process, can be made precise in the following way. There is a random point $\xi \in X$ such that the distribution of $X - \xi$ is the Palm distribution $\mathbb{P}^0$ of X. This is a consequence of Thorisson's results on shift-coupling (see [386, Lemma 11.7]). Holroyd and Peres [349] give an explicit construction of ξ, depending only on X.

6. The approach to Palm measures through disintegration of the Campbell measure can be pursued also in the non-stationary setting and produces a family $\mathbb{P}^x$, $x \in \mathbb{R}^d$, of probability measures $\mathbb{P}^x$ which can be interpreted as conditional distributions of the random measure (or point process) X, given that x is a point in (the support of) X. This set-up is used, for example, in Daley and Vere–Jones [194] and in

Kallenberg [385]. A rather general theory of disintegration with applications to Palm measures, without the usual explicit topological requirements and for more general group operations, is presented in Kallenberg [387].

7. The notion of Palm distributions can be extended to that of n-point (or n-fold) Palm distributions. General inversion formulas for these, in the spirit of Theorem 3.3.4, were treated by Hanisch [320].

3.4 Palm Distributions – General Approach

The Palm distribution $\mathbb{P}^0$ of a random measure X on $\mathbb{R}^d$, which we introduced in the last section, is a probability measure on $(\mathsf{M}, \mathcal{M})$. This is motivated by the invariance properties with respect to translations which we exploited and by the fact that the translation group operates naturally on M. In this section we describe a slightly more abstract approach, which is useful to derive an important exchange formula for Palm distributions.

We now assume that the basic probability space $(\Omega, \mathbf{A}, \mathbb{P})$ is supplied with an additional structure, namely a group $\mathcal{T} = \{T_x : x \in \mathbb{R}^d\}$ of automorphisms. Here, a map $T : \Omega \to \Omega$ is called an **automorphism** if T is bijective and T and T^{-1} are measurable and leave the probability measure $\mathbb{P}$ invariant (that is, satisfy $\mathbb{P}(TA) = \mathbb{P}(T^{-1}A) = \mathbb{P}(A)$ for all $A \in \mathbf{A}$). The group structure of $\mathcal{T}$ is enforced by requiring that $T_x T_y = T_{x+y}$ for all $x, y \in \mathbb{R}^d$, thus $\mathcal{T}$ with the composition is an abelian group (the neutral element is T_0, the identity map). $\mathcal{T}$ is also called a **flow** on $(\Omega, \mathbf{A}, \mathbb{P})$ (parameterized by $\mathbb{R}^d$), and the quadruple $(\Omega, \mathbf{A}, \mathbb{P}, \mathcal{T})$ is called a **dynamical system** (we shall discuss such dynamical systems further in connection with ergodic limits in Section 9.3). Within this framework, a random measure X on $\mathbb{R}^d$ is called **stationary** or **adapted** (to the flow $\mathcal{T}$) if

$$X(\omega, B + x) = X(T_x \omega, B) \qquad \text{for } \omega \in \Omega,\ x \in \mathbb{R}^d,\ B \in \mathcal{B}.$$

As in Section 3.3, we obtain the following results. Since the proofs are nearly identical, we only emphasize instances where there are essential differences.

Lemma 3.4.1. *Let X be a random measure on $\mathbb{R}^d$, and let $f : \mathbb{R}^d \times \Omega \to \mathbb{R}$ be a nonnegative $\mathcal{B} \otimes \mathbf{A}$-measurable function. Then the mappings*

$$\omega \mapsto \int_{\mathbb{R}^d} f(x, \omega)\, X(\omega, \mathrm{d}x) \quad \textit{and} \quad \omega \mapsto \int_{\mathbb{R}^d} f(x, T_x \omega)\, X(\omega, \mathrm{d}x)$$

are measurable.

The measures μ and $\mathbf{C}$ are now defined on $\mathcal{B} \otimes \Omega$ by

$$\mu(\widetilde{A}) := \int_\Omega \int_{\mathbb{R}^d} \mathbf{1}_{\widetilde{A}}(x, T_x \omega)\, X(\omega, \mathrm{d}x)\, \mathbb{P}(\mathrm{d}\omega)$$

and

$$\mathbf{C}(\widetilde{A}) := \int_\Omega \int_{\mathbb{R}^d} \mathbf{1}_{\widetilde{A}}(x,\omega)\, X(\omega, \mathrm{d}x)\, \mathbb{P}(\mathrm{d}\omega).$$

Now, $\mathbf{C}$ is the image measure of μ under $(x,\omega) \mapsto (x, T_{-x}\,\omega)$, and μ is the image measure of $\mathbf{C}$ under $(x,\omega) \mapsto (x, T_x\,\omega)$. The measure $\mathbf{C}$ is called the **Campbell measure** of X. It fulfills

$$\int_\Omega \int_{\mathbb{R}^d} f(x,\omega)\, X(\omega, \mathrm{d}x)\, \mathbb{P}(\mathrm{d}\omega) = \int_{\mathbb{R}^d \times \Omega} f(x,\omega)\, \mathbf{C}(\mathrm{d}(x,\omega))$$

for nonnegative measurable functions $f : \mathbb{R}^d \times \Omega \to \mathbb{R}$.

Theorem 3.4.1. *Let X be a stationary random measure on $\mathbb{R}^d$ with intensity $\gamma > 0$. Then*

$$\mu = \gamma\lambda \otimes \mathbb{P}^0$$

with a (uniquely determined) probability measure $\mathbb{P}^0$ on Ω. For $B \in \mathcal{B}$ and $A \in \mathbf{A}$,

$$\mathbf{C}(B \times A) = \gamma \int_B \mathbb{P}^0(T_x A)\, \lambda(\mathrm{d}x).$$

The measure $\mathbb{P}^0$ is called the **Palm distribution** of X.

Theorem 3.4.2. *Let X be a stationary random measure on $\mathbb{R}^d$ with intensity $\gamma > 0$. For an arbitrary Borel set $B \in \mathcal{B}$ with $\lambda(B) = 1$ and for $A \in \mathbf{A}$,*

$$\gamma \mathbb{P}^0(A) = \int_\Omega \int_{\mathbb{R}^d} \mathbf{1}_B(x) \mathbf{1}_A(T_x\,\omega)\, X(\omega, \mathrm{d}x)\, \mathbb{P}(\mathrm{d}\omega).$$

Theorem 3.4.3 (Refined Campbell theorem). *Let X be a stationary random measure on $\mathbb{R}^d$ with intensity $\gamma > 0$, and let $f : \mathbb{R}^d \times \Omega \to \mathbb{R}$ be a nonnegative measurable function. Then $\omega \to \int_{\mathbb{R}^d} f(x,\omega)\, X(\omega, \mathrm{d}x)$ is measurable, and*

$$\int_\Omega \int_{\mathbb{R}^d} f(x,\omega)\, X(\omega, \mathrm{d}x)\, \mathbb{P}(\mathrm{d}\omega) = \gamma \int_{\mathbb{R}^d} \int_\Omega f(x, T_{-x}\,\omega)\, \mathbb{P}^0(\mathrm{d}\omega)\, \lambda(\mathrm{d}x).$$

The general framework described so far reduces to the situation in Section 3.3 if we consider a stationary random measure X (in the sense of Section 3.3) and take as $(\Omega, \mathbf{A}, \mathbb{P})$ the canonical space $(\mathsf{M}, \mathcal{M}, \mathbb{P}_X)$ where the automorphisms T_x are the translations t_{-x}, $x \in \mathbb{R}^d$. The advantages of the abstract approach become apparent if we now consider two stationary random measures X, Y.

Theorem 3.4.4. *Let X be a stationary random measure on $\mathbb{R}^d$ with intensity $\gamma > 0$ and Palm distribution $\mathbb{P}^0$, and let Y be a further stationary random measure on $\mathbb{R}^d$. Then, for a nonnegative measurable function $g : \mathbb{R}^d \times \mathbb{R}^d \times \Omega \to \mathbb{R}$, we have*

$$\int_\Omega \int_{\mathbb{R}^d} \int_{\mathbb{R}^d} g(x, y, T_x\,\omega)\, Y(\omega, \mathrm{d}y)\, X(\omega, \mathrm{d}x)\, \mathbb{P}(\mathrm{d}\omega)$$
$$= \gamma \int_\Omega \int_{\mathbb{R}^d} \int_{\mathbb{R}^d} g(x, x+y, \omega)\, Y(\omega, \mathrm{d}y)\, \lambda(\mathrm{d}x)\, \mathbb{P}^0(\mathrm{d}\omega).$$

Proof. We apply Theorem 3.4.3 to the (measurable) function

$$f(x, \omega) := \int_{\mathbb{R}^d} g(x, x+y, \omega)\, Y(\omega, \mathrm{d}y)$$

and get

$$\gamma \int_\Omega \int_{\mathbb{R}^d} \int_{\mathbb{R}^d} g(x, x+y, \omega)\, Y(\omega, \mathrm{d}y)\, \lambda(\mathrm{d}x)\, \mathbb{P}^0(\mathrm{d}\omega)$$
$$= \gamma \int_\Omega \int_{\mathbb{R}^d} f(x, \omega)\, \lambda(\mathrm{d}x)\, \mathbb{P}^0(\mathrm{d}\omega)$$
$$= \int_\Omega \int_{\mathbb{R}^d} f(x, T_x\,\omega)\, X(\omega, \mathrm{d}x)\, \mathbb{P}(\mathrm{d}\omega)$$
$$= \int_\Omega \int_{\mathbb{R}^d} \int_{\mathbb{R}^d} g(x, x+y, T_x\,\omega)\, Y(T_x\,\omega, \mathrm{d}y)\, X(\omega, \mathrm{d}x)\, \mathbb{P}(\mathrm{d}\omega)$$
$$= \int_\Omega \int_{\mathbb{R}^d} \int_{\mathbb{R}^d} g(x, y, T_x\,\omega)\, Y(\omega, \mathrm{d}y)\, X(\omega, \mathrm{d}x)\, \mathbb{P}(\mathrm{d}\omega),$$

as asserted. □

We use Theorem 3.4.4 to establish a useful exchange formula for Palm distributions.

Theorem 3.4.5 (Exchange formula of Neveu). *Let X, Y be stationary random measures on $\mathbb{R}^d$ with intensities $\gamma_X, \gamma_Y > 0$ and Palm distributions $\mathbb{P}^{0,X}, \mathbb{P}^{0,Y}$. Then, for a nonnegative measurable function $f : \mathbb{R}^d \times \Omega \to \mathbb{R}$, we have*

$$\gamma_X \int_\Omega \int_{\mathbb{R}^d} f(y, T_y\,\omega)\, Y(\omega, \mathrm{d}y)\, \mathbb{P}^{0,X}(\mathrm{d}\omega)$$
$$= \gamma_Y \int_\Omega \int_{\mathbb{R}^d} f(-x, \omega)\, X(\omega, \mathrm{d}x)\, \mathbb{P}^{0,Y}(\mathrm{d}\omega).$$

Proof. Using Theorem 3.4.2, the stationarity and Fubini's theorem, we obtain

$$\gamma_X \int_\Omega \int_{\mathbb{R}^d} f(y, T_y\,\omega)\, Y(\omega, \mathrm{d}y)\, \mathbb{P}^{0,X}(\mathrm{d}\omega)$$
$$= \int_\Omega \int_{\mathbb{R}^d} \int_{\mathbb{R}^d} \mathbf{1}_B(x) f(y-x, T_y\,\omega)\, X(\omega, \mathrm{d}x)\, Y(\omega, \mathrm{d}y)\, \mathbb{P}(\mathrm{d}\omega),$$

where $B \in \mathcal{B}$ is a set with $\lambda(B) = 1$. Now we apply Theorem 3.4.4 with

$$g(y, x, \omega) := \mathbf{1}_B(x) f(y - x, \omega)$$

(and the roles of X, Y interchanged) and get the assertion after another application of Fubini's theorem. □

Notes for Section 3.4

1. The use of abstract flows in the set-up for Palm measures appears in Geman and Horowitz [256], Mecke [477] and Neveu [582].

2. As we remarked in the Notes for Section 3.3, Palm-type formulas can be established for stationary random measures with not necessarily finite intensity γ by using the Palm measure ρ^0 (which equals $\gamma\mathbb{P}^0$ for finite γ). Even more generally, in the framework underlying this section, a Palm theory has been developed in which the basic probability measure $\mathbb{P}$ is replaced by a σ-finite measure (see the pioneering paper by Mecke [476]).

3. As another generalization, the Palm theory has been extended to random measures and point processes in a (suitable) topological group G (which is then also the parameter group for the automorphisms T_x, $x \in G$). The role of the Lebesgue measure in the formulas is then played by the Haar measure on G (again, see Mecke [476]).

4. For a stationary point process X, Thorisson [757] (see also Thorisson [758]) showed that the Palm distribution $\mathbb{P}^0$ of X is invariant under bijective point-shifts. If X_0 is the point process distributed as $\mathbb{P}^0$, a **point-shift** is a mapping picking a point of $X_0 \setminus \{0\}$ by some unbiased rule. Thorisson asked whether this **point-stationarity** characterizes Palm measures. A positive answer was given by Heveling and Last [339, 341] (see also Heveling [337]). General invariance properties of Palm measures as well as an extension of the concept of point-stationarity to general random measures are discussed in Last and Thorisson [435].

3.5 Marked Point Processes

The construction of Palm distributions in the previous two sections made use of the fact that the random measures are defined on $E = \mathbb{R}^d$, which at the same time is the group of transformations to which stationarity refers. This set-up can be extended to the more general situation where the basic space E is a product, with $\mathbb{R}^d$ being one of the factors.

Thus, we assume now that $E = \mathbb{R}^d \times M$, where M is a locally compact space with a countable base and E carries the product topology. For a translation t_x of $\mathbb{R}^d$, with translation vector x, and for $(y, m) \in E$ we define

$$t_x(y, m) := (t_x y, m),$$

letting translations operate on the first component only. The mapping $t_x : (y, m) \mapsto t_x(y, m)$ is continuous and hence measurable, and for a measure η on E we define $\eta + x$ as the image measure under t_x. Together with X, also $X + x$ is a point process in $\mathbb{R}^d \times M$.

Definition 3.5.1. *A* **marked point process** *in $\mathbb{R}^d$ with mark space M is a simple point process X in $\mathbb{R}^d \times M$ with intensity measure Θ satisfying*

$$\Theta(C \times M) < \infty \qquad \text{for all } C \in \mathcal{C}. \tag{3.12}$$

We remark that by means of the mapping

$$\begin{aligned} e : \mathsf{N}(\mathbb{R}^d) &\to \quad \mathsf{N}_s(\mathbb{R}^d \times \mathbb{N}) \\ \eta \quad &\mapsto \sum_{x \in \operatorname{supp} \eta} \delta_{(x, \eta(\{x\}))} \end{aligned}$$

every point process in $\mathbb{R}^d$ can be represented as a marked point process with mark space $\mathbb{N}$, and thus as a simple point process.

If X is a marked point process in $\mathbb{R}^d$, its image under the projection $(y, m) \mapsto y$ is an ordinary point process X^0 in $\mathbb{R}^d$. The process X^0 is called the **unmarked point process** or **ground process** of X. The interpretation of the pair (y, m) is that of a point y in $\mathbb{R}^d$ to which a mark $m \in M$ is attached. The motivation for introducing marked point processes comes from applications where a mark is used to provide additional information about the point. In simple cases, the marks will be real numbers or tuples of numbers, but they can also be more complicated objects, for example compact sets, as in the next chapter.

Obviously, for a marked point process X, the unmarked point process X^0 need not be simple. In particular, this can happen for a Poisson process X. However, if we speak of a **marked Poisson process** X, in the following, we assume that X is a Poisson process and that X^0 is simple (and therefore a Poisson process, too).

The main interest in the following lies in marked point processes X in $\mathbb{R}^d$ which are **stationary**, in the sense that

$$X \stackrel{\mathcal{D}}{=} X + x$$

for all $x \in \mathbb{R}^d$.

For stationary marked point processes, the intensity measure has a useful decomposition.

Theorem 3.5.1. *If X is a stationary marked point process in $\mathbb{R}^d$ with mark space M and intensity measure $\Theta \neq 0$, then*

$$\Theta = \gamma \lambda \otimes \mathbb{Q}$$

with a number $0 < \gamma < \infty$ and a (uniquely determined) probability measure $\mathbb{Q}$ on M.

Proof. Let $A \in \mathcal{B}(M)$. Putting

$$\mu_A(B) := \Theta(B \times A) \qquad \text{for } B \in \mathcal{B},$$

we have defined a locally finite measure μ_A on $\mathbb{R}^d$. It is translation invariant, since

$$\mu_A(t_x B) = \mathbb{E}X(t_x B \times A) = \mathbb{E}t_{-x}X(B \times A) = \mu_A(B)$$

for $x \in \mathbb{R}^d$, and hence $\mu_A = c(A)\lambda$ with $0 \le c(A) < \infty$. From $c(A) = \Theta(C^d \times A)$ it follows that c is a measure, and for $\gamma := c(M)$ we have $0 < \gamma < \infty$. Putting $\mathbb{Q} := \gamma^{-1}c$, we obtain

$$\Theta(B \times A) = \gamma(\lambda \otimes \mathbb{Q})(B \times A)$$

for $B \in \mathcal{B}$, $A \in \mathcal{B}(M)$, which implies the assertion. □

The number γ is again called the **intensity** of the marked point process X, and the probability measure $\mathbb{Q}$ is called the **mark distribution**. This name is plausible, since the Campbell theorem gives

$$\mathbb{Q}(A) = \frac{1}{\gamma}\mathbb{E} \sum_{(y,m)\in X} \mathbf{1}_B(y)\mathbf{1}_A(m)$$

for $B \in \mathcal{B}$ with $\lambda(B) = 1$ and $A \in \mathcal{B}(M)$.

If the marked point process X is stationary, then the unmarked process X^0 is a stationary ordinary point process, and γ is its intensity. A possible construction (and simulation) of stationary marked point processes with given data $\gamma, \mathbb{Q}$ consists in first generating a stationary point process X^0 in $\mathbb{R}^d$ with intensity γ and then equipping the points $y \in X^0$ independently with random marks m distributed according to $\mathbb{Q}$. However, in this way one obtains only a special class of marked point processes. These independently marked point processes will be considered later in this section.

To construct Palm distributions of stationary marked point processes, we proceed in a similar way to Section 3.3, with $\mathbb{R}^d$ replaced by $\mathbb{R}^d \times M$. Let X be a stationary marked point process in $\mathbb{R}^d$ with mark space M. We define the **Campbell measure** of X by

$$\mathbf{C}(\widetilde{A}) := \mathbb{E} \sum_{(y,m)\in X} \mathbf{1}_{\widetilde{A}}(y, m, X)$$

for $\widetilde{A} \in \mathcal{B} \otimes \mathcal{B}(M) \otimes \mathcal{N}_s(\mathbb{R}^d \times M)$. In particular, we have

$$\mathbf{C}(B \times A \times N) = \mathbb{E}[X(B \times A)\mathbf{1}_N(X)]$$

for $B \in \mathcal{B}$, $A \in \mathcal{B}(M)$ and $N \in \mathcal{N}_s(\mathbb{R}^d \times M)$. The following theorem comprises the assertions corresponding to Theorems 3.3.1 and 3.3.2. The proof is analogous (observing the assumption (3.12)).

Theorem 3.5.2. *Let X be a stationary marked point process in $\mathbb{R}^d$ with mark space M and intensity $\gamma > 0$. Then there exists a uniquely determined probability measure $\mathbb{P}^0$ on $M \times \mathsf{N}_s(\mathbb{R}^d \times M)$ such that*

$$\mathbf{C}(B \times A \times N) = \gamma \int_B \mathbb{P}^0(A \times (N - y))\, \lambda(\mathrm{d}y) \tag{3.13}$$

for $B \in \mathcal{B}$, $A \in \mathcal{B}(M)$ and $N \in \mathcal{N}_s(\mathbb{R}^d \times M)$, and

$$\gamma \mathbb{P}^0(\widetilde{A}) = \mathbb{E} \sum_{(y,m) \in X} \mathbf{1}_B(y) \mathbf{1}_{\widetilde{A}}(m, X - y)$$

for all $B \in \mathcal{B}$ with $\lambda(B) = 1$ and all $\widetilde{A} \in \mathcal{B}(M) \otimes \mathcal{N}_s(\mathbb{R}^d \times M)$.

The probability measure $\mathbb{P}^0$ is called the **Palm distribution** of the marked point process X. Similarly to Section 3.3, we can interpret $\mathbb{P}^0(A \times \cdot)$, for $A \in \mathcal{B}(M)$, as the conditional distribution of X under the condition that at the origin there is a point of X (more precisely, of X^0) with mark in A.

Because of

$$\gamma \mathbb{P}^0(A \times \mathsf{N}_s(\mathbb{R}^d \times M)) = \mathbf{C}(B \times A \times \mathsf{N}_s(\mathbb{R}^d \times M)) = \Theta(B \times A) = \gamma \mathbb{Q}(A)$$

for $A \in \mathcal{B}(M)$ and $B \in \mathcal{B}$ with $\lambda(B) = 1$, the mark distribution $\mathbb{Q}$ is the projection of the Palm distribution $\mathbb{P}^0$ to the first component.

Again there is a refined Campbell theorem, with an analogous proof.

Theorem 3.5.3 (Refined Campbell theorem). *Let X be a stationary marked point process in $\mathbb{R}^d$ with mark space M and intensity $\gamma > 0$. Let $f : \mathbb{R}^d \times M \times \mathsf{N}_s(\mathbb{R}^d \times M) \to \mathbb{R}$ be a nonnegative measurable function. Then $\sum_{(y,m) \in X} f(y, m, X)$ is measurable, and*

$$\begin{aligned} &\mathbb{E} \sum_{(y,m) \in X} f(y, m, X) \\ &= \gamma \int_{\mathbb{R}^d} \int_{M \times \mathsf{N}_s(\mathbb{R}^d \times M)} f(y, m, \eta + y)\, \mathbb{P}^0(\mathrm{d}(m, \eta))\, \lambda(\mathrm{d}y). \end{aligned}$$

As shown by (3.13), also in the present situation the Palm distribution $\mathbb{P}^0$ can be obtained by disintegration (and subsequent normalization) from the Campbell measure $\mathbf{C}$. We can go yet one step further and disintegrate also $\mathbb{P}^0$ (with respect to the mark distribution $\mathbb{Q}$). For this, we observe that for fixed $N \in \mathcal{N}_s(\mathbb{R}^d \times M)$ the measure $\mathbb{P}^0(\cdot \times N)$ on M is absolutely continuous with respect to $\mathbb{Q}$. In fact, if $\mathbb{Q}(A) = 0$ for some $A \in \mathcal{B}(M)$, then Theorem 3.5.1 shows that for $B \in \mathcal{B}$ with $\lambda(B) = 1$ we have $\Theta(B \times A) = 0$ and therefore, by Theorem 3.5.2,

$$\gamma \mathbb{P}^0(A \times N) \leq \gamma \mathbb{P}^0(A \times \mathsf{N}_s(\mathbb{R}^d \times M)) = \gamma \mathbb{Q}(A) = 0.$$

Hence, the Radon–Nikodym theorem provides a density g_N^0 of $\mathbb{P}^0(\cdot \times N)$ with respect to $\mathbb{Q}$.

By means of the bijective mapping

$$i_s : \mathsf{N}_s(\mathbb{R}^d \times M) \to \mathcal{F}_{\ell f}(\mathbb{R}^d \times M)$$

we can consider the measure $\mathbb{P}^0$ as a probability measure on $M \times \mathcal{F}(\mathbb{R}^d \times M)$ (cf. Theorem 3.1.1). This space is locally compact and has a countable base, since M and $\mathcal{F}(\mathbb{R}^d \times M)$ have these properties, the first by assumption and the second by Theorem 12.2.1. Thus, the topological assumptions for the application of a known existence and uniqueness theorem for regular conditional probabilities are satisfied (see, for example, Kallenberg [386, Th. 6.3]). By this result, there exists a regular version $m \mapsto g_N^0(m)$ of the density found above, that is, for $\mathbb{Q}$-almost all $m \in M$,

$$\mathbb{P}^{0,m}(N) := g_N^0(m)$$

defines a probability measure on $\mathcal{F}(\mathbb{R}^d \times M)$ (concentrated on $\mathcal{F}_{\ell f}(\mathbb{R}^d \times M)$). To obtain a uniform presentation in this section, we now interpret $\mathbb{P}^{0,m}$ again as a measure on $\mathcal{N}_s(\mathbb{R}^d \times M)$. We call $(\mathbb{P}^{0,m})_{m \in M}$ a **regular family**. (The mapping $(m, N) \mapsto \mathbb{P}^{0,m}(N)$ has the properties by which one usually defines transition probabilities or Markov kernels.) The following theorem collects what we have obtained.

Theorem 3.5.4. *There exists a regular family $(\mathbb{P}^{0,m})_{m \in M}$ of probability measures on $\mathcal{N}_s(\mathbb{R}^d \times M)$ which satisfies*

$$\mathbb{P}^0(A \times N) = \int_A \mathbb{P}^{0,m}(N)\, \mathbb{Q}(\mathrm{d}m)$$

for all $A \in \mathcal{B}(M)$ and $N \in \mathcal{N}_s(\mathbb{R}^d \times M)$. The family $(\mathbb{P}^{0,m})_{m \in M}$ is uniquely determined, up to a set of $\mathbb{Q}$-measure zero.

We call also $\mathbb{P}^{0,m}$ a **Palm distribution**, and we interpret it as the distribution of X under the condition $(0, m) \in X$, that is, under the condition that at the origin there is a point of X^0 with mark m.

Combining Theorems 3.5.4 and 3.5.3, we obtain a further version of the refined Campbell theorem (using Fubini's theorem for kernels).

Theorem 3.5.5 (Refined Campbell theorem). *Let X be a stationary marked point process in $\mathbb{R}^d$ with mark space M and intensity $\gamma > 0$. Let $f : \mathbb{R}^d \times M \times \mathsf{N}_s(\mathbb{R}^d \times M) \to \mathbb{R}$ be a nonnegative measurable function. Then $\sum_{(y,m) \in X} f(y, m, X)$ is measurable, and*

$$\mathbb{E} \sum_{(y,m) \in X} f(y, m, X)$$
$$= \gamma \int_{\mathbb{R}^d} \int_M \int_{\mathsf{N}_s(\mathbb{R}^d \times M)} f(y, m, \eta + y)\, \mathbb{P}^{0,m}(\mathrm{d}\eta)\, \mathbb{Q}(\mathrm{d}m)\, \lambda(\mathrm{d}y).$$

A third type of Palm distribution for stationary marked point processes X arises in the context of the previous section. To explain this, let $(\Omega, \mathbf{A}, \mathbb{P}, \mathcal{T})$ be the dynamical system given by the canonical space $(\mathsf{N}_s(\mathbb{R}^d \times M), \mathcal{N}_s(\mathbb{R}^d \times M), \mathbb{P}_X)$ and the class $\mathcal{T}$ of translations t_{-x}, $x \in \mathbb{R}^d$, as they were introduced at the beginning of this section. Then, the unmarked process X^0 is a stationary random measure, which has a **Palm distribution** $\mathbb{P}^{0,X^0}$. It describes the distribution of X under the condition that there is a point of X^0 at the origin (with no information on the mark). Obviously, we have

$$\mathbb{P}^{0,X^0} = \mathbb{P}^0(M \times \cdot).$$

Although $\mathbb{P}^{0,X^0}$ carries less information than $\mathbb{P}^0$, it still determines the mark distribution $\mathbb{Q}$. Namely, for $x \in \mathbb{R}^d$ and $\eta \in \mathsf{N}_s(\mathbb{R}^d \times M)$, let us define $\zeta_x(\eta) \in M$ as m if $\eta(\{(x,m)\}) > 0$, and as some fixed mark m_0 if there is no such m, that is, if $\eta(\{x\} \times M) = 0$. Then

$$\zeta_x : \mathsf{N}_s(\mathbb{R}^d \times M) \to M$$

is a measurable mapping, as follows by arguments similar to the proof of Lemma 3.1.3. Since

$$\mathbb{Q}(A) = \mathbb{P}^{0,X^0}(\{\eta \in \mathsf{N}_s(\mathbb{R}^d \times M) : \eta(\{(0,m)\}) > 0,\ \zeta_0(\eta) \in A\})$$

and since $\mathbb{P}^{0,X^0}$ is concentrated on

$$\{\eta \in \mathsf{N}_s(\mathbb{R}^d \times M) : \eta(\{(0,m)\}) > 0\},$$

$\mathbb{Q}$ is the image measure of $\mathbb{P}^{0,X^0}$ under ζ_0.

Independently Marked Point Processes

We turn now to the investigation of an important subclass of marked point processes. The notion of mark distribution of a marked point process is in general meaningless without additional structural information, such as stationarity, but it does have a meaning under suitable independence assumptions. By Lemma 3.1.3, a marked point process X in $\mathbb{R}^d$ with mark space M can be represented in the form

$$X = \sum_{i=1}^{\tau} \delta_{(\xi_i,\mu_i)}, \tag{3.14}$$

where $(\xi_i, \mu_i)_{i\in\mathbb{N}}$ is a sequence of random variables in $\mathbb{R}^d \times M$ and $\tau := X(\mathbb{R}^d \times M) = X^0(\mathbb{R}^d)$.

Definition 3.5.2. *The marked point process X is* **independently marked** *if it has a representation* (3.14) *where the random marks $\mu_1, \mu_2, \dots$ are independently and identically distributed and are independent of $((\xi_i)_{i\in\mathbb{N}}, \tau)$. The distribution $\mathbb{Q}$ of the μ_i is then called the* **mark distribution** *of X.*

The following theorem shows that the mark distribution of an independently marked point process X does not depend on the special representation (3.14) and that in the stationary case it coincides with the mark distribution defined by Theorem 3.5.1.

Theorem 3.5.6. *Let X be an independently marked point process in $\mathbb{R}^d$ with intensity measure Θ and mark distribution $\mathbb{Q}$. Then*

$$\Theta = \vartheta \otimes \mathbb{Q},$$

where ϑ is the intensity measure of the unmarked point process X^0.

Proof. Let M be the mark space of X. Put $X(\mathbb{R}^d \times M) =: \tau \in \mathbb{N}_0 \cup \{\infty\}$. For $B \in \mathcal{B}(\mathbb{R}^d)$ and $A \in \mathcal{B}(M)$ the assumed independence properties yield (employing a representation (3.14))

$$\begin{aligned}
\Theta(B \times A) &= \mathbb{E} \sum_{i=1}^{\tau} \delta_{(\xi_i, \mu_i)}(B \times A) \\
&= \sum_{k \in \mathbb{N}_0 \cup \{\infty\}} \mathbb{E} \left[\mathbf{1}_{\{k\}}(\tau) \sum_{i=1}^{k} \delta_{(\xi_i, \mu_i)}(B \times A) \right] \\
&= \sum_{k \in \mathbb{N}_0 \cup \{\infty\}} \sum_{i=1}^{k} \mathbb{E} \left[\mathbf{1}_{\{k\}}(\tau) \mathbf{1}_B(\xi_i) \mathbf{1}_A(\mu_i) \right] \\
&= \sum_{k \in \mathbb{N}_0 \cup \{\infty\}} \sum_{i=1}^{k} \mathbb{E} \left[\mathbf{1}_{\{k\}}(\tau) \mathbf{1}_B(\xi_i) \right] \mathbb{E}[\mathbf{1}_A(\mu_i)] \\
&= \mathbb{Q}(A) \sum_{k \in \mathbb{N}_0 \cup \{\infty\}} \mathbb{E} \left[\mathbf{1}_{\{k\}}(\tau) \sum_{i=1}^{k} \delta_{\xi_i}(B) \right] \\
&= \mathbb{Q}(A) \vartheta(B) = (\vartheta \otimes \mathbb{Q})(B \times A),
\end{aligned}$$

from which the assertion follows. □

Of particular interest is the case where the unmarked point process X^0 is a Poisson process.

Theorem 3.5.7. *Let X be an independently marked point process in $\mathbb{R}^d$, and assume that the unmarked process X^0 is a Poisson process. Then X is a Poisson process.*

Proof. Let M be the mark space of X. Let X' be a Poisson process in $\mathbb{R}^d \times M$ with intensity measure $\vartheta \otimes \mathbb{Q}$, where ϑ is the intensity measure of X^0 and $\mathbb{Q}$ is the mark distribution of X. We shall show that

$$\mathbb{P}(X(C) = 0) = \mathbb{P}(X'(C) = 0) \tag{3.15}$$

holds for all $C \in \mathcal{C}(\mathbb{R}^d \times M)$. By Theorem 3.1.1, this implies $X \stackrel{\mathcal{D}}{=} X'$ and thus the assertion. To prove (3.15), we assume a representation (3.14) and exploit the independence properties. This yields

$$\begin{aligned}
\mathbb{P}(X(C)=0) &= \mathbb{P}\left(\sum_{i=1}^{\tau} \delta_{(\xi_i,\mu_i)}(C)=0\right) \\
&= \sum_{k\in\mathbb{N}_0\cup\{\infty\}} \mathbb{P}(\tau=k,\ \mathbf{1}_C(\xi_i,\mu_i)=0 \text{ for } i=1,\dots,k) \\
&= \sum_{k\in\mathbb{N}_0\cup\{\infty\}} \mathbb{P}\left(\tau=k,\ \prod_{i=1}^{k}(1-\mathbf{1}_C(\xi_i,\mu_i))=1\right) \\
&= \sum_{k\in\mathbb{N}_0\cup\{\infty\}} \mathbb{E}\left[\mathbf{1}_{\{k\}}(\tau)\prod_{i=1}^{k}(1-\mathbf{1}_C(\xi_i,\mu_i))\right] \\
&= \sum_{k\in\mathbb{N}_0\cup\{\infty\}} \mathbb{E}\left[\mathbf{1}_{\{k\}}(\tau)\prod_{i=1}^{k}\int_M (1-\mathbf{1}_C(\xi_i,m))\,\mathbb{Q}(\mathrm{d}m)\right] \\
&= \mathbb{E}\prod_{i=1}^{\tau}\int_M (1-\mathbf{1}_C(\xi_i,m))\,\mathbb{Q}(\mathrm{d}m).
\end{aligned}$$

Now Theorem 3.2.4, applied to X^0, gives

$$\begin{aligned}
\mathbb{P}(X(C)=0) &= \exp\left(-\int_{\mathbb{R}^d}\int_M \mathbf{1}_C(x,m)\,\mathbb{Q}(\mathrm{d}m)\,\vartheta(\mathrm{d}x)\right) \\
&= \exp\left(-\vartheta\otimes\mathbb{Q}(C)\right) \\
&= \mathbb{P}(X'(C)=0),
\end{aligned}$$

which completes the proof. □

The previous theorem shows that independent marking, applied to a Poisson process X^0 in $\mathbb{R}^d$, yields a marked point process X which is a Poisson process. However, by Theorem 3.5.6, this procedure yields only marked point processes for which the intensity measure has a product form. For marked Poisson processes X with a general intensity measure Θ, the unmarked process X^0 is again a Poisson process, but X cannot be obtained from X^0 by independent marking (the marks are position-dependent). In the stationary case, though, the situation is different.

Theorem 3.5.8. *Let X be a stationary Poisson process in $\mathbb{R}^d \times M$ with intensity measure Θ satisfying* (3.12). *Then X is independently marked.*

Proof. We assume $\Theta \neq 0$, since for $\Theta = 0$ there is nothing to show. Condition (3.12) being satisfied, X is a stationary marked point process in $\mathbb{R}^d$ with mark space M. Theorem 3.5.1 shows that

$$\Theta = \gamma\lambda \otimes \mathbb{Q}$$

with $0 < \gamma < \infty$ and a probability measure $\mathbb{Q}$ on M.

For $\eta \in \mathsf{N}_s(\mathbb{R}^d \times M)$ we put

$$\eta^0 := \sum_{(x,m)\in\eta} \delta_x$$

and define

$$\mathsf{N}_s^0(\mathbb{R}^d \times M) := \{\eta \in \mathsf{N}_s(\mathbb{R}^d \times M) : \eta^0 \text{ simple}, \eta(\mathbb{R}^d \times M) = \infty,$$
$$\eta(C \times M) < \infty \text{ for all } C \in \mathcal{C}(\mathbb{R}^d)\}.$$

Then $\mathsf{N}_s^0(\mathbb{R}^d \times M)$ is measurable. Now we modify the argument in the proof of Lemma 3.1.3. In that proof, we choose $E = \mathbb{R}^d$ and define the sets A_i^k, the functions j_i, and the linear order $\prec$ on $\mathbb{R}^d$ in the same way. Then we define, for $p \in \mathbb{N}$, a mapping $\zeta_p : \mathsf{N}_s^0(\mathbb{R}^d \times M) \to \mathbb{R}^d \times M$ in the following way. Let $\eta \in \mathsf{N}_s^0(\mathbb{R}^d \times M)$, and let (x, m) be an atom of η. All atoms (x', m') of η with $x' \prec x$, $x' \neq x$, lie in the set $\bigcup_{i=1}^{j_1(x)} A_i^1 \times M$, hence their number is finite, say $p - 1$. We define $\zeta_p(\eta) := (x, m)$. If this is done for all atoms (x, m) of η, then ζ_p is defined for all $p \in \mathbb{N}$. The measurability of ζ_p is obtained in a similar way to the proof of Lemma 3.1.3.

Setting $\zeta_i(\eta) =: (x_i, m_i)$ for $\mathsf{N}_s^0(\mathbb{R}^d \times M)$, we have defined a measurable mapping

$$\begin{aligned} \Xi : \mathsf{N}_s^0(\mathbb{R}^d \times M) &\to (\mathbb{R}^d \times M)^{\mathbb{N}} \\ \eta \qquad &\mapsto (x_i, m_i)_{i\in\mathbb{N}}, \end{aligned}$$

from which we get the measurable mapping

$$\begin{aligned} \Xi' : \mathsf{N}_s^0(\mathbb{R}^d \times M) &\to \quad (\mathbb{R}^d)^{\mathbb{N}} \times M^{\mathbb{N}} \\ \eta \qquad &\mapsto ((x_i)_{i\in\mathbb{N}}, (m_i)_{i\in\mathbb{N}}). \end{aligned}$$

For the stationary Poisson process X in $\mathbb{R}^d \times M$, the mappings ζ_p, Ξ, Ξ' are defined $\mathbb{P}_X$-almost surely. Hence, we can define $(\xi_i, \mu_i) := \zeta_i \circ X$ for $i \in \mathbb{N}$ and thus obtain a sequence $(\xi_i, \mu_i)_{i\in\mathbb{N}}$ of random variables with

$$X = \sum_{i=1}^{\infty} \delta_{(\xi_i, \mu_i)}.$$

We assert that the marks $\mu_1, \mu_2, \dots$ are independently and uniformly distributed and that they are independent of $(\xi_i)_{i\in\mathbb{N}}$. For the proof, we note that the joint distribution

$$\mathbb{P}_{((\xi_i)_{i\in\mathbb{N}}, (\mu_i)_{i\in\mathbb{N}})}$$

is the image of $\mathbb{P}_X$ under Ξ'. Let $(\nu_i)_{i\in\mathbb{N}}$ be a sequence of independent, identically distributed random variables in M with distribution $\mathbb{Q}$ and independent of $(\xi_i)_{i\in\mathbb{N}}$. Then

$$\widetilde{X} := \sum_{i=1}^{\infty} \delta_{(\xi_i, \nu_i)}$$

is an independently marked stationary Poisson process which has intensity measure Θ and hence satisfies

$$\mathbb{P}_X = \mathbb{P}_{\widetilde{X}},$$

by Theorem 3.2.1. Since $\widetilde{X}$ is independently marked, we have

$$\mathbb{P}_{((\xi_i)_{i\in\mathbb{N}}, (\nu_i)_{i\in\mathbb{N}})} = \overline{\mathbb{P}} \otimes \mathbb{Q}^{\mathbb{N}}$$

with a probability measure $\overline{\mathbb{P}}$ on $(\mathbb{R}^d)^{\mathbb{N}}$. Since $\mathbb{P}_{((\xi_i)_{i\in\mathbb{N}}, (\nu_i)_{i\in\mathbb{N}})}$ is the image of $\mathbb{P}_{\widetilde{X}}$ under Ξ', we also have

$$\mathbb{P}_{((\xi_i)_{i\in\mathbb{N}}, (\mu_i)_{i\in\mathbb{N}})} = \overline{\mathbb{P}} \otimes \mathbb{Q}^{\mathbb{N}},$$

which implies what we have stated. □

In the preceding proof, the stationarity was only needed in so far as it implies the product form of the intensity measure. Therefore, a generalization of Theorem 3.5.8 can be obtained whenever the intensity measure is of the form $\Theta = \vartheta \otimes \mathbb{Q}$ with suitable ϑ.

For stationary marked point processes, there is also a version of Theorem 3.3.5, Slivnyak's theorem. For that, we again use the map $i_s : \mathsf{N}_s(\mathbb{R}^d \times M) \to \mathcal{F}_{\ell f}(\mathbb{R}^d \times M)$ to interpret measures on $\mathsf{N}_s(\mathbb{R}^d \times M)$ also as measures on $\mathcal{F}(\mathbb{R}^d \times M)$. Corresponding to the convention that translations on $\mathbb{R}^d \times M$ affect only the first component, we define $A + x$ for $A \in \mathcal{F}(\mathbb{R}^d \times M)$ and $x \in \mathbb{R}^d$ by $\{(y + x, m) : (y, m) \in A\}$.

Theorem 3.5.9 (Theorem of Slivnyak). *Let X be a stationary marked point process in $\mathbb{R}^d$ with intensity $\gamma > 0$ and with mark space M and mark distribution $\mathbb{Q}$ and let $\mathbb{P}^{0,m}$, $m \in M$, be the family of Palm distributions of X. Then, X is a Poisson process if and only if, for $\mathbb{Q}$-almost all $m \in M$, we have*

$$\mathbb{P}^{0,m}(A) = \mathbb{P}(X \cup \{(0, m)\} \in A)$$

for all $A \in \mathcal{B}(\mathcal{F}(\mathbb{R}^d \times M))$.

Proof. Though the proof is analogous to that of Theorem 3.3.5, we present it for the reader's convenience. From Theorems 3.2.5 and 3.5.1 we obtain that X is a Poisson process if and only if

$$\begin{aligned}
&\mathbb{E} \sum_{(x,m)\in X} g(X, x, m) \\
&= \gamma \int_M \int_{\mathbb{R}^d} \mathbb{E}\, g(X \cup \{(x, m)\}, x, m)\, \lambda(\mathrm{d}x)\, \mathbb{Q}(\mathrm{d}m) \\
&= \gamma \int_M \int_{\mathbb{R}^d} \int_{\mathsf{N}_s(\mathbb{R}^d \times M)} g((\eta \cup \{(0, m)\}) + x, x, m)\, \mathbb{P}_X(\mathrm{d}\eta)\, \lambda(\mathrm{d}x)\, \mathbb{Q}(\mathrm{d}m)
\end{aligned}$$

holds for all measurable functions $g : \mathsf{N}_s(\mathbb{R}^d \times M) \times \mathbb{R}^d \times M \to [0, \infty)$. Applying the refined Campbell theorem (Theorem 3.5.5) to the left side and using Fubini's theorem, we obtain that X is a Poisson process if and only if

$$\int_M \int_{\mathbb{R}^d} \int_{\mathsf{N}_s(\mathbb{R}^d \times M)} g(\eta + x, x, m)\, \mathbb{P}^{0,m}(\mathrm{d}\eta)\, \lambda(\mathrm{d}x)\, \mathbb{Q}(\mathrm{d}m)$$

$$= \int_M \int_{\mathbb{R}^d} \int_{\mathsf{N}_s(\mathbb{R}^d \times M)} g((\eta \cup \{(0,m)\}) + x, x, m)\, \mathbb{P}_X(\mathrm{d}\eta)\, \lambda(\mathrm{d}x)\, \mathbb{Q}(\mathrm{d}m)$$

holds for all g. Choosing $g(\eta, x, m) := \mathbf{1}_A(\eta - x)\mathbf{1}_{C\times B}(x, m)$ with $A \in \mathcal{B}(\mathcal{F}(\mathbb{R}^d \times M))$, $C \in \mathcal{B}(\mathbb{R}^d)$ and $B \in \mathcal{B}(M)$, we see that the latter equality is equivalent to

$$\int_B \mathbb{P}(X \cup \{(0,m)\} \in A)\, \mathbb{Q}(\mathrm{d}m) = \int_B \mathbb{P}^{0,m}(A)\, \mathbb{Q}(\mathrm{d}m)$$

for all $B \in \mathcal{B}(M)$ and $A \in \mathcal{B}(\mathcal{F}(\mathbb{R}^d \times M))$. Since the regular family in Theorem 3.5.4 is $\mathbb{Q}$-almost surely uniquely determined, the assertion follows. □

Up to now, Poisson processes and Cox processes were the only non-trivial examples of (stationary) point processes in $\mathbb{R}^d$ which we discussed. Although the Poisson process $X_{\gamma\lambda}$ yields the basis for most of the more sophisticated geometric models which we shall investigate later, we want to mention two other (classes of) stationary point processes in $\mathbb{R}^d$. They are obtained as variations of $X_{\gamma\lambda}$, with the help of marked processes.

A **cluster process** X in $\mathbb{R}^d$ is generated by a marked point process $\widetilde{X}$, where the mark space M is the subset $\mathsf{N}_{sf} \subset \mathsf{N}_s$ of simple finite counting measures on $\mathbb{R}^d$, by superposition of the translated marks,

$$X = \sum_{(x,\eta)\in\widetilde{X}} \sum_{y\in\eta} \delta_{x+y} = \sum_{(x,\eta)\in\widetilde{X}} (\eta + x).$$

Under appropriate conditions on the size of the marks, which guarantee that X is locally finite, X is a point process. For $(x, \eta) \in \widetilde{X}$, the point x is called a parent point of X, whereas the points $x+y$, $y \in \eta$, are called daughter points. They form a 'cluster around' x, therefore x is often called the center of the cluster. If $0 \in \eta$, the parent points appear in the cluster process, but this is not part of the definition. Although the intensity measure of the marked point process $\widetilde{X}$ is locally finite (by our general assumption), the intensity measure Θ of the cluster process X need not have this property, so it has to be added as an additional requirement. If $\widetilde{X}$ is stationary (with intensity $\widetilde{\gamma} > 0$), the cluster process X is stationary and has intensity

$$\gamma = \mathbb{E} \sum_{z\in X} \mathbf{1}_{C^d}(z)$$

$$
\begin{aligned}
&= \mathbb{E} \sum_{(x,\eta)\in\widetilde{X}} \sum_{y\in\eta} \mathbf{1}_{C^d}(x+y) \\
&= \widetilde{\gamma} \int_{\mathsf{N}_{sf}} \int_{\mathbb{R}^d} \sum_{y\in\eta} \mathbf{1}_{C^d}(x+y)\, \lambda(\mathrm{d}x)\, \mathbb{Q}(\mathrm{d}\eta) \\
&= \widetilde{\gamma} \int_{\mathsf{N}_{sf}} \sum_{y\in\eta} 1\, \mathbb{Q}(\mathrm{d}\eta) \\
&= \widetilde{\gamma} n_c,
\end{aligned}
$$

where n_c is the mean cluster size (mean number of points in the typical cluster).

If $\widetilde{X}$ is an independently marked Poisson process, the cluster process X is called a **Neyman–Scott process**. By Theorem 3.5.8, a (stationary) Neyman–Scott process arises if $\widetilde{X}$ is a stationary Poisson process. A special Neyman–Scott process is the **Matérn cluster process**. It has Poissonian clusters, that is, its mark distribution is the distribution of a second stationary Poisson process Y restricted to the ball RB^d. Here, the intensity μ of Y and the cluster radius $R > 0$ are additional parameters. A Neyman–Scott process X is simple, due to the following argument. Let ϑ be the intensity measure of the (Poisson) process of parent points and let V_0 denote the Euler characteristic (see Section 14.2). From Corollary 3.2.4 and the Campbell theorem, we obtain

$$
\begin{aligned}
&\mathbb{E} \sum_{(x,\eta),(x',\eta')\in\widetilde{X}^2_{\neq}} V_0((\eta+x)\cap(\eta'+x')) \\
&= \widetilde{\gamma}^2 \int_{\mathsf{N}_{sf}} \int_{\mathsf{N}_{sf}} \int_{\mathbb{R}^d} \int_{\mathbb{R}^d} V_0((\eta+x)\cap(\eta'+x'))\, \vartheta(\mathrm{d}x)\, \vartheta(\mathrm{d}x')\, \mathbb{Q}(\mathrm{d}\eta)\, \mathbb{Q}(\mathrm{d}\eta') \\
&= 0,
\end{aligned}
$$

since the inner integral vanishes (ϑ has no atoms). Hence, almost surely, all clusters $\eta + x$ and $\eta' + x'$ in X with $x \neq x'$ are disjoint, and thus X is simple.

In contrast to this, the definition of a general cluster process given above allows multiple points. If we identify $\eta \in \mathsf{N}_{sf}$ with its support, then

$$
X = \bigcup_{(x,\eta)\in\widetilde{X}} (\eta + x)
$$

defines a simple point process which is also called a cluster process. For Neyman–Scott processes, both definitions coincide. Since, in this latter interpretation, clusters are finite sets, cluster processes appear as union sets of special particle processes, as they will be discussed in Section 4.1. Neyman–Scott processes are then special cases of Boolean models (see Section 4.3).

For the second model which we present here, we also assume a special setup. We start with a stationary Poisson process Y in $\mathbb{R}^d$ and generate a **hard**

core process X by deleting some points of Y, such that the remaining points in X have a minimal distance from each other. The process X is determined by the hard core distance $c > 0$ and by the thinning procedure used. Several methods are popular here. The simplest one consists in deleting all pairs of points $x, y \in Y$, $x \neq y$, with distance $d(x, y) < c$. The resulting process is called the **Matérn process (first kind)**. For the **Matérn process (second kind)** X, we start with a stationary marked Poisson process $\widetilde{X}$ with intensity $\widetilde{\gamma}$ and mark space $[0, 1]$ (with uniform mark distribution). For each pair $(x_1, w_1), (x_2, w_2) \in \widetilde{X}^2_{\neq}$ with $d(x_1, x_2) < c$, we delete the point $x_i \in \widetilde{X}_0$ with the higher weight w_i. The undeleted points form the point process X. In order to calculate the intensity γ of X, we apply Theorems 3.2.5 and 3.5.1 and get

$$\begin{aligned}
\gamma &= \mathbb{E} \sum_{z \in X} \mathbf{1}_{C^d}(z) \\
&= \mathbb{E} \sum_{(x,w) \in \widetilde{X}} \mathbf{1}_{C^d}(x) \prod_{(y,v) \in \widetilde{X} \setminus \{(x,w)\},\, y \in x + cB^d} \mathbf{1}_{[w,1]}(v) \\
&= \widetilde{\gamma} \int_0^1 \int_{\mathbb{R}^d} \mathbf{1}_{C^d}(x)\, \mathbb{E} \prod_{(y,v) \in \widetilde{X},\, y \in x + cB^d} \mathbf{1}_{[w,1]}(v)\, \lambda(\mathrm{d}x)\, \mathrm{d}w \\
&= \widetilde{\gamma} \int_0^1 \mathbb{P}(\widetilde{X}(cB^d \times [0, w]) = 0)\, \mathrm{d}w \\
&= \widetilde{\gamma} \int_0^1 \exp\{-\widetilde{\gamma} c^d \kappa_d\, w\}\, \mathrm{d}w \\
&= \frac{1}{c^d \kappa_d} \left(1 - \mathrm{e}^{-\widetilde{\gamma} c^d \kappa_d}\right).
\end{aligned}$$

In a similar way, the intensity γ of the Matérn process (first kind) can be derived as

$$\gamma = \widetilde{\gamma}\, \mathrm{e}^{-\widetilde{\gamma} c^d \kappa_d},$$

where $\widetilde{\gamma}$ is the intensity of the original Poisson process Y before thinning.

Notes for Section 3.5

1. Since marked point processes can be subsumed under the general theory of point processes, as (special) point processes on product spaces, they are treated in most of the monographs on point processes. Thorough presentations are found in Matthes, Kerstan and Mecke [465], König and Schmidt [423] and – with a view to applications in stochastic geometry – in Stoyan, Kendall and Mecke [743].

2. Poisson cluster processes also appear as infinitely divisible point processes which have a regularity property (see Daley and Vere–Jones [196, sect. 10.2]).

3.6 Point Processes of Closed Sets

Now we consider point processes where the 'points' are themselves nonempty, closed subsets of $\mathbb{R}^d$. Thus, the basic space in this section is $E = \mathcal{F}'(\mathbb{R}^d)$. We often write $\mathcal{F}'(\mathbb{R}^d) = \mathcal{F}'$ again, and also symbols such as $\mathsf{N}, \mathcal{N}$ etc. are mostly used with suppression of their argument E. Let X be a point process in $\mathcal{F}'$. Its intensity measure Θ is a measure on $\mathcal{F}'$; according to our convention in Section 3.1, it is always assumed to be locally finite. By Lemma 2.3.1, local finiteness of Θ is equivalent to $\Theta(\mathcal{F}_C) < \infty$ for all $C \in \mathcal{C}$.

As for ordinary and marked point processes in $\mathbb{R}^d$, also for point processes X of closed sets, invariance properties play an important role. X is called **stationary** if $X + x \stackrel{\mathcal{D}}{=} X$ for all $x \in \mathbb{R}^d$, and it is **isotropic** if $\vartheta X \stackrel{\mathcal{D}}{=} X$ for all $\vartheta \in SO_d$. The following assertion follows directly from the definition of the intensity measure Θ.

Lemma 3.6.1. *If X is a stationary point process in $\mathcal{F}'$, then its intensity measure Θ is translation invariant. If X is isotropic, then Θ is rotation invariant.*

For Poisson processes, the converse assertion is also true. For stationary point processes in $\mathcal{F}'$ with Poisson counting variables, we give a condition ensuring that they are simple and thus Poisson processes.

Theorem 3.6.1. *A Poisson process X in $\mathcal{F}'$ is stationary (isotropic) if and only if its intensity measure Θ is translation invariant (rotation invariant).*

A point process X in $\mathcal{F}'$ with Poisson counting variables, with translation invariant intensity measure Θ and satisfying $X(\{\mathbb{R}^d\}) = 0$ a.s., is a stationary Poisson process.

Proof. Let X be a Poisson process in $\mathcal{F}'$ with intensity measure Θ. For $g \in G_d$ (the motion group), gX has intensity measure $g\Theta$. By the uniqueness of Poisson processes with given intensity measure (Theorem 3.2.1), $gX \stackrel{\mathcal{D}}{=} X$ is equivalent to $g\Theta = \Theta$. This yields the first assertion.

For the second assertion, we need only show that, under the assumptions, Θ has no atoms. Let $\{F\}$, $F \in \mathcal{F}'$, be an atom of Θ. Then also every translate $\{F\} + x$ with $x \in \mathbb{R}^d$ is an atom. If $F \neq \mathbb{R}^d$, then there is a compact set C and there are infinitely many $x \in \mathbb{R}^d$ so that the sets $F + x$ are all distinct and satisfy $(F + x) \cap C \neq \emptyset$. This gives $\Theta(\mathcal{F}_C) = \infty$, a contradiction. Thus, only $F = \mathbb{R}^d$ is possible. But then $\mathbb{P}(X(\{\mathbb{R}^d\}) > 0) > 0$, which contradicts the assumption. Hence, Θ has no atoms. □

Let X be a point process in $\mathcal{F}'$, and let $A \in \mathcal{F}'$ be a fixed closed set. We define $X \cap A$ by

$$(X \cap A)(\omega) := \sum_{F_i \cap A \neq \emptyset} \delta_{F_i \cap A} \qquad \text{if} \quad X(\omega) = \sum \delta_{F_i}.$$

The mapping $\alpha : F \mapsto F \cap A$ from $\mathcal{F}'$ in $\mathcal{F}$ is measurable, by Theorem 12.2.6. For $G \in \mathcal{B}(\mathcal{F}')$ and $k \in \mathbb{N}_0$,

$$(X \cap A)^{-1}(\mathsf{N}_{G,k}) = X^{-1}(\mathsf{N}_{\alpha^{-1}(G),k}),$$

hence, by Lemma 3.1.5, $X \cap A$ is a point process in $\mathcal{F}'$ (though possibly with intensity measure zero). We call $X \cap A$ the **section process** of X with the set A.

Starting from point processes in $\mathcal{F}'$, one obtains random sets by taking unions. For a point process X in $\mathcal{F}'$ we put

$$Z_X(\omega) := \bigcup_{F \in \operatorname{supp} X(\omega)} F \qquad \text{for } \omega \in \Omega.$$

Theorem 3.6.2. *The union set Z_X of a point process in $\mathcal{F}'$ is a random closed set. If X is stationary (isotropic), then Z_X is stationary (isotropic).*

Proof. We have to show that Z_X is closed. For that, let $\omega \in \Omega$. Let $(x_j)_{j \in \mathbb{N}}$ be a convergent sequence with $x_j \in Z_X(\omega)$ and $x_j \to x$. The set $C := \{x, x_1, x_2, \dots\}$ is compact. Since $X(\omega)$ is locally finite and $\mathcal{F}_C$ is compact, we have $X(\omega, \mathcal{F}_C) := X(\omega)(\mathcal{F}_C) < \infty$, hence only finitely many sets $F_1, \dots, F_k$ from $\operatorname{supp} X(\omega)$ meet the set C. Thus $x_j \in \bigcup_{i=1}^k F_i$ for $j \in \mathbb{N}$ and hence $x \in \bigcup_{i=1}^k F_i \subset Z_X(\omega)$. The set $Z_X(\omega)$ being closed, we have indeed defined a map $Z_X : \Omega \to \mathcal{F}$.

To show that Z_X is measurable, let $C \in \mathcal{C}$. Then $Z_X(\omega) \in \mathcal{F}^C$ is equivalent to $X(\omega, \mathcal{F}^C) = 0$. From the definition of $\mathcal{N}$ it follows that $\{X(\mathcal{F}^C) = 0\}$ is measurable, hence $\{Z_X \in \mathcal{F}^C\}$ is measurable. The measurability of Z_X now follows from Lemma 2.1.1.

The remaining assertions are clear. ⊓

In particular, the union set of a point process in $\mathcal{F}'$ with Poisson counting variables is a random closed set. With the aid of the results from Section 2.3, the random closed sets generated in this way can be characterized as follows.

Theorem 3.6.3. *For a random closed Z set in $\mathbb{R}^d$, the following conditions* (a), (b), (c) *are equivalent:*

(a) *Z is (up to equivalence) the union set Z_X of a point process X in $\mathcal{F}'$ with Poisson counting variables.*
(b) *There is a locally finite measure Θ on $\mathcal{F}'$ with*

$$T_Z(C) = 1 - e^{-\Theta(\mathcal{F}_C)}, \qquad C \in \mathcal{C}.$$

(c) *Z is infinitely divisible and has no fixed points.*

If (a) *and* (b) *are satisfied, then Θ is the intensity measure of X.*

Proof. First we show the equivalence of (a) and (b). If Z is the union set of a point process X in $\mathcal{F}'$ with Poisson counting variables and if $C \in \mathcal{C}$, then

$$\begin{aligned} T_Z(C) &= \mathbb{P}(Z \cap C \neq \emptyset) = \mathbb{P}(X(\mathcal{F}_C) > 0) \\ &= 1 - \mathbb{P}(X(\mathcal{F}_C) = 0) = 1 - e^{-\Theta(\mathcal{F}_C)}, \end{aligned}$$

where Θ is the intensity measure of X.

If, conversely,

$$T_Z(C) = 1 - e^{-\Theta(\mathcal{F}_C)}, \qquad C \in \mathcal{C},$$

holds with a locally finite measure Θ, then, as shown in the proof of Theorem 3.2.1, there exists a point process X in $\mathcal{F}'$ satisfying (3.2) with intensity measure Θ. From the first part already proved we have

$$T_{Z_X}(C) = 1 - e^{-\Theta(\mathcal{F}_C)} = T_Z(C), \qquad C \in \mathcal{C},$$

thus $T_{Z_X} = T_Z$. By Theorem 2.1.3, this implies $Z_X \stackrel{\mathcal{D}}{=} Z$.

Suppose that (b) holds, that is,

$$T_Z(C) = 1 - e^{-\Theta(\mathcal{F}_C)}, \qquad C \in \mathcal{C}.$$

For $m \in \mathbb{N}$, we define

$$T^{(m)} := 1 - (1 - T_Z)^{1/m},$$

thus

$$T^{(m)}(C) = 1 - e^{-\Theta(\mathcal{F}_C)/m} = 1 - e^{-\Theta_m(\mathcal{F}_C)}, \qquad C \in \mathcal{C},$$

with $\Theta_m := \Theta/m$. To the locally finite measure Θ_m there exists, again by Theorem 3.2.1, a point process $X^{(m)}$ satisfying (3.2) with intensity measure Θ_m. Its union set $Z^{(m)}$ has the capacity functional $T^{(m)}$, as was proved above. By Lemma 2.3.2(a), Z is infinitely divisible. Since Θ is locally finite,

$$\mathbb{P}(x \in Z) = T_Z(\{x\}) = 1 - e^{-\Theta(\mathcal{F}_{\{x\}})} < 1,$$

hence Z has no fixed points. Thus (c) holds.

The implication (c) $\Rightarrow$ (b) is the assertion of Theorem 2.3.2. □

We cannot conclude, in general, that the point process with Poisson counting variables appearing in Theorem 3.6.3 is simple. In the stationary case, however, this conclusion is possible, under a mild extra condition.

Theorem 3.6.4. *For a stationary random closed set Z in $\mathbb{R}^d$ satisfying $Z \neq \mathbb{R}^d$ almost surely, the following conditions* (a), (b), (c) *are equivalent:*

(a) *Z is (equivalent to) the union set of a Poisson process X in $\mathcal{F}'$.*
(b) *There is a locally finite measure without atoms on $\mathcal{F}'$ with*

$$T_Z(C) = 1 - e^{-\Theta(\mathcal{F}_C)}, \qquad C \in \mathcal{C}.$$

(c) *Z is infinitely divisible and has no fixed points.*

If (a) *and* (b) *are satisfied, then* Θ *is the intensity measure of* X*, and* Θ *is translation invariant.*

Proof. That (a) implies (b) follows from Theorem 3.6.3. For the converse, it remains to show that the point process X satisfying (3.2) with intensity measure Θ, which exists by Theorem 3.6.3, is simple. Since Z is stationary and satisfies

$$T_Z(C) = 1 - e^{\Theta(\mathcal{F}_C)},$$

we have $(t_x\Theta)(\mathcal{F}_C) = \Theta(\mathcal{F}_C)$ for $C \in \mathcal{C}$. By Lemma 2.3.1, this implies $t_x\Theta = \Theta$, thus Θ is translation invariant. The relation $\mathbb{P}(Z = \mathbb{R}^d) = 0$ implies $\mathbb{P}(X(\mathbb{R}^d) > 0) = 0$, hence X is simple, by Theorem 3.6.1. $\square$

General assumption. All point processes appearing in the following are assumed to be simple, except in the cases where the opposite is explicitly stated. Thus, a point process in $\mathcal{F}'$ can be viewed as a random locally finite collection of closed sets. However, the interpretation as a counting measure is maintained where it simplifies assertions and notation.

Note for Section 3.6

Theorems 3.6.3 and 3.6.4 are due to Matheron [462].

4

Geometric Models

Having laid the general foundations in the previous chapters, we now study geometric processes in $\mathbb{R}^d$ and the random sets derived from them. By geometric processes we understand point processes of closed sets which are concentrated on geometrically distinguished subclasses of $\mathcal{F}'$. In particular, we consider particle processes and flat processes. **Particle processes** are point processes in the subset $\mathcal{C}'$ of nonempty compact sets. Special processes, in general more tractable, are obtained if only particles from the convex ring $\mathcal{R}$ or even the class $\mathcal{K}$ of convex bodies are admitted. A k-**flat process** is a point process in $\mathcal{F}'$ whose intensity measure is concentrated on the space $A(d,k)$ of k-dimensional flats (planes, affine subspaces) of $\mathbb{R}^d$.

We begin with the investigation of particle processes. For these we introduce, in the stationary case, intensities, grain distributions, and densities of functionals in various representations. Special cases are fiber and surface processes; they are treated after the flat processes, in the fifth section. The second section establishes a connection between particle processes and marked point processes. In particular, we introduce the **germ-grain processes**, where compact sets serve as marks. The **germ-grain models** of the third section, which are generated from germ-grain processes, are important examples of random closed sets. An especially tractable subclass are the **Boolean models**, derived from Poisson processes. In the fourth section we treat flat processes. Of particular interest are the processes arising from flat processes by intersections, either with a fixed plane or by intersecting fixed numbers of the flats in the process. Some assertions about the intensities and the directional distributions of these derived processes are obtained, mainly in the case of Poisson processes. The sixth section is concerned with a set-valued parameter, which can be attached to different processes of geometric objects. This is Matheron's 'Steiner compact set', which we call here the **associated zonoid**. It permits us to obtain, among other results, several geometric inequalities for particle or flat processes. In some cases, these can be used to characterize processes with specific extremal properties.

We remind the reader of two general assumptions that we have made. The first one (at the end of Section 3.6) is that *all point processes considered from now on are simple*, except when the opposite is explicitly stated. The second assumption (at the end of Section 3.1) is that *only point processes with locally finite intensity measures are admitted.*

4.1 Particle Processes

By a **particle process** in $\mathbb{R}^d$ we understand a point process in $\mathcal{F}' = \mathcal{F}'(\mathbb{R}^d)$ that is concentrated on the subset $\mathcal{C}'$ of nonempty compact sets, that is, the intensity measure Θ of which satisfies $\Theta(\mathcal{F}' \setminus \mathcal{C}') = 0$. In particular, a point process in $\mathcal{F}'$ whose intensity measure is concentrated on $\mathcal{R}' = \mathcal{R} \setminus \{\emptyset\}$ or $\mathcal{K}' = \mathcal{K} \setminus \{\emptyset\}$, is called a **particle process in** $\mathcal{R}$, respectively, **in** $\mathcal{K}$, in the latter case also a **process of convex particles**. The local finiteness of the intensity measure Θ of a particle process is, by Lemma 2.3.1, equivalent to

$$\Theta(\mathcal{F}_C) < \infty \qquad \text{for all } C \in \mathcal{C}. \tag{4.1}$$

The assumption (4.1) is essential for many later consequences. This is one reason for the fact that we did not define a particle process as a point process in the space $(\mathcal{C}', \delta)$ (with the Hausdorff metric); local finiteness of an intensity measure Θ in this case would only mean that $\Theta(\mathcal{F}^{C^c}) < \infty$ for $C \in \mathcal{C}$.

Nevertheless, it is convenient in the following, when we work with the set $\mathcal{C}'$, to equip it with the Hausdorff metric δ. In particular, continuity of functions on $\mathcal{C}'$ will refer to the Hausdorff metric. Although this continuity differs from continuity with respect to the topology of $\mathcal{F}$, for measurability there is no difference (see Theorem 2.4.1).

The intensity measure of a stationary particle process has a useful decomposition, obtained, roughly speaking, by factoring out the translations. For this, we need a **center function**, and we choose here the mapping

$$c : \mathcal{C}' \to \mathbb{R}^d$$

that associates with each $C \in \mathcal{C}'$ the **circumcenter** $c(C)$ of C. By definition, this is the center of the (uniquely determined) smallest ball containing C. We denote this ball by $B(C)$ and call it the **circumball** and its radius $r(C)$ the **circumradius** of C.

Lemma 4.1.1. *The mapping c is continuous on $\mathcal{C}'$.*

Proof. Let $r(C)$ denote the radius of $B(C)$, for $C \in \mathcal{C}'$. We show first that r is continuous. Let $C_i \to C$ be a convergent sequence in $\mathcal{C}'$. Every accumulation point of the sequence $(B(C_i))_{i\in\mathbb{N}}$ is a ball containing C. This implies $r(C) \le \liminf r(C_i)$. Conversely, for given $\epsilon > 0$, almost all C_i are contained in the ball $B(C) + \epsilon B^d$, hence $\limsup r(C_i) \le r(C) + \epsilon$. For $\epsilon \to 0$ we obtain $r(C) = \lim r(C_i)$.

Next, we show that $B(C_i) \to B(C)$. The sequence of balls $B(C_i)$ is bounded, hence we can assume without loss of generality that it converges, say $B(C_i) \to B$. The limit body B is a ball containing C, and since $r(C_i) \to r(C)$, it has radius $r(C)$. Since the circumball is unique, we have $B = B(C)$.

Finally, $B(C_i) \to B(C)$ implies $c(C_i) \to c(C)$. □

We put

$$\mathcal{C}_0 := \{C \in \mathcal{C}' : c(C) = 0\}$$

and call $\mathcal{C}_0$ the **grain space** (for particle processes). This grain space may also be considered as the set of all translation classes in $\mathcal{C}'$. The set $\mathcal{C}_0$ is closed in $\mathcal{C}'$ and hence (by Lemma 2.1.2) a Borel set in $\mathcal{F}$. Similarly, we define the subsets $\mathcal{K}_0 := \mathcal{C}_0 \cap \mathcal{K}'$, and $\mathcal{R}_0 := \mathcal{C}_0 \cap \mathcal{R}'$. For a subset $B \subset \mathbb{R}^d$ we put

$$\mathcal{C}_c(B) := \{C \in \mathcal{C}' : c(C) \in B\}.$$

The mapping

$$\begin{aligned} \Phi : \mathbb{R}^d \times \mathcal{C}_0 \to & \quad \mathcal{C}' \\ (x, C) & \mapsto x + C \end{aligned}$$

is a homeomorphism, by Lemma 4.1.1 and Theorem 12.3.5.

Theorem 4.1.1. *Let X be a stationary particle process in $\mathbb{R}^d$ with intensity measure $\Theta \neq 0$. Then there exist a number $\gamma \in (0, \infty)$ and a probability measure $\mathbb{Q}$ on $\mathcal{C}_0$ such that*

$$\Theta = \gamma \Phi(\lambda \otimes \mathbb{Q}). \tag{4.2}$$

The number γ and the measure $\mathbb{Q}$ are uniquely determined.

Proof. Let $\widetilde{\Theta} := \Phi^{-1}(\Theta)$ be the image measure of Θ on $\mathbb{R}^d \times \mathcal{C}_0$. We first show a finiteness property. For the unit cube $C^d = [0,1]^d$, we put $C_0^d := [0,1)^d = C^d \setminus \partial^+ C^d$, where

$$\partial^+ C^d := \{x = (x^1, \dots, x^d) \in C^d : \max_{1 \le i \le d} x^i = 1\}$$

is the **upper right boundary** of C^d. Let $(z_i)_{i \in \mathbb{N}}$ be an enumeration of $\mathbb{Z}^d$. We have

$$\begin{aligned} \widetilde{\Theta}(C_0^d \times \mathcal{C}_0) &= \Theta(\{C \in \mathcal{C}' : c(C) \in C_0^d\}) \\ &\le \sum_{i=1}^{\infty} \Theta(\{C \in \mathcal{C}' : C \cap (C_0^d + z_i) \neq \emptyset,\ c(C) \in C_0^d\}) \\ &= \sum_{i=1}^{\infty} \Theta(\{C \in \mathcal{C}' : C \cap C_0^d \neq \emptyset,\ c(C) \in C_0^d - z_i\}) \\ &= \Theta(\{C \in \mathcal{C}' : C \cap C_0^d \neq \emptyset\}) \le \Theta(\mathcal{F}_{C^d}) < \infty, \end{aligned}$$

due to the translation invariance of Θ (which follows from the stationarity of X) and (4.1).

Now we can copy the proof of Theorem 3.5.1, to obtain a representation

$$\widetilde{\Theta} = \gamma(\lambda \otimes \mathbb{Q})$$

with $\gamma \in (0, \infty)$ and a probability measure $\mathbb{Q}$ on $\mathcal{C}_0$. This proves (4.2). The uniqueness is trivial. □

Note that Theorem 4.1.1 shows that for Θ-integrable functions f on $\mathcal{C}'$ we have

$$\int_{\mathcal{C}'} f \, \mathrm{d}\Theta = \gamma \int_{\mathcal{C}_0} \int_{\mathbb{R}^d} f(C + x) \, \lambda(\mathrm{d}x) \, \mathbb{Q}(\mathrm{d}C), \tag{4.3}$$

which will be used frequently.

We call γ the **intensity** and $\mathbb{Q}$ the **grain distribution** of the stationary particle process X. A random set with distribution $\mathbb{Q}$ is called the **typical grain** of X. If X is isotropic, then $\mathbb{Q}$ is rotation invariant (since $c(\vartheta C) = \vartheta c(C)$ for $C \in \mathcal{C}'$ and $\vartheta \in SO_d$), but the converse is generally false. Unless explicitly stated otherwise, we occasionally allow also stationary particle processes with $\Theta = 0$; in this case we define $\gamma = 0$, the grain distribution $\mathbb{Q}$ is not defined, and $\gamma\mathbb{Q}$ has to be read as the zero measure.

For later applications, it will be necessary to admit other center functions, besides the circumcenter. If c is replaced by a measurable mapping $z : \mathcal{C}' \to \mathbb{R}^d$ satisfying $z(tC) = tz(C)$ for $C \in \mathcal{C}'$ and every translation t, then again a decomposition (4.2) is obtained, with different $\mathbb{Q}$. However, isotropy of X is reflected in rotation invariance of $\mathbb{Q}$ only if z has the additional property $z(\vartheta C) = \vartheta z(C)$ for $\vartheta \in SO_d$. (An example, different from the circumcenter, with this property is provided by the Steiner point of the convex hull.) Such center functions play a role if a particle process is to be represented as a marked point process; this is explained in Section 4.2.

It must be noted that the assumed local finiteness of the intensity measure on $\mathcal{F}'$ has the consequence that not every probability measure on $\mathcal{C}_0$ can occur as the grain distribution $\mathbb{Q}$ of a stationary particle process. The following theorem clarifies this.

Theorem 4.1.2. *The probability measure $\mathbb{Q}$ on $\mathcal{C}_0$ is the grain distribution of some stationary particle process if and only if*

$$\int_{\mathcal{C}_0} V_d(C + rB^d) \, \mathbb{Q}(\mathrm{d}C) < \infty \quad \text{for some (or all) } r > 0. \tag{4.4}$$

This is equivalent to the $\mathbb{Q}$-integrability of the dth power of the circumradius, and in the case of a process of convex particles it is equivalent to the $\mathbb{Q}$-integrability of the intrinsic volumes $V_1, \dots, V_d$.

If $\mathbb{Q}$ satisfies (4.4) and if $\gamma > 0$ is given, then there exists (up to stochastic equivalence) precisely one stationary Poisson particle process X in $\mathbb{R}^d$ with intensity γ and grain distribution $\mathbb{Q}$. The process X is isotropic if and only if $\mathbb{Q}$ is rotation invariant.

Proof. If Θ is the intensity measure of the stationary particle process X with intensity $\gamma > 0$ and grain distribution $\mathbb{Q}$, then (4.3) gives, for $K \in \mathcal{C}$,

$$\begin{aligned}\Theta(\mathcal{F}_K) &= \gamma \int_{\mathcal{C}_0} \int_{\mathbb{R}^d} \mathbf{1}_{\mathcal{F}_K}(C + x)\, \lambda(\mathrm{d}x)\, \mathbb{Q}(\mathrm{d}C) \\ &= \gamma \int_{\mathcal{C}_0} V_d(-C + K)\, \mathbb{Q}(\mathrm{d}C),\end{aligned}$$

since

$$\mathbf{1}_{\mathcal{F}_K}(C + x) = 1 \Leftrightarrow (C + x) \cap K \neq \emptyset \Leftrightarrow x \in -C + K.$$

Hence, the local finiteness of Θ implies, in particular, that

$$\int_{\mathcal{C}_0} V_d(C + rB^d)\, \mathbb{Q}(\mathrm{d}C) < \infty \tag{4.5}$$

for all $r > 0$. If (4.5) is satisfied for one number $r > 0$, then

$$\int_{\mathcal{C}_0} V_d(C + K)\, \mathbb{Q}(\mathrm{d}C) < \infty$$

holds for all $K \in \mathcal{C}$, since K can be covered by finitely many translates of rB^d. The remaining equivalences follow from

$$V_d(C + B^d) \leq 2^d \kappa_d \max\{r(C)^d, 1\}$$

and, in the case of a process of convex particles, from the Steiner formula (14.5).

Suppose, conversely, that $\mathbb{Q}$ satisfies (4.4) and that $\gamma > 0$ is given. Then the measure Θ defined by (4.2) is locally finite and translation invariant. By Theorems 3.2.1 and 3.6.1, there exists a Poisson process, unique up to equivalence, with intensity measure Θ. It is stationary, and if $\mathbb{Q}$ is rotation invariant, it is also isotropic. □

The intuitive meaning of the intensity and the grain distribution of a stationary particle process will become clearer by the representations given below, as special cases of the next theorem. With this theorem, we turn to a refined quantitative description of particle processes, which we begin with the definition of densities for geometric functionals. For stationary particle processes, Theorem 4.1.1 opens an easy way of introducing mean values of geometric quantities.

Let $\varphi : \mathcal{C}' \to \mathbb{R}$ be a translation invariant, measurable function, and let X be a stationary particle process with intensity $\gamma > 0$ and grain distribution $\mathbb{Q}$. If φ is nonnegative or $\mathbb{Q}$-integrable, we define the **φ-density** of X by

$$\overline{\varphi}(X) := \gamma \int_{\mathcal{C}_0} \varphi \, \mathrm{d}\mathbb{Q}. \tag{4.6}$$

Remark. We emphasize that $\overline{\varphi}(X)$ is defined here as the mean value of φ with respect to the grain distribution $\mathbb{Q}$, *multiplied by the intensity* γ. That the factor γ has been included in the definition, simplifies many formulas, but must be observed when these formulas are compared with other literature.

For nonnegative φ, it is permitted in (4.6) that $\overline{\varphi}$ (and thus also the limit in Theorem 4.1.3(b)) is infinite.

The following theorem gives different representations of the φ-density, and it also justifies this name.

Theorem 4.1.3. *Let X be a stationary particle process in $\mathbb{R}^d$ with grain distribution $\mathbb{Q}$, and let $\varphi : \mathcal{C}' \to \mathbb{R}$ be a translation invariant measurable function which is nonnegative or $\mathbb{Q}$-integrable.*

(a) *For all $B \in \mathcal{B}(\mathbb{R}^d)$ with $0 < \lambda(B) < \infty$,*

$$\overline{\varphi}(X) = \frac{1}{\lambda(B)} \mathbb{E} \sum_{C \in X,\, c(C) \in B} \varphi(C).$$

(b) *For all $W \in \mathcal{K}$ with $V_d(W) > 0$,*

$$\overline{\varphi}(X) = \lim_{r \to \infty} \frac{1}{V_d(rW)} \mathbb{E} \sum_{C \in X,\, C \subset rW} \varphi(C).$$

(c) *If*

$$\int_{\mathcal{C}_0} |\varphi(C)| V_d(C + B^d)\, \mathbb{Q}(\mathrm{d}C) < \infty,$$

then

$$\overline{\varphi}(X) = \lim_{r \to \infty} \frac{1}{V_d(rW)} \mathbb{E} \sum_{C \in X,\, C \cap rW \neq \emptyset} \varphi(C)$$

for all $W \in \mathcal{K}$ with $V_d(W) > 0$.

Proof. (a) From Campbell's theorem (Theorem 3.1.2) and (4.3), together with the translation invariance of φ, we get

$$\begin{aligned}
\mathbb{E} \sum_{C \in X,\, c(C) \in B} \varphi(C) &= \mathbb{E} \sum_{C \in X} \mathbf{1}_B(c(C)) \varphi(C) \\
&= \gamma \int_{\mathcal{C}_0} \int_{\mathbb{R}^d} \mathbf{1}_B(c(C + x)) \varphi(C)\, \lambda(\mathrm{d}x)\, \mathbb{Q}(\mathrm{d}C) \\
&= \gamma \lambda(B) \int_{\mathcal{C}_0} \varphi(C)\, \mathbb{Q}(\mathrm{d}C) \\
&= \lambda(B) \overline{\varphi}(X).
\end{aligned}$$

(b) As above, we get

$$\mathbb{E} \sum_{C \in X,\, C \subset rW} \varphi(C) = \gamma \int_{\mathcal{C}_0} \varphi(C) \lambda(\{x \in \mathbb{R}^d : C + x \subset rW\})\, \mathbb{Q}(\mathrm{d}C).$$

We may assume that 0 is in the interior of W. Then there is a nonnegative measurable function $\rho : \mathcal{C}_0 \to \mathbb{R}$ such that $C' \subset \rho(C')W$ for all $C' \in \mathcal{C}_0$. For given $C \in \mathcal{C}_0$, suppose that $r > \rho(C)$. For $x \in (r - \rho(C))W$ we have $C + x \subset rW$. It follows that

$$\left(1 - \frac{\rho(C)}{r}\right)^d \le \frac{\lambda(\{x \in \mathbb{R}^d : C + x \subset rW\})}{V_d(rW)} \le 1.$$

For $r \to \infty$, the left side converges monotonically to 1, hence the monotone convergence theorem proves the assertion if φ is nonnegative. The dominated convergence theorem gives the result if φ is integrable.

(c) We have

$$\frac{1}{V_d(rW)} \mathbb{E} \sum_{C \in X,\, C \cap rW \neq \emptyset} \varphi(C) = \frac{\gamma}{V_d(W)} \int_{\mathcal{C}_0} \varphi(C) V_d\left(W - \frac{1}{r}C\right) \mathbb{Q}(\mathrm{d}C).$$

For $C \in \mathcal{C}_0$, let B be a ball with $-C \subset B$, then $W - \frac{1}{r}C \subset W + \frac{1}{r}B$ and hence $V_d(W) \le V_d(W - \frac{1}{r}C) \le V_d(W + \frac{1}{r}B)$. From the continuity of the volume functional on $\mathcal{K}$ it follows that

$$V_d\left(W - \frac{1}{r}C\right) \to V_d(W) \qquad \text{for } r \to \infty.$$

Let $y_1, \dots, y_m \in -C$ be points with $(W + y_i) \cap (W + y_j) = \emptyset$ for $i \neq j$, and assume that m is maximal. For $x \in -C$ there exists i with $(W+x) \cap (W+y_i) \neq \emptyset$, thus $x \in W - W + y_i$. This shows that $-C \subset \bigcup_i (W - W + y_i)$. We may assume that $0 \in W$. For $r \ge 1$, we get $W - \frac{1}{r}C \subset \bigcup_i \left[2W - W + \frac{1}{r}y_i\right]$ and, therefore, $V_d(W - \frac{1}{r}C) \le mV_d(2W - W)$. From $mV_d(W) \le V_d(W - C)$ we obtain

$$V_d\left(W - \frac{1}{r}C\right) \le b(W) V_d(W - C)$$

with a constant $b(W)$ that does not depend on C or r.

There are finitely many vectors $t_1, \dots, t_n \in \mathbb{R}^d$ such that $W \subset \bigcup_{i=1}^n (B^d + t_i)$. This yields $W - C \subset \bigcup_{i=1}^n (B^d - C + t_i)$, hence $V_d(W - C) \le nV_d(B^d - C) = nV_d(B^d + C)$ and thus

$$\int_{\mathcal{C}_0} |\varphi(C)| V_d(W - C)\, \mathbb{Q}(\mathrm{d}C) < \infty.$$

The assertion now follows from the dominated convergence theorem. □

For additive functionals φ, further representations of the φ-density will be given in Theorem 9.2.2. The most important such functionals will be the intrinsic volumes of convex bodies.

As a special case of Theorem 4.1.3, we may choose

$$\varphi(C) := \mathbf{1}_A(C - c(C)) \qquad \text{with } A \in \mathcal{B}(\mathcal{C}_0).$$

If B is as in Theorem 4.1.3, we get

$$\gamma \mathbb{Q}(A) = \frac{1}{\lambda(B)} \, \mathbb{E} \sum_{C \in X,\, c(C) \in B} \mathbf{1}_A(C - c(C)),$$

in particular

$$\gamma = \frac{1}{\lambda(B)} \, \mathbb{E} \sum_{C \in X,\, c(C) \in B} 1. \tag{4.7}$$

Thus, the intensity γ can be interpreted as the expected number of particles per unit volume.

Further we obtain, with W is as in Theorem 4.1.3,

$$\gamma \mathbb{Q}(A) = \lim_{r \to \infty} \frac{1}{V_d(rW)} \, \mathbb{E} \sum_{C \in X,\, C \subset rW} \mathbf{1}_A(C - c(C)) \tag{4.8}$$

$$= \lim_{r \to \infty} \frac{1}{V_d(rW)} \, \mathbb{E} \sum_{C \in X,\, C \cap rW \neq \emptyset} \mathbf{1}_A(C - c(C)). \tag{4.9}$$

If X is a stationary particle process in $\mathbb{R}^d$, then the points $c(C)$, $C \in X$, almost surely generate an ordinary point process X^0 in $\mathbb{R}^d$ (the necessary finiteness condition follows from the proof of Theorem 4.1.1). The corresponding assertion is also true in the non-stationary case if the particles are convex; however, for non-convex particles, the measure $\sum_{C \in X} \delta_{c(C)}$ is not necessarily locally finite a.s. For a stationary (simple) particle process X, the point process X^0 need no longer be simple. We shall see, however, that for a stationary Poisson particle process the point process X^0 is always simple (and thus a stationary Poisson process, too).

In the case of a stationary particle process X the intensity γ is, according to (4.7), also the intensity of the stationary point process X^0. One might interpret this as a construction of X, starting from the ordinary point process X^0 and attaching random compact sets with distribution $\mathbb{Q}$ to the points (regarding multiplicities). However, this idea is not precise, since the random $\mathcal{C}_0$-sets corresponding to different points of X^0 need not be independent. The following example may be instructive.

In $\mathbb{R}^2$, let s_h be a horizontal and s_v a vertical unit segment, both with center 0, and consider the system of segments $s_h + z$, $s_v + z'$, $z, z' \in \mathbb{Z}^2$, where z has even and z' has odd sum of coordinates. Applying to this system the translation by a random vector, uniformly distributed in $[0,1]^2$, we obtain a stationary particle process for which $\mathbb{Q}$ is concentrated on the set $\{s_h, s_v\}$ (associating probability 1/2 to either segment). In this case, the particle, s_h

or s_v, attached to one point of the ordinary point process X^0, completely determines all the other particles of the realization.

The situation is different for Poisson processes.

Theorem 4.1.4. *Let X be a stationary Poisson particle process in $\mathbb{R}^d$ with intensity $\gamma > 0$ and grain distribution $\mathbb{Q}$. Then the ordinary point process X^0 is a (stationary) Poisson process, and the following holds.*

To every $B \in \mathcal{C}$ with $\lambda(B) > 0$ and every $k \in \mathbb{N}$ there exist random points $\xi_1, \dots, \xi_k$ in B with distribution $\lambda \llcorner B/\lambda(B)$ and random closed sets $Z_1, \dots, Z_k$ in $\mathcal{C}_0$ with distribution $\mathbb{Q}$ such that $\xi_1, \dots, \xi_k, Z_1, \dots, Z_k$ are independent and

$$\mathbb{P}(X \llcorner \mathcal{C}_c(B) \in \cdot \mid X(\mathcal{C}_c(B)) = k) = \mathbb{P}\left(\sum_{i=1}^{k} \delta_{\xi_i + Z_i} \in \cdot\right).$$

In other words, a stationary Poisson process X in $\mathcal{C}'$ can be generated (and also simulated) by taking an ordinary stationary Poisson process X^0 with intensity γ and adding to every $x \in X^0$ independently a random closed set Z_x with distribution $\mathbb{Q}$,

$$X = \sum_{x \in X^0} \delta_{x + Z_x}.$$

Proof. Since

$$\mathbb{P}(X^0(A) = k) = \mathbb{P}(X(\mathcal{C}_c(A)) = k), \qquad A \in \mathcal{B}(\mathbb{R}^d),$$

X^0 is a stationary Poisson process in $\mathbb{R}^d$ with intensity measure

$$\Theta_c(A) = \Theta(\mathcal{C}_c(A)) = \gamma\lambda(A), \qquad A \in \mathcal{B}(\mathbb{R}^d);$$

here we have applied Theorem 4.1.1 to the set $\mathcal{C}_c(A)$. In particular, if B is compact and $\lambda(B) > 0$, then $0 < \Theta(\mathcal{C}_c(B)) < \infty$, hence we can apply Theorem 3.2.2(b) and obtain independent random closed sets $\widetilde{Z}_1, \dots, \widetilde{Z}_k$ with

$$\mathbb{P}(X \llcorner \mathcal{C}_c(B) \in \cdot \mid X(\mathcal{C}_c(B)) = k) = \mathbb{P}\left(\sum_{i=1}^{k} \delta_{\widetilde{Z}_i} \in \cdot\right);$$

here each $\widetilde{Z}_i$ has distribution

$$\mathbb{P}_{\widetilde{Z}_i} = \frac{\Theta \llcorner \mathcal{C}_c(B)}{\Theta(\mathcal{C}_c(B))}.$$

From $\Theta \llcorner \mathcal{C}_c(B) = \Phi((\lambda \llcorner B) \otimes \gamma\mathbb{Q})$ and $\Theta(\mathcal{C}_c(B)) = \gamma\lambda(B)$ it follows that

$$\mathbb{P}_{\widetilde{Z}_i} = \Phi\left(\frac{\lambda \llcorner B}{\lambda(B)} \otimes \mathbb{Q}\right).$$

Hence, if we define Z_i, ξ_i by $\Phi^{-1} \circ \widetilde{Z}_i =: (\xi_i, Z_i)$, we obtain independent random variables Z_i (random closed sets with distribution $\mathbb{Q}$) and ξ_i (random points with distribution $\lambda \llcorner B/\lambda(B)$), and we have

$$\widetilde{Z}_i = \Phi(\xi_i, Z_i) = \xi_i + Z_i,$$

therefore also

$$\mathbb{P}\left(\sum_{i=1}^{k} \delta_{\widetilde{Z}_i} \in \cdot\right) = \mathbb{P}\left(\sum_{i=1}^{k} \delta_{\xi_i + Z_i} \in \cdot\right).$$

This completes the proof. □

Marked Particle Processes

In Chapter 10 it will be useful to have limit relations, analogous to those of Theorem 4.1.3, for marked particle processes. By a **marked particle process** we understand a simple point process in $\mathcal{C}' \times M$, where M denotes the mark space, as in Section 3.5. For the intensity measure Θ we assume, corresponding to (3.12), that

$$\Theta(C \times M) < \infty \qquad \text{for all } C \in \mathcal{C}(\mathcal{F}').$$

Stationarity again means invariance of the distribution $\mathbb{P}_X$ under translations, where these, as in the case of marked point processes, affect only the first component. Let X be a stationary marked particle process with intensity measure $\Theta \neq 0$. Using the mapping

$$\begin{aligned} \Phi : \mathbb{R}^d \times \mathcal{C}_0 \times M &\to \quad \mathcal{C}' \times M \\ (x, C, m) \quad &\mapsto (x + C, m), \end{aligned}$$

we obtain, in analogy to Theorem 4.1.1, a decomposition

$$\Theta = \gamma \Phi(\lambda \otimes \mathbb{Q})$$

where $\mathbb{Q}$ is now a probability measure on $\mathcal{C}_0 \times M$; it is called the **grain-mark distribution** of X. With this decomposition, Campbell's theorem and the analog of (4.3) read as follows. For every nonnegative measurable function f on $\mathcal{C}' \times M$,

$$\begin{aligned} \mathbb{E} \sum_{(C,m) \in X} f(C, m) &= \int_{\mathcal{C}' \times M} f \, \mathrm{d}\Theta \\ &= \gamma \int_{\mathcal{C}_0 \times M} \int_{\mathbb{R}^d} f(C + x, m) \, \lambda(\mathrm{d}x) \, \mathbb{Q}(\mathrm{d}(C, m)). \end{aligned}$$

Theorem 4.1.5. *Let X be a stationary marked particle process in $\mathbb{R}^d$ with grain-mark distribution $\mathbb{Q}$, and let $\varphi : \mathcal{C}' \times M \to \mathbb{R}$ be a measurable function* *which is translation invariant in the first variable and either nonnegative or $\mathbb{Q}$-integrable. Then the φ-density defined by*

$$\overline{\varphi}(X) := \gamma \int_{\mathcal{C}_0 \times M} \varphi \, \mathrm{d}\mathbb{Q}$$

has the representations

$$\overline{\varphi}(X) = \frac{1}{\lambda(B)} \mathbb{E} \sum_{(C,m)\in X,\, c(C)\in B} \varphi(C,m) \tag{4.10}$$

for $B \in \mathcal{B}(\mathbb{R}^d)$ with $0 < \lambda(B) < \infty$,

$$\overline{\varphi}(X) = \lim_{r\to\infty} \frac{1}{V_d(rW)} \mathbb{E} \sum_{(C,m)\in X,\, C\subset rW} \varphi(C,m) \tag{4.11}$$

for $W \in \mathcal{K}$ with $V_d(W) > 0$, and

$$\overline{\varphi}(X) = \lim_{r\to\infty} \frac{1}{V_d(rW)} \mathbb{E} \sum_{(C,m)\in X,\, C\cap rW\neq\emptyset} \varphi(C,m), \tag{4.12}$$

if, in addition,

$$\int_{\mathcal{C}_0 \times M} |\varphi(C,m)| V_d(C + B^d)\, \mathbb{Q}(\mathrm{d}(C,m)) < \infty$$

is satisfied.

The proof is obtained by the obvious modification of that of Theorem 4.1.3.

Note for Section 4.1

Theorem 4.1.1 shows, as did also Theorems 3.3.1 and 3.5.1, how the assumption of stationarity leads to a decomposition of the intensity measure, where the Lebesgue measure appears as a factor. Such a factorization of measures with partial invariance properties on (local) product spaces, where a Haar measure appears as one factor, was raised by Ambartzumian [35] to a basic principle of stochastic geometry.

4.2 Germ-grain Processes

We recall the convention, agreed upon in Section 3.1, that simple counting measures are often identified with their supports, so that, for example, for $\eta \in \mathsf{N}_s(\mathcal{C}')$ the notations $\eta(\{C\}) = 1$ and $C \in \eta$ are employed synonymously.

In the previous section, the decomposition of the intensity measure of a stationary particle process X in $\mathbb{R}^d$ was based on a representation of the particles $C \in X$ in the form $C = C_0 + x$ with $x := c(C)$ and $C_0 := C - c(C)$. Of course, the formulation of Theorem 4.1.1 is strongly reminiscent of the corresponding Theorem 3.5.1 on marked point processes. One can, in fact, identify the stationary particle process X with the marked point process

$$\Phi^{-1}(X) = \{(x, C) \in \mathbb{R}^d \times \mathcal{C}_0 : C + x \in X\}.$$

Here the mark space is $\mathcal{C}_0$, and the grain distribution $\mathbb{Q}$ becomes the mark distribution. Before we make use of this connection and apply the results of Section 3.5 on Palm distributions of marked point processes to particle processes, we want to clarify the role that the choice of the circumcenter as a center of the particles plays here.

Generally, we understand by a **center function** a measurable map $z : \mathcal{C}' \to \mathbb{R}^d$ which is compatible with translations, that is, satisfies

$$z(C + x) = z(C) + x \qquad \text{for all } x \in \mathbb{R}^d.$$

Examples of center functions, besides the circumcenter c, are the center of gravity, if only particles with positive Lebesgue measure are considered, or the Steiner point of the convex hull. These center functions are also equivariant under rotations (that is, $z(\vartheta C) = \vartheta z(C)$ for $\vartheta \in SO_d$). The following examples do not have this property. For $C \in \mathcal{C}'(\mathbb{R}^2)$ (the definition can be extended to $d \geq 2$), the **lower tangent point** of C is defined by $\tilde{z}(C) = (z^1, z^2)$ with

$$\begin{aligned} z^2 &:= \min\{x^2 : (x^1, x^2) \in C\} \\ z^1 &:= \min\{x^1 : (x^1, z^2) \in C\}, \end{aligned}$$

where x^1, x^2 are the coordinates of x. Further, the **left lower corner** of C is defined by

$$z'(C) := \left(\min_{x \in C} x^1, \min_{x \in C} x^2 \right)$$

(in general, $z'(C) \notin C$). These center functions are applied in certain estimation procedures. As in the example of the center of gravity, it is sometimes convenient to allow center functions that are defined only on measurable subclasses of $\mathcal{C}'$ which are closed under translations.

If X is a particle process and z is a center function, then

$$X^z := \sum_{C \in X} \delta_{z(C)}$$

is a random counting measure on $\mathbb{R}^d$ which, however, need neither be simple nor locally finite. In the stationary case, local finiteness is ensured. Thus, the following connection between stationary particle processes and marked point processes can be established.

Theorem 4.2.1. *Let X be a stationary particle process in $\mathbb{R}^d$, and let z be a center function. Then X^z is a stationary point process in $\mathbb{R}^d$, and*

$$X_z := \sum_{C \in X} \delta_{(z(C), C - z(C))}$$

is a stationary marked point process with mark space $\mathcal{C}'$. The intensities of X, X^z and X_z are the same. The mark distribution $\mathbb{Q}_z$ of X_z is the image of the grain distribution $\mathbb{Q}$ of X under the mapping $C \mapsto C - z(C)$.

In particular, the grain distribution $\mathbb{Q}$ is the mark distribution of X_c.

Proof. To show the measurability of X_z, it suffices by Lemma 3.1.5 to verify that $\{X_z(A) = k\}$ is measurable for all $A \in \mathcal{B}(\mathbb{R}^d \times \mathcal{C}')$ and all $k \in \mathbb{N}_0$. Let A and k be given. The function

$$\begin{aligned} \varphi : \mathcal{C}' \to \quad & \mathbb{R}^d \times \mathcal{C}' \\ C \mapsto & (z(C), C - z(C)) \end{aligned}$$

is measurable, since z is measurable. Hence, $\{X_z(A) = k\} = \{X(\varphi^{-1}(A)) = k\}$ is measurable.

As in the proof of Theorem 4.1.1 (where we replace $\mathcal{C}_0$ by $\mathcal{C}'$ and c by z) we obtain

$$\mathbb{E} X_z(C_0^d \times \mathcal{C}') \le \Theta(\mathcal{F}_{C^d}) < \infty,$$

where Θ is the intensity measure of X. This gives $\mathbb{E} X_z(C \times \mathcal{C}') < \infty$ for every compact set $C \in \mathcal{C}$; thus the measure X_z is a.s. locally finite. Hence, X^z is a point process with locally finite intensity measure, and X_z satisfies (3.12) and is, therefore, a marked point process in $\mathbb{R}^d$.

For $t \in \mathbb{R}^d$, the definition of the operation of the translation group on $\mathbb{R}^d \times \mathcal{C}'$ and the compatibility of z with translations give

$$\begin{aligned} X_z + t &= \sum_{C \in X} \delta_{(z(C)+t, C - z(C))} \\ &= \sum_{C \in X} \delta_{(z(C+t), C+t-z(C+t))} \\ &= \sum_{C \in X+t} \delta_{(z(C), C - z(C))} \\ &= (X+t)_z. \end{aligned}$$

Since X and $X + t$ have the same distribution, the same holds for X_z and $X_z + t$, which means that X_z is stationary. From this it follows that also X^z is stationary.

Let $\mathbb{Q}_z$ be the mark distribution of X_z and let γ_z be its intensity. Denoting by γ and $\mathbb{Q}$ the intensity and the grain distribution of X, from Theorems 3.5.1, 3.1.2 and 4.1.1 we get, for $B \in \mathcal{B}(\mathbb{R}^d)$ and $A \in \mathcal{B}(\mathcal{C}')$,

$$\begin{aligned}
\gamma_z \lambda \otimes \mathbb{Q}_z(B \times A) &= \mathbb{E} X_z(B \times A) \\
&= \mathbb{E} \sum_{C \in X} \mathbf{1}_B(z(C)) \mathbf{1}_A(C - z(C)) \\
&= \gamma \int_{\mathcal{C}_0} \int_{\mathbb{R}^d} \mathbf{1}_B(z(C+x)) \mathbf{1}_A(C + x - z(C+x)) \, \lambda(\mathrm{d}x) \, \mathbb{Q}(\mathrm{d}C) \\
&= \gamma \lambda(B) \int_{\mathcal{C}_0} \mathbf{1}_A(C - z(C)) \, \mathbb{Q}(dC) \\
&= \gamma(\lambda \otimes f_z(\mathbb{Q}))(B \times A)
\end{aligned}$$

with $f_z : \mathcal{C}_0 \to \mathcal{C}'$ defined by $f_z(C) = C - z(C)$. The special case $B = C^d$ and $A = \mathcal{C}'$ gives $\gamma = \gamma_z$, which is also the intensity of X^z. Now it follows that $\mathbb{Q}_z$ is the image measure of $\mathbb{Q}$ under the mapping f_z. □

Some properties of the mark distribution, for example rotation invariance, depend essentially on the choice of the center function. If X is isotropic and z is equivariant under rotations, then the mark distribution, too, is rotation invariant.

By Theorem 4.2.1, to every stationary particle process X there corresponds a whole family of marked point processes with mark space $\mathcal{C}'$; every center function z generates an element X_z of this family. The special choice $z = c$ yields the canonical model X_c of X with $X^z = X^0$, which we mostly use in the following. If X is a stationary Poisson process, a corresponding assertion holds for X_z.

Theorem 4.2.2. *Let X be a stationary Poisson particle process in $\mathbb{R}^d$, and let z be a center function. Then X_z is an independently marked stationary Poisson process.*

Proof. We define

$$\begin{aligned}
\varphi : \mathcal{C}' \to \quad & \mathbb{R}^d \times \mathcal{C}_0 \\
C \mapsto \ & (z(C), C - z(C)).
\end{aligned}$$

As we have seen in the proof of Theorem 4.2.1,

$$\{X_z(A) = k\} = \{X(\varphi^{-1}(A)) = k\}$$

holds for Borel sets $A \in \mathcal{B}(\mathbb{R}^d \times \mathcal{C}_0)$ and $k \in \mathbb{N}_0$. Hence, X_z is a stationary Poisson process in $\mathbb{R}^d \times \mathcal{C}_0$. The assertion now follows from Theorem 3.5.8. □

We return to the general situation and now consider, conversely, a marked point process $\widetilde{X}$ with mark space $\mathcal{C}'$. Then

$$X := \sum_{(x,C) \in \widetilde{X}} \delta_{x+C} \tag{4.13}$$

defines a particle process X, if the local finiteness of the counting measures on the right side (and of the intensity measure) is guaranteed. In this case, we call $\widetilde{X}$ a **germ-grain process**. The intuitive idea behind this is that the points x of the pairs $(x, C) \in \widetilde{X}$ are the 'germs' and the compact sets $x + C$ are the 'grains'. The process (4.13) is called the **particle process generated by** $\widetilde{X}$. If, in particular, $\widetilde{X}$ is stationary, then in analogy to Theorem 4.1.2 one finds that for the local finiteness of the intensity measure of X the condition

$$\int_{\mathcal{C}'} V_d(C + B^d)\, \mathbb{Q}(dC) < \infty \tag{4.14}$$

on the mark distribution $\mathbb{Q}$ of $\widetilde{X}$ is necessary and sufficient. In the stationary case, the mark distribution $\mathbb{Q}$ is also called the **distribution of the typical grain**, and every random closed set Z_0 with distribution $\mathbb{Q}$ is called the **typical grain** (or **primary grain**) of $\widetilde{X}$.

If $\widetilde{X}$ is an independently marked point process with mark space $\mathcal{C}'$, then even without stationarity it is possible to work with the mark distribution $\mathbb{Q}$, as in Section 3.5. Also in this case, a random closed set Z_0 with distribution $\mathbb{Q}$ is called the **typical grain** of $\widetilde{X}$. The finiteness condition (4.1) for the intensity measure Θ of the particle process X generated by (4.13) can now be rewritten in the following way. For $C \in \mathcal{C}'$ we have

$$\begin{aligned}\Theta(\mathcal{F}_C) &= \int_{\mathbb{R}^d} \int_{\mathcal{C}'} \mathbf{1}_{\mathcal{F}_C}(x + K)\, \mathbb{Q}(\mathrm{d}K)\, \vartheta(\mathrm{d}x) \\ &= \int_{\mathbb{R}^d} T_{Z_0}(C - x)\, \vartheta(\mathrm{d}x).\end{aligned}$$

Here, ϑ is the intensity measure of the unmarked process X^0, and T_{Z_0} is the capacity functional of the typical grain Z_0 of $\widetilde{X}$. Hence, for the particle process X, (4.1) is equivalent to

$$\int_{\mathbb{R}^d} T_{Z_0}(C - x)\, \vartheta(\mathrm{d}x) < \infty \qquad \text{for } C \in \mathcal{C}'. \tag{4.15}$$

If $\widetilde{X}$ is stationary, this is again equivalent to (4.14). An independently marked point process $\widetilde{X}$ satisfying (4.15) is called an **independent germ-grain process**.

If, for an independent germ-grain process $\widetilde{X}$, the germ process X^0 is a Poisson process, and thus $\widetilde{X}$, according to Theorem 3.5.7, is a Poisson process in $\mathbb{R}^d \times \mathcal{C}'$, with intensity measure $\vartheta \otimes \mathbb{Q}$, then the generated particle process X is the image $\sigma(\widetilde{X})$ of $\widetilde{X}$ under the mapping

$$\begin{aligned}\sigma : \mathbb{R}^d \times \mathcal{C}' &\to \quad \mathcal{C}' \\ (x, C) &\mapsto x + C\end{aligned}$$

and hence is also a Poisson process (with intensity measure $\Theta = \sigma(\vartheta \otimes \mathbb{Q})$). Since ϑ has no atoms, the same holds true for Θ.

We continue these considerations in the next section, where we treat random closed sets which arise from independent germ-grain processes by taking union sets of the generated particle processes.

Now we apply the results of Section 3.5 to particle processes.

Theorem 4.2.3. *Let X be a stationary particle process in $\mathbb{R}^d$ with intensity $\gamma > 0$, and let z be a center function. Then there is a (uniquely determined) probability measure $\mathbb{P}^0$ on $\mathcal{C}' \times \mathsf{N}_s(\mathcal{C}')$ such that*

$$\gamma \mathbb{P}^0(A) = \mathbb{E} \sum_{C \in X} \mathbf{1}_B(z(C)) \mathbf{1}_A(C - z(C), X - z(C))$$

for all $A \in \mathcal{B}(\mathcal{C}') \otimes \mathcal{N}_s(\mathcal{C}')$ and all $B \in \mathcal{B}(\mathbb{R}^d)$ with $\lambda(B) = 1$.

If $f : \mathbb{R}^d \times \mathcal{C}' \times \mathsf{N}_s(\mathcal{C}') \to \mathbb{R}$ is a nonnegative measurable function, then $\sum_{C \in X} f(z(C), C - z(C), X)$ is measurable and

$$\mathbb{E} \sum_{C \in X} f(z(C), C - z(C), X)$$
$$= \gamma \int_{\mathbb{R}^d} \int_{\mathcal{C}' \times \mathsf{N}_s(\mathcal{C}')} f(x, C, \eta + x) \, \mathbb{P}^0(\mathrm{d}(C, \eta)) \, \lambda(\mathrm{d}x).$$

Proof. We apply Theorem 3.5.2 to the marked point process X_z with mark space $\mathcal{C}'$. More precisely, if $A \in \mathcal{B}(\mathcal{C}') \otimes \mathcal{N}_s(\mathcal{C}')$ is given, we apply it to the set $\widetilde{A} := (\mathrm{id} \times \psi)(A)$, where $\psi : \mathsf{N}_s(\mathcal{C}') \to \mathsf{N}_s(\mathbb{R}^d \times \mathcal{C}')$ is defined by

$$\psi(\eta) := \sum \delta_{(z(C_i), C_i - z(C_i))}, \qquad \text{if } \eta = \sum \delta_{C_i}.$$

If the measure that results from Theorem 3.5.2 is denoted by $\widetilde{\mathbb{P}}^0$, then $\mathbb{P}^0(A) = \widetilde{\mathbb{P}}^0(\widetilde{A})$ yields the required measure. The second part of Theorem 4.2.3 follows from Theorem 3.5.3. $\square$

Theorem 4.2.4. *Let X be a stationary particle process in $\mathbb{R}^d$ with intensity $\gamma > 0$, let z be a center function. Let $\mathcal{C}_{z,0} := \{C \in \mathcal{C}' : z(C) = 0\}$ denote the mark space of the marked point process X_z, and let $\mathbb{Q}$ be the mark distribution of X_z. Then there exists a ($\mathbb{Q}$-a.s. uniquely determined) regular family $(\mathbb{P}^{0,C})_{C \in \mathcal{C}_{z,0}}$ of probability measures on $\mathcal{N}_s(\mathcal{C}')$ with*

$$\mathbb{P}^0(B \times A) = \int_B \mathbb{P}^{0,C}(A) \, \mathbb{Q}(\mathrm{d}C)$$

for $B \in \mathcal{B}(\mathcal{C}_{z,0})$ and $A \in \mathcal{N}_s(\mathcal{C}')$.

If $f : \mathbb{R}^d \times \mathcal{C}_{z,0} \times \mathsf{N}_s(\mathcal{C}') \to \mathbb{R}$ is a nonnegative measurable function, then $\sum_{C \in X} f(z(C), C - z(C), X)$ is measurable, and

$$\mathbb{E} \sum_{C \in X} f(z(C), C - z(C), X)$$
$$= \gamma \int_{\mathbb{R}^d} \int_{\mathcal{C}_{z,0}} \int_{\mathsf{N}_s(\mathcal{C}')} f(x, C, \eta + x) \, \mathbb{P}^{0,C}(\mathrm{d}\eta) \, \mathbb{Q}(\mathrm{d}C) \, \lambda(\mathrm{d}x).$$

Proof. Let $\widetilde{\mathbb{P}}^0$ be the measure obtained in the proof of Theorem 4.2.3. By Theorem 3.5.4, there exists a ($\mathbb{Q}$-a.s. uniquely determined) regular family $(\widetilde{\mathbb{P}}^{0,C})_{C \in \mathcal{C}_{z,0}}$ of probability measures on $\mathsf{N}_s(\mathbb{R}^d \times \mathcal{C}_{z,0})$ with

$$\widetilde{\mathbb{P}}^0(B \times A) = \int_B \widetilde{\mathbb{P}}^{0,C}(A) \, \mathbb{Q}(\mathrm{d}C)$$

for all $B \in \mathcal{B}(\mathcal{C}_{z,0})$ and $A \in \mathcal{N}_s(\mathbb{R}^d \times \mathcal{C}_{z,0})$. Defining $\mathbb{P}^{0,C}$ as the image measure of $\widetilde{\mathbb{P}}^{0,C}$ under the mapping

$$\begin{aligned} \mathsf{N}(\mathbb{R}^d \times \mathcal{C}_{z,0}) &\to \quad \mathsf{N}_s(\mathcal{C}'), \\ \widetilde{\eta} \quad &\mapsto \sum_{(x,C) \in \widetilde{\eta}} \delta_{x+C} \end{aligned}$$

we obtain the assertion. □

Now we consider sections with a fixed k-dimensional plane $S \in G(d,k)$. Let $\widetilde{X}$ be a stationary (but otherwise arbitrary) germ-grain process in $\mathbb{R}^d$. With it, we can associate in a natural way the section process

$$\widetilde{X} \cap S := \sum_{(x,C) \in \widetilde{X},\, (x+C) \cap S \neq \emptyset} \delta_{(x_S, (x^S + C) \cap S)},$$

where $x = x_S + x^S$ with $x_S \in S$ and $x^S \in S^\perp$ is the orthogonal decomposition. Thus, the germs of $\widetilde{X} \cap S$ arise by orthogonally projecting to S those germs of $\widetilde{X}$ for which the corresponding grain has nonempty intersection with S. Observe that

$$(x + C) \cap S = x_S + [(x^S + C) \cap S]. \tag{4.16}$$

If we assume in addition that $\widetilde{X} \cap S$ is simple and that the condition corresponding to (3.12) is satisfied, then $\widetilde{X} \cap S$ is a germ-grain process in the space S (which we can identify with $\mathbb{R}^k$) with mark space $\mathcal{C}'(S)$; the marked process $\widetilde{X} \cap S$ is stationary in S. For the particle process generated by $\widetilde{X}$,

$$X := \sum_{(x,C) \in \widetilde{X}} \delta_{x+C},$$

the section process $X \cap S$ was already defined in Section 3.6. Because of (4.16), $X \cap S$ coincides with the particle process generated by $\widetilde{X} \cap S$.

Suppose now that γ is the intensity and $\mathbb{Q}$ is the mark distribution of $\widetilde{X}$, and let $\gamma_{\widetilde{X} \cap S}$, $\mathbb{Q}_{\widetilde{X} \cap S}$ be the corresponding parameters for the section process

$\widetilde{X} \cap S$. Then, for $B \in \mathcal{B}(S)$ and $A \in \mathcal{B}(\mathcal{C}'(S))$, by Theorem 3.5.1 and the Campbell theorem we have

$$\begin{aligned}
&\gamma_{\widetilde{X}\cap S}\lambda_S(B)\mathbb{Q}_{\widetilde{X}\cap S}(A) \\
&= \mathbb{E} \sum_{(x,C)\in\widetilde{X}} \mathbf{1}_B(X_S)\mathbf{1}_A((x^S + C)\cap S) \\
&= \gamma \int_{\mathcal{C}'}\int_{\mathbb{R}^d} \mathbf{1}_B(x_S)\mathbf{1}_A((x^S + C)\cap S)\,\lambda(\mathrm{d}x)\,\mathbb{Q}(\mathrm{d}C) \\
&= \gamma \int_{\mathcal{C}'}\int_{S^\perp}\int_S \mathbf{1}_B(y)\mathbf{1}_A((z + C)\cap S)\,\lambda_S(\mathrm{d}y)\,\lambda_{S^\perp}(\mathrm{d}z)\,\mathbb{Q}(\mathrm{d}C) \\
&= \gamma\lambda_S(B)\int_{\mathcal{C}'}\int_{S^\perp} \mathbf{1}_A((z + C)\cap S)\,\lambda_{S^\perp}(\mathrm{d}z)\,\mathbb{Q}(\mathrm{d}C).
\end{aligned}$$

Hence, setting

$$M_S(A) := \int_{\mathcal{C}'}\int_{S^\perp} \mathbf{1}_A((z + C)\cap S)\,\lambda_{S^\perp}(\mathrm{d}z)\,\mathbb{Q}(\mathrm{d}C),$$

we obtain

$$\gamma_{\widetilde{X}\cap S} = \gamma M_S(\mathcal{C}'(S)) = \gamma \int_{\mathcal{C}'} \lambda_{S^\perp}(C|S^\perp)\,\mathbb{Q}(\mathrm{d}C), \tag{4.17}$$

where $C|S^\perp$ is the image of C under the orthogonal projection to $S^\perp$, and

$$\mathbb{Q}_{\widetilde{X}\cap S}(A) = M_S(A)/M_S(\mathcal{C}'(S)), \tag{4.18}$$

if $M_S(\mathcal{C}'(S)) \neq 0$. Thus, the mark distribution of the section process depends only on the mark distribution of the original process. More explicit results for $\gamma_{\widetilde{X}\cap S}$ can be obtained for stationary and isotropic processes of convex particles (a general result of this type is Theorem 9.4.8).

Notes for Section 4.2

1. Generalized center functions. As a generalization of the notion of center function z, a **generalized center function** z maps each particle C in a particle collection η to a point which may depend not only on C but also on the other particles in η. Formally, z is a measurable mapping from $\mathcal{C}' \circ \mathsf{N}_s(\mathcal{C}') := \{(C,\eta) \in \mathcal{C}' \times \mathsf{N}_s(\mathcal{C}') : C \in \eta\}$ to $\mathbb{R}^d$, which is compatible with translations,

$$\mathsf{z}(C + x, \eta + x) = \mathsf{z}(C,\eta) + x \qquad \text{for all } x \in \mathbb{R}^d.$$

Every center function z defines a generalized center function $\langle z\rangle$ by $\langle z\rangle(C,\eta) := z(C)$. In generalization of Theorem 4.2.1, the following holds (see Schneider and Weil [717, Satz 4.3.1]).

Let X be a stationary particle process in $\mathbb{R}^d$, and let z be a generalized center function. Then

$$X^{\mathsf{z}} := \sum_{C \in X} \delta_{\mathsf{z}(C,X)}$$

is a stationary point process in $\mathbb{R}^d$, *and*

$$X_{\mathsf{z}} := \sum_{C \in X} \delta_{(\mathsf{z}(C,X), C-\mathsf{z}(C,X))}$$

is a stationary marked point process with mark space $\mathcal{C}'$. *The intensities of* X, X^{z} *and* X_{z} *are the same.*

Generalized center functions occur, for example, in connection with Voronoi mosaics. If A is a locally finite set in $\mathbb{R}^d$ such that the Voronoi cells $C(x, A)$, $x \in A$, are all bounded (see Section 10.2), then, for the corresponding Voronoi mosaic $\mathsf{m} := \{C(x, A) : x \in A\}$, the mapping $\mathsf{z} : (C, \eta) \mapsto x$, if $\eta = \mathsf{m}$ and $C = C(x, A)$, (and $\mathsf{z}(C, \eta) := c(C)$ otherwise) is a generalized center function.

2. The section formulas (4.17) and (4.18) can be found in Stoyan [740].

3. The interpretation of germ-grain processes as processes of points around which grains have grown randomly already indicates a temporal aspect which can, and has been, pursued further. Motivated by applications to the growth of crystals, tumor cells and other growing structures, various spatio-temporal models have been developed. Starting with the realization of a spatial point process X, one can, for example, let balls grow around the points of X with constant or random speed, in a dependent way or independently, at the same time or at different, random times. The growth can be stopped or modified, according to different rules, when the growing balls touch, or the growing balls can overlap, penetrate or get deformed to form a tessellation of space. Finally, also the underlying point process X may vary in time, for example as a birth-and-death process. Examples of this kind are the Stienen model (compare Note 9 to Section 10.2), the lilypond model (lilypond growth protocol, see Daley and Last [193], Heveling and Last [340]), the dead leaves model (see Serra [729, pp. 508–511], Cowan and Tsang [184]), the Johnson–Mehl tessellation model (see Møller [552]) and the general class of crystallization processes investigated by Capasso and co-workers [157], [158], [159], [515].

Some of the mentioned spatio-temporal models produce random systems of non-overlapping balls, others can be modified to do so, for example by thinning. There are current efforts by statisticians and physicists to generate random packings of balls with high volume density (Torquato [759], Stoyan and Schlather [745], Döge et al. [205]).

4.3 Germ-grain Models, Boolean Models

In Theorem 3.6.2 we have seen that for a point process X in $\mathcal{F}'$ the union set

$$Z_X := \bigcup_{F \in X} F$$

is a random closed set, and in Theorem 3.6.3 we have characterized those Z_X for which X has Poisson distributed counting variables. Now we study the random closed sets arising as the union sets of particle processes. We shall be particularly interested in the random closed sets resulting from special germ-grain processes.

It is easy to see that a given random closed set Z can always be represented as the union set of a particle process X. In the following, we describe a construction which has the advantage that invariance properties of Z carry over to X. If Z is a random $\mathcal{S}$-set, then it is even possible to choose the particles of X as convex bodies. However, in order to ensure in this case the local finiteness of the intensity measure of X, we need an integrability assumption on the random $\mathcal{S}$-set Z. To formulate it, we define

$$N(K) := \min\left\{m \in \mathbb{N} : K = \bigcup_{i=1}^{m} K_i \text{ with } K_i \in \mathcal{K}\right\} \qquad \text{for } K \in \mathcal{R}',$$

and $N(\emptyset) := 0$.

Lemma 4.3.1. *The function $N : \mathcal{R} \to \mathbb{N}_0$ is measurable.*

Proof. By Theorem 2.4.1, it is sufficient to show that N is semicontinuous with respect to the Hausdorff metric. Let $M_j, M \in \mathcal{R}$ be sets with $M_j \to M$ (in the Hausdorff metric) as $j \to \infty$. We assert that

$$N(M) \le \liminf N(M_j). \tag{4.19}$$

Suppose this were false. Going over to a subsequence, we can assume that there exists a number $m \in \mathbb{N}$ with

$$N(M_j) = m < N(M) \qquad \text{for } j \in \mathbb{N},$$

thus

$$M_j = \bigcup_{i=1}^{m} K_j^{(i)} \qquad \text{with } K_j^{(i)} \in \mathcal{K}'.$$

Since the sets M_j and hence also the sets $K_j^{(i)}$ are uniformly bounded, there exists a subsequence $(j_k)_{k\in\mathbb{N}}$ such that

$$K_{j_k}^{(i)} \to K^{(i)} \qquad \text{as } k \to \infty, \qquad i = 1, \dots, m,$$

with $K^{(i)} \in \mathcal{K}'$. Theorem 12.3.5 gives

$$M_{j_k} = \bigcup_{i=1}^{m} K_{j_k}^{(i)} \to \bigcup_{i=1}^{m} K^{(i)},$$

thus $M = \bigcup_{i=1}^{m} K^{(i)}$, and hence $N(M) \le m$, a contradiction. This completes the proof of (4.19) and thus of the lemma. □

Now we can prove the announced representation result.

Theorem 4.3.1. *To every random closed set Z in $\mathbb{R}^d$ there exists a simple particle process X with $Z = Z_X$ and such that $X \stackrel{\mathcal{D}}{=} gX$ for all rigid motions $g \in G_d$ for which $Z \stackrel{\mathcal{D}}{=} gZ$. In particular, X is stationary (isotropic) if Z is stationary (isotropic).*

If Z is a random $\mathcal{S}$-set with $\mathbb{E}N(Z \cap C) < \infty$ for all $C \in \mathcal{C}'$, then X can in addition be chosen so that all particles are convex.

Proof. The decomposition of a random closed set Z into compact particles is easier to achieve than that of a random $\mathcal{S}$-set into convex bodies. Therefore, we restrict ourselves in the proof to this more difficult situation. The decomposition into compact particles can be done in a similar way if the mapping ψ used below is replaced by $\widetilde{\psi} : C \mapsto \delta_C$, $C \in \mathcal{C}'$ (and $\widetilde{\psi}(\emptyset) = 0$).

From the proof of Theorem 14.4.4 we get the existence of a measurable map $\psi : \mathcal{R} \to \mathsf{N}_s(\mathcal{K}')$ with

$$\psi(C) = \sum_{i=1}^{N(C)} \delta_{K_i}, \qquad C = \bigcup_{i=1}^{N(C)} K_i$$

for $C \neq \emptyset$, and $\psi(\emptyset) = 0$.

As before, we denote by C_0^d the half-open unit cube. With an enumeration $(z_k)_{k\in\mathbb{N}}$ of $\mathbb{Z}^d$ and with $C_{0k}^d := C_0^d + z_k$, we put

$$\Psi(Z) := \sum_{k=1}^{\infty} \left[\psi\left(\mathrm{cl}\,(Z \cap C_{0k}^d) - z_k\right) + z_k\right]. \tag{4.20}$$

Then $\Psi(Z)$ is a simple point process in $\mathcal{K}'$ with a locally finite intensity measure. In fact, we have

$$\mathbb{E}\,\Psi(Z)(\mathcal{F}_C) \leq p(C)\mathbb{E}N(Z \cap w(C)) < \infty$$

for $C \in \mathcal{C}'$, where $p(C)$ denotes the number of cubes $C^d + z_k$, $k \in \mathbb{N}$, with $C \cap (C^d + z_k) \neq \emptyset$, and $w(C)$ is the union of these cubes.

Obviously, we have $Z = \bigcup_{K\in\Psi(Z)} K$ and $t_{-z}\Psi(t_z Z) = \Psi(Z)$ for all $z \in \mathbb{Z}^d$ (to achieve this invariance, and the simplicity, Ψ has been defined by (4.20)). To obtain the stronger invariance properties as required, the construction has yet to be modified. For a motion $g \in G_d$, we define $\Psi_g(Z) := g\Psi(g^{-1}Z)$. We put

$$G_d^0 := \left\{g = \vartheta t_x \in G_d : \vartheta \in SO_d,\ x \in C_0^d\right\}$$

and denote by μ^0 the Haar measure on the motion group G_d, restricted to the relatively compact set G_d^0 and normalized to a probability measure. Let ξ be a random motion, independent of Z and with distribution μ^0. We define

$$X := \Psi_\xi(Z).$$

As shown above, X is a point process in $\mathcal{K}'$ with $Z = Z_X$.

Now suppose that $g_0 \in G_d$ is a motion with $Z \stackrel{\mathcal{D}}{=} g_0 Z$. Then

$$g_0 X = g_0 \xi \Psi\left(\xi^{-1} g_0^{-1} g_0 Z\right) = \Psi_{g_0 \xi}(g_0 Z),$$

hence, for all $A \in \mathcal{N}_s(\mathcal{K}')$,

$$\mathbb{P}(g_0 X \in A) = \int_{G_d^0} \mathbb{P}(\Psi_{g_0 g}(Z) \in A)\, \mu^0(\mathrm{d}g),$$

by the independence of Z and ξ and the g_0-invariance of Z. For $g_0 = \vartheta_0 t_{x_0}$, $g = \vartheta t_x$ we have

$$g_0 g = \vartheta_0 \vartheta t_{x + \vartheta^{-1} x_0},$$

hence (using the decomposition (13.8) of the invariant measure on G_d)

$$\mathbb{P}(g_0 X \in A) = \int_{SO_d} \int_{C_0^d} \mathbb{P}\left(\Psi_{\vartheta_0 \vartheta t_{x+\vartheta^{-1} x_0}}(Z) \in A\right) \lambda(\mathrm{d}x)\, \nu(\mathrm{d}\vartheta).$$

If $\vartheta \in SO_d$ is fixed, to each $x \in C_0^d$ there exists a unique representation

$$x + \vartheta^{-1} x_0 = y(x) + z(x)$$

with $y(x) \in C_0^d$ and $z(x) \in \mathbb{Z}^d$. If x varies in C_0^d, the norm of $z(x)$ remains bounded, hence $z(x)$ attains only finitely many values $z_1, \dots, z_r \in \mathbb{Z}^d$. For $D_i := \{x \in C_0^d : z(x) = z_i\}$, $i = 1, \dots, r$, we then have

$$C_0^d = \bigcup_{i=1}^r D_i,$$

and the sets D_i are pairwise disjoint. Consider the mapping $\varphi : x \mapsto y(x)$ on C_0^d. On each D_i, it is a translation. For $x, x' \in C_0^d$ with $y(x) = y(x')$ we have $x - x' = z(x) - z(x') \in \mathbb{Z}^d$, hence $x = x'$. Thus φ is injective. This map is also surjective, since to each $y \in C_0^d$ there exists a decomposition $y - \vartheta^{-1} x_0 = x - z$, $x \in C_0^d$, $z \in \mathbb{Z}^d$. This gives $x + \vartheta^{-1} x_0 = y + z$, hence $y = y(x)$. Thus φ is a bijection onto C_0^d, which leaves λ invariant. Therefore, we obtain

$$\begin{aligned}
&\int_{C_0^d} \mathbb{P}\left(\Psi_{\vartheta_0 \vartheta t_{x+\vartheta^{-1} x_0}}(Z) \in A\right) \lambda(\mathrm{d}x) \\
&= \int_{C_0^d} \mathbb{P}\left(\Psi_{\vartheta_0 \vartheta t_{y(x)+z(x)}}(Z) \in A\right) \lambda(\mathrm{d}x) \\
&= \int_{C_0^d} \mathbb{P}\left(\vartheta_0 \vartheta \Psi_{t_{y(x)+z(x)}}(\vartheta^{-1} \vartheta_0^{-1} Z) \in A\right) \lambda(\mathrm{d}x) \\
&= \int_{C_0^d} \mathbb{P}\left(\vartheta_0 \vartheta \Psi_{t_{y(x)}}(\vartheta^{-1} \vartheta_0^{-1} Z) \in A\right) \lambda(\mathrm{d}x) \\
&= \int_{C_0^d} \mathbb{P}(\vartheta_0 \vartheta \Psi_{t_x}(\vartheta^{-1} \vartheta_0^{-1} Z) \in A)\, \lambda(\mathrm{d}x) \\
&= \int_{C_0^d} \mathbb{P}(\Psi_{\vartheta_0 \vartheta t_x}(Z) \in A)\, \lambda(\mathrm{d}x).
\end{aligned}$$

The rotation invariance of the measure ν now yields

$$\begin{aligned}\mathbb{P}(g_0 X \in A) &= \int_{SO_d} \int_{C_0^d} \mathbb{P}(\Psi_{\vartheta t_x}(Z) \in A)\, \lambda(\mathrm{d}x)\, \nu(\mathrm{d}\vartheta) \\ &= \int_{G_d} \mathbb{P}(\Psi_g(Z) \in A)\, \mu^0(\mathrm{d}g) \\ &= \mathbb{P}(X \in A)\end{aligned}$$

and thus $g_0 X \stackrel{\mathcal{D}}{=} X$. □

Due to this theorem, in particular every stationary random closed set Z is the union set of a stationary germ-grain process $\widetilde{X}$. Moreover, in the case of a random $\mathcal{S}$-set satisfying the finiteness condition of Theorem 4.3.1 the grains are convex. Here, the union set $Z_{\widetilde{X}}$ of a germ-grain process $\widetilde{X}$ is defined as the union set

$$Z_{\widetilde{X}} := \bigcup_{(x,C)\in\widetilde{X}} (x + C)$$

of the particle process induced by $\widetilde{X}$ according to (4.13).

To obtain more accessible models, we now consider random sets $Z = Z_{\widetilde{X}}$ arising from an independent germ-grain process. Such a random closed set is called a **germ-grain model**. If the mark distribution of $\widetilde{X}$ is concentrated on $\mathcal{K}'$, we call $Z_{\widetilde{X}}$ a **germ-grain model with convex grains**. For a germ-grain model Z, the capacity functional T_Z can be expressed in terms of the process X^0 of germs and the capacity functional of the typical grain Z_0. In fact, for $C \in \mathcal{C}$ we have

$$T_Z(C) = 1 - \mathbb{E} \prod_{x\in X^0} (1 - T_{Z_0}(C - x)). \tag{4.21}$$

To prove this, we choose a suitable representation

$$\widetilde{X} = \sum_{i=1}^{\tau} \delta_{(\xi_i, Z_i)}, \qquad \tau = \widetilde{X}(\mathbb{R}^d \times \mathcal{C}'),$$

and then argue as follows.

$$\begin{aligned}1 - T_Z(C) &= \mathbb{P}\left(\bigcup_{i=1}^{\tau} (\xi_i + Z_i) \cap C = \emptyset\right) \\ &= \mathbb{P}(\xi_i \notin C - Z_i,\ i = 1, \dots, \tau) \\ &= \mathbb{P}\left(\prod_{i=1}^{\tau} (1 - \mathbf{1}_{C - Z_i}(\xi_i)) = 1\right) \\ &= \mathbb{E} \prod_{i=1}^{\tau} (1 - \mathbf{1}_{C - Z_i}(\xi_i))\end{aligned}$$

$$= \mathbb{E} \prod_{i=1}^{\tau} \left(1 - \int_{C'} \mathbf{1}_{C-K}(\xi_i)\, \mathbb{Q}(\mathrm{d}K)\right)$$

$$= \mathbb{E} \prod_{x \in X^0} (1 - T_{Z_0}(C - x)).$$

Particularly accessible are those germ-grain models Z for which the process of germs is a Poisson process. They are called **Boolean models (Boolean models with convex grains** if the generating germ-grain process $\widetilde{X}$ has convex grains). These random closed sets are the most tractable ones for applications. The Boolean model

$$Z := \bigcup_{(x,C) \in \widetilde{X}} (x + C) \tag{4.22}$$

is (up to stochastic equivalence) determined by the intensity measure ϑ of the Poisson process of germs and by the distribution $\mathbb{Q}$ of the typical grain. For that reason, we also write $Z =: Z(\vartheta, \mathbb{Q})$.

Formulas for Boolean models are the subject of Section 9.1.

Now let Z be a stationary Boolean model, that is, a stationary random closed set that is generated, according to (4.22), by an independent germ-grain process $\widetilde{X}$ with a Poisson germ process X^0. Then the generated particle process

$$X := \sum_{(x,C) \in \widetilde{X}} \delta_{x+C},$$

is a Poisson process, too, as remarked before Theorem 4.2.3. Hence, Z is the union set of a (by Theorem 3.6.4) stationary Poisson particle process X. From Theorem 4.2.2 we obtain also a converse, and thus the following result.

Theorem 4.3.2. *The stationary Boolean models are precisely the union sets of stationary Poisson particle processes.*

If Z is a stationary Boolean model and X is a generating Poisson particle process, then the intensity measure of X is uniquely determined by the distribution of Z, by Theorem 3.6.3 and Lemma 2.3.1. It is translation invariant, and by Theorem 4.1.1 the intensity γ (assumed positive) and the grain distribution $\mathbb{Q}$ of X are uniquely determined. By Theorem 4.1.2, also the corresponding particle process is uniquely determined. However, this does not hold for the corresponding marked processes. If, besides $Z = Z(\gamma\lambda, \mathbb{Q})$ one also has $Z = Z(\gamma\lambda, \mathbb{Q}')$, then $\mathbb{Q}'$ is in general distinct from the grain distribution $\mathbb{Q}$ of X. Yet, it is true that $\mathbb{Q}$ is the image of $\mathbb{Q}'$ under the mapping $\pi_c : C \mapsto C - c(C)$. This follows from Theorem 4.2.1. Thus, the generation of a stationary Boolean model by an independent germ-grain process with a Poisson germ process can be achieved in different ways. It is, however, always possible to choose the 'canonical' generating process X_c.

If a stationary Boolean model is intersected with a plane S, then in S one obtains again a Boolean model, stationary with respect to S. In fact, for a Poisson particle process X, the section process $X \cap S$, too, is a Poisson process, as follows immediately from (3.2).

For a point process of lower-dimensional sets, it may be possible to obtain certain quantities of X from studying the union set Z. In general this will be difficult, due to overlappings. In particular, for Poisson processes in $\mathcal{C}$, $\mathcal{R}$ or $\mathcal{K}$ and with full-dimensional particles, such overlappings occur with positive probability, due to Theorem 4.1.4. On the other hand, for a stationary Boolean model Z, which is the union set of a stationary Poisson particle process X, this process X is already uniquely determined, as just remarked. Therefore, all characteristic parameters of X, for example the intensity, must be obtainable from quantities of Z. We shall study this phenomenon in greater detail in Section 9.1.

Notes for Section 4.3

1. Theorem 4.3.1 is due to Weil and Wieacker [805].

2. General germ-grain models were introduced by Hanisch [319] and further studied by Heinrich [324] and others. In Hanisch [319] one finds, for example, formula (4.21).

3. The Boolean model was, after a few precursors (see Cressie [185, p. 753]), at first mainly studied by the Fontainebleau school; this is reflected in the books of Matheron [462] and Serra [729]. The book by Hall [317] contains a detailed discussion of qualitative and quantitative properties of the Boolean model and more general germ-grain models, in particular with a view to covering and connectivity properties. Meesters and Roy [509] study Boolean models in the framework of percolation theory. Statistical methods for Boolean models are treated by Molchanov [546].

4. Quermass-interaction models. Starting with a Poisson process Y of convex particles with a finite intensity measure Θ and the corresponding Boolean model Z, Kendall, van Lieshout and Baddeley [398] defined quermass-interaction processes Y' and their union sets Z'. The particle process Y' is supposed to be absolutely continuous to Y with density

$$p(\mathbf{y}) = \alpha\beta^{n(\mathbf{y})} \exp\left[-\sum_{j=0}^{d} \gamma_j V_j(U(\mathbf{y}))\right].$$

Here, $\mathbf{y} = \{K_1, \dots, K_n\}$ is a (finite) realization of Y with $n(\mathbf{y}) = n$, $\beta > 0$ and $\gamma_j \in \mathbb{R}$ are model parameters, α is a normalizing constant and $U(\mathbf{y}) = \bigcup_{i=1}^n K_i$.

The particles of Y' are no longer independent, in general, but satisfy a Markov property. The main question discussed in [398] is whether Y' is stable in the sense of Ruelle, a property which implies integrability of the density p. The results mostly concern the planar case with particles being disks or convex polygons.

4.4 Processes of Flats

In this section, we study processes of flats. A **process** of k-**flats**, or k-**flat process**, in $\mathbb{R}^d$ is a point process in the space $A(d,k)$ of k-flats (k-dimensional planes) in $\mathbb{R}^d$, where $k \in \{1,\dots,d-1\}$, and thus a point process in $\mathcal{F}'$ with intensity measure concentrated on $A(d,k)$. For $k=1$, we also speak of a **line process**, and for $k=d-1$, of a **hyperplane process**.

For stationary k-flat processes, there is again a decomposition of the intensity measure. The proof is not quite as simple as for the analogous results in Theorems 3.3.1 and 3.5.1, since $A(d,k)$ can, for $k<d-1$, only locally be represented as a product space with a Euclidean factor.

The Grassmannian $G(d,k)$ of k-dimensional linear subspaces of $\mathbb{R}^d$ is a subset of $A(d,k)$ and is closed in $\mathcal{F}$ and $\mathcal{F}'$. For $L \in G(d,k)$, recall that λ_L is the k-dimensional Lebesgue measure on L. The (continuous) mapping

$$\pi_0 : \bigcup_{k=1}^{d-1} A(d,k) \to \bigcup_{k=1}^{d-1} G(d,k)$$

associates with every plane its translate through 0.

Theorem 4.4.1. *Let Θ be a locally finite, translation invariant measure on $A(d,k)$. Then there exists a uniquely determined finite measure Θ_0 on $G(d,k)$ such that*

$$\Theta(A) = \int_{G(d,k)} \int_{L^\perp} \mathbf{1}_A(L+x)\,\lambda_{L^\perp}(\mathrm{d}x)\,\Theta_0(\mathrm{d}L) \tag{4.23}$$

for every Borel set $A \in \mathcal{B}(A(d,k))$.

Proof. Let $U \in G(d,d-k)$, and define

$$G_U := \{L \in G(d,k) : \dim(L \cap U) = 0\}$$

and $A_U := \{L+x : L \in G_U,\ x \in U\}$. The mapping

$$\begin{aligned} \varphi : G_U \times U &\to \ A_U \\ (L,x) &\mapsto L+x \end{aligned}$$

is a homeomorphism. Let $A \subset G_U$ be a Borel set. For Borel sets $B \subset U$, let

$$\eta(B) := \Theta(\varphi(A \times B)).$$

Then η is a locally finite, translation invariant measure on U and thus a multiple of the Lebesgue measure λ_U. Denoting the factor by $\rho(A)$, we have

$$\Theta(\varphi(A \times B)) = \rho(A)\lambda_U(B).$$

Evidently, ρ is a finite measure on G_U. Thus,

$$\varphi^{-1}(\Theta)(A \times B) = (\rho \otimes \lambda_U)(A \times B),$$

which gives $\varphi^{-1}(\Theta) = \rho \otimes \lambda_U$ and, therefore, $\Theta \llcorner A_U = \varphi(\rho \otimes \lambda_U)$. Hence, for every nonnegative measurable function f on $A(d,k)$ we have

$$\int_{A_U} f \, \mathrm{d}\Theta = \int_{G_U} \int_U f(L+x) \, \lambda_U(\mathrm{d}x) \, \rho(\mathrm{d}L).$$

For given $L \in G_U$, let $\Pi_L : U \to L^\perp$ denote the orthogonal projection to the orthogonal complement of L. It is bijective, since $L \in G_U$. Therefore, $\Pi_L(\lambda_U) = a(L)\lambda_{L^\perp}$, with a factor $a(L) > 0$ that depends only on L. Further, $f(L+x) = f(L + \Pi_L(x))$. This yields

$$\int_U f(L+x) \, \lambda_U(\mathrm{d}x) = a(L) \int_{L^\perp} f(L+x) \, \lambda_{L^\perp}(\mathrm{d}x).$$

Defining a measure Θ_U on G_U by $a(L)\rho(\mathrm{d}L) =: \Theta_U(\mathrm{d}L)$, we have

$$\int_{A_U} f \, \mathrm{d}\Theta = \int_{G_U} \int_{L^\perp} fL + x) \, \lambda_{L^\perp}(\mathrm{d}x) \, \Theta_U(\mathrm{d}L).$$

We interpret Θ_U as a measure on all of $G(d,k)$, with $\Theta_U(G(d,k) \setminus G_U) = 0$, and then have

$$\int_{A_U} f \, \mathrm{d}\Theta = \int_{G(d,k)} \int_{L^\perp} f(L+x) \, \lambda_{L^\perp}(\mathrm{d}x) \, \Theta_U(\mathrm{d}L).$$

Every set G_U, $U \in G(d, d-k)$, is open in $G(d,k)$, hence there are finitely many subspaces $U_1, \dots, U_m \in G(d, d-k)$ with $G(d,k) = \bigcup_{i=1}^m G_{U_i}$. The sets A_{U_i}, $i = 1, \dots, m$, cover $A(d,k)$ and are invariant under translations. The translation invariant Borel sets defined by $A_k := A_{U_k} \setminus (A_1 \cup \ldots \cup A_{k-1})$, $k = 1, \dots, m$, form a disjoint covering of $A(d,k)$. Since $\Theta \llcorner A_i$ is translation invariant, the measure $\Theta_i := (\Theta \llcorner A_i)_{U_i}$, defined as above, satisfies

$$\int_{A_i} f \, \mathrm{d}\Theta = \int_{G(d,k)} \int_{L^\perp} f(L+x) \, \lambda_{L^\perp}(\mathrm{d}x) \, \Theta_i(\mathrm{d}L).$$

Therefore, the measure $\Theta_0 := \Theta_1 + \ldots + \Theta_m$ satisfies (4.23).

From (4.23) we obtain, for $A \in \mathcal{B}(G(d,k))$,

$$\Theta_0(A) = \frac{1}{\kappa_{d-k}} \Theta\left(\mathcal{F}_{B^d} \cap \pi_0^{-1}(A)\right). \tag{4.24}$$

From (4.24) it is obvious that Θ_0 is finite and uniquely determined. □

Applying the preceding theorem to intensity measures, we immediately obtain the following result.

Theorem 4.4.2. *Let X be a stationary k-flat process in $\mathbb{R}^d$ with intensity measure $\Theta \neq 0$. Then there are a number $\gamma \in (0, \infty)$ and a probability measure $\mathbb{Q}$ on $G(d,k)$ with*

$$\int_{A(d,k)} f \, \mathrm{d}\Theta = \gamma \int_{G(d,k)} \int_{L^\perp} f(L+x) \, \lambda_{L^\perp}(\mathrm{d}x) \, \mathbb{Q}(\mathrm{d}L) \tag{4.25}$$

for all nonnegative measurable functions f on $A(d,k)$. Here γ and $\mathbb{Q}$ are uniquely determined by Θ.

We call γ the **intensity** and $\mathbb{Q}$ the **directional distribution** of the stationary flat process X. If X is moreover isotropic, then $\mathbb{Q}$ is rotation invariant, as follows from the uniqueness. By Theorem 13.2.11, there is only one normalized rotation invariant measure on $G(d,k)$, the Haar measure ν_k.

Occasionally (for example, when sections are considered) we have to allow flat processes with $\Theta = 0$; for these we define $\gamma = 0$.

The interpretation of γ and $\mathbb{Q}$ is clear from (4.25), since for $A \in \mathcal{B}(G(d,k))$ this gives

$$\gamma \mathbb{Q}(A) = \frac{1}{\kappa_{d-k}} \mathbb{E} X \left(\mathcal{F}_{B^d} \cap \pi_0^{-1}(A) \right), \tag{4.26}$$

in particular

$$\gamma = \frac{1}{\kappa_{d-k}} \, \mathbb{E} X \left(\mathcal{F}_{B^d} \right) \tag{4.27}$$

and

$$\mathbb{Q}(A) = \frac{\mathbb{E} X \left(\mathcal{F}_{B^d} \cap \pi_0^{-1}(A) \right)}{\mathbb{E} X (\mathcal{F}_{B^d})}. \tag{4.28}$$

The representation (4.28) explains why the measure $\mathbb{Q}$ is called the directional distribution of X.

For a further interpretation of the intensity γ, we need a measurability result.

Lemma 4.4.1. *Let X be a point process in $A(d,k)$. Then*

$$\sum_{E \in X} \lambda_E(A) \tag{4.29}$$

is measurable for all $A \in \mathcal{B}(\mathbb{R}^d)$.

Proof. It is sufficient to consider the case $A \subset mB^d$, $m \in \mathbb{N}$. First let A be compact, and assume that $E_i \to E$ in $G(d,k)$. Then there exist rotations $g_i \in SO_d$, $i \in \mathbb{N}$, converging to the identity for $i \to \infty$, and such that $g_i^{-1} E = E_i$. Using the representation

$$\lambda_{E_i}(A) = \int_E \mathbf{1}_{g_i A}(x) \, \lambda_E(\mathrm{d}x),$$

one shows as in the proof of Theorem 12.3.6 that the function $E \mapsto \lambda_E(A)$ is upper semicontinuous and thus measurable. By Theorem 3.1.2, also (4.29) is measurable. Now let $\mathcal{A}$ be the system of all Borel sets $A \subset mB^d$ for which (4.29) is measurable. We have shown that $\mathcal{A}$ contains all compact subsets of mB^d. Evidently, $\mathcal{A}$ is closed under disjoint countable unions and relative complements. Since $\mathcal{C}$ is $\cap$-stable, $\mathcal{A}$ contains the σ-algebra generated by the compact sets in mB^d, and, therefore, all Borel sets in mB^d. $\square$

The measurability being shown, we can define

$$\varphi_X(A) := \mathbb{E} \sum_{E \in X} \lambda_E(A), \qquad A \in \mathcal{B}(\mathbb{R}^d),$$

and thus obtain a locally finite measure φ_X. If X is stationary, φ_X is translation invariant and, therefore, of the form $\varphi_X = \alpha\lambda$ with a number $\alpha \in [0,\infty)$. The following theorem shows that this constant is precisely the intensity γ.

Theorem 4.4.3. *Let X be a stationary k-flat process in $\mathbb{R}^d$ with intensity γ. Then*

$$\mathbb{E} \sum_{E \in X} \lambda_E = \gamma\lambda.$$

Proof. Using the Campbell theorem and Theorem 4.4.2, we get

$$\begin{aligned}
\mathbb{E} \sum_{E \in X} \lambda_E(A) &= \int_{A(d,k)} \lambda_E(A)\, \Theta(\mathrm{d}E) \\
&= \gamma \int_{G(d,k)} \int_{L^\perp} \lambda_{L+x}(A)\, \lambda_{L^\perp}(\mathrm{d}x)\, \mathbb{Q}(\mathrm{d}L) \\
&= \gamma \int_{G(d,k)} \lambda(A)\, \mathbb{Q}(\mathrm{d}L) \\
&= \gamma\lambda(A),
\end{aligned}$$

as stated. $\square$

Further interpretations of the intensity will be obtained in Section 9.4. In particular, formula (9.33) provides $k+1$ such interpretations.

Processes of k-flats satisfying Poisson assumptions again have particular properties. From Theorems 3.2.1 and 3.6.1, the following is immediately clear.

Theorem 4.4.4. *Let $\gamma \in (0,\infty)$ and let $\mathbb{Q}$ be a probability measure on $G(d,k)$. Then there is (up to equivalence) precisely one stationary Poisson k-flat process X in $\mathbb{R}^d$ with intensity γ and directional distribution $\mathbb{Q}$. The process X is isotropic if and only if $\mathbb{Q} = \nu_k$.*

In the next theorem, we collect some consequences of the independence properties of Poisson k-flat processes. We say that two linear subspaces L, L' of $\mathbb{R}^d$ are in **general position** if

$$\mathrm{lin}\,(L \cup L') = \mathbb{R}^d \quad \text{or} \quad \dim\,(L \cap L') = 0.$$

Two k-planes E, E' are said to be in general position if their direction spaces $\pi_0(E), \pi_0(E')$ are in general position.

Theorem 4.4.5. *Let X be a stationary Poisson k-flat process in $\mathbb{R}^d$.*

(a) *If $k < d/2$, then a.s. any two k-flats of X are disjoint.*

(b) *If the directional distribution $\mathbb{Q}$ of X has no atoms, then a.s. any two k-flats of the process X are not translates of each other.*

(c) *If the directional distribution of X is absolutely continuous with respect to the invariant measure ν_k, then a.s. any two k-planes of the process X are in general position.*

Proof. Let $A \in \mathcal{B}(A(d,k)^2)$. From Theorem 3.1.3, Corollary 3.2.4, Theorem 4.1.2 we get

$$\begin{aligned}\mathbb{E} \sum_{(E_1,E_2)\in X^2_{\neq}} \mathbf{1}_A(E_1,E_2) &= \int_{A(d,k)^2} \mathbf{1}_A \,\mathrm{d}\Lambda^{(2)} \\ &= \int_{A(d,k)} \int_{A(d,k)} \mathbf{1}_A(E_1,E_2)\, \Theta(\mathrm{d}E_1)\, \Theta(\mathrm{d}E_2) \\ &= \gamma^2 \int_{G(d,k)} \int_{G(d,k)} \int_{L_2^\perp} \int_{L_1^\perp} \mathbf{1}_A(L_1 + x_1, L_2 + x_2) \\ &\quad \times \lambda_{L_1^\perp}(\mathrm{d}x_1)\, \lambda_{L_2^\perp}(\mathrm{d}x_2)\, \mathbb{Q}(\mathrm{d}L_1)\, \mathbb{Q}(\mathrm{d}L_2).\end{aligned}$$

To prove (a), suppose that $k < d/2$ and choose

$$A := \{(E_1,E_2) \in A(d,k)^2 : E_1 \cap E_2 \neq \emptyset\}.$$

For fixed k-flats $L_1 \in G(d,k)$, $E_2 \in A(d,k)$, the integral

$$\int_{L_1^\perp} \mathbf{1}_A(L_1 + x_1, E_2)\, \lambda_{L_1^\perp}(\mathrm{d}x_1)$$

gives the $(d-k)$-dimensional Lebesgue measure of the image of E_2 under the orthogonal projection to $L_1^\perp$, which is zero. We deduce that

$$\mathbb{E} \sum_{(E_1,E_2)\in X^2_{\neq}} \mathbf{1}_A(E_1,E_2) = 0,$$

and from this the assertion (a) follows.

To prove (b), let $m \in \mathbb{N}$ and

$$A := \{(E_1,E_2) \in A(d,k)^2 : E_i \cap mB^d \neq \emptyset,\ i = 1,2,\ E_1 \text{ is a translate of } E_2\}.$$

Then we get

$$\begin{aligned}
&\mathbb{E} \sum_{(E_1,E_2)\in X^2_{\neq}} \mathbf{1}_A(E_1,E_2) \\
&\le \left(\gamma m^{d-k}\kappa_{d-k}\right)^2 \int_{G(d,k)} \int_{G(d,k)} \mathbf{1}_A(L_1,L_2)\,\mathbb{Q}(\mathrm{d}L_1)\,\mathbb{Q}(\mathrm{d}L_2) \\
&= \left(\gamma m^{d-k}\kappa_{d-k}\right)^2 \int_{G(d,k)} \mathbb{Q}(\{L_2\})\,\mathbb{Q}(\mathrm{d}L_2) \\
&= 0,
\end{aligned}$$

since $\mathbb{Q}$ has no atoms. Assertion (b) follows, since $m \in \mathbb{N}$ was arbitrary.

To prove (c), suppose that $\mathbb{Q}$ has a density f with respect to ν_k. For $m \in \mathbb{N}$ we choose

$$\begin{aligned}
A := \{(E_1,E_2) \in A(d,k)^2 : E_i \cap mB^d \neq \emptyset,\ i=1,2,& \\
E_1, E_2 \text{ not in general position}\}.&
\end{aligned}$$

As above, we obtain similarly

$$\begin{aligned}
&\mathbb{E} \sum_{(E_1,E_2)\in X^2_{\neq}} \mathbf{1}_A(E_1,E_2) \\
&\le \left(\gamma m^{d-k}\kappa_{d-k}\right)^2 \int_{G(d,k)} \int_{A(L_2)} f(L_1)\,\nu_k(\mathrm{d}L_1) f(L_2)\,\nu_k(\mathrm{d}L_2) \\
&= 0,
\end{aligned}$$

since the set $A(L_2) := \{L_1 \in G(d,k) : (L_1,L_2) \in A\}$ satisfies $\nu_k(A(L_2)) = 0$, as can be deduced from Lemma 13.2.1. Since $m \in \mathbb{N}$ was arbitrary, assertion (c) follows. □

When considering section processes in the following, we shall also meet j-flat processes for $j = 0$. These can be considered as ordinary point processes. Namely, we identify every one-pointed set $\{x\}$ with x, observing that the mapping $\{x\} \mapsto x$ maps the subspace $\{\{x\} : x \in \mathbb{R}^d\}$ of $\mathcal{F}'$ homeomorphically to $\mathbb{R}^d$.

Sections with a Fixed Plane

We turn to sections with fixed planes, a topic which, at least in small dimensions, is important for applications. Let X be a stationary k-flat process in $\mathbb{R}^d$ ($k \in \{1,\dots,d-1\}$), and let S be a fixed $(d-k+j)$-flat with $0 \le j \le k-1$. We recall the definition of the section process,

$$(X \cap S)(\omega) := \sum_{E\in X,\, E\cap S\neq\emptyset} \delta_{E\cap S}.$$

As we shall see below, $X \cap S$ is a j-flat process. Its realizations lie in the section plane S and it is stationary with respect to S. Therefore, in the following $X \cap S$ is considered as a stationary j-flat process in S. The question arises how intensity and directional distribution of $X \cap S$ are related to the corresponding parameters of X. This will now be investigated.

Because of the stationarity of the process X it is no restriction to assume that $S \in G(d, d-k+j)$. The nonempty intersections $E \cap S$, $E \in X$, can be r-flats with $r \in \{j, \dots, \min(k, d-k+j)\}$; we show, however, that almost surely they are j-flats. Let

$$A := \{E \in A(d,k) : \dim (E \cap S) > j\}$$

(with the usual convention that $\dim \emptyset := -1$). By (4.25),

$$\mathbb{E}X(A) = \Theta(A) = \gamma \int_{G(d,k)} \int_{L^\perp} \mathbf{1}_A(L + x)\, \lambda_{L^\perp}(\mathrm{d}x)\, \mathbb{Q}(\mathrm{d}L).$$

If $\mathbf{1}_A(L+x) = 1$, then L and S span only a proper subspace U of $\mathbb{R}^d$, and we have $x \in U$ and $\dim (L^\perp \cap U) < \dim L^\perp$. This gives

$$\mathbb{E}X(A) \le \gamma \int_{G(d,k)} \lambda_{L^\perp}(L^\perp \cap U)\, \mathbb{Q}(\mathrm{d}L) = 0.$$

Hence, almost surely we have $\dim (E \cap S) = j$ or $E \cap S = \emptyset$ for $E \in X$. Therefore, $X \cap S$ is a j-flat process in S (which may have intensity 0, though). Similarly we obtain that $X \cap S$ is a.s. simple. In fact, if a j-flat in S is generated as the intersection of two distinct k-flats E_1, E_2 with S, then $E_1 \cap E_2$ is an i-flat with $j \le i \le k-1$. For given i, we consider all i-flats which are the intersection of two flats of X (counting every such i-flat only once, even if it is generated in two different ways). In this way, a process Y_i of i-flats is obtained (the measurability is not difficult to prove). The process Y_i is stationary. By the argument used above and because of $i \le k-1$, the flats of Y_i intersect the plane S a.s. in planes of dimension less than j. This shows that $X \cap S$ is a.s. simple.

First we consider now the case $\dim S = d-k$, where $X \cap S$ is an ordinary point process in S. In the next theorem, we determine the intensity of this point process. For the subspace determinant $[\cdot,\cdot]$ occurring in the following we refer to Section 14.1.

Theorem 4.4.6. *Let $k \in \{1, \dots, d-1\}$, and let X be a stationary k-flat process in $\mathbb{R}^d$ with intensity γ and directional distribution $\mathbb{Q}$. Let $S \in G(d, d-k)$, and let $\gamma_{X \cap S}$ be the intensity of the point process $X \cap S$. Then*

$$\gamma_{X \cap S} = \gamma \int_{G(d,k)} [S, L]\, \mathbb{Q}(\mathrm{d}L).$$

Proof. Let B^{d-k} be the unit ball in S. By the definition of the intensity of the point process $X \cap S$,

$$\begin{aligned}\kappa_{d-k}\gamma_{X\cap S} &= \mathbb{E}(X \cap S)(\mathcal{F}_{B^{d-k}}) \\ &= \mathbb{E}X(\mathcal{F}_{B^{d-k}}) = \Theta(\mathcal{F}_{B^{d-k}}) \\ &= \gamma \int_{G(d,k)} \int_{L^\perp} \mathbf{1}_{\mathcal{F}_{B^{d-k}}}(L + x)\, \lambda_{L^\perp}(\mathrm{d}x)\, \mathbb{Q}(\mathrm{d}L) \\ &= \gamma \int_{G(d,k)} \lambda_{L^\perp}(B^{d-k}|L^\perp)\, \mathbb{Q}(\mathrm{d}L).\end{aligned}$$

Here $B^{d-k}|L^\perp$ is the image of B^{d-k} under the orthogonal projection to $L^\perp$. The $(d-k)$-volume of this image is given by $\lambda_S(B^{d-k})[S,L]$, from which the assertion follows. □

In the cases $k = 1$ and $k = d-1$ it is convenient to replace the directional distribution $\mathbb{Q}$ by the **spherical directional distribution** φ. This is the measure on the unit sphere S^{d-1} which, for a set $A \in \mathcal{B}(S^{d-1})$ without antipodal points, is defined by

$$\varphi(A) := \frac{1}{2}\mathbb{Q}(\{L(u) : u \in A\}) \qquad \text{if } k = 1,$$

for $L(u) := \mathrm{lin}\{u\}$, respectively

$$\varphi(A) := \frac{1}{2}\mathbb{Q}(\{u^\perp : u \in A\}) \qquad \text{if } k = d-1. \tag{4.30}$$

(The factor $\frac{1}{2}$ appears here since $L(u) = L(-u)$ and $u^\perp = (-u)^\perp$.) By additivity, φ is then defined for all $A \in \mathcal{B}(S^{d-1})$. Thus, φ is an even probability measure on S^{d-1}. Writing $\gamma_X(u) := \gamma_{X\cap L(u)}$ if $k = 1$, respectively $\gamma_X(u) := \gamma_{X\cap u^\perp}$ if $k = d-1$, we then have

$$\gamma_X(u) = \gamma \int_{S^{d-1}} |\langle u, v\rangle|\, \varphi(\mathrm{d}v). \tag{4.31}$$

The right side of (4.31) defines the support function of a centrally symmetric convex body, which can be associated with the measure $\gamma\varphi$. This body belongs to the class of zonoids. Such associated zonoids will be studied and applied in Section 4.6.

A corresponding uniqueness theorem (Theorem 14.3.4) shows that the function γ_X in (4.31) uniquely determines the measure $\gamma\varphi$ (and therefore also γ and φ). In particular, for a stationary Poisson line or hyperplane process X, the distribution $\mathbb{P}_X$ is uniquely determined by the **section intensities** $\gamma_{X\cap S}$, $S \in G(d, d-1)$, respectively $S \in G(d, 1)$ (see also Section 4.6). For $1 < k < d-1$, however, a stationary Poisson k-flat process X is in general not uniquely determined by the section intensities $\gamma_{X\cap S}$, $S \in G(d, d-k)$; see Note 2 of this section.

Now we consider also the case of higher-dimensional section planes S, where we obtain in S an intersection process of j-flats with $j > 0$. Let X be a stationary k-flat process, and let $S \in G(d, d-k+j)$, with $j \in \{1, \ldots, k-1\}$, be a fixed plane. As shown above, $X \cap S$ is a.s. a j-flat process. Its intensity measure, $\Theta_{X \cap S}$, is concentrated on the space

$$G(S,j) := \{L \in G(d,j) : L \subset S\}.$$

Theorem 4.4.7. *Let $k \in \{2, \ldots, d-1\}$, and let X be a stationary k-flat process in $\mathbb{R}^d$ with intensity γ and directional distribution $\mathbb{Q}$. Let $j \in \{1, \ldots, k-1\}$ and $S \in G(d, d-k+j)$; let $\gamma_{X \cap S}$ be the intensity and $\mathbb{Q}_{X \cap S}$ the directional distribution of the j-flat process $X \cap S$. Then, for $A \in \mathcal{B}(G(d,j))$,*

$$\gamma_{X \cap S} \mathbb{Q}_{X \cap S}(A) = \gamma \int_{G(d,k)} \mathbf{1}_A(L \cap S)[L, S] \, \mathbb{Q}(\mathrm{d}L).$$

(If $\gamma_{X \cap S} = 0$, then $\mathbb{Q}_{X \cap S}$ is not defined, and the expression $\gamma_{X \cap S} \mathbb{Q}_{X \cap S}$ has to be read as the zero measure.)

Proof. Let $P \in \mathcal{B}(A(d,j))$. By Campbell's theorem and Theorem 4.4.2,

$$\begin{aligned}
\Theta_{X \cap S}(P) &= \mathbb{E}(X \cap S)(P) = \mathbb{E} \sum_{E \in X} \mathbf{1}_P(E \cap S) \\
&= \int_{A(d,k)} \mathbf{1}_P(E \cap S) \, \Theta(\mathrm{d}E) \\
&= \gamma \int_{G(d,k)} \int_{L^\perp} \mathbf{1}_P((L + x) \cap S) \, \lambda_{L^\perp}(\mathrm{d}x) \, \mathbb{Q}(\mathrm{d}L).
\end{aligned}$$

The intensity measure $\Theta_{X \cap S}$ is concentrated on the j-flats in S and is invariant under the translations of S into itself. By (4.24) (applied in S) and the definition of intensity and directional distribution, for $A \in \mathcal{B}(G(d,j))$ and $B_S := B^d \cap S$ we get

$$\begin{aligned}
&\gamma_{X \cap S} \mathbb{Q}_{X \cap S}(A) \\
&= \frac{1}{\kappa_{d-k}} \Theta_{X \cap S}(\mathcal{F}_{B_S} \cap \pi_0^{-1}(A)) \\
&= \frac{\gamma}{\kappa_{d-k}} \int_{G(d,k)} \int_{L^\perp} \mathbf{1}_{\mathcal{F}_{B_S} \cap \pi_0^{-1}(A)}((L + x) \cap S) \, \lambda_{L^\perp}(\mathrm{d}x) \, \mathbb{Q}(\mathrm{d}L) \\
&= \frac{\gamma}{\kappa_{d-k}} \int_{G(d,k)} \mathbf{1}_A(L \cap S) \lambda_{d-k}(B_S | L^\perp) \, \mathbb{Q}(\mathrm{d}L),
\end{aligned}$$

since $(L + x) \cap S \in \mathcal{F}_{B_S} \cap \pi_0^{-1}(A)$ obviously holds if and only if $L \cap S \in A$ and $x \in B_S | L^\perp$. If T denotes the orthogonal complement of $L \cap S$ in S, then

$$B_S | L^\perp = (B_S | T) | L^\perp = B_T | L^\perp.$$

The orthogonal projection from T to $L^\perp$ has the absolute determinant $[L,S]$, hence

$$\lambda_{d-k}(B_S|L^\perp) = \kappa_{d-k}[L,S].$$

This yields the assertion. $\square$

Intersection Processes

By intersecting flats in a k-flat process among themselves, we obtain new lower-dimensional flat processes. We shall now study such intersection processes in the case of stationary Poisson flat processes. In particular, we are interested in how the intensity and the directional distribution of an intersection process depend on the data of the original process. We restrict ourselves to two cases: intersecting k-tuples of hyperplanes, or intersecting pairs of r-flats, where $r \geq d/2$. In some cases we shall be able to obtain sharp inequalities between the intensities of the intersection process and the original process; this will be explained in Section 4.6.

First we consider hyperplane processes. It is convenient to represent hyperplanes in the form

$$H(u,\tau) := \{x \in \mathbb{R}^d : \langle x,u\rangle = \tau\} \tag{4.32}$$

with a unit vector $u \in S^{d-1}$ and a number $\tau \in \mathbb{R}$. Every hyperplane $H \in A(d,d-1)$ has two such representations. Instead of $H(u,0)$, we shall write $u^\perp$ again.

Let X be a stationary hyperplane process in $\mathbb{R}^d$ with intensity $\gamma \neq 0$ and directional distribution $\mathbb{Q}$. Using the spherical directional distribution φ introduced by (4.30), the decomposition of the intensity measure Θ given by Theorem 4.4.2 can be written in the form

$$\int_{A(d,d-1)} f \,\mathrm{d}\Theta = \gamma \int_{S^{d-1}} \int_{-\infty}^{\infty} f(H(u,\tau))\,\mathrm{d}\tau\, \varphi(\mathrm{d}u). \tag{4.33}$$

Let $k \in \{2,\dots,d\}$. For every realization of X, we consider the intersection of any k hyperplanes in the process which are in general position. We want to show that in this way we obtain a stationary $(d-k)$-flat process X_k; we shall call this the **intersection process of order** k of the process X. For $P \in \mathcal{B}(A(d,d-k))$, define the function $f_P : A(d,d-1)^k \to \mathbb{R}$ by

$$f_P(H_1,\dots,H_k) := \begin{cases} 1, & \text{if } H_1 \cap \ldots \cap H_k \in P, \\ 0 & \text{else.} \end{cases} \tag{4.34}$$

The set of all $(H_1,\dots,H_k) \in A(d,d-1)^k$ with $\dim(H_1 \cap \ldots \cap H_k) = d-k$ is open, and on this set the mapping $(H_1,\dots,H_k) \mapsto H_1 \cap \ldots \cap H_k$ is continuous. Hence, f_P is measurable. By Theorem 3.1.3, the function

$$X_k(P) := \frac{1}{k!} \sum_{(H_1,\dots,H_k)\in X^k_{\neq}} f_P(H_1,\dots,H_k)$$

is measurable. If P is compact, there exists a ball that is hit by all $(d-k)$-flats in P and hence also by all hyperplanes $H_1,\dots,H_k$ with $f_P(H_1,\dots,H_k)=1$. It follows that $X_k(P)$ is a.s. finite. Thus, X_k is a point process in $A(d,d-k)$. Obviously, it is stationary, but it may have intensity zero and need not be simple. If X is a stationary Poisson hyperplane process, then almost surely either the intersection $H_1 \cap \ldots \cap H_k$ is empty or $H_1,\dots,H_k$ are in general position, as follows by the method used in the proof of Theorem 4.4.5. Therefore, X_k is a.s. simple. That X_k is not a Poisson process, in general, is already seen in the case $d=2$, $k=2$, since for a stationary Poisson point process in $\mathbb{R}^2$ a.s. no three points are collinear.

In the following, for vectors $u_1,\dots,u_m \in \mathbb{R}^d$, $m \le d$, we denote by $\nabla_m(u_1,\dots,u_m)$ the m-dimensional volume of the parallelepiped spanned by $u_1,\dots,u_m$.

Theorem 4.4.8. *Let X be a stationary Poisson hyperplane process in $\mathbb{R}^d$ with intensity $\gamma \neq 0$ and spherical directional distribution φ. Let $k \in \{2,\dots,d\}$, and let X_k be the intersection process of order k of X. Then the intensity γ_k and the directional distribution $\mathbb{Q}_k$ of X_k are given by*

$$\gamma_k \mathbb{Q}_k(A)$$
$$= \frac{\gamma^k}{k!} \int_{S^{d-1}} \cdots \int_{S^{d-1}} \mathbf{1}_A(u_1^\perp \cap \ldots \cap u_k^\perp) \nabla_k(u_1,\dots,u_k)\, \varphi(\mathrm{d}u_1)\cdots\varphi(\mathrm{d}u_k)$$

for $A \in \mathcal{B}(G(d,d-k))$.

(For $k=d$, $\mathbb{Q}_k(A)$ and $\mathbf{1}_A(u_1^\perp \cap \ldots \cap u_k^\perp)$ have to be omitted from the formula. If $\gamma_k = 0$, the measure $\mathbb{Q}_k$ is not defined; then $\gamma_k\mathbb{Q}_k$ has to be read as the zero measure.)

Proof. Let Θ_k be the intensity measure of the intersection process X_k (the subsequent proof also yields that Θ_k is locally finite). For $P \in \mathcal{B}(A(d,d-k))$ let f_P be the function defined by (4.34). Then

$$\begin{aligned}\Theta_k(P) &= \mathbb{E}X_k(P)\\ &= \frac{1}{k!}\mathbb{E} \sum_{(H_1,\dots,H_k)\in X^k_{\neq}} f_P(H_1,\dots,H_k)\\ &= \frac{1}{k!} \int_{A(d,d-1)^k} f_P \,\mathrm{d}\Lambda^{(k)},\end{aligned}$$

by Theorem 3.1.3. Here $\Lambda^{(k)} = \Theta^k$ by Corollary 3.2.4. Together with (4.33) this gives

$$
\begin{aligned}
k!\Theta_k(P) &= \int_{A(d,d-1)} \cdots \int_{A(d,d-1)} f_P(H_1, \dots, H_k)\, \Theta(\mathrm{d}H_1) \cdots \Theta(\mathrm{d}H_k) \\
&= \gamma^k \int_{S^{d-1}} \cdots \int_{S^{d-1}} \int_{-\infty}^{\infty} \cdots \int_{-\infty}^{\infty} f_P(H(u_1, \tau_1), \dots, H(u_k, \tau_k)) \\
&\quad \times \mathrm{d}\tau_1 \cdots \mathrm{d}\tau_k\, \varphi(\mathrm{d}u_1) \cdots \varphi(\mathrm{d}u_k).
\end{aligned}
$$

Let $A \in \mathcal{B}(G(d, d-k))$ and choose $P := \mathcal{F}_{B^d} \cap \pi_0^{-1}(A)$. By (4.26),

$$
\begin{aligned}
k!\gamma_k \mathbb{Q}_k(A) &= \frac{k!}{\kappa_k} \Theta_k(\mathcal{F}_{B^d} \cap \pi_0^{-1}(A)) \\
&= \frac{\gamma^k}{\kappa_k} \int_{S^{d-1}} \cdots \int_{S^{d-1}} \int_{-\infty}^{\infty} \cdots \int_{-\infty}^{\infty} f_P(H(u_1, \tau_1), \dots, H(u_k, \tau_k)) \\
&\quad \times \mathrm{d}\tau_1 \cdots \mathrm{d}\tau_k\, \varphi(\mathrm{d}u_1) \cdots \varphi(\mathrm{d}u_k),
\end{aligned}
$$

where

$$
\begin{aligned}
& f_P(H(u_1, \tau_1), \dots, H(u_k, \tau_k)) \\
&= \mathbf{1}_A(u_1^\perp \cap \ldots \cap u_k^\perp) \mathbf{1}_{\mathcal{F}_{B^d}}(H(u_1, \tau_1) \cap \ldots \cap H(u_k, \tau_k)).
\end{aligned}
$$

For the computation of the integral

$$
I_k := \int_{-\infty}^{\infty} \cdots \int_{-\infty}^{\infty} \mathbf{1}_{\mathcal{F}_{B^d}}(H(u_1, \tau_1) \cap \ldots \cap H(u_k, \tau_k))\, \mathrm{d}\tau_1 \cdots \mathrm{d}\tau_k,
$$

we first assume that $u_1, \dots, u_k$ are linearly independent. Let $k = d$. For $\tau := (\tau_1, \dots, \tau_d)$ let $T(\tau)$ be the intersection point of the hyperplanes $H(u_1, \tau_1), \dots, H(u_d, \tau_d)$. Then I_d is the d-dimensional Lebesgue measure of the set $T^{-1}(B^d)$. The mapping T is injective, and its inverse is given by $T^{-1}(x) = (\langle x, u_1 \rangle, \dots, \langle x, u_d \rangle)$; the Jacobian of T^{-1} is $\nabla_d(u_1, \dots, u_d)$. Therefore,

$$
I_d = \kappa_d \nabla_d(u_1, \dots, u_d).
$$

For $k < d$ we obtain

$$
I_k = \kappa_k \nabla_k(u_1, \dots, u_k),
$$

by applying the obtained result in the space $\operatorname{lin}\{u_1, \dots, u_k\}$. Thus we get

$$
\begin{aligned}
& \int_{-\infty}^{\infty} \cdots \int_{-\infty}^{\infty} f_P(H(u_1, \tau_1), \dots, H(u_k, \tau_k))\, \mathrm{d}\tau_1 \cdots \mathrm{d}\tau_k \\
&= \mathbf{1}_A(u_1^\perp \cap \ldots \cap u_k^\perp) \kappa_k \nabla_k(u_1, \dots, u_k).
\end{aligned}
$$

This equation also holds if $u_1, \dots, u_k$ are linearly dependent, since in that case both sides are zero. This completes the proof. □

Now we consider a stationary process X of r-flats, where $d/2 \leq r \leq d-1$. In every realization of X, we take the intersection of any two flats in the process which are in general position. By similar arguments to those used for hyperplanes, we see that we obtain in this way a stationary process of $(2r-d)$-flats. We denote it by X_2 and call it the **intersection process of order** 2 of X.

Theorem 4.4.9. *Let $d/2 \leq r \leq d-1$, let X be a stationary Poisson process of r-flats in $\mathbb{R}^d$ with intensity $\gamma \neq 0$ and directional distribution $\mathbb{Q}$. Let X_2 be the intersection process of order 2 of X. Then the intensity γ_2 and the directional distribution $\mathbb{Q}_2$ of X_2 are given by*

$$\gamma_2 \mathbb{Q}_2(A) = \frac{\gamma^2}{2} \int_{G(d,r)} \int_{G(d,r)} \mathbf{1}_A(E \cap F)[E,F]\, \mathbb{Q}(\mathrm{d}E)\, \mathbb{Q}(\mathrm{d}F)$$

for $A \in \mathcal{B}(G(d,2r-d))$.

(If $\gamma_2 = 0$, the measure $\mathbb{Q}_2$ is not defined; then $\gamma_2\mathbb{Q}_2$ has to be read as the zero measure.)

Proof. Let Θ_2 be the intensity measure of X_2. For $A \in \mathcal{B}(G(d,2r-d))$ we obtain, similarly to the proof of Theorem 4.4.8,

$$\begin{aligned}
&\gamma_2 \mathbb{Q}_2(A) \\
&= \frac{1}{\kappa_{2(d-r)}} \Theta_2(\mathcal{F}_{B^d} \cap \pi_0^{-1}(A)) \\
&= \frac{\gamma^2}{2\kappa_{2(d-r)}} \int_{G(d,r)} \int_{G(d,r)} \int_{E^\perp} \int_{F^\perp} \mathbf{1}_A(E \cap F) \mathbf{1}_{\mathcal{F}_{B^d}}((E+x) \cap (F+y)) \\
&\quad \times \lambda_{F^\perp}(\mathrm{d}y)\, \lambda_{E^\perp}(\mathrm{d}x)\, \mathbb{Q}(\mathrm{d}E)\, \mathbb{Q}(\mathrm{d}F).
\end{aligned}$$

In the integral

$$I(x) := \int_{F^\perp} \mathbf{1}_{\mathcal{F}_{B^d}}((E+x) \cap (F+y))\, \lambda_{F^\perp}(\mathrm{d}y),$$

the integrand is equal to 1 if and only if $y \in (B^d \cap (E+x))|F^\perp$. As in the proof of Theorem 4.4.7 (observing that $B^d \cap (E+x)$ is now a ball of radius $\sqrt{1-\|x\|^2}$), we obtain

$$I(x) = \kappa_{d-r}[E,F](1-\|x\|^2)^{(d-r)/2}.$$

This gives

$$\begin{aligned}
&\int_{E^\perp} \int_{F^\perp} \mathbf{1}_{\mathcal{F}_{B^d}}((E+x) \cap (F+y))\, \lambda_{F^\perp}(\mathrm{d}y)\, \lambda_{E^\perp}(\mathrm{d}x) \\
&= \kappa_{d-r}[E,F] \int_{B^d \cap E^\perp} (1-\|x\|^2)^{(d-r)/2}\, \lambda_{E^\perp}(\mathrm{d}x) \\
&= \kappa_{2(d-r)}[E,F].
\end{aligned}$$

This yields the assertion. □

The Proximity of Non-intersecting Poisson Flats

The considered intersection densities of stationary Poisson hyperplane processes are examples of real parameters that describe the geometric behavior of such processes and are not determined by the intensity alone. We now suggest a similar parameter for r-flat processes, where $r < d/2$. For these, we cannot work with intersections. The proposed parameter is a means to measure how close flats in general position of the process approach each other, in the mean. (If the directional distribution of a Poisson r-flat process is absolutely continuous, then by Theorem 4.4.5 almost surely any two flats in the process are in general position.)

Let $1 \leq r < d/2$ and $E_1, E_2 \in A(d,r)$. If E_1, E_2 are in general position, there are uniquely determined points $x_1 \in E_1$ and $x_2 \in E_2$ such that

$$d(E_1, E_2) := ||x_1 - x_2|| = \inf\{||y_1 - y_2|| : y_1 \in E_1,\ y_2 \in E_2\}.$$

We call the point

$$m(E_1, E_2) := \frac{1}{2}(x_1 + x_2)$$

the **midpoint** of E_1 and E_2.

Let X be a stationary r-flat process in $\mathbb{R}^d$, where $1 \leq r < d/2$. For every realization of X we take the midpoint $m(E_1, E_2)$ of any two flats E_1, E_2 of the realization which are in general position and satisfy $d(E_1, E_2) \leq 1$. (The bound 1 for the distance is only chosen for convenience; for a Poisson process, a different bound would result in an additional factor in (4.35).) In this way, we obtain a stationary point process in $\mathbb{R}^d$, the **midpoint process** of X. Its intensity is denoted by $\pi(X)$ and called the **proximity** of the flat process X. (Here $\pi(X) = 0$ is possible, for example, if the directional distribution of X is degenerate.)

Theorem 4.4.10. *Let $1 \leq r < d/2$, and let X be a stationary Poisson r-flat process in $\mathbb{R}^d$ with intensity $\gamma > 0$ and directional distribution $\mathbb{Q}$. Then the proximity of X is given by*

$$\pi(X) = \frac{1}{2}\kappa_{d-2r}\gamma^2 \int_{G(d,r)} \int_{G(d,r)} [E, F]\, \mathbb{Q}(\mathrm{d}E)\, \mathbb{Q}(\mathrm{d}F). \tag{4.35}$$

Proof. For $E_1, E_2 \in A(d,r)$, define

$$g(E_1, E_2) := \begin{cases} 1, & \text{if } E_1, E_2 \text{ are in general position,} \\ & \quad d(E_1, E_2) \leq 1 \text{ and } m(E_1, E_2) \in B^d, \\ 0 & \text{otherwise.} \end{cases}$$

Since the proximity $\pi(X)$ is the intensity of the midpoint process of X, it is given by the expectation

$$\pi(X) = \frac{1}{2\kappa_d}\mathbb{E}\sum_{(E_1,E_2)\in X^2_{\neq}} g(E_1,E_2).$$

By Theorems 3.1.3, 4.4.2 and Corollary 3.2.4, we obtain

$$\begin{aligned}\pi(X) &= \frac{1}{2\kappa_d}\int_{A(d,r)}\int_{A(d,r)} g(E_1,E_2)\,\Theta(\mathrm{d}E_1)\,\Theta(\mathrm{d}E_2)\\ &= \frac{\gamma^2}{2\kappa_d}\int_{G(d,r)}\int_{G(d,r)}\int_{E^\perp}\int_{F^\perp} g(E+x,F+y)\\ &\quad\times\lambda_{F^\perp}(\mathrm{d}y)\,\lambda_{E^\perp}(\mathrm{d}x)\,\mathbb{Q}(\mathrm{d}E)\,\mathbb{Q}(\mathrm{d}F).\end{aligned}$$

We compute the inner double integral

$$I(E,F) := \int_{E^\perp}\int_{F^\perp} g(E+x,F+y)\,\lambda_{F^\perp}(\mathrm{d}y)\,\lambda_{E^\perp}(\mathrm{d}x)$$

for two fixed subspaces $E,F\in G(d,r)$ in general position.

Let $E+F =: V$ and $U := V^\perp$. Vectors $x\in E^\perp$ and $y\in F^\perp$ have unique decompositions

$$\begin{aligned} x &= x_1+x_2, \quad x_1\in E^\perp\cap V,\ x_2\in U,\\ y &= y_1+y_2, \quad y_1\in F^\perp\cap V,\ y_2\in U,\end{aligned}$$

which gives

$$I(E,F) = \int_U\int_U J(E,F,x_2,y_2)\,\lambda_U(\mathrm{d}x_2)\,\lambda_U(\mathrm{d}y_2)$$

with

$$\begin{aligned}&J(E,F,x_2,y_2)\\ &= \int_{E^\perp\cap V}\int_{F^\perp\cap V} g(E+x_1+x_2,F+y_1+y_2)\,\lambda_{F^\perp\cap V}(\mathrm{d}y_1)\,\lambda_{E^\perp\cap V}(\mathrm{d}x_1).\end{aligned}$$

To compute this double integral, let $z\in V$ be the intersection point of $E+x_1$ and $F+y_1$. The distance of $E+x_1+x_2$ and $F+y_1+y_2$ is realized by the points $z+x_2$ and $z+y_2$, hence $d(E+x_1+x_2,F+y_1+y_2) = \|x_2-y_2\|$ and $m(E+x_1+x_2,F+y_1+y_2) = z+(x_2+y_2)/2$. Thus, $J(E,F,x_2,y_2)=0$ if $\|x_2-y_2\|>1$. Assume that $\|x_2-y_2\|\le 1$. Then $g(E+x_1+x_2,F+y_1+y_2)=1$ if and only if $z+(x_2+y_2)/2\in B^d$. The set $V\cap(B^d-(x_2+y_2)/2)$ is a $2r$-dimensional ball with radius $(1-\|(x_2+y_2)/2\|^2)^{1/2}$. It follows that

$$J(E,F,x_2,y_2) = \kappa_{2r}(1-\|(x_2+y_2)/2\|^2)^r[E,F],$$

if $\|x_2 - y_2\| \le 1$ and $\|(x_2+y_2)/2\| \le 1$, and 0 otherwise. This yields

$$\pi(X) = \frac{\kappa_{2r}}{2\kappa_d}\gamma^2 \int_{G(d,r)} \int_{G(d,r)} [E,F]\, \mathbb{Q}(\mathrm{d}E)\, \mathbb{Q}(\mathrm{d}F) \cdot K$$

with

$$K := \int_{U^2} \mathbf{1}\{\|x_2 - y_2\| \le 1\} \mathbf{1}\{\|(x_2+y_2)/2\| \le 1\} \\ \times (1 - \|(x_2+y_2)/2\|^2)^r \, \lambda_U^2(\mathrm{d}(x_2,y_2)).$$

The substitution $x_2 - y_2 = u$, $(x_2+y_2)/2 = v$ allows us to compute this integral, which completes the proof. □

Remark. Let X be as in Theorem 4.4.10. Let $\mathbb{Q}^\perp$ be the image measure of $\mathbb{Q}$ under the mapping $L \mapsto L^\perp$ from $G(d,r)$ to $G(d,d-r)$. There is a stationary Poisson $(d-r)$-flat process $X^\perp$ with directional distribution $\mathbb{Q}^\perp$ and intensity γ. A comparison of Theorems 4.4.9 and 4.4.10 shows that the second intersection density $\gamma_2(X^\perp)$ of the process $X^\perp$ and the proximity $\pi(X)$ of the process X are related by

$$\frac{1}{2}\kappa_{d-2r}\gamma_2(X^\perp) = \pi(X).$$

Therefore, inequalities for the second intersection density of a stationary Poisson flat process, as they are treated in Section 4.6, can be transferred to the proximity.

Notes for Section 4.4

1. Flat processes, in particular under Poisson assumptions, were first studied intensively by Miles [521, 523] and Matheron [460, 461, 462]. In the book by Matheron [462] one finds most of the results of Section 4.4, though partially with different proofs. For example, Theorem 4.4.1 appears there (p. 66) with a proof involving an extension of conditional probabilities, whereas we have preferred to give a direct and more elementary proof.

2. In the discussion following Theorem 4.4.6, we have mentioned the result (first pointed out by Matheron) that the distribution $\mathbb{P}_X$ of a stationary Poisson k-flat process X is uniquely determined by the section intensities $\gamma_{X\cap S}$, $S \in G(d,d-k)$, if either $k=1$ or $k=d-1$. That there is no corresponding uniqueness result for $1<k<d-1$, was shown by Goodey and Howard [271]. Sections with planes S of dimension $d-k+j$, $j \in \{1,\dots,k-1\}$, raise at least two questions: whether the section intensities $\gamma_{X\cap S}$, or whether the intensity measures of $X \cap S$, $S \in G(d,d-k+j)$, are sufficient to determine the distribution of the Poisson k-flat process X. These questions were answered partially by Goodey and Howard [271, 272] and completely by Goodey, Howard and Reeder [273].

The distribution $\mathbb{P}_X$ of a stationary Poisson hyperplane process X is, more generally, determined by the section intensities $\gamma_{X\cap S}$, $S \in G(d,r)$, for fixed

$r \in \{1, \dots, d-1\}$. Corresponding inversion formulas (for the intensity measure of X) are discussed in Spodarev [733, 734] (in a purely analytic setting, more general inversion formulas are treated by Rubin [653]).

3. Theorem 4.4.10, together with Theorem 4.6.6, is found in Schneider [699], though with a factor 1/2 missing.

4. In analogy to the idea of proximity, Spodarev [733, 735] introduced the **rose of neighborhood** γ_{kr} of a stationary k-flat process X, as a function on $G(d,r)$ where $k+r < d$. For $S \in G(d,r)$, $\gamma_{kr}(S)$ is the intensity of the (stationary) process of points in S arising as projections of midpoints $m(E,S), E \in X$, with distance $d(E,S) \leq 1$ (say). Relating X to a 'dual' process X' of $(d-k)$-flats, $\gamma_{kr}(S)$ transforms into the section intensity $\gamma_{X' \cap S^\perp}$ of X'. Therefore, the uniqueness, respectively non-uniqueness, results for section intensities (see Note 2 above) carry over to the roses of neighborhood.

For a similar situation, a process X of k-flats and a fixed r-plane S with $r+s < d$, Hug, Last and Weil [360] discussed the question whether distance measurements from S to (the union set of) the flats in X suffice to determine the directional distribution of X. Their results also hold for non-stationary processes X (see the Notes to Section 11.3).

5. The complementary theorem of Miles (see Note 5 of Section 3.2), in its versions for Poisson flat processes due to Miles [523] and to Møller and Zuyev [555], was considerably extended by Baumstark and Last [86]. They considered stationary Poisson processes of k-flats ($k \in \{0, \dots, d-1\}$) in $\mathbb{R}^d$ and obtained that the integral geometric contents of several closed sets constructed on such processes have conditional Gamma distributions.

4.5 Surface Processes

After studying processes of k-dimensional flats, it is a natural next step to consider processes of k-dimensional surfaces. Since unbounded surfaces can be represented as unions of countably many bounded surfaces, we may restrict ourselves to the latter. In particular, we shall consider particle processes where the particles are compact surfaces. For example, a **surface process** in $\mathbb{R}^3$ is obtained if the particles are almost surely two-dimensional surfaces, and a particle process consisting of curves is a **curve process** or **fiber process**, etc. The technical requirements for a theory of particle processes of k-dimensional surfaces depend very much on the generality of the notion of k-surface that is employed. A suitable general concept is that of a $\mathcal{H}^k$-rectifiable closed set. We refer to Section 14.5 for the definition of $\mathcal{H}^k$-rectifiable sets, and to Zähle [823] for a proof of the fact that the system $\mathcal{X}^{(k)}$ of $\mathcal{H}^k$-rectifiable closed sets in $\mathbb{R}^d$ is a measurable subset of $\mathcal{F}$. Therefore, a k-surface process can be defined as a point process in $\mathcal{F}$ the intensity measure of which is concentrated on $\mathcal{X}^{(k)}$. The treatment of such processes, however, requires methods from geometric measure theory, which are outside the scope of this book. For that reason, in the following we treat, with complete proofs, only an elementary version of surface processes, namely special processes in the convex ring $\mathcal{R}$. For this, we consider k-dimensional surfaces ($k = 1, \dots, d-1$) that can be represented as finite unions of k-dimensional compact convex sets, for example, polyhedral surfaces of dimension k.

For $k \in \{1, \dots, d-1\}$, we denote by $\mathcal{K}^{(k)}$ the set of all convex sets $K \in \mathcal{K}$ of dimension k and by $\mathcal{R}^{(k)} \subset \mathcal{R}$ the set of all finite unions of elements from $\mathcal{K}^{(k)}$. Elements of $\mathcal{R}^{(k)}$ are briefly called **k-surfaces** in the following. Obvious modifications of the proof of Theorem 2.4.2 show that $\mathcal{K}^{(k)}$ and $\mathcal{R}^{(k)}$ are Borel subsets of $\mathcal{F}$. By a **k-surface process** in $\mathbb{R}^d$ we understand a particle process with intensity measure concentrated on $\mathcal{R}^{(k)}$. This elementary case is sufficient for demonstrating the typical questions and results about surface processes. The extension to more general models then requires no principally new ideas, but is technically more involved. The methods and results from [823] allow us to obtain the results below with the system $\mathcal{R}^{(k)}$ of elementary k-surfaces replaced by the system $\mathcal{X}^{(k)}$ of $\mathcal{H}^k$-rectifiable closed sets, but this is not carried out here.

Let X be a stationary k-surface process with intensity measure $\Theta \neq 0$. According to Theorem 4.1.1, this process has an intensity γ and a grain distribution $\mathbb{Q}$. The intensity γ has to be distinguished from the k-volume density or specific k-volume. The latter is the intensity of the induced random k-volume measure (and is, therefore, by some authors called the 'intensity' of X). It can be introduced as follows. First we note that for $C \in \mathcal{R}^{(k)}$ we have

$$V_k(C) = \mathcal{H}^k(C), \tag{4.36}$$

where V_k is the additive extension of the kth intrinsic volume to the convex ring $\mathcal{R}$ (see Sections 14.2 and 14.4) and $\mathcal{H}^k$ is the k-dimensional Hausdorff measure. This follows by additivity, since (4.36) is true for $C \in \mathcal{K}^{(k)}$. The function $C \mapsto V_k(C)$, $C \in \mathcal{R}^{(k)}$, which we call the k-volume, is measurable by Theorem 14.4.4 (the additive extension of V_k is measurable on $\mathcal{R}$). By (4.36) it is nonnegative.

According to (4.6), the k-volume density of X is defined by

$$\overline{V}_k(X) := \gamma \int_{\mathcal{C}_0} V_k \, d\mathbb{Q}. \tag{4.37}$$

We call $\overline{V}_k(X)$ the **specific k-volume** of X (the possibility of $\overline{V}_k(X) = \infty$ is not excluded).

We define a random measure η by

$$\eta := \sum_{C \in X} \mathcal{H}^k \llcorner C.$$

Almost surely η is locally finite, since X is a particle process, and each particle $C \in \mathcal{R}^{(k)}$ satisfies $\mathcal{H}^k(C) < \infty$. The following theorem shows that the specific k-volume, if finite, can be interpreted as the intensity of the stationary random measure η.

Theorem 4.5.1. *Let X be a stationary k-surface process in $\mathbb{R}^d$ with $\overline{V}_k(X) < \infty$. Then*

$$\sum_{C\in X} \mathcal{H}^k \llcorner C$$

is a stationary random measure, and $\overline{V}_k(X)$ is its intensity, that is,

$$\overline{V}_k(X) = \frac{1}{\lambda(A)} \mathbb{E} \sum_{C\in X} \mathcal{H}^k(C \cap A) \tag{4.38}$$

for all $A \in \mathcal{B}(\mathbb{R}^d)$ with $0 < \lambda(A) < \infty$.

Proof. Let $A \in \mathcal{B}(\mathbb{R}^d)$ be given. We define

$$f(C) := (\mathcal{H}^k \llcorner C)(A) = \mathcal{H}^k(A \cap C) \qquad \text{for } C \in \mathcal{R}^{(k)}.$$

Assume, first, that A is compact and that $K_i, K \in \mathcal{K}^{(k)}$ satisfy $K_i \to K$ in the Hausdorff metric. Let $E \in A(d,k)$ be the plane with $K \subset E$. There exist rigid motions g_i, converging to the identity, such that $g_iK_i \subset E$ and $g_iK_i \to K$. For every $x \in \mathbb{R}^d$,

$$\limsup \mathbf{1}_{g_i(A\cap K_i)}(x) \le \mathbf{1}_{A\cap K}(x).$$

As in the proof of Theorem 12.3.6, we get

$$\begin{aligned}\mathcal{H}^k(A \cap K) &= \int_E \mathbf{1}_{A\cap K}(x)\, \mathcal{H}^k(\mathrm{d}x) \ge \int_E \limsup \mathbf{1}_{g_i(A\cap K_i)}(x)\, \mathcal{H}^k(\mathrm{d}x) \\ &\ge \limsup \int_E \mathbf{1}_{g_i(A\cap K_i)}(x)\, \mathcal{H}^k(\mathrm{d}x) = \limsup \mathcal{H}^k(A \cap K_i).\end{aligned}$$

Thus, on $\mathcal{K}^{(k)}$ the function f is upper semicontinuous and, therefore, measurable. Modifying the proof of Theorem 14.4.4, we see that f is measurable on $\mathcal{R}^{(k)}$. Since this holds for all compact sets A, it holds for all Borel sets A. Now the Campbell theorem shows that $\sum_{C\in X}(\mathcal{H}^k \llcorner C)(A)$ is measurable. It follows that $\sum_{C\in X} \mathcal{H}^k \llcorner C$ is a random measure; clearly it is stationary. Campbell's theorem further shows that

$$\begin{aligned}\mathbb{E} \sum_{C\in X} (\mathcal{H}^k \llcorner C)(A) &= \gamma \int_{\mathcal{C}_0} \int_{\mathbb{R}^d} \mathcal{H}^k(A \cap (C + x))\, \lambda(\mathrm{d}x)\, \mathbb{Q}(\mathrm{d}C) \\ &= \gamma \int_{\mathcal{C}_0} V_k(C)\lambda(A)\, \mathbb{Q}(\mathrm{d}C) \\ &= \overline{V}_k(X)\lambda(A),\end{aligned}$$

where Theorem 5.2.1 (with $\alpha := \mathcal{H}^k \llcorner C$) and (4.37) were used. This proves (4.38). Now the assumption $\overline{V}_k(X) < \infty$ implies that $\sum_{C\in X} \mathcal{H}^k \llcorner C$ has locally finite intensity measure and intensity $\overline{V}_k(X)$. □

A k-surface has, at $\mathcal{H}^k$-almost every point, a k-dimensional tangent plane. For a k-surface process, this leads to the notion of its directional distribution. Let $C \in \mathcal{R}^{(k)}$, and let $C = \bigcup_{i=1}^m C_i$ be a representation with $C_i \in \mathcal{K}^{(k)}$ for $i = 1, \dots, m$. The set of all $y \in C$ lying in some C_i and some C_j, where C_i and C_j have different affine hulls, is of $\mathcal{H}^k$-measure zero. For the remaining $y \in C$, we can choose i with $y \in C_i$ and then define the **tangent plane** $T_y C$ of C at y as the linear subspace of $\mathbb{R}^d$ which is parallel to the affine hull of C_i. Thus, at $\mathcal{H}^k$-almost all $y \in C$, the tangent plane is uniquely determined.

Theorem 4.5.2. *Let X be a stationary k-surface process in $\mathbb{R}^d$ with specific k-volume satisfying $0 < \overline{V}_k(X) < \infty$. Then there is a unique probability measure $\mathbb{T}$ on the Grassmannian $G(d,k)$ satisfying*

$$\mathbb{E} \sum_{C \in X} \int_{B \cap C} \mathbf{1}_A(T_y C)\, \mathcal{H}^k(\mathrm{d}y) = \overline{V}_k(X) \lambda(B) \mathbb{T}(A)$$

for all $B \in \mathcal{B}(\mathbb{R}^d)$ with $0 < \lambda(B) < \infty$ and all $A \in \mathcal{B}(G(d,k))$.

Proof. From the Campbell theorem and from Theorem 5.2.1 we obtain

$$\begin{aligned}
&\mathbb{E} \sum_{C \in X} \int_{B \cap C} \mathbf{1}_A(T_y C)\, \mathcal{H}^k(\mathrm{d}y) \\
&= \gamma \int_{\mathcal{C}_0} \int_{\mathbb{R}^d} \int_{B \cap (C+x)} \mathbf{1}_A(T_y(C+x))\, \mathcal{H}^k(\mathrm{d}y)\, \lambda(\mathrm{d}x)\, \mathbb{Q}(\mathrm{d}C) \\
&= \gamma \int_{\mathcal{C}_0} \int_{\mathbb{R}^d} \int_{(B-x) \cap C} \mathbf{1}_A(T_y C)\, \mathcal{H}^k(\mathrm{d}y)\, \lambda(\mathrm{d}x)\, \mathbb{Q}(\mathrm{d}C) \\
&= \gamma \lambda(B) \int_{\mathcal{C}_0} \int_C \mathbf{1}_A(T_y C)\, \mathcal{H}^k(\mathrm{d}y)\, \mathbb{Q}(\mathrm{d}C).
\end{aligned}$$

The mapping

$$A \mapsto \gamma \int_{\mathcal{C}_0} \int_C \mathbf{1}_A(T_y C)\, \mathcal{H}^k(\mathrm{d}y)\, \mathbb{Q}(\mathrm{d}C), \qquad A \in \mathcal{B}(G(d,k)),$$

is a measure η with $\eta(G(d,k)) = \overline{V}_k(X)$. Defining $\mathbb{T} := \eta / \overline{V}_k(X)$, we obtain the assertion. The uniqueness is clear. □

For later use, we note that

$$\overline{V}_k(X) \mathbb{T}(A) = \gamma \int_{\mathcal{C}_0} \int_C \mathbf{1}_A(T_y C)\, \mathcal{H}^k(\mathrm{d}y)\, \mathbb{Q}(\mathrm{d}C). \tag{4.39}$$

We call the probability measure $\mathbb{T}$ the **directional distribution** of the k-surface process X (another common name is **rose of directions**, in particular for fiber processes). The directional distribution can be interpreted as the distribution of the tangent plane in a typical point of the surface process.

Now we consider section processes derived from k-surface processes. Let X be a stationary k-surface process with positive, finite specific k-volume, and let $S \in G(d, d-k+j)$ be a $(d-k+j)$-plane, where $0 \le j \le k-1$. In Section 3.6 we have defined the section process $X \cap S$. It is a particle process in S. Because of the elementary notion of k-surface that we employ, it is not difficult to show that $X \cap S$ is almost surely a j-surface process in S. We do not carry out the proof here, since very similar arguments were already employed when we treated processes of flats (before Theorem 4.4.6). It is clear that the j-surface process $X \cap S$ is stationary in S.

Theorem 4.5.3. *Let X be a stationary k-surface process in $\mathbb{R}^d$ with positive, finite specific k-volume $\overline{V}_k(X)$ and with directional distribution $\mathbb{T}$. Let $S \in G(d, d-k+j)$, $0 \le j \le k-1$, and let $\overline{V}_j(X \cap S)$ be the specific j-volume of the section process $X \cap S$. Then*

$$\overline{V}_j(X \cap S) = \overline{V}_k(X) \int_{G(d,k)} [S, L] \, \mathbb{T}(\mathrm{d}L).$$

Proof. Let $A \subset S$ be a compact set with $\lambda_S(A) = 1$. By Theorem 4.5.1 and Campbell's theorem,

$$\begin{aligned} \overline{V}_j(X \cap S) &= \mathbb{E} \sum_{C \in X} \mathcal{H}^j(C \cap A) \\ &= \gamma \int_{\mathcal{C}_0} \int_{\mathbb{R}^d} \mathcal{H}^j((C+x) \cap A) \, \lambda(\mathrm{d}x) \, \mathbb{Q}(\mathrm{d}C). \end{aligned}$$

Let $C = \bigcup_{i=1}^m C_i$ with $C_i \in \mathcal{K}^{(k)}$. By the inclusion–exclusion principle (with the notation used in (14.48)) we have

$$\begin{aligned} \int_{\mathbb{R}^d} \mathcal{H}^j((C+x) \cap A) \, \lambda(\mathrm{d}x) &= \sum_{v \in S(m)} (-1)^{|v|-1} \int_{\mathbb{R}^d} \mathcal{H}^j((C_v + x) \cap A) \, \lambda(\mathrm{d}x) \\ &= \sum_{v \in S(m)} (-1)^{|v|-1} [S, \operatorname{aff} C_v] V_k(C_v) \\ &= \sum_{v \in S(m)} (-1)^{|v|-1} \int_{C_v} [S, T_y C_v] \, \mathcal{H}^k(\mathrm{d}y) \\ &= \int_C [S, T_y C] \, \mathcal{H}^k(\mathrm{d}y) \end{aligned}$$

(observe that $V_k(C_v) = 0$ if $\dim C_v < k$). This yields

$$\begin{aligned} \overline{V}_j(X \cap S) &= \gamma \int_{\mathcal{C}_0} \int_C [S, T_y C] \, \mathcal{H}^k(\mathrm{d}y) \, \mathbb{Q}(\mathrm{d}C) \\ &= \overline{V}_k(X) \int_{G(d,k)} [S, L] \, \mathbb{T}(\mathrm{d}L), \end{aligned}$$

where we have used (4.39), extended from indicator functions to nonnegative measurable functions on $G(d,k)$. □

In the cases $k = 1$ (fiber processes) and $k = d-1$ (hypersurface processes) it is again convenient (as after Theorem 4.4.6) to interpret the directional distribution as an even measure on the sphere S^{d-1}. For a unit vector $u \in S^{d-1}$, $L(u)$ denotes the one-dimensional linear subspace spanned by u, and $u^\perp$ is the $(d-1)$-dimensional linear subspace orthogonal to u. Corresponding to a directional distribution $\mathbb{T}$ we define a spherical directional distribution φ by setting, for a set $A \in \mathcal{B}(S^{d-1})$ without pairs of antipodal points,

$$\varphi(A) := \frac{1}{2}\mathbb{T}(\{L(u) : u \in A\}) \qquad \text{if } k = 1$$

and

$$\varphi(A) := \frac{1}{2}\mathbb{T}(\{u^\perp : u \in A\}) \qquad \text{if } k = d-1.$$

For the specific 0-volumes (intersection point densities) of the section processes found in Theorem 4.5.3, we now obtain, for $v \in S^{d-1}$,

$$\overline{V}_0(X \cap v^\perp) = \overline{V}_1(X) \int_{S^{d-1}} |\langle u, v\rangle| \, \varphi(\mathrm{d}u) \qquad \text{if } k = 1 \tag{4.40}$$

and

$$\overline{V}_0(X \cap L(v)) = \overline{V}_{d-1}(X) \int_{S^{d-1}} |\langle u, v\rangle| \, \varphi(\mathrm{d}u) \qquad \text{if } k = d-1. \tag{4.41}$$

Note for Section 4.5

The investigation of the directional distribution (also called 'rose of directions') of fiber and surface processes was initiated in papers by Mecke and Stoyan, beginning with [501], which was generalized by Mecke and Nagel [495]. Pohlmann, Mecke and Stoyan [606] treated stereological formulas for stationary surface processes. For a very general investigation of fiber and surface processes (using Hausdorff rectifiable sets), we refer to Zähle [822].

4.6 Associated Convex Bodies

For a stationary particle process X in $\mathbb{R}^d$ and a suitable translation invariant function φ on $\mathcal{C}_0$, the φ-density $\overline{\varphi}$ was defined in Section 4.1 by

$$\overline{\varphi}(X) := \gamma \int_{\mathcal{C}_0} \varphi \, \mathrm{d}\mathbb{Q}.$$

This procedure is not restricted to real-valued functions φ. In particular, on the space of convex bodies, there are some geometrically meaningful translation invariant mappings into spaces of functions or measures which can be employed. In this way one can associate with a particle process, besides intensities of real-valued functionals, also measures or convex bodies as describing parameters. Similar procedures are possible for other geometric processes, such as processes of flats, fibers, or surfaces, or even for certain random closed sets. One motivation for this comes from the fact that associated measures or convex bodies contain more information than real-valued parameters, and may yet be accessible to estimation procedures. Another reason for introducing auxiliary convex bodies lies in the observation that sometimes the application of results from convex geometry to associated auxiliary bodies leads to results, for example to solutions of extremal problems, which otherwise would be out of reach. Such results from convex geometry are applied in this section; they are collected in Section 14.3, with references to sources where proofs can be found.

Processes of Convex Particles

First we consider a stationary process X of convex particles in $\mathbb{R}^d$ with intensity $\gamma > 0$ and grain distribution $\mathbb{Q}$.

Since a convex body K is determined by its support function $h(K, \cdot)$, defined by

$$h(K, u) := \max\{\langle x, u \rangle : x \in K\}, \qquad u \in \mathbb{R}^d,$$

it appears natural to consider the density of the functional $h(\cdot, u)$ for $u \in \mathbb{R}^d$. However, the support function is not translation invariant. This is remedied by introducing the **centered support function**, by

$$h^*(K, u) := h(K, u) - \langle s(K), u \rangle = h(K - s(K), u), \tag{4.42}$$

where $s(K)$ is the Steiner point of K (see (14.28)). We have

$$h^*(K + x, \cdot) = h^*(K, \cdot) \qquad \text{for } x \in \mathbb{R}^d$$

and $h^*(K, \cdot) \geq 0$. From (14.7) and (14.28) we obtain an estimate of the form $h^*(K, u) \leq c(d)V_1(K)\|u\|$ with a constant $c(d)$. Since V_1 is $\mathbb{Q}$-integrable by Theorem 4.1.2, $h^*(\cdot, u)$ is $\mathbb{Q}$-integrable. Hence, we can define

$$\overline{h}(X, u) := \gamma \int_{\mathcal{K}_0} h^*(K, u)\, \mathbb{Q}(\mathrm{d}K) \qquad \text{for } u \in \mathbb{R}^d.$$

Obviously, the function $\overline{h}(X, \cdot)$ is again convex and positively homogeneous, hence it is the support function of a uniquely determined convex body. We denote this body by $M(X)$ and call it the **mean body** of the particle process X.

In a similar way, the surface area measure $S_{d-1}(K, \cdot)$ (see (14.22)) can be employed. For $A \in \mathcal{B}(S^{d-1})$, the function $S_{d-1}(\cdot, A)$ is measurable and translation invariant. Further, $0 \leq S_{d-1}(K, \cdot) \leq S_{d-1}(K, S^{d-1}) = 2V_{d-1}(K)$, where V_{d-1} is one of the intrinsic volumes (see Section 14.2). Since V_{d-1} is $\mathbb{Q}$-integrable by Theorem 4.1.2, we can define

$$\overline{S}_{d-1}(X,A) := \gamma \int_{\mathcal{K}_0} S_{d-1}(K,A)\,\mathbb{Q}(\mathrm{d}K) \tag{4.43}$$

for $A \in \mathcal{B}(S^{d-1})$. By monotone convergence, $\overline{S}_{d-1}(X,\cdot)$ is a measure.

We indicate how this measure-valued parameter can be interpreted in the case where the particles of the process X are a.s. of dimension d. From the process X we then also obtain a hypersurface process, by replacing each particle by its boundary. For such a hypersurface process, a directional distribution can be defined, similarly to Section 4.1. In contrast to the case $k = d-1$ of Theorem 4.5.2, we now consider an oriented directional distribution, taking into account that for the boundary of a d-dimensional convex body one can distinguish between an outer and an inner normal direction. For the boundary hypersurface $\mathrm{bd}\,K$ of a convex body it is convenient to describe the direction of a tangent hyperplane by its outer normal vector. For $\mathcal{H}^{d-1}$-almost all $y \in \mathrm{bd}\,K$, the outer unit normal vector $n_K(y)$ of K at y is uniquely determined. For a Borel set $A \subset S^{d-1}$, we have $S_{d-1}(K,A) = \mathcal{H}^{d-1}(n_K^{-1}(A))$. For $A \in \mathcal{B}(S^{d-1})$ and $B \in \mathcal{B}(\mathbb{R}^d)$ with $\lambda(B) < \infty$, the mapping $K \mapsto \mathcal{H}^{d-1}(B \cap n_K^{-1}(A))$ is measurable (as follows from Schneider [695, Theorem 4.2.1]), and the Campbell theorem together with (4.3) gives

$$\begin{aligned}
&\mathbb{E} \sum_{K\in X} \mathcal{H}^{d-1}\left(B \cap n_K^{-1}(A)\right) \\
&= \gamma \int_{\mathcal{K}_0} \int_{\mathbb{R}^d} \mathcal{H}^{d-1}\left(B \cap n_{K+x}^{-1}(A)\right) \lambda(\mathrm{d}x)\,\mathbb{Q}(\mathrm{d}K) \\
&= \gamma \int_{\mathcal{K}_0} \int_{\mathbb{R}^d} \mathcal{H}^{d-1}\left((B-x) \cap n_K^{-1}(A)\right) \lambda(\mathrm{d}x)\,\mathbb{Q}(\mathrm{d}K) \\
&= \gamma \int_{\mathcal{K}_0} \lambda(B)\mathcal{H}^{d-1}\left(n_K^{-1}(A)\right) \mathbb{Q}(\mathrm{d}K) \\
&= \gamma\lambda(B) \int_{\mathcal{K}_0} S_{d-1}(K,A)\,\mathbb{Q}(\mathrm{d}K),
\end{aligned}$$

where Theorem 5.2.1 was used. Thus, for $B \in \mathcal{B}(\mathbb{R}^d)$ with $\lambda(B) = 1$ we have

$$\overline{S}_{d-1}(X,A) = \mathbb{E} \sum_{K\in X} \mathcal{H}^{d-1}\left(B \cap n_K^{-1}(A)\right).$$

For this reason, the normalized measure $\overline{S}_{d-1}(X,\cdot)/2\overline{V}_{d-1}(X)$ can be interpreted as the **distribution of the normal vector in a typical boundary point** of the particle process X.

The measure $\overline{S}_{d-1}(X,\cdot)$ is called the **mean normal measure** of X (also in the case where the particles are not necessarily d-dimensional).

Starting from the measure-valued parameter $\overline{S}_{d-1}(X,\cdot)$, we now associate two convex bodies with the particle process X. This requires a preliminary consideration.

For a convex body K and for $u \in \mathbb{R}^d \setminus \{0\}$, we denote by $V_{d-1}(K|u^\perp)$ the $(d-1)$-dimensional volume of the orthogonal projection of K to $u^\perp$. The density of the function $K \mapsto V_{d-1}(K|u^\perp)$ for the particle process X is denoted by $\overline{V}_{d-1}(X|u^\perp)$, thus

$$\overline{V}_{d-1}(X|u^\perp) = \gamma \int_{\mathcal{K}_0} V_{d-1}(K|u^\perp)\,\mathbb{Q}(\mathrm{d}K).$$

With Fubini's theorem for kernels and with (14.41), for unit vectors $u \in S^{d-1}$ we get

$$\begin{aligned}\overline{V}_{d-1}(X|u^\perp) &= \frac{\gamma}{2}\int_{\mathcal{K}_0}\int_{S^{d-1}} |\langle u,v\rangle|\, S_{d-1}(K,\mathrm{d}v)\,\mathbb{Q}(\mathrm{d}K)\\ &= \frac{1}{2}\int_{S^{d-1}} |\langle u,v\rangle|\, \overline{S}_{d-1}(X,\mathrm{d}v). \end{aligned} \tag{4.44}$$

From the Campbell theorem and Theorem 4.1.2, we get for $r > 0$

$$\begin{aligned}&\mathbb{E} \sum_{K\in X,\, c(K)\in rB^d} V_{d-1}(K|u^\perp)\\ &= \gamma \int_{\mathcal{K}_0}\int_{\mathbb{R}^d} \mathbf{1}_{rB^d}(c(K+x)) V_{d-1}((K+x)|u^\perp)\,\lambda(\mathrm{d}x)\,\mathbb{Q}(\mathrm{d}K)\\ &= \kappa_d r^d\, \overline{V}_{d-1}(X|u^\perp).\end{aligned}$$

Thus, $\overline{V}_{d-1}(X|u^\perp) = 0$ holds if and only if

$$\sum_{K\in X} V_{d-1}(K|u^\perp) = 0$$

almost surely. If there exists a vector $u \in S^{d-1}$ with this property, we say that the particle process X is **degenerate**.

We assume now that X is not degenerate. Then (4.44) shows that the measure $\overline{S}_{d-1}(X,\cdot)$ is not concentrated on a great subsphere. Since

$$\int_{S^{d-1}} u\, S_{d-1}(K,\mathrm{d}u) = 0$$

always holds, we also have

$$\int_{S^{d-1}} u\, \overline{S}_{d-1}(X,\mathrm{d}u) = 0.$$

By the Theorem of Minkowski (Theorem 14.3.1), there exists a uniquely determined convex body $B(X) \in \mathcal{K}_0$ with

$$S_{d-1}(B(X),\cdot) = \overline{S}_{d-1}(X,\cdot). \tag{4.45}$$

We call $B(X)$ the **Blaschke body** of the particle process X. (The name reflects the fact that the addition of surface area measures induces the so-called Blaschke addition of the corresponding convex bodies.)

For a convex body K, we denote by Π_K its projection body (see Section 14.3, in particular (14.40)). The projection body of the Blaschke body, that is,

$$\Pi_X := \Pi_{B(X)}, \tag{4.46}$$

is called the **associated zonoid** of the particle process X. (The name refers to the fact that projection bodies belong to the special class of convex bodies known as zonoids; these are precisely the bodies which can be approximated by vector sums of line segments.) Using (14.40), (14.41), (4.44), (4.45), we get

$$\begin{aligned} h(\Pi_X,u) &= \frac{1}{2}\int_{S^{d-1}} |\langle u,v\rangle|\, \overline{S}_{d-1}(X,\mathrm{d}v) \\ &= \overline{V}_{d-1}(X|u^\perp) \\ &= \gamma \int_{\mathcal{K}_0} h(\Pi_K,u)\,\mathbb{Q}(\mathrm{d}K). \end{aligned} \tag{4.47}$$

Thus, the support function of the associated zonoid has a simple geometric meaning: on unit vectors, it represents the density of the projection volume in the direction of the vector. Moreover, Π_X can be interpreted as the mean projection body of the particle process X.

Rewriting (4.47) in the form

$$h(\Pi_X,u) = \frac{\gamma}{2}\int_{\mathcal{K}_0}\int_{S^{d-1}} |\langle u,v\rangle|\, S_{d-1}(K,\mathrm{d}v)\,\mathbb{Q}(\mathrm{d}K),$$

integrating over S^{d-1} with respect to the spherical Lebesgue measure, and observing $S_{d-1}(K,S^{d-1}) = 2V_{d-1}(K)$ as well as (14.7), we obtain the identity

$$V_1(\Pi_X) = 2\overline{V}_{d-1}(X), \tag{4.48}$$

The intrinsic volume V_1 appearing here is essentially the mean width; hence, the identity says that the mean width of the associated zonoid is, up to a constant factor, the surface area density of the particle process X.

Next we show how further geometric quantities of the particle process X are related to the associated zonoid Π_X. First we determine

$$f(u) := \mathbb{E}\sum_{K\in X} \operatorname{card}\left([0,u]\cap \operatorname{bd} K\right),$$

the expected number of points in which the segment with endpoints 0 and $u \in \mathbb{R}^d \setminus \{0\}$ meets the boundaries of the bodies of the particle process.

Thus, for a unit vector u, the value $f(u)$ gives the intensity $\gamma_{L(u)}$ of the point process that is generated by intersecting the hypersurface process induced by the boundaries of the particles with the line $L(u)$ (as considered similarly in Theorem 4.5.3 for a different class of surface processes). With the Campbell theorem and the decomposition (4.2) we obtain

$$\begin{aligned} f(u) &= \gamma \int_{\mathcal{K}_0} \int_{\mathbb{R}^d} \operatorname{card}\left([0,u] \cap \operatorname{bd}(K+x)\right) \lambda(\mathrm{d}x)\, \mathbb{Q}(\mathrm{d}K) \\ &= 2\gamma \int_{\mathcal{K}_0} \|u\| V_{d-1}(K|u^\perp)\, \mathbb{Q}(\mathrm{d}K) \\ &= 2\|u\| \overline{V}_{d-1}(X|u^\perp) \\ &= 2h(\Pi_X, u). \end{aligned}$$

Thus, we have

$$h(\Pi_X, u) = \frac{1}{2} \mathbb{E} \sum_{K \in X} \operatorname{card}\left([0,u] \cap \operatorname{bd} K\right) = \frac{1}{2} \|u\| \gamma_{L(u)}, \tag{4.49}$$

which provides a further interpretation of the support function of the associated zonoid. In particular, for the intersection intensity we obtain from (4.47) the formula

$$\gamma_{L(u)} = \int_{S^{d-1}} |\langle u, v\rangle|\, \overline{S}_{d-1}(X, \mathrm{d}v). \tag{4.50}$$

Since $\overline{S}_{d-1}(X,\cdot)/2\overline{V}_{d-1}(X)$ is a probability measure, this equation is analogous to (4.41). However, it must be observed that X in (4.41) is a hypersurface process, whereas in (4.50) it is a process of convex particles (for a convex body K, $2V_{d-1}(K)$ is the surface area).

Applications to Boolean Models

The associated zonoid is particularly useful when dealing with stationary Boolean models with convex grains. Therefore, we assume now in addition that X is a Poisson process. We still assume that X is nondegenerate, that is, it satisfies $\overline{V}_{d-1}(X|u^\perp) \neq 0$ for all $u \in S^{d-1}$. In this case, also the Boolean model $Z = Z_X$ is called nondegenerate (this property depends only on Z, since Z determines the particle process X up to equivalence). Hence, the Boolean model Z is degenerate if and only if there is a direction u such that the orthogonal projection of Z to $u^\perp$ a.s. has Lebesgue measure zero.

For $F \in \mathcal{F}$ and $x \in \mathbb{R}^d$, we write

$$S_x(F) := \{y \in \mathbb{R}^d : [x,y] \cap F = \emptyset\}$$

for the region visible from x; here F is regarded as opaque. The set $S_x(F)$ is open and star-shaped with respect to x; it is empty if $x \in F$. For the stationary Boolean model $Z = Z_X$, the conditional expectation

$$\overline{V}_s(Z) := \mathbb{E}(\lambda(S_0(Z)) \mid 0 \notin Z)$$

is called the **mean visible volume** outside Z (note that we always have $\mathbb{P}(0 \notin Z) > 0$, by (9.5)). The measurability of the function $\lambda(S_0(Z))$ and of the function $(u, \omega) \mapsto s_u(Z(\omega))$ used below follows from the measurability of the set

$$\left\{(\omega, u, \alpha) \in \Omega \times S^{d-1} \times \mathbb{R}_0^+ : [0, \alpha u] \cap Z(\omega) = \emptyset\right\}.$$

The quantity $\overline{V}_s(Z)$ is a further simple parameter which, besides volume and surface area density, can be used for the description of a Boolean model. (Here, volume and surface area density refer to the underlying particle process X; a connection with corresponding parameters of the union set Z_X will be established later in Section 9.1.)

First we observe that also the visible domain $S_0(Z)$ can itself be averaged in a natural way, namely by averaging its radial function $\rho(S_0(Z), \cdot)$. For $u \in S^{d-1}$,

$$s_u(Z) := \rho(S_0(Z), u) = \sup\{\alpha \geq 0 : [0, \alpha u] \cap Z = \emptyset\}$$

defines the **visibility range** from 0 in direction u. For $r \geq 0$, we have

$$\mathbb{P}(s_u(Z) \leq r \mid 0 \notin Z) = H_l^{(u)}(r) = 1 - \mathrm{e}^{-r\overline{V}_{d-1}(X|u^\perp)},$$

as will be proved in Theorem 9.1.1. Thus, the visibility range $s_u(Z)$ has (under the condition $0 \notin Z$) an exponential distribution with parameter $\overline{V}_{d-1}(X|u^\perp)$ (which is positive, since X was assumed to be nondegenerate). Therefore, the kth moment of the visibility range $s_u(Z)$ is equal to $k!\overline{V}_{d-1}(X|u^\perp)^{-k}$; in particular, the expectation is $\overline{V}_{d-1}(X|u^\perp)^{-1}$. We define the **mean visible region $\overline{K}_s$** outside Z as the star-shaped set with radial function

$$\rho(\overline{K}_s, \cdot) = \mathbb{E}(\rho(S_0(Z), \cdot) \mid 0 \notin Z),$$

thus

$$\overline{K}_s = \{\alpha u : u \in S^{d-1}, 0 \leq \alpha \leq \mathbb{E}(s_u(Z) \mid 0 \notin Z)\}.$$

Because of

$$\rho(\overline{K}_s, u) = \overline{V}_{d-1}(X|u^\perp)^{-1} = h(\Pi_X, u)^{-1}$$

for $u \in S^{d-1}$, the set $\overline{K}_s$ is the polar body of the associated zonoid, which in the following will be denoted by Π_X^o. In particular, it follows that the mean visible region is convex.

The volume of the visible region $S_0(Z)$ is given by

$$V_d(S_0(Z)) = \frac{1}{d} \int_{S^{d-1}} s_u(Z)^d \, \sigma(\mathrm{d}u),$$

where σ denotes spherical Lebesgue measure. Therefore, for the mean visible volume outside Z we obtain

$$\begin{aligned}
\overline{V}_s(Z) &= \mathbb{E}(\lambda(S_0(Z)) \mid 0 \notin Z) \\
&= \frac{1}{d}\int_{S^{d-1}} \mathbb{E}(s_u(Z)^d \mid 0 \notin Z)\,\sigma(\mathrm{d}u) \\
&= (d-1)! \int_{S^{d-1}} \overline{V}_{d-1}(X|u^\perp)^{-d}\,\sigma(\mathrm{d}u) \\
&= d! V_d(\Pi_X^o).
\end{aligned}$$

We resume this as a theorem.

Theorem 4.6.1. *Let $Z = Z_X$ be a nondegenerate stationary Boolean model with convex grains in $\mathbb{R}^d$. The mean visible region outside Z is the polar body Π_X^o of the associated zonoid of X; the mean visible volume outside Z is given by*

$$\overline{V}_s(Z) = (d-1)! \int_{S^{d-1}} \overline{V}_{d-1}(X|u^\perp)^{-d}\,\sigma(\mathrm{d}u) = d! V_d(\Pi_X^o). \tag{4.51}$$

We are now in a position to establish a few sharp inequalities between different parameters of the Boolean model Z_X, respectively of the corresponding particle process X. From (4.48) and (14.43) we obtain the inequality

$$\overline{V}_s(Z) \geq d!\kappa_d \left(\frac{\kappa_{d-1}}{d\kappa_d} 2\overline{V}_{d-1}(X)\right)^{-d}. \tag{4.52}$$

Here, equality holds if and only if the associated zonoid Π_X is a ball. This occurs, for instance, if the density $\overline{S}_{d-1}(X,\cdot)$ of the surface area measure is rotation invariant (and, hence, the Blaschke body is a ball). Therefore, we can formulate the following result.

Theorem 4.6.2. *Let $Z = Z_X$ be a nondegenerate stationary Boolean model with convex grains in $\mathbb{R}^d$, generated by a Poisson particle process X with given surface area density. The mean visible volume outside Z_X is minimal if the process is isotropic.*

This raises the question whether there also exists an upper estimate of the mean visible volume $\overline{V}_s(Z)$ in terms of a functional density of X. In terms of the surface area density this is not possible, as can be shown by examples. A suitable functional for such an estimate is given by $V_d^{1-1/d}$ (which is of the same degree of homogeneity as the surface area). For this, we employ the Blaschke body $B(X)$. Applying successively (14.23), (4.45), (14.23), (14.30) and making use of mixed volumes (see Section 14.2) we obtain

$$\begin{aligned}
V_d(B(X)) &= \frac{1}{d}\int_{S^{d-1}} h(B(X),u)\, S_{d-1}(B(X),\mathrm{d}u) \\
&= \frac{\gamma}{d}\int_{\mathcal{K}_0}\int_{S^{d-1}} h(B(X),u)\, S_{d-1}(K,\mathrm{d}u)\,\mathbb{Q}(\mathrm{d}K)
\end{aligned}$$

$$= \gamma \int_{\mathcal{K}_0} V(B(X), K, \dots, K) \, \mathbb{Q}(\mathrm{d}K)$$

$$\geq V_d(B(X))^{1/d} \gamma \int_{\mathcal{K}_0} V_d(K)^{1-1/d} \, \mathbb{Q}(\mathrm{d}K),$$

hence

$$V_d(B(X))^{1-1/d} \geq \overline{V_d^{1-1/d}}(X).$$

From (4.51), (4.46), (14.44) we now obtain, as a counterpart to (4.52), the inequality

$$\overline{V}_s(Z) \leq d! \left(\frac{\kappa_{d-1}}{\kappa_d} \overline{V_d^{1-1/d}}(X) \right)^{-d}. \tag{4.53}$$

Here, equality holds if and only if the grain distribution $\mathbb{Q}$ is concentrated on a set of homothetic ellipsoids. This follows from the available information about the equality cases in the inequalities (14.30) and (14.44).

As a further parameter for a geometric description of a Poisson particle process X we introduce the intersection density of the boundaries. For a bounded Borel set $B \subset \mathbb{R}^d$, let $s(X, B)$ be the number of points in B arising as intersection points of the boundaries of any d distinct bodies of the process. The **intersection density** of X is the number $\gamma_d(X)$ satisfying

$$\mathbb{E}\, s(X, B) = \gamma_d(X) \lambda(B)$$

for all bounded Borel sets B. In order to show its existence and to compute it, we use Theorem 3.1.3, Corollary 3.2.4, and Theorem 4.1.1 and obtain

$$\mathbb{E}\, s(X, B) = \frac{1}{d!} \mathbb{E} \sum_{(K_1, \dots, K_d) \in X^d_{\neq}} \operatorname{card} (B \cap \operatorname{bd} K_1 \cap \dots \cap \operatorname{bd} K_d)$$

$$= \frac{1}{d!} \int_{\mathcal{K}^d} \operatorname{card} (B \cap \operatorname{bd} K_1 \cap \dots \cap \operatorname{bd} K_d) \, \Lambda^{(d)}(\mathrm{d}(K_1, \dots, K_d))$$

$$= \frac{\gamma^d}{d!} \int_{\mathcal{K}_0} \cdots \int_{\mathcal{K}_0} I(K_1, \dots, K_d) \, \mathbb{Q}(\mathrm{d}K_1) \cdots \mathbb{Q}(\mathrm{d}K_d)$$

with

$$I(K_1, \dots, K_d) :=$$

$$\int_{\mathbb{R}^d} \cdots \int_{\mathbb{R}^d} \operatorname{card} (B \cap \operatorname{bd} (K_1 + x_1) \cap \dots \cap \operatorname{bd} (K_d + x_d)) \, \lambda(\mathrm{d}x_1) \cdots \lambda(\mathrm{d}x_d).$$

We abbreviate $B \cap \operatorname{bd} (K_1 + x) =: F_x$ and $\operatorname{bd} K_i =: F_i$ for $i = 2, \dots, d$. Using the body Π_{F_x} given by (5.31) and applying Theorem 5.4.4, (14.21) and Theorem 5.2.1, we get

$$
\begin{aligned}
&I(K_1,\dots,K_d)\\
&= \int_{\mathbb{R}^d}\cdots\int_{\mathbb{R}^d} \mathrm{card}\,(F_x\cap(F_2+x_2)\cap\ldots\cap(F_d+x_d))\,\lambda(\mathrm{d}x_2)\cdots\lambda(\mathrm{d}x_d)\,\lambda(\mathrm{d}x)\\
&= d!\int_{\mathbb{R}^d} V(\Pi_{F_x},\Pi_{K_2},\dots,\Pi_{K_d})\,\lambda(\mathrm{d}x)\\
&= (d-1)!\int_{\mathbb{R}^d}\int_{S^{d-1}} h(\Pi_{F_x},u)\,S(\Pi_{K_2},\dots,\Pi_{K_d},\mathrm{d}u)\,\lambda(\mathrm{d}x)\\
&= (d-1)!\int_{S^{d-1}}\int_{\mathbb{R}^d} h(\Pi_{F_x},u)\,\lambda(\mathrm{d}x)\,S(\Pi_{K_2},\dots,\Pi_{K_d},\mathrm{d}u)\\
&= (d-1)!\int_{S^{d-1}} h(\Pi_{K_1},u)\,S(\Pi_{K_2},\dots,\Pi_{K_d},\mathrm{d}u)\,\lambda(B)\\
&= d!V(\Pi_{K_1},\Pi_{K_2},\dots,\Pi_{K_d})\lambda(B).
\end{aligned}
$$

Thus, we obtain

$$
\mathbb{E}\,s(X,B) = \gamma^d\int_{\mathcal{K}_0}\cdots\int_{\mathcal{K}_0} V(\Pi_{K_1},\dots,\Pi_{K_d})\,\mathbb{Q}(\mathrm{d}K_1)\cdots\mathbb{Q}(\mathrm{d}K_d)\,\lambda(B).
$$

Here we have, by (14.21) and (4.47),

$$
\begin{aligned}
&\gamma\int_{\mathcal{K}_0} V(\Pi_{K_1},\dots,\Pi_{K_d})\,\mathbb{Q}(\mathrm{d}K_1)\\
&= \frac{\gamma}{d}\int_{\mathcal{K}_0}\int_{S^{d-1}} h(\Pi_{K_1},u)\,S(\Pi_{K_2},\dots,\Pi_{K_d},\mathrm{d}u)\,\mathbb{Q}(\mathrm{d}K_1)\\
&= \frac{1}{d}\int_{S^{d-1}} h(\Pi_X,u)\,S(\Pi_{K_2},\dots,\Pi_{K_d},\mathrm{d}u)\\
&= V(\Pi_X,\Pi_{K_2},\dots,\Pi_{K_d}).
\end{aligned}
$$

Repeating this procedure, we finally get

$$
\mathbb{E}\,s(X,B) = V_d(\Pi_X)\lambda(B),
$$

and thus the existence of the intersection density, together with the representation

$$
\gamma_d(X) = V_d(\Pi_X). \tag{4.54}
$$

From (4.54), (4.48) and (14.31) we obtain *a sharp inequality between the intersection density and the surface area density*, namely

$$
\gamma_d(X) \le \kappa_d\left(\frac{2\kappa_{d-1}}{d\kappa_d}\overline{V}_{d-1}(X)\right)^d. \tag{4.55}
$$

Here equality holds if and only if the associated zonoid Π_X is a ball, which occurs, for example, if the Poisson particle process X is isotropic.

It is intuitively plausible that a large intersection density indicates that much overlapping of particles occurs and that, therefore, the mean visible volume must be small. This intuition is indeed precise, in so far as the product $\gamma_d(X)\overline{V}_s(X)$ does not depend on the intensity of the process. For this quantity, we are able to establish sharp inequalities.

Theorem 4.6.3. *Let $Z = Z_X$ be a nondegenerate stationary Boolean model with convex grains in $\mathbb{R}^d$. The intersection density and the mean visible volume satisfy the inequalities*

$$4^d \leq \gamma_d(X)\overline{V}_s(Z) \leq d!\kappa_d^2. \tag{4.56}$$

On the right side, equality holds if the process X is isotropic. On the left side, equality holds if and only if the particles of X are almost surely parallelepipeds with edges of d fixed directions.

Proof. The inequalities follow from (4.51), (4.54), and (14.45). On the right side, equality holds if and only if the associated zonoid Π_X is an ellipsoid, thus in particular if the process is isotropic. On the left side, equality holds if and only if Π_X is a parallelepiped. This is equivalent to the existence of d linearly independent vectors $v_1, \ldots, v_d \in S^{d-1}$ such that the measure $\overline{S}_{d-1}(X, \cdot)$ is concentrated on $\{\pm v_i : i = 1, \ldots, d\}$. By (4.43), this holds if and only if for $\mathbb{Q}$-almost all $K \in \mathcal{K}_0$ the measure $S_{d-1}(K, \cdot)$ is concentrated on $\{\pm v_i : i = 1, \ldots, d\}$, hence if K is a parallelepiped with facet normal vectors $\pm v_i$. We conclude that equality in the left inequality of (4.56) holds if and only if the particles of X are almost surely parallelepipeds whose facet normals are parallel to d fixed directions. The facet normals determine also the directions of the edges. □

Processes of Flats

We turn now to processes of flats and want to show how associated zonoids can be utilized for them. We describe a general construction, which is not only applicable to flat processes, but also, for instance, to fiber and surface processes. We begin with a finite Borel measure τ on the space $G(d,k)$ of k-dimensional linear subspaces of $\mathbb{R}^d$, $k \in \{1, \ldots, d-1\}$. There exists a convex body $\Pi^k(\tau)$ with support function

$$h(\Pi^k(\tau), \cdot) = \frac{1}{2} \int_{G(d,k)} h(L^\perp \cap B^d, \cdot)\, \tau(\mathrm{d}L). \tag{4.57}$$

That this is indeed a support function, is clear, since the integrand is a support function. Since $L^\perp \cap B^d$ is a ball (of dimension $d-k$) and thus a zonoid, $\Pi^k(\tau)$ is a zonoid, too. We have $h(L^\perp \cap B^d, u) = [L^\perp, u^\perp] = [L, L(u)]$ for $u \in S^{d-1}$, hence also

$$h(\Pi^k(\tau), u) = \frac{1}{2} \int_{G(d,k)} [L, L(u)] \, \tau(\mathrm{d}L), \qquad u \in S^{d-1}. \tag{4.58}$$

Now we consider, first, a stationary hyperplane process X in $\mathbb{R}^d$. Let $\gamma > 0$ be its intensity and $\mathbb{Q}$ its directional distribution. We put

$$\Pi_X := \Pi^{d-1}(\gamma\mathbb{Q})$$

and call Π_X the **associated zonoid** of the hyperplane process X. By (4.58), we have

$$h(\Pi_X, u) = \frac{\gamma}{2} \int_{S^{d-1}} |\langle u, v\rangle| \, \varphi(\mathrm{d}v), \qquad u \in \mathbb{R}^d, \tag{4.59}$$

where φ is the spherical directional distribution of X. According to Theorem 14.3.4, Π_X determines the measure $\gamma\varphi$ uniquely, hence also the intensity γ and the spherical directional distribution φ of X are uniquely determined by Π_X. In particular, a Poisson process X is isotropic if and only if Π_X is a ball. From Theorem 4.4.1 we get the following.

Theorem 4.6.4. *For every centered zonoid $Z \subset \mathbb{R}^d$ there is up to equivalence precisely one stationary Poisson hyperplane process X with associated zonoid Z.*

The support function of the associated zonoid is again connected with intersection densities. As in Section 4.4, let $\gamma_{X\cap L(u)}$ denote the intensity of the point process $X \cap L(u)$. By (4.59) and (4.34) we have

$$2h(\Pi_X, u) = \|u\|\gamma_{X\cap L(u)} = \mathbb{E}X(\mathcal{F}_{[0,u]}) \qquad \text{for } u \in \mathbb{R}^d. \tag{4.60}$$

From (4.60) we see immediately how to obtain the associated zonoid of a section process. For an r-dimensional linear subspace $S \in G(d,r)$ with $r \in \{1, \dots, d-1\}$, let $X \cap S$ be the section process (see Section 4.4). Its associated zonoid $\Pi_{X\cap S}$ is defined as a convex body in S. For $u \in S$ we have, by (4.60),

$$2h(\Pi_{X\cap S}, u) = \mathbb{E}(X \cap S)(\mathcal{F}_{[0,u]}) = \mathbb{E}X(\mathcal{F}_{[0,u]}) = 2h(\Pi_X, u).$$

For the orthogonal projection $\Pi_X|S$, we have $h(\Pi_X|S, u) = h(\Pi_X, u)$ for $u \in S$, hence

$$\Pi_{X\cap S} = \Pi_X|S. \tag{4.61}$$

This means *that the associated zonoid of the section process $X \cap S$ is the orthogonal projection of the associated zonoid of X to the linear subspace S.*

Now we assume, in particular, that X is a stationary Poisson hyperplane process with intensity $\gamma > 0$. With the aid of the associated zonoid, we can obtain information on the intersection processes of X of higher order. In Section 4.4 we have defined, for $k \in \{2, \dots, d\}$, the intersection process of order k of X, as the $(d-k)$-flat process X_k that is obtained if one takes the intersection of any k hyperplanes of the process X which are in general position. Similar intersection processes can be formed more generally for stationary surface processes. As a different special case, we have previously considered the intersection point density γ_d of the boundary hypersurfaces of a stationary process of convex particles. Now, for the stationary Poisson hyperplane process X, let γ_k be the intensity and $\mathbb{Q}_k$ the directional distribution of the intersection process X_k. By Theorem 4.4.8 we then have, for $A \in \mathcal{B}(G(d, d-k))$,

$$\gamma_k \mathbb{Q}_k(A)$$
$$= \frac{\gamma^k}{k!} \int_{S^{d-1}} \cdots \int_{S^{d-1}} \mathbf{1}_A \left(u_1^\perp \cap \ldots \cap u_k^\perp\right) \nabla_k(u_1, \ldots, u_k)\, \varphi(\mathrm{d}u_1) \cdots \varphi(\mathrm{d}u_k),$$

where φ is the spherical directional distribution of X. The associated zonoid Π_X of X is given by

$$h(\Pi_X, u) = \int_{S^{d-1}} |\langle u, v \rangle|\, \rho(\mathrm{d}v) \qquad \text{for } u \in \mathbb{R}^d,$$

where $\rho := \gamma\varphi/2$. Hence, if $\rho_{(k)}$ is the kth projection generating measure of Π_X, as defined by (14.36), then, for $A \in \mathcal{B}(G(d, d-k))$,

$$\gamma_k \mathbb{Q}_k(A) = \kappa_k \int_{G(d,k)} \mathbf{1}_A(L^\perp)\, \rho_{(k)}(\mathrm{d}L) = \kappa_k \rho_{(k)}^\perp(A),$$

where $\rho_{(k)}^\perp$ is the image measure of $\rho_{(k)}$ under the mapping $L \mapsto L^\perp$ from $G(d,k)$ to $G(d, d-k)$. Therefore, we have

$$\gamma_k \mathbb{Q}_k = \kappa_k \rho_{(k)}^\perp, \tag{4.62}$$

saying that *the intensity measure of the intersection process of order k of X is determined by the kth projection generating measure of the associated zonoid Π_X*.

The intensity γ_k of the intersection process of order k of X is called the kth **intersection density** of X. Here, $\gamma_1 = \gamma$. By the definition of the intersection processes and by Theorem 4.4.3, the kth intersection density is given by

$$\gamma_k = \frac{1}{d!\kappa_d} \mathbb{E} \sum_{(H_1,\ldots,H_k) \in X^k} \lambda_{d-k}^*(H_1 \cap \ldots \cap H_k \cap B^d).$$

Here $\lambda_{d-k}^*(A)$ is the $(d-k)$-dimensional volume of A if $\dim A = d-k$, and is zero otherwise. For the intersection densities, we can obtain inequalities. They are based on the fact that $\gamma_k = \kappa_k \rho_{(k)}(G(d,k))$ by (4.62) and hence, by (14.37),

$$\gamma_k = V_k(\Pi_X); \tag{4.63}$$

thus, *the kth intersection density is the kth intrinsic volume of the associated zonoid*. In particular, the dth intersection density, which is the density of the intersection points generated by X, is nothing but the volume of the associated zonoid. An analog of this fact is (4.54); both equations are special cases of a corresponding result for general stationary hypersurface processes (with the k-volume density instead of γ_k). Also (4.55) and (4.56) can be generalized in this sense.

Now, from (14.31) we obtain the inequality

$$\left(\frac{\kappa_{d-j}}{\binom{d}{j}}\gamma_j\right)^k \geq \kappa_d^{k-j}\left(\frac{\kappa_{d-k}}{\binom{d}{k}}\gamma_k\right)^j \tag{4.64}$$

for $1 \leq j < k \leq d$. If $\gamma_j > 0$, then equality in (4.64) holds if and only if Π_X is a ball. By the uniqueness theorem 14.3.4, this holds if and only if the spherical directional distribution φ of X is the normalized spherical Lebesgue measure, hence, if and only if the Poisson hyperplane process X is isotropic. The case $\gamma_j = 0$ occurs, by (4.63), if and only if $\dim \Pi_X < j$, hence if and only if the spherical directional distribution φ is concentrated on $S^{d-1} \cap L$ for some subspace $L \in G(d, j-1)$. An equivalent condition is that the hyperplanes of the process almost surely contain a translate of the $(d+1-j)$-dimensional plane $L^\perp$.

We formulate the special case $j = 1$ as a theorem.

Theorem 4.6.5. *The kth intersection density, $k \in \{2, \dots, d\}$, of a stationary Poisson hyperplane process of intensity $\gamma > 0$ in $\mathbb{R}^d$ satisfies the inequality*

$$\gamma_k \leq \frac{\binom{d}{k}\kappa_{d-1}^k}{d^k \kappa_{d-k}\kappa_d^{k-1}}\gamma^k.$$

Equality holds if and only if the process is isotropic.

Thus, the isotropic processes are characterized here by an extremal property of isoperimetric type: for given intensity, they have maximal intersection densities.

If X_1 and X_2 are independent stationary Poisson hyperplane processes, then their superposition $X_1 + X_2$ is also a stationary Poisson process, with intensity measure $\Theta_1 + \Theta_2$, if Θ_i is the intensity measure of X_i. It follows that the associated zonoids also add:

$$\Pi_{X_1+X_2} = \Pi_{X_1} + \Pi_{X_2}.$$

If $\gamma_k(X)$ denotes the kth intersection density of X, then (4.63) and (14.32) yield the inequality

$$\gamma_k(X_1 + X_2)^{1/k} \geq \gamma_k(X_1)^{1/k} + \gamma_k(X_2)^{1/k}, \tag{4.65}$$

for $k = 2, \dots, d$. Equality in (4.65) holds at least if the hyperplane processes X_1 and X_2 have the same directional distribution, since then their associated zonoids are homothetic.

We can also derive a sharp estimate for the proximity (defined before Theorem 4.4.10) of a Poisson line process.

Theorem 4.6.6. *Let X be a stationary Poisson line process of given intensity $\gamma > 0$. The proximity $\pi(X)$ attains its maximum if and only if X is isotropic.*

Proof. This follows from the remark after Theorem 4.4.10 and the case $k = 2$ of Theorem 4.6.5. □

For a k-flat process X and a fixed $(d-k)$-plane S, we have considered in Section 4.4 the section process $X \cap S$. There it was mentioned that a stationary Poisson k-flat process is uniquely determined, up to stochastic equivalence, by its intersection densities $\gamma_{X \cap S}$, $S \in G(d, d-k)$, if either $k = 1$ or $k = d-1$, but not in the cases $1 < k < d-1$. However, for kth-order intersection processes of stationary Poisson hyperplane processes, there exists a corresponding uniqueness result. An even stronger assertion is expressed by the subsequent theorem. Here it has to be observed that

$$\gamma_{X_k \cap S} = \gamma_k(X \cap S).$$

A stationary hyperplane process is **nondegenerate** if the hyperplanes of the process are not almost surely parallel to a fixed line.

Theorem 4.6.7. *Let X be a nondegenerate stationary Poisson hyperplane process of intensity γ in $\mathbb{R}^d$, let $r \in \{1, \dots, d-1\}$ and $k \in \{1, \dots, r\}$. Then X is uniquely determined (up to stochastic equivalence) by the kth intersection densities $\gamma_k(X \cap S)$ of the section processes $X \cap S$, $S \in G(d, r)$.*

Proof. By (4.63) and (4.61) we have

$$\gamma_k(X \cap S) = V_k(\Pi_{X \cap S}) = V_k(\Pi_X | S)$$

for $S \in G(d, r)$. Since X is nondegenerate, $\dim \Pi_X \geq d$, as was remarked earlier. By a theorem from convex geometry (Aleksandrov's Projection Theorem; see Gardner [244, Theorem 3.3.6]), the convex body Π_X, which is centrally symmetric with respect to 0, is uniquely determined by the intrinsic volumes $V_k(\Pi_X | S)$, $S \in G(d, r)$. Now the assertion follows from Theorem 4.6.4. □

Flat Processes Hitting Convex Bodies

Now we consider more general flat processes. Let X be a stationary k-flat process of intensity $\gamma > 0$ in $\mathbb{R}^d$. We suppose that a convex 'test body' $K \in \mathcal{K}'$ is hit by the flats of the process, and we want to measure in different ways how intensively it is hit. We could, for example, be interested in deciding which shape a convex body of given volume must have so that in the mean it is hit by as few flats as possible. This depends on how the intensity of hitting is measured. For instance, if we use the k-dimensional volume of the intersections as a measure, then Theorem 4.4.3 gives the answer

$$\mathbb{E} \sum_{E \in X} V_k(K \cap E) = \gamma V_d(K),$$

saying that the left side is independent of the shape of K. Instead of the k-dimensional volume of the intersections, we could ask for the number of nonempty intersections or, more generally, using the jth intrinsic volume V_j, $j \in \{0, \dots, k\}$, ask for

$$\mathbb{E} \sum_{E \in X} V_j(K \cap E). \tag{4.66}$$

The number of nonempty intersections is included here, for $j = 0$. If X is in addition isotropic, then Theorem 9.4.8, to be proved later, gives the result

$$\mathbb{E} \sum_{E \in X} V_j(K \cap E) = \gamma c_{j,d}^{k,d-k+j} V_{d+j-k}(K),$$

with certain constants $c_{j,d}^{k,d-k+j}$. For $j < k$ and given positive volume, the functional $V_{d+j-k}(K)$ attains its minimum if and only if K is a ball (cf. (14.31)). Here the assumption of isotropy cannot be deleted; without it, the quantity (4.66) will not only depend on the intrinsic volume $V_{d+j-k}(K)$, but the shape of K will play an essential role. To see this, at least in some special cases, we first compute the expectation (4.66). Let Θ be the intensity measure and $\mathbb{Q}$ the directional distribution of X. From the Campbell theorem and Theorem 4.4.2 we get

$$\begin{aligned} \mathbb{E} \sum_{E \in X} V_j(K \cap E) &= \int_{A(d,k)} V_j(K \cap E)\, \Theta(\mathrm{d}E) \\ &= \gamma \int_{G(d,k)} \int_{L^\perp} V_j(K \cap (L + x))\, \lambda_{L^\perp}(\mathrm{d}x)\, \mathbb{Q}(\mathrm{d}L). \end{aligned}$$

Using the integral geometric formula (6.39), we obtain

$$\begin{aligned} &\mathbb{E} \sum_{E \in X} V_j(K \cap E) \\ &= \frac{\binom{d}{k-j}}{\kappa_{k-j}} \gamma \int_{G(d,k)} V(K[d+j-k], (L \cap B^d)[k-j])\, \mathbb{Q}(\mathrm{d}L), \end{aligned} \tag{4.67}$$

where the integrand is a mixed volume. For $j = 0$, the integral geometric formula is not needed, and one obtains directly

$$\mathbb{E} \sum_{E \in X} V_0(K \cap E) = \gamma \int_{G(d,k)} V_{d-k}(K|L^\perp)\, \mathbb{Q}(\mathrm{d}L). \tag{4.68}$$

A further treatment of the integral (4.67) has only been successful in special cases. First we consider the case $j = k - 1$, that is, the surface area of the k-dimensional sections $K \cap E$. If $S_{d-1}(K, \cdot)$ denotes the surface area measure of K, then formula (14.23) for mixed volumes and (4.57) give

$$\begin{aligned}&\int_{G(d,k)} V(K,\dots,K,L\cap B^d)\,\mathbb{Q}(\mathrm{d}L)\\&=\frac{1}{d}\int_{G(d,k)}\int_{S^{d-1}} h(L\cap B^d,u)\,S_{d-1}(K,\mathrm{d}u)\,\mathbb{Q}(\mathrm{d}L)\\&=\frac{1}{d}\int_{S^{d-1}}\int_{G(d,k)} h(L\cap B^d,u)\,\mathbb{Q}(\mathrm{d}L)\,S_{d-1}(K,\mathrm{d}u).\end{aligned}$$

We define a zonoid $\Pi_k(\mathbb{Q})$ by

$$h(\Pi_k(\mathbb{Q}),u)=\frac{1}{2}\int_{G(d,k)} h(L\cap B^d,u)\,\mathbb{Q}(\mathrm{d}L).$$

Thus,

$$\Pi_k(\mathbb{Q})=\Pi^{d-k}(\mathbb{Q}^\perp),$$

where $\mathbb{Q}^\perp$ is the image measure of $\mathbb{Q}$ under the mapping $L\mapsto L^\perp$ from $G(d,k)$ to $G(d,d-k)$. Then we put

$$\Pi^X:=\gamma\Pi_k(\mathbb{Q}).$$

We obtain

$$\begin{aligned}&\int_{G(d,k)} V(K,\dots,K,L\cap B^d)\,\mathbb{Q}(\mathrm{d}L)\\&=\int_{G(d,k)}\frac{1}{d}\int_{S^{d-1}} h(L\cap B^d,u)\,S_{d-1}(K,\mathrm{d}u)\,\mathbb{Q}(\mathrm{d}L)\\&=\frac{2}{d}\int_{S^{d-1}} h(\Pi_k(\mathbb{Q}),u)\,S_{d-1}(K,\mathrm{d}u)\\&=2V(\Pi_k(\mathbb{Q}),K,\dots,K),\end{aligned}$$

hence

$$\mathbb{E}\sum_{E\in X} V_{k-1}(K\cap E)=dV(\Pi^X,K,\dots,K).$$

From Minkowski's inequality (14.30), we now deduce the following extremal property.

Theorem 4.6.8. *Let X be a stationary k-flat process of intensity $\gamma>0$ in $\mathbb{R}^d$, and let K be a convex body of given positive volume. Then the expected value*

$$\mathbb{E}\sum_{E\in X} V_{k-1}(K\cap E)$$

is minimal if and only if K is homothetic to the zonoid Π^X.

In the case of a line process ($k = 1$), the quantity $\mathbb{E}\sum_{E\in X} V_{k-1}(K\cap E)$ is just the expected number of lines hitting the body K.

Further information on the expected number of hitting k-planes is available in the case $k = d-1$. Let X be a stationary hyperplane process of intensity $\gamma > 0$ and with spherical directional distribution φ. We assume that X is nondegenerate, so that the hyperplanes of the process are not almost surely parallel to a fixed line. Under this assumption, the measure φ is not concentrated on a great subsphere, and it follows that $V_d(\Pi_X) > 0$. Since φ is an even measure, it follows from Theorem 14.3.1 that there exists a uniquely determined convex body $B(X)$, centrally symmetric with respect to 0, for which

$$S_{d-1}(B(X),\cdot) = \gamma\varphi.$$

We call $B(X)$ the **Blaschke body** of the hyperplane process X. Thus, by (4.59) we have

$$\Pi_X = \Pi_{B(X)},$$

in analogy to (4.46).

Now from (4.68) and (14.23), we obtain

$$\begin{aligned}\mathbb{E}\sum_{E\in X} V_0(K\cap E) &= \gamma\int_{G(d,d-1)} V_1(K|L^\perp)\,\mathbb{Q}(\mathrm{d}L)\\ &= \gamma\int_{S^{d-1}} [h(K,u)+h(K,-u)]\,\varphi(\mathrm{d}u)\\ &= 2\int_{S^{d-1}} h(K,u)\,S_{d-1}(B(X),\mathrm{d}u)\\ &= 2dV(K,B(X),\dots,B(X)). \qquad (4.69)\end{aligned}$$

Again, we can apply Minkowski's inequality (14.30), and deduce the following result.

Theorem 4.6.9. *Let X be a nondegenerate stationary hyperplane process of intensity $\gamma > 0$ in $\mathbb{R}^d$, and let $K \in \mathcal{K}$ be a convex body with given volume $V_d(K) > 0$. The expected number of hyperplanes of the process X hitting the convex body K is minimal if and only if K is homothetic to the Blaschke body $B(X)$ of X.*

Notes for Section 4.6

1. The associated zonoid of a stationary process X of convex particles was introduced here as the projection body of the Blaschke body of X; this is equivalent to the original definition. The construction of the Blaschke body requires Minkowski's existence theorem (Theorem 14.3.1). This existence theorem was first used in Schneider [686] for associating a convex body with a directional distribution (of finitely many random hyperplanes in that case) and then applying results from convex geometry. Extensive use of this association in the case of random hypersurfaces was made by Wieacker [817, 818]. The Blaschke body $B(X)$ of a particle process X was introduced in Weil [796]; that paper also provides information on the mean body $M(X)$. In Weil [797], Blaschke bodies were also suggested and investigated for random closed sets with values in the extended convex ring. For Boolean models, for example, this paper established a connection between the Blaschke body and the contact distribution function.

2. The associated zonoid of a stationary Poisson hyperplane process was introduced by Matheron [461, 462], under the name of 'Steiner compact set'. The book by Matheron [462] has already the formulas (4.61) and (4.63), and in principle also (4.62). Associated zonoids for random hyperplanes were also used in Schneider [685, 686]. Generalization and systematic application of associated zonoids then followed in the work of Wieacker [816, 817, 818]; see also Sections 6 and 7 in the survey article of Weil and Wieacker [806] and Section 6 of Schneider and Wieacker [720]. Wieacker has introduced different types of associated zonoids, and he has applied them to random surfaces, surface processes, k-flat processes, particle processes, and random mosaics. In Wieacker [817] one finds, for example, the assertions (4.48), (4.49), Theorem 4.6.1 (essentially), (4.52), (4.54), (4.55), (4.56) (the right-hand inequality), and a generalization of Theorem 4.6.5. For extensions and supplements (such as (4.56) (left side) and (4.53), see Schneider [689]; that paper treats Poisson processes of convex cylinders, which includes as special cases flat processes and processes of convex particles. Compare also inequality (10.52) and Note 4 for Section 10.4.

3. Inequality (4.64) and with it Theorem 4.6.5 are due to Thomas [756]. Similar arguments, but with different interpretations, appear in Schneider [686, 697]. Alternative proofs for special cases were found by Mecke [480, 484].

Theorem 4.6.5 raises the question which stationary Poisson k-flat processes of given (positive) intensity, where $d/2 \leq k < d-1$, have maximal second intersection density. According to Theorem 4.4.9, this amounts to finding the maximum of the integral

$$\int_{G(d,k)} \int_{G(d,k)} [E,F]\, \mathbb{Q}(\mathrm{d}E)\, \mathbb{Q}(\mathrm{d}F)$$

over all probability measures $\mathbb{Q}$ on $G(d,k)$. For $k = d-1$, the maximum is attained precisely by the rotation invariant probability measures, by Theorem 4.6.5. For $k < d-1$, however, the maximum is not attained by invariant measures, as was discovered by Mecke and Thomas [504] (see also Mecke [489]). Mecke [486, 487] was able to determine explicitly the extremal measures for $d = 2k$. Keutel [401] has completely settled the case where $k < d-2$ and $d-k$ divides d. The general case is still open.

Theorem 4.6.6 goes back, in principle, to Janson and Kallenberg [378], though with a different approach.

Theorems 4.6.8 and 4.6.9 are special cases of considerably more general assertions in Wieacker [818]. Theorem 4.6.7 and related results were first published in Schneider and Weil [717].

Part II

Integral Geometry

5

Averaging with Invariant Measures

As soon as stochastic geometry deals with structures satisfying invariance properties with respect to some group, such as stationarity or isotropy in Euclidean spaces, there arises the need for a theory allowing averaging with respect to invariant measures. Integral geometry in the sense of Blaschke and Santaló is perfectly made for obtaining such averaging formulas. In this chapter we develop the basic tools, namely intersection formulas for fixed and moving geometric objects, where suitable geometric quantities of the intersections are integrated with respect to invariant measures. Basic facts about invariant measures on locally compact topological groups and homogeneous spaces, as far as they are needed for our purposes, are collected in the Appendix in Chapter 13.

The main purpose of Section 5.1 is the calculation of general kinematic integrals of the form

$$\int_{G_d} \varphi(K \cap gM)\, \mu(\mathrm{d}g) \tag{5.1}$$

for convex bodies K, M in $\mathbb{R}^d$. Here G_d is the motion group of $\mathbb{R}^d$, and the integration is with respect to its Haar measure μ. Such integrals are called 'kinematic', since one imagines M as moving and one averages the functional φ over all intersections of the moving set with the fixed set K. The integral (5.1) takes a simple form if the functional φ satisfies two natural assumptions, additivity and continuity. This result is known as Hadwiger's general integral geometric theorem (Theorem 5.1.2). The assumptions on φ are satisfied, in particular, by the intrinsic volumes V_j. A brief introduction to these important functionals from convex geometry is given in Section 14.2. For the intrinsic volumes, Hadwiger's general theorem reduces to the classical principal kinematic formula for convex bodies.

If the moving convex body is replaced by a moving flat, one is led to the Crofton formulas, giving explicit expressions for the integrals

$$\int_{A(d,k)} V_j(K \cap E)\, \mu_k(\mathrm{d}E),$$

where μ_k is the invariant measure on the affine Grassmannian $A(d,k)$.

The intrinsic volumes have local versions, the support measures or (generalized) curvature measures. These are introduced in Section 14.2, by means of a local Steiner formula. The purpose of the two subsequent sections is a derivation of the principal kinematic formula for curvature measures. Section 5.2 treats only integrations over the translation group, in a more general fashion with a view to later applications, and Section 5.3 then deals with the additional integrations over the rotation group. In each case, formulas for intrinsic volumes result by specialization.

Section 5.4 leaves the domain of convex or polyconvex sets and studies translative, kinematic and Crofton formulas for Hausdorff rectifiable sets and the Hausdorff measures of their intersections.

5.1 The Kinematic Formula for Additive Functionals

We make use of the homogeneous spaces and invariant measures of Euclidean geometry, as introduced in the Appendix. We assume that the reader is familiar with these, either from Chapter 13 or from other sources. We recall and collect here only the basic notation.

We denote by SO_d the group of proper (that is, orientation-preserving) rotations of $\mathbb{R}^d$. Being a compact group, it carries a unique rotation invariant (Borel) probability measure, which we denote by ν. The group of (proper) rigid motions of $\mathbb{R}^d$ is denoted by G_d. Let μ be its invariant (or Haar) measure, normalized so that

$$\mu(\{g \in G_d : gx \in B^d\}) = \kappa_d$$

for $x \in \mathbb{R}^d$. More explicitly, the mapping

$$\begin{aligned} \gamma : \mathbb{R}^d \times SO_d \to \quad & G_d \\ (x, \vartheta) \quad & \mapsto t_x \circ \vartheta, \end{aligned}$$

where t_x is the translation by the vector x, is a homeomorphism, and μ is the image measure of the product measure $\lambda \otimes \nu$ under γ.

The Grassmannian $G(d,q)$ of q-dimensional linear subspaces of $\mathbb{R}^d$, $q \in \{0,\dots,d\}$, is a compact homogeneous space with respect to the rotation group SO_d. It carries a unique rotation invariant probability measure, which we denote by ν_q. The affine Grassmannian $A(d,q)$, the space of q-flats in $\mathbb{R}^d$, is a locally compact homogeneous space with respect to the motion group and carries a locally finite motion invariant measure. We denote it by μ_q and normalize it so that

$$\mu_q(\{E \in A(d,q) : E \cap B^d \neq \emptyset\}) = \kappa_{d-q}.$$

More explicitly, we may choose a fixed subspace $L_q \in G(d,q)$, denote its orthogonal complement by $L_q^\perp$, and define mappings

$$\begin{aligned} \beta_q : SO_d &\to G(d,q) \\ \vartheta &\mapsto \vartheta L_q \end{aligned}$$

and

$$\begin{aligned} \gamma_q : L_q^\perp \times SO_d &\to \quad A(d,q) \\ (x,\vartheta) &\mapsto \vartheta(L_q + x). \end{aligned} \tag{5.2}$$

These maps are continuous and surjective. Now, ν_q is the image measure of the invariant measure ν under β_q, and μ_q is the image measure of the product measure $\lambda_{L_q^\perp} \otimes \nu$ under γ_q, where $\lambda_{L_q^\perp}$ denotes the $(d-q)$-dimensional Lebesgue measure on $L_q^\perp$.

A basic task involving these invariant measures consists in the evaluation of integrals such as

$$\int_{G_d} \varphi(K \cap gM)\, \mu(\mathrm{d}g) \qquad \text{and} \qquad \int_{A(d,k)} \varphi(K \cap E)\, \mu_k(\mathrm{d}E),$$

for suitable sets K, M and functions φ. A typical simple case arises if K and M are convex bodies and $\varphi = \chi$, the Euler characteristic. Since $\chi(K) = 1$ for nonempty convex bodies K and $\chi(\emptyset) = 0$, we have

$$\int_{G_d} \chi(K \cap gM)\, \mu(\mathrm{d}g) = \mu(\{g \in G_d : K \cap gM \neq \emptyset\}),$$

which is the total invariant measure of all rigid motions g for which the body gM hits (that is, has nonempty intersection with) the body K. In order to get an idea of what an explicit computation will involve, we first consider the special case where M is a ball of radius $r > 0$. By the representation of the invariant measure μ described above, we then have

$$\begin{aligned} \int_{G_d} \chi(K \cap grB^d)\, \mu(\mathrm{d}g) &= \int_{SO_d} \int_{\mathbb{R}^d} \chi(K \cap (\vartheta r B^d + x))\, \lambda(\mathrm{d}x)\, \nu(\mathrm{d}\vartheta) \\ &= V_d(K + rB^d), \end{aligned}$$

since $K \cap (\vartheta r B^d + x) \neq \emptyset$ if and only if x lies in the parallel body

$$K + rB^d = \{k + b : k \in K,\ b \in rB^d\}.$$

The **Steiner formula** of convex geometry (see (14.5)) tells us that

$$V_d(K + rB^d) = \sum_{j=0}^{d} r^{d-j} \kappa_{d-j} V_j(K),$$

where $V_0, \dots, V_d$ are the **intrinsic volumes**. Once it is known (and this is not difficult to prove) that $V_d(K + rB^d)$ is a polynomial in r, the Steiner formula can serve to define the intrinsic volumes. For these functionals and their properties, as well as for a local version of the Steiner formula, we refer to Section 14.3 and the literature quoted there.

We see already from this special case, $M = rB^d$, that for the computation of the integrals $\int_{G_d} \chi(K \cap gM)\,\mu(\mathrm{d}g)$ the intrinsic volumes must play an essential role. It is a remarkable fact that no further functions are needed for the general case: the integrals

$$\int_{G_d} \chi(K \cap gM)\,\mu(\mathrm{d}g) \qquad \text{and} \qquad \int_{A(d,k)} \chi(K \cap E)\,\mu_k(\mathrm{d}E)$$

can be expressed in terms of the intrinsic volumes of K and M. These results will be obtained as special cases of formulas involving more general functions φ in the integrands. The essential property of these integrand functions, which makes explicit formulas possible, is their additivity. Generally, a function φ on $\mathcal{K}'$ with values in an abelian group is **additive** if

$$\varphi(K \cup M) + \varphi(K \cap M) = \varphi(K) + \varphi(M)$$

for all $K, M \in \mathcal{K}'$ with $K \cup M \in \mathcal{K}'$. For an additive function φ on $\mathcal{K}'$, one always extends the definition by $\varphi(\emptyset) := 0$. A reader not familiar with additive functionals on convex bodies is advised to have a look at Section 14.4. We shall make essential use of Hadwiger's characterization theorem for the intrinsic volumes, which is proved in that section.

To obtain these formulas for more general integrands, we begin with computing the integral

$$\psi(K) := \int_{A(d,k)} V_j(K \cap E)\,\mu_k(\mathrm{d}E) \tag{5.3}$$

for convex bodies $K \in \mathcal{K}'$, where V_j is the jth intrinsic volume, $j \in \{0, \dots, d\}$. (Recall that V_j is additive, and that we have defined $V_j(\emptyset) = 0$.) Equation (5.3) defines a functional ψ on $\mathcal{K}'$. Since the intrinsic volume V_j is additive, invariant under rigid motions, and continuous, it is not difficult to show that the functional ψ is additive, motion invariant and continuous (for the continuity, compare the argument used in the proof of Theorem 5.1.2 below). Therefore, Hadwiger's characterization theorem (Theorem 14.4.6) yields a representation

$$\psi(K) = \sum_{r=0}^{d} c_r V_r(K), \qquad K \in \mathcal{K}',$$

with constant coefficients $c_0, \dots, c_d$. Here only one coefficient is different from zero, due to the homogeneity property

$$\psi(\alpha K) = \alpha^{d-k+j} \psi(K)$$

for $\alpha > 0$; this property follows from the representation

$$\psi(K) = \int_{G(d,k)} \int_{L^\perp} V_j(K \cap (L + x))\, \lambda_{d-k}(\mathrm{d}x)\, \nu_k(\mathrm{d}L).$$

Since V_r is homogeneous of degree r, we see that $c_r = 0$ for $r \neq d - k + j$, hence

$$\int_{A(d,k)} V_j(K \cap E)\, \mu_k(\mathrm{d}E) = cV_{d-k+j}(K)$$

with some constant c. In order to determine this constant, we take for K the unit ball B^d. For $\epsilon \geq 0$, the Steiner formula gives

$$\sum_{j=0}^{d} \epsilon^{d-j} \kappa_{d-j} V_j(B^d) = V_d(B^d + \epsilon B^d) = (1+\epsilon)^d \kappa_d = \sum_{j=0}^{d} \epsilon^{d-j} \binom{d}{j} \kappa_d,$$

hence

$$V_j(B^d) = \binom{d}{j} \frac{\kappa_d}{\kappa_{d-j}} \qquad \text{for } j = 0, \dots, d.$$

In the following, we make use of the fact that the intrinsic volume V_j of a convex body does not depend on the dimension of the space in which the body is embedded. Choosing $L \in G(d,k)$, we obtain

$$\begin{aligned}
cV_{d-k+j}(B^d) &= \int_{A(d,k)} V_j(B^d \cap E)\, \mu_k(\mathrm{d}E) \\
&= \int_{SO_d} \int_{L^\perp} V_j(B^d \cap \vartheta(L + x))\, \lambda_{d-k}(\mathrm{d}x)\, \nu(\mathrm{d}\vartheta) \\
&= \int_{L^\perp \cap B^d} (1 - \|x\|^2)^{j/2} V_j(B^d \cap L)\, \lambda_{d-k}(\mathrm{d}x) \\
&= \binom{k}{j} \frac{\kappa_k}{\kappa_{k-j}} \int_{L^\perp \cap B^d} (1 - \|x\|^2)^{j/2}\, \lambda_{d-k}(\mathrm{d}x).
\end{aligned}$$

Introducing polar coordinates, we transform the latter integral into a Beta integral and obtain

$$c = \binom{k}{j} \frac{\kappa_k \kappa_{d-k+j}}{V_{d-k+j}(B^d) \kappa_{k-j} \kappa_j} = c_j^k c_d^{d-k+j}.$$

Here we have denoted by

$$c_j^k := \frac{k!\kappa_k}{j!\kappa_j} \tag{5.4}$$

a frequently occurring constant. By using the identity

$$m!\kappa_m = 2^m \pi^{\frac{m-1}{2}} \Gamma\left(\frac{m+1}{2}\right),$$

it can also be put in the form

$$c_j^k = \frac{\Gamma\left(\frac{k+1}{2}\right)}{\Gamma\left(\frac{j+1}{2}\right)}.$$

To simplify later expressions, we also introduce the notation

$$c_{s_1,\dots,s_k}^{r_1,\dots,r_k} := \prod_{i=1}^{k} c_{s_i}^{r_i} = \prod_{i=1}^{k} \frac{r_i!\kappa_{r_i}}{s_i!\kappa_{s_i}}. \tag{5.5}$$

With these notations, we have obtained the following result.

Theorem 5.1.1. *Let $K \in \mathcal{K}'$ be a convex body. For $k \in \{1,\dots,d-1\}$ and $j \le k$ the* **Crofton formula**

$$\int_{A(d,k)} V_j(K \cap E)\, \mu_k(\mathrm{d}E) = c_{j,d}^{k,d-k+j} V_{d-k+j}(K) \tag{5.6}$$

holds.

The special case $j = 0$ of (5.6) gives

$$V_m(K) = c_{m,d-m}^{0,d} \int_{A(d,d-m)} \chi(K \cap E)\, \mu_{d-m}(\mathrm{d}E) \tag{5.7}$$

and thus provides an integral geometric interpretation of the intrinsic volumes: $V_m(K)$ is, up to a normalizing factor, the invariant measure of the set of $(d-m)$-flats intersecting K.

Using the explicit representation of the measure μ_{d-m} and the fact that the map $L \mapsto L^\perp$ transforms ν_{d-m} into ν_m, we can rewrite the representation (5.7) as

$$V_m(K) = c_{m,d-m}^{0,d} \int_{G(d,m)} \lambda_m(K|L)\, \nu_m(\mathrm{d}L), \tag{5.8}$$

where $K|L$ denotes the image of K under orthogonal projection to the subspace L. The special case $m = 1$ shows that V_1, up to a factor, is the mean width.

When we consider in the following a fixed and a moving convex body, we shall often have to exclude the touching positions. We say that the convex bodies K and M **touch** if $K \cap M \neq \emptyset$, but K and M can be separated weakly by a hyperplane. The following lemma is useful.

Lemma 5.1.1. *Let $K, M \in \mathcal{K}'$ be convex bodies, and let $(K_i)_{i\in\mathbb{N}}$, $(M_i)_{i\in\mathbb{N}}$ be sequences in $\mathcal{K}'$ with $K_i \to K$ and $M_i \to M$ for $i \to \infty$. Then the following holds:*

(a) *If $K \cap M = \emptyset$, then $K_i \cap M_i = \emptyset$ for all sufficiently large i.*

(b) *If $K \cap M \neq \emptyset$ and K and M do not touch, then $K_i \cap M_i \to K \cap M$ for $i \to \infty$.*

Proof. Assertion (a) follows immediately from the definition of convergence with respect to the Hausdorff metric.

To prove (b), let $x \in K \cap M$. We put $x_i := p(K_i \cap M_i, x)$ (the point in $K_i \cap M_i$ nearest to x, see Section 14.2) for those i for which $K_i \cap M_i \neq \emptyset$. We claim that x_i is defined for almost all i and that $x_i \to x$ for $i \to \infty$. Suppose this were false. Then there exists a ball B with center x such that $B \cap K_i \cap M_i = \emptyset$ holds for infinitely many i. For sufficiently large i we have $B \cap K_i \neq \emptyset$, since $K_i \to K$ and $x \in K$. By a standard separation theorem, for each such i there exists a hyperplane separating $B \cap K_i$ and M_i. A suitable subsequence of this sequence of hyperplanes converges to a hyperplane H; this hyperplane separates $B \cap K$ and M. Since $x \in K \cap M$, we have $x \in H$, hence H separates also K and M. This contradicts the assumption that K and M do not touch. It follows that $x_i \to x$ for $i \to \infty$.

Let $x_{i_j} \in K_{i_j} \cap M_{i_j}$ for some increasing sequence $(i_j)_{j \in \mathbb{N}}$, and assume that $x_{i_j} \to y$ for $j \to \infty$. Then $y \in K \cap M$.

The assertion $K_i \cap M_i \to K \cap M$ now follows from Theorem 12.2.2 (together with Theorem 12.3.4). □

From Hadwiger's characterization theorem, we now deduce a general kinematic formula, involving a functional on convex bodies that need not have any invariance property; crucial is the additivity of this functional. (Recall that an additive functional φ on $\mathcal{K}'$ is always extended to $\mathcal{K}$, by $\varphi(\emptyset) := 0$.)

Theorem 5.1.2 (Hadwiger's general integral geometric theorem). *If $\varphi : \mathcal{K}' \to \mathbb{R}$ is additive and continuous, then*

$$\int_{G_d} \varphi(K \cap gM)\,\mu(\mathrm{d}g) = \sum_{k=0}^{d} \varphi_{d-k}(K) V_k(M) \tag{5.9}$$

for $K, M \in \mathcal{K}'$, where the coefficients $\varphi_{d-k}(K)$ are given by

$$\varphi_{d-k}(K) = \int_{A(d,k)} \varphi(K \cap E)\,\mu_k(\mathrm{d}E).$$

Proof. The μ-integrability of the integrand in (5.9) is seen as follows. For $K, M \in \mathcal{K}'$, let $G_d(K, M)$ be the set of all motions $g \in G_d$ for which K and gM touch. It is not difficult to check that $\gamma(x, \vartheta) \in G_d(K, M)$ if and only if $x \in \mathrm{bd}\,(K - \vartheta M)$ and, hence, that $\mu(G_d(K, M)) = 0$.

Let $g \in G_d \setminus G_d(K, M)$, and let $(M_j)_{j \in \mathbb{N}}$ be a sequence in $\mathcal{K}'$ converging to M. Then $gM_j \to gM$ and hence $K \cap gM_j \to K \cap gM$, by Lemma 5.1.1, thus $\varphi(K \cap gM_j) \to \varphi(K \cap gM)$ for $j \to \infty$. It follows that the map $g \mapsto \varphi(K \cap gM)$ is continuous outside a closed set of μ-measure zero. Moreover, the continuous function φ is bounded on the compact set $\{L \in \mathcal{K}' : L \subset K\}$, and

$$\mu(\{g \in G_d : \varphi(K \cap gM) \neq 0\}) \leq \mu(\{g \in G_d : K \cap gM \neq \emptyset\}) < \infty.$$

This shows the μ-integrability of the function $g \mapsto \varphi(K \cap gM)$.

Now we fix a convex body $K \in \mathcal{K}'$ and define

$$\psi(M) := \int_{G_d} \varphi(K \cap gM)\,\mu(\mathrm{d}g) \qquad \text{for } M \in \mathcal{K}'.$$

Then $\psi : \mathcal{K}' \to \mathbb{R}$ is additive and motion invariant. The foregoing consideration together with the dominated convergence theorem shows that ψ is continuous. By Hadwiger's characterization theorem, there exist constants $\varphi_0(K), \dots, \varphi_d(K)$ such that

$$\psi(M) = \sum_{i=0}^{d} \varphi_{d-i}(K) V_i(M)$$

for all $M \in \mathcal{K}'$. We have to determine the coefficients $\varphi_{d-i}(K)$.

Let $k \in \{0, \dots, d\}$, and choose $L_k \in G(d,k)$. Let $C \subset L_k$ be a k-dimensional unit cube with center 0, and let $r > 0$. Then

$$\psi(rC) = \sum_{i=0}^{d} \varphi_{d-i}(K) V_i(rC) = \sum_{i=0}^{k} \varphi_{d-i}(K) r^i V_i(C).$$

On the other hand, using the rotation invariance of λ, we get

$$\begin{aligned}
&\psi(rC)\\
&= \int_{G_d} \varphi(K \cap grC)\,\mu(\mathrm{d}g)\\
&= \int_{SO_d} \int_{\mathbb{R}^d} \varphi(K \cap (\vartheta rC + x))\,\lambda(\mathrm{d}x)\,\nu(\mathrm{d}\vartheta)\\
&= \int_{SO_d} \int_{L_k^\perp} \int_{L_k} \varphi(K \cap (\vartheta rC + \vartheta x_1 + \vartheta x_2))\,\lambda_k(\mathrm{d}x_1)\,\lambda_{d-k}(\mathrm{d}x_2)\,\nu(\mathrm{d}\vartheta)\\
&= \int_{SO_d} \int_{L_k^\perp} \int_{L_k} \varphi(K \cap [\vartheta r(C + x_1) + \vartheta x_2]) r^k\,\lambda_k(\mathrm{d}x_1)\,\lambda_{d-k}(\mathrm{d}x_2)\,\nu(\mathrm{d}\vartheta).
\end{aligned}$$

Comparison gives

$$\begin{aligned}
&\varphi_{d-k}(K)\\
&= \lim_{r \to \infty} \int_{SO_d} \int_{L_k^\perp} \int_{L_k} \varphi(K \cap [\vartheta r(C + x_1) + \vartheta x_2])\,\lambda_k(\mathrm{d}x_1)\,\lambda_{d-k}(\mathrm{d}x_2)\,\nu(\mathrm{d}\vartheta).
\end{aligned}$$

For $r \to \infty$, we have

$$\varphi(K \cap [\vartheta r(C + x_1) + \vartheta x_2]) \to \begin{cases} \varphi(K \cap \vartheta(L_k + x_2)) & \text{if } 0 \in \operatorname{relint}(C + x_1),\\ 0 & \text{if } 0 \notin C + x_1. \end{cases}$$

Hence, the dominated convergence theorem yields

$$\varphi_{d-k}(K) = \int_{SO_d} \int_{L_k^\perp} \varphi(K \cap \vartheta(L_k + x_2))\lambda_k(C)\,\lambda_{d-k}(\mathrm{d}x_2)\,\nu(\mathrm{d}\vartheta)$$
$$= \int_{A(d,k)} \varphi(K \cap E)\,\mu_k(\mathrm{d}E),$$

as asserted. □

In Theorem 5.1.2 we can choose for φ, in particular, the intrinsic volume V_j. In this case, the Crofton formula (5.6) gives

$$(V_j)_{d-k}(K) = \int_{A(d,k)} V_j(K \cap E)\,\mu_k(\mathrm{d}E) = c_{j,d}^{k,d-k+j} V_{d-k+j}(K).$$

Hence, we obtain the following result.

Theorem 5.1.3. *Let $K, M \in \mathcal{K}'$ be convex bodies, and let $j \in \{0, \ldots, d\}$. Then the* **principal kinematic formula**

$$\int_{G_d} V_j(K \cap gM)\,\mu(\mathrm{d}g) = \sum_{k=j}^{d} c_{j,d}^{k,d-k+j} V_k(K) V_{d-k+j}(M) \tag{5.10}$$

holds.

We note that the special case $j = 0$, or

$$\int_{G_d} \chi(K \cap gM)\,\mu(\mathrm{d}g) = \sum_{k=0}^{d} c_{0,d}^{k,d-k} V_k(K) V_{d-k}(M),$$

gives the total measure of the set of rigid motions bringing M into a hitting position with K.

Hadwiger's general formula can be iterated, that is, extended to a finite number of moving convex bodies.

Theorem 5.1.4. *Let $\varphi : \mathcal{K}' \to \mathbb{R}$ be additive and continuous, and let $K_0, K_1, \ldots, K_k \in \mathcal{K}'$, $k \geq 1$, be convex bodies. Then*

$$\int_{G_d} \cdots \int_{G_d} \varphi(K_0 \cap g_1 K_1 \cap \ldots \cap g_k K_k)\,\mu(\mathrm{d}g_1) \cdots \mu(\mathrm{d}g_k)$$
$$= \sum_{\substack{r_0, \ldots, r_k = 0 \\ r_0 + \ldots + r_k = kd}}^{d} c_{d-r_0}^{d} \varphi_{r_0}(K_0) \prod_{i=1}^{k} c_d^{r_i} V_{r_i}(K_i),$$

where the coefficients are given by (5.4).

The specialization $\varphi = V_j$ yields the following.

Theorem 5.1.5 (Iterated kinematic formula). *Let $K_0, K_1, \ldots, K_k \in \mathcal{K}'$, $k \geq 1$, be convex bodies, and let $j \in \{0, \ldots, d\}$. Then*

$$\int_{G_d} \cdots \int_{G_d} V_j(K_0 \cap g_1 K_1 \cap \ldots \cap g_k K_k)\, \mu(\mathrm{d}g_1) \cdots \mu(\mathrm{d}g_k)$$

$$= \sum_{\substack{m_0,\ldots,m_k=j \\ m_0+\ldots+m_k=kd+j}}^{d} c_j^d \prod_{i=0}^{k} c_d^{m_i} V_{m_i}(K_i).$$

Proof. We prove Theorem 5.1.4. The proof proceeds by induction with respect to k. Theorem 5.1.2 is the case $k = 1$. Suppose that $k \geq 1$ and that the assertion of Theorem 5.1.4, and hence that of Theorem 5.1.5, has been proved for $k+1$ convex bodies. Let $K_0, \ldots, K_{k+1} \in \mathcal{K}'$. Using Fubini's theorem twice, the invariance of the measure μ, and Theorem 5.1.2 followed by Theorem 5.1.5 for $k+1$ convex bodies, we obtain

$$\int_{G_d} \cdots \int_{G_d} \varphi(K_0 \cap g_1 K_1 \cap \ldots \cap g_{k+1} K_{k+1})\, \mu(\mathrm{d}g_1) \cdots \mu(\mathrm{d}g_{k+1})$$

$$= \int_{G_d} \cdots \int_{G_d} \left[\int_{G_d} \varphi(K_0 \cap g_1(K_1 \cap g_2 K_2 \cap \ldots \cap g_{k+1} K_{k+1}))\, \mu(\mathrm{d}g_1) \right]$$
$$\times\, \mu(\mathrm{d}g_2) \cdots \mu(\mathrm{d}g_{k+1})$$

$$= \int_{G_d} \cdots \int_{G_d} \sum_{j=0}^{d} \varphi_{d-j}(K_0) V_j(K_1 \cap g_2 K_2 \cap \ldots \cap g_{k+1} K_{k+1})$$
$$\times\, \mu(\mathrm{d}g_2) \cdots \mu(\mathrm{d}g_{k+1})$$

$$= \sum_{j=0}^{d} c_j^d \varphi_{d-j}(K_0) \sum_{\substack{m_0,\ldots,m_k=j \\ m_0+\ldots+m_k=kd+j}}^{d} c_d^{m_0} \cdots c_d^{m_k} V_{m_0}(K_1) \cdots V_{m_k}(K_{k+1})$$

$$= \sum_{\substack{r_0,\ldots,r_{k+1}=0 \\ r_0+\ldots+r_{k+1}=(k+1)d}}^{d} c_{d-r_0}^d \varphi_{r_0}(K_0) c_d^{r_1} \cdots c_d^{r_{k+1}} V_{r_1}(K_1) \cdots V_{r_{k+1}}(K_{k+1}).$$

This completes the proof. □

Remark on renormalization. The preceding formulas suggest renormalization of the intrinsic volumes, by putting

$$\widetilde{V}_j := c_d^j V_j,$$

and also of the invariant measures on the affine Grassmannians, by putting

$$\widetilde{\mu}_k := c_k^d \mu_k.$$

Then the Crofton formula (5.6) becomes

$$\int_{A(d,k)} \widetilde{V}_j(K \cap E)\, \widetilde{\mu}_k(\mathrm{d}E) = \widetilde{V}_{d-k+j}(K).$$

Hadwiger's general integral geometric theorem reads

$$\int_{G_d} \varphi(K \cap gM)\, \mu(\mathrm{d}g) = \sum_{k=0}^{d} \widetilde{\varphi}_{d-k}(K)\widetilde{V}_k(M)$$

with

$$\widetilde{\varphi}_{d-k}(K) = \int_{A(d,k)} \varphi(K \cap E)\, \widetilde{\mu}_k(\mathrm{d}E).$$

In particular,

$$\widetilde{(\widetilde{V}_j)}_{d-k} = \widetilde{V}_{d-k+j}.$$

The principal kinematic formula (5.10) simplifies to

$$\int_{G_d} \widetilde{V}_j(K \cap gM)\, \mu(\mathrm{d}g) = \sum_{k=j}^{d} \widetilde{V}_k(K)\widetilde{V}_{d-k+j}(M). \tag{5.11}$$

Theorem 5.1.4 becomes

$$\int_{G_d} \cdots \int_{G_d} \varphi(K_0 \cap g_1 K_1 \cap \ldots \cap g_k K_k)\, \mu(\mathrm{d}g_1) \cdots \mu(\mathrm{d}g_k)$$

$$= \sum_{\substack{r_0,\ldots,r_k=0 \\ r_0+\ldots+r_k=kd}}^{d} \widetilde{\varphi}_{r_0}(K_0) \prod_{i=1}^{k} \widetilde{V}_{r_i}(K_i),$$

and the iterated kinematic formula attains the form

$$\int_{G_d} \cdots \int_{G_d} \widetilde{V}_j(K_0 \cap g_1 K_1 \cap \ldots \cap g_k K_k)\, \mu(\mathrm{d}g_1) \cdots \mu(\mathrm{d}g_k)$$

$$= \sum_{\substack{m_0,\ldots,m_k=j \\ m_0+\ldots+m_k=kd+j}}^{d} \prod_{i=0}^{k} \widetilde{V}_{m_i}(K_i).$$

Although these simplifications increase the elegance of the formulas, we retain the original normalization of the intrinsic volumes, since a different normalization might lead to confusion in several other instances.

Remark on extension to the convex ring. All the integral geometric formulas of this section remain valid if the involved convex bodies are replaced by polyconvex sets, that is, finite unions of convex bodies, and the involved additive functionals are replaced by their additive extensions to the convex ring. The simple principle of such extensions will be explained at the end of Section 5.2.

Notes for Section 5.1

1. Integral geometry as a subject of its own was first presented in two booklets by Blaschke in 1935 and 1937; a third edition [107] appeared in 1955 (see also vol. 2 of Blaschke's Collected Works [108]). The earlier development and its connection with geometric probability are subsumed in the book by Deltheil [202]. Introductions to integral geometry, from distinctly different points of view, were given by Santaló [659], Hadwiger [307, ch. 6], Stoka [738]. The standard source on integral geometry is the monograph by Santaló [662]. It stresses the applications to geometric probability. In a similar spirit is the book by Ren [635]. The book by Voss [772] describes integral geometry as a tool for stereology and image reconstruction.

The survey by Schneider and Wieacker [720] emphasizes the relations to convex geometry, and the article by Hug and Schneider [369] surveys integral geometric intersection formulas.

Combinatorial aspects of integral geometry are in the foreground of the original approaches in the books by Ambartzumian [34] and by Klain and Rota [416].

2. The principal kinematic formula (5.10) is a central result of classical integral geometry. For easier comparison with older literature, we write the special case $j = 0$ in terms of the Euler characteristic $\chi = V_0$ and the quermassintegrals, which are defined by (14.6). It then takes the form

$$\int_{G_d} \chi(K \cap gM)\,\mu(\mathrm{d}g) = \frac{1}{\kappa_d}\sum_{k=0}^{d}\binom{d}{k} W_k(K)W_{d-k}(M). \tag{5.12}$$

Here K and M can be arbitrary polyconvex sets. Often, only (5.12) is called the **principal kinematic formula**. It goes back to Blaschke and to Santaló, under different assumptions on the sets occurring in it. Hints to the origins can be found in the work of Blaschke [106], Hadwiger [307], Santaló [662].

When comparing with this literature, one has to observe that Santaló and Hadwiger normalize the invariant measure on the rotation group so that SO_d has total measure given by

$$c_d := \frac{d!}{2}\kappa_1 \cdots \kappa_d.$$

Under this normalization, the right side of (5.12) attains the additional factor c_d, and in Santaló's work the further factor 2, since Santaló integrates also over the improper rigid motions. (The constant O_k that often occurs in Santaló's work is given by $(k+1)\kappa_{k+1}$.)

If K is a convex body whose boundary is a regular (twice continuously differentiable) hypersurface, then

$$dW_i(K) = \int_{\mathrm{bd}\,K} H_{i-1}\,\mathrm{d}S =: M_{i-1}(\mathrm{bd}\,K) \qquad \text{for } i = 1, \ldots, d,$$

where H_{i-1} denotes the $(i-1)$th normalized elementary symmetric function of the principal curvatures of $\mathrm{bd}\,K$ (and M_{i-1} is the notation used by Santaló). With this interpretation of the functionals W_i as curvature integrals, equation (5.12) holds also if K and M are non-convex domains of $\mathbb{R}^d$ with boundary hypersurfaces of class C^2. This is the differential-geometric version of the principal kinematic formula. It goes back to Chern and Yien [171] and was proved with greater care by Chern [169]; see also Santaló [662, pp. 262 ff]. There (p. 269) one also finds a differential-geometric version of the formula

$$\int_{G_d} W_i(K \cap gM)\, \mu(\mathrm{d}g) = \frac{1}{\kappa_d} \sum_{k=d-i}^{d} C_{ki} W_{i+k-d}(K) W_{d-k}(M)$$

with

$$C_{ki} := \binom{i}{d-k} \frac{\kappa_k \kappa_i \kappa_{2d-k-i}}{\kappa_{d-k}\kappa_{d-i}\kappa_{k+i-d}}$$

($i = 1, \dots, d$). For elements of the convex ring, this is formula (5.10), rewritten in terms of the quermassintegrals W_i. Further kinematic intersection formulas in a differential-geometric version, valid for lower-dimensional compact differentiable submanifolds without boundary, are due to Chern [170]; see also Chapter V in the book by Sulanke and Wintgen [749]. This book, in contrast to [662], also provides the technical foundations that are required for using the elegant machinery of differential forms in the derivation of integral geometric formulas.

Also for the Crofton formulas, there are differential-geometric versions; one finds them in the quoted books of Sulanke and Wintgen and of Santaló.

A common generalization of kinematic formulas for convex bodies and for smooth compact submanifolds is Federer's extension to sets of positive reach; see Note 1 for Section 5.3.

3. The approach to integral geometric formulas for convex bodies that uses the axiomatic characterization of the intrinsic volumes, goes in principle back to W. Blaschke. It came into full force only when Hadwiger had proved his characterization theorem (Theorem 14.4.6). Hadwiger's general integral geometric theorem (Theorem 5.1.2) was proved in this way in [306, 307]; no other proof is currently known.

4. Hadwiger's general integral geometric theorem provides a kinematic formula for arbitrary additive continuous functions on convex bodies. For integrations over the translation group, an analogous result can be proved for simply additive functions. Let φ be a continuous real function on $\mathcal{K}'$ which is a simple valuation, that is, additive and satisfying $\varphi(K) = 0$ for convex bodies of dimension less than d. Then

$$\int_{\mathbb{R}^d} \varphi(K \cap (M + x))\, \lambda(\mathrm{d}x) = \varphi(K) V_d(M) + \int_{S^{d-1}} f_{K,\varphi}(u)\, S_{d-1}(M, \mathrm{d}u)$$

for convex bodies $K, M \in \mathcal{K}'$, where the function $f_{K,\varphi} : S^{d-1} \to \mathbb{R}$ is given by

$$f_{K,\varphi}(u) = \int_{-h(K,-u)}^{h(K,u)} \varphi(K \cap H^-(u,\tau))\, \mathrm{d}\tau - \varphi(K) h(K,u).$$

Here $h(K,\cdot)$ is the support function of K and $H^-(u,\tau)$ is the closed halfspace $\{x \in \mathbb{R}^d : \langle x, u \rangle \le \tau\}$; the measure $S_{d-1}(M,\cdot)$ is the surface area measure of M (see Section 14.2). This formula was proved by Schneider [708].

5. Alesker's work on valuations (see [22] for a survey) has also shed new light on kinematic formulas and their generalizations. Part of his work extends Hadwiger's characterization theorem and its integral geometric applications. Let G be a compact subgroup of the orthogonal group O_d acting transitively on the unit sphere S^{d-1}. Let $\mathbf{Val}^G$ denote the vector space of continuous, translation invariant and G-invariant real valuations on $\mathcal{K}$. Alesker [18] has shown that $\mathbf{Val}^G$ has finite dimension. In [19], he provided explicit bases for the case of $G = U(n)$ (where $\mathbb{C}^n$ is identified with $\mathbb{R}^{2n}$), thus establishing a unitary counterpart to Hadwiger's characterization theorem (with a much deeper proof, though). For the case of $SU(2)$, see Alesker [21]. Further, Alesker [20] has introduced a multiplication for continuous, translation invariant valuations. With this, $\mathbf{Val}^G$ becomes a graded algebra over $\mathbb{R}$, satisfying the Poincaré duality. The structure of this algebra was determined for the case $G = SO_d$ by Alesker [20], and for $G = U(d)$ by Fu [239].

Applications to kinematic formulas, where the role of the rotation group in the classical case is now played by a group G as above, were investigated by Alesker [19], Fu [239], Bernig and Fu [95, 96], Bernig [92].

6. Iterations of the principal kinematic formula, as in Theorem 5.1.5, were used, for instance, by Streit [747].

7. As explained in the remark on renormalization, the intrinsic volumes and the invariant measures on the Grassmannians can be renormalized so that the principal kinematic formula takes the particularly simple form (5.11), where all the coefficients of the bilinear expression are equal to one. Some authors have elaborated upon this fact from a structural point of view; see Nijenhuis [585] and Fu [239].

Questions of normalization with desirable properties are also an issue in the book by Klain and Rota [416].

5.2 Translative Integral Formulas

Our next major aim is an extension of the principal kinematic formula (5.10) to the curvature measures Φ_m, which are introduced in Section 14.2. Thus, we want to compute the integral

$$\begin{aligned}&\int_{G_d} \Phi_j(K \cap gM, A \cap gB)\, \mu(\mathrm{d}g)\\ &= \int_{SO_d} \int_{\mathbb{R}^d} \Phi_j(K \cap (\vartheta M + x), A \cap (\vartheta B + x))\, \lambda(\mathrm{d}x)\, \nu(\mathrm{d}\vartheta)\end{aligned}$$

for convex bodies K, M and Borel sets $A, B \in \mathcal{B}(\mathbb{R}^d)$ (recall that $\Phi_j(\emptyset, \cdot) = 0$, by definition). The result will be stated in Theorem 5.3.2. In this section we study only the inner integral

$$\int_{\mathbb{R}^d} \Phi_j(K \cap (M + x), A \cap (B + x))\, \lambda(\mathrm{d}x). \tag{5.13}$$

Integrals of this type can be considered to extend over the translation group of $\mathbb{R}^d$ and are therefore known as **translative integrals**. The computation of (5.13) is the first step towards a direct proof of the kinematic formula for curvature measures, but is also of independent interest, in view of later applications to non-isotropic stochastic models.

The integral (5.13) is easily computed for $j = d$. This is a special case of a simple but often useful integral geometric formula with respect to the translation group, which can be obtained without much effort (and was already used in Sections 4.5 and 4.6). It is quite general and is an immediate consequence of the translation and inversion invariance of the Lebesgue measure.

Theorem 5.2.1. *If α is a σ-finite measure on $\mathbb{R}^d$ and if $A, B \in \mathcal{B}(\mathbb{R}^d)$, then*

$$\int_{\mathbb{R}^d} \alpha(A \cap (B+t))\, \lambda(\mathrm{d}t) = \alpha(A)\lambda(B).$$

Proof. Fubini's theorem gives

$$\begin{aligned}\int_{\mathbb{R}^d} \alpha(A \cap (B+t))\, \lambda(\mathrm{d}t) &= \int_{\mathbb{R}^d}\int_{\mathbb{R}^d} \mathbf{1}_A(x)\mathbf{1}_{B+t}(x)\, \alpha(\mathrm{d}x)\, \lambda(\mathrm{d}t)\\ &= \int_{\mathbb{R}^d} \mathbf{1}_A(x) \int_{\mathbb{R}^d} \mathbf{1}_{-B+x}(t)\, \lambda(\mathrm{d}t)\, \alpha(\mathrm{d}x)\\ &= \int_{\mathbb{R}^d} \mathbf{1}_A(x)\lambda(-B+x)\, \alpha(\mathrm{d}x)\\ &= \alpha(A)\lambda(B),\end{aligned}$$

as asserted. □

The formulas

$$\int_{\mathbb{R}^d} V_d(K \cap (M+x))\, \lambda(\mathrm{d}x) = V_d(K)V_d(M) \tag{5.14}$$

and

$$\int_{\mathbb{R}^d} V_{d-1}(K \cap (M+x))\, \lambda(\mathrm{d}x) = V_{d-1}(K)V_d(M) + V_d(K)V_{d-1}(M) \tag{5.15}$$

for convex bodies $K, M \in \mathcal{K}'$ are special cases of Theorem 5.2.1. The second is obtained by applying the theorem twice, taking for the measure α the $(d-1)$-dimensional Hausdorff measure, restricted to the boundary of one of the bodies. We do not carry this out here, since we shall give a detailed proof of the much more general Theorem 5.2.3.

The corresponding translative formulas for the intrinsic volumes V_j with $j < d-1$ are no longer so simple as (5.14) and (5.15). This is already seen from the case $j = 0$. Since $V_0(K \cap (M+x)) = 1$ is equivalent to $K \cap (M+x) \neq \emptyset$ and hence to $x \in K - M$, we have

$$\int_{\mathbb{R}^d} V_0(K \cap (M+x))\, \lambda(\mathrm{d}x) = V_d(K-M).$$

Similarly to the case of the Steiner formula, the volume of a sum $K + \epsilon L$, $\epsilon \geq 0$, for convex bodies $K, L \in \mathcal{K}'$, can be expanded into a polynomial in ϵ (see (14.17)). In this way, one obtains

$$\int_{\mathbb{R}^d} V_0(K \cap (M + x))\, \lambda(\mathrm{d}x) = \sum_{k=0}^{d} \binom{d}{k} V(\underbrace{K, \dots, K}_{k}, \underbrace{-M, \dots, -M}_{d-k}), \tag{5.16}$$

where V on the right side denotes a mixed volume. We see that the result involves functionals that depend on K and M simultaneously. It is in general not possible to separate the roles of K and M, as was the case with the principal kinematic formula (5.10). There, the resulting bilinear form owes its existence to the further integration over the rotation group. The occurrence of simultaneous functionals is typical for translative integral geometry.

Our proof of translative and kinematic formulas for curvature measures is prepared by a measurability lemma. We recall from the proof of Theorem 5.1.2 that the set $G_d(K, M) = \{g \in G_d : K \text{ and } M \text{ touch}\}$ satisfies $\mu(G_d(K, M)) = 0$.

Lemma 5.2.1. *Let $K, M \in \mathcal{K}'$ and $A, B \in \mathcal{B}(\mathbb{R}^d)$, let $j \in \{0, \dots, d\}$. The mapping*

$$x \mapsto \Phi_j(K \cap (M + x), A \cap (B + x)), \qquad x \in \mathbb{R}^d,$$

is measurable on $\mathbb{R}^d \setminus \mathrm{bd}\,(K - M)$, where $\lambda(\mathrm{bd}\,(K - M)) = 0$.

The mapping

$$g \mapsto \Phi_j(K \cap gM, A \cap gB), \qquad g \in G_d,$$

is measurable on $G_d \setminus G_d(K, M)$, where $\mu(G_d(K, M)) = 0$.

Proof. It suffices to prove the second assertion, since the proof of the first one is analogous. For fixed $(x, \vartheta) \in \mathbb{R}^d \times SO_d$, we define

$$\begin{aligned} T_{x,\vartheta} : \mathbb{R}^d &\to \quad \mathbb{R}^d \times \mathbb{R}^d \\ y &\mapsto (y, \vartheta^{-1}(y - x)) \end{aligned}$$

and the image measure

$$\varphi^{(j)}(x, \vartheta, K, M, \cdot) := T_{x,\vartheta}(\Phi_j(K \cap (\vartheta M + x), \cdot)).$$

Then $\varphi^{(j)}(x, \vartheta, K, M, \cdot)$ is a finite measure on $\mathbb{R}^d \times \mathbb{R}^d$, and

$$\varphi^{(j)}(x, \vartheta, K, M, A \times B) = \Phi_j(K \cap (\vartheta M + x), A \cap (\vartheta B + x))$$

for $A, B \in \mathcal{B}(\mathbb{R}^d)$. By the transformation formula for integrals,

$$\begin{aligned} &\int_{\mathbb{R}^d \times \mathbb{R}^d} f(y, z)\, \varphi^{(j)}(x, \vartheta, K, M, \mathrm{d}(y, z)) \\ &= \int_{\mathbb{R}^d} f(y, \vartheta^{-1}(y - x))\, \Phi_j(K \cap (\vartheta M + x), \mathrm{d}y) \end{aligned}$$

for $f \in \mathbf{C}(\mathbb{R}^d \times \mathbb{R}^d)$. By Lemma 5.1.1, the mapping $(x, \vartheta) \mapsto K \cap (\vartheta M + x)$ is continuous outside the set $\gamma^{-1}(G_d(K, M))$, hence, by Theorem 14.2.2(c) the mapping $(x, \vartheta) \mapsto \Phi_j(K \cap (\vartheta M + x), \cdot)$ is continuous (with respect to the weak topology) on $(\mathbb{R}^d \times SO_d) \setminus \gamma^{-1}(G_d(K, M))$. For $f \in \mathbf{C}(\mathbb{R}^d \times \mathbb{R}^d)$ and for $(x_i, \vartheta_i) \to (x_0, \vartheta_0) \notin \gamma^{-1}(G_d(K, M))$ we deduce that

$$\int_{\mathbb{R}^d} f(y, \vartheta_i^{-1}(y - x_i))\, \Phi_j(K \cap (\vartheta_i M + x_i), \mathrm{d}y)$$

$$\to \int_{\mathbb{R}^d} f(y, \vartheta_0^{-1}(y - x_0))\, \Phi_j(K \cap (\vartheta_0 M + x_0), \mathrm{d}y)$$

(since $\Phi_j(K \cap (\vartheta_i M + x_i), \cdot)$ vanishes outside a suitable compact set independent of i, and since f is uniformly continuous on any compact set). Therefore, the mapping

$$(x, \vartheta) \mapsto \int_{\mathbb{R}^d \times \mathbb{R}^d} f(y, z)\, \varphi^{(j)}(x, \vartheta, K, M, \mathrm{d}(y, z))$$

is continuous on $\mathbb{R}^d \setminus \gamma^{-1}(G_d(K, M))$. As shown in Lemma 12.1.1, this implies the measurability of the mapping

$$(x, \vartheta) \mapsto \varphi^{(j)}(x, \vartheta, K, M, A \times B) = \Phi_j(K \cap (\vartheta M + x), A \cap (\vartheta B + x))$$

on $\mathbb{R}^d \setminus \gamma^{-1}(G_d(K, M))$, for arbitrary $A, B \in \mathcal{B}(\mathbb{R}^d)$. □

In the following, we shall have to use the subspace determinant $[L_1, \dots, L_k]$, which is introduced in Section 14.1. We extend its definition as follows. If $A_1, \dots, A_k \subset \mathbb{R}^d$ are nonempty subsets, we denote by $L(A_i)$ the linear subspace which is a translate of the affine hull of A_i, and we write

$$[A_1, \dots, A_k] := [L(A_1), \dots, L(A_k)]$$

if the latter is defined.

First we investigate a translative formula for polytopes. For the external angles, we refer to (14.10). For polytopes $K, M \in \mathcal{P}'$ and for faces F of K and G of M we define a **common external angle** by

$$\gamma(F, G; K, M) := \gamma(F \cap (G + x), K \cap (M + x)),$$

where $x \in \mathbb{R}^d$ is chosen so that

$$\operatorname{relint} F \cap \operatorname{relint}(G + x) \neq \emptyset.$$

Obviously, this definition does not depend on the special choice of x.

Two faces F and G of a polytope are said to be in **special position** if the linear subspaces $L(F)$ and $L(G)$ parallel to F and G are in special position, that is, satisfy

$$L(F) \cap L(G) \neq \{0\} \quad \text{and} \quad \text{lin}\,(L(F) \cup L(G)) \neq \mathbb{R}^d.$$

For a face F of a polytope, the measure λ_F is defined by

$$\lambda_F(A) := \lambda_{\dim F}(A \cap F) \qquad \text{for } A \in \mathcal{B}(\mathbb{R}^d).$$

Theorem 5.2.2. *If $K, M \in \mathcal{P}'$ are polytopes, $A, B \in \mathcal{B}(\mathbb{R}^d)$ are Borel sets and $j \in \{0, \ldots, d\}$, then*

$$\int_{\mathbb{R}^d} \Phi_j(K \cap (M + x), A \cap (B + x))\, \lambda(\mathrm{d}x) = \sum_{k=j}^{d} \Phi_k^{(j)}(K, M; A \times B)$$

with finite measures $\Phi_k^{(j)}(K, M; \cdot)$ on $\mathbb{R}^d \times \mathbb{R}^d$, which are defined by

$$\Phi_k^{(j)}(K, M; \cdot) := \sum_{F \in \mathcal{F}_k(K)} \sum_{G \in \mathcal{F}_{d-k+j}(M)} \gamma(F, G; K, M)[F, G] \lambda_F \otimes \lambda_G$$

$(k = j, \ldots, d)$. In particular,

$$\Phi_j^{(j)}(K, M; A \times B) = \Phi_j(K, A) \Phi_d(M, B),$$
$$\Phi_d^{(j)}(K, M; A \times B) = \Phi_d(K, A) \Phi_j(M, B).$$

Proof. Let

$$I := \int_{\mathbb{R}^d} \Phi_j(K \cap (M + x), A \cap (B + x))\, \lambda(\mathrm{d}x).$$

By Lemma 5.2.1, this is well defined. The representation (14.13) gives

$$I = \int_{\mathbb{R}^d} \sum_{F' \in \mathcal{F}_j(K \cap (M+x))} \gamma(F', K \cap (M + x)) \lambda_{F'}(A \cap (B + x))\, \lambda(\mathrm{d}x).$$

The faces $F' \in \mathcal{F}_j(K \cap (M + x))$ are precisely the j-dimensional sets of the form $F' = F \cap (G + x)$, where $F \in \mathcal{F}_k(K)$ and $G \in \mathcal{F}_i(M)$ for suitable $k, i \in \{j, \ldots, d\}$. For the computation of the integral I, only those vectors x are relevant that together with $F \cap (G + x) \neq \emptyset$ for a pair F, G satisfy $\text{relint}\, F \cap \text{relint}\,(G + x) \neq \emptyset$, since the remaining vectors x form a set of Lebesgue measure zero. Moreover, the pairs F, G for which $k + i < d$ or which are in special position, do not contribute to the integral, since for these we have

$$\lambda(\{x \in \mathbb{R}^d : F \cap (G + x) \neq \emptyset\}) = \lambda(F - G) = 0.$$

In the other cases, $\dim F' = \dim F + \dim G - d$, hence $k + i = d + j$. Thus, we obtain

$$I = \sum_{k=j}^{d} \sum_{F \in \mathcal{F}_k(K)} \sum_{G \in \mathcal{F}_{d-k+j}(M)}$$

$$\int_{\mathbb{R}^d} \gamma(F \cap (G+x), K \cap (M+x)) \lambda_{F \cap (G+x)}(A \cap (B+x)) \, \lambda(\mathrm{d}x)$$

$$= \sum_{k=j}^{d} \sum_{F \in \mathcal{F}_k(K)} \sum_{G \in \mathcal{F}_{d-k+j}(M)} \gamma(F, G; K, M) J(F, G)$$

with

$$J(F, G) := \int_{\mathbb{R}^d} \lambda_{F \cap (G+x)}(A \cap (B+x)) \, \lambda(\mathrm{d}x).$$

For the computation of $J(F, G)$ we suppose, without loss of generality, that

$$0 \in L_1 := \operatorname{aff} F \cap \operatorname{aff} G.$$

Let

$$L_2 := L_1^{\perp} \cap \operatorname{aff} F, \qquad L_3 := L_1^{\perp} \cap \operatorname{aff} G,$$

and let $\lambda_j, \lambda_{k-j}, \lambda_{d-k}$ be the Lebesgue measure on L_1, L_2, L_3, respectively. Then $\mathbb{R}^d = L_1 \oplus L_2 \oplus L_3$, and $x \in \mathbb{R}^d$ can uniquely be written in the form $x = x_1 + x_2 + x_3$ with $x_i \in L_i$ for $i = 1, 2, 3$. Writing $A' := A \cap F$, $B' := B \cap G$, we have

$$J(F, G) = [F, G] \int_{L_3} \int_{L_2} \int_{L_1} \lambda_{F \cap (G + x_1 + x_2 + x_3)}(A' \cap (B' + x_1 + x_2 + x_3))$$
$$\times \lambda_j(\mathrm{d}x_1) \, \lambda_{k-j}(\mathrm{d}x_2) \, \lambda_{d-k}(\mathrm{d}x_3).$$

Since

$$(A' \cap (B' + x_1 + x_2 + x_3)) - x_2 = (A' - x_2) \cap (B' + x_1 + x_3) \subset L_1,$$

we obtain

$$\int_{L_1} \lambda_{F \cap (G + x_1 + x_2 + x_3)}(A' \cap (B' + x_1 + x_2 + x_3)) \, \lambda_j(\mathrm{d}x_1)$$
$$= \int_{L_1} \lambda_j((A' - x_2) \cap (B' + x_3 + x_1)) \, \lambda_j(\mathrm{d}x_1)$$
$$= \lambda_j((A' - x_2) \cap L_1) \lambda_j((B' + x_3) \cap L_1),$$

by Theorem 5.2.1. Fubini's theorem yields

$$\int_{L_2} \lambda_j((A' - x_2) \cap L_1) \, \lambda_{k-j}(\mathrm{d}x_2) = \lambda_j \otimes \lambda_{k-j}(A') = \lambda_F(A)$$

and

$$\int_{L_3} \lambda_j((B' + x_3) \cap L_1)\, \lambda_{d-k}(\mathrm{d}x_3) = \lambda_j \otimes \lambda_{d-k}(B') = \lambda_G(B).$$

Altogether this gives

$$J(F, G) = [F, G] \lambda_F(A) \lambda_G(B),$$

and thus the representation of the measure $\Phi_k^{(j)}(K, M; \cdot)$ as stated in the theorem.

In the special case $k = j$ we have

$$\begin{aligned}
&\sum_{F \in \mathcal{F}_k(K)} \sum_{G \in \mathcal{F}_{d-k+j}(M)} \gamma(F, G; K, M)[F, G] \lambda_F \otimes \lambda_G \\
&= \sum_{F \in \mathcal{F}_j(K)} \gamma(F, M; K, M)[F, M] \lambda_F \otimes \lambda_M \\
&= \sum_{F \in \mathcal{F}_j(K)} \gamma(F, K) \lambda_F \otimes \lambda_M \\
&= \Phi_j(K, \cdot) \otimes \Phi_d(M, \cdot).
\end{aligned}$$

Similarly, for $k = d$ we obtain the measure $\Phi_d(K, \cdot) \otimes \Phi_j(M, \cdot)$. □

Corollary 5.2.1. *If $K, M \in \mathcal{P}'$ are polytopes and $j \in \{0, \dots, d\}$, then*

$$\begin{aligned}
&\int_{\mathbb{R}^d} V_j(K \cap (M + x))\, \lambda(\mathrm{d}x) \\
&= V_j(K) V_d(M) + \sum_{k=j+1}^{d-1} V_k^{(j)}(K, M) + V_d(K) V_j(M),
\end{aligned}$$

where

$$V_k^{(j)}(K, M) := \sum_{F \in \mathcal{F}_k(K)} \sum_{G \in \mathcal{F}_{d-k+j}(M)} \gamma(F, G; K, M)[F, G] V_k(F) V_{d-k+j}(G).$$

Theorem 5.2.2 and Corollary 5.2.1 will now be extended, by means of approximation, to arbitrary convex bodies $K, M \in \mathcal{K}'$. In contrast to the case of polytopes, for the measures $\Phi_k^{(j)}(K, M; \cdot)$ and functionals $V_k^{(j)}(K, M)$ that occur, no simple explicit representations are known in the general case.

Theorem 5.2.3. *For convex bodies $K, M \in \mathcal{K}'$ and for $j \in \{0, \dots, d\}$, there exist finite measures $\Phi_{j+1}^{(j)}(K, M; \cdot), \dots, \Phi_{d-1}^{(j)}(K, M; \cdot)$ on $\mathbb{R}^d \times \mathbb{R}^d$, concentrated on $\mathrm{bd}\, K \times \mathrm{bd}\, M$, such that*

$$\begin{aligned}
&\int_{\mathbb{R}^d} \Phi_j(K \cap (M + x), A \cap (B + x))\, \lambda(\mathrm{d}x) \qquad (5.17) \\
&= \Phi_j(K, A) \Phi_d(M, B) + \sum_{k=j+1}^{d-1} \Phi_k^{(j)}(K, M; A \times B) + \Phi_d(K, A) \Phi_j(M, B)
\end{aligned}$$

for all $A, B \in \mathcal{B}(\mathbb{R}^d)$. *In particular,*

$$\int_{\mathbb{R}^d} V_j(K \cap (M+x))\, \lambda(\mathrm{d}x) = V_j(K)V_d(M) + \sum_{k=j+1}^{d-1} V_k^{(j)}(K,M) + V_d(K)V_j(M)$$

with $V_k^{(j)}(K,M) := \Phi_k^{(j)}(K,M;\mathbb{R}^d \times \mathbb{R}^d)$.

The measure $\Phi_k^{(j)}(K,M;\cdot)$ *depends continuously on* $K, M \in \mathcal{K}'$ *and is homogeneous of degree* k *in* K *and of degree* $d-k+j$ *in* M. *It is additive in each of its first two arguments. For polytopes* K, M, *the measure* $\Phi_k^{(j)}(K,M;\cdot)$ *coincides with the one appearing in Theorem* 5.2.2.

Proof. As was already verified in the proof of Theorem 5.2.2, the integrand on the left side of (5.17) is measurable for λ-almost all x, hence the integral in (5.17) is well defined. We now first remark that equality (5.17) is equivalent to

$$\int_{\mathbb{R}^d}\int_{\mathbb{R}^d} f(x, x-y)\, \Phi_j(K \cap (M+y), \mathrm{d}x)\, \lambda(\mathrm{d}y)$$

$$= \sum_{k=j}^{d} \int_{\mathbb{R}^d \times \mathbb{R}^d} f(x,y)\, \Phi_k^{(j)}(K,M;\mathrm{d}(x,y)) \tag{5.18}$$

for all continuous functions f on $\mathbb{R}^d \times \mathbb{R}^d$, provided that the measures $\Phi_k^{(j)}(K,M;\cdot)$ exist; here we have written

$$\Phi_j^{(j)}(K,M;\cdot) := \Phi_j(K,\cdot) \otimes \Phi_d(M,\cdot),$$
$$\Phi_d^{(j)}(K,M;\cdot) := \Phi_d(K,\cdot) \otimes \Phi_j(M,\cdot).$$

In fact, if (5.17) holds, then (5.18) is true for $f = \mathbf{1}_{A\times B}$, hence (5.18) follows for elementary functions and then, by a standard argument, for integrable functions. If (5.18) holds, then (5.17) is obtained for compact sets A, B, since $\mathbf{1}_{A\times B}$ is in this case the limit of a decreasing sequence of continuous functions, and for arbitrary Borel sets it then follows since both sides, as functions of A and B, are measures.

By Theorem 5.2.2, formulas (5.17) and (5.18) are valid if K and M are polytopes.

For convex bodies $K, M \in \mathcal{K}'$ and for a continuous function f on $\mathbb{R}^d \times \mathbb{R}^d$ we now define

$$J(f,K,M) := \int_{\mathbb{R}^d}\int_{\mathbb{R}^d} f(x, x-y)\, \Phi_j(K \cap (M+y), \mathrm{d}x)\, \lambda(\mathrm{d}y).$$

We show that $J(f,K,M)$ depends continuously on K and M. For this, let $K_i \to K$, $M_i \to M$ be convergent sequences in $\mathcal{K}'$. From Lemma 5.1.1 and Theorem 14.2.2(c) we infer the weak convergence

$$\Phi_j(K_i \cap (M_i + y), \cdot) \xrightarrow{w} \Phi_j(K \cap (M + y), \cdot)$$

and, therefore, the pointwise convergence

$$\int_{\mathbb{R}^d} f(x, x-y)\, \Phi_j(K_i \cap (M_i + y), \mathrm{d}x) \to \int_{\mathbb{R}^d} f(x, x-y)\, \Phi_j(K \cap (M + y), \mathrm{d}x)$$

for $i \to \infty$, for all $y \notin \operatorname{bd}(K - M)$. From this we deduce that

$$\begin{aligned}
&\lim_{i\to\infty} J(f, K_i, M_i) \\
&= \int_{\mathbb{R}^d} \left(\lim_{i\to\infty} \int_{\mathbb{R}^d} f(x, x-y)\, \Phi_j(K_i \cap (M_i + y), \mathrm{d}x) \right) \lambda(\mathrm{d}y) \\
&= \int_{\mathbb{R}^d} \int_{\mathbb{R}^d} f(x, x-y)\, \Phi_j(K \cap (M + y), \mathrm{d}x)\, \lambda(\mathrm{d}y) \\
&= J(f, K, M).
\end{aligned}$$

Here we have applied the dominated convergence theorem. This is legitimate, since we can find a λ-integrable function of y dominating

$$\left| \int_{\mathbb{R}^d} f(x, x-y)\, \Phi_j(K_i \cap (M_i + y), \mathrm{d}x) \right| \tag{5.19}$$

for all i. To see this, we choose a ball rB^d, $r > 0$, containing all K_i, M_i and hence also K and M, and denote by $\|f\|_r$ the maximum of the continuous function f on $rB^d \times rB^d$. Then

$$\left| \int_{\mathbb{R}^d} f(x, x-y)\, \Phi_j(K_i \cap (M_i + y), \mathrm{d}x) \right| \le \|f\|_r V_j(K_i \cap (M_i + y)).$$

The monotonicity of the intrinsic volumes gives

$$V_j(K_i \cap (M_i + y)) \le V_j(K_i) \mathbf{1}_{K_i - M_i}(y),$$

and this yields the required function dominating (5.19).

For $r, s > 0$ we now define a continuous mapping $D_{r,s}$ from $\mathbb{R}^d \times \mathbb{R}^d$ into itself by

$$D_{r,s}(x, y) := \left(\frac{x}{r}, \frac{y}{s} \right) \qquad \text{for } x, y \in \mathbb{R}^d.$$

If K and M are polytopes, (5.18) gives

$$\begin{aligned}
D_{r,s} J(f, K, M) &:= \int_{\mathbb{R}^d} \int_{\mathbb{R}^d} f\left(\frac{x}{r}, \frac{x-y}{s} \right) \Phi_j(K \cap (M + y), \mathrm{d}x)\, \lambda(\mathrm{d}y) \\
&= \sum_{k=j}^{d} \int_{\mathbb{R}^d \times \mathbb{R}^d} f\left(\frac{x}{r}, \frac{y}{s} \right) \Phi_k^{(j)}(K, M; \mathrm{d}(x, y)) \\
&= \sum_{k=j}^{d} \int_{\mathbb{R}^d \times \mathbb{R}^d} f(x, y)\, D_{r,s}\left(\Phi_k^{(j)}(K, M; \cdot) \right) (\mathrm{d}(x, y)).
\end{aligned}$$

For the polytopes rK and sM, the image measure $D_{r,s}\left(\Phi_k^{(j)}(rK,sM;\cdot)\right)$ can be determined by means of the formula in Theorem 5.2.2; this yields

$$D_{r,s}\left(\Phi_k^{(j)}(rK,sM;\cdot)\right) = r^k s^{d-k+j}\Phi_k^{(j)}(K,M;\cdot).$$

For given convex bodies K, M we now choose polytopes K_i, M_i ($i \in \mathbb{N}$) so that $K_i \to K$ and $M_i \to M$ for $i \to \infty$. Then it follows that

$$D_{r,s}J(f,rK_i,sM_i) \to D_{r,s}J(f,rK,sM)$$

for every continuous function f on $\mathbb{R}^d \times \mathbb{R}^d$ and all $r, s > 0$. As we have just seen,

$$\begin{aligned} &D_{r,s}J(f,rK_i,sM_i) \\ &= \sum_{k=j}^{d} r^k s^{d-k+j} \int_{\mathbb{R}^d\times\mathbb{R}^d} f(x,y)\,\Phi_k^{(j)}(K_i,M_i;\mathrm{d}(x,y)). \end{aligned} \tag{5.20}$$

We deduce the convergence of the coefficients

$$\int_{\mathbb{R}^d\times\mathbb{R}^d} f(x,y)\,\Phi_k^{(j)}(K_i,M_i;\mathrm{d}(x,y))$$

in the polynomial (5.20) and thus the weak convergence of the measures

$$\Phi_k^{(j)}(K_i,M_i;\cdot), \qquad k = j,\dots,d,$$

for $i \to \infty$. The limits, denoted by $\Phi_k^{(j)}(K,M;\cdot)$, $k = j,\dots,d$, are again finite measures, satisfying

$$D_{r,s}J(f,rK,sM) = \sum_{k=j}^{d} r^k s^{d-k+j} \int_{\mathbb{R}^d\times\mathbb{R}^d} f(x,y)\,\Phi_k^{(j)}(K,M;\mathrm{d}(x,y)), \tag{5.21}$$

from which we see that they are independent of the approximating sequences $(K_i)_{i\in\mathbb{N}}$, $(M_i)_{i\in\mathbb{N}}$. For $r = s = 1$ we obtain (5.18).

From the polynomial expansion (5.21), we also deduce that $\Phi_k^{(j)}(K,M;\cdot)$ depends continuously on K and M. That $\Phi_{j+1}^{(j)}(K,M;\cdot),\dots,\Phi_{d-1}^{(j)}(K,M;\cdot)$ are concentrated on $\operatorname{bd} K \times \operatorname{bd} M$ is a consequence of Theorem 5.2.2, if K and M are polytopes, and for general convex bodies K, M it is obtained by approximation with polytopes. The stated homogeneity properties are obvious for polytopes, and for the general case they follow by approximation. The additivity of $\Phi_k^{(j)}$ in any of its first two arguments follows immediately from the expansion (5.17), if one uses the additivity of Φ_j in its first argument and then compares summands of equal degrees of homogeneity. □

We supplement the definition by $\Phi_k^{(j)}(K, M; \cdot) = 0$ if $K = \emptyset$ or $M = \emptyset$. In Section 6.4 we shall extend $\Phi_m^{(j)}(K, M; \cdot)$ to more than two convex bodies; these functions are then called mixed measures.

Additive Extension

The integral geometric formulas obtained so far are not restricted to convex bodies, but can be extended to sets of the convex ring $\mathcal{R}$, by means of additivity. First we note that the curvature measure Φ_j, as a function of its first argument, has an additive extension to $\mathcal{R}$. This follows from Groemer's extension theorem (Theorem 14.4.2), since Φ_j is additive on $\mathcal{K}'$ and is continuous as a map from $\mathcal{K}'$ into the vector space of finite signed measures on $\mathbb{R}^d$ with the weak topology. The extension is denoted by the same symbol. In a similar way, the function $\Phi_k^{(j)}$ can be extended. First we fix a convex body $M \in \mathcal{K}'$. By the same argument as just used, $\Phi_k^{(j)}(\cdot, M; \cdot)$ as a function of its first argument has an additive extension to the convex ring $\mathcal{R}$; we denote it by the same symbol. Next, we fix a polyconvex set $K \in \mathcal{R}$. We choose a representation $K = K_1 \cup \ldots \cup K_m$ with convex bodies $K_i \in \mathcal{K}'$. From the representation

$$\Phi_k^{(j)}(K, \cdot; \cdot) = \sum_{v \in S(m)} (-1)^{|v|-1} \Phi_k^{(j)}(K_v, \cdot; \cdot)$$

it follows that $\Phi_k^{(j)}(K, \cdot; \cdot)$, as a function of its second argument, is additive and continuous, hence it has an additive extension to $\mathcal{R}$. In this way, $\Phi_k^{(j)}(K, M; \cdot)$ is defined for all $K, M \in \mathcal{R}$ and is additive in each of its first two arguments.

Now both sides of the formula (5.17) make sense for arbitrary polyconvex sets $K, M \in \mathcal{R}$. Suppose, first, that M is convex. As a function of K, both sides are additive, and they are equal if K is convex. By the inclusion–exclusion principle, two additive functions coinciding on $\mathcal{K}'$ also coincide on $\mathcal{R}$. Thus, (5.17) remains true if K is a polyconvex set. In the same way, M can be replaced by a polyconvex set.

Theorem 5.2.4. *The translative formula* (5.17) *holds for polyconvex sets* $K, M \in \mathcal{R}$.

The investigation of translative integral geometry will be continued in Section 6.4.

The Notes for this section are included in those for Section 5.3.

5.3 The Principal Kinematic Formula for Curvature Measures

As mentioned at the beginning of the previous section, our aim is the derivation of a formula for the kinematic integral

$$\int_{G_d} \Phi_j(K \cap gM, A \cap gB)\,\mu(\mathrm{d}g)$$
$$= \int_{SO_d} \int_{\mathbb{R}^d} \Phi_j(K \cap (\vartheta M + x), A \cap (\vartheta B + x))\,\lambda(\mathrm{d}x)\,\nu(\mathrm{d}\vartheta). \quad (5.22)$$

The measurability of the integrand was proved in Lemma 5.2.1. If K and M are polytopes, we can apply Theorem 5.2.2 and obtain for the right side of (5.22) the expression

$$\Phi_j(K,A)\Phi_d(M,B) + \sum_{k=j+1}^{d-1} \sum_{F\in\mathcal{F}_k(K)} \sum_{G\in\mathcal{F}_{d-k+j}(M)} \lambda_F(A)\lambda_G(B)$$
$$\times \int_{SO_d} \gamma(F,\vartheta G; K, \vartheta M)[F,\vartheta G]\,\nu(\mathrm{d}\vartheta) + \Phi_d(K,A)\Phi_j(M,B). \quad (5.23)$$

The integral over the rotation group occurring here is evaluated in the next theorem.

Theorem 5.3.1. *If $K, M \in \mathcal{P}'$ are polytopes, $j \in \{0,\dots,d-2\}$, $k \in \{j+1,\dots,d-1\}$, $F \in \mathcal{F}_k(K)$ and $G \in \mathcal{F}_{d-k+j}(M)$, then*

$$\int_{SO_d} \gamma(F,\vartheta G; K,\vartheta M)[F,\vartheta G]\,\nu(\mathrm{d}\vartheta) = c_{j,d}^{k,d-k+j}\gamma(F,K)\gamma(G,M), \quad (5.24)$$

where the constant is given by (5.5).

Proof. Let $\vartheta \in SO_d$ be a rotation for which F and ϑG are not in special position. By Lemma 13.2.1, only such rotations need to be considered for the computation of the integral in (5.24). By definition,

$$\gamma(F,\vartheta G; K,\vartheta M) = \gamma(F \cap (\vartheta G + x), K \cap (\vartheta M + x))$$

with a suitable vector $x \in \mathbb{R}^d$. Denoting by $N(P,F)$ the normal cone of P at a relatively interior point of F, we see from the definition of the external angle that

$$\gamma(F,\vartheta G; K,\vartheta M) = \frac{\sigma_{d-1-j}(N(K\cap(\vartheta M + x), F \cap (\vartheta G + x)) \cap S^{d-1})}{\sigma_{d-1-j}(L \cap S^{d-1})},$$

where $L \in G(d, d-j)$ is the subspace totally orthogonal to $F \cap (\vartheta G + x)$. Since

$$N(K\cap(\vartheta M + x), F\cap(\vartheta G + x)) = N(K,F) + \vartheta N(M,G)$$

(see Schneider [695, Theorem 2.2.1]), we have to consider the integral

$$\int_{SO_d} \sigma_{d-j-1}((N(K,F) + \vartheta N(M,G)) \cap S^{d-1})[F,\vartheta G]\,\nu(\mathrm{d}\vartheta).$$

More generally, we denote by L_1 and L_2 the orthogonal spaces of F and G, respectively. Noting that $[F, \vartheta G] = [L_1^\perp, \vartheta L_2^\perp] = [L_1, \vartheta L_2]$, we define

$$I(A, B) := \int_{SO_d} \sigma_{d-j-1}((\breve{A} + \vartheta \breve{B}) \cap S^{d-1})[L_1, \vartheta L_2]\, \nu(\mathrm{d}\vartheta)$$

for arbitrary Borel sets $A \subset L_1 \cap S^{d-1}$ and $B \subset L_2 \cap S^{d-1}$, where

$$\breve{A} := \{\alpha x : x \in A,\, \alpha \geq 0\}$$

denotes the cone generated by A. Concerning the measurability of the integrand, we observe the following. The function $\vartheta \mapsto [L_1, \vartheta L_2]$ is continuous. Let U be the set of all rotations $\vartheta \in SO_d$ for which L_1 and ϑL_2 are not in special position; then $\nu(SO_d \setminus U) = 0$ by Lemma 13.2.1. Since $\dim L_1 + \dim L_2 = d - j \leq d$, the sum $L_1 + \vartheta L_2$ is direct if $\vartheta \in U$, hence $\breve{A} + \vartheta \breve{B}$ is a Borel set. For $\vartheta \in U$, all sets $\breve{A} + \vartheta \breve{B}$ are images of a fixed one under linear transformations of $\mathbb{R}^d$. Using this fact, it is not difficult to show that the map

$$\vartheta \mapsto \sigma_{d-j-1}((\breve{A} + \vartheta \breve{B}) \cap S^{d-1})$$

is measurable on U.

For fixed $B \in \mathcal{B}(L_2 \cap S^{d-1})$ we now set

$$\omega(A) := I(A, B) \qquad \text{for } A \in \mathcal{B}(L_1 \cap S^{d-1}).$$

If $\bigcup_{i=1}^\infty A_i$ is a disjoint union of sets $A_i \in \mathcal{B}(L_1 \cap S^{d-1})$, then

$$\left(\bigcup_{i=1}^\infty \breve{A}_i + \vartheta \breve{B} \right) \cap S^{d-1} = \bigcup_{i=1}^\infty \left((\breve{A}_i + \vartheta \breve{B}) \cap S^{d-1} \right)$$

for $\vartheta \in U$, and this is again a disjoint union up to a set of σ_{d-j-1}-measure zero. It follows that

$$\sigma_{d-j-1} \left(\left(\bigcup_{i=1}^\infty \breve{A}_i + \vartheta \breve{B} \right) \cap S^{d-1} \right) = \sum_{i=1}^\infty \sigma_{d-j-1} \left((\breve{A}_i + \vartheta \breve{B}) \cap S^{d-1} \right)$$

for $\vartheta \in U$, hence

$$\omega \left(\bigcup_{i=1}^\infty A_i \right) = \sum_{i=1}^\infty \omega(A_i)$$

by the monotone convergence theorem. Thus ω is a finite measure on $L_1 \cap S^{d-1}$. Let $\rho \in SO_d$ be a rotation mapping L_1 into itself and fixing every point of $L_1^\perp$. Then

$$\rho \breve{A} + \vartheta \breve{B} = \rho(\breve{A} + \rho^{-1} \vartheta \breve{B})$$

and

$$[L_1, \vartheta L_2] = [\rho L_1, \vartheta L_2] = [L_1, \rho^{-1} \vartheta L_2],$$

hence

$$\begin{aligned}\omega(\rho A) &= \int_{SO_d} \sigma_{d-j-1}((\rho\breve{A} + \vartheta\breve{B}) \cap S^{d-1})[L_1, \vartheta L_2]\, \nu(\mathrm{d}\vartheta) \\ &= \int_{SO_d} \sigma_{d-j-1}((\breve{A} + \rho^{-1}\vartheta\breve{B}) \cap S^{d-1})[L_1, \rho^{-1}\vartheta L_2]\, \nu(\mathrm{d}\vartheta) \\ &= \omega(A).\end{aligned}$$

By the uniqueness of spherical Lebesgue measure (a special case of Theorem 13.1.3), ω is a constant multiple of σ_{d-k-1} on $L_1 \cap S^{d-1}$. Similarly we obtain for fixed $A \in \mathcal{B}(L_1 \cap S^{d-1})$ that $I(A, \cdot)$ is a constant multiple of σ_{k-j-1} on $L_2 \cap S^{d-1}$. Altogether this yields a representation

$$I(A, B) = \alpha(L_1, L_2)\sigma_{d-k-1}(A)\sigma_{k-j-1}(B)$$

for all $A \in \mathcal{B}(L_1 \cap S^{d-1})$, $B \in \mathcal{B}(L_2 \cap S^{d-1})$, where $\alpha(L_1, L_2)$ is a constant that depends only on L_1 and L_2. The choice $A = L_1 \cap S^{d-1}$ and $B = L_2 \cap S^{d-1}$, together with the invariance properties of the functional I resulting from its definition, shows that $\alpha(L_1, L_2)$ does, in fact, depend only on the dimensions d, j, k.

In particular, this gives

$$I(N(K, F) \cap S^{d-1}, N(M, G) \cap S^{d-1}) = \alpha_{djk}\gamma(F, K)\gamma(G, M)$$

with a constant $\alpha_{djk} > 0$ and thus the assertion of Theorem 5.3.1, up to the determination of α_{djk}. We insert (5.24) into (5.22), using (5.23) for the right side. If we choose for K an r-polytope with $r \in \{j+1, \dots, d-1\}$, for M a $(d-r+j)$-polytope, and $A = B = \mathbb{R}^d$, then the result must coincide with formula (5.10). From this we conclude that $\alpha_{djr} = c_{j,d}^{r,d-r+j}$. This completes the proof. □

Corollary 5.3.1. *If $K, M \in \mathcal{K}'$ are convex bodies, $A, B \in \mathcal{B}(\mathbb{R}^d)$ are Borel sets and if $j \in \{0, \dots, d-2\}$, $k \in \{j+1, \dots, d-1\}$, then*

$$\int_{SO_d} \Phi_k^{(j)}(K, \vartheta M; A \times \vartheta B)\, \nu(\mathrm{d}\vartheta) = c_{j,d}^{k,d-k+j} \Phi_k(K, A)\Phi_{d-k+j}(M, B). \quad (5.25)$$

Proof. If K, M are polytopes, the definition of $\Phi_k^{(j)}(K, M; \cdot)$ and formula (5.24) show that

$$\begin{aligned}&\int_{SO_d} \Phi_k^{(j)}(K, \vartheta M; A \times \vartheta B)\, \nu(\mathrm{d}\vartheta) \\ &= \sum_{F \in \mathcal{F}_k(K)} \sum_{G \in \mathcal{F}_{d-k+j}(M)} \lambda_F(A)\lambda_G(B) \int_{SO_d} \gamma(F, \vartheta G; K, \vartheta M)[F, \vartheta G]\, \nu(\mathrm{d}\vartheta)\end{aligned}$$

$$= c_{j,d}^{k,d-k+j} \left(\sum_{F \in \mathcal{F}_k(K)} \lambda_F(A)\gamma(F,K) \right) \left(\sum_{G \in \mathcal{F}_{d-k+j}(M)} \lambda_G(B)\gamma(G,M) \right)$$
$$= c_{j,d}^{k,d-k+j} \Phi_k(K,A)\Phi_{d-k+j}(M,B).$$

Approximation by polytopes yields (5.25) for general convex bodies $K, M \in \mathcal{K}'$. For this, we first have to verify the measurability of the integrand. It is obtained from the weak continuity of the measures $\Phi_k^{(j)}(K, \vartheta M; \cdot)$, established in Theorem 5.2.3, and from the identity

$$\int_{SO_d} \int_{\mathbb{R}^d \times \mathbb{R}^d} f(x, \vartheta^{-1} y)\, \Phi_k^{(j)}(K, \vartheta M; \mathrm{d}(x,y))\, \nu(\mathrm{d}\vartheta)$$
$$= c_{j,d}^{k,d-k+j} \int_{\mathbb{R}^d} \int_{\mathbb{R}^d} f(x,y)\, \Phi_k(K, \mathrm{d}x)\, \Phi_{d-k+j}(M, \mathrm{d}y),$$

valid for all $f \in \mathbf{C}(\mathbb{R}^d \times \mathbb{R}^d)$. The latter identity is equivalent to (5.25); this is seen as in the proof of Theorem 5.2.3. Also the final limit procedure is analogous to that in the proof of Theorem 5.2.3; one uses the weak continuity of the involved measures (Theorems 14.2.2 and 5.2.3) and the dominated convergence theorem. □

We can now state the main result of this section.

Theorem 5.3.2 (Local principal kinematic formula). *If $K, M \in \mathcal{R}$ are polyconvex sets, $A, B \in \mathcal{B}(\mathbb{R}^d)$ are Borel sets and if $j \in \{0, \dots, d\}$, then*

$$\int_{G_d} \Phi_j(K \cap gM, A \cap gB)\, \mu(\mathrm{d}g) = \sum_{k=j}^{d} c_{j,d}^{k,d-k+j} \Phi_k(K,A)\Phi_{d-k+j}(M,B).$$

Proof. The extension to polyconvex sets K, M was explained at the end of Section 5.2. □

Remark. The **extended convex ring** $\mathcal{S}$ is defined as the system of all subsets of $\mathbb{R}^d$ that intersect every convex body in a union of finitely many convex bodies. Since the curvature measures are locally determined (Theorem 14.2.3), they can be extended to sets of the extended convex ring, as long as the involved Borel sets remain bounded. Hence, Theorem 5.3.2 can be extended in the same sense.

In the principal kinematic formula, curvature measures of the intersection of a fixed and a moving set from the convex ring are integrated over all rigid motions. Here the moving compact set can be replaced by a moving affine subspace, and the integration can be carried out with respect to the corresponding invariant measure. One can derive such formulas directly from the principal kinematic formula.

Theorem 5.3.3 (Local Crofton formula). *If $K \in \mathcal{R}^d$ is a polyconvex set, $A \in \mathcal{B}(\mathbb{R}^d)$ is a Borel set and $q \in \{0, \dots, d\}$, $j \in \{0, \dots, q\}$, then*

$$\int_{A(d,q)} \Phi_j(K \cap E, A \cap E)\, \mu_q(\mathrm{d}E) = c_{j,d}^{q,d-q+j} \Phi_{d-q+j}(K, A).$$

Proof. We may assume that $K \in \mathcal{K}'$; the extension to $K \in \mathcal{R}$ is then achieved as explained at the end of Section 5.2. We fix $L_q \in G(d,q)$ and use the map γ_q defined by (5.2), then $\mu_q = \gamma_q(\lambda_{d-q} \otimes \nu)$. Let C be a q-dimensional unit cube in L_q. Since $L_q \in \mathcal{S}$, C is bounded and A can be replaced by the bounded set $A \cap K$, the remark after the proof of Theorem 5.3.2 shows that

$$J := \int_{G_d} \Phi_j(L_q \cap gK, C \cap gA)\, \mu(\mathrm{d}g) = \sum_{k=j}^{d} c_{j,d}^{k,d-k+j} \Phi_k(L_q, C) \Phi_{d-k+j}(K, A).$$

Now

$$\Phi_k(L_q, C) = \begin{cases} \lambda_q(C) & \text{for } k = q, \\ 0 & \text{for } k \neq q, \end{cases}$$

hence

$$J = c_{j,d}^{q,d-q+j} \Phi_{d-q+j}(K, A).$$

On the other hand,

$$\begin{aligned} J &= \int_{SO_d} \int_{\mathbb{R}^d} \Phi_j(L_q \cap (\vartheta K + x), C \cap (\vartheta A + x))\, \lambda(\mathrm{d}x)\, \nu(\mathrm{d}\vartheta) \\ &= \int_{SO_d} \int_{L_q^\perp} \int_{L_q} \Phi_j(L_q \cap (\vartheta K + x_1 + x_2), C \cap (\vartheta A + x_1 + x_2))\, \lambda_q(\mathrm{d}x_2) \\ &\quad \times \lambda_{d-q}(\mathrm{d}x_1)\, \nu(\mathrm{d}\vartheta). \end{aligned}$$

To compute the inner integral, we put

$$\Phi_j(L_q \cap (\vartheta K + x_1), \cdot) =: \phi, \qquad \vartheta A + x_1 =: A';$$

then

$$\begin{aligned} &\int_{L_q} \Phi_j(L_q \cap (\vartheta K + x_1 + x_2), C \cap (\vartheta A + x_1 + x_2))\, \lambda_q(\mathrm{d}x_2) \\ &= \int_{L_q} \phi((C - x_2) \cap A')\, \lambda_q(\mathrm{d}x_2) \\ &= \phi(A') \lambda_q(C) \\ &= \Phi_j(L_q \cap (\vartheta K + x_1), L_q \cap (\vartheta A + x_1)), \end{aligned}$$

where Theorem 5.2.1 was used. This yields

$$\begin{aligned} J &= \int_{SO_d} \int_{L_q^\perp} \Phi_j(L_q \cap (\vartheta K + x_1), L_q \cap (\vartheta A + x_1))\, \lambda_{d-q}(\mathrm{d}x_1)\, \nu(\mathrm{d}\vartheta) \\ &= \int_{SO_d} \int_{L_q^\perp} \Phi_j(K \cap \vartheta(L_q + x), A \cap \vartheta(L_q + x))\, \lambda_{d-q}(\mathrm{d}x)\, \nu(\mathrm{d}\vartheta) \\ &= \int_{A(d,q)} \Phi_j(K \cap E, A \cap E)\, \mu_q(\mathrm{d}E), \end{aligned}$$

where we have used the motion covariance of the curvature measures and the inversion invariance of λ_{d-q} and ν. The two representations obtained for J prove the assertion. □

The case $j = 0$,

$$\Phi_{d-q}(K, A) = c_{q,d-q}^{d,0} \int_{A(d,q)} \Phi_0(K \cap E, A \cap E)\, \mu_q(\mathrm{d}E),$$

gives an interpretation of the measure $\Phi_{d-q}(K, A)$: up to a numerical factor, it is the mean value of $\Phi_0(K \cap E, A \cap E)$, where the mean is taken over the intersections with q-flats. The Gaussian curvature measure Φ_0 has a simple intuitive interpretation, as mentioned in Section 14.2.

Notes for Sections 5.2 and 5.3

1. A general local principal kinematic formula, which coincides with Theorem 5.3.2 in the case of convex bodies, was first obtained by Federer [228]. He proved it for sets of positive reach and for their curvature measures, which he introduced for this purpose. The generality of the admissible point sets requires deeper techniques from geometric measure theory. Using such techniques, in particular Martina Zähle studied new approaches to curvature measures and to integral geometric formulas valid for them; see Zähle [824, 825, 826, 827], Rother and Zähle [649].

There have been several successful attempts to define curvature measures and to obtain kinematic and Crofton formulas in very general situations, where strong singularities are permitted. We refer here to Fu [236, 237, 238], Bröcker and Kuppe [122], Bernig and Bröcker [94, 93], Rataj and Zähle [620, 621].

In contrast to this trend to deep generalizations, it has been our aim in this book to follow an approach to local integral geometric formulas for convex bodies and sets of the convex ring that needs only elementary measure-theoretic and geometric arguments, and which (we hope) is more in the spirit of the integral geometry of Blaschke and Hadwiger. Different approaches of this kind are also found in Schneider [676, 680].

In deriving the Crofton formula (Theorem 5.3.3) from the local principal kinematic formula, we followed Federer [228].

2. In order to extend the curvature measures additively to the convex ring, we have referred here to Groemer's extension theorem. For the support measures, and thus for the curvature measures, a more explicit construction of an additive extension to polyconvex sets is found in Schneider [679] and in Section 4.4 of [695]. It is based on an extension of the local Steiner formula for polyconvex sets, with the Lebesgue measure replaced by the integral of the multiplicity function that arises from additive extension of the indicator function of a parallel set. See also Note 3 of Section 14.4.

3. A more general version of Theorem 5.2.1 is Theorem 13.1.4. We refer to Note 2 of Section 13.1 for some references.

Translative integral geometry was first investigated by Blaschke [106] and Berwald and Varga [98]; see Schneider and Weil [715] for further references. From the latter paper, we essentially took the proofs of Theorems 5.2.2 and 5.3.1, and thus of the local principal kinematic formula, Theorem 5.3.2. A first version of Theorem 5.2.3 appeared in Weil [786]. A better understanding of the mixed measures $\Phi_k^{(j)}(K,M;\cdot)$ of Theorem 5.2.3 is desirable. Results concerning the total measures $\Phi_k^{(j)}(K,M;\mathbb{R}^d\times\mathbb{R}^d) =: V_k^{(j)}(K,M)$ were found by Goodey and Weil [277], Weil [790, 791, 800].

For further information on translative integral geometry, we refer to the Notes for Section 6.4.

4. Kinematic formulas for support measures. The curvature measures, for which we have proved the local principal kinematic formula and the Crofton formula, are specializations of the support measures Ξ_m introduced in Section 14.2. There are also versions of these formulas for support measures. They require that the intersection of Borel sets in $\mathbb{R}^d$ be replaced by a suitable law of composition for subsets of $\Sigma = \mathbb{R}^d \times S^{d-1}$, which is adapted to intersections of convex bodies. For $A, B \subset \Sigma$, let

$$A \wedge B := \{(x,u) \in \Sigma :\ \text{there are } u_1, u_2 \in S^{d-1} \text{ with}$$
$$(x,u_1) \in A,\ (x,u_2) \in B,\ u \in \operatorname{pos}\{u_1,u_2\}\},$$

where $\operatorname{pos}\{u_1,u_2\} := \{\lambda_1 u_1 + \lambda_2 u_2 : \lambda_1,\lambda_2 \ge 0\}$ is the positive hull of $\{u_1,u_2\}$. Now for convex bodies $K, M \in \mathcal{K}'$, Borel sets $A \subset \operatorname{Nor} K$ and $B \subset \operatorname{Nor} M$ (where Nor denotes the generalized normal bundle, see Section 14.2), and for $j \in \{0,\dots,d-2\}$, the formula

$$\int_{G_d} \Xi_j(K \cap gM, A \wedge gB)\,\mu(\mathrm{d}g) = \sum_{k=j+1}^{d-1} c_{j,d}^{k,d-k+j}\,\Xi_k(K,A)\Xi_{d-k+j}(M,B) \quad (5.26)$$

holds (for $j = d-1$, both sides would give 0).

For a q-flat $E \in A(d,q)$, $q \in \{1,\dots,d-1\}$, one defines

$$A \wedge E := \{(x,u) \in \Sigma :\ \text{there are } u_1, u_2 \in S^{d-1} \text{ with}$$
$$(x,u_1) \in A,\ x \in E,\ u_2 \in E^\perp,\ u \in \operatorname{pos}\{u_1,u_2\}\},$$

where $E^\perp$ is the linear subspace totally orthogonal to E. Then the local Crofton formula has the following extension. Let $K \in \mathcal{K}'$ be a convex body, $k \in \{1,\dots,d-1\}$, $j \in \{0,\dots,k-1\}$, and let $A \subset \operatorname{Nor} K$ be a Borel set. Then

$$\int_{A(d,k)} \Xi_j(K \cap E, A \wedge E)\,\mu_k(\mathrm{d}E) = c_{j,d}^{k,d-k+j}\,\Xi_{d-k+j}(K,A).$$

These results are due to Glasauer [266], under an additional assumption in the case of (5.26). A common boundary point x of the convex bodies K, M is said to be 'exceptional' if the linear hulls of the normal cones of K and M at x have a non-zero intersection. Glasauer assumed that the set of rigid motions g for which K and gM have some exceptional common boundary point, is of Haar measure zero. He conjectured that this assumption is always satisfied. This was proved by Schneider [700]. An alternative proof of a more general result appears in Zähle [831].

5. A local version of Hadwiger's general integral geometric theorem. The local principal kinematic formula together with the local Crofton formula (Theorems 5.3.2 and 5.3.3) can be extended in the same way as the principal kinematic formula and the Crofton formula (Theorems 5.1.3 and 5.1.1) are extended by Hadwiger's general integral geometric theorem. This abstract version of (5.17) reads as follows.

Theorem. *Let* $\Lambda : \mathcal{K}' \times \mathcal{B}(\mathbb{R}^d) \to \mathbb{R}$ *be a mapping with the following properties:*

(a) $\Lambda(K, \cdot)$ *is a finite positive measure concentrated on* K, *for all* $K \in \mathcal{K}'$.

(b) *The map* $K \mapsto \Lambda(K, \cdot)$ *is additive and weakly continuous.*

(c) *If* $K, M \in \mathcal{K}'$, $A \subset \mathbb{R}^d$ *is open and* $K \cap A = M \cap A$, *then* $\Lambda(K, B) = \Lambda(M, B)$ *for all Borel sets* $B \subset A$.

Then, for $K, M \in \mathcal{K}'$, $A, B \in \mathcal{B}(\mathbb{R}^d)$ *and* $j \in \{0, \dots, d\}$, *the formula*

$$\int_{G_d} \Lambda(K \cap gM, A \cap gB)\, \mu(\mathrm{d}g) = \sum_{k=0}^{d} \Lambda_{d-k}(K, A) \Phi_k(M, B)$$

(*with* $\Lambda(\emptyset, \cdot) := 0$) *holds, where*

$$\Lambda_{d-k}(K, B) := \int_{A(d,k)} \Lambda(K \cap E, B)\, \mu_k(\mathrm{d}E).$$

This was proved by Schneider [696]. An analog in spherical space and a simpler proof in Euclidean space were given by Glasauer [264]. Examples of mappings Λ satisfying the above properties are the relative curvature measures introduced in Schneider [696]. Also (5.26) admits an abstract generalization in the spirit of Hadwiger's general integral geometric theorem; see Glasauer [268], Theorem 7.

6. Tensor valuations. The intrinsic volumes and their local versions arise from the notion of volume, through the Steiner formula. Replacement of the volume by vectorial or higher rank tensorial moments leads to tensor-valued valuations on convex bodies and raises the question whether their properties and their role in integral geometry extend those of the intrinsic volumes. To explain this, we denote by $\mathbb{T}^p$ the vector space of symmetric tensors of rank p over $\mathbb{R}^d$ (we identify $\mathbb{R}^d$ with its dual space, using the scalar product, so that no distinction between covariant and contravariant tensors is necessary). If $p \in \mathbb{N}$ and $x \in \mathbb{R}^d$, we write x^p for the p-fold tensor product $x \otimes \cdots \otimes x$, and we put $x^0 := 1$. For symmetric tensors a and b, their symmetric product is denoted by ab. For $K \in \mathcal{K}'$ and $p \in \mathbb{N}_0$, let

$$\Psi_p(K) := \frac{1}{p!} \int_K x^p\, \lambda(\mathrm{d}x).$$

The Steiner formula extends to a polynomial expansion

$$\Psi_p(K + \epsilon B^d) = \sum_{k=0}^{d+p} \epsilon^{d+p-k} \kappa_{d+p-k} V_k^{(p)}(K) \tag{5.27}$$

for $\epsilon > 0$, with $V_k^{(p)}(K) \in \mathbb{T}^p$. Each function $V_k^{(p)} : \mathcal{K}' \to \mathbb{T}^p$ is additive, continuous and isometry covariant, which means that $V_k^{(p)}(\vartheta K) = \vartheta V_k^{(p)}(K)$ for every rotation $\vartheta \in SO_d$ and that $V_k^{(p)}(K+t)$ is a (tensor) polynomial in $t \in \mathbb{R}^d$ of degree p. The known facts in the case $p = 0$ suggest the following questions: (a) Is an additive, continuous, isometry covariant function $f : \mathcal{K}' \to \mathbb{T}^p$ necessarily a linear combination of $V_0^{(p)}, \dots, V_{d+p}^{(p)}$? (b) Do the coefficients $V_k^{(p)}$ satisfy kinematic and Crofton formulas? For $p = 0$, positive answers were given in this chapter. For $p = 1$, both questions were answered affirmatively by Hadwiger and Schneider [312]. For $p > 1$, however, the situation is different. One has to consider more general tensor valuations, defined by

$$\Phi_{m,r,s}(K) := \frac{1}{r!s!} \frac{\omega_{d-m}}{\omega_{d-m+s}} \int_{\Sigma} x^r u^s \, \Xi_m(K, \mathrm{d}(x,u))$$

for $K \in \mathcal{K}'$ and integers $r, s \geq 0$, $0 \leq m \leq d-1$ (the factors before the integral turn out to be convenient). They were introduced (via a polytopal approach) by McMullen [472]. Besides these tensor functions $\Phi_{m,r,s} : \mathcal{K}' \to \mathbb{T}^{r+s}$, one also needs the metric tensor $G \in \mathbb{T}^2$ of $\mathbb{R}^d$. The functions $G^q \Phi_{m,r,s}$ and $G^q \Psi_p$ ($q \in \mathbb{N}_0$) are called **basic tensor valuations**. Answering a question posed by McMullen [472], Alesker [17], based on his earlier work in [16], proved the following extension of Hadwiger's characterization theorem:

Theorem. *If $p \in \mathbb{N}_0$ and if $f : \mathcal{K}' \to \mathbb{T}^p$ is an additive, continuous, isometry covariant function, then f is a linear combination of the functions $G^q \Phi_{m,r,s}$ (with $2q + r + s = p$) and the functions $G^q \Psi_r$ (with $2q + r = p$).*

McMullen [472] had already discovered a set of nontrivial linear relations between the basic tensor valuations. Therefore, Alesker's result yielded a generating system, but not a basis or the dimension of the vector space of continuous, isometry covariant tensor valuations of fixed rank. This remaining problem was settled by Hug, Schneider and Schuster [374], who proved that the relations between the basic tensor valuations discovered by McMullen are essentially the only ones.

By Alesker's result, the coefficients $V_k^{(p)}$ appearing in the Steiner polynomial (5.27) are linear combinations of basic tensor valuations. Question (b) above should, therefore, be modified, asking whether the functions $\Phi_{m,r,s}$ and Ψ_p satisfy kinematic and Crofton formulas. Unlike in the cases of rank zero or one, the characterization theorem does not seem useful for obtaining integral geometric formulas, due to the linear relations between the basic tensor valuations; hence, direct computations are required. It is sufficient to derive Crofton formulas, since then Hadwiger's general integral geometric theorem, which in the case of tensor functions can be applied coordinate-wise, immediately yields kinematic formulas. For dimension two and rank one or two, kinematic formulas were already obtained by Müller [567] (except for $\Phi_{0,1,1}$, in our notation), who took up a suggestion of Blaschke. An investigation for all dimensions and ranks was begun by Schneider [701] and continued by Schneider and Schuster [713]. This led, in particular, to a complete set of Crofton and kinematic formulas in two and three dimensions. The higher-dimensional case turned out to be intricate; it was settled by Hug, Schneider and Schuster [375].

7. Non-intersecting sets: distances. All the integral geometric results considered up to now in this chapter concern the intersection of a fixed and a moving set. For convex sets, there are also kinematic formulas involving relations between non-intersecting sets. One possibility consists in taking distances into account. The distance $d(K,L)$ of a compact set $K \subset \mathbb{R}^d$ and a closed set $L \subset \mathbb{R}^d$, $K, L \neq \emptyset$, is defined by

$$d(K,L) := \min\{\|x-y\| : x \in K,\ y \in L\}.$$

Let $f : [0,\infty) \to [0,\infty)$ be a measurable function satisfying $f(0) = 0$ and

$$m_k(f) := \frac{1}{k!}\int_0^\infty f(r)r^k \,\mathrm{d}r < \infty \qquad \text{for } k = 0,\dots,d-1.$$

Then, for convex bodies $K, M \in \mathcal{K}'$, the kinematic formula

$$\int_{G_d} f(d(K,gM))\,\mu(\mathrm{d}g) = \sum_{j=0}^{d-1}\sum_{k=0}^{d-j-1} c_{d,0}^{d-j,d-k} m_{d-j-1-k}(f) V_j(K) V_k(M)$$

holds. This can be generalized in various directions. To give one example, suppose that for the convex bodies K, M with $K \cap M = \emptyset$ there is a unique pair $x \in K$, $y \in M$ with $\|x - y\| = d(K,M)$. Then one can define $p(K,M) := x$. One can show that $p(K,gM)$ exists for μ-almost all $g \in G_d$ with $K \cap gM = \emptyset$. If $f : (0,\infty) \times \mathrm{bd}\,K \times \mathrm{bd}\,M \to \mathbb{R}$ is a measurable function for which the integral

$$\int_{K\cap gM=\emptyset} f(d(K,gM), p(K,gM), g^{-1}p(gM,K))\,\mu(\mathrm{d}g)$$

is finite, then this integral can be expressed in terms of integrals of curvature measures of K and M. Similarly, one can treat kinematic integrals involving functions of the unit vector pointing from K to gM. Further, the moving convex body can be replaced by a moving flat.

For the special case where M is one-pointed, a related formula is given by Theorem 14.3.3.

Contributions to this area are due to Hadwiger [309, 310], Bokowski, Hadwiger and Wills [111], Schneider [675], Groemer [291], Weil [779, 780, 782]. We refer also to Section 4 of the survey article by Schneider and Wieacker [720].

Translative formulas for non-intersecting convex bodies in suitable general position have been studied by Kiderlen and Weil [409]; the results involve mixed curvature measures. Hug, Last and Weil [358] give a quite general translative formula, allowing also non-Euclidean distances and using relative support measures (a special case is Theorem 14.3.3). A corresponding version for flats is contained in Hug, Last and Weil [360].

8. Non-intersecting sets: convex hulls. Glasauer [267] found a new type of kinematic formulas, involving the convex hull of a fixed and a moving convex body. Since convex hulls with a freely moving convex body are not uniformly bounded, the results can only be of the type of weighted limits. Let $K \vee M$ denote the convex hull of $K \cup M$. A typical result of Glasauer concerns the mixed volumes with fixed convex bodies $K_{j+1},\dots,K_d$ and states that

$$\lim_{r\to\infty}\frac{1}{r^{d+1}}\frac{(d+1)\kappa_d}{\kappa_{d-1}}\int_{\{g\in G_d:\ gM\subset rB^d\}} V(K\vee gM[j], K_{j+1},\dots,K_d)\,\mu(\mathrm{d}g)$$

$$= \sum_{k=0}^{j-1} V(K[k], B^d[j-k], K_{j+1},\dots,K_d) V(M[j-k-1], B^d[d-j+k+1]).$$

This is a special case of Theorem 3 of Glasauer [267]. He has considerably more general results, for not necessarily invariant measures, and with mixed area measures instead of mixed volumes. For $K_{j+1} = \ldots = K_d = B^d$, the formula reduces to one for intrinsic volumes. For this result, there is also a local version, which is 'dual' to formula (5.26). It involves a law of composition for subsets of Σ which is adapted to the convex hull operation for pairs of convex bodies. For $A, B \subset \Sigma$, let

$$\begin{aligned} A \vee B := \; & \{(x,u) \in \Sigma : \text{ there are } x_1, x_2 \in \mathbb{R}^d \text{ with} \\ & \langle x_1 - x_2, u \rangle = 0,\ (x_1,u) \in A,\ (x_2,u) \in B,\ x \in \operatorname{conv}\{x_1,x_2\}\}. \end{aligned}$$

Now suppose that $K, M \in \mathcal{K}'$, $A \subset \operatorname{Nor} K$ and $B \subset \operatorname{Nor} M$ are Borel sets, and $j \in \{0, \ldots, d-1\}$. Then Glasauer [268] proved (with different notation) that

$$\begin{aligned} & \lim_{r\to\infty} \frac{1}{r^{d+1}} \int_{\{g \in G_d:\, gM \subset rB^d\}} \Xi_j(K \vee gM, A \vee gB)\, \mu(\mathrm{d}g) \\ & = \sum_{k=0}^{j-1} \beta_{djk} \Xi_k(K,A) \Xi_{j-k-1}(M,B), \end{aligned}$$

with explicit constants β_{djk}. The proof requires the following regularity result. A common supporting hyperplane H of the convex bodies K, M (leaving K and M on the same side) is said to be exceptional if the affine hulls of the sets $H \cap K$ and $H \cap M$ have a nonempty intersection or contain parallel lines. Then the set of all rigid motions g for which K and gM have some exceptional common supporting hyperplane is of Haar measure zero. This was conjectured by Glasauer and proved by Schneider [700].

9. Dual quermassintegrals. The principal kinematic formula for convex bodies involves the intrinsic volumes, which belong to the Brunn–Minkowski theory. There are analogs in the dual Brunn–Minkowski theory. This analogy becomes clearer in terms of the quermassintegrals $W_0, \ldots, W_n$ (see (14.6)). Equivalent to (5.8) is the formula

$$W_{d-i}(K) = \frac{\kappa_d}{\kappa_i} \int_{G(d,i)} \lambda_i(K|L)\, \nu_i(\mathrm{d}L)$$

for $i = 0, \ldots, d$. In terms of the quermassintegrals, the principal kinematic formula has the form (5.12). Let $K \subset \mathbb{R}^d$ be a star body (a compact set, star-shaped with respect to 0, with continuous radial function). The **dual quermassintegrals** $\widetilde{W}_0, \ldots, \widetilde{W}_d$ are defined by

$$\widetilde{W}_{d-i}(K) = \frac{\kappa_d}{\kappa_i} \int_{G(d,i)} \lambda_i(K \cap L)\, \nu_i(\mathrm{d}L).$$

For $g \in G_d$, let N_g denote the segment joining 0 and $g0$. Zhang [833] has proved the kinematic formula

$$\int_{G_d} \chi(K \cap gM \cap N_g)\, \mu(\mathrm{d}g) = \frac{1}{\kappa_d} \sum_{i=0}^{d} \binom{d}{i} \widetilde{W}_i(K) \widetilde{W}_{d-i}(M)$$

for star bodies $K, M \subset \mathbb{R}^d$, which is formally very similar to (5.12).

10. Striking combinatorial analogs of the kinematic formula in the context of finite lattices were found by Klain [415]; see also Klain and Rota [416, p. 29].

11. Kinematic formulas for boundaries of convex bodies. Let $K, M \subset \mathbb{R}^d$ be convex bodies with nonempty interiors, and let $\partial K, \partial L$ denote their boundaries. The following two kinematic formulas, involving intersections of two convex surfaces or of a convex surface and a convex body, were conjectured by Firey (see Problem 18 in the collection of Gruber and Schneider [298]):

$$\int_{G_d} \chi(\partial K \cap g\partial M)\, \mu(\mathrm{d}g) = \frac{1+(-1)^d}{\kappa_d} \sum_{k=0}^{d-1} \binom{d}{k} (1-(-1)^k) W_{d-k}(K) W_k(M), \quad (5.28)$$

$$\int_{G_d} \chi(\partial K \cap gM)\, \mu(\mathrm{d}g) = \frac{1}{\kappa_d} \sum_{k=0}^{d-1} \binom{d}{k} (1-(-1)^{d-k}) W_{d-k}(K) W_k(M). \quad (5.29)$$

For polytopes, these formulas can easily be verified. However, there is no simple approximation argument to extend the results to general convex bodies. In Hug and Schätzle [368], Firey's conjecture was confirmed by proving the following more general translative versions of (5.28) and (5.29):

$$\int_{\mathbb{R}^d} \chi(\partial K \cap (\partial M + x))\, \lambda(\mathrm{d}x)$$
$$= (1+(-1)^d) \sum_{k=0}^{d-1} \binom{d}{k} \left(V_k(K, -M) + (-1)^{k-1} V_k(K, M) \right),$$

where $V_k(K, L)$ denotes the mixed volume of k copies of K and $d-k$ copies of L, and

$$\int_{\mathbb{R}^d} \chi(\partial K \cap (M + x))\, \lambda(\mathrm{d}x) = \sum_{k=0}^{d-1} \binom{d}{k} \left(V_k(K, -M) + (-1)^{d-k-1} V_k(K, M) \right).$$

From these formulas, (5.28) and (5.29) are obtained if one replaces M by ϑM, integrates over all $\vartheta \in SO_d$ with respect to the invariant measure, and then applies [695, formula (5.3.25)]. In fact, Firey's original question was already answered implicitly by a result of Fu [238], which, however, does not cover the translative case.

In Hug, Mani–Levitska and Schätzle [363], these integral geometric results are extended further, to lower-dimensional sets. Furthermore, iterated formulas are established concerning intersections of several convex bodies, which then are applied to obtain formulas of stochastic geometry. Defining intrinsic volumes for intersections of convex surfaces in a suitable way by a Crofton type expression, integral formulas for such functionals are also derived.

12. Further information on kinematic and Crofton formulas is contained in the survey article by Hug and Schneider [369].

13. A Gaussian kinematic formula. Taylor [754] obtains an analog of the Steiner formula and Weyl tube formula, with Lebesgue measure replaced by Gaussian measure. This is then applied in an analog of the principal kinematic formula, expressing the expected Euler characteristic of excursion sets for certain random fields. For the geometry of random fields, see Adler [1] and Adler and Taylor [2].

5.4 Intersection Formulas for Submanifolds

The integral geometric formulas considered so far all refer to intersections of a fixed and a moving set, and these sets, with the exception of Theorem 5.2.1, were either convex bodies or affine subspaces. Certain applications to stochastic geometry or stereology, dealing with fiber or surface processes, require intersection formulas for submanifolds of various dimensions and for Hausdorff measures of their intersections. In the present section we describe such results. The technical requirements for such a treatment depend on the generality of the notion of k-dimensional surface that is used. For the most elementary notion, polyhedral surfaces, the results stated below are easily obtained from the results previously established and by the methods used in this book. However, already smooth surfaces would require different methods. The more general k-surfaces, for which the results will be formulated, need notions and techniques from geometric measure theory. Since this is outside the scope of this book, we present only the results and give references to complete proofs (including the measurability considerations omitted here).

Some notions from geometric measure theory, which are used in the following, are collected in Section 14.5 of the Appendix. In this section, we do not aim at the greatest generality, but prefer simpler formulations which are sufficient for our applications.

Let $k \in \{0, \dots, d\}$. Recall that a subset $M \subset \mathbb{R}^d$ is **k-rectifiable** if it is the image of some bounded subset of $\mathbb{R}^k$ under some Lipschitz map. The set M is **countably k-rectifiable** if it is the union of countably many k-rectifiable sets. By a **k-surface** we understand, in this section, a countably k-rectifiable Borel set M with $\mathcal{H}^k(M) < \infty$, where $\mathcal{H}^k$ denotes the k-dimensional Hausdorff measure.

A trivial case of the translative integrals we want to consider is obtained if we take a k-dimensional convex body K and a $(d-k)$-dimensional convex body M. In that case, we immediately get

$$\int_{\mathbb{R}^d} \mathcal{H}^0(K \cap (M+t))\, \lambda(\mathrm{d}t) = [K, M]\mathcal{H}^k(K)\mathcal{H}^{d-k}(M).$$

The first theorem of this section is a generalization of this simple formula to k-surfaces.

Let M be a k-surface. Then there exist k-dimensional C^1-submanifolds $N_1, N_2, \dots$ such that $\mathcal{H}^k(M \setminus \bigcup_{i\in\mathbb{N}} N_i) = 0$. Let $T_x N_i$ denote the tangent space of N_i at $x \in N_i$ (considered as a subspace of $\mathbb{R}^d$). For a Borel set $A \in \mathcal{B}(G(d,k))$, we define

$$\tau_M(A) := \mathcal{H}^k\left(\bigcup_{i\in\mathbb{N}} \{x \in M \cap N_i : T_x N_i \in A\}\right).$$

This defines a finite measure τ_M on $G(d,k)$, which depends only on M.

Theorem 5.4.1. *Let $n \in \{1, \dots, d-1\}$, and let M_i be a k_i-surface, for $i = 0, \dots, n$, with $k := k_0 + \dots + k_n \geq nd$. Then*

$$\int_{\mathbb{R}^d} \cdots \int_{\mathbb{R}^d} \mathcal{H}^{k-nd}(M_0 \cap (M_1 + t_1) \cap \dots \cap (M_n + t_n))\, \lambda(\mathrm{d}t_1) \cdots \lambda(\mathrm{d}t_n)$$

$$= \int_{G(d,k_n)} \cdots \int_{G(d,k_0)} [L_0, \dots, L_n]\, \tau_{M_0}(\mathrm{d}L_0) \cdots \tau_{M_n}(\mathrm{d}L_n).$$

For the proof, we refer to Wieacker [816]. He has a more general result, for $(\mathcal{H}^{k_i}, k_i)$-rectifiable subsets M_i, but in that case, an additional assumption on the product $M_0 \times \dots \times M_n$ is required.

Our first conclusion from Theorem 5.4.1 is a kinematic formula for two surfaces.

Theorem 5.4.2. *Let $k_0, k_1 \in \{1, \dots, d-1\}$ be numbers with $k_0 + k_1 \geq d$, let M_i be a k_i-surface, for $i = 0, 1$. Then*

$$\int_{G_d} \mathcal{H}^{k_0+k_1-d}(M_0 \cap gM_1)\, \mu(\mathrm{d}g) = c^{k_0,k_1}_{k_0+k_1-d,d}\, \mathcal{H}^{k_0}(M_0)\mathcal{H}^{k_1}(M_1).$$

This theorem holds, more generally, for $\mathcal{H}^{k_i}$-rectifiable subsets M_i, $i = 0, 1$, if it is assumed that $M_0 \times M_1$ is $(\mathcal{H}^{k_0+k_1}, k_0 + k_1)$-rectifiable; see Zähle [823]. We also refer to this paper for the necessary measurability considerations and the proof that $M_0 \cap gM_1$ is $(\mathcal{H}^{k_0+k_1-d}, k_0 + k_1 - d)$-rectifiable for μ-almost all $g \in G_d$. Here we show only how the formula of Theorem 5.4.2 follows from that of Theorem 5.4.1.

Proof. Making use of the obvious facts that $\tau_{\vartheta M} = \vartheta(\tau_M)$ for $\vartheta \in SO_d$ and that the integral $\int_{SO_d} [L_0, \vartheta L_1]\, \nu(\mathrm{d}\vartheta)$ is invariant under rotations of L_0, we obtain

$$\int_{G_d} \mathcal{H}^{k_0+k_1-d}(M_0 \cap gM_1)\, \mu(\mathrm{d}g)$$

$$= \int_{SO_d} \int_{\mathbb{R}^d} \mathcal{H}^{k_0+k_1-d}(M_0 \cap (\vartheta M_1 + t))\, \lambda(\mathrm{d}t)\, \nu(\mathrm{d}\vartheta)$$

$$= \int_{SO_d} \int_{G(d,k_0)} \int_{G(d,k_1)} [L_0, L_1]\, \tau_{\vartheta M_1}(\mathrm{d}L_1)\, \tau_{M_0}(\mathrm{d}L_0)\, \nu(\mathrm{d}\vartheta)$$

$$= \int_{SO_d} \int_{G(d,k_0)} \int_{G(d,k_1)} [L_0, \vartheta L_1]\, \tau_{M_1}(\mathrm{d}L_1)\, \tau_{M_0}(\mathrm{d}L_0)\, \nu(\mathrm{d}\vartheta)$$

$$= c\, \mathcal{H}^{k_0}(M_0)\mathcal{H}^{k_1}(M_1)$$

with a constant c. Its value is obtained from the principal kinematic formula (5.10), if we choose for M_i a convex body of dimension k_i ($i = 0, 1$) and observe that then $V_j(M_0 \cap gM_1) = \mathcal{H}^{k_0+k_1-d}(K \cap gM)$, $V_k(M_0) = 0$ for $k > k_0$ and $V_{k_0+k_1-k}(M_1) = 0$ for $k < k_0$. □

From this kinematic formula, we can deduce a Crofton formula.

Theorem 5.4.3. *Let $k, q \in \{1, \dots, d-1\}$ be numbers with $k + q \geq d$, let M be a k-surface. Then*

$$\int_{A(d,q)} \mathcal{H}^{k+q-d}(M \cap E)\, \mu_q(\mathrm{d}E) = c^{k,q}_{k+q-d,d}\, \mathcal{H}^k(M).$$

Proof. The proof is similar to that of Theorem 5.3.3, but simpler. Choose $L_q \in G(d,q)$ and a q-dimensional unit cube $C \subset L_q$. By Theorem 5.4.2,

$$J := \int_{G_d} \mathcal{H}^{k+q-d}(C \cap gM)\, \mu(\mathrm{d}g) = c^{k,q}_{k+q-d,d}\, \mathcal{H}^k(M).$$

On the other hand,

$$\begin{aligned} J &= \int_{SO_d} \int_{\mathbb{R}^d} \mathcal{H}^{k+q-d}(C \cap (\vartheta M + t))\, \lambda(\mathrm{d}t)\, \nu(\mathrm{d}\vartheta) \\ &= \int_{SO_d} \int_{L_q^\perp} \int_{L_q} \mathcal{H}^{k+q-d}(C \cap (\vartheta M + t_1 + t_2))\, \lambda_q(\mathrm{d}t_2)\, \lambda_{d-q}(\mathrm{d}t_1)\, \nu(\mathrm{d}\vartheta) \end{aligned}$$

and

$$\begin{aligned} &\int_{L_q} \mathcal{H}^{k+q-d}(C \cap (\vartheta M + t_1 + t_2))\, \lambda_q(\mathrm{d}t_2) \\ &= \int_{L_q} \mathcal{H}^{k+q-d}((C - t_2) \cap (\vartheta M + t_1))\, \lambda_q(\mathrm{d}t_2) \\ &= \mathcal{H}^{k+q-d}(L_q \cap (\vartheta M + t_1)) \end{aligned}$$

by Theorem 5.2.1 (the σ-finiteness condition is satisfied for almost all ϑ). This gives

$$\begin{aligned} J &= \int_{SO_d} \int_{L_q^\perp} \mathcal{H}^{k+q-d}(L_q \cap (\vartheta M + t_1))\, \lambda_{d-q}(\mathrm{d}t_1)\, \nu(\mathrm{d}\vartheta) \\ &= \int_{SO_d} \mathcal{H}^{k+q-d}(M \cap \vartheta(L_q + t))\, \lambda_{d-q}(\mathrm{d}t)\, \nu(\mathrm{d}\vartheta) \\ &= \int_{A(d,q)} \mathcal{H}^{k+q-d}(M \cap E)\, \mu_q(\mathrm{d}E), \end{aligned}$$

which completes the proof. □

Finally, we consider the special case of Theorem 5.4.1 where each k_i is equal to $d-1$. If M is a $(d-1)$-surface, it is convenient to replace the measure τ_M by the even measure σ_M on the unit sphere which for $A \in \mathcal{B}(S^{d-1})$ without antipodal points is defined by

$$\sigma_M(A) := \frac{1}{2}\tau_M(\{u^\perp : u \in A\}).$$

We define an auxiliary convex body Π_M, a zonoid, by its support function

$$h(\Pi_M, u) := \frac{1}{2}\int_{S^{d-1}} |\langle u, v\rangle| \,\sigma_M(\mathrm{d}v), \qquad u \in S^{d-1}. \tag{5.30}$$

If $K \in \mathcal{K}'$ is a convex body, then $\sigma_{\mathrm{bd}\, K} = (1/2)[S_{d-1}(K,\cdot)+S_{d-1}(-K,\cdot)]$ and, therefore,

$$\Pi_{\mathrm{bd}\, K} = \Pi_K,$$

where Π_K is the projection body of K, introduced in (14.40), (14.41).

If $m \in \{2,\dots,d\}$ and M_i is a $(d-1)$-surface ($i = 1,\dots,m$), then the formula of Theorem 5.4.1 can be written as follows (recall the definition of ∇_m before Theorem 4.4.8).

$$\int_{\mathbb{R}^d} \cdots \int_{\mathbb{R}^d} \mathcal{H}^{d-m}(M_1 \cap (M_2+t_2) \cap \ldots \cap (M_m+t_m))\, \lambda(\mathrm{d}t_2)\cdots\lambda(\mathrm{d}t_m)$$
$$= \int_{S^{d-1}} \cdots \int_{S^{d-1}} \nabla_m(u_1,\dots,u_m)\, \sigma_{M_1}(\mathrm{d}u_1)\cdots\sigma_{M_m}(\mathrm{d}u_m)$$
$$= \frac{d!}{(d-m)!\kappa_{d-m}} V(\Pi_{M_1},\dots,\Pi_{M_m}, B^d,\dots,B^d),$$

where the right side is a mixed volume. The last equality follows from (14.34), observing the factor 1/2 in (5.30).

We state the last result as a theorem.

Theorem 5.4.4. *Let $m \in \{2,\dots,d\}$, and let M_i be a $(d-1)$-surface, for $i = 1,\dots,m$. Then*

$$\int_{\mathbb{R}^d} \cdots \int_{\mathbb{R}^d} \mathcal{H}^{d-m}(M_1 \cap (M_2+t_2) \cap \ldots \cap (M_m+t_m))\, \lambda(\mathrm{d}t_2)\cdots\lambda(\mathrm{d}t_m)$$
$$= \frac{d!}{(d-m)!\kappa_{d-m}} V(\Pi_{M_1},\dots,\Pi_{M_m}, B^d,\dots,B^d).$$

By a **convex hypersurface** we understand any $(d-1)$-surface of the form $F = B \cap \mathrm{bd}\, K$, where $K \in \mathcal{K}$ is a convex body with interior points and $B \in \mathcal{B}(\mathbb{R}^d)$ is a Borel set. For such a convex hypersurface, the body defined by (5.30) can be represented by

$$h(\Pi_F, u) := \frac{1}{2}\int_F |\langle u, n_K(x)\rangle| \,\mathcal{H}^{d-1}(\mathrm{d}x), \qquad u \in S^{d-1}. \tag{5.31}$$

Here $n_K(x)$ denotes the outer unit normal vector of K at $x \in \mathrm{bd}\, K$; it is uniquely determined $\mathcal{H}^{d-1}$-almost everywhere on $\mathrm{bd}\, K$. The integral in (5.31) depends only on F and not on K.

Notes for Section 5.4

1. The techniques of geometric measure theory that are needed for the general versions of the results of this section are found in the book by Federer [229].

The first general versions of Theorems 5.4.2 and 5.4.3 are due to Federer [227].

Applications to random processes of Hausdorff rectifiable closed sets were investigated by Zähle [822].

2. The translative formulas of Theorems 5.4.1 and 5.4.4 appear in Wieacker [816]. He has extended the approach considerably and has studied various applications to stochastic geometry; see [817, 818].

3. Crofton formulas in Minkowski spaces and projective Finsler spaces. The particular case of Theorem 5.4.3, where M is a k-surface and $q = d - k$ is the complementary dimension, reduces to

$$\int_{A(d,d-k)} \operatorname{card}(M \cap E)\, \mu_{d-k}(\mathrm{d}E) = \alpha_{dk} \mathcal{H}^k(M) \tag{5.32}$$

with a constant α_{dk}. This formula provides a beautiful interpretation of the k-dimensional area $\mathcal{H}^k(M)$: it is, up to a normalizing factor, the invariant measure of the $(d-k)$-flats hitting M, weighted by the number of hits. This motivates the following reverse question. If some other notion of k-dimensional area, denoted by vol_k, is given, does there exist a measure (or a signed measure) η_{d-k} on $A(d, d-k)$ such that

$$\int_{A(d,d-k)} \operatorname{card}(M \cap E)\, \eta_{d-k}(\mathrm{d}E) = \mathrm{vol}_k(M) \tag{5.33}$$

holds for all k-surfaces M (or at least for a nontrivial subclass, such as polyhedral surfaces)? This question has been studied in various degrees of generality, in particular, for Minkowski spaces and for projective Finsler spaces. A **Minkowski space** is a finite-dimensional real normed vector space. A **Finsler metric** on an open convex subset C of $\mathbb{R}^d$ is (here) a continuous function $F : C \times \mathbb{R}^d \to [0, \infty)$ such that $F(x, \cdot)$ is a norm on $\mathbb{R}^d$ for each $x \in C$. In the following, the pair (C, F) is called a **Finsler space**, and it is called smooth if F is of class C^∞ on $C \times \mathbb{R}^d \setminus \{0\}$ and the unit sphere of the norm $F(x, \cdot)$ is quadratically convex (has positive curvatures), for each $x \in C$. In a Finsler space, there is a canonical notion of curve length (and an induced metric), denoted by vol_1. The Finsler space (C, F) is called **projective** if line segments are shortest curves connecting their endpoints. The classical examples of projective Finsler spaces are Minkowski spaces and the Hilbert geometries in bounded open convex sets.

In a Finsler space, for $k > 1$ there are many different possibilities of defining a reasonable notion of k-dimensional area, but no canonical one (see Álvarez and Thompson [31] for a survey). Two such notions are particularly natural and important from a geometric point of view. These are the **Busemann k-area**, which is defined by the k-dimensional Hausdorff measure coming from the induced metric, and the **Holmes–Thompson k-area**, which is defined via the symplectic volume. For a more detailed introduction, we refer to Schneider [712, pp. 165–177].

For the existence of Crofton type formulas, it has turned out that the Holmes–Thompson area is the right area notion to be used. Let vol_k denote the k-dimensional Holmes–Thompson area. It was observed, with different degrees of generality, by Busemann [143], El-Ekhtiar [216] and Schneider and Wieacker [721] that for vol_{d-1}

in a Minkowski space there always exists a translation invariant measure η_1 on $A(d,1)$ so that (5.33) holds. In order that (5.33) hold for vol_1 with a translation invariant measure η_{d-1}, it is necessary and sufficient that the Minkowski space be hypermetric. (A metric space (S,δ) is **hypermetric** if $\sum_{i,j=1}^k \delta(p_i,p_j)N_iN_j \le 0$ holds for $k \ge 2$, all $p_1,\dots,p_k \in S$ and all integers $N_1,\dots,N_k$ with $\sum_{i=1}^k N_i = 1$.) This is equivalent to the condition that the unit ball of the dual Minkowski space is a zonoid. If this assumption is satisfied, then there are translation invariant measures η_j on $A(d,j)$ such that the general Crofton type formula

$$\int_{A(d,j)} \mathrm{vol}_{k+j-d}(M \cap E)\, \eta_j(\mathrm{d}E) = \alpha_{nkj}\mathrm{vol}_k(M) \tag{5.34}$$

holds for all $k \in \{1,\dots,d\}$, $j \in \{d-k,\dots,d-1\}$ and for all k-surfaces M. This was proved by Schneider and Wieacker [721, Th. 7.3]. For general (not necessarily smooth) hypermetric projective Finsler spaces, the existence of measures η_{d-k} so that (5.33) holds at least for k-dimensional compact convex sets M was established by Schneider [705] (for $k = d-1$, the assumption 'hypermetric' can be deleted). The proof yields merely the existence; an explicit construction for the line measure η_1 in polytopal Hilbert geometries is described in Schneider [711].

For smooth projective Finsler spaces, general investigations on Crofton densities have been undertaken by Gelfand and Smirnov [255] and by Álvarez, Gelfand and Smirnov [30], in part related to Hilbert's fourth problem and to symplectic geometry. Subsequent work by Álvarez and Fernandes [26, 27, 28, 29] and the thesis of Fernandes [231] use double fibrations and the Gelfand transform as a unifying approach to integral geometric intersection formulas and obtain, in particular, Crofton type formulas (with signed measures) for Holmes–Thompson areas of smooth submanifolds in smooth projective Finsler spaces. The first of these papers makes use of the symplectic structure on the space of geodesics of a projective Finsler space. Later it turned out that the methods applied by Schneider and Wieacker [721] for the case of hypermetric Minkowski spaces (where they yield measures η_j) can be adapted to the case of smooth projective Finsler spaces (where they yield signed measures). In this way, a very general version of the Crofton formula (5.34) was obtained, namely for $k = 1,\dots,d$, $j = d-k,\dots,d-1$ and for Holmes–Thompson areas of $(\mathcal{H}^k,k)$-rectifiable Borel sets M in smooth projective Finsler spaces (where the local unit spheres need not be quadratically convex); see Schneider [706].

The special role that the Holmes–Thompson area plays in connection with Crofton type formulas can be illuminated from other sides. Following Busemann, one can define a general notion of Minkowskian $(d-1)$-area by a few natural axioms. It was shown by Schneider [698] that there exist Minkowski spaces for which, among all Minkowskian $(d-1)$-areas, only the multiples of the Holmes–Thompson area allow a Crofton formula (5.33) for $k = d-1$ with a translation invariant measure η_1. For the Busemann area, the picture is not clear. Let us say that, for a Minkowski space $S = (\mathbb{R}^d, \|\cdot\|)$, the Busemann area is **integral geometric** if (5.33) holds for S and for the Busemann $(d-1)$-area with a translation invariant measure η_1, and at least for all $(d-1)$-dimensional compact convex sets M. The following was shown by Schneider [703], for $d \ge 3$. Every neighborhood (in the sense of the Banach–Mazur distance) of the Euclidean space ℓ_2^d contains Minkowski spaces for which the Busemann area is not integral geometric, as well as spaces (different from ℓ_2^d) for which the Busemann area is integral geometric. If d is sufficiently large, then a full neighborhood of the Minkowski space ℓ_∞^d consists of Minkowski spaces for which the Busemann area is not integral geometric. We conjecture that it is generically true (that is, for a dense open subset of the space of all d-dimensional Minkowski spaces) that the Busemann area is not integral geometric. In the preceding counterexamples, non-smoothness properties of the unit ball of the Minkowski space play a role. On the other hand, Álvarez and Berck [25] have constructed smooth projective Finsler spaces in which there is no Crofton formula for the Busemann area, not even with a signed measure.

Additional and more detailed information can be found in the survey of Schneider [710].

6

Extended Concepts of Integral Geometry

In this chapter, we derive further integral geometric formulas for convex bodies. They are related to the principal kinematic formula, either directly or indirectly. As in the latter formula, we have a fixed and a moving set, but in the two subsequent sections we do not consider intersections of both; we form sums of convex bodies or projections of convex bodies to subspaces. First we treat rotation means of Minkowski sums, which will later (Section 8.5) be applied to touching probabilities. The global version is an immediate consequence of the principal kinematic formula; the local version will be proved by techniques similar to those in Sections 5.2 and 5.3. From the formulas for rotation means of sums we deduce projection formulas.

In Section 6.3, we admit (infinite) convex cylinders as moving sets. For these, we derive a local kinematic formula, and we also obtain a formula that combines sections with projections.

Section 6.4 is devoted to a continuation of translative integral geometry. We treat iterated translative formulas, which involve a more general series of mixed measures, and consider rotation means and results of Crofton type for the mixed measures. The integral formulas for mixed measures and their global versions, the mixed functionals, also yield kinematic formulas for certain mixed volumes and for projection functions and support functions of convex bodies.

Section 6.5 provides an introduction to the integral geometry of spherically convex sets in the spherical space S^{d-1}.

6.1 Rotation Means of Minkowski Sums

In this section, we are interested in mean value formulas for the Minkowski sum of a fixed and a moving convex body. The functions to be integrated are again intrinsic volumes and curvature measures. Since $V_j(K + (\vartheta M + x))$, for example, does not depend on x, only the rotations of M are relevant, hence we shall be interested in the integral

$$\int_{SO_d} V_j(K + \vartheta M)\,\nu(\mathrm{d}\vartheta).$$

In order to illustrate the connection with the principal kinematic formula, we first prove the global version of the **rotational mean value formula.**

Theorem 6.1.1. *If $K, M \in \mathcal{K}'$ are convex bodies and if $j \in \{0, \dots, d\}$, then*

$$\int_{SO_d} V_j(K + \vartheta M)\,\nu(\mathrm{d}\vartheta) = \sum_{k=0}^{j} c_{d-j,d}^{d+k-j,d-k} V_k(K) V_{j-k}(M).$$

Proof. First we consider the case $j = d$. We have

$$\int_{SO_d} V_d(K + \vartheta M)\,\nu(\mathrm{d}\vartheta) = \int_{SO_d} \int_{\mathbb{R}^d} \mathbf{1}_{K+\vartheta M}(x)\,\lambda(\mathrm{d}x)\,\nu(\mathrm{d}\vartheta).$$

The relation $x \in K + \vartheta M$ is equivalent to $K \cap (\vartheta M' + x) \neq \emptyset$, where $M' := -M$. Hence, we obtain

$$\begin{aligned}\int_{SO_d} V_d(K + \vartheta M)\,\nu(\mathrm{d}\vartheta) &= \int_{SO_d} \int_{\mathbb{R}^d} V_0(K \cap (\vartheta M' + x))\,\lambda(\mathrm{d}x)\,\nu(\mathrm{d}\vartheta)\\ &= \int_{G_d} V_0(K \cap gM')\,\mu(\mathrm{d}g)\\ &= \sum_{k=0}^{d} c_{0,d}^{k,d-k} V_k(K) V_{d-k}(M),\end{aligned}$$

where we have used the principal kinematic formula (Theorem 5.1.3) and the fact that $V_j(M') = V_j(M)$ for $j = 0, \dots, d$.

Now we replace K by $K + \epsilon B^d$ with $\epsilon > 0$ and apply the Steiner formula (14.16), to obtain

$$\begin{aligned}&\sum_{j=0}^{d} \epsilon^{d-j} \kappa_{d-j} \int_{SO_d} V_j(K + \vartheta M)\,\nu(\mathrm{d}\vartheta)\\ &= \int_{SO_d} V_d((K + \vartheta M) + \epsilon B^d)\,\nu(\mathrm{d}\vartheta)\\ &= \int_{SO_d} V_d((K + \epsilon B^d) + \vartheta M)\,\nu(\mathrm{d}\vartheta)\\ &= \sum_{m=0}^{d} c_{0,d}^{m,d-m} V_m(K + \epsilon B^d) V_{d-m}(M)\\ &= \sum_{m=0}^{d} \sum_{k=0}^{m} \epsilon^{m-k} \frac{1}{(m-k)!} c_{0,d}^{m,d-k} V_k(K) V_{d-m}(M).\end{aligned}$$

Putting $m = d + k - j$ and changing the order of summation, we get the double sum

$$\sum_{j=0}^{d} \epsilon^{d-j} \kappa_{d-j} \sum_{k=0}^{j} c_{d-j,d}^{d+k-j,d-k} V_k(K) V_{j-k}(M).$$

Comparing the coefficients, we obtain the assertion for all $j \in \{0, \ldots, d\}$. □

We want to extend the previous theorem to curvature measures, that is, replace the integrand $V_j(K + \vartheta M)$ by $\Phi_j(K + \vartheta M, A + \vartheta B)$. Evidently, this requires a restriction to the cases $j < d$ and to Borel sets $A \subset K$, $B \subset M$, contained in the respective bodies. Even under this assumption, $A + \vartheta B$ is in general not a Borel set, so that $\Phi_j(K + \vartheta M, A + \vartheta B)$ would not be defined. However, it will be sufficient to know the following.

Lemma 6.1.1. *Let $K, M \in \mathcal{K}'$, $A, B \in \mathcal{B}(\mathbb{R}^d)$ and $A \subset K$, $B \subset M$. For ν-almost all $\vartheta \in SO_d$ the set*

$$(A + \vartheta B) \cap \mathrm{bd}\,(K + \vartheta M)$$

is a Borel set, hence $\Phi_j(K + \vartheta M, A + \vartheta B)$ is defined for $j = 0, \ldots, d-1$.

Proof. For $x \in \mathrm{bd}\,(K + \vartheta M)$ there is a representation $x = y + z$ with $y \in K$, $z \in \vartheta M$. The points x, y, z lie in parallel supporting hyperplanes of $K + \vartheta M$, K, and ϑM, respectively; in particular, $y \in \mathrm{bd}\, K$ and $z \in \mathrm{bd}\, \vartheta M$. Suppose there is another representation $x = y_1 + z_1$ with $y_1 \in K$ and $z_1 \in \vartheta M$, then $y - y_1 = z_1 - z$, and the segments $\overline{yy_1}$, $\overline{z_1 z}$ satisfy $\overline{yy_1} \subset \mathrm{bd}\, K$ and $\overline{z_1 z} \subset \mathrm{bd}\, \vartheta M$. Hence the bodies K and ϑM contain parallel segments lying in parallel supporting hyperplanes. A theorem from the theory of convex bodies (see Schneider [695, Theorem 2.3.10]) says that for ν-almost all $\vartheta \in SO_d$ this does not occur. Hence, for these ϑ the representation $x = y + z$ with $y \in K$ and $z \in \vartheta M$ is unique for each $x \in \mathrm{bd}\,(K + \vartheta M)$. Putting

$$\pi_1(K, M, \vartheta, x) := y, \qquad \pi_2(K, M, \vartheta, x) := \vartheta^{-1} z,$$

we obtain mappings

$$\pi_1(K, M, \vartheta, \cdot) : \ \mathrm{bd}\,(K + \vartheta M) \to \mathrm{bd}\, K,$$
$$\pi_2(K, M, \vartheta, \cdot) : \ \mathrm{bd}\,(K + \vartheta M) \to \mathrm{bd}\, M.$$

From the compactness of the bodies K, M it follows easily that the mapping

$$\pi := \pi_1(K, M, \vartheta, \cdot) \times \pi_2(K, M, \vartheta, \cdot) : \ \mathrm{bd}\,(K + \vartheta M) \to \mathrm{bd}\, K \times \mathrm{bd}\, M$$

is continuous. Hence, for Borel sets $A \subset K$, $B \subset M$ the set

$$(A + \vartheta B) \cap \mathrm{bd}\,(K + \vartheta M) = \pi^{-1}(A \times B)$$

is a Borel set, too. □

In proving the local version of Theorem 6.1.1, we proceed similarly to the case of the principal kinematic formula, so we first consider polytopes. We say that two polytopes $K, M \in \mathcal{P}'$ are in **general relative position** if for any two faces F of K and G of M the linear subspaces $L(F), L(G)$ parallel to aff F, aff G, respectively, are in general position.

Theorem 6.1.2. *If $K, M \in \mathcal{K}'$ are convex bodies, $A, B \in \mathcal{B}(\mathbb{R}^d)$ are Borel sets satisfying $A \subset K$ and $B \subset M$, and if $j \in \{0, \dots, d-1\}$, then*

$$\int_{SO_d} \Phi_j(K + \vartheta M, A + \vartheta B)\, \nu(\mathrm{d}\vartheta)$$
$$= \sum_{k=0}^{j} c_{d-j,d}^{d+k-j,d-k} \Phi_k(K, A) \Phi_{j-k}(M, B). \tag{6.1}$$

Proof. The measurability of the integrand will be verified in the course of the proof. First we consider the case $j = d - 1$.

Let K, M be d-dimensional polytopes. By Lemmas 13.2.1 and 6.1.1, there is a Borel set $D_{K,M} \subset SO_d$ with $\nu(D_{K,M}) = 1$ such that K and ϑM are in general relative position and $(A + \vartheta B) \cap \mathrm{bd}\,(K + \vartheta M)$ is a Borel set if $\vartheta \in D_{K,M}$. Let $\vartheta \in D_{K,M}$. Since $K + \vartheta M$ is a polytope, we have

$$\Phi_{d-1}(K + \vartheta M, A + \vartheta B) = \sum_{F' \in \mathcal{F}_{d-1}(K+\vartheta M)} \gamma(F', K + \vartheta M) \lambda_{F'}(A + \vartheta B).$$

Because of $\vartheta \in D_{K,M}$, each facet $F' \in \mathcal{F}_{d-1}(K + \vartheta M)$ is of the form $F' = F + \vartheta G$ with $F \in \mathcal{F}_k(K)$ and $G \in \mathcal{F}_{d-1-k}(M)$, for some $k \in \{0, \dots, d-1\}$. For faces $F \in \mathcal{F}_k(K)$ and $G \in \mathcal{F}_{d-1-k}(M)$, we put $L_1 := (\mathrm{aff}\, F)^\perp$, $L_2 := (\mathrm{aff}\, G)^\perp$; then $L_1 \cap \vartheta L_2$ is of dimension one. The external angle $\gamma(F + \vartheta G, K + \vartheta M)$ is zero if $F + \vartheta G$ is not a face of $K + \vartheta M$; otherwise it is equal to $1/2$, and this happens if and only if

$$N(K, F) \cap \vartheta N(M, G) \cap S^{d-1} \neq \emptyset.$$

For arbitrary subsets $U \subset L_1$, $V \subset L_2$, the intersection $U \cap \vartheta V \cap S^{d-1}$ is either empty or one-pointed or two-pointed; we put

$$I(U, V, \vartheta) := \frac{1}{2} \mathrm{card}\,(U \cap \vartheta V \cap S^{d-1}).$$

If $F + \vartheta G$ is a face of $K + \vartheta M$, then

$$(A + \vartheta B) \cap (F + \vartheta G) = (A \cap F) + \vartheta (B \cap G).$$

Since the sum $F + \vartheta G$ is direct, we obtain

$$\lambda_{F+\vartheta G}(A + \vartheta B) = [F, \vartheta G] \lambda_F(A) \lambda_G(B).$$

This yields

$$\begin{aligned}&\Phi_{d-1}(K+\vartheta M, A+\vartheta B)\\&=\sum_{k=0}^{d-1}\sum_{F\in\mathcal{F}_k(K)}\sum_{\mathcal{F}_{d-1-k}(M)}\lambda_F(A)\lambda_G(B)I(N(K,F),N(M,G),\vartheta)[F,\vartheta G].\end{aligned}$$

If faces F, G are given, we now define

$$J(U,V) := \int_{SO_d} I(U,V,\vartheta)[F,\vartheta G]\,\nu(\mathrm{d}\vartheta)$$

for arbitrary Borel sets $U \subset L_1 \cap S^{d-1}$, $V \subset L_2 \cap S^{d-1}$. The measurability of the integrand is easily verified. This also yields the measurability of the integrand in (6.1) for the case where K and M are polytopes. Similarly to the proof of Theorem 5.3.1 one now proves the equality

$$J(U,V) = \alpha_{dk}\,\sigma_{d-k-1}(U)\,\sigma_k(V)$$

with a certain constant $\alpha_{dk} > 0$. This gives

$$\begin{aligned}&\int_{SO_d}\Phi_{d-1}(K+\vartheta M, A+\vartheta B)\,\nu(\mathrm{d}\vartheta)\\&=\sum_{k=0}^{d-1}\sum_{F\in\mathcal{F}_k(K)}\sum_{G\in\mathcal{F}_{d-k-1}(M)}\alpha'_{dk}\gamma(F,K)\gamma(G,M)\lambda_F(A)\lambda_G(B)\\&=\sum_{k=0}^{d-1}\alpha'_{dk}\Phi_k(K,A)\Phi_{d-k-1}(M,B).\end{aligned}$$

If $A = K$ and $B = M$, this formula must coincide with the corresponding one in Theorem 6.1.1, hence $\alpha'_{dk} = c_{1,d}^{k+1,d-k}$. Thus the proof of the case $j = d-1$ of (6.1) for d-dimensional polytopes K, M is complete.

Now let K, M be arbitrary d-dimensional convex bodies. Without loss of generality, we assume $0 \in \operatorname{int} K \cap \operatorname{int} M$. Then 0 is an inner point of $K+\vartheta M$, for all rotations $\vartheta \in SO_d$. By Lemma 6.1.1, there is a Borel set $D_{K,M} \subset SO_d$ with $\nu(D_{K,M}) = 1$ such that for $\vartheta \in D_{K,M}$ there exist the mappings

$$\begin{aligned}\pi_1(K,M,\vartheta,\cdot):\ &\operatorname{bd}(K+\vartheta M) \to \operatorname{bd} K,\\ \pi_2(K,M,\vartheta,\cdot):\ &\operatorname{bd}(K+\vartheta M) \to \operatorname{bd} M\end{aligned}$$

introduced in the proof of the lemma. Let $\vartheta \in D_{K,M}$. We extend the domain of $\pi_1(K,M,\vartheta,\cdot)$ and $\pi_2(K,M,\vartheta,\cdot)$ to all of $\mathbb{R}^d$. Since $0 \in \operatorname{int}(K+\vartheta M)$, to each $x \in \mathbb{R}^d$ there exist $\alpha \geq 0$ and $\overline{x} \in \operatorname{bd}(K+\vartheta M)$ with $x = \alpha\overline{x}$. We set

$$\pi_k(K,M,\vartheta,x) := \alpha\pi_k(K,M,\vartheta,\overline{x}) \qquad \text{for } k=1,2.$$

Evidently the mappings $\pi_k(K, M, \vartheta, \cdot) : \mathbb{R}^d \to \mathbb{R}^d$ thus defined are continuous ($k = 1, 2$). Let $\varphi(K, M, \vartheta, \cdot)$ be the image measure of $\Phi_{d-1}(K + \vartheta M, \cdot)$ under the map

$$\pi(K, M, \vartheta, \cdot) := \pi_1(K, M, \vartheta, \cdot) \times \pi_2(K, M, \vartheta, \cdot).$$

Then $\varphi(K, M, \vartheta, \cdot)$ is a finite Borel measure on $\mathbb{R}^d \times \mathbb{R}^d$, and for $A, B \in \mathcal{B}(\mathbb{R}^d)$ with $A \subset K$ and $B \subset M$, we have

$$\varphi(K, M, \vartheta, A \times B) = \Phi_{d-1}(K + \vartheta M, A + \vartheta B). \tag{6.2}$$

By the transformation rule for integrals,

$$\begin{aligned} &\int_{\mathbb{R}^d \times \mathbb{R}^d} f(x, y)\, \varphi(K, M, \vartheta, \mathrm{d}(x, y)) \\ &= \int_{\mathbb{R}^d} f(\pi(K, M, \vartheta, z))\, \Phi_{d-1}(K + \vartheta M, \mathrm{d}z) \end{aligned} \tag{6.3}$$

for all continuous functions f on $\mathbb{R}^d \times \mathbb{R}^d$.

Now let $\vartheta \in D_{K,M}$ and let $(\vartheta_i)_{i \in \mathbb{N}}$ be a sequence in $D_{K,M}$ converging to ϑ. We show that $\varphi(K, M, \vartheta_i, \cdot)$ converges weakly to $\varphi(K, M, \vartheta, \cdot)$ if $i \to \infty$. We can choose a convex body $C \in \mathcal{K}^d$ with $K + \vartheta_i M \subset C$ for all $i \in \mathbb{N}$. Let $f \in \mathbf{C}(\mathbb{R}^d \times \mathbb{R}^d)$. The function f is uniformly continuous on $C \times C$, hence for given $\epsilon > 0$ there is δ with $|f(x, y) - f(x', y')| < \epsilon$ for all $x, y, x', y' \in C$ with $\|x - x'\| + \|y - y'\| < 2\delta$. It is easy to see that $\pi_k(K, M, \vartheta_i, \cdot)$ converges to $\pi_k(K, M, \vartheta, \cdot)$, uniformly on C, for $k = 1, 2$. We infer that $\|\pi_k(K, M, \vartheta_i, z) - \pi_k(K, M, \vartheta, z)\| < \delta$ for all $z \in C$, almost all $i \in \mathbb{N}$, and $k = 1, 2$. Together with (6.3) this gives

$$\begin{aligned} &\left| \int_{\mathbb{R}^d \times \mathbb{R}^d} f \, \mathrm{d}\varphi(K, M, \vartheta_i, \cdot) - \int_{\mathbb{R}^d \times \mathbb{R}^d} f \, \mathrm{d}\varphi(K, M, \vartheta, \cdot) \right| \\ &\le \int_{\mathbb{R}^d} |f(\pi(K, M, \vartheta_i, z)) - f(\pi(K, M, \vartheta, z))| \, \Phi_{d-1}(K + \vartheta_i M, \mathrm{d}z) \\ &\quad + \left| \int_{\mathbb{R}^d} f(\pi(K, M, \vartheta, z)) \, (\Phi_{d-1}(K + \vartheta_i M, \mathrm{d}z) - \Phi_{d-1}(K + \vartheta M, \mathrm{d}z)) \right| \\ &< a\epsilon \end{aligned}$$

for almost all $i \in \mathbb{N}$, with a constant a not depending on i. Here we have used the fact that $\Phi_{d-1}(K + \vartheta_i M, \mathbb{R}^d)$ is bounded by a constant depending only on C (similarly to the proof of Theorem 5.2.3); further, the weak convergence $\Phi_{d-1}(K + \vartheta_i M, \cdot) \xrightarrow{w} \Phi_{d-1}(K + \vartheta M, \cdot)$ and the continuity of the function $f(\pi(K, M, \vartheta, \cdot))$ were applied.

The weak convergence thus established shows that for each $f \in \mathbf{C}(\mathbb{R}^d \times \mathbb{R}^d)$ the mapping

$$\vartheta \mapsto \int_{\mathbb{R}^d \times \mathbb{R}^d} f \, \mathrm{d}\varphi(K, M, \vartheta, \cdot)$$

is continuous on $D_{K,M}$. By Lemma 12.1.1 this implies the measurability of the mapping

$$\vartheta \mapsto \varphi(K, M, \vartheta, U)$$

on $D_{K,M}$, for all $U \in \mathcal{B}(\mathbb{R}^d \times \mathbb{R}^d)$. In particular, for $A, B \in \mathcal{B}(\mathbb{R}^d)$ with $A \subset K$ and $B \subset M$ we obtain from (6.2) the measurability of the map

$$\vartheta \mapsto \Phi_{d-1}(K + \vartheta M, A + \vartheta B)$$

on $D_{K,M}$ and hence the measurability ν-almost everywhere of the integrand in (6.1), if $j = d - 1$.

Putting

$$\varphi(K, M, \cdot) := \int_{SO_d} \varphi(K, M, \vartheta, \cdot)\, \nu(\mathrm{d}\vartheta),$$

we now obtain a finite measure on $\mathbb{R}^d \times \mathbb{R}^d$ satisfying

$$\int_{\mathbb{R}^d \times \mathbb{R}^d} f \,\mathrm{d}\varphi(K, M, \cdot) = \int_{SO_d} \int_{\mathbb{R}^d} f(\pi(K, M, \vartheta, z))\, \Phi_{d-1}(K + \vartheta M, \mathrm{d}z)\, \nu(\mathrm{d}\vartheta)$$

for $f \in \mathbf{C}(\mathbb{R}^d \times \mathbb{R}^d)$. We consider convergent sequences $K_i \to K$ and $M_i \to M$ of convex bodies K_i, M_i with $0 \in \operatorname{int} K_i \cap \operatorname{int} M_i$, and we put $D := \bigcap_{i=1}^{\infty} D_{K_i,M_i} \cap D_{K,M}$. As before, we see that for $\vartheta \in D$ the functions $f(\pi(K_i, M_i, \vartheta, \cdot))$ converge for $i \to \infty$, uniformly on every compact set, and we deduce in a similar way that

$$\int_{\mathbb{R}^d} f(\pi(K_i, M_i, \vartheta, z))\, \Phi_{d-1}(K_i + \vartheta M_i, \mathrm{d}z)$$
$$\to \int_{\mathbb{R}^d} f(\pi(K, M, \vartheta, z))\, \Phi_{d-1}(K + \vartheta M, \mathrm{d}z)$$

for $i \to \infty$. The dominated convergence theorem yields

$$\int_{SO_d} \int_{\mathbb{R}^d} f(\pi(K_i, M_i, \vartheta, z))\, \Phi_{d-1}(K_i + \vartheta M_i, \mathrm{d}z)\, \nu(\mathrm{d}\vartheta)$$
$$\to \int_{SO_d} \int_{\mathbb{R}^d} f(\pi(K, M, \vartheta, z))\, \Phi_{d-1}(K + \vartheta M, \mathrm{d}z)\, \nu(\mathrm{d}\vartheta)$$

and thus the weak convergence $\varphi(K_i, M_i, \cdot) \xrightarrow{w} \varphi(K, M, \cdot)$ for $i \to \infty$.

Obviously, the assertion of the theorem for $j = d - 1$ is equivalent to

$$\int_{\mathbb{R}^d \times \mathbb{R}^d} f(x) g(y)\, \varphi(K, M, \mathrm{d}(x, y))$$
$$= \sum_{k=0}^{d-1} c_{1,d}^{k+1,d-k} \int_{\mathbb{R}^d} f \,\mathrm{d}\Phi_k(K, \cdot) \int_{\mathbb{R}^d} g \,\mathrm{d}\Phi_{d-k-1}(M, \cdot)$$

for all $f,g \in \mathbf{C}(\mathbb{R}^d)$. Since we have proved the assertion for d-dimensional polytopes, we can use the latter equality, where both sides depend continuously on K and M, to extend it by approximation to arbitrary d-dimensional convex bodies K, M.

The extension to convex bodies without interior points and to $j < d-1$ is now achieved by an application of the local Steiner formula of Theorem 14.2.4. We assume first that $M \in \mathcal{K}'$ still has interior points, while $K \in \mathcal{K}'$ may be arbitrary. The assertion to be proved holds for $j = d-1$ and for the bodies $K + \epsilon B^d$ and M, where $\epsilon > 0$ is arbitrary. Using Theorem 14.2.4 twice, we therefore obtain the measurability of the integrand in (6.1) and the equalities

$$\sum_{j=0}^{d-1} \epsilon^{d-1-j} \frac{1}{(d-1-j)!} c_1^{d-j} \int_{SO_d} \Phi_j(K + \vartheta M, A + \vartheta B)\, \nu(\mathrm{d}\vartheta)$$

$$= \int_{SO_d} \Phi_{d-1}(K + \epsilon B^d + \vartheta M, A + \epsilon S^{d-1} + \vartheta B)\, \nu(\mathrm{d}\vartheta)$$

$$= \sum_{r=0}^{d-1} c_{1,d}^{r+1,d-r} \Phi_r(K + \epsilon B^d, A + \epsilon S^{d-1}) \Phi_{d-r-1}(M, B)$$

$$= \sum_{r=0}^{d-1} c_{1,d}^{r+1,d-r} \sum_{k=0}^{r} \epsilon^{r-k} \frac{1}{(r-k)!} c_{d-r}^{d-k} \Phi_k(K, A) \Phi_{d-r-1}(M, B)$$

$$= \sum_{j=0}^{d-1} \epsilon^{d-1-j} \frac{1}{(d-1-j)!} c_1^{d-j} \sum_{k=0}^{j} c_{d-j,d}^{d+k-j,d-k} \Phi_k(K, A) \Phi_{j-k}(M, B).$$

Comparing the coefficients, we obtain the assertion for the bodies K and M. Analogously, M can be replaced by an arbitrary convex body. □

Notes for Section 6.1

1. Theorem 6.1.1 goes back, with a different proof, to Hadwiger [307, p. 231]. A local version of this mean value formula under Minkowski addition was first proved by Schneider [673], though not for the curvature measures Φ_j, but for the area measures Ψ_j. Weil [780] used a result of Schneider [677] on curvature measures to prove Theorem 6.1.2. A simpler proof and a generalization appear in Schneider [688].

2. The rotation formula (6.1) has an extension to support measures. It involves an operation for sets of support elements which is adapted to the Minkowski addition of convex bodies. For sets $A, B \subset \Sigma = \mathbb{R}^d \times S^{d-1}$ we define

$$A * B := \{(x+y, u) \in \Sigma : (x,u) \in A,\ (y,u) \in B\}.$$

This operation combines the behaviors of sets of boundary points and of normal vectors of convex bodies under addition, in the following way. If $A \subset \operatorname{Nor} K$ and

$B \subset \operatorname{Nor} M$, then $A * B \subset \operatorname{Nor}(K + M)$, and for $A_1, A_2 \subset \mathbb{R}^d$ and $B_1, B_2 \subset S^{d-1}$ we have

$$(A_1 \times B_1) * (A_2 \times B_2) = (A_1 + A_2) \times (B_1 \cap B_2).$$

The following result holds for convex bodies $K, M \in \mathcal{K}'$, Borel sets $A \subset \operatorname{Nor} K$, $B \subset \operatorname{Nor} M$, and for $j = 0, \dots, d-1$:

$$\int_{SO_d} \Xi_j(K + \vartheta M, A * \vartheta B)\, \nu(\mathrm{d}\vartheta) = \sum_{k=0}^{j} c_{d-j,d}^{d+k-j,d-k}\, \Xi_k(K, A) \Xi_{j-k}(M, B). \tag{6.4}$$

Special cases are (6.1) and the formula

$$\int_{SO_d} \Psi_j(K + \vartheta M, A \cap \vartheta B)\, \nu(\mathrm{d}\vartheta) = \sum_{k=0}^{j} c_{d-j,d}^{d+k-j,d-k} \Psi_k(K, A) \Psi_{j-k}(M, B) \tag{6.5}$$

for Borel sets $A, B \subset S^{d-1}$.

Formula (6.4) was proved by Schneider [688]; the special case (6.5) was obtained earlier by Schneider [672].

3. For special pairs of convex bodies $K, M \in \mathcal{K}'$, a counterpart to Theorem 6.1.1 holds with the sum $K + \vartheta M$ replaced by the Minkowski difference $K \ominus \vartheta M$. One says that M **rolls freely** in K if for each rotation $\vartheta \in SO_d$ and each point $x \in \operatorname{bd} K$ there is a vector t such that $x \in \vartheta M + t \subset K$, equivalently, if each rotation image ϑM is a summand of K (see Schneider [695, p. 150]). If M rolls freely in K, then

$$\int_{SO_d} V_d(K \ominus \vartheta M)\, \nu(\mathrm{d}\vartheta) = \sum_{k=0}^{d} (-1)^{d-k} c_{d,0}^{d-k,k} V_k(K) V_{d-k}(M). \tag{6.6}$$

In fact, Theorem 6.1.1 together with (14.20) yields

$$\begin{aligned}
\sum_{k=0}^{d} c_{d,0}^{d-k,k} V_k(K) V_{d-k}(M) \epsilon^{d-k} &= \int_{SO_d} V_d(K + \epsilon\vartheta M)\, \nu(\mathrm{d}\vartheta) \\
&= \sum_{k=0}^{d} \epsilon^{d-k} \binom{d}{k} \int_{SO_d} V(K[k], \vartheta M[d-k])\, \nu(\mathrm{d}\vartheta).
\end{aligned}$$

Comparing the coefficients, we obtain

$$\binom{d}{k} \int_{SO_d} V(K[k], \vartheta M[d-k])\, \nu(\mathrm{d}\vartheta) = c_{d,0}^{d-k,k} V_k(K) V_{d-k}(M) \tag{6.7}$$

for $k = 0, \dots, d$. Since $(K \ominus M) + M = K$, the symmetry and linearity properties of mixed volumes imply

$$\begin{aligned}
V_d(K \ominus M) &= V(K \ominus M, \dots, K \ominus M) \\
&= V(K \ominus M, \dots, K \ominus M, K) - V(K \ominus M, \dots, K \ominus M, M) \\
&= \dots = \sum_{k=0}^{d} (-1)^{d-k} \binom{d}{k} V(K[k], M[d-k]),
\end{aligned}$$

and similarly for ϑM instead of M. Together with (6.7) this yields (6.6).

4. Containment measures. While the principal kinematic formula expresses the hitting measure $\mu(\{g \in G_d : gM \cap K \neq \emptyset\})$ of two convex bodies $K, M \in \mathcal{K}'$ in terms of intrinsic volumes, there is in general no simple expression for the **containment measure** (also called **inclusion measure**)

$$I(M,K) := \mu(\{g \in G_d : gM \subset K\}).$$

An exception is the case of the previous note: if M rolls freely in K, then $\{t \in \mathbb{R}^d : \vartheta M + t \subset K\} = K \ominus \vartheta M$, hence $I(M,K) = \int_{SO_d} V_d(K \ominus \vartheta M)\, \nu(\mathrm{d}\vartheta)$.

For results on containment measures, in particular for the case where M is a segment, we refer to Santaló [664], Ren [635], Zhang [832], the survey by Zhang and Zhou [834], and the literature quoted there.

6.2 Projection Formulas

A further familiar operation for convex bodies is the projection to a subspace. For a subspace $L \in G(d,q)$, recall that $A|L$ is the image of the set $A \subset \mathbb{R}^d$ under the orthogonal projection to L. From the results of the last sections we shall now derive **projection formulas**.

Theorem 6.2.1. *If $K \in \mathcal{K}'$ is a convex body, $A \in \mathcal{B}(\mathbb{R}^d)$ is a Borel set satisfying $A \subset K$, and if $q \in \{1,\dots,d-1\}$, $j \in \{0,\dots,q-1\}$, then*

$$\int_{G(d,q)} \Phi_j(K|L, A|L)\, \nu_q(\mathrm{d}L) = c_{d,q-j}^{q,d-j} \Phi_j(K,A). \tag{6.8}$$

Proof. Let $L_q \in G(d,q)$ be fixed. By the definition of ν_q,

$$\int_{G(d,q)} \Phi_j(K|L, A|L)\, \nu_q(\mathrm{d}L) = \int_{SO_d} \Phi_j(K|\vartheta L_q, A|\vartheta L_q)\, \nu(\mathrm{d}\vartheta).$$

Let M be a unit cube in $L_q^\perp$ and $B := \operatorname{relint} M$, then

$$\Phi_k(M,B) = \begin{cases} 1 \text{ for } k = d-q, \\ 0 \text{ for } k \neq d-q. \end{cases} \tag{6.9}$$

Let $\vartheta \in SO_d$ be chosen in such a way that K and ϑM do not contain parallel segments in parallel supporting hyperplanes. We consider the local parallel set

$$U_\epsilon(K,A) := \{x \in \mathbb{R}^d : \|x - p(K,x)\| \le \epsilon,\ p(K,x) \in A\}.$$

For $\epsilon > 0$ we have

$$\begin{aligned} &U_\epsilon(K + \vartheta M, (A + \vartheta B) \cap \operatorname{bd}(K + \vartheta M)) \\ &= \{z \in U_\epsilon(K, A') : z - p(K,z) \in \vartheta L_q\} + \vartheta B \end{aligned}$$

with $A' := \{a \in A : a|\vartheta L_q \in \operatorname{relbd}(K|\vartheta L_q)\}$. In fact, if $y = p(K + \vartheta M, x)$ and $y \in (A + \vartheta B) \cap \operatorname{bd}(K + \vartheta M)$, then $y = a + \vartheta b$ with $a \in A$, $b \in B$. There is a supporting hyperplane H to $K + \vartheta M$ through y. Since ϑb lies in a supporting hyperplane of ϑM parallel to H and since $b \in \operatorname{relint} M$, we obtain $\vartheta L_q^\perp + y \subset H$ and thus $a \in A'$. The argument can be reversed.

Trivially we have

$$(A|\vartheta L_q) \cap \operatorname{relbd}(K|\vartheta L_q) = A'|\vartheta L_q. \tag{6.10}$$

This set is a Borel set, since the orthogonal projection to ϑL_q, restricted to the points that are projected to $\operatorname{relbd}(K|\vartheta L_q)$, is a homeomorphism, by the choice of ϑ. Fubini's theorem now gives

$$\lambda(U_\epsilon(K + \vartheta M, (A + \vartheta B) \cap \operatorname{bd}(K + \vartheta M))) = \lambda_q(U_\epsilon^{(q)}(K|\vartheta L_q, A'|\vartheta L_q)),$$

where $U_\epsilon^{(q)}$ is a local parallel set in ϑL_q. Using the local Steiner formula (14.12) and (6.10), we obtain

$$\sum_{i=0}^{d-1} \epsilon^{d-i}\kappa_{d-i}\Phi_i(K + \vartheta M, A + \vartheta B) = \sum_{j=0}^{q-1} \epsilon^{q-j}\kappa_{q-j}\Phi_j(K|\vartheta L_q, A|\vartheta L_q),$$

hence

$$\Phi_j(K|\vartheta L_q, A|\vartheta L_q) = \Phi_{d-q+j}(K + \vartheta M, A + \vartheta B)$$

for $j = 0, \dots, q-1$ and

$$\Phi_i(K + \vartheta M, A + \vartheta B) = 0$$

for $i = 0, \dots, d-q-1$. This entails the measurability, up to a set of ν-measure zero, of the mapping $\vartheta \mapsto \Phi_j(K|\vartheta L_q, A|\vartheta L_q)$, and from Theorem 6.1.2 and (6.9) we then obtain

$$\begin{aligned}
&\int_{SO_d} \Phi_j(K|\vartheta L_q, A|\vartheta L_q)\, \nu(\mathrm{d}\vartheta) \\
&= \int_{SO_d} \Phi_{d-q+j}(K + \vartheta M, A + \vartheta B)\, \nu(\mathrm{d}\vartheta) \\
&= \sum_{r=0}^{d+j-q} c_{d,q-j}^{q+r-j,d-r} \Phi_r(K, A)\Phi_{d-q+j-r}(M, B) \\
&= c_{d,q-j}^{q,d-j} \Phi_j(K, A),
\end{aligned}$$

as asserted. □

Theorem 6.2.1 implies a projection formula for the intrinsic volumes V_j, $j = 0, \dots, q-1$. This formula holds for V_q, too.

Theorem 6.2.2. *If $K \in \mathcal{K}'$ and $q \in \{1, \dots, d-1\}$, $j \in \{0, \dots, q\}$, then*

$$\int_{G(d,q)} V_j(K|L)\, \nu_q(\mathrm{d}L) = c_{d,q-j}^{q,d-j} V_j(K).$$

Proof. Only the case $j = q$ needs to be proved. Fix $L_q \in G(d,q)$ and let B^{d-q} be the unit ball in $L_q^\perp$. For $\epsilon > 0$, Theorem 6.1.1 gives

$$\int_{SO_d} V_d(\vartheta K + \epsilon B^{d-q})\, \nu(\mathrm{d}\vartheta) = \sum_{k=0}^{d} c_{d,0}^{k,d-k} V_k(K) V_{d-k}(B^{d-q}) \epsilon^{d-k}.$$

The coefficient of ϵ^{d-q} is $c_{d,0}^{q,d-q} V_q(K)\kappa_{d-q}$. On the other hand, Fubini's theorem gives

$$V_d(\vartheta K + \epsilon B^{d-q}) = \int_{L_q} V_{d-q}((\vartheta K \cap (L_q^\perp + x)) + \epsilon B^{d-q})\, \lambda_q(\mathrm{d}x).$$

Applying the Steiner formula (14.5) in $L_q^\perp + x$, we obtain on the right side a polynomial in ϵ, where the coefficient of ϵ^{d-q} is equal to

$$\int_{L_q} \kappa_{d-q} V_0(\vartheta K \cap (L_q^\perp + x))\, \lambda_q(\mathrm{d}x) = \kappa_{d-q} V_q(\vartheta K|L_q) = \kappa_{d-q} V_q(K|\vartheta^{-1} L_q).$$

Integrating over SO_d (observing the invariance of ν) and comparing the coefficients, we obtain the assertion. □

Choosing $q = j$ in Theorem 6.2.2, we get

$$V_j(K) = c_{j,d-j}^{0,d} \int_{G(d,j)} V_j(K|L)\, \nu_j(\mathrm{d}L), \tag{6.11}$$

which is known as **Kubota's formula**. The special case $j = d-1$ of (6.11) yields the representation

$$\begin{aligned} S(K) = 2V_{d-1}(K) &= \frac{d\kappa_d}{\kappa_{d-1}} \int_{G(d,d-1)} V_{d-1}(K|L)\, \nu_{d-1}(\mathrm{d}L) \\ &= \frac{1}{\kappa_{d-1}} \int_{S^{d-1}} V_{d-1}(K|u^\perp)\, \sigma(\mathrm{d}u) \end{aligned} \tag{6.12}$$

for the surface area $S(K)$ of K. The latter equation is called **Cauchy's surface area formula**. For $j = 1$, (6.11) reduces to

$$V_1(K) = \frac{d\kappa_d}{2\kappa_{d-1}} \int_{G(d,1)} V_1(K|L)\, \nu_1(\mathrm{d}L).$$

Since $V_1(K|L)$ is the width of K in direction L, the integral

$$\int_{G(d,1)} V_1(K|L)\,\nu_1(\mathrm{d}L)$$

is the mean width, $b(K)$, of K. Hence we obtain formula (14.7),

$$V_1(K) = \frac{d\kappa_d}{2\kappa_{d-1}}\, b(K).$$

Notes for Section 6.2

1. The projection formulas of Theorem 6.2.2 are classical results of the integral geometry of convex bodies; a special case was already known to Cauchy. Local versions are found in Schneider [673] and Weil [780]. The reduction to the rotation formula for sums, which is used in the proof of Theorem 6.2.1, was noted in Schneider [688].

2. The projection formula (6.8) has an extension to support measures. For a set $A \subset \Sigma$ and a linear subspace of $\mathbb{R}^d$ we define

$$A|L := \{(x|L, u) : (x, u) \in A,\ u \in L\}.$$

Let $K \in \mathcal{K}'$ be a convex body and $A \subset \operatorname{Nor} K$ a Borel set. For $q \in \{1, \dots, d-1\}$, $j \in \{0, \dots, q-1\}$, the formula

$$\int_{G(d,q)} \Xi'_j(K|L, A|L)\,\nu_q(\mathrm{d}L) = c_{d,q-j}^{q,d-j}\,\Xi_j(K, A)$$

holds, where Ξ'_j denotes the jth support measure with respect to L. In a different, but equivalent formulation, this is Theorem 4.5.10 in Schneider [695].

3. An extension of the projection formula (6.8) to polyconvex sets was treated in Schneider [693]; here suitable multiplicities of tangential projections have to be taken into account.

6.3 Cylinders and Thick Sections

As we have seen, the Crofton formulas can be deduced from the principal kinematic formula, and the Cauchy–Kubota formulas are consequences of the rotation formulas for Minkowski sums. This shows that integral geometric formulas for convex bodies on one side and for affine subspaces on the other side are closely connected. This connection will become even more evident when we now consider cylinders and prove a common generalization of the principal kinematic formula and the Crofton formula.

By a (convex) **cylinder** C in $\mathbb{R}^d$ we understand a set of the form $C = M + L$ with $L \in G(d,q)$, $q \in \{0, \dots, d-1\}$, and $M \in \mathcal{K}'$, $M \subset L^\perp$. The linear subspace L is called the **direction space** of the cylinder C, and M is its **base**. Also the images gC of C under $g \in G_d$ are called cylinders, but C will always be of the standard form as described (with fixed $L \in G(d,q)$).

Since C is a closed convex set, the curvature measures $\Phi_0(C,\cdot),\dots,\Phi_d(C,\cdot)$ are well defined. They are finite on bounded Borel sets and have a special form. In the following, we identify $\mathbb{R}^d$ with $L^\perp \times L$. We denote by λ_{d-q} and λ_q, respectively, the Lebesgue measures in $L^\perp$ and L.

Lemma 6.3.1. *The curvature measures of the cylinder C satisfy*

$$\Phi_j(C,\cdot) = \begin{cases} \Phi_{j-q}(M,\cdot)\otimes\lambda_q & \text{for } q \le j \le d,\\ 0 & \text{for } 0 \le j < q. \end{cases}$$

Proof. We can assume that the base M is a polytope; the general case then follows by approximation, due to the weak continuity of the mapping $C \mapsto \Phi_j(C\cap K,\cdot)$, for each $K \in \mathcal{K}'$. Since C is polyhedral in that case, the representation (14.13) of the curvature measures of polytopes gives

$$\Phi_j(C,\cdot) = \sum_{F\in\mathcal{F}_j(C)} \gamma(F,C)\lambda_F.$$

Since $C = M + L$, we have $\mathcal{F}_j(C) = \emptyset$ for $j < q$, thus $\Phi_j(C,\cdot) = 0$. For $j \ge q$,

$$\mathcal{F}_j(C) = \{F + L : F \in \mathcal{F}_{j-q}(M)\},$$

hence in this case we get

$$\Phi_j(C,\cdot) = \sum_{F\in\mathcal{F}_{j-q}(M)} \gamma(F+L, M+L)\lambda_{F+L}.$$

Together with $\gamma(F+L,M+L) = \gamma(F,M)$ and $\lambda_{F+L} = \lambda_F \otimes \lambda_q$, this yields

$$\Phi_j(C,\cdot) = \left(\sum_{F\in\mathcal{F}_{j-q}(M)} \gamma(F,M)\lambda_F\right)\otimes\lambda_q = \Phi_{j-q}(M,\cdot)\otimes\lambda_q,$$

as stated. □

In analogy to the principal kinematic formula and the Crofton formula, we now consider intersections of a fixed convex body and a moving cylinder. The principal kinematic formula involves an integration over the motion group. Although the motion group has infinite invariant measure, the integrals remain finite, since for $K, M \in \mathcal{K}$ the relation $K \cap gM \neq \emptyset$ holds only for the motions g from a suitable compact set. However, for a convex body K with inner points and for a cylinder C with $q > 0$, the set of rigid motions g with $K \cap gC \neq \emptyset$ has infinite measure. In the case of the Crofton formula, which concerns the case $\dim M = 0$, the integration was therefore with respect to the invariant measure μ_q on the space $A(d,q)$ of q-flats. In a similar way, we can interpret the set of cylinders congruent to C as a homogeneous space, on which we can introduce an invariant measure. Implicitly, this has been done in the following theorem where, though, we work directly with a suitable representation of this invariant measure.

Theorem 6.3.1 (Local kinematic formula for cylinders). *Suppose that* $q \in \{0, \dots, d-1\}$ *and* $j \in \{0, \dots, d\}$. *Let* $K \in \mathcal{K}'$ *be a convex body, let* C *be a cylinder with direction space* $L \in G(d,q)$ *and base* M, *and let* $A, B \in \mathcal{B}(\mathbb{R}^d)$ *be Borel sets with* $B \subset L^\perp$. *Then*

$$\int_{SO_d} \int_{L^\perp} \Phi_j(K \cap \vartheta(C + x), A \cap \vartheta(B + L + x))\, \lambda_{d-q}(\mathrm{d}x)\, \nu(\mathrm{d}\vartheta)$$

$$= \sum_{k=j}^{N(d,j,q)} c_{j,d}^{k,d-k+j} \Phi_k(K, A) \Phi_{d-k+j-q}(M, B)$$

with $N(d,j,q) := \min\{d, d+j-q\}$.

Proof. First we note that

$$\int_{SO_d} \int_{L^\perp} \Phi_j(K \cap \vartheta(C + x), A \cap \vartheta(B + L + x))\, \lambda_{d-q}(\mathrm{d}x)\, \nu(\mathrm{d}\vartheta)$$

$$= \int_{SO_d} \int_{L^\perp} \Phi_j(\vartheta K \cap (C + x), \vartheta A \cap (B + L + x))\, \lambda_{d-q}(\mathrm{d}x)\, \nu(\mathrm{d}\vartheta).$$

Since $\{\vartheta K : \vartheta \in SO_d\}$ is bounded, there exists a compact set $B' \subset L$ (with $\lambda_q(B') > 0$) such that

$$\Phi_j(\vartheta K \cap (C + x), \vartheta A \cap (B + L + x))$$

$$= \Phi_j(\vartheta K \cap (C + x), \vartheta A \cap (B + B' + x))$$

for all $x \in L^\perp$ and all $\vartheta \in SO_d$. From Theorem 5.3.2 and Lemma 6.3.1 we get

$$\int_{SO_d} \int_{L^\perp} \int_L \Phi_j(\vartheta K \cap (C + x), \vartheta A \cap (B + B' + x + y))$$

$$\times \lambda_q(\mathrm{d}y)\, \lambda_{d-q}(\mathrm{d}x)\, \nu(\mathrm{d}\vartheta)$$

$$= \sum_{k=j}^{d} c_{j,d}^{k,d-k+j} \Phi_k(K, A) \Phi_{d-k+j}(C, B + B')$$

$$= \sum_{k=j}^{N(d,j,q)} c_{j,d}^{k,d-k+j} \Phi_k(K, A) \Phi_{d-k+j-q}(M, B) \lambda_q(B').$$

On the other hand, with $K' := \vartheta K - x$ and $A' := \vartheta A - x$ we have

$$\int_L \Phi_j(\vartheta K \cap (C + x), \vartheta A \cap (B + B' + x + y))\, \lambda_q(\mathrm{d}y)$$

$$= \int_L \Phi_j(K' \cap C, A' \cap (B + B' + y))\, \lambda_q(\mathrm{d}y)$$

$$
\begin{aligned}
&= \int_L \int_{\mathbb{R}^d} \mathbf{1}_{A'}(u)\mathbf{1}_{B+B'}(u-y)\,\Phi_j(K'\cap C,\mathrm{d}u)\,\lambda_q(\mathrm{d}y) \\
&= \int_{\mathbb{R}^d} \int_L \mathbf{1}_{B+B'}(u-y)\,\lambda_q(\mathrm{d}y)\mathbf{1}_{A'}(u)\,\Phi_j(K'\cap C,\mathrm{d}u) \\
&= \int_{\mathbb{R}^d} \mathbf{1}_{B+L}(u)\lambda_q(B')\mathbf{1}_{A'}(u)\,\Phi_j(K'\cap C,\mathrm{d}u) \\
&= \Phi_j(K'\cap C, A'\cap(B+L))\lambda_q(B') \\
&= \Phi_j(\vartheta K\cap(C+x), \vartheta A\cap(B+L+x))\lambda_q(B').
\end{aligned}
$$

Dividing by $\lambda_q(B')$, we obtain the assertion. □

As special cases, Theorem 6.3.1 contains both, the principal kinematic formula (Theorem 5.3.2) and the Crofton formula (Theorem 5.3.3). The former is obtained for $q = 0$ (thus $L = \{0\}$), and the latter for $M = B = \{0\}$ and $j \le q$, since then

$$
\Phi_{d-k+j-q}(M,B) = \begin{cases} 1 & \text{for } d-k+j-q=0, \\ 0 & \text{else.} \end{cases}
$$

The global version of Theorem 6.3.1 (that is, $A = B = \mathbb{R}^d$) results in a cylinder formula for intrinsic volumes.

Corollary 6.3.1 (Principal kinematic formula for cylinders). *Let $q \in \{0,\dots,d-1\}$ and $j \in \{0,\dots,d\}$. If $K \in \mathcal{K}'$ is a convex body and C is a cylinder with direction space $L \in G(d,q)$ and base M, then*

$$
\begin{aligned}
&\int_{SO_d}\int_{L^\perp} V_j(K\cap\vartheta(C+x))\,\lambda_{d-q}(\mathrm{d}x)\,\nu(\mathrm{d}\vartheta) \\
&\quad = \sum_{k=j}^{N(d,j,q)} c_{j,d}^{k,d-k+j} V_k(K)V_{d-k+j-q}(M).
\end{aligned}
$$

Especially for cylinders, there is a further operation besides section and projection – combining section and projection. Namely, for K and C as above, the intersection $K\cap\vartheta(C+x)$ can be projected orthogonally to the direction space ϑL of $\vartheta(C+x)$. In a special case, such a combination appears in certain applications. For example, microscopical sections, as they are treated in stereology by means of integral geometric methods, have a non-zero thickness. Therefore, a microscopical section is not an intersection with a plane, but with a cylinder $C = M + L$, where L is a plane and $M \subset L^\perp$ is a segment. Only the projection $(K\cap C)|L$ is observable. For such projections of sections with cylinders we state a general integral geometric formula, restricted to the global case.

Theorem 6.3.2 (Projected thick sections). *Let $q \in \{0,\dots,d-1\}$ and $j \in \{0,\dots,q\}$. If $K \in \mathcal{K}'$ is a convex body and C is a cylinder with direction space $L \in G(d,q)$ and base M, then*

$$\int_{SO_d} \int_{L^\perp} V_j((K \cap \vartheta(C+x))|\vartheta L)\, \lambda_{d-q}(\mathrm{d}x)\, \nu(\mathrm{d}\vartheta)$$
$$= \sum_{k=j}^{d+j-q} c_{j,q-j,d}^{k,d-k,q} V_k(K) V_{d-k+j-q}(M).$$

Proof. First, the double integral in the assertion is again written in the form

$$I_j := \int_{SO_d} \int_{L^\perp} V_j(((\vartheta K + x) \cap (M+L))|L)\, \lambda_{d-q}(\mathrm{d}x)\, \nu(\mathrm{d}\vartheta).$$

Since

$$((\vartheta K + x) \cap (M+L))|L = (\vartheta K - M + x) \cap L,$$

we get

$$I_j = \int_{SO_d} \int_{L^\perp} V_j((\vartheta K - M + x) \cap L)\, \lambda_{d-q}(\mathrm{d}x)\, \nu(\mathrm{d}\vartheta).$$

We put $\vartheta K - M =: C$ and $B^d \cap L =: B^q$ and let $\epsilon > 0$. Using Fubini's theorem, the Steiner formula (14.5) and the invariance properties of the Lebesgue measure, we obtain

$$\begin{aligned} V_d(C + \epsilon B^q) &= \int_{L^\perp} V_q((C + \epsilon B^q) \cap (L+y))\, \lambda_{d-q}(\mathrm{d}y) \\ &= \int_{L^\perp} V_q((C \cap (L+y)) + \epsilon B^q)\, \lambda_{d-q}(\mathrm{d}y) \\ &= \int_{L^\perp} \sum_{j=0}^{q} \epsilon^{q-j} \kappa_{q-j} V_j(C \cap (L+y))\, \lambda_{d-q}(\mathrm{d}y) \\ &= \sum_{j=0}^{q} \epsilon^{q-j} \kappa_{q-j} \int_{L^\perp} V_j((C+x) \cap L)\, \lambda_{d-q}(\mathrm{d}x). \end{aligned}$$

Inserting $C = \vartheta K - M$ and integrating over SO_d, we get

$$\begin{aligned} \sum_{j=0}^{q} \epsilon^{q-j} \kappa_{q-j} I_j &= \int_{SO_d} V_d(\vartheta K - M + \epsilon B^q)\, \nu(\mathrm{d}\vartheta) \\ &= \sum_{k=0}^{d} c_{0,d}^{d-k,k} V_k(K) V_{d-k}(-M + \epsilon B^q) \\ &= \sum_{k=0}^{d} c_{0,d}^{d-k,k} V_k(K) \sum_{r=0}^{d-k} V_r(-M) V_{d-k-r}(B^q) \epsilon^{d-k-r} \end{aligned}$$

$$= \sum_{j=0}^{q} \sum_{k=j}^{d+j-q} c_{0,d}^{d-k,k} V_k(K) V_{d-k+j-q}(M) V_{q-j}(B^q) \epsilon^{q-j}.$$

Here we have used Theorem 6.1.1 and Lemma 14.2.1. Since

$$V_{q-j}(B^q) = \binom{q}{j} \frac{\kappa_q}{\kappa_j},$$

a comparison of the coefficients yields the assertion. □

Notes for Section 6.3

1. Kinematic formulas for cylinders were treated by Santaló [662, p. 270 ff]. The local kinematic formula for a fixed convex body and a moving cylinder (Theorem 6.3.1) was proved in Schneider [680].

2. Theorem 6.3.2 and its proof are taken from Schneider [681].

6.4 Translative Integral Geometry, Continued

Our proof of the local principal kinematic formula, Theorem 5.3.2, was preceded by a translative version, Theorem 5.2.3. This translative formula involves a series of mixed measures $\Phi_k^{(j)}(K, M; \cdot)$, which are measures on the product space $\mathbb{R}^d \times \mathbb{R}^d$, depending homogeneously (of degrees k and $d-k+j$, respectively) and additively on the convex bodies K and M. In the following, we continue the investigation of translative formulas and consider iterations, rotation means and Crofton-type results for the mixed measures. In contrast to the iterated kinematic formula of Theorem 5.1.5, the iteration of the translative formula of Theorem 5.2.3 involves new functions at each iteration step. Altogether, a series of **mixed measures** is required, which depend on an increasing number of convex bodies. The total mixed measures define **mixed functionals**, which generalize the intrinsic volumes of one body and the mixed volumes of two convex bodies. We start with the definition of the mixed measures.

In order to simplify the presentation within this section, we frequently abbreviate the translate $A + x$ of a set A by A^x.

For polytopes $P_1, \ldots, P_k$ and faces F_i of P_i $(i = 1, \ldots, k)$ with

$$\sum_{i=1}^{k} \dim F_i \geq (k-1)d,$$

we define the **common external angle** $\gamma(F_1, \ldots, F_k; P_1, \ldots, P_k)$ by

$$\gamma(F_1, \ldots, F_k; P_1, \ldots, P_k) := \gamma(F_1 \cap F_2^{x_2} \cap \ldots \cap F_k^{x_k}, P_1 \cap P_2^{x_2} \cap \ldots \cap P_k^{x_k}),$$

where $x_2, \dots, x_k \in \mathbb{R}^d$ are chosen so that the sets $F_1, F_2^{x_2}, \dots, F_k^{x_k}$ have relatively interior points in common. The common external angle does not depend on the choice of $x_2, \dots, x_k$.

Definition 6.4.1. *Let*

$$k \in \mathbb{N}, \quad j \in \{0, \dots, d\}, \quad m_1, \dots, m_k \in \{j, \dots, d\}, \\ j = \textstyle\sum_{i=1}^k m_i - (k-1)d. \tag{6.13}$$

For polytopes $K_1, \dots, K_k \in \mathcal{P}'$, the **mixed measure** $\Phi^{(j)}_{m_1,\dots,m_k}(K_1, \dots, K_k; \cdot)$ *is the measure on $(\mathbb{R}^d)^k$ defined by*

$$\begin{aligned} &\Phi^{(j)}_{m_1,\dots,m_k}(K_1, \dots, K_k; A_1 \times \dots \times A_k) \\ &:= \sum_{F_1 \in \mathcal{F}_{m_1}(K_1)} \cdots \sum_{F_k \in \mathcal{F}_{m_k}(K_k)} \gamma(F_1, \dots, F_k; K_1, \dots, K_k) \\ &\quad \times [F_1, \dots, F_k] \lambda_{F_1}(A_1) \cdots \lambda_{F_k}(A_k) \end{aligned} \tag{6.14}$$

for $A_1, \dots, A_k \in \mathcal{B}(\mathbb{R}^d)$.

Convention. Integers k, j and k-tuples $(m_1, \dots, m_k)$ occurring in this section are always assumed to satisfy (6.13).

Obviously, the case $k = 1$ of (6.14) reduces to the representations of the curvature measures of polytopes given by (14.13), thus

$$\Phi^{(j)}_j(K; \cdot) = \Phi_j(K, \cdot).$$

The case $k = 2$ reduces to the mixed measures introduced in Theorem 5.2.2.

First we collect the essential properties of the mixed measures and state the iterated translative formula.

Theorem 6.4.1. *The mixed measure $\Phi^{(j)}_{m_1,\dots,m_k}(K_1, \dots, K_k; \cdot)$ depends continuously on the polytopes $K_1, \dots, K_k$ (in the weak topology). It has a (unique) continuous extension to arbitrary convex bodies $K_1, \dots, K_k \in \mathcal{K}'$. The extended measures have the following properties, valid for all $K_1, \dots, K_k \in \mathcal{K}'$ and $A_1, \dots, A_k \in \mathcal{B}(\mathbb{R}^d)$.*

(a) **Symmetry:**

$$\begin{aligned} &\Phi^{(j)}_{m_{i_1},\dots,m_{i_k}}(K_{i_1}, \dots, K_{i_k}; A_{i_1} \times \dots \times A_{i_k}) \\ &= \Phi^{(j)}_{m_1,\dots,m_k}(K_1, \dots, K_k; A_1 \times \dots \times A_k) \end{aligned}$$

for every permutation $(i_1, \dots, i_k)$ of $(1, \dots, k)$.

(b) **Decomposability:**

$$\Phi^{(j)}_{m_1,\dots,m_{k-1},d}(K_1,\dots,K_{k-1},K_k;\cdot)$$
$$= \Phi^{(j)}_{m_1,\dots,m_{k-1}}(K_1,\dots,K_{k-1};\cdot) \otimes (\lambda \llcorner K_k).$$

For $m_1,\dots,m_k < d$, *the measure* $\Phi^{(j)}_{m_1,\dots,m_k}(K_1,\dots,K_k;\cdot)$ *is concentrated on* $\mathrm{bd}\, K_1 \times \ldots \times \mathrm{bd}\, K_k$.

(c) **Homogeneity:**

$$\Phi^{(j)}_{m_1,\dots,m_k}(\alpha K_1, K_2,\dots,K_k;\alpha A_1 \times A_2 \times \ldots \times A_k)$$
$$= \alpha^{m_1} \Phi^{(j)}_{m_1,\dots,m_k}(K_1,\dots,K_k;A_1 \times \ldots \times A_k)$$

for $\alpha \geq 0$.

(d) **Additivity:** *The measure* $\Phi^{(j)}_{m_1,\dots,m_k}(K_1,\dots,K_k;\cdot)$ *is additive in each of its arguments* $K_1,\dots,K_k$.

(e) **Local determination:** *The measure* $\Phi^{(j)}_{m_1,\dots,m_k}(K_1,\dots,K_k;\cdot)$ *is locally determined, that is, for an open set* $U \subset (\mathbb{R}^d)^k$ *and for* $M_1,\dots,M_k \in \mathcal{K}'$ *with* $K_1 \times \ldots \times K_k \cap U = M_1 \times \ldots \times M_k \cap U$, *we have*

$$\Phi^{(j)}_{m_1,\dots,m_k}(K_1,\dots,K_k;\cdot) = \Phi^{(j)}_{m_1,\dots,m_k}(M_1,\dots,M_k;\cdot)$$

on U.

The following **iterated translative formula** *holds:*

$$\int_{(\mathbb{R}^d)^{k-1}} \Phi_j(K_1 \cap K_2^{x_2} \cap \ldots \cap K_k^{x_k}, A_1 \cap A_2^{x_2} \cap \ldots \cap A_k^{x_k})\, \lambda^{k-1}(\mathrm{d}(x_2,\dots,x_k))$$
$$= \sum_{\substack{m_1,\dots,m_k=j \\ m_1+\ldots+m_k=(k-1)d+j}}^{d} \Phi^{(j)}_{m_1,\dots,m_k}(K_1,\dots,K_k;A_1 \times \ldots \times A_k). \tag{6.15}$$

Proof. Concerning (6.15), the measurability of the integrand on the left side follows from the obvious extension of Lemma 5.2.1. We now show first that (6.15) holds for polytopes $K_1,\dots,K_k$, by using induction on k. For $k=1$, (6.15) is trivial, and for $k=2$ it reduces to Theorem 5.2.2. For $k \geq 3$, the induction hypothesis, Theorem 5.2.2 and Lemma 14.1.1 yield

$$\int_{(\mathbb{R}^d)^{k-1}} \Phi_j(K_1 \cap K_2^{x_2} \cap \ldots \cap K_k^{x_k}, A_1 \cap A_2^{x_2} \cap \ldots \cap A_k^{x_k})\, \lambda^{k-1}(\mathrm{d}(x_2,\dots,x_k))$$
$$= \sum_{\substack{m_1,\dots,m_{k-2},m=j \\ m_1+\ldots+m_{k-2}+m=(k-2)d+j}}^{d} \int_{\mathbb{R}^d} \Phi^{(j)}_{m_1,\dots,m_{k-2},m}(K_1,\dots,K_{k-2},K_{k-1} \cap K_k^{x};$$

$$A_1 \times \ldots \times A_{k-2} \times (A_{k-1} \cap A_k^x))\, \lambda(\mathrm{d}x)$$

$$= \sum_{\substack{m_1,\ldots,m_{k-2},m=j \\ m_1+\ldots+m_{k-2}+m=(k-2)d+j}}^{d} \sum_{F_1\in\mathcal{F}_{m_1}(K_1)} \cdots \sum_{F_{k-2}\in\mathcal{F}_{m_{k-2}}(K_{k-2})}$$

$$\int_{\mathbb{R}^d} \sum_{F\in\mathcal{F}_m(K_{k-1}\cap K_k^x)} \gamma(F_1,\ldots,F_{k-2},F;K_1,\ldots,K_{k-2},K_{k-1}\cap K_k^x)$$

$$\times [F_1,\ldots,F_{k-2},F]\lambda_{F_1}(A_1)\cdots\lambda_{F_{k-2}}(A_{k-2})\lambda_F(A_{k-1}\cap A_k^x)\,\lambda(\mathrm{d}x)$$

$$= \sum_{\substack{m_1,\ldots,m_k=j \\ m_1+\ldots+m_k=(k-1)d+j}}^{d} \sum_{F_1\in\mathcal{F}_{m_1}(K_1)} \cdots \sum_{F_k\in\mathcal{F}_{m_k}(K_k)} \lambda_{F_1}(A_1)\cdots\lambda_{F_{k-2}}(A_{k-2})$$

$$\times \int_{\mathbb{R}^d} \gamma(F_1,\ldots,F_{k-2},F_{k-1}\cap F_k^x;K_1,\ldots,K_{k-2},K_{k-1}\cap K_k^x)$$

$$\times [F_1,\ldots,F_{k-2},F_{k-1}\cap F_k^x]\lambda_{F_{k-1}\cap F_k^x}(A_{k-1}\cap A_k^x)\,\lambda(\mathrm{d}x)$$

$$= \sum_{\substack{m_1,\ldots,m_k=j \\ m_1+\ldots+m_k=(k-1)d+j}}^{d} \sum_{F_1\in\mathcal{F}_{m_1}(K_1)} \cdots \sum_{F_k\in\mathcal{F}_{m_k}(K_k)} \gamma(F_1,\ldots,F_k;K_1,\ldots,K_k)$$

$$\times [F_1,\ldots,F_{k-2},L(F_{k-1})\cap L(F_k)]\lambda_{F_1}(A_1)\cdots\lambda_{F_{k-2}}(A_{k-2})$$

$$\times \int_{\mathbb{R}^d} \lambda_{F_{k-1}\cap F_k^x}(A_{k-1}\cap A_k^x)\,\lambda(\mathrm{d}x)$$

$$= \sum_{\substack{m_1,\ldots,m_k=j \\ m_1+\ldots+m_k=(k-1)d+j}}^{d} \sum_{F_1\in\mathcal{F}_{m_1}(K_1)} \cdots \sum_{F_k\in\mathcal{F}_{m_k}(K_k)} \gamma(F_1,\ldots,F_k;K_1,\ldots,K_k)$$

$$\times [F_1,\ldots,F_{k-2},L(F_{k-1})\cap L(F_k)][F_{k-1},F_k]$$

$$\times \lambda_{F_1}(A_1)\cdots\lambda_{F_{k-2}}(A_{k-2})\lambda_{F_{k-1}}(A_{k-1})\lambda_{F_k}(A_k)$$

$$= \sum_{\substack{m_1,\ldots,m_k=j \\ m_1+\ldots+m_k=(k-1)d+j}}^{d} \sum_{F_1\in\mathcal{F}_{m_1}(K_1)} \cdots \sum_{F_k\in\mathcal{F}_{m_k}(K_k)} \gamma(F_1,\ldots,F_k;K_1,\ldots,K_k)$$

$$\times [F_1,\ldots,F_k]\lambda_{F_1}(A_1)\cdots\lambda_{F_k}(A_k)$$

$$= \sum_{\substack{m_1,\ldots,m_k=j \\ m_1+\ldots+m_k=(k-1)d+j}}^{d} \Phi^{(j)}_{m_1,\ldots,m_k}(K_1,\ldots,K_k;A_1\times\ldots\times A_k).$$

The integral formula (6.15) is thus established for polytopes.

We now extend (6.15), and thus the mixed measures, to arbitrary convex bodies, employing approximation by polytopes. For this purpose, we first remark that (6.15), for all Borel sets $A_1,\ldots,A_k$, is equivalent to

$$\int_{(\mathbb{R}^d)^{k-1}} \int_{\mathbb{R}^d} f(x_1, x_1 - x_2, \ldots, x_1 - x_k)\, \Phi_j(K_1 \cap K_2^{x_2} \cap \ldots \cap K_k^{x_k}, \mathrm{d}x_1) \\ \times \lambda^{k-1}(\mathrm{d}(x_2, \ldots, x_k)) \tag{6.16}$$

$$= \sum_{\substack{m_1,\ldots,m_k=j \\ m_1+\ldots+m_k=(k-1)d+j}}^{d} \int_{(\mathbb{R}^d)^k} f(x_1, \ldots, x_k)\, \Phi^{(j)}_{m_1,\ldots,m_k}(K_1, \ldots, K_k; \mathrm{d}(x_1, \ldots, x_k))$$

for all continuous functions f on $(\mathbb{R}^d)^k$ (provided that the mixed measures exist). For $k = 2$, this equivalence is explained at the beginning of the proof of Theorem 5.2.3; the general case follows analogously. Hence, (6.15) and (6.16) are valid if $K_1, \ldots, K_k$ are polytopes. As in the proof of Theorem 5.2.3, we consider the functional

$$J(f, K_1, \ldots, K_k) \\ := \int_{(\mathbb{R}^d)^{k-1}} \int_{\mathbb{R}^d} f(x_1, x_1 - x_2, \ldots, x_1 - x_k)\, \Phi_j(K_1 \cap K_2^{x_2} \cap \ldots \cap K_k^{x_k}, \mathrm{d}x_1) \\ \times \lambda^{k-1}(\mathrm{d}(x_2, \ldots, x_k))$$

and obtain that J depends continuously on $K_1, \ldots, K_k$. For $r_1, \ldots, r_k > 0$, we define a continuous mapping $D_{r_1,\ldots,r_k}$ from $(\mathbb{R}^d)^k$ into itself by

$$D_{r_1,\ldots,r_k}(x_1, \ldots, x_k) := \left(\frac{x_1}{r_1}, \ldots, \frac{x_k}{r_k}\right) \qquad \text{for } x_1, \ldots, x_k \in \mathbb{R}^d.$$

For polytopes $K_1, \ldots, K_k$, relation (6.16) and the definition of the mixed measures imply

$$D_{r_1,\ldots,r_k} J(f, r_1 K_1, \ldots, r_k K_k) \\ := \int_{(\mathbb{R}^d)^{k-1}} \int_{\mathbb{R}^d} f\left(\frac{x_1}{r_1}, \frac{x_1 - x_2}{r_2}, \ldots, \frac{x_1 - x_k}{r_k}\right) \\ \times \Phi_j(r_1 K_1 \cap (r_2 K_2)^{x_2} \cap \ldots \cap (r_k K_k)^{x_k}, \mathrm{d}x_1)\, \lambda^{k-1}(\mathrm{d}(x_2, \ldots, x_k))$$

$$= \sum_{\substack{m_1,\ldots,m_k=j \\ m_1+\ldots+m_k=(k-1)d+j}}^{d} \\ \int_{(\mathbb{R}^d)^k} f\left(\frac{x_1}{r_1}, \ldots, \frac{x_k}{r_k}\right) \Phi^{(j)}_{m_1,\ldots,m_k}(r_1 K_1, \ldots, r_2 K_k; \mathrm{d}(x_1, \ldots, x_k))$$

$$= \sum_{\substack{m_1,\ldots,m_k=j \\ m_1+\ldots+m_k=(k-1)d+j}}^{d} r_1^{m_1} \cdots r_k^{m_k} \\ \times \int_{(\mathbb{R}^d)^k} f(x_1, \ldots, x_k)\, \Phi^{(j)}_{m_1,\ldots,m_k}(K_1, \ldots, K_k; \mathrm{d}(x_1, \ldots, x_k)).$$

For arbitrary convex bodies $K_1, \ldots, K_k$, we choose sequences of polytopes $K_{1i}, \ldots K_{ki}$ ($i \in \mathbb{N}$) such that $K_{1i} \to K_1, \ldots, K_{ki} \to K_k$ for $i \to \infty$. Then

$$D_{r_1,\ldots,r_k} J(f, r_1 K_{1i}, \ldots, r_k K_{ki}) \to D_{r_1,\ldots,r_k} J(f, r_1 K_1, \ldots, r_k K_k)$$

for every continuous function f on $(\mathbb{R}^d)^k$ and all $r_1, \ldots, r_k > 0$. From the polynomial expansion just established, we deduce the convergence of the coefficients

$$\int_{(\mathbb{R}^d)^k} f(x_1, \ldots, x_k)\, \Phi^{(j)}_{m_1,\ldots,m_k}(K_{1i}, \ldots, K_{ki}; \mathrm{d}(x_1, \ldots, x_k))$$

and thus the weak convergence of the measures

$$\Phi^{(j)}_{m_1,\ldots,m_k}(K_{1i}, \ldots, K_{ki}; \cdot)$$

for $i \to \infty$. The limits, denoted by $\Phi^{(j)}_{m_1,\ldots,m_k}(K_1, \ldots, K_k; \cdot)$, are again finite measures, satisfying

$$\begin{aligned} & D_{r_1,\ldots,r_k} J(f, r_1 K_1, \ldots, r_k K_k) \\ & = \sum_{\substack{m_1,\ldots,m_k=j \\ m_1+\ldots+m_k=(k-1)d+j}}^{d} r_1^{m_1} \cdots r_k^{m_k} \\ & \quad \times \int_{(\mathbb{R}^d)^k} f(x_1, \ldots, x_k)\, \Phi^{(j)}_{m_1,\ldots,m_k}(K_1, \ldots, K_k; \mathrm{d}(x_1, \ldots, x_k)), \end{aligned} \tag{6.17}$$

from which we see that they are independent of the approximating sequences $(K_{1i})_{i\in\mathbb{N}}, \ldots, (K_{ki})_{i\in\mathbb{N}}$. For $r_1 = \ldots = r_k = 1$, we obtain (6.16).

Thus, mixed measures for arbitrary bodies $K_1, \ldots, K_k$ are defined which fulfill (6.15). Moreover, properties (a), (b) and (c), which follow for polytopes $K_1, \ldots, K_k$ from the definition, transfer to general convex bodies by means of approximation and an application of (6.17). Also, (6.17) shows that $\Phi^{(j)}_{m_1,\ldots,m_k}(K_1, \ldots, K_k; \cdot)$ depends continuously on the bodies $K_1, \ldots, K_k$, and (d) can be deduced from the corresponding additivity property of curvature measures, similarly to the proof of Theorem 5.2.3. To prove (e), suppose that its assumptions are satisfied. Without loss of generality, we may assume that $U = U_1 \times \ldots \times U_k$ with open sets $U_1, \ldots, U_k \subset \mathbb{R}^d$. Then, for $r_1, \ldots, r_k > 0$ and $x_2, \ldots, x_k \in \mathbb{R}^d$, the set

$$r_1 K_1 \cap (r_2 K_2)^{x_2} \cap \ldots \cap (r_k K_k)^{x_k} \cap r_1 U_1 \cap (r_2 U_2)^{x_2} \cap \ldots \cap (r_k U_k)^{x_k}$$

remains the same if K_i is replaced by M_i, $i = 1, \ldots, k$. Since, by Theorem 14.2.3, the curvature measures are locally determined, the value

$$\Phi_j(r_1 K_1 \cap (r_2 K_2)^{x_2} \cap \ldots \cap (r_k K_k)^{x_k}, r_1 A_1 \cap (r_2 A_2)^{x_2} \cap \ldots \cap (r_k A_k)^{x_k}),$$

for Borel sets $A_i \subset U_i$, does not change if K_i is replaced by M_i. Let f be a continuous function on $(\mathbb{R}^d)^k$ with support in U. Then the case $r_1 = \ldots = r_k = 1$ shows that the left side of (6.16) does not change if K_i is replaced by M_i. More generally, we obtain

$$D_{r_1,\ldots,r_k} J(f, r_1K_1, \ldots, r_kK_k) = D_{r_1,\ldots,r_k} J(f, r_1M_1, \ldots, r_kM_k).$$

Therefore, the right side of (6.17) does not change if K_i is replaced by M_i. This yields the assertion. □

For the total mixed measures, we introduce the notation

$$V^{(j)}_{m_1,\ldots,m_k}(K_1, \ldots, K_k) := \Phi^{(j)}_{m_1,\ldots,m_k}(K_1, \ldots, K_k; (\mathbb{R}^d)^k),$$

and we call these the **mixed functionals**. In particular,

$$V^{(j)}_j(K) = V_j(K),$$

and the case $k = 2$ reduces to the mixed functionals introduced in Theorem 5.2.3. If $K_1, \ldots, K_k$ are polytopes, then

$$\begin{aligned} &V^{(j)}_{m_1,\ldots,m_k}(K_1, \ldots, K_k) \\ &= \sum_{F_1\in\mathcal{F}_{m_1}(K_1)} \cdots \sum_{F_k\in\mathcal{F}_{m_k}(K_k)} \gamma(F_1, \ldots, F_k; K_1, \ldots, K_k) \\ &\quad \times [F_1, \ldots, F_k] V_{m_1}(F_1) \cdots V_{m_k}(F_k). \end{aligned} \tag{6.18}$$

Results on mixed measures contain results on mixed functionals as special cases. In the sequel, we therefore concentrate on mixed measures and mention mixed functionals only when their behavior deviates from that of mixed measures.

In view of the decomposability property (b) we can, in large parts of the following, concentrate on the case $k \le d$.

Since mixed measures are locally determined, we can extend them to unbounded closed convex sets $K_1, \ldots, K_k$. We shall use this extension, in particular, for linear or affine subspaces and for closed halfspaces. The representation (6.14) remains valid for polyhedral sets. It is important to note that also the integral geometric formulas for mixed measures obtained in this section extend in the same way. In fact, any unbounded convex set K_i in such a formula, with bounded corresponding Borel set A_i, can be replaced by the intersection of K_i with a cube (say) that contains A_i in its interior. This replacement does not affect the values of the involved mixed measures.

The next theorem collects some of the integral geometric formulas that hold for mixed measures.

Theorem 6.4.2. *For $k \in \mathbb{N}$, convex bodies $K_1, \ldots, K_k \in \mathcal{K}'$ and Borel sets $A_1, \ldots, A_k \in \mathcal{B}(\mathbb{R}^d)$, the mixed measures satisfy the* **translative formula**

$$\int_{\mathbb{R}^d} \Phi^{(j)}_{m_1,\dots,m_{k-2},m}(K_1,\dots,K_{k-2},K_{k-1}\cap K_k^x;$$
$$A_1\times\ldots\times A_{k-2}\times(A_{k-1}\cap A_k^x))\,\lambda(\mathrm{d}x)$$
$$= \sum_{\substack{m_{k-1},m_k=m\\ m_{k-1}+m_k=d+m}}^{d} \Phi^{(j)}_{m_1,\dots,m_k}(K_1,\dots,K_k;A_1\times\ldots\times A_k), \tag{6.19}$$

the **rotation formula**

$$\int_{SO_d} \Phi^{(j)}_{m_1,\dots,m_{k-1},m}(K_1,\dots,K_{k-1},\vartheta K_k;$$
$$A_1\times\ldots\times A_{k-1}\times\vartheta A_k)\,\nu(\mathrm{d}\vartheta) \tag{6.20}$$
$$= c^{m,d-m+j}_{d,j}\Phi^{(d-m+j)}_{m_1,\dots,m_{k-1}}(K_1,\dots,K_{k-1};A_1\times\ldots\times A_{k-1})\Phi_m(K_k,A_k),$$

and the **principal kinematic formula**

$$\int_{G_d} \Phi^{(j)}_{m_1,\dots,m_{k-2},m}(K_1,\dots,K_{k-2},K_{k-1}\cap gK_k;$$
$$A_1\times\ldots\times A_{k-2}\times(A_{k-1}\cap gA_k))\,\mu(\mathrm{d}g)$$
$$= \sum_{r=m}^{d} c^{d-r+m,j-m+r}_{d,j}\Phi^{(j-m+r)}_{m_1,\dots,m_{k-2},r}(K_1,\dots,K_{k-1};A_1\times\ldots\times A_{k-1})$$
$$\times\,\Phi_{d-r+m}(K_k,A_k). \tag{6.21}$$

Proof. It is sufficient to prove the results for polytopes; the general case then follows by approximation, using arguments similar to those of the previous proof.

For polytopes, (6.19) was obtained during the proof of (6.15). In the case of (6.20), we use the definition of the mixed measures, Lemma 14.1.1, and Theorem 5.3.1 to get

$$\int_{SO_d} \Phi^{(j)}_{m_1,\dots,m_{k-1},m}(K_1,\dots,K_{k-1},\vartheta K_k;A_1\times\ldots\times A_{k-1}\times\vartheta A_k)\,\nu(\mathrm{d}\vartheta)$$
$$= \sum_{F_1\in\mathcal{F}_{m_1}(K_1)}\cdots\sum_{F_{k-1}\in\mathcal{F}_{m_{k-1}}(K_{k-1})}\sum_{F\in\mathcal{F}_m(K_k)} \lambda_{F_1}(A_1)\cdots\lambda_{F_{k-1}}(A_{k-1})\lambda_F(A_k)$$
$$\times\int_{SO_d}\gamma(F_1,\dots,F_{k-1},\vartheta F;K_1,\dots,K_{k-1},\vartheta K_k)[F_1,\dots,F_{k-1},\vartheta F]\,\nu(\mathrm{d}\vartheta)$$
$$= \sum_{F_1\in\mathcal{F}_{m_1}(K_1)}\cdots\sum_{F_{k-1}\in\mathcal{F}_{m_{k-1}}(K_{k-1})}\sum_{F\in\mathcal{F}_m(K_k)} \lambda_{F_1}(A_1)\cdots\lambda_{F_{k-1}}(A_{k-1})\lambda_F(A_k)$$
$$\times[F_1,\dots,F_{k-1}]\int_{SO_d}\gamma(G,\vartheta F;M,\vartheta K_k)[G,\vartheta F]\,\nu(\mathrm{d}\vartheta)$$

$$= c_{d,j}^{m,d-m+j} \sum_{F_1 \in \mathcal{F}_{m_1}(K_1)} \cdots \sum_{F_{k-1} \in \mathcal{F}_{m_{k-1}}(K_{k-1})} [F_1, \dots, F_{k-1}]$$

$$\times \gamma(G, M) \lambda_{F_1}(A_1) \cdots \lambda_{F_{k-1}}(A_{k-1}) \sum_{F \in \mathcal{F}_m(K_k)} \gamma(F, K_k) \lambda_F(A_k)$$

$$= c_{d,j}^{m,d-m+j} \Phi_{m_1,\dots,m_{k-1}}^{(d-m+j)}(K_1, \dots, K_{k-1}; A_1 \times \dots \times A_{k-1}) \Phi_m(K_k, A_k),$$

where $G := F_1 \cap F_2^{x_2} \cap \dots \cap F_{k-1}^{x_{k-1}}$, $M := K_1 \cap K_2^{x_2} \cap \dots \cap K_{k-1}^{x_{k-1}}$ and $x_2, \dots, x_{k-1}$ are suitably chosen vectors.

Finally, (6.21) follows immediately by combining (6.19) and (6.20). □

Further formulas can be obtained by iteration. In particular, there is an iterated translative formula for mixed measures.

We present next a Crofton formula for the mixed measures, first in a translative version and then in its kinematic form.

Theorem 6.4.3. *For convex bodies $K_1, \dots, K_k \in \mathcal{K}'$, Borel sets $A_1, \dots, A_k \in \mathcal{B}(\mathbb{R}^d)$, a subspace $L \in G(d,q)$ with $q \in \{m, \dots, d-1\}$, and any Borel set $A_L \subset L$ with $\lambda_q(A_L) = 1$, the mixed measures satisfy the* **translative Crofton formula**

$$\int_{L^\perp} \Phi_{m_1,\dots,m_{k-1},m}^{(j)}(K_1, \dots, K_{k-1}, K_k \cap L^x;$$
$$A_1 \times \dots \times A_{k-1} \times (A_k \cap L^x))\, \lambda_{d-q}(\mathrm{d}x)$$
$$= \Phi_{m_1,\dots,m_{k-1},d-q+m,q}^{(j)}(K_1, \dots, K_k, L; A_1 \times \dots \times A_k \times A_L) \quad (6.22)$$

and the **Crofton formula**

$$\int_{A(d,q)} \Phi_{m_1,\dots,m_{k-1},m}^{(j)}(K_1, \dots, K_{k-1}, K_k \cap E;$$
$$A_1 \times \dots \times A_{k-1} \times (A_k \cap E))\, \mu_q(\mathrm{d}E)$$
$$= c_{d,j}^{q,d-q+j} \Phi_{m_1,\dots,m_{k-1},d-q+m}^{(d-q+j)}(K_1, \dots, K_k; A_1 \times \dots \times A_k). \quad (6.23)$$

Proof. For the proof of (6.22) we may again concentrate on the case of polytopes. Moreover, we can assume that the faces of the polytope K_k and the subspace L are in general position. This implies that the m-dimensional faces of $K_k \cap L^x$, for $x \in L^\perp$, are intersections $F \cap L^x$ of $(d-q+m)$-dimensional faces F of K_k, at least for those x for which the sets intersect at relatively interior points.

Using this observation, we can proceed, in large parts, similarly to the proof of (6.15) and get

$$\int_{L^\perp} \Phi_{m_1,\dots,m_{k-1},m}^{(j)}(K_1, \dots, K_{k-1}, K_k \cap L^x;$$

$$
\begin{aligned}
&A_1 \times \ldots \times A_{k-1} \times (A_k \cap L^x))\, \lambda_{d-q}(\mathrm{d}x) \\
&= \sum_{F_1 \in \mathcal{F}_{m_1}(K_1)} \cdots \sum_{F_{k-1} \in \mathcal{F}_{m_{k-1}}(K_{k-1})} \lambda_{F_1}(A_1) \cdots \lambda_{F_{k-1}}(A_{k-1}) \\
&\quad \times \int_{L^\perp} \sum_{F \in \mathcal{F}_m(K_k \cap L^x)} \gamma(F_1, \ldots, F_{k-1}, F; K_1, \ldots, K_{k-1}, K_k \cap L^x) \\
&\quad \times [F_1, \ldots, F_{k-1}, F] \lambda_F(A_k \cap L^x)\, \lambda_{d-q}(\mathrm{d}x) \\
&= \sum_{F_1 \in \mathcal{F}_{m_1}(K_1)} \cdots \sum_{F_{k-1} \in \mathcal{F}_{m_{k-1}}(K_{k-1})} \lambda_{F_1}(A_1) \cdots \lambda_{F_{k-1}}(A_{k-1}) \\
&\quad \times \sum_{F \in \mathcal{F}_{d-q+m}(K_k)} \int_{L^\perp} \gamma(F_1, \ldots, F_{k-1}, F \cap L^x; K_1, \ldots, K_{k-1}, K_k \cap L^x) \\
&\quad \times [F_1, \ldots, F_{k-1}, F \cap L^x] \lambda_{F \cap L^x}(A_k \cap L^x)\, \lambda_{d-q}(\mathrm{d}x) \\
&= \sum_{F_1 \in \mathcal{F}_{m_1}(K_1)} \cdots \sum_{F_{k-1} \in \mathcal{F}_{m_{k-1}}(K_{k-1})} \sum_{F \in \mathcal{F}_{d-q+m}(K_k)} \\
&\quad \gamma(F_1, \ldots, F_{k-1}, F, L; K_1, \ldots, K_{k-1}, K_k, L) [F_1, \ldots, F_{k-1}, L(F) \cap L] \\
&\quad \times \lambda_{F_1}(A_1) \cdots \lambda_{F_{k-1}}(A_{k-1}) \int_{L^\perp} \lambda_{F \cap L^x}(A_k \cap L^x)\, \lambda_{d-q}(\mathrm{d}x) \\
&= \sum_{F_1 \in \mathcal{F}_{m_1}(K_1)} \cdots \sum_{F_{k-1} \in \mathcal{F}_{m_{k-1}}(K_{k-1})} \sum_{F \in \mathcal{F}_{d-q+m}(K_k)} \\
&\quad \gamma(F_1, \ldots, F_{k-1}, F, L; K_1, \ldots, K_{k-1}, K_k, L) [F_1, \ldots, F_{k-1}, L(F) \cap L] \\
&\quad \times [F, L] \lambda_{F_1}(A_1) \cdots \lambda_{F_{k-1}}(A_{k-1}) \lambda_F(A_k) \\
&= \Phi^{(j)}_{m_1, \ldots, m_{k-1}, d-q+m, q}(K_1, \ldots, K_k, L; A_1 \times \ldots \times A_k \times A_L).
\end{aligned}
$$

The Crofton formula (6.23) is a direct consequence of (6.22) and the rotation formula (6.20). □

Remark on extension to the convex ring. Since the mixed measure $\Phi^{(j)}_{m_1,\ldots,m_k}$ is additive and weakly continuous in each of its first k arguments, it has a unique additive extension to the convex ring. As in Section 5.1, the integral geometric formulas for mixed measures obtained so far in this section remain valid if the involved convex bodies are replaced by polyconvex sets. The arguments explained at the end of Section 5.2 can easily be adapted to the present situation.

By specializing some of the integral geometric formulas, we obtain useful information about mixed measures and functionals. For that, we assume one of the bodies to be the unit ball $B^d \subset \mathbb{R}^d$. If we put $K_k = B^d$ and $A_k = \mathbb{R}^d$ in (6.20) and insert the value of $V_m(B^d)$ given by (14.8), then we obtain

$$\Phi^{(j)}_{m_1,\dots,m_{k-1},m}(K_1,\dots,K_{k-1},B^d;A_1\times\dots\times A_{k-1}\times\mathbb{R}^d)$$
$$=\frac{1}{m!}c^{m,d-m+j}_{j,d-m}\Phi^{(d-m+j)}_{m_1,\dots,m_{k-1}}(K_1,\dots,K_{k-1};A_1\times\dots\times A_{k-1}). \quad (6.24)$$

The following result is a consequence of (6.24).

Theorem 6.4.4. *For $K_1,\dots,K_k\in\mathcal{K}'$ and $A_1,\dots,A_k\in\mathcal{B}(\mathbb{R}^d)$, the mixed measures satisfy the* **reduction property**

$$\Phi^{(j)}_{m_1,\dots,m_k}(K_1,\dots,K_k;A_1\times\dots\times A_k)$$
$$=\frac{1}{\kappa_{d-j}}\Phi^{(0)}_{m_1,\dots,m_k,d-j}(K_1,\dots,K_k,B^d;A_1\times\dots\times A_k\times\mathbb{R}^d)$$
$$=\left(\frac{2}{\kappa_{d-1}}\right)^j\frac{1}{j!\kappa_j}\Phi^{(0)}_{m_1,\dots,m_k,d-1,\dots,d-1}(K_1,\dots,K_k,\underbrace{B^d,\dots,B^d}_{j};$$
$$A_1\times\dots\times A_k\times(\mathbb{R}^d)^j).$$

Proof. The first equation is obtained from (6.24) if m,j,k are replaced by $d-j,0,k$, respectively. For the second, we put $m=d-1$ and replace $k-1$ by k and j by $j-1$ in (6.24). This gives

$$\Phi^{(j-1)}_{m_1,\dots,m_k,d-1}(K_1,\dots,K_k,B^d;A_1\times\dots\times A_k\times\mathbb{R}^d)$$
$$=\frac{1}{(d-1)!}c^{j,d-1}_{j-1,1}\Phi^{(j)}_{m_1,\dots,m_k}(K_1,\dots,K_k;A_1\times\dots\times A_k).$$

The assertion is now obtained by j-fold iteration. □

It follows from this result that all mixed measures can be reduced to the (series of) measures $\Phi^{(0)}_{m_1,\dots,m_k}(K_1,\dots,K_k;\cdot)$, where $k\in\{1,\dots,d\}$ and $m_1,\dots,m_k\in\{1,\dots,d-1\}$ satisfy $m_1+\dots+m_k=(k-1)d$.

As a consequence of the Crofton formula, we note a connection between the mixed functionals $V^{(j)}_{k,d-k+j}(K,M)$ of two convex bodies K,M and mixed volumes. It already follows from (5.16) and Corollary 5.2.1 that

$$V^{(0)}_{k,d-k}(K,M)=\binom{d}{k}V(K[k],-M[d-k]). \quad (6.25)$$

Combining (6.25) with the Crofton formula (6.23), we immediately get a representation of the mixed functionals $V^{(j)}_{k,d-k+j}(K,M)$ as Crofton-type integrals of mixed volumes.

Theorem 6.4.5. *Let $K,M\in\mathcal{K}'$. If $j\in\{0,\dots,d-2\}$ and $k\in\{j+1,\dots,d-1\}$, then*

$$V^{(j)}_{k,d-k+j}(K,M) \quad (6.26)$$
$$=\binom{d}{k-j}c^{d,0}_{j,d-j}\int_{A(d,d-j)}V((K\cap E)[k-j],-M[d-k+j])\,\mu_{d-j}(\mathrm{d}E).$$

Our next aim is the derivation of a translative integral formula for support functions. It can be deduced from a translative formula for special mixed measures. As in (4.42), we replace the support function $h(K,\cdot)$ by its centered version $h^*(K,\cdot)$. Here, a continuous function f on S^{d-1} is **centered** if

$$\int_{S^{d-1}} f(u)u\,\sigma(\mathrm{d}u) = 0.$$

The centered support function of K is invariant under translations of K. The following lemma establishes a connection between the centered support function and a special mixed measure. We use the notation

$$u^+ := \{x \in \mathbb{R}^d : \langle x, u\rangle \geq 0\}$$

for the closed halfspace with 0 in the boundary and inner normal vector $u \in S^{d-1}$.

Lemma 6.4.1. *Let $P \in \mathcal{P}'$ be a polytope, and let $u \in S^{d-1}$. Then*

$$h^*(P,u) = \sum_{F\in\mathcal{F}_1(P)} \gamma(F,u^\perp;P,u^+)[F,u^\perp]\lambda_1(F). \tag{6.27}$$

Let $K \in \mathcal{K}'$, and let $A_{u^\perp} \subset u^\perp$ be a Borel set with $\lambda_{d-1}(A_{u^\perp}) = 1$. Then

$$h^*(K,u) = \Phi^{(0)}_{1,d-1}(K,u^+;\mathbb{R}^d \times A_{u^\perp}). \tag{6.28}$$

Proof. For a vertex e of P, we do not distinguish between the vector e and the corresponding 0-face $\{e\}$. We use the relations

$$\sum_{e\in\mathcal{F}_0(P)} \gamma(e,P) = \chi(P) = 1,$$

which is obvious, and

$$\sum_{e\in\mathcal{F}_0(P)} \gamma(e,P)e = s(K),$$

which is given by (14.29). Writing

$$u_t^+ := \{x \in \mathbb{R}^d : \langle x,u\rangle \geq t\}, \qquad u_t^\perp := \{x \in \mathbb{R}^d : \langle x,u\rangle = t\}$$

for $u \in S^{d-1}$ and $t \in \mathbb{R}$, and choosing a number c with $P \subset u_c^+$, we get

$$\begin{aligned}
&h(P,u) - c\\
&= \int_c^\infty \chi(P\cap u_t^+)\,\mathrm{d}t\\
&= \int_c^\infty \sum_{e\in\mathcal{F}_0(P\cap u_t^+)} \gamma(e,P\cap u_t^+)\,\mathrm{d}t
\end{aligned}$$

$$= \int_c^\infty \sum_{e \in \mathcal{F}_0(P)} \gamma(e,P) \mathbf{1}\{\langle e,u\rangle \geq t\}\,\mathrm{d}t$$

$$+ \int_c^\infty \sum_{F \in \mathcal{F}_1(P)} \gamma(F,u^\perp;P,u^+)\chi(F \cap u_t^\perp)\,\mathrm{d}t$$

$$= \sum_{e \in \mathcal{F}_0(P)} \gamma(e,P)(\langle e,u\rangle - c) + \sum_{F \in \mathcal{F}_1(P)} \gamma(F,u^\perp;P,u^+)[F,u^\perp]\lambda_1(F)$$

$$= \langle s(P),u\rangle - c + \sum_{F \in \mathcal{F}_1(P)} \gamma(F,u^\perp;P,u^+)[F,u^\perp]\lambda_1(F).$$

This proves (6.27), and (6.28) for polytopes follows from (6.14) (extended to polyhedral sets).

For a polytope P, the definition (6.14) implies

$$\Phi_{1,d-1}^{(0)}(P,u^+;\cdot) = \left(\sum_{F \in \mathcal{F}_1(P)} \gamma(F,u^\perp;P,u^+)[F,u^\perp]\,\lambda_F \right) \otimes \lambda_{u^\perp}.$$

Using the weak continuity of the mixed measures, we conclude by approximation that also $\Phi_{1,d-1}^{(0)}(K,u^+;\cdot)$ for $K \in \mathcal{K}'$ is a product measure with $\lambda_{u^\perp}$ as second factor. Now (6.28) follows by approximation and continuity. □

We state a translative formula for centered support functions. Here we restrict ourselves to the case of two convex bodies; the extension to $k \geq 2$ bodies presents no additional difficulties.

Theorem 6.4.6. *For convex bodies $K,M \in \mathcal{K}'$, there exist continuous functions $h_1^*(K,M;\cdot),\dots,h_d^*(K,M;\cdot)$ on S^{d-1} such that*

$$\int_{\mathbb{R}^d} h^*(K \cap M^x,\cdot)\,\lambda(\mathrm{d}x) = \sum_{k=1}^d h_k^*(K,M;\cdot), \tag{6.29}$$

where $h_1^*(K,M;\cdot) = h^*(K,\cdot)V_d(M)$ and $h_d^*(K,M;\cdot) = V_d(K)h^*(M,\cdot)$. The function $h_k^*(K,M;\cdot)$ is centered and symmetric, in the sense that

$$h_k^*(K,M;\cdot) = h_{d+1-k}^*(M,K;\cdot),$$

it depends continuously on $K,M \in \mathcal{K}'$ and is homogeneous of degree k in K and of degree $d+1-k$ in M. Moreover, it is additive in each of its arguments K and M.

For polytopes K,M, we have

$$h_k^*(K,M;u) \tag{6.30}$$
$$= \sum_{F \in \mathcal{F}_k(K)} \sum_{G \in \mathcal{F}_{d+1-k}(M)} \gamma(F,G,u^\perp;K,M,u^+)[F,G,u^\perp]\lambda_k(F)\lambda_{d-k+1}(G).$$

Proof. Let $K, M \in \mathcal{K}'$, let $u \in S^{d-1}$. By (6.28),

$$h^*(K \cap M^x, u) = \Phi^{(0)}_{d-1,1}(u^+, K \cap M^x; A_{u^\perp} \times \mathbb{R}^d).$$

The translative formula (6.19) with $k = 3$, $K_1 = u^+$, $K_2 = K$, $K_3 = M$, $A_1 = A_{u^\perp}$, $A_2 = A_3 = \mathbb{R}^d$, $j = 0$, $m_1 = d-1$ and $m = 1$ gives (6.29) with

$$h^*_k(K, M; u) := \Phi^{(0)}_{d-1,k,d+1-k}(u^+, K, M; A_{u^\perp} \times \mathbb{R}^d \times \mathbb{R}^d) \tag{6.31}$$

for $k = 1, \dots, d$. The representation (6.30) in the case of polytopes follows from (6.14).

From (6.31) and the known properties of mixed measures we immediately obtain the assertions about symmetry, continuity in K, M, homogeneity, and additivity of $h^*_k(K, M; u)$. By the homogeneity property, we have

$$\int_{\mathbb{R}^d} h^*(rK \cap M^x, \cdot)\, \lambda(\mathrm{d}x) = \sum_{k=1}^{d} r^k h^*_k(K, M; \cdot)$$

for all $r \geq 0$. Inserting $r = 1, \dots, d$ and solving the resulting system of equations, we get a representation

$$h^*_k(K, M; \cdot) = \sum_{n=1}^{d} a_{kn} \int_{\mathbb{R}^d} h^*(nK \cap M^x, \cdot)\, \lambda(\mathrm{d}x)$$

with coefficients a_{kn} independent of K and M. From this representation, we see that $h^*_k(K, M; \cdot)$ is a continuous function. Inserting u, multiplying by u, integrating over S^{d-1}, and using Fubini's theorem, we also see that $h^*(K, M; \cdot)$ is centered. □

We next derive a kinematic formula for support functions. For fixed $u \in \mathbb{R}^d$, the function $K \mapsto h^*(K, u)$ satisfies the assumptions of Theorem 5.1.2, hence we get

$$\int_{G_d} h^*(K \cap gM, \cdot)\, \mu(\mathrm{d}g) = \sum_{k=1}^{d} \left(\int_{A(d,k)} h^*(K \cap E, \cdot)\, \mu_k(\mathrm{d}E) \right) V_k(M) \tag{6.32}$$

(observing that $h^*(\{x\}, \cdot) = 0$). The coefficient of $V_k(M)$ is evidently a support function. We define, for $k \in \{1, \dots, d-1\}$, the kth **mean section body** $M_k(K)$ of a convex body $K \in \mathcal{K}'$ by

$$h(M_k(K), \cdot) := \int_{A(d,k)} h^*(K \cap E, \cdot)\, \mu_k(\mathrm{d}E).$$

We complement the definition by setting $h(M_d(K), \cdot) := h^*(K, \cdot)$, that is, $M_d(K) = K - s(K)$, and $h(M_0(K), \cdot) := 0$, thus $M_0(K) = \{0\}$. To obtain a connection with mixed measures, we choose $M = B^d$ in (6.32). Applying (6.29), (6.31) and Theorem 6.4.4, we get, for $u \in \mathbb{R}^d$,

$$h(M_k(K),u)V_k(B^d) = \Phi^{(0)}_{d+1-k,k,d-1}(K,B^d,u^+;\mathbb{R}^d \times \mathbb{R}^d \times A_{u^\perp})$$
$$= \kappa_k \Phi^{(d-k)}_{d+1-k,d-1}(K,u^+;\mathbb{R}^d \times A_{u^\perp}). \tag{6.33}$$

If K is a polytope, an explicit form of the latter expression is obtained from (6.14). We collect the obtained results in the following theorem.

Theorem 6.4.7. *If $K, M \in \mathcal{K}'$, then*

$$\int_{G_d} h^*(K \cap gM, \cdot)\,\mu(\mathrm{d}g) = \sum_{k=1}^{d} h(M_k(K),\cdot)V_k(M),$$

where

$$h(M_k(K),u) = c^{d-k,k}_{d,0}\Phi^{(d-k)}_{d+1-k,d-1}(K,u^+;\mathbb{R}^d \times A_{u^\perp}).$$

If K is a polytope, then

$$h(M_k(K),u) = c^{d-k,k}_{d,0} \sum_{F \in \mathcal{F}_{d+1-k}(K)} \gamma(F,u^\perp)[F,u^\perp]\lambda_{d+1-k}(F). \tag{6.34}$$

Finally, we use some of the obtained information on mixed measures to derive a kinematic and a Crofton formula for projection functions, that is, volumes of projections of convex bodies. Let $j \in \{1,\dots,d-1\}$ and $L \in G(d,j)$. For $K \in \mathcal{K}'$, the j-dimensional volume of the orthogonal projection $K|L$ defines the jth **projection function** $L \mapsto V_j(K|L)$. From (14.19) and (6.25) we have

$$V_j(K|L) = \binom{d}{j} V(K[j], B_{L^\perp}[d-j]) = V^{(0)}_{j,d-j}(K,B_{L^\perp})$$

where $B_{L^\perp} \subset L^\perp$ is a ball with $\lambda_{d-j}(B_{L^\perp}) = 1$. Therefore, (6.21) and (6.23) yield, for $K, M \in \mathcal{K}'$,

$$\int_{G_d} V_j((K \cap gM)|L)\,\mu(\mathrm{d}g) = \sum_{i=j}^{d} c^{d-i+j,i-j}_{d,0} V^{(i-j)}_{i,d-j}(K,B_{L^\perp})V_{d-i+j}(M),$$

$$\int_{A(d,d-i+j)} V_j((K \cap E)|L)\,\mu_{d-i+j}(\mathrm{d}E) = c^{d-i+j,i-j}_{d,0} V^{(i-j)}_{i,d-j}(K,B_{L^\perp}).$$

(Of course, if one of the two results is known, the other one can also be deduced from Theorem 5.1.2.) The mixed functionals appearing here can be expressed as Radon transforms of the projection function. The **Radon transform** R_{ij} : $\mathbf{C}(G(d,i)) \to \mathbf{C}(G(d,j))$ is defined by

$$(R_{ij}f)(L) := \int_{G(L,i)} f(M)\, \nu_i^L(\mathrm{d}M), \qquad L \in G(d,j). \tag{6.35}$$

In the following, we assume $i \in \{j+1,\dots,d\}$.

Using the symmetry of the mixed functionals together with (6.26) and denoting by $c_1, c_2, \dots$ constants depending only on d, i, j, we get

$$\begin{aligned} V_{i,d-j}^{(i-j)}(K, B_{L^\perp}) &= V_{d-j,i}^{(i-j)}(B_{L^\perp}, K) \\ &= c_1 \int_{A(d,d-i+j)} V((B_{L^\perp} \cap E)[d-i], K[i])\, \mu_{d-i+j}(\mathrm{d}E). \end{aligned}$$

The integrand, as a function of E, depends only on $E \cap L^\perp$. Therefore, we can use the integral geometric identity

$$\int_{A(d,d-i+j)} f(E \cap L^\perp)\, \mu_{d-i+j}(\mathrm{d}E) = c_2 \int_{A(L^\perp,d-i)} f(F)\, \mu_{d-i}^{L^\perp}(\mathrm{d}F),$$

which holds for all nonnegative measurable functions f on $A(L^\perp, d-i)$. Here $A(L^\perp, d-i)$ is the space of $(d-i)$-flats contained in $L^\perp$, and $\mu_{d-i}^{L^\perp}$ is its invariant measure (see Section 13.2). To prove the identity, we note that its left side, applied to indicator functions of Borel sets in $A(L^\perp, d-i)$, defines a measure on $A(L^\perp, d-i)$, which is locally finite and invariant under rigid motions of $L^\perp$ into itself and hence is a multiple of the invariant measure $\mu_{d-i}^{L^\perp}$. Thus, we obtain

$$\begin{aligned} & V_{i,d-j}^{(i-j)}(K, B_{L^\perp}) \\ &= c_3 \int_{A(L^\perp,d-i)} V((B_{L^\perp} \cap F)[d-i], K[i])\, \mu_{d-i}^{L^\perp}(\mathrm{d}F) \\ &= c_3 \int_{G(L^\perp,d-i)} \int_{H^\perp \cap L^\perp} V((B_{L^\perp} \cap H^x)[d-i], K[i])\, \lambda_{i-j}(\mathrm{d}x)\, \nu_{d-i}^{L^\perp}(\mathrm{d}H). \end{aligned}$$

Here we have

$$\int_{H^\perp \cap L^\perp} V((B_{L^\perp} \cap H^x)[d-i], K[i])\, \lambda_{i-j}(\mathrm{d}x) = c_4 V_i(K|H^\perp),$$

which follows from (14.19), since $B_{L^\perp} \cap H^x$, if not empty, is homothetic to $B_{L^\perp} \cap H$, with homothety factor depending only on $\|x\|$. We deduce that

$$\begin{aligned} V_{i,d-j}^{(i-j)}(K, B_{L^\perp}) &= c_5 \int_{G(L^\perp,d-i)} V_i(K|H^\perp)\, \nu_{d-i}^{L^\perp}(\mathrm{d}H) \\ &= c_5 \int_{G(L,i)} V_i(K|M)\, \nu_i^L(\mathrm{d}M) \\ &= c_5 (R_{ij} V_i(K|\cdot))(L). \end{aligned}$$

Since (6.24) implies

$$V_{i,d-j}^{(i-j)}(B^d, B_{L^\perp}) = \kappa_i c_{d-i,i-j}^{d-j,0},$$

we conclude that $c_5 = c_{d-i,i-j}^{d-j,0}$. We have obtained the following result.

Theorem 6.4.8. *Let $K, M \in \mathcal{K}'$. If $j \in \{1, \dots, d-1\}$ and $L \in G(d,j)$, then the* **principal kinematic formula for projection functions**,

$$\int_{G_d} V_j((M \cap gK)|L)\,\mu(\mathrm{d}g) = \sum_{i=j}^{d} c_{d,d-i}^{d-i+j,d-j}(R_{ij}V_i(K|\cdot))(L)V_{d-i+j}(M),$$

and the **Crofton formula for projection functions**,

$$\int_{A(d,d-i+j)} V_j((K \cap E)|L)\,\mu_{d-i+j}(\mathrm{d}E) = c_{d,d-i}^{d-i+j,d-j}(R_{ij}V_i(K|\cdot))(L),$$

hold.

Notes for Section 6.4

1. An iterated translative integral formula in the plane was first derived by Miles [529].

The iterated translation formula for curvature measures was proved in Weil [792] and applied to non-isotropic Poisson particle processes and Boolean models. Shorter surveys are given in [790] and [791]. The presentation in this section follows closely the one in Weil [800].

2. Extensions of translative integral formulas to sets of positive reach have been studied by Rataj and Zähle, using methods of geometric measure theory. First, a translative formula for support measures of sets with positive reach was proved in [617]. An iterated version was obtained by Rataj [613]. Various extensions and supplements were provided by Rataj [614], Hug [355], Zähle [831], Rataj and Zähle [618], [619]. Translative Crofton formulas for support measures were treated by Rataj [615].

The iterated translative integral formula for support measures can be written in the form

$$\begin{aligned}
&\int_{\mathbb{R}^d} \cdots \int_{\mathbb{R}^d} \int_{\mathbb{R}^d \times \mathbb{R}^d} h(x, x - x_2, \dots, x - x_k, u) \\
&\times \Xi_j(K_1 \cap K_2^{x_2} \cap \dots \cap K_k^{x_k}, \mathrm{d}(x,u))\,\lambda(\mathrm{d}x_2) \cdots \lambda(\mathrm{d}x_k) \\
&= \sum_{\substack{m_1,\dots,m_k=j \\ m_1+\dots+m_k=(k-1)d+j}}^{d} \int_{(\mathbb{R}^d)^{k+1}} h(x_1, \dots, x_k, u) \\
&\quad \times \Xi_{m_1,\dots,m_k}^{(j)}(K_1, \dots, K_k; \mathrm{d}(x_1, \dots, x_k, u))
\end{aligned}$$

for nonnegative measurable functions h on $(\mathbb{R}^d)^{k+1}$, with certain **mixed support measures** $\Xi^{(j)}_{m_1,\dots,m_k}(K_1,\dots,K_k;\cdot)$ on $(\mathbb{R}^d)^{k+1}$. In Rataj's version for sets of positive reach, the mixed support measures are expressed as currents evaluated at specially chosen differential forms. For closed convex sets, Hug [355] has a more general version for relative support measures, as well as more explicit expressions for the mixed support measures, which imply, in particular, representations of special mixed measures by Goodey and Weil [280].

3. In Schneider [702], the mixed functionals $V^{(0)}_{m_1,\dots,m_k}(K_1,\dots,K_k)$ of convex bodies are embedded in a wider theory, together with the mixed volumes. For polytopes, more general representations of type (6.18) (for $j=0$) are obtained.

4. The reduction property in Theorem 6.4.4 can be generalized to lower-dimensional balls. The following result was proved in Weil [800]. It also provides a kind of **exchangeability**, since the role of the subspace L and the unit ball B_L in L can be exchanged. Let $K_1,\dots,K_k \in \mathcal{K}'$ and $A_1,\dots,A_k \in \mathcal{B}(\mathbb{R}^d)$. For $q \in \{j,\dots,d\}$ and $m \in \{j,\dots,q\}$, let $L \in G(d,q)$ and let B_L be the unit ball in L. Then

$$\begin{aligned} &\Phi^{(j)}_{m_1,\dots,m_k,m}(K_1,\dots,K_k,B_L;A_1\times\dots\times A_k\times L)\\ &= \frac{1}{m!\kappa_q} c^{m,q-m+j}_{j,q-m} \Phi^{(q-m+j)}_{m_1,\dots,m_k,q}(K_1,\dots,K_k,L;A_1\times\dots\times A_k\times B_L)\\ &= \frac{1}{m!\kappa_q} c^{m,q-m+j}_{j,q-m} \Phi^{(q-m+j)}_{m_1,\dots,m_k,q}(K_1,\dots,K_k,B_L;A_1\times\dots\times A_k\times L). \end{aligned} \tag{6.36}$$

Replacing m and j by $q-j$ and 0, we obtain

$$\begin{aligned} &\Phi^{(j)}_{m_1,\dots,m_k,q}(K_1,\dots,K_k,L;A_1\times\dots\times A_k\times B_L)\\ &= \Phi^{(j)}_{m_1,\dots,m_k,q}(K_1,\dots,K_k,B_L;A_1\times\dots\times A_k\times L)\\ &= \frac{\kappa_q}{\kappa_{q-j}} \Phi^{(0)}_{m_1,\dots,m_k,q-j}(K_1,\dots,K_k,B_L;A_1\times\dots\times A_k\times L). \end{aligned} \tag{6.37}$$

For $q=d$ (and using the reduction property of mixed measures), formula (6.36) reduces to (6.24) and (6.37) yields the first formula in Theorem 6.4.4.

For the mixed functionals, (6.37) implies

$$V^{(j)}_{m_1,\dots,m_k,q}(K_1,\dots,K_k,B_L) = \frac{\kappa_q}{\kappa_{q-j}} V^{(0)}_{m_1,\dots,m_k,q-j}(K_1,\dots,K_k,B_L). \tag{6.38}$$

5. Translative Crofton formula for mixed volumes. Combining (6.25) with the translative Crofton formula in Theorem 6.4.3, we obtain a translative integral formula for mixed volumes of convex bodies K, M (see Weil [800]).

Let $j \in \{1,\dots,d-1\}$, $q \in \{j,\dots,d-1\}$ and $L \in G(d,q)$, then

$$\int_{L^\perp} V((K\cap L^x)[j], M[d-j])\,\lambda_{d-q}(\mathrm{d}x) = \frac{1}{\binom{d}{j}\kappa_q} V^{(0)}_{d-q+j,d-j,q}(K,-M,B_L).$$

For $M = B^d$, and using (6.25) and (6.38), a translative Crofton formula for intrinsic volumes results,

$$\int_{L^\perp} V_j(K\cap L^x)\,\lambda_{d-q}(\mathrm{d}x) = \frac{\binom{d}{q-j}}{\kappa_{q-j}} V(K[d-q+j], B_L[q-j]), \tag{6.39}$$

which was first proved in Schneider [681].

6. Formulas for halfspaces. Crofton-type formulas, where the moving k-flat is replaced by a moving halfspace, were also discussed in Weil [800]. Let $C_{u^\perp}$ be a unit cube in $u^\perp$.

The following formula is the analog of Theorem 6.4.3 (for $q = d-1$) and is proved in the same way:

$$\begin{aligned}\int_{-\infty}^{\infty} &\Phi^{(j)}_{m_1,\dots,m_{k-1},m}(K_1,\dots,K_{k-1},K\cap(u^+)^{ru};\\ &A_1\times\dots\times A_{k-1}\times(A\cap(u^\perp)^{ru}))\,\mathrm{d}r\\ &=\Phi^{(j)}_{m_1,\dots,m_{k-1},m+1,d-1}(K_1,\dots,K_{k-1},K,u^+;\\ &\quad A_1\times\dots\times A_{k-1}\times A\times C_{u^\perp}).\end{aligned} \tag{6.40}$$

Combining Theorem 6.4.3 and (6.40), a quite general result for halfspaces of the form $L^{u,+} := L\cap u^+, L\in G(d,q), u\in S^{d-1}\cap L$, with bounding flat $L^u := L\cap u^\perp$, can be deduced:

$$\begin{aligned}&\Phi^{(j)}_{m_1,\dots,m_k,q-1}(K_1,\dots,K_k,L^{u,+};A_1\times\dots\times A_k\times C_{L^u})\\ &=\Phi^{(j)}_{m_1,\dots,m_k,q,d-1}(K_1,\dots,K_k,L,u^+;A_1\times\dots\times A_k\times C_L\times C_{u^\perp}).\end{aligned}$$

7. Mean section body. The kth mean section body $M_k(K)$ of a convex body K was introduced and investigated in Goodey and Weil [279]. In particular, the representation (6.34) was proved there.

A special case of (6.26) is worth mentioning. If $k = j+1$, then

$$V^{(j)}_{j+1,d-1}(K,M) = dc^{d,0}_{j,d-j}\int_{A(d,d-j)} V((K\cap E)[1],-M[d-1])\,\mu_{d-j}(\mathrm{d}E).$$

The linearity properties of the mixed volume imply that the latter integral equals the mixed volume $V(M_{d-j}(K)[1],-M[d-1])$, thus

$$V^{(j)}_{j+1,d-1}(K,M) = dc^{d,0}_{j,d-j}V(M_{d-j}(K)[1],-M[d-1]).$$

8. Spherical integral representations. For mixed volumes $V(K[1],M[d-1])$ of two convex bodies K,M the spherical integral representation (14.23), namely

$$V(K[1],M[d-1]) = \frac{1}{d}\int_{S^{d-1}} h^*(K,u)\,S(M,\mathrm{d}u),$$

is classical. It involves the support function $h(K,\cdot)$ of K (here replaced by its centered version $h^*(K,\cdot)$) and the surface area measure $S(M,\cdot) := S_{d-1}(M,\cdot)$ of M. Since

$$V(K[1],M[d-1]) = \frac{1}{d}V^{(0)}_{1,d-1}(K,-M),$$

one can ask for extensions to mixed functionals of more than two bodies. The following result is obtained in Weil [800] for convex bodies $K_1,\dots,K_k$, $M_1,\dots,M_i$:

$$\Phi^{(j)}_{m_1,\ldots,m_k,d-1,\ldots,d-1}(K_1,\ldots,K_k,-M_1,\ldots,-M_i;A_1\times\ldots\times A_k\times(\mathbb{R}^d)^i)$$
$$=\int_{S^{d-1}}\cdots\int_{S^{d-1}}\Phi^{(j)}_{m_1,\ldots,m_k,d-1,\ldots,d-1}(K_1,\ldots,K_k,u_1^+,\ldots,u_i^+;$$
$$A_1\times\ldots\times A_k\times C_{u_1^\perp}\times\ldots\times C_{u_i^\perp})\,S(M_1,\mathrm{d}u_1)\cdots S(M_i,\mathrm{d}u_i). \tag{6.41}$$

For $k=1$, (6.41) again implies the formulas (6.28) and (6.33). The latter representation was proved by Goodey and Weil [281] (in correction of an erroneous statement from [795]).

If there are no bodies K_j in (6.41), then a formula for $V^{(j)}_{d-1,\ldots,d-1}(M_1,\ldots,M_{d-j})$ results. Since

$$\Phi^{(j)}_{d-1,\ldots,d-1}(u_1^+,\ldots,u_{d-j}^+;C_{u_1^\perp}\times\ldots\times C_{u_{d-j}^\perp})$$
$$=\frac{1}{(d-j)\kappa_{d-j}}\sigma_{d-j-1}(\mathrm{co}(u_1,\ldots,u_{d-j}))|\det(u_1,\ldots,u_{d-j})|,$$

where co denotes the spherical convex hull, we obtain

$$V^{(j)}_{d-1,\ldots,d-1}(M_1,\ldots,M_{d-j})$$
$$=\frac{1}{(d-j)\kappa_{d-j}}\int_{S^{d-1}}\cdots\int_{S^{d-1}}\sigma_{d-j-1}(\mathrm{co}(u_1,\ldots,u_{d-j}))$$
$$\times|\det(u_1,\ldots,u_{d-j})|\,S(M_1,\mathrm{d}u_1)\cdots S(M_{d-j},\mathrm{d}u_{d-j}). \tag{6.42}$$

From the case $j=d-2$ of (6.42) (or from (6.33)), we get a representation of $h(M_2(K),\cdot)$ due to Goodey and Weil [279],

$$h(M_2(K),u)=\frac{1}{2\pi}c^{2,d-2}_{d,0}\int_{S^{d-1}}\alpha(u,v)\sin\alpha(u,v)\,S(-K,\mathrm{d}v).$$

Here, $\alpha(u,v)\in[0,\pi]$ denotes the (smaller) angle between $u,v\in S^{d-1}$.

9. Centrally symmetric bodies. For smooth centrally symmetric bodies, representations of mixed measures in terms of the projection generating measures are possible. If M is a generalized zonoid (a centrally symmetric body, for which (14.33) holds with a signed measure ρ), the signed measure $\rho_{(j)}$ introduced by (14.36) exists and satisfies (14.38). For convex bodies $K_1,\ldots,K_k\in\mathcal{K}$ and generalized zonoids $M_1,\ldots,M_i$, the following was shown in Weil [800]:

$$\Phi^{(j)}_{m_1,\ldots,m_k,r_1,\ldots,r_i}(K_1,\ldots,K_k,M_1,\ldots,M_i;A_1\times\ldots\times A_k\times(\mathbb{R}^d)^i)$$
$$=2^{\sum_{j=1}^i r_j}\int_{G(d,r_1)}\cdots\int_{G(d,r_i)}\Phi^{(j)}_{m_1,\ldots,m_k,r_1,\ldots,r_i}(K_1,\ldots,K_k,L_1,\ldots,L_i;$$
$$A_1\times\ldots\times A_k\times C_{L_1}\times\ldots\times C_{L_i})\,\rho_{(r_1)}(M_1,\mathrm{d}L_1)\cdots\rho_{(r_i)}(M_i,\mathrm{d}L_i). \tag{6.43}$$

For $k=0$, (6.43) yields Theorem 10.1 in Weil [792]:

$$V^{(j)}_{r_1,\ldots,r_i}(M_1,\ldots,M_i) \tag{6.44}$$
$$=\frac{2^{(i-1)d+j}}{r_1!\cdots r_i!}\int_{G(d,r_1)}\cdots\int_{G(d,r_i)}[L_1,\ldots,L_i]\,\rho_{(r_1)}(M_1,\mathrm{d}L_1)\cdots\rho_{(r_i)}(M_i,\mathrm{d}L_i).$$

In the case $r_1 = \ldots = r_i = d-1$, (6.44) implies an iterated variant of (6.40) for centrally symmetric bodies $M_1, \ldots, M_i$ where the halfspaces are replaced by their bounding hyperplanes (see Weil [800], for details). As special cases, for centrally symmetric convex bodies K, M, the following formulas result:

$$V^{(j)}_{j+1,d-1}(K,M) = \frac{1}{2}\int_{S^{d-1}} \Phi^{(j)}_{j+1,d-1}(K,u^\perp;\mathbb{R}^d \times C_{u^\perp})\, S(M,\mathrm{d}u),$$

$$h(M_{d-j}(K),u) = \frac{c^{j,d-j}_{d,0}}{2}\Phi^{(j)}_{j+1,d-1}(K,u^\perp;\mathbb{R}^d \times C_{u^\perp}).$$

10. Support functions. The translative formula for support functions of Theorem 6.4.6 and the kinematic formula of Theorem 6.4.7 were proved by Weil [795]. The approach to (6.29) that is presented here comes from Schneider [708].

Since the left side of (6.29) defines a support function and the summands on the right side have different degrees of homogeneity, one may conjecture that the mixed functions $h^*_k(K,M;\cdot)$ are support functions, too. This was indeed proved by Goodey and Weil [281]. A simpler approach and an extension to mixed functions of more than two convex bodies are found in Schneider [709].

The case of two convex bodies can also be formulated as follows. If $K, M \in \mathcal{K}'$, then the translative integral

$$\int_{\mathbb{R}^d} h^*(K \cap M^x, \cdot)\, \lambda(\mathrm{d}x)$$

defines the support function of a convex body $T(K,M)$, called the **translation mixture** of K and M. There exists a polynomial expansion

$$T(rK, sM) = \sum_{k=1}^{d} r^k s^{d+1-k} T_k(K,M)$$

with convex bodies $T_k(K,M)$, called the **mixed bodies** of K and M. For the case of polytopes K, M, the vertices and edges of $T_k(K,M)$ were explicitly determined in [709].

Applications of the integral geometric formulas for support functions to stochastic geometry appear in Weil [793, 798].

11. Projection functions. Kinematic and Crofton formulas for projection functions were first studied by Goodey and Weil [278]. Theorem 6.4.8 in its present form appears in Goodey, Schneider and Weil [275].

6.5 Spherical Integral Geometry

Large parts of integral geometry in Euclidean spaces can be extended, in a suitable way, to spaces of constant curvature. In this section, we treat basic facts of the integral geometry of convex bodies in spherical space, since this is of some relevance for stochastic geometry. The approach will be similar to the Euclidean case: for (spherically) convex bodies we introduce generalized curvature measures via a Steiner formula, and integral geometric intersection formulas involving curvature measures are proved for polytopes, using characterization theorems, and then extended to general convex bodies. Our presentation owes much to the work of Glasauer [264, 265], which we follow in several aspects and details. We shall be rather brief at points where the procedure is in an obvious way similar to the Euclidean case. On the other hand, some geometric facts of spherical geometry are proved here instead of deferring them to the Appendix, since they are needed only in this section.

The spherical space to be considered is the unit sphere S^{d-1} of $\mathbb{R}^d$. The usual metric in S^{d-1} is denoted by d_s, thus $d_s(x,y) = \arccos\langle x,y\rangle$ for $x,y \in S^{d-1}$. It induces the trace topology from $\mathbb{R}^d$ on S^{d-1}, and topological notions in S^{d-1} refer to this topology. For points $x,y \in S^{d-1}$ with $d_s(x,y) < \pi$, the set $[x,y] := S^{d-1} \cap \operatorname{pos}\{x,y\}$ is the unique **spherical segment** joining x and y. A **spherically convex body** in S^{d-1} is the intersection of S^{d-1} with a closed convex cone different from $\{0\}$ in $\mathbb{R}^d$. In the present section, we shall mostly say 'convex' instead of 'spherically convex'. The convex bodies in S^{d-1} are precisely the nonempty closed subsets that contain with any two points of spherical distance less than π also the spherical segment joining them. The set of all convex bodies in S^{d-1} is denoted by $\mathcal{K}_s$. It is equipped with the Hausdorff metric induced by the metric d_s. Note that S^{d-1} is an isolated point of $\mathcal{K}_s$. For $K \in \mathcal{K}_s$, we denote by $\breve{K} := \operatorname{pos} K$ the cone with $K = S^{d-1} \cap \breve{K}$. The correspondence $K \leftrightarrow \breve{K}$ is quite useful for the study of spherically convex bodies. The **dimension** of $K \in \mathcal{K}_s$ is defined as $\dim K := \dim \breve{K} - 1$. The **relative interior** of K, denoted by relint K, is the interior of K relative to $S^{d-1} \cap \operatorname{lin} K$.

A distinguished subset of $\mathcal{K}_s$ is the set $\mathcal{S}_k$ of k-dimensional great subspheres, which are the intersections of S^{d-1} with $(k+1)$-dimensional linear subspaces of $\mathbb{R}^d$, $k = 0, \dots, d-1$. We write $\mathcal{S}_\bullet := \bigcup_{k=0}^{d-1} \mathcal{S}_k$, and we often say 'subsphere' instead of 'great subsphere'. A set $K \in \mathcal{K}_s$ is called a **proper convex body** if it is contained in an open hemisphere, equivalently, if the cone $\breve{K}$ is pointed (does not contain a line). We write $\mathcal{K}_s^p$ for the set of all proper convex bodies.

Let $K, M \in \mathcal{K}_s$. We denote by

$$K \vee M := S^{d-1} \cap \operatorname{pos}(K \cup M)$$

the **spherically convex hull** of K and M. For $K \vee \{x\}$ we write $K \vee x$, and $x \vee y := [x,y]$ if x and y are not antipodal. The set

$$K^* := \{x \in S^{d-1} : \langle x,y\rangle \le 0 \text{ for all } y \in K\}$$

is the **polar body** of K; thus K^* is the intersection of S^{d-1} with the dual cone of $\breve{K}$. It is again in $\mathcal{K}_s$. Further, $(K^*)^* = K$ and

$$(K \vee M)^* = K^* \cap M^*, \qquad (K \cap M)^* = K^* \vee M^*. \tag{6.45}$$

The polar body of K can also be represented as

$$K^* := \{x \in S^{d-1} : d_s(K, x) \geq \pi/2\}.$$

If $S \in \mathcal{S}_k$ for $k \in \{0, \dots, d-2\}$, then $S^* = S^{d-1} \cap (\operatorname{lin} S)^\perp \in \mathcal{S}_{d-k-2}$.

Let $K \in \mathcal{K}_s$ and $x \in S^{d-1}$. If $0 \leq d_s(K, x) < \pi/2$, there is a unique point in K that is nearest to x; we denote it by $p_s(K, x)$. This defines the **nearest-point map** or **metric projection** $p_s(K, \cdot)$. If $x \notin K$, we define $u_s(K, x) = u(\breve{K}, x)$ (see Section 14.2 for the latter); note that $u_s(K, x) = p_s(K^*, x)$. For $x \in \operatorname{bd} K$,

$$N_s(K, x) := \{y \in K^* : \langle x, y \rangle = 0\}$$

is the set of outer (unit) normal vectors to K at x. Note that $\operatorname{pos} N_s(K, x) = N(\breve{K}, x)$ is the normal cone of $\breve{K}$ at x (as introduced in Section 14.2), but $N_s(K, x)$ consists of unit vectors. A pair (x, u) with $x \in \operatorname{bd} K$ and $u \in N_s(K, x)$ is called a **support element** of K. It is easy to see that

$$(x, u) \text{ is a support element of } K \quad \Leftrightarrow \quad (u, x) \text{ is a support element of } K^*.$$

The set of all support elements of K, denoted by $\operatorname{Nor} K$, is a closed subset of the product space $\Sigma_s := S^{d-1} \times S^{d-1}$.

A convex body $P \in \mathcal{K}_s$ is a (spherical) **polytope** if the cone $\breve{P}$ is polyhedral, that is, an intersection of finitely many closed halfspaces with 0 in the boundary. The set of polytopes in S^{d-1} is denoted by $\mathcal{P}_s$. Let $P \in \mathcal{P}_s$. A k-**face** of P is a set $F = S^{d-1} \cap \breve{F}$, where $\breve{F}$ is a $(k+1)$-face of $\breve{P}$, $k \in \{0, \dots, d-1\}$. The set of all k-faces of P is denoted by $\mathcal{F}_k(P)$, and we write $\mathcal{F}_\bullet(P) := \bigcup_{k=0}^{d-1} \mathcal{F}_k(P)$.

Let F be a k-face of P. The set $N_s(P, x)$ is the same for all $x \in \operatorname{relint} F$ and is denoted by $N_s(P, F)$. The **internal angle** $\beta(0, \breve{F})$ of the cone $\breve{F}$ at 0 is defined by

$$\beta(0, \breve{F}) := \frac{\sigma_k(F)}{\omega_{k+1}},$$

and the **external angle** of P at F by

$$\gamma(F, P) := \gamma(\breve{F}, \breve{P}) := \frac{\sigma_{d-k-2}(N_s(P, F))}{\omega_{d-k-1}}.$$

The proof of a local Steiner formula for spherical polytopes will rest on the following lemma. It is a spherical counterpart to the representation of Lebesgue measure in $\mathbb{R}^d$ as the product of the Lebesgue measures on a subspace and its orthogonal complement.

Lemma 6.5.1. *Let $S \in \mathcal{S}_k$, where $k \in \{0, \dots, d-2\}$, and let $f : S^{d-1} \to \mathbb{R}$ be a nonnegative measurable function. Then*

$$\int_{S^{d-1}} f \, d\sigma = \int_S \int_{S^* \vee v} \sin^k(d_s(S^*, u)) f(u) \, \sigma_{d-k-1}(du) \, \sigma_k(dv).$$

Proof. We extend f to $\mathbb{R}^d$ by $\bar{f}(x) := \|x\|^{-d+1} f(x/\|x\|)$ for $0 < \|x\| < 1$, and $\bar{f}(x) = 0$ otherwise. Let $L := \operatorname{lin} S$. Using spherical coordinates, we obtain

$$\begin{aligned}
\int_{S^{d-1}} f \, \mathrm{d}\sigma &= \int_{S^{d-1}} \int_0^1 (f(x)/t^{d-1}) t^{d-1} \, \mathrm{d}t \, \sigma(\mathrm{d}x) \\
&= \int_{\mathbb{R}^d} \bar{f} \, \mathrm{d}\lambda \\
&= \int_L \int_{L^\perp} \bar{f}(x+y) \, \lambda_{L^\perp}(\mathrm{d}y) \, \lambda_L(\mathrm{d}x) \\
&= \int_S \left[\int_0^1 t^k \int_{L^\perp} \bar{f}(tv+y) \, \lambda_{L^\perp}(\mathrm{d}y) \, \mathrm{d}t \right] \sigma_k(\mathrm{d}v).
\end{aligned}$$

Recall that $d(\cdot,\cdot)$ denotes the Euclidean distance. With $tv + y =: w = \tau u$, $\|u\| = 1$, we have $t = d(L^\perp, w) = \tau \sin(d_s(S^*, u))$. Hence, the integral in brackets is equal to

$$\begin{aligned}
&\int_{\operatorname{pos}(L^\perp \cup \{v\})} d(L^\perp, w)^k \bar{f}(w) \, \lambda_{d-k}(\mathrm{d}w) \\
&= \int_{S^* \vee v} \int_0^1 \tau^{d-k-1} \tau^k \sin^k(d_s(S^*, u)) \bar{f}(\tau u) \, \mathrm{d}\tau \, \sigma_{d-k-1}(\mathrm{d}u) \\
&= \int_{S^* \vee v} \sin^k(d_s(S^*, u)) f(u) \, \sigma_{d-k-1}(\mathrm{d}u).
\end{aligned}$$

This yields the assertion. □

For $K \in \mathcal{K}_s$, the **local parallel set** of K, determined by a Borel set $A \subset \Sigma_s$ and a number $0 < \epsilon < \pi/2$, is defined by

$$M_\epsilon(K, A) := \{x \in S^{d-1} : d_s(K, x) \leq \epsilon, \, (p_s(K, x), u_s(K, x)) \in A\}.$$

Theorem 6.5.1 (Local spherical Steiner formula). *For $K \in \mathcal{K}_s$, there exist uniquely determined finite measures $\Theta_0(K, \cdot), \ldots, \Theta_{d-2}(K, \cdot)$ on Σ_s such that the following holds. If $A \in \mathcal{B}(\Sigma_s)$ and $0 < \epsilon < \pi/2$, then*

$$\sigma(M_\epsilon(K, A)) = \sum_{m=0}^{d-2} g_{d,m}(\epsilon) \Theta_m(K, A)$$

with

$$g_{d,m}(\epsilon) := \omega_{m+1} \omega_{d-m-1} \int_0^\epsilon \cos^m \varphi \sin^{d-m-2} \varphi \, \mathrm{d}\varphi, \qquad 0 \leq \epsilon \leq \pi/2.$$

If $P \in \mathcal{P}_s$, then

$$\Theta_m(P, A) = \frac{1}{\omega_{m+1} \omega_{d-m-1}} \sum_{F \in \mathcal{F}_m(P)} \int_F \int_{N_s(P,F)} \mathbf{1}_A(x, u) \, \sigma_{d-m-2}(\mathrm{d}u) \, \sigma_m(\mathrm{d}x).$$

Proof. First let $P \in \mathcal{P}_s$, let $m \in \{0, \dots, d-2\}$ and $F \in \mathcal{F}_m(P)$. Put $S := S^{d-1} \cap \operatorname{lin} F$ and $U := p_s(P, \cdot)^{-1}(\operatorname{relint} F)$. Let $f \geq 0$ be a measurable function on S^{d-1}. We apply Lemma 6.5.1, first to S^{d-1} and its subsphere S, then to the sphere $S^* \vee \{-x, x\}$, where $x \in S$, and its subsphere S^*. This gives

$$\begin{aligned}
\int_U f \, d\sigma &= \int_S \int_{S^* \vee x} \mathbf{1}_U(u) f(u) \sin^m(d_s(S^*, u)) \, \sigma_{d-m-1}(du) \, \sigma_m(dx) \\
&= \int_S \int_{S^*} \int_{\{-x,x\} \vee v} \mathbf{1}_{U \cap (S^* \vee x)}(z) f(z) \sin^m(d_s(S^*, z)) \\
&\quad \times \sin^{d-m-2}(d_s(\{-x, x\}, z)) \, \sigma_1(dz) \, \sigma_{d-m-2}(dv) \, \sigma_m(dx) \\
&= \int_S \int_{S^*} \int_{[x,v]} \mathbf{1}_U(z) f(z) \cos^m(d_s(S, z)) \sin^{d-m-2}(d_s(S, z)) \, \sigma_1(dz) \\
&\quad \times \sigma_{d-m-2}(dv) \, \sigma_m(dx) \\
&= \int_F \int_{N_s(P,F)} \int_0^{\pi/2} f(x \cos\varphi + v \sin\varphi) \cos^m \varphi \sin^{d-m-2} \varphi \, d\varphi \\
&\quad \times \sigma_{d-m-2}(dv) \, \sigma_m(dx).
\end{aligned}$$

The choice $f = \mathbf{1}_{M_\epsilon(P,A)}$ yields

$$\begin{aligned}
&\sigma(M_\epsilon(P, A) \cap p_s(P, \cdot)^{-1}(\operatorname{relint} F)) \\
&= \int_F \int_{N_s(P,F)} \mathbf{1}_A(x, v) \, \sigma_{d-m-2}(dv) \, \sigma_m(dx) \int_0^\epsilon \cos^m \varphi \sin^{d-m-2} \varphi \, d\varphi.
\end{aligned}$$

Similarly to Euclidean space, $\operatorname{bd} P = \bigcup_{m=0}^{d-2} \bigcup_{F \in \mathcal{F}_m(P)} \operatorname{relint} F$ is a disjoint union, hence we get

$$\begin{aligned}
\sigma(M_\epsilon(P, A)) &= \sum_{m=0}^{d-2} \sum_{F \in \mathcal{F}_m(P)} \int_F \int_{N_s(P,F)} \mathbf{1}_A(x, v) \, \sigma_{d-m-2}(dv) \, \sigma_m(dx) \\
&\quad \times \int_0^\epsilon \cos^m \varphi \sin^{d-m-2} \varphi \, d\varphi \\
&= \sum_{m=0}^{d-2} g_{d,m}(\epsilon) \Theta_m(P, A),
\end{aligned}$$

if $\Theta_m(P, A)$ is defined as shown in the theorem. We observe that the functions $g_{d,0}, \dots, g_{d,d-2}$ are linearly independent on $(0, \pi/2)$. The remaining parts of the proof (measurability, extension to general convex bodies, uniqueness) are so similar to the Euclidean case (which is treated in Schneider [695, sect. 4.1, 4.2]) that we omit them. □

We add a remark to the Steiner formula. Applying it with $K = S \in \mathcal{S}_i$ for $i \in \{0, \dots, d-2\}$ and $A = \Sigma_s$, and observing that $\lim_{\epsilon \to \pi/2} \sigma(M_\epsilon(S, \Sigma_s)) = \omega_d$ and

$$\Theta_m(S, \Sigma_s) = \delta_{im} \qquad \text{for } S \in \mathcal{S}_i \tag{6.46}$$

(where δ_{im} denotes the Kronecker symbol), we find that $g_{d,i}(\pi/2) = \omega_d$ (which can, of course, also be obtained from the definition). If P is a polytope and $F \in \mathcal{F}_m(P)$, then

$$\mathrm{cl} \bigcup_{0<\epsilon<\pi/2} M_\epsilon(P, \Sigma_s) \cap p_s(P, \cdot)^{-1}(\mathrm{relint}\, F) = F \vee N_s(P, F),$$

hence

$$\frac{\sigma(F \vee N_s(P,F))}{\omega_d} = \frac{\sigma_m(F)}{\omega_{m+1}} \frac{\sigma_{d-m-2}(N_s(P,F))}{\omega_{d-m-1}} = \frac{1}{\omega_{m+1}} \gamma(F,P)\sigma_m(F). \tag{6.47}$$

The polytopes $F \vee N_s(P,F)$, $F \in \mathcal{F}_m(P)$, $m = 0, \dots, d-2$, together with P and P^*, tile the sphere S^{d-1}, that is, they cover it and have pairwise no common interior points. It follows that

$$\sum_{m=0}^{d-2} \frac{1}{\omega_{m+1}} \sum_{F \in \mathcal{F}_m(P)} \gamma(F,P)\sigma_m(F) + \frac{1}{\omega_d}\sigma(P) + \frac{1}{\omega_d}\sigma(P^*) = 1, \tag{6.48}$$

a fact which will later become important.

We call $\Theta_m(K, \cdot)$ the *m*th **support measure** or **generalized curvature measure** of K. The chosen normalization has a simplifying effect in later formulas. The following theorem collects the main properties of the support measures.

Theorem 6.5.2. *For $m = 0, \dots, d-2$, the mapping $\Theta_m : \mathcal{K}_s \times \mathcal{B}(\Sigma_s) \to \mathbb{R}$ has the following properties:*

(a) **Rotation covariance:** $\Theta_m(\vartheta K, \vartheta A) = \Theta_m(K, A)$ *for* $\vartheta \in SO_d$, *where* $\vartheta A := \{(\vartheta x, \vartheta u) : (x,u) \in A\}$,
(b) **Weak continuity:** $K_j \to K$ *(in the Hausdorff metric on* $\mathcal{K}_s$*) implies* $\Theta_m(K_j, \cdot) \xrightarrow{w} \Theta_m(K, \cdot)$,
(c) $\Theta_m(\cdot, A)$ *is* **additive**, *for each fixed* $A \in \mathcal{B}(\Sigma_s)$,
(d) $\Theta_m(\cdot, A)$ *is* **measurable**, *for each fixed* $A \in \mathcal{B}(\Sigma_s)$.

This theorem is analogous to Theorem 14.2.2, whose proof can be found in Schneider [695]. In the spherical case, the proof is very similar, so that we omit it here.

There is no Euclidean counterpart to the following nice behavior of the support measures under polarity.

Theorem 6.5.3. *If $K \in \mathcal{K}_s$ and $A \in \mathcal{B}(\Sigma_s)$, then*

$$\Theta_m(K, A) = \Theta_{d-m-2}(K^*, A^{-1})$$

for $m \in \{0, \dots, d-2\}$, where $A^{-1} := \{(u,x) \in \Sigma_s : (x,u) \in A\}$.

Proof. By the weak continuity of the support measures and the continuity of the polarity $K \mapsto K^*$ (which is easy to see), it suffices to prove this for the case where K is a polytope P. The assertion then follows from the explicit representation of $\Theta_m(P, A)$ given in Theorem 6.5.1 and the fact that $F \in \mathcal{F}_m(P)$ and $N_s(P, F) =: G$ implies $G \in \mathcal{F}_{d-m-2}(P^*)$ and $N_s(P^*, G) = F$; the latter is again easy to see. □

As a marginal measure of the mth support measure, we obtain the mth **curvature measure**, by

$$\phi_m(K, A) := \Theta_m(K, A \times S^{d-1}), \qquad A \in \mathcal{B}(S^{d-1}).$$

We supplement the definition by

$$\phi_{d-1}(K, A) := \frac{1}{\omega_d}\sigma(K \cap A), \qquad A \in \mathcal{B}(S^{d-1}).$$

Theorem 6.5.2 (together with properties of the spherical Lebesgue measure) implies that the curvature measure ϕ_m, $m \in \{0, \dots, d-1\}$, is rotation covariant, in the sense that $\phi_m(\vartheta K, \vartheta A) = \phi(K, A)$ for $K \in \mathcal{K}_s$ and $A \in \mathcal{B}(S^{d-1})$, weakly continuous, and additive and measurable in its first argument. Further, it follows easily from the definition that $\phi_m(K, \cdot)$ is concentrated on K and that $\phi_m(K, \cdot)$ is **locally determined**, in the sense that $K_1, K_2 \in \mathcal{K}_s$ and $K_1 \cap B = K_2 \cap B$ for an open set $B \subset S^{d-1}$ implies $\phi(K_1, A) = \phi(K_2, A)$ for all $A \in \mathcal{B}(B)$. These properties can be defined similarly for mappings $\psi : \mathcal{K}_s \times \mathcal{B}(S^{d-1}) \to \mathbb{R}$ and play a role in the following characterization theorem.

Theorem 6.5.4. *Let $\psi : \mathcal{P}_s \times \mathcal{B}(S^{d-1}) \to \mathbb{R}$ be a mapping which is rotation covariant, locally determined, additive in its first argument, and such that $\psi(P, \cdot)$ is a finite measure concentrated on P, for all $P \in \mathcal{P}_s$. Then there are constants $c_0, \dots, c_{d-1} \geq 0$ such that*

$$\psi(P, \cdot) = \sum_{m=0}^{d-1} c_m \phi_m(P, \cdot)$$

for all $P \in \mathcal{P}_s$.

Proof. Let $k \in \{0, \dots, d-2\}$. First let $S_k \in \mathcal{S}_k$, and let $\mathcal{P}(S_k^*)$ be the set of all polytopes contained in S_k^*. Let $Q \in \mathcal{P}(S_k^*) \cup \{\emptyset\}$. For $Q = \emptyset$, we define $Q^* := S^{d-1}$. The mapping $A \mapsto \psi(S_k \vee Q, A)$, $A \in \mathcal{B}(S_k)$, is a finite measure which is invariant under all rotations that map S_k into itself and fix S_k^* pointwise. By the uniqueness of the spherical Lebesgue measure, there exists a constant $c(S_k, Q) \geq 0$ with

$$\psi(S_k \vee Q, A) = c(S_k, Q)\sigma_k(A), \qquad A \in \mathcal{B}(S_k). \tag{6.49}$$

We write Q^o for the polar body of Q with respect to S_k^* as surrounding sphere, thus $Q^o = Q^* \cap S_k^*$; in particular, $Q^o = S_k^*$ if $Q = \emptyset$ (one has to keep in mind that Q^o depends on S_k^*). Choosing $A = S_k$, we put

$$f(Q) := c(S_k, Q^o) = \frac{1}{\omega_{k+1}} \psi(S_k \vee Q^o, S_k) \qquad \text{for } Q \in \mathcal{P}(S_k^*) \cup \{\emptyset\}.$$

The function f is nonnegative and invariant under the rotations of S_k^* into itself. If $Q_1 \cup Q_2$ is convex, then

$$S_k \vee (Q_1 \cup Q_2)^o = (S_k \vee Q_1^o) \cap (S_k \vee Q_2^o),$$
$$S_k \vee (Q_1 \cap Q_2)^o = (S_k \vee Q_1^o) \cup (S_k \vee Q_2^o).$$

From the additivity of ψ in its first argument it follows that f is additive. Let $Q \in \mathcal{P}(S_k^*)$ be a polytope with $\dim Q < d-k-2$. Then there exists $S_0 \in \mathcal{S}_0$ with $S_0 \subset Q^o$. Putting $S_{k+1} := S_k \vee S_0$, we have $S_{k+1} \in \mathcal{S}_{k+1}$ and

$$S_k \vee Q^o = S_{k+1} \vee (Q^* \cap S_{k+1}^*).$$

Therefore, using (6.49) with k replaced by $k+1$,

$$\begin{aligned} \omega_{k+1} f(Q) &= \psi(S_k \vee Q^o, S_k) = \psi(S_{k+1} \vee (Q^* \cap S_{k+1}^*), S_k) \\ &= c(S_{k+1}, Q^* \cap S_{k+1}^*)\sigma_{k+1}(S_k) = 0. \end{aligned}$$

This shows that the mapping f satisfies the assumptions of Theorem 14.4.7, with S^{d-1} replaced by S_k^*. It follows that

$$f(Q) = c(S_k)\sigma_{d-k-2}(Q)$$

with a constant $c(S_k) \geq 0$. By the rotation covariance of ψ, this constant depends only on k; we put $c(S_k) =: b_k$. Thus we have $c(S_k, Q^o) = b_k \sigma_{d-k-2}(Q)$ and hence $c(S_k, Q) = b_k \sigma_{d-k-2}(Q^o)$. Altogether, we arrive at

$$\psi(S_k \vee Q, A) = b_k \sigma_{d-k-2}(Q^o)\sigma_k(A)$$

for $Q \in \mathcal{P}(S_k^*)$ and $A \in \mathcal{B}(S_k)$.

Now let $P \in \mathcal{P}_s$, $F \in \mathcal{F}_k(P)$ for some $k \in \{0, \dots, d-2\}$, and $A \in \mathcal{B}(S^{d-1})$ with $A \subset \operatorname{relint} F$. With $S_k := S^{d-1} \cap \operatorname{lin} F$ and $Q := (S_k \vee P) \cap S_k^*$ we have $N_s(P, F) = P^* \cap S_k^* = Q^* \cap S_k^* = Q^o$. A sufficiently small open neighborhood B of A satisfies $P \cap B = (S_k \vee Q) \cap B$. Since ψ is locally determined, we get

$$\begin{aligned} \psi(P, A) &= \psi(S_k \vee Q, A) = b_k \sigma_{d-k-2}(Q^o)\sigma_k(A) \\ &= b_k \sigma_{d-k-2}(N_s(P, F))\sigma_k(A). \end{aligned}$$

Finally, let $P \in \mathcal{P}_s$ and $A \in \mathcal{B}(S^{d-1})$. Then

$$A = (A \setminus P) \cup (A \cap \operatorname{int} P) \cup \bigcup_{k=0}^{d-2} \bigcup_{F \in \mathcal{F}_k(P)} A \cap \operatorname{relint} F$$

is a disjoint union. Since $\psi(P,\cdot)$ is concentrated on P, we get

$$\begin{aligned}
&\psi(P,A)\\
&=\psi(P,A\cap \operatorname{int} P)+\sum_{k=0}^{d-2}\sum_{F\in\mathcal{F}_k(P)} b_k\sigma_{d-k-2}(N_s(P,F))\sigma_k(A\cap \operatorname{relint} F)\\
&=\psi(P,A\cap \operatorname{int} P)+\sum_{k=0}^{d-2} c_k\phi_k(P,A),
\end{aligned}$$

with $c_k := b_k\omega_{k+1}\omega_{d-k-1}$. Since ψ is locally determined, we have $\psi(P,A\cap \operatorname{int} P)=\psi(S^{d-1},A\cap \operatorname{int} P)$. Here $\psi(S^{d-1},\cdot)$ is a rotation invariant finite measure and hence proportional to σ, thus $\psi(P,A\cap \operatorname{int} P)=b_{d-1}\sigma(A\cap \operatorname{int} P)=c_{d-1}\phi_{d-1}(P,A)$ with a constant $c_{d-1}\geq 0$. This completes the proof. □

Before applying Theorem 6.5.4 to the proof of a kinematic formula for curvature measures, we consider the total curvature measures. We write

$$v_m(K) := \phi_m(K,S^{d-1}), \qquad m=0,\dots,d-1.$$

The functional v_m is called the mth **(spherical) intrinsic volume**. For a polytope P, we obtain from Theorem 6.5.1 the representation

$$v_m(P)=\frac{1}{\omega_{m+1}}\sum_{F\in\mathcal{F}_m(P)}\gamma(F,P)\sigma_m(F)=\sum_{G\in\mathcal{F}_{m+1}(\breve{P})}\beta(0,G)\gamma(G,\breve{P}). \quad (6.50)$$

The duality relation of Theorem 6.5.3 gives

$$v_m(K)=v_{d-m-2}(K^*), \qquad m=0,\dots,d-2. \quad (6.51)$$

It is consistent with this to supplement the definition by

$$v_{-1}(K) := v_{d-1}(K^*). \quad (6.52)$$

With this definition, the intrinsic volumes satisfy two linear relations, which also have no counterpart in Euclidean space.

Theorem 6.5.5. *For $K\in\mathcal{K}_s$,*

$$\sum_{i=-1}^{d-1} v_i(K)=1, \quad (6.53)$$

and if $K\in\mathcal{K}_s\setminus\mathcal{S}_\bullet$, then

$$\sum_{i=-1}^{d-1}(-1)^i v_i(K)=0, \quad (6.54)$$

hence also

$$\sum_{i=0}^{\lfloor\frac{d-1}{2}\rfloor} v_{2i}(K)=\frac{1}{2}. \quad (6.55)$$

Proof. Relation (6.53) is just (6.48). For the proof of (6.54), it is convenient to consider first a pointed polyhedral cone $C \subset \mathbb{R}^d$ with interior points and its dual cone

$$C^* := \{y \in \mathbb{R}^d : \langle x, y\rangle \le 0 \text{ for all } X \in C\}.$$

If $F \in \mathcal{F}_m(C)$ for some $m \in \{0, \dots, d\}$, then we define $\widehat{F} := N(C, F)$ (normal cone of C at F) and observe that $\widehat{F} \in \mathcal{F}_{d-m}(C^*)$. The cone $F + (-\widehat{F})$ has dimension d. We denote by U the union of all faces of dimensions less than $d-1$ of all the cones $F + (-\widehat{F})$, $F \in \mathcal{F}_\bullet(C)$, and assert that

$$\eta(x) := \sum_{F \in \mathcal{F}_\bullet(C)} (-1)^{\dim F} \mathbf{1}_{F+(-\widehat{F})}(x) = 0 \qquad \text{for all } x \in \mathbb{R}^d \setminus U. \tag{6.56}$$

Since C and C^* can be separated weakly by a hyperplane, there is a point $y \in \mathbb{R}^d$ with $y \notin C \cup (-C^*)$, hence with $\eta(y) = 0$. Let $x \in \mathbb{R}^d \setminus U$. The point x can be joined to y by a polygonal path in $\mathbb{R}^d \setminus U$. Hence, it suffices to show that η is constant along this path, and for this it is sufficient to show that η does not change when entering some $(d-1)$-face of a cone $F + (-\widehat{F})$.

Let H be a $(d-1)$-face of some cone $F + (-\widehat{F})$, $F \in \mathcal{F}_\bullet(C)$. Being a facet of the direct sum $F + (-\widehat{F})$, H is the direct sum, $H = F_1 + (-G)$, of a face $F_1 \in \mathcal{F}_k(C)$, for some $k \in \{0, \dots, d-1\}$, and a face $-G \in \mathcal{F}_{d-1-k}(-\widehat{F})$. There is a face F_2 of C with $G = \widehat{F_2}$. From $\dim F_1 + \dim G = d - 1$ and $\dim G = d - \dim F_2$ it follows that $\dim F_2 = k+1$. From $F_1 \subset F$ and $\widehat{F_2} \subset \widehat{F}$, hence $F_1 \subset F \subset F_2$, it follows that either $F = F_1$ or $F = F_2$.

Since $F_1 \subset H$, $F_2 \not\subset H$, F_1 is a face of F_2 and $\dim F_1 = \dim F_2 - 1$, it follows that the cone F_2 lies in one of the closed halfspaces bounded by $\operatorname{lin} H$. Similarly, $\widehat{F_1}$ lies in one of the closed halfspaces bounded by $\operatorname{lin} H$. Let u be the unit normal vector of $\operatorname{lin} H$ pointing into the halfspace not containing F_2; note that $u \in (\operatorname{lin} \widehat{F_2})^\perp = \operatorname{lin} F_2$. There exists $x \in F_1$ with $x - u \in F_2 \setminus F_1$. Since $x - u \in C$, but $x - u \notin F_1$, there exists $y \in \widehat{F_1}$ with $\langle x - u, y\rangle < 0$, hence with $\langle u, y\rangle > 0$. Thus, $\widehat{F_1}$ lies in the halfspace not containing F_2. Since $F_1 \subset H$ and $\widehat{F_2} \subset H$, we conclude that $F_1 + (-\widehat{F_1})$ and $F_2 + (-\widehat{F_2})$ lie on the same side of the hyperplane $\operatorname{lin} H$.

As a consequence, when entering the facet H from $\operatorname{int}(F + (-\widehat{F})) \setminus U$, the changes in the contributions to the function η coming from $F_1 + (-\widehat{F_1})$ and from $F_2 + (-\widehat{F_2})$ cancel each other. Should part of the facet H also belong to some other cone $G + (-\widehat{G})$, $G \in \mathcal{F}_\bullet(C)$, the same argument applies. In this way, relation (6.56) is proved.

If now $P \in \mathcal{K}_s^p$ is a $(d-1)$-dimensional polytope, we can apply relation (6.56) to the cone $C = \breve{P}$ and obtain

$$\mathbf{1}_{-P^*}(x) + \sum_{i=0}^{d-1} (-1)^{i+1} \sum_{F \in \mathcal{F}_i(P)} \mathbf{1}_{F \vee (-N_s(P,F))}(x) = 0$$

for σ-almost all $x \in S^{d-1}$. Integrating this relation over the unit sphere and using the reflection invariance of σ, we obtain (6.54), in view of (6.47) and (6.50). The extension to general convex bodies which are not subspheres follows by approximation. Relation (6.55) follows from (6.53) and (6.54). $\square$

The **spherical convex ring** $\mathcal{R}_s$ is defined as the system of all finite unions of spherically convex bodies in S^{d-1}, including the empty set $\emptyset$. Groemer's extension theorem 14.4.2, with the obvious adaptation to S^{d-1}, shows that every continuous additive functional on $\mathcal{K}_s^p$ with values in a topological vector space has a continuous extension to $\mathcal{R}_s$. The function χ defined by $\chi(K) := 1$ for $K \in \mathcal{K}_s^p$ is additive and hence has such an extension, which is also denoted by χ and called the **Euler characteristic**. Since the intrinsic volumes, too, have additive extensions to $\mathcal{R}_s$, relation (6.55) generalizes to

$$2 \sum_{i=0}^{\lfloor \frac{d-1}{2} \rfloor} v_{2i}(K) = \chi(K)$$

for $K \in \mathcal{R}_s$. This is a version of the **spherical Gauss–Bonnet theorem**. (Note that, in contrast to the Euclidean case, v_0 is *not* the Euler characteristic.)

Returning to the curvature measures, we prove an integral geometric intersection formula.

Theorem 6.5.6 (Spherical kinematic formula). *If $K, M \in \mathcal{R}_s$ and $A, B \in \mathcal{B}(S^{d-1})$, then*

$$\int_{SO_d} \phi_j(K \cap \vartheta M, A \cap \vartheta B)\, \nu(\mathrm{d}\vartheta) = \sum_{k=j}^{d-1} \phi_k(K, A)\phi_{d-1-k+j}(M, B) \qquad (6.57)$$

for $j = 0, \dots, d-1$.

Proof. For convex bodies $K, M \in \mathcal{K}_s$ we define

$$T(K, M) := \{\vartheta \in SO_d : K \text{ and } M \text{ touch}\},$$

where K and M are said to **touch** if $K \cap M \neq \emptyset$ but the cones $\breve{K}, \breve{M}$ can be separated weakly by a hyperplane. Similarly to the proof of Lemma 5.2.1 one shows that the mapping

$$\vartheta \mapsto \phi_j(K \cap \vartheta M, A \cap \vartheta B), \qquad \vartheta \in SO_d,$$

is measurable on $SO_d \setminus T(K, M)$ and hence coincides almost everywhere on SO_d with a measurable mapping if $\nu(T(K, M)) = 0$. The proof of the latter fact is not so straightforward as that for the Euclidean counterpart (see the beginning of Theorem 5.1.2). For polytopes P, Q, the relation $\nu(T(P, Q)) = 0$ is easily deduced from Lemma 13.2.1. Therefore, we first prove the kinematic formula for polytopes. This is used to prove $\nu(T(K,M)) = 0$ for general convex bodies, which then allows us to extend the kinematic formula to this case.

Let $j \in \{0,\dots,d-1\}$. The left side of (6.57) is well defined if K, M are polytopes. We fix $Q \in \mathcal{P}_s$ and an open set $B \subset S^{d-1}$ and put

$$\psi(P,A) := \int_{SO_d} \phi_j(P \cap \vartheta Q, A \cap \vartheta B)\, \nu(\mathrm{d}\vartheta),$$

for $P \in \mathcal{P}_s$ and $A \in \mathcal{B}(S^{d-1})$. It is easy to check that ψ satisfies the assumptions of Theorem 6.5.4, hence there exist constants $c_0(Q,B),\dots,c_{d-1}(Q,B) \geq 0$ such that

$$\int_{SO_d} \phi_j(P \cap \vartheta Q, A \cap \vartheta B)\, \nu(\mathrm{d}\vartheta) = \sum_{k=0}^{d-1} c_k(Q,B)\phi_k(P,A)$$

for all $P \in \mathcal{P}_s$ and all Borel sets $A \subset S^{d-1}$. Since, by (6.46), $\phi_m(S_k, S^{d-1}) = \delta_{km}$ for $S_k \in \mathcal{S}_k$, we obtain

$$c_k(Q,B) = \int_{SO_d} \phi_j(S_k \cap \vartheta Q, \vartheta B)\, \nu(\mathrm{d}\vartheta) \tag{6.58}$$

for $k = 0,\dots,d-1$. Admitting arbitrary Borel sets B in (6.58), we can again apply Theorem 6.5.4 and deduce that

$$c_k(Q,B) = \int_{SO_d} \phi_j(S_k \cap \vartheta Q, \vartheta B)\, \nu(\mathrm{d}\vartheta) = \sum_{i=0}^{d-1} b_{ik}\phi_i(Q,B)$$

with constants $b_{ik} \geq 0$. Here we choose $Q = S_m \in \mathcal{S}_m$ and $B = S^{d-1}$. It follows from Lemma 13.2.1 that either $S_k \cap \vartheta S_m = \emptyset$ for ν-almost all ϑ or $S_k \cap \vartheta S_m \in \mathcal{S}_{k+m-d+1}$ for ν-almost all ϑ, hence

$$b_{mk} = \begin{cases} 1, \text{ if } m = d-1-k+j, \\ 0 \text{ else.} \end{cases}$$

Thus we get $c_k(Q,B) = 0$ for $k = 0,\dots,j-1$ and $c_k(Q,B) = \phi_{d-1-k+j}(Q,B)$ for $k = j,\dots,d-1$. We conclude that

$$\int_{SO_d} \phi_j(P \cap \vartheta Q, A \cap \vartheta B)\, \nu(\mathrm{d}\vartheta) = \sum_{k=j}^{d-1} \phi_k(P,A)\phi_{d-1-k+j}(Q,B) \tag{6.59}$$

for $P, Q \in \mathcal{P}_s$, $A \in \mathcal{B}(S^{d-1})$ and open sets B. Since both sides define measures if B varies, (6.59) holds for arbitrary Borel sets B.

We want to replace P in (6.59) by a general convex body K. Since subspheres are polytopes, we may assume that K is not a subsphere. We can choose polytopes $P_1, P_2 \in \mathcal{P}_s$ which are not subspheres and satisfy $P_1 \subset K \subset P_2$. Then

$$T(K,Q) \subset (\{\vartheta \in SO_d : P_2 \cap \vartheta Q \neq \emptyset\} \setminus \{\vartheta \in SO_d : P_1 \cap \vartheta Q \neq \emptyset\}) \cup T(P_1, Q).$$

Since P_r is not a subsphere ($r = 1, 2$), it is easy to check that $P_r \cap \vartheta Q$ is not a subsphere for almost all ϑ, hence for almost all ϑ with $P_r \cap \vartheta Q \neq \emptyset$ we have

$$2 \sum_{i=0}^{\lfloor \frac{d-1}{2} \rfloor} v_{2i}(P_r \cap \vartheta Q) = 1$$

by (6.55). Now formula (6.59) with $A = B = \mathbb{R}^d$ gives

$$\begin{aligned} \nu(\{\vartheta \in SO_d : P_r \cap \vartheta Q \neq \emptyset\}) &= \int_{SO_d} 2 \sum_{i=0}^{\lfloor \frac{d-1}{2} \rfloor} v_{2i}(P_r \cap \vartheta Q) \, \nu(\mathrm{d}\vartheta) \\ &= 2 \sum_{i=0}^{\lfloor \frac{d-1}{2} \rfloor} \sum_{k=2i}^{d-1} v_k(P_r) v_{d-1-k+2i}(Q) \end{aligned}$$

for $r = 1, 2$ and hence

$$\nu(T(K,Q)) \le 2 \sum_{i=0}^{\lfloor \frac{d-1}{2} \rfloor} \sum_{k=2i}^{d-1} [v_k(P_2) - v_k(P_1)] v_{d-1-k+2i}(Q).$$

Since P_1 and P_2 can be chosen arbitrarily close to K in the Hausdorff metric and since the spherical intrinsic volumes are continuous (by Theorem 6.5.2(b)), we conclude that $\nu(T(K,Q)) = 0$.

Now we can conclude, as before, that $\vartheta \mapsto \phi_j(K \cap \vartheta Q, A \cap \vartheta B)$ coincides almost everywhere on SO_d with a measurable function, hence the left side of our next assertion,

$$\int_{SO_d} \phi_j(K \cap \vartheta Q, A \cap \vartheta B) \, \nu(\mathrm{d}\vartheta) = \sum_{k=j}^{d-1} \phi_k(K, A) \phi_{d-1-k+j}(Q, B), \qquad (6.60)$$

is well defined. To prove (6.60), we proceed similarly to the Euclidean case, see Theorem 5.2.3. Assertion (6.60) is equivalent to

$$\begin{aligned} &\int_{SO_d} \int_{S^{d-1}} f(x) g(\vartheta^{-1} x) \, \phi_j(K \cap \vartheta Q, \mathrm{d}x) \, \nu(\mathrm{d}\vartheta) \\ &= \sum_{k=j}^{d-1} \int_{S^{d-1}} f \, \mathrm{d}\phi_k(K, \cdot) \int_{S^{d-1}} g \, \mathrm{d}\phi_{d-1-k+j}(Q, \cdot) \end{aligned}$$

for all continuous functions $f, g : S^{d-1} \to \mathbb{R}$. The proof of this relation and hence of (6.60) is now completed by approximating K by polytopes, using the weak continuity of the curvature measures and the bounded convergence theorem.

Due to the inversion invariance of the measure ν, the result (6.60) can be written in the form

$$\int_{SO_d} \phi_j(Q \cap \vartheta K, B \cap \vartheta A)\, \nu(\mathrm{d}\vartheta) = \sum_{k=j}^{d-1} \phi_k(Q, B)\phi_{d-1-k+j}(K, A),$$

valid for polytopes Q and convex bodies K. As before, the polytope Q can now similarly be replaced by a general convex body.

The final extension to the spherical convex ring $\mathcal{R}_s$ is also similar to the Euclidean case. □

As a particular case of the spherical kinematic formula, we note its global version

$$\int_{SO_d} v_j(K \cap \vartheta M)\, \nu(\mathrm{d}\vartheta) = \sum_{k=j}^{d-1} v_k(K) v_{d-1-k+j}(M), \tag{6.61}$$

for $K, M \in \mathcal{R}_s$ and $j = 0, \dots, d-1$. Due to the duality relations (6.45), (6.51), (6.52), we obtain a dual kinematic formula for convex bodies by applying (6.61) to polar bodies. The result is

$$\int_{SO_d} v_j(K \vee \vartheta M)\, \nu(\mathrm{d}\vartheta) = \sum_{k=-1}^{j} v_k(K) v_{j-k-1}(M)$$

for $K, M \in \mathcal{K}_s$ and $j = -1, \dots, d-2$.

In contrast to the Euclidean case, where V_0 is the Euler characteristic, we must use (6.55) to obtain the integral

$$\int_{SO_d} \chi(K \cap \vartheta M)\, \nu(\mathrm{d}\vartheta) = 2 \sum_{k=0}^{\lfloor \frac{d-1}{2} \rfloor} \sum_{i=2k}^{d-1} v_i(K) v_{d-1-i+2k}(M)$$

for $K, M \in \mathcal{R}_s$.

The case of (6.57) where one of the sets is a subsphere deserves special attention. To have a concise notation, for $q \in \{0, \dots, d-1\}$ we choose $S_0 \in \mathcal{S}_q$ and denote by τ_q the image measure of ν under the map $\vartheta \mapsto \vartheta S_0$ from SO_d to $\mathcal{S}_q$. Thus, τ_q is the uniquely determined rotation invariant probability measure on $\mathcal{S}_q$. With this notation, we get the **spherical Crofton formula**

$$\int_{\mathcal{S}_q} \phi_j(K \cap S, A \cap S)\, \tau_q(\mathrm{d}S) = \phi_{d-1-q+j}(K, A)$$

for $K \in \mathcal{R}_s$, $A \in \mathcal{B}(S^{d-1})$, $q \in \{0, \dots, d-1\}$ and $j \in \{0, \dots, q\}$.

Defining

$$U_j(K) := \frac{1}{2} \int_{\mathcal{S}_{d-1-j}} \chi(K \cap S)\, \tau_{d-1-j}(\mathrm{d}S) \tag{6.62}$$

for $K \in \mathcal{R}_s$ and $j \in \{0, \dots, d-1\}$, we further obtain

$$U_j(K) = \sum_{k=0}^{\lfloor \frac{d-1-j}{2} \rfloor} v_{j+2k}(K) \tag{6.63}$$

If K in (6.62) is a convex body and not a subsphere, then for almost all $S \in \mathcal{S}_{d-1-j}$ the intersection $K \cap S$ is not a subsphere, hence $\chi(K \cap S) = 1$ if $K \cap S \neq \emptyset$. Therefore, $2U_j(K)$ is the total invariant measure of the set of all $(d-1-j)$-dimensional subspheres hitting the convex body K.

Notes for Section 6.5

1. A general source for integral geometry in spaces of constant curvature, from the differential geometric viewpoint, is the book by Santaló [662] and the literature quoted there, in particular Santaló [658].

Steiner formulas in spaces of constant curvature, in a differential geometric setting, were studied by Herglotz [336], Allendoerfer [23], Santaló [657]. A very general local version, for sets of positive reach, is due to Kohlmann [422]. The approach followed here is taken from Glasauer [264]. For differential geometric proofs of the spherical Gauss–Bonnet formula, we refer to Allendoerfer and Weil [24], Santaló [660, 661].

2. Since the linear relations between intrinsic volumes in Theorem 6.5.5 are special cases of the Steiner formula and the Gauss–Bonnet formula in spherical space, they appeared first, with differential geometric proofs, in the relevant literature quoted above. For spherical polytopes, in an equivalent version for polyhedral cones in $\mathbb{R}^d$, McMullen [469] has given interesting new proofs, more combinatorial in nature. The proof of (6.54) given here is based on a note by McMullen [470], which expands his remark at the beginning of §3 in [469].

3. The proof of Theorem 6.5.6, the spherical kinematic formula for curvature measures, is modeled after the proof given for its Euclidean counterpart by Schneider [676, Th. (6.1)]. The presentation given here follows the one by Glasauer [264]. This work contains many more results of spherical integral geometry, among them an abstract version of the kinematic formula and a version for support measures.

4. For spherical polytopes, or rather their spanned polyhedral cones, the functionals U_j were studied by Grünbaum [300], under the name of **Grassmann angles**.

5. In contrast to the Euclidean case, the spherical intrinsic volumes are in general not monotone under set inclusion. We restrict ourselves here to the set $\mathcal{K}_s \setminus \mathcal{S}_\bullet$. Clearly v_{d-1} is increasing under set inclusion, and so is v_{d-2}, being equal to U_{d-2}. By duality, v_{-1} and v_0 are decreasing. For $j \in \{1, \dots, d-3\}$, however, the functional v_j is not monotone. This follows, for example, by considering spherical balls B_r with spherical radius r, $0 \le r \le \pi/2$. From the Steiner formula one sees that

$$v_j(B_r) = \frac{\omega_d}{\omega_{j+1}\omega_{d-1-j}} \binom{d-2}{j} \cos^{d-2-j} r \sin^j r,$$

which is not a monotone function of r.

On the other hand, the functionals $U_0, \dots, U_{d-1}$ are increasing, as follows immediately from their definition. They may as well be considered as spherical analogs of the Euclidean intrinsic volumes, sharing with them the integral geometric interpretation as total measures of intersecting flats, respectively subspheres, of a suitable dimension. There is still another series of functionals which can be considered as counterparts to the Euclidean intrinsic volumes. Let $q \in \{0, \dots, d-1\}$ and $S \in \mathcal{S}_q$. The **spherical projection** of $K \in \mathcal{K}_s$ to S is defined by $K|S := S \cap (K \vee S^*)$. Then the function defined by

$$W_j(K) := \frac{1}{\omega_{j+1}} \int_{S_j} \sigma_j(K|S)\, \tau_j(\mathrm{d}S)$$

for $K \in \mathcal{K}_s$ can be expressed in terms of intrinsic volumes. The relation

$$W_j(K) = \sum_{k=j}^{d-1} v_k(K)$$

was proved by Glasauer [264]. Clearly, also W_j is increasing. Thus, in spherical space, there are three series of functionals which, with some reason, can be considered as counterparts to the Euclidean intrinsic volumes. All functionals v_j, U_j, W_j are nonnegative, additive, continuous, and rotation invariant. The U_j and W_j are linear combinations of the v_j with nonnegative coefficients, and they are increasing under set inclusion.

It is a longstanding (and repeatedly asked) open question whether Hadwiger's characterization theorem 14.4.6 has a spherical counterpart. For example, if a function $\varphi : \mathcal{K}_s \to \mathbb{R}$ is additive, continuous and rotation invariant, must it be of the form $\varphi = \sum_{i=0}^{d-1} c_i v_i$ with constant coefficients $c_0, \dots, c_{d-1}$? An affirmative answer to the question posed in Note 6 of Section 14.4 would be an essential step towards a solution. As a variant, one might ask whether a function $\varphi : \mathcal{K}_s \setminus \mathcal{S}_\bullet \to \mathbb{R}$ which is additive, rotation invariant and increasing must be a nonnegative linear combination of the functions U_j or W_j.

6. Motivated by the Euclidean case, one may ask for inequalities existing between the functionals v_j, U_j, W_j. For example, among all convex bodies $K \in \mathcal{K}_s^p$ of given positive volume $v_{d-1}(K)$, which ones are extremal for one of the functionals? Only the following nontrivial cases seem to be known. The minimum of v_{d-2} is attained if and only if K is a ball (the classical isoperimetric problem in spherical space). The maximum of $v_{-1}(K) = v_{d-1}(K^*)$ (and, because of $U_1(K) = \frac{1}{2} - v_{d-1}(K^*)$; also the minimum of $U_1(K)$) is attained if and only if K is a ball. The latter result, which can be considered as a spherical counterpart to the Blaschke–Santaló inequality, was proved by Gao, Hug and Schneider [243].

7

Integral Geometric Transformations

Mean value formulas with respect to invariant measures, as treated in the preceding two chapters, are a central topic of integral geometry. Another one is transformation formulas for integrals over various spaces of geometric objects. The need for such results in stochastic geometry can be demonstrated by simple examples. Consider, for instance, two independent, identically distributed random hyperplanes in $\mathbb{R}^d$. Suppose the distribution is such that the intersection of the two hyperplanes is almost surely a $(d-2)$-flat. What is the distribution of this random $(d-2)$-flat? Or, consider $k \leq d$ independent, identically distributed random points in $\mathbb{R}^d$, and suppose their distribution is such that they almost surely span a $(k-1)$-flat. What is its distribution? In the cases where the original distributions are derived from invariant measures (by restriction, for example), the answers can be obtained from simple cases of the transformation formulas of this chapter. Generally, these transformation formulas relate integrations over tuples of flats, with respect to invariant measures, to integrations over other sets of flats (or other geometric objects) that are obtained by geometric operations, such as intersection or span. As an example, consider the integral of a function depending on d points. It may happen that the function depends only on the hyperplane spanned (almost everywhere) by the points. Then it may have a simplifying effect to integrate first over the d-tuples of points lying in a fixed hyperplane, and then over all hyperplanes. In principle, the required transformation formulas are just versions of the transformation rule for multiple integrals under differentiable mappings. However, since the mappings are defined by geometric operations, the Jacobians have geometric interpretations, and therefore direct geometric arguments are often simpler and more perspicuous than the use of special parametrizations.

The transformation formulas to be proved have various applications in stochastic geometry, for example in the investigation of convex hulls of random points (Chapter 8), the study of random mosaics (Chapter 10), or in the foundations of stereology. We do not aim at presenting the integral geometric transformation formulas in their greatest generality, but rather give typical and basic examples. This will be done in Sections 7.2 and 7.3. The first section provides simple rules for invariant measures on flag spaces.

7.1 Flag Spaces

In this section, we consider pairs of linear or affine subspaces, one contained in the other.

Let $p, q \in \{0, \ldots, d\}$, and let $L \in G(d,p)$ be a fixed p-dimensional linear subspace. We denote by $G(L,q)$ the space of all q-dimensional linear subspaces contained in L if $q \leq p$, respectively containing L if $q > p$. In a similar way, for an affine subspace $E \in A(d,p)$, the space $A(E,q)$ of q-flats contained in E, respectively containing E, is defined. These spaces are described in detail in Section 13.2. There also the invariant measures ν_q^L on $G(L,q)$ and μ_q^E on $A(E,q)$ are constructed. These measures will be used in the following.

Now we turn to spaces of pairs of linear subspaces or flats. For $0 \leq p, q \leq d$ with $p \neq q$ we define

$$G(d,p,q) := \{(L,M) \in G(d,p) \times G(d,q) : L \subset M\}, \qquad \text{if } p < q,$$

$$G(d,p,q) := \{(L,M) \in G(d,p) \times G(d,q) : L \supset M\}, \qquad \text{if } p > q,$$

and

$$A(d,p,q) := \{(E,F) \in A(d,p) \times A(d,q) : E \subset F\}, \qquad \text{if } p < q,$$

$$A(d,p,q) := \{(E,F) \in A(d,p) \times A(d,q) : E \supset F\}, \qquad \text{if } p > q.$$

In an obvious way, these definitions could be extended to more than two linear or affine subspaces. Spaces of the type $G(d,p,q)$ or $A(d,p,q)$ are called **flag spaces**. The flag space $G(d,p,q)$, for example, is evidently a homogeneous SO_d-space. Defining

$$\begin{aligned} \beta_{p,q} : SO_d &\to G(d,p,q), \\ \vartheta &\mapsto (\vartheta L_p, \vartheta L_q) \end{aligned}$$

where $(L_p, L_q) \in G(d,p,q)$ is arbitrary but fixed, and

$$\nu_{p,q} := \beta_{p,q}(\nu),$$

we obtain a rotation invariant probability measure $\nu_{p,q}$ on $G(d,p,q)$. Thus, by definition,

$$\int_{G(d,p,q)} f \, \mathrm{d}\nu_{p,q} = \int_{SO_d} f(\vartheta L_p, \vartheta L_q) \, \nu(\mathrm{d}\vartheta) \tag{7.1}$$

for every nonnegative measurable function f on $G(d,p,q)$. We shall first show that this measure can be computed, as one might expect, by iterated integrations over p- and q-dimensional subspaces.

Theorem 7.1.1. *If $0 \le p < q \le d-1$ and if $f : G(d,p,q) \to \mathbb{R}$ is a nonnegative measurable function, then*

$$\int_{G(d,p,q)} f \, \mathrm{d}\nu_{p,q} = \int_{G(d,q)} \int_{G(M,p)} f(L,M)\, \nu_p^M(\mathrm{d}L)\, \nu_q(\mathrm{d}M)$$
$$= \int_{G(d,p)} \int_{G(L,q)} f(L,M)\, \nu_q^L(\mathrm{d}M)\, \nu_p(\mathrm{d}L).$$

Proof. Measurability follows, for example, from the fact that the mapping $(M,B) \mapsto \nu_p^M(B)$, $M \in G(d,q)$, $B \in \mathcal{B}(G(d,p))$, is a kernel; for this, see Lemma 13.2.2. Let $(L_p, L_q) \in G(d,p,q)$. In the subsequent chain of equalities we use, in this order, the definition of ν_q as the image measure of ν under β_q, the invariance property (13.12), the definition of $\nu_p^{L_q}$, Fubini's theorem, the equality $L_q = \rho L_q$ for $\rho \in SO(L_q)$, and the right invariance of ν. We obtain

$$\int_{G(d,q)} \int_{G(M,p)} f(L,M)\, \nu_p^M(\mathrm{d}L)\, \nu_q(\mathrm{d}M)$$
$$= \int_{SO_d} \int_{G(\vartheta L_q,p)} f(L,\vartheta L_q)\, \nu_p^{\vartheta L_q}(\mathrm{d}L)\, \nu(\mathrm{d}\vartheta)$$
$$= \int_{SO_d} \int_{G(L_q,p)} f(\vartheta L',\vartheta L_q)\, \nu_p^{L_q}(\mathrm{d}L')\, \nu(\mathrm{d}\vartheta)$$
$$= \int_{SO_d} \int_{SO(L_q)} f(\vartheta\rho L_p,\vartheta L_q)\, \nu_{L_q}(\mathrm{d}\rho)\, \nu(\mathrm{d}\vartheta)$$
$$= \int_{SO(L_q)} \int_{SO_d} f(\vartheta\rho L_p,\vartheta L_q)\, \nu(\mathrm{d}\vartheta)\, \nu_{L_q}(\mathrm{d}\rho)$$
$$= \int_{SO(L_q)} \int_{SO_d} f(\vartheta L_p,\vartheta L_q)\, \nu(\mathrm{d}\vartheta)\, \nu_{L_q}(\mathrm{d}\rho)$$
$$= \int_{SO_d} f(\vartheta L_p,\vartheta L_q)\, \nu(\mathrm{d}\vartheta).$$

In an analogous manner (though with a difference since $p < q$) we get

$$\int_{G(d,p)} \int_{G(L,q)} f(L,M)\, \nu_q^L(\mathrm{d}M)\, \nu_p(\mathrm{d}L)$$
$$= \int_{SO_d} \int_{G(\vartheta L_p,q)} f(\vartheta L_p,M)\, \nu_q^{\vartheta L_p}(\mathrm{d}M)\, \nu(\mathrm{d}\vartheta)$$
$$= \int_{SO_d} \int_{G(L_p,q)} f(\vartheta L_p,\vartheta M')\, \nu_q^{L_p}(\mathrm{d}M')\, \nu(\mathrm{d}\vartheta)$$
$$= \int_{SO_d} \int_{SO(L_p^\perp)} f(\vartheta L_p,\vartheta\rho L_q)\, \nu_{L_p^\perp}(\mathrm{d}\rho)\, \nu(\mathrm{d}\vartheta)$$

$$= \int_{SO(L_p^\perp)} \int_{SO_d} f(\vartheta L_p, \vartheta \rho L_q)\, \nu(\mathrm{d}\vartheta)\, \nu_{L_p^\perp}(\mathrm{d}\rho)$$

$$= \int_{SO(L_p^\perp)} \int_{SO_d} f(\vartheta L_p, \vartheta L_q)\, \nu(\mathrm{d}\vartheta)\, \nu_{L_p^\perp}(\mathrm{d}\rho)$$

$$= \int_{SO_d} f(\vartheta L_p, \vartheta L_q)\, \nu(\mathrm{d}\vartheta).$$

This, together with (7.1), completes the proof of Theorem 7.1.1. □

Remark. Let $0 \le p < q \le d-1$. The special case $f(L, M) = \mathbf{1}_A(L)$ with $A \in \mathcal{B}(G(d,p))$ in Theorem 7.1.1 yields

$$\nu_p(A) = \int_{G(d,q)} \nu_p^M(A)\, \nu_q(\mathrm{d}M), \tag{7.2}$$

and similarly one has

$$\nu_q(A) = \int_{G(d,p)} \nu_q^L(A)\, \nu_p(\mathrm{d}L) \tag{7.3}$$

for $A \in \mathcal{B}(G(d,q))$.

Remark. Let $p, q \in \{0, \dots, d-1\}$. For the Radon transform

$$R_{pq} : \mathbf{C}(G(d,p)) \to \mathbf{C}(G(d,q))$$

defined by (6.35), Theorem 7.1.1 implies the symmetry relation

$$\int_{G(d,q)} (R_{pq} f) g \,\mathrm{d}\nu_q = \int_{G(d,p)} f(R_{qp} g)\,\mathrm{d}\nu_p$$

for $f \in \mathbf{C}(G(d,p))$ and $g \in \mathbf{C}(G(d,q))$.

Theorem 7.1.1 can be generalized. For example, let integers $r < p < q < d$ or $q < p < r < d$ be given. For $L_0 \in G(d,r)$ and $L_2 \in G(d,q)$, let $G(L_0, L_2, p)$ denote the space of all p-dimensional subspaces L_1 with $L_0 \subset L_1 \subset L_2$ if $r < p < q$, respectively with $L_2 \subset L_1 \subset L_0$ if $q < p < r$. It carries a unique probability measure $\nu_p^{L_0, L_2}$ which is invariant under the rotations that map L_0 into itself and L_2 into itself. Let $L_0 \in G(d,r)$ be fixed. If $f : G(d,p,q) \to \mathbb{R}$ is a nonnegative measurable function, then

$$\int_{G(L_0,p)} \int_{G(L_1,q)} f(L_1, L_2)\, \nu_q^{L_1}(\mathrm{d}L_2)\, \nu_p^{L_0}(\mathrm{d}L_1)$$

$$= \int_{G(L_0,q)} \int_{G(L_0,L_2,p)} f(L_1, L_2)\, \nu_p^{L_0,L_2}(\mathrm{d}L_1)\, \nu_q^{L_0}(\mathrm{d}L_2). \tag{7.4}$$

This can be proved along similar lines to above.

A result analogous to Theorem 7.1.1 is valid for affine subspaces. It can be deduced from this theorem.

Theorem 7.1.2. *If* $0 \le p < q \le d-1$ *and if* $f : A(d,p,q) \to \mathbb{R}$ *is a nonnegative measurable function, then*

$$\int_{A(d,q)} \int_{A(F,p)} f(E,F)\, \mu_p^F(\mathrm{d}E)\, \mu_q(\mathrm{d}F)$$
$$= \int_{A(d,p)} \int_{A(E,q)} f(E,F)\, \mu_q^E(\mathrm{d}F)\, \mu_p(\mathrm{d}E). \tag{7.5}$$

Proof. For measurability, we again refer to Lemma 13.2.2. In the subsequent chain of equations we use (13.9), (13.13), Theorem 7.1.1, (13.14), again (13.9), and several times the theorem of Fubini. In this way, we get

$$\int_{A(d,q)} \int_{A(F,p)} f(E,F)\, \mu_p^F(\mathrm{d}E)\, \mu_q(\mathrm{d}F)$$
$$= \int_{G(d,q)} \int_{L^\perp} \int_{A(L+t,p)} f(E, L+t)\, \mu_p^{L+t}(\mathrm{d}E)\, \lambda_{d-q}(\mathrm{d}t)\, \nu_q(\mathrm{d}L)$$
$$= \int_{G(d,q)} \int_{L^\perp} \int_{G(L,p)} \int_{M^\perp \cap L} f(M+x+t, L+t)$$
$$\times \lambda_{q-p}(\mathrm{d}x)\, \nu_p^L(\mathrm{d}M)\, \lambda_{d-q}(\mathrm{d}t)\, \nu_q(\mathrm{d}L)$$
$$= \int_{G(d,q)} \int_{G(L,p)} \int_{L^\perp} \int_{M^\perp \cap L} f(M+x+t, L+x+t)$$
$$\times \lambda_{q-p}(\mathrm{d}x)\, \lambda_{d-q}(\mathrm{d}t)\, \nu_p^L(\mathrm{d}M)\, \nu_q(\mathrm{d}L)$$
$$= \int_{G(d,q)} \int_{G(L,p)} \int_{M^\perp} f(M+z, L+z)\, \lambda_{d-p}(\mathrm{d}z)\, \nu_p^L(\mathrm{d}M)\, \nu_q(\mathrm{d}L)$$
$$= \int_{G(d,p)} \int_{G(M,q)} \int_{M^\perp} f(M+z, L+z)\, \lambda_{d-p}(\mathrm{d}z)\, \nu_q^M(\mathrm{d}L)\, \nu_p(\mathrm{d}M)$$
$$= \int_{G(d,p)} \int_{M^\perp} \int_{G(M,q)} f(M+z, L+z)\, \nu_q^M(\mathrm{d}L)\, \lambda_{d-p}(\mathrm{d}z)\, \nu_p(\mathrm{d}M)$$
$$= \int_{G(d,p)} \int_{M^\perp} \int_{A(M+z,q)} f(M+z, F)\, \mu_q^{M+z}(\mathrm{d}F)\, \lambda_{d-p}(\mathrm{d}z)\, \nu_p(\mathrm{d}M)$$
$$= \int_{A(d,p)} \int_{A(E,q)} f(E,F)\, \mu_q^E(\mathrm{d}F)\, \mu_p(\mathrm{d}E).$$

This completes the proof of Theorem 7.1.2. □

Remark. Analogously to (7.2) one obtains, for $0 \le p < q \le d-1$, a representation of the invariant measure μ_p in the form

$$\mu_p(A) = \int_{A(d,q)} \mu_p^F(A)\, \mu_q(\mathrm{d}F),$$

for $A \in \mathcal{B}(A(d,p))$. There is, however, no representation corresponding to (7.3), because the measure μ_p^F, where $F \in A(d,q)$ and $p < q$, is not finite.

Note for Section 7.1

Theorems 7.1.1 and 7.1.2 can also be deduced from the essential uniqueness of invariant measures on homogeneous spaces (Theorem 13.3.1).

7.2 Blaschke–Petkantschin Formulas

We recall that for $E \in A(d,q)$ we denote by λ_E the q-dimensional Lebesgue measure on E, considered as a measure on all of $\mathbb{R}^d$, thus

$$\lambda_E(A) = \lambda_q(A \cap E) \qquad \text{for } A \in \mathcal{B}(\mathbb{R}^d).$$

For several applications one needs integral geometric transformations of a kind for which the following is a typical example. Suppose we have to integrate a function of q-tuples of points in $\mathbb{R}^d$, where $q \in \{1,\dots,d-1\}$, with respect to the product measure λ^q. In some cases it may simplify the computation to integrate first over the q-tuples of points in a fixed q-dimensional linear subspace L, with respect to the product measure λ_L^q, and then to integrate over all linear subspaces L, with respect to the invariant measure ν_q on $G(d,q)$. The case $q=1$ corresponds essentially to the well-known computation of a volume integral in terms of polar coordinates. The Jacobian appearing in the general transformation formula has a simple geometric meaning. A similar transformation formula exists for affine, instead of linear, subspaces. Results of this type are called **Blaschke–Petkantschin formulas**. We prepare the proof of these formulas by a lemma which extends the polar coordinate formula.

We denote by $d(x,L)$ the distance of the point $x \in \mathbb{R}^d$ from the subspace $L \subset \mathbb{R}^d$.

Lemma 7.2.1. *If $r \in \{0,\dots,d-1\}$ and $L \in G(d,r)$ is a fixed linear subspace, then*

$$\int_{\mathbb{R}^d} f \,\mathrm{d}\lambda = \frac{\omega_{d-r}}{2} \int_{G(L,r+1)} \int_M f d(\cdot,L)^{d-r-1} \,\mathrm{d}\lambda_M \, \nu_{r+1}^L(\mathrm{d}M)$$

for every nonnegative measurable function f on $\mathbb{R}^d$.

Proof. We denote by L_u the positive hull of L and a vector u. Using spherical coordinates in $L^\perp$ and Fubini's theorem, we obtain

$$\begin{aligned}&\int_{\mathbb{R}^d} f(x)\,\lambda(\mathrm{d}x)\\ &= \int_L \int_{L^\perp} f(x_0+x_1)\lambda_{L^\perp}(\mathrm{d}x_1)\,\lambda_L(\mathrm{d}x_0)\end{aligned}$$

$$
\begin{aligned}
&= \int_L \int_0^\infty \int_{S^{d-1}\cap L^\perp} f(x_0+\rho u)\rho^{d-r-1}\,\sigma_{d-r-1}(\mathrm{d}u)\,\mathrm{d}\rho\,\lambda_L(\mathrm{d}x_0)\\
&= \int_{S^{d-1}\cap L^\perp} \int_L \int_0^\infty f(x_0+\rho u)\rho^{d-r-1}\,\mathrm{d}\rho\,\lambda_L(\mathrm{d}x_0)\,\sigma_{d-r-1}(\mathrm{d}u)\\
&= \int_{S^{d-1}\cap L^\perp} \int_{L_u} f(x)d(x,L)^{d-r-1}\,\lambda_{L_u}(\mathrm{d}x)\,\sigma_{d-r-1}(\mathrm{d}u)\\
&= \frac{\omega_{d-r}}{2}\int_{G(L,r+1)}\int_M f(x)d(x,L)^{d-r-1}\,\lambda_M(\mathrm{d}x)\,\nu^L_{r+1}(\mathrm{d}M).
\end{aligned}
$$

This was the assertion. □

We recall (from Section 4.4 or Section 14.1) that for $q \in \{1,\ldots,d\}$ and $x_1,\ldots,x_q \in \mathbb{R}^d$ we denote by $\nabla_q(x_1,\ldots,x_q)$ the q-dimensional volume of the parallelepiped spanned by the vectors $x_1,\ldots,x_q$. For $q+1$ points $x_0,x_1,\ldots,x_q \in \mathbb{R}^d$,

$$
\Delta_q(x_0,\ldots,x_q) := \frac{1}{q!}\nabla_q(x_1-x_0,\ldots,x_q-x_0) \tag{7.6}
$$

is the q-dimensional volume of the convex hull of $\{x_0,\ldots,x_q\}$.

The following result is known as the **linear Blaschke–Petkantschin formula**.

Theorem 7.2.1. *If $q \in \{1,\ldots,d\}$ and if $f : (\mathbb{R}^d)^q \to \mathbb{R}$ is a nonnegative measurable function, then*

$$
\int_{(\mathbb{R}^d)^q} f\,\mathrm{d}\lambda^q = b_{dq}\int_{G(d,q)}\int_{L^q} f\nabla_q^{d-q}\,\mathrm{d}\lambda_L^q\,\nu_q(\mathrm{d}L) \tag{7.7}
$$

with

$$
b_{dq} := \frac{\omega_{d-q+1}\cdots\omega_d}{\omega_1\cdots\omega_q}. \tag{7.8}
$$

Proof. The subsequent proof, which is adapted from Miles [525], proceeds by induction. For $q=1$, the assertion reduces to Lemma 7.2.1 (case $r=0$) and hence is true. We assume that the assertion has been proved for some $q \geq 1$ and all dimensions d. In the inductive step we make use of the fact that for $x_1,\ldots,x_q \in L \in G(d,q)$ and $x_{q+1} \in \mathbb{R}^d$ one has

$$
\nabla_{q+1}(x_1,\ldots,x_{q+1}) = \nabla_q(x_1,\ldots,x_q)d(x_{q+1},L). \tag{7.9}
$$

Below we abbreviate $(x_1,\ldots,x_q)$ by $\mathbf{x}$. First we use, besides Fubini's theorem, the induction hypothesis and Lemma 7.2.1, to obtain

$$
I := \int_{(\mathbb{R}^d)^{q+1}} f\,\mathrm{d}\lambda^{q+1}
$$

$$= \int_{\mathbb{R}^d} \int_{(\mathbb{R}^d)^q} f(\mathbf{x}, x)\, \lambda^q(\mathrm{d}\mathbf{x})\, \lambda(\mathrm{d}x)$$

$$= b_{dq} \int_{\mathbb{R}^d} \int_{G(d,q)} \int_{L^q} f(\mathbf{x}, x) \nabla_q(\mathbf{x})^{d-q}\, \lambda_L^q(\mathrm{d}\mathbf{x})\, \nu_q(\mathrm{d}L)\, \lambda(\mathrm{d}x)$$

$$= b_{dq} \int_{G(d,q)} \int_{L^q} \nabla_q(\mathbf{x})^{d-q} \int_{\mathbb{R}^d} f(\mathbf{x}, x)\, \lambda(\mathrm{d}x)\, \lambda_L^q(\mathrm{d}\mathbf{x})\, \nu_q(\mathrm{d}L)$$

$$= \frac{b_{dq}\omega_{d-q}}{2} \int_{G(d,q)} \int_{L^q} \nabla_q(\mathbf{x})^{d-q} \int_{G(L,q+1)} \int_M f(\mathbf{x}, x)\, d(x, L)^{d-q-1}$$
$$\times\, \lambda_M(\mathrm{d}x)\, \nu_{q+1}^L(\mathrm{d}M)\, \lambda_L^q(\mathrm{d}\mathbf{x})\, \nu_q(\mathrm{d}L).$$

Applying Theorem 7.1.1 for interchanging the integrations over q- and $(q+1)$-dimensional subspaces and then using (7.9), we get

$$I = \frac{b_{dq}\omega_{d-q}}{2} \int_{G(d,q+1)} \int_{G(M,q)} \int_{L^q} \int_M f(\mathbf{x}, x) \nabla_q(\mathbf{x})^{d-q}\, d(x, L)^{d-q-1}$$
$$\times\, \lambda_M(\mathrm{d}x)\, \lambda_L^q(\mathrm{d}\mathbf{x})\, \nu_q^M(\mathrm{d}L)\, \nu_{q+1}(\mathrm{d}M)$$
$$= \frac{b_{dq}\omega_{d-q}}{2} \int_{G(d,q+1)} \int_M \int_{G(M,q)} \int_{L^q} f(\mathbf{x}, x) \nabla_{q+1}(\mathbf{x}, x)^{d-q-1} \nabla_q(\mathbf{x})$$
$$\times\, \lambda_L^q(\mathrm{d}\mathbf{x})\, \nu_q^M(\mathrm{d}L)\, \lambda_M(\mathrm{d}x)\, \nu_{q+1}(\mathrm{d}M).$$

Now we apply the induction hypothesis again, to a q-fold integration over the $(q+1)$-dimensional space M and the function $f(\cdot, x)\nabla_{q+1}(\cdot, x)^{d-q-1}$. We obtain

$$I = \frac{b_{dq}\omega_{d-q}}{2b_{(q+1)q}} \int_{G(d,q+1)} \int_M \int_{M^q} f(\mathbf{x}, x) \nabla_{q+1}(\mathbf{x}, x)^{d-q-1}$$
$$\times\, \lambda_M^q(\mathrm{d}\mathbf{x})\, \lambda_M(\mathrm{d}x)\, \nu_{q+1}(\mathrm{d}M)$$
$$= b_{d(q+1)} \int_{G(d,q+1)} \int_{M^{q+1}} f \nabla_{q+1}^{d-q-1}\, \mathrm{d}\lambda_M^{q+1}\, \nu_{q+1}(\mathrm{d}M),$$

which is the assertion for a $(q+1)$-fold integration. □

Before proceeding further, we want to explain in which situations we talk of a formula of 'Blaschke–Petkantschin type'; thus, we try to describe the common feature of these transformations. The starting point is an integration over a product (possibly with one factor only) of measure spaces of geometric objects (points or flats, as a rule), mostly homogeneous spaces with their invariant measures. Almost everywhere, the integration variable, which is a tuple of geometric objects, determines a new geometric object (for example, by span or intersection). We call this new object the 'pivot'. The initial integration is then decomposed into an outer and an inner integration. The outer integration space is the space of all possible pivots, with a natural measure; often it is a homogeneous space. For a given pivot, the inner integration space consists of the tuples of the initial integration space which determine precisely this pivot; as a rule, it is a product of homogeneous spaces.

Lemma 7.2.1 was already of this type. The initial integration is over $\mathbb{R}^d$. The integration variable $x \in \mathbb{R}^d$ determines (almost everywhere) the $(q+1)$-subspace which is spanned by x and the fixed q-subspace L. This $(q+1)$-subspace is the pivot. The outer integration space is the space $G(L, q+1)$ of all $(q+1)$-subspaces containing L. For M in this space, the inner integration space is equal to M. In the case of Theorem 7.2.1, the initial integration is over $(\mathbb{R}^d)^q$, and the pivot is the q-subspace spanned by the integration variable $(x_1, \dots, x_q) \in (\mathbb{R}^d)^q$. Hence, the outer integration space is the Grassmannian $G(d,q)$ of all q-subspaces. For $L \in G(d,q)$, the inner integration space is the product L^q.

There are also extensions of formulas of Blaschke–Petkantschin type where the pivot is not uniquely determined by the integration variable, but only associated with it in some way. For example, in the situation of Theorem 7.2.1, a pivot associated with $(x_1, \dots, x_q)$ could be a subspace of fixed dimension $s \geq q$ containing $x_1, \dots, x_q$, or the span of $x_1, \dots, x_q$ and of a fixed subspace. One can also combine both possibilities; this gives the following generalization of the linear Blaschke–Petkantschin formula. Here we denote by $\nabla_{q,r}(x_1, \dots x_q, L_0)$ the $(q+r)$-dimensional volume spanned by the vectors $x_1, \dots, x_q \in \mathbb{R}^d$ and an orthonormal basis of the subspace $L_0 \in G(d,r)$.

Theorem 7.2.2. *Let $q \geq 1$, $r \geq 0$ and s be integers with $q + r \leq s \leq d$, let $L_0 \in G(d,r)$. If $f : (\mathbb{R}^d)^q \to \mathbb{R}$ is a nonnegative measurable function, then*

$$\int_{(\mathbb{R}^d)^q} f \,\mathrm{d}\lambda^q = \frac{b_{(d-r)q}}{b_{(s-r)q}} \int_{G(L_0,s)} \int_{L^q} f \nabla_{q,r}(\cdot, L_0)^{d-s} \,\mathrm{d}\lambda_L^q \, \nu_s^{L_0}(\mathrm{d}L). \tag{7.10}$$

Proof. First we consider the case $s = q + r$, that is, the formula

$$\int_{(\mathbb{R}^d)^q} f \,\mathrm{d}\lambda^q = b_{(d-r)q} \int_{G(L_0,q+r)} \int_{L^q} f \nabla_{q,r}(\cdot, L_0)^{d-q-r} \,\mathrm{d}\lambda_L^q \, \nu_{q+r}^{L_0}(\mathrm{d}L). \tag{7.11}$$

Its proof proceeds by induction, in a similar manner to Theorem 7.2.1. The case of $q = 1$ is again provided by Lemma 7.2.1. In the induction step, one applies Lemma 7.2.1 to a fixed subspace L of dimension $q + r$ and then uses the interchange formula (7.4) instead of Theorem 7.1.1. After observing that

$$\nabla_{q+1,r}(x_1, \dots, x_{q+1}, L_0) = \nabla_{q,r}(x_1, \dots, x_q, L_0) d(x_{q+1}, L) \quad \text{for } L_0 \subset L$$

and applying Fubini's theorem, the induction hypothesis is applied to a q-fold integration over a $(q+r+1)$-dimensional subspace M containing L_0. Apart from these changes, the proof is the same as before. Thus, the formula (7.11) is proved.

To prove (7.10), we assume that $q + r \leq s \leq d$ and start with the integral

$$I := \int_{G(L_0,s)} \int_{L^q} f \, \mathrm{d}\lambda_L^q \, \nu_s^{L_0}(\mathrm{d}L).$$

We apply (7.11) to the integral over L^q; here $\dim L = s$ and $L_0 \subset L$. Then we use the interchange formula (7.4) and Fubini's theorem. This yields

$$\begin{aligned} I &= b_{(s-r)q} \int_{G(L_0,s)} \int_{G(L_0,L,q+r)} \int_{M^q} f \nabla_{q,r}(\cdot, L_0)^{s-q-r} \\ &\quad \times \mathrm{d}\lambda_M^q \, \nu_{q+r}^{L_0,L}(\mathrm{d}M) \, \nu_s^{L_0}(\mathrm{d}L) \\ &= b_{(s-r)q} \int_{G(L_0,q+r)} \int_{G(M,s)} \int_{M^q} f \nabla_{q,r}(\cdot, L_0)^{s-q-r} \, \mathrm{d}\lambda_M^q \, \nu_s^M(\mathrm{d}L) \, \nu_{q+r}^{L_0}(\mathrm{d}M) \\ &= b_{(s-r)q} \int_{G(L_0,q+r)} \int_{M^q} \int_{G(M,s)} f \nabla_{q,r}(\cdot, L_0)^{s-q-r} \, \nu_s^M(\mathrm{d}L) \, \mathrm{d}\lambda_M^q \, \nu_{q+r}^{L_0}(\mathrm{d}M) \\ &= b_{(s-r)q} \int_{G(L_0,q+r)} \int_{M^q} f \nabla_{q,r}(\cdot, L_0)^{s-q-r} \, \mathrm{d}\lambda_M^q \, \nu_{q+r}^{L_0}(\mathrm{d}M) \\ &= \frac{b_{(s-r)q}}{b_{(d-r)q}} \int_{(\mathbb{R}^d)^q} f \nabla_{q,r}(\cdot, L_0)^{s-d} \, \mathrm{d}\lambda^q, \end{aligned}$$

by another application of (7.11). Replacing f by $f \nabla_{q,r}(\cdot, L_0)^{d-s}$, we obtain the assertion. □

The original Blaschke–Petkantschin formula is a source of a series of further integral geometric transformations, of which we shall give some examples. First we derive transformation formulas for integrals over tuples of linear subspaces. The cases where the sum of the dimensions of the subspaces is at most d or larger than d have to be distinguished.

In the subsequent theorem, the initial integration space is $\prod_{i=1}^q G(d, r_i)$, with $\sum_{i=1}^q r_i =: p \le d$, and the pivot determined by a q-tuple of subspaces is their linear span. Correspondingly, the outer integration space is $G(d,p)$, and for $L \in G(d,p)$, the inner integration space is $\prod_{i=1}^q G(L, r_i)$.

In the following, we shall have to use the subspace determinant $[\cdot, \dots, \cdot]$ defined in Section 14.1. If $q \in \mathbb{N}$, $r_1, \dots, r_q \in \{1, \dots, d-1\}$ and $(L_1, \dots, L_q) \in G(d, r_1) \times \ldots \times G(d, r_q)$, we write

$$[L_1, \dots, L_q] =: [L_1, \dots, L_q]_{\mathbf{r}},$$

where $\mathbf{r} := (r_1, \dots, r_q)$ serves as a multi-index. If L_0 is a fixed linear subspace, we also write

$$[L_1, \dots, L_q, L_0] =: [L_1, \dots, L_q, L_0]_{\mathbf{r}}.$$

Thus, for $\mathbf{r} := (r_1, \dots, r_q)$, the functions $[\cdot, \dots, \cdot]_{\mathbf{r}}$ and $[\cdot, \dots, \cdot, L_0]_{\mathbf{r}}$ are both defined on $G(d, r_1) \times \ldots \times G(d, r_q)$.

Theorem 7.2.3. *Let* $r_1, \dots, r_q \in \{1, \dots, d-1\}$ *be integers with* $r_1 + \dots + r_q =: p \leq d$, *and put* $\mathbf{r} := (r_1, \dots, r_q)$. *If* $f : G(d, r_1) \times \dots \times G(d, r_q) \to \mathbb{R}$ *is a nonnegative measurable function, then*

$$\int_{G(d,r_1)\times\dots\times G(d,r_q)} f \, \mathrm{d}(\nu_{r_1} \otimes \dots \otimes \nu_{r_q})$$
$$= b \int_{G(d,p)} \int_{G(L,r_1)\times\dots\times G(L,r_q)} f \, [\cdot, \dots, \cdot]_{\mathbf{r}}^{d-p} \, \mathrm{d}(\nu_{r_1}^L \otimes \dots \otimes \nu_{r_q}^L) \, \nu_p(\mathrm{d}L)$$

with

$$b := b_{dp} \prod_{j=1}^{q} \frac{b_{pr_j}}{b_{dr_j}}. \tag{7.12}$$

Proof. We begin with a preparatory remark. If $r \in \{1, \dots, d-1\}$ and if $h : G(d, r) \to \mathbb{R}$ is a nonnegative measurable function, then

$$\int_{\mathbb{R}^d} \cdots \int_{\mathbb{R}^d} h(\mathrm{lin}\{x_1, \dots, x_r\}) \prod_{j=1}^{r} \mathbf{1}_{B^d}(x_j) \, \lambda(\mathrm{d}x_1) \cdots \lambda(\mathrm{d}x_r)$$
$$= \kappa_d^r \int_{G(d,r)} h(L) \, \nu_r(\mathrm{d}L).$$

In fact, choosing for h the indicator function of a Borel set, one can use the left side to define a finite measure on $G(d, r)$. Since it is rotation invariant, it must be a multiple of the invariant measure ν_r. The factor can then be determined by choosing $h = 1$.

Now we define, almost everywhere on $(\mathbb{R}^d)^p$, a function g by

$$g(x_1^1, \dots, x_{r_1}^1, \dots, x_1^q, \dots, x_{r_q}^q)$$
$$:= f(\mathrm{lin}\{x_1^1, \dots, x_{r_1}^1\}, \dots, \mathrm{lin}\{x_1^q, \dots, x_{r_q}^q\}) \prod_{j=1}^{q} \prod_{i=1}^{r_j} \mathbf{1}_{B^d}(x_i^j).$$

Applying Fubini's theorem and q times the preceding remark, we obtain

$$I := \int_{(\mathbb{R}^d)^p} g \, \mathrm{d}\lambda^p = \kappa_d^p \int_{G(d,r_1)\times\dots\times G(d,r_q)} f \, \mathrm{d}(\nu_{r_1} \otimes \dots \otimes \nu_{r_q}).$$

On the other hand, Theorem 7.2.1 gives

$$I = b_{dp} \int_{G(d,p)} \int_{L^p} g \nabla_p^{d-p} \, \mathrm{d}\lambda_L^p \, \nu_p(\mathrm{d}L).$$

We abbreviate $(x_{r_1}^j, \dots, x_{r_j}^j) =: \mathbf{x}_j$ and $\mathrm{lin}\{x_{r_1}^j, \dots, x_{r_j}^j\} =: \mathrm{lin}\, \mathbf{x}_j$ for $j = 1, \dots, q$, then

$$I = b_{dp} \int_{G(d,p)} \int_{L^{r_1}} \cdots \int_{L^{r_q}} g(\mathbf{x}_1, \ldots, \mathbf{x}_q) \nabla_p(\mathbf{x}_1, \ldots, \mathbf{x}_q)^{d-p} \\ \times \lambda_L^{r_q}(\mathrm{d}\mathbf{x}_1) \cdots \lambda_L^{r_1}(\mathrm{d}\mathbf{x}_q)\, \nu_p(\mathrm{d}L).$$

From the definitions of $[\cdot, \ldots, \cdot]_{\mathbf{r}}$ and ∇_r it follows that

$$\nabla_p(\mathbf{x}_1, \ldots, \mathbf{x}_q) = \nabla_{r_1}(\mathbf{x}_1) \cdots \nabla_{r_q}(\mathbf{x}_q)[\operatorname{lin} \mathbf{x}_1, \ldots, \operatorname{lin} \mathbf{x}_q]_{\mathbf{r}}.$$

We insert this in the last integrand. Then, for fixed $L \in G(d,p)$, we use Theorem 7.2.1 (with $(\mathbb{R}^d, q)$ replaced by (L, r_1)) to transform the integration involving $\mathbf{x}_1$ into an integration over $G(L, r_1)$ and, for fixed $L_1 \in G(L, r_1)$, an integration with respect to the measure $\lambda_{L_1}^{r_1}$. The integral

$$\int_{L_1^{r_1}} \mathbf{1}_{(B^d \cap L_1)^{r_1}}(\mathbf{x}_1) \nabla_{r_1}(\mathbf{x}_1)^{d-r_1}\, \lambda_{L_1}^{r_1}(\mathrm{d}\mathbf{x}_1) = I(r_1, r_1, d - r_1)$$

occurring here can be evaluated by means of Theorem 8.2.2. In a similar way the integrations involving $\mathbf{x}_j$, $j = 2, \ldots, q$, are treated. Now the assertion follows. □

In the following generalization of Theorem 7.2.3, a fixed subspace is given, and the pivot determined by a tuple of subspaces is the linear span of these and the given one.

Theorem 7.2.4. *Let $r_1, \ldots, r_q \in \{1, \ldots, d-1\}$ and $r_0 \in \{0, \ldots, d-1\}$ be integers with*

$$r_1 + \ldots + r_q =: p \le d - r_0;$$

put $\mathbf{r} := (r_1, \ldots, r_q)$. Let $L_0 \in G(d, r_0)$ be a fixed subspace. If $f : G(d, r_1) \times \ldots \times G(d, r_q) \to \mathbb{R}$ is a nonnegative measurable function, then

$$\int_{G(d,r_1) \times \ldots \times G(d,r_q)} f \, \mathrm{d}(\nu_{r_1} \otimes \ldots \otimes \nu_{r_q}) \\ = c \int_{G(L_0, p+r_0)} \int_{G(L,r_1) \times \ldots \times G(L,r_q)} f\, [\cdot, \ldots, \cdot, L_0]_{\mathbf{r}}^{d-p-r_0} \\ \times \mathrm{d}(\nu_{r_1}^L \otimes \ldots \otimes \nu_{r_q}^L)\, \nu_{p+r_0}^{L_0}(\mathrm{d}L)$$

with

$$c := b_{(d-r_0)p} \prod_{j=1}^{q} \frac{b_{(p+r_0)r_j}}{b_{dr_j}}.$$

Proof. The proof is the obvious extension of the previous one. Instead of applying Theorem 7.2.1 first, we employ (7.11), with (q, r) replaced by (p, r_0). After using the identity

$$\nabla_{p,r_0}(\mathbf{x}_1, \ldots, \mathbf{x}_q, L_0) = \nabla_{r_1}(\mathbf{x}_1) \cdots \nabla_{r_q}(\mathbf{x}_q)[\operatorname{lin} \mathbf{x}_1, \ldots, \operatorname{lin} \mathbf{x}_q, L_0]_{\mathbf{r}},$$

the rest of the proof is the same. □

The preceding theorems have counterparts where linear spans are replaced by intersections. In that case, we consider linear subspaces $L_1, \dots, L_q \subset \mathbb{R}^d$ with $\sum_{i=1}^q \dim L_i \geq (q-1)d$. In Section 14.1 we define $[L_1, \dots, L_q] = [L_1^\perp, \dots, L_q^\perp]$. In particular, if $L_1, \dots, L_q$ are hyperplanes through 0 and if u_i is a unit normal vector of L_i for $i = 1, \dots, q$, then $[L_1, \dots, L_q]$ is the q-dimensional volume of the parallelepiped spanned by $u_1, \dots, u_q$, also denoted by $\nabla_q(u_1, \dots, u_q)$. As before, we write $[L_1, \dots, L_q] = [L_1, \dots, L_q]_{\mathbf{s}}$ with $\mathbf{s} := (s_1, \dots, s_q)$ if $\dim L_i = s_i$, $i = 1, \dots, q$.

In the next theorem, the initial integration space is $\prod_{i=1}^q G(d, s_i)$, with $\sum_{i=1}^q s_i \geq (q-1)d$, and the pivot determined by a q-tuple of subspaces is their intersection.

Theorem 7.2.5. *Let $s_1, \dots, s_q \in \{1, \dots, d-1\}$ be integers satisfying*

$$s_1 + \ldots + s_q - (q-1)d =: m \geq 0;$$

put $\mathbf{s} := (s_1, \dots, s_q)$. If $f : G(d, s_1) \times \ldots \times G(d, s_q) \to \mathbb{R}$ is a nonnegative measurable function, then

$$\int_{G(d,s_1)\times\ldots\times G(d,s_q)} f \, \mathrm{d}(\nu_{s_1} \otimes \ldots \otimes \nu_{s_q})$$
$$= \bar{b} \int_{G(d,m)} \int_{G(L,s_1)\times\ldots\times G(L,s_q)} f \, [\cdot, \dots, \cdot]_{\mathbf{s}}^m \, \mathrm{d}(\nu_{s_1}^L \otimes \ldots \otimes \nu_{s_q}^L) \, \nu_m(\mathrm{d}L)$$

with

$$\bar{b} := b_{d(d-m)} \prod_{j=1}^q \frac{b_{(d-m)(d-s_j)}}{b_{d(d-s_j)}}. \tag{7.13}$$

Proof. We put $d - s_j =: r_j$ and $\mathbf{r} := (r_1, \dots, r_q)$. For $M_j \in G(d, r_j)$, $j = 1, \dots, q$, we set

$$f^\perp(M_1, \dots, M_q) := f(M_1^\perp, \dots, M_q^\perp).$$

By Theorem 7.2.3,

$$\int_{G(d,r_1)\times\ldots\times G(d,r_q)} f^\perp \, \mathrm{d}(\nu_{r_1} \otimes \ldots \otimes \nu_{r_q})$$
$$= b \int_{G(d,d-m)} \int_{G(L,r_1)\times\ldots\times G(L,r_q)} f^\perp [\cdot, \dots, \cdot]_{\mathbf{r}}^m \, \mathrm{d}(\nu_{r_1}^L \otimes \ldots \otimes \nu_{r_q}^L) \, \nu_{d-m}(\mathrm{d}L)$$

with b given by (7.12). Now we observe that the mapping $L \mapsto L^\perp$ maps the space $G(d,k)$ to $G(d, d-k)$ and transforms the measure ν_k into ν_{d-k} Moreover, for a fixed subspace M, it maps the space $G(M,k)$ onto $G(M^\perp, d-k)$ and transforms the measure ν_k^M into $\nu_{d-k}^{M^\perp}$, as follows from the uniqueness of these invariant measures. Hence, the last equation is equivalent to the assertion. □

In the same way, one obtains from Theorem 7.2.4 the following generalization of the preceding result. Here a fixed subspace is given, and the pivot determined by a tuple of subspaces is the intersection of these and the given one.

Theorem 7.2.6. *Let $s_1, \ldots, s_q \in \{1, \ldots, d-1\}$ and $s_0 \in \{1, \ldots, d\}$ be integers with*

$$s_1 + \ldots + s_q - (q-1)d =: m \geq d - s_0;$$

put $\mathbf{s} := (s_1, \ldots, s_q)$. Let $L_0 \in G(d, s_0)$ be a fixed subspace. If $f : G(d, s_1) \times \ldots \times G(d, s_q) \to \mathbb{R}$ is a nonnegative measurable function, then

$$\int_{G(d,s_1)\times\ldots\times G(d,s_q)} f \,\mathrm{d}(\nu_{s_1} \otimes \ldots \otimes \nu_{s_q})$$
$$= \bar{c} \int_{G(L_0,m+s_0-d)} \int_{G(L,s_1)\times\ldots\times G(L,s_q)} f\,[\cdot, \ldots, \cdot, L_0]_{\mathbf{s}}^{m+s_0-d}$$
$$\times \,\mathrm{d}(\nu_{s_1}^L \otimes \ldots \otimes \nu_{s_q}^L)\, \nu_{m+s_0-d}^{L_0}(\mathrm{d}L)$$

with

$$\bar{c} := b_{s_0(d-m)} \prod_{j=1}^{q} \frac{b_{(2d-m-s_0)(d-s_j)}}{b_{d(d-s_j)}}.$$

Now we turn to affine transformation formulas, with affine subspaces instead of linear subspaces. First we derive the **affine Blaschke–Petkantschin formula**. Here the initial integration is over $(\mathbb{R}^d)^{q+1}$, and the pivot is the q-flat affinely spanned (almost everywhere) by the integration variable $(x_0, \ldots, x_q) \in (\mathbb{R}^d)^{q+1}$. The outer integration space is the affine Grassmannian $A(d,q)$, and for $E \in A(d,q)$, the inner integration space is the product E^{q+1}. Recall that $\Delta_q(x_0, \ldots, x_q)$, as defined by (7.6), denotes the q-dimensional volume of the simplex with vertices $x_0, \ldots, x_q$.

Theorem 7.2.7. *If $q \in \{1, \ldots, d\}$ and if $f : (\mathbb{R}^d)^{q+1} \to \mathbb{R}$ is a nonnegative measurable function, then*

$$\int_{(\mathbb{R}^d)^{q+1}} f \,\mathrm{d}\lambda^{q+1} = b_{dq} (q!)^{d-q} \int_{A(d,q)} \int_{E^{q+1}} f \Delta_q^{d-q} \,\mathrm{d}\lambda_E^{q+1}\, \mu_q(\mathrm{d}E) \tag{7.14}$$

with b_{dq} given by (7.8).

Proof. We apply Theorem 7.2.1 and several times the theorem of Fubini:

$$\int_{(\mathbb{R}^d)^{q+1}} f \,\mathrm{d}\lambda^{q+1}$$
$$= \int_{\mathbb{R}^d} \int_{(\mathbb{R}^d)^q} f(x_0, y_1 + x_0, \ldots, y_q + x_0)\, \lambda^q(\mathrm{d}(y_1, \ldots, y_q))\, \lambda(\mathrm{d}x_0)$$

$$= b_{dq} \int_{\mathbb{R}^d} \int_{G(d,q)} \int_{L^q} f(x_0, y_1 + x_0, \ldots, y_q + x_0) \nabla_q(y_1, \ldots, y_q)^{d-q}$$
$$\times \lambda_L^q(\mathrm{d}(y_1, \ldots, y_q))\, \nu_q(\mathrm{d}L)\, \lambda(\mathrm{d}x_0)$$
$$= b_{dq} \int_{G(d,q)} \int_{L^\perp} \int_L \int_{L^q} f(z + t, y_1 + z + t, \ldots, y_q + z + t)$$
$$\times \nabla_q(y_1, \ldots, y_q)^{d-q}\, \lambda_L^q(\mathrm{d}(y_1, \ldots, y_q))\, \lambda_L(\mathrm{d}z)\, \lambda_{L^\perp}(\mathrm{d}t)\, \nu_q(\mathrm{d}L)$$
$$= b_{dq}(q!)^{d-q} \int_{G(d,q)} \int_{L^\perp} \int_{(L+t)^{q+1}} f(x_0, \ldots, x_q)$$
$$\times \Delta_q(x_0, \ldots, x_q)^{d-q}\, \lambda_{L+t}^{q+1}(\mathrm{d}(x_0, \ldots, x_q))\, \lambda_{L^\perp}(\mathrm{d}t)\, \nu_q(\mathrm{d}L)$$
$$= b_{dq}(q!)^{d-q} \int_{A(d,q)} \int_{E^{q+1}} f(x_0, \ldots, x_q) \Delta_q(x_0, \ldots, x_q)^{d-q}$$
$$\times \lambda_E^{q+1}(\mathrm{d}(x_0, \ldots, x_q))\, \mu_q(\mathrm{d}E).$$

Here we have used (13.9). □

Postponing the treatment of affine spans of flats of small dimensions, we now consider the affine counterpart to Theorem 7.2.5. Here, the pivot determined by a q-tuple of flats of large dimensions is their intersection. For an affine subspace E we denote by E^0 the linear subspace parallel to E. For $E_1, \ldots, E_q \subset \mathbb{R}^d$ with $\dim E_i = s_i$ we put $\mathbf{s} := (s_1, \ldots, s_q)$ and

$$[E_1, \ldots, E_q]_{\mathbf{s}} := [E_1^0, \ldots, E_q^0]_{\mathbf{s}},$$

provided the right side is defined.

Theorem 7.2.8. *Let $s_1, \ldots, s_q \in \{1, \ldots, d-1\}$ be integers satisfying*

$$s_1 + \ldots + s_q - (q-1)d =: m \geq 0;$$

put $\mathbf{s} := (s_1, \ldots, s_q)$. If $f : A(d, s_1) \times \ldots \times A(d, s_q) \to \mathbb{R}$ is a nonnegative measurable function, then

$$\int_{A(d,s_1)\times\ldots\times A(d,s_q)} f \,\mathrm{d}(\mu_{s_1} \otimes \ldots \otimes \mu_{s_q})$$
$$= \bar{b} \int_{A(d,m)} \int_{A(E,s_1)\times\ldots\times A(E,s_q)} f\, [\cdot, \ldots, \cdot]_{\mathbf{s}}^{m+1}\, \mathrm{d}(\mu_{s_1}^E \otimes \ldots \otimes \mu_{s_q}^E)\, \mu_m(\mathrm{d}E)$$

with $\bar{b}$ given by (7.13).

Proof. By (13.9) we can write

$$I := \int_{A(d,s_1)\times\ldots\times A(d,s_q)} f \,\mathrm{d}(\mu_{s_1} \otimes \ldots \otimes \mu_{s_q}) \tag{7.15}$$
$$= \int_{G(d,s_1)\times\ldots\times G(d,s_q)} J(L_1, \ldots, L_q)\, (\nu_{s_1} \otimes \ldots \otimes \nu_{s_q})(\mathrm{d}(L_1, \ldots, L_q))$$

with

$$J(L_1,\dots,L_q)$$
$$=\int_{L_1^\perp\times\dots\times L_q^\perp} f(L_1+t_1,\dots,L_q+t_q)\,(\lambda_{L_1^\perp}\otimes\dots\otimes\lambda_{L_q^\perp})(\mathrm{d}(t_1,\dots,t_q)).$$

Let $L_j\in G(d,s_j)$, $j=1,\dots,q$ and assume, without loss of generality (by Lemma 13.2.1), that these subspaces are in general position. We put $L_1\cap\dots\cap L_q=:L$. For $(t_1,\dots,t_q)\in L_1^\perp\times\dots\times L_q^\perp$ we have

$$(L_1+t_1)\cap\dots\cap(L_q+t_q)=L+\xi(t_1,\dots,t_q)$$

with a unique vector $\xi(t_1,\dots,t_q)\in L^\perp$. This defines a linear map

$$\xi: L_1^\perp\times\dots\times L_q^\perp\to L^\perp.$$

If $\pi_j: L^\perp\to L_j^\perp$ denotes the orthogonal projection, then the inverse map ξ^{-1} is given by $\xi^{-1}(x)=(\pi_1(x),\dots,\pi_q(x))$. Choosing in each space $L_j^\perp$ an orthonormal basis and applying a linear map from $L_1^\perp\times\dots\times L_q^\perp$ to $L^\perp$ that maps the union of these bases to an orthonormal basis of $L^\perp$, we see that

$$J(L_1,\dots,L_q)=[L_1,\dots,L_q]_{\mathbf{s}}\int_{L^\perp} f(L_1+x,\dots,L_q+x)\,\lambda_{L^\perp}(\mathrm{d}x).$$

We insert this in (7.15) and use Theorem 7.2.5. In the subsequent integrals we have $L=L_1\cap\dots\cap L_q$ up to sets of measure zero.

$$\begin{aligned}
I=&\int_{G(d,s_1)\times\dots\times G(d,s_q)}[L_1,\dots,L_q]_{\mathbf{s}}\int_{L^\perp} f(L_1+x,\dots,L_q+x)\,\lambda_{L^\perp}(\mathrm{d}x)\\
&\times(\nu_{s_1}\otimes\dots\otimes\nu_{s_q})(\mathrm{d}(L_1,\dots,L_q))\\
=&\,\bar{b}\int_{G(d,m)}\int_{G(L,s_1)\times\dots\times G(L,s_q)}\int_{L^\perp} f(L_1+x,\dots,L_q+x)[L_1,\dots,L_q]_{\mathbf{s}}^{m+1}\\
&\times\lambda_{L^\perp}(\mathrm{d}x)\,(\nu_{s_1}^L\otimes\dots\otimes\nu_{s_q}^L)(\mathrm{d}(L_1,\dots,L_q))\,\nu_m(\mathrm{d}L)\\
=&\,\bar{b}\int_{G(d,m)}\int_{L^\perp}\int_{G(L,s_1)\times\dots\times G(L,s_q)} f(L_1+x,\dots,L_q+x)[L_1,\dots,L_q]_{\mathbf{s}}^{m+1}\\
&\times(\nu_{s_1}^L\otimes\dots\otimes\nu_{s_q}^L)(\mathrm{d}(L_1,\dots,L_q))\,\lambda_{L^\perp}(\mathrm{d}x)\,\nu_m(\mathrm{d}L)\\
=&\,\bar{b}\int_{G(d,m)}\int_{L^\perp}\int_{A(L+x,s_1)\times\dots\times A(L+x,s_q)} f(E_1,\dots,E_q)[E_1,\dots,E_q]_{\mathbf{s}}^{m+1}\\
&\times(\mu_{s_1}^{L+x}\otimes\dots\otimes\mu_{s_q}^{L+x})(\mathrm{d}(E_1,\dots,E_q))\,\lambda_{L^\perp}(\mathrm{d}x)\,\nu_m(\mathrm{d}L)\\
=&\,\bar{b}\int_{A(d,m)}\int_{A(E,s_1)\times\dots\times A(E,s_q)} f(E_1,\dots,E_q)[E_1,\dots,E_q]_{\mathbf{s}}^{m+1}\\
&\times(\mu_{s_1}^{E}\otimes\dots\otimes\mu_{s_q}^{E})(\mathrm{d}(E_1,\dots,E_q))\,\mu_m(\mathrm{d}E).
\end{aligned}$$

Here we have used (13.14) and (13.9). □

In the case of flats of small dimensions, we consider only two flats. Let $E_1 \in A(d,r)$, $E_2 \in A(d,s)$ be flats with dimensions satisfying $r+s \le d-1$. We assume that they are in general position, that is, the dimension of their affine span is equal to $r+s+1$. Under this assumption, the distance between E_1 and E_2 is realized by unique points $x_1 \in E_1$, $x_2 \in E_2$, and the line F through x_1 and x_2 is orthogonal to both, E_1 and E_2. We call F the **ortholine** of E_1 and E_2 and denote the distance $\|x_1 - x_2\|$ by $D(E_1,E_2)$.

In the following theorem, the pivot determined by a pair of flats of small dimensions is their affine span.

Theorem 7.2.9. *Let $r_1, r_2 \in \{0,\dots,d-1\}$ be integers satisfying $r_1+r_2+1 =: p \le d$; put $\mathbf{r} := (r_1,r_2)$. If $f : A(d,r_1) \times A(d,r_2) \to \mathbb{R}$ is a nonnegative measurable function, then*

$$\int_{A(d,r_1)\times A(d,r_2)} f \,\mathrm{d}(\mu_{r_1} \otimes \mu_{r_2})$$

$$= b \int_{A(d,p)} \int_{A(E,r_1)\times A(E,r_2)} f D^{d-p} [\cdot,\cdot]_{\mathbf{r}}^{d-p} \,\mathrm{d}(\mu_{r_1}^E \otimes \mu_{r_2}^E)\, \mu_p(\mathrm{d}E)$$

with b given by (7.12) for $q=2$.

Proof. It is sufficient to prove the assertion for a function f for which there exists a ball B with $f(E_1,E_2)=0$ if $E_j \cap \operatorname{int} B = \emptyset$ for at least one $j \in \{1,2\}$. If this is established, then the general case follows with an application of the monotone convergence theorem.

For $\mathbf{x}_j := (x_0^j,\dots,x_{r_j}^j) \in (\mathbb{R}^d)^{r_j+1}$ we write $\operatorname{aff}\{x_0^j,\dots,x_{r_j}^j\} =: \operatorname{aff}\mathbf{x}_j$, and we define

$$g(\mathbf{x}_1,\mathbf{x}_2) := \prod_{j=1}^{2} \mathbf{1}_{B^{r_j+1}}(\mathbf{x}_j) \lambda_{\operatorname{aff}\mathbf{x}_j}(B)^{-r_j-1} \quad \text{if } \prod_{j=1}^{2} \lambda_{\operatorname{aff}\mathbf{x}_j}(B) \neq 0,$$

and $g(\mathbf{x}_1,\mathbf{x}_2) := 0$ otherwise. To each of the two integrals in

$$I := (r_1!)^{r_1-d}(r_2!)^{r_2-d} \int_{(\mathbb{R}^d)^{r_1+1}} \int_{(\mathbb{R}^d)^{r_2+1}} f(\operatorname{aff}\mathbf{x}_1, \operatorname{aff}\mathbf{x}_2)$$

$$\times\, g(\mathbf{x}_1,\mathbf{x}_2) \Delta_{r_1}(\mathbf{x}_1)^{r_1-d} \Delta_{r_2}(\mathbf{x}_2)^{r_2-d}\, \lambda^{r_2+1}(\mathrm{d}\mathbf{x}_2)\, \lambda^{r_1+1}(\mathrm{d}\mathbf{x}_1)$$

we apply the affine Blaschke–Petkantschin formula (7.14). This gives

$$I = \prod_{j=1}^{2} b_{dr_j} \int_{A(d,r_1)} \int_{A(d,r_2)} f(E_1,E_2)\, \mu_{r_2}(\mathrm{d}E_2)\, \mu_{r_1}(\mathrm{d}E_1).$$

On the other hand, we can view I as an integral over $(\mathbb{R}^d)^{p+1}$ with respect to the measure λ^{p+1} and apply (7.14) to this. The result can be written as

$$I = b_{dp} \int_{A(d,p)} \int_{E^{r_1+1}} \int_{E^{r_2+1}} f(\operatorname{aff} \mathbf{x}_1, \operatorname{aff} \mathbf{x}_2) g(\mathbf{x}_1, \mathbf{x}_2)$$
$$\times (r_1!)^{r_1-d} (r_2!)^{r_2-d} (p!)^{d-p} \Delta_{r_1}(\mathbf{x}_1)^{r_1-d} \Delta_{r_2}(\mathbf{x}_2)^{r_2-d}$$
$$\times \Delta_p(\mathbf{x}_1, \mathbf{x}_2)^{d-p} \, \lambda_E^{r_2+1}(\mathrm{d}\mathbf{x}_2) \, \lambda_E^{r_1+1}(\mathrm{d}\mathbf{x}_1) \, \mu_p(\mathrm{d}E).$$

Here we employ the (easily established) fact that

$$p!\Delta_p(\mathbf{x}_1, \mathbf{x}_2) = r_1! r_2! \Delta_{r_1}(\mathbf{x}_1) \Delta_{r_2}(\mathbf{x}_2) D(\operatorname{aff} \mathbf{x}_1, \operatorname{aff} \mathbf{x}_2)[\operatorname{aff} \mathbf{x}_1, \operatorname{aff} \mathbf{x}_2].$$

We insert this and then apply (7.14) to the two inner integrals over E^{r_j+1}, $j = 1, 2$. This immediately yields the assertion of the theorem. □

In the following theorem, we restrict ourselves to flats of small dimensions which affinely span the whole space. The pivot determined by a pair (E_1, E_2) of flats in general position will now be the triple (F, x_1, x_2), consisting of the ortholine F of E_1, E_2 and the points x_1, x_2 where F intersects the flats. For a given triple (F, x_1, x_2), the inner integration space is in effect (though written in a more convenient way) the space $A(x_1, F^\perp + x_1, r_1) \times A(x_2, F^\perp + x_2, r_2)$. Here $A(x, F^\perp + x, s)$ denotes the space of s-flats through x and contained in $F^\perp + x$ (recall that $F^\perp := (F^0)^\perp$ is a linear subspace).

Theorem 7.2.10. *Let $r_1, r_2 \in \{0, \dots, d-2\}$ be integers satisfying $r_1 + r_2 = d - 1$. If $f : A(d, r_1) \times A(d, r_2) \to \mathbb{R}$ is a nonnegative measurable function, then*

$$\int_{A(d,r_1)\times A(d,r_2)} f \, \mathrm{d}(\mu_{r_1} \otimes \mu_{r_2})$$
$$= b \int_{A(d,1)} \int_{F^2} \int_{G(F^\perp,r_1)\times G(F^\perp,r_2)} f(L_1 + x_1, L_2 + x_2)[L_1, L_2]^2$$
$$\times (\nu_{r1}^{F^\perp} \otimes \nu_{r2}^{F^\perp})(\mathrm{d}(L_1, L_2)) \, \lambda_F^2(\mathrm{d}(x_1, x_2)) \, \mu_1(\mathrm{d}F)$$

with b given by (7.12) *for $q = 2$.*

Proof. In the proof, we use repeatedly Fubini's theorem and (13.9). We apply Theorem 7.2.3 and then go over to orthogonal complements:

$$I := \int_{A(d,r_1)\times A(d,r_2)} f \, \mathrm{d}(\mu_{r_1} \otimes \mu_{r_2})$$
$$= \int_{G(d,r_1)\times G(d,r_2)} \int_{L_1^\perp \times L_2^\perp} f(L_1 + x_1, L_2 + x_2)$$
$$\times (\lambda_{L_1^\perp} \otimes \lambda_{L_2^\perp})(\mathrm{d}(x_1, x_2)) \, (\nu_{r_1} \otimes \nu_{r_2})(\mathrm{d}(L_1, L_2))$$
$$= b \int_{G(d,d-1)} \int_{G(H,r_1)\times G(H,r_2)} \int_{L_1^\perp \times L_2^\perp} f(L_1 + x_1, L_2 + x_2)$$

$$\times\, [L_1, L_2]\, (\lambda_{L_1^\perp} \otimes \lambda_{L_2^\perp})(\mathrm{d}(x_1, x_2))(\nu_{r_1}^H \otimes \nu_{r_2}^H)(\mathrm{d}(L_1, L_2))\, \nu_{d-1}(\mathrm{d}H)$$

$$= b \int_{G(d,1)} \int_{G(L^\perp, r_1) \times G(L^\perp, r_2)} \int_{L_1^\perp \times L_2^\perp} f(L_1 + x_1, L_2 + x_2)$$

$$\times\, [L_1, L_2]\, (\lambda_{L_1^\perp} \otimes\ \lambda_{L_2^\perp})(\mathrm{d}(x_1, x_2))\, (\nu_{r_1}^{L^\perp} \otimes \nu_{r_2}^{L^\perp})(\mathrm{d}(L_1, L_2))\, \nu_1(\mathrm{d}L).$$

Using the direct sum decomposition $L_j^\perp = L \oplus (L_j^\perp \cap L^\perp)$ and writing $x_j = y_j + z_j$ with $y_j \in L$ and $z_j \in L_j^\perp \cap L^\perp$, we obtain

$$I = b \int_{G(d,1)} \int_{G(L^\perp, r_1) \times G(L^\perp, r_2)} \int_{L^2} \int_{(L_1^\perp \cap L^\perp) \times (L_2^\perp \cap L^\perp)}$$

$$f(L_1 + y_1 + z_1, L_2 + y_2 + z_2)[L_1, L_2]\, (\lambda_{L_1^\perp \cap L^\perp} \otimes \lambda_{L_2^\perp \cap L^\perp})(\mathrm{d}(z_1, z_2))$$

$$\times\, \lambda_L^2(\mathrm{d}(y_1, y_2))\, (\nu_{r_1}^{L^\perp} \otimes \nu_{r_2}^{L^\perp})(\mathrm{d}(L_1, L_2))\, \nu_1(\mathrm{d}L)$$

$$= b \int_{G(d,1)} \int_{L^2} \left\{ \int_{A(L^\perp, r_1) \times A(L^\perp, r_2)} f(E_1 + y_1, E_2 + y_2) \right.$$

$$\left. \times\, [E_1, E_2]\, (\mu_{r_1}^{L^\perp} \otimes \mu_{r_2}^{L^\perp})(\mathrm{d}(E_1, E_2)) \right\} \lambda_L^2(\mathrm{d}(y_1, y_2))\, \nu_1(\mathrm{d}L).$$

To the integral in braces we apply Theorem 7.2.8 (in $L^\perp$); here $m = 0$ (and thus $\bar{b} = 1$), so that $A(L^\perp, m)$ is identified with $L^\perp$. This gives

$$I = b \int_{G(d,1)} \int_{L^2} \left\{ \int_{L^\perp} \int_{G(L^\perp, r_1) \times G(L^\perp, r_2)} f(L_1 + t + y_1, L_2 + t + y_2) \right.$$

$$\left. \times\, [L_1, L_2]^2\, (\nu_{r_1}^{L^\perp} \otimes \nu_{r_2}^{L^\perp})(\mathrm{d}(L_1, L_2))\, \lambda_{L^\perp}(\mathrm{d}t) \right\} \lambda_L^2(\mathrm{d}(y_1, y_2))\, \nu_1(\mathrm{d}L)$$

$$= b \int_{G(d,1)} \int_{L^\perp} \int_{(L+t)^2} \int_{G(L^\perp, r_1) \times G(L^\perp, r_2)} f(L_1 + y_1, L_2 + y_2)$$

$$\times\, [L_1, L_2]^2\, (\nu_{r_1}^{L^\perp} \otimes \nu_{r_2}^{L^\perp})(\mathrm{d}(L_1, L_2))\, \lambda_{L+t}^2(\mathrm{d}(y_1, y_2))\lambda_{L^\perp}(\mathrm{d}t)\, \nu_1(\mathrm{d}L)$$

$$= b \int_{A(d,1)} \int_{F^2} \int_{G(F^\perp, r_1) \times G(F^\perp, r_2)} f(L_1 + y_1, L_2 + y_2)$$

$$\times\, [L_1, L_2]^2\, (\nu_{r_1}^{F^\perp} \otimes \nu_{r_2}^{F^\perp})(\mathrm{d}(L_1, L_2))\, \lambda_F^2(\mathrm{d}(y_1, y_2))\, \mu_1(\mathrm{d}F).$$

This completes the proof. □

In Theorem 7.2.10 we have assumed that $r_1 + r_2 + 1$, the dimension of the affine span (if the flats are in general position) is equal to d. If this dimension is less than d, we obtain the corresponding result by first applying Theorem 7.2.9 and then transforming the inner integral by means of Theorem 7.2.10.

Notes for Section 7.2

1. In this note we give an interesting alternative proof of the Blaschke–Petkantschin formula of Theorem 7.2.1. This proof, which is due to Møller [550], is based on a uniqueness result for relatively invariant measures. The method is also briefly described in Barndorff–Nielsen, Blæsild and Eriksen [79, pp. 59–60]. In the following, we use notation and results from Section 13.3.

Second proof of Theorem 7.2.1. To prove (7.7), we need evidently consider only linearly independent q-tuples $(x_1,\dots,x_q)$. We denote by $U \subset (\mathbb{R}^d)^q$ the subspace of linearly independent q-tuples. In the following, we consider the elements of $\mathbb{R}^d$ as column vectors (with respect to some fixed basis) and, correspondingly, $(x_1,\dots,x_q)$ as a (d,q)-matrix; then U is the space of real (d,q)-matrices of rank q. Let $\mathcal{GL}(q)$ be the group of regular (q,q)-matrices with the standard topology; it is locally compact. The same holds true for the direct product $G := \mathcal{SO}(d) \times \mathcal{GL}(q)$, where $\mathcal{SO}(d)$ denotes the group of orthogonal (d,d)-matrices with determinant one. By

$$((D,M),(x_1,\dots,x_q)) \mapsto D(x_1,\dots,x_q)M^t =: (D,M).(x_1,\dots,x_q)$$

for $(D,M) \in G$ and $(x_1,\dots,x_q) \in U$, where M^t denotes the transpose of the matrix M, we define a transitive operation of G on U. It is not difficult to verify that U, with this operation, becomes a homogeneous G-space. Recall that (Sec. 13.3), if a group G operates on a set U, one defines $(g.f)(u) := f(g^{-1}u)$, $u \in U$, for a function f on U.

Now we define two positive linear functionals I_1, I_2 on $\mathbf{C}_c(U)$ by

$$I_1(f) := \int_{\mathbb{R}^d} \cdots \int_{\mathbb{R}^d} f(x_1,\dots,x_q)\,\lambda(\mathrm{d}x_1)\cdots\lambda(\mathrm{d}x_q),$$

$$I_2(f) := \int_{G(d,q)} \int_L \cdots \int_L f(x_1,\dots,x_q)\nabla_q(x_1,\dots,x_q)^{d-q} \times \lambda_L(\mathrm{d}x_1)\cdots\lambda_L(\mathrm{d}x_q)\,\nu_q(\mathrm{d}L)$$

for $f \in \mathbf{C}_c(U)$. For $(D,M) \in G$ we get

$$\begin{aligned} I_1((D,M).f) &= \int_{\mathbb{R}^d} \cdots \int_{\mathbb{R}^d} f(D^{-1}(x_1,\dots,x_q)M^{-t})\,\lambda(\mathrm{d}x_1)\cdots\lambda(\mathrm{d}x_q) \\ &= |\det M|^d I_1(f), \end{aligned}$$

because λ is rotation invariant, and the linear map $(x_1,\dots,x_q) \mapsto (x_1,\dots,x_q)M$ from the space of (d,q)-matrices into itself has determinant $(\det M)^d$. Further, for $L \in G(d,q)$ we get

$$\begin{aligned} &\int_L \cdots \int_L f(D^{-1}(x_1,\dots,x_q)M^{-t})\nabla_q(x_1,\dots,x_q)^{d-q}\,\lambda_L(\mathrm{d}x_1)\cdots\lambda_L(\mathrm{d}x_q) \\ &= |\det M|^q \int_L \cdots \int_L f(D^{-1}(x_1,\dots,x_q))\nabla_q((x_1,\dots,x_q)M^t)^{d-q} \\ &\quad \times \lambda_L(\mathrm{d}x_1)\cdots\lambda_L(\mathrm{d}x_q) \\ &= |\det M|^d \int_{\vartheta^{-1}L} \cdots \int_{\vartheta^{-1}L} f(x_1,\dots,x_q)\nabla_q(x_1,\dots,x_q)^{d-q} \\ &\quad \times \lambda_{\vartheta^{-1}L}(\mathrm{d}x_1)\cdots\lambda_{\vartheta^{-1}L}(\mathrm{d}x_q), \end{aligned}$$

where $\vartheta \in SO_d$ is the rotation defined by D. The rotation invariance of ν_q now implies

$$I_2((D,M).f) = |\det M|^d I_2(f).$$

Thus the integrals I_1 and I_2 are relatively invariant with the same multiplier. From Theorem 13.3.1 it follows that $I_1 = cI_2$ with a constant c. The value of this constant can be obtained from Theorem 8.2.2. □

2. A general reference for the formulas of Section 7.2 is Santaló's book [662]. His proofs, as much of the original literature, use differential forms. Another flexible tool for obtaining integral geometric transformation formulas is Federer's coarea formula. In contrast to this, our aim was here to give more elementary and geometric proofs, based either on direct integration procedures or invariance arguments.

Results of Blaschke–Petkantschin type can in principle be traced back to Lebesgue [437], who used the transformation rule for multiple integrals to give new proofs for results of Crofton. After an influential lecture course by Herglotz [335] in Göttingen on geometric probabilities, and papers by Blaschke [105] and Varga [762], a systematic and general investigation of such integral geometric transformation formulas was undertaken by Petkantschin [601]. In special forms, most of the results of this section appear already in that paper. The usefulness of Blaschke–Petkantschin type formulas in stochastic geometry was emphasized by Miles. In [531], he gave new proofs and extensions of some results going back to Petkantschin, for example, of Theorem 7.2.2 above. In the style of the present chapter, though less generally, Blaschke–Petkantschin formulas were presented in Schneider and Weil [716].

For a recent application (in particular of Theorem 7.2.3) outside stochastic geometry, we mention E. Milman [540].

3. There are more general versions of Lemma 7.2.1, for integrations over $A(d,q)$ instead of $\mathbb{R}^d$; see Petkantschin [601, formula (49)]. A special case, where the given linear subspace is of dimension zero, reads as follows. Let $q < r \le d$. For the pivot associated with $E \in A(d,q)$ one can choose a linear r-subspace containing E. Then the outer integration space is $G(d,r)$, and for $L \in G(d,r)$, the inner integration space is $A(L,q)$. The resulting formula is

$$\int_{A(d,q)} f \,\mathrm{d}\mu_q = c \int_{G(d,r)} \int_{A(L,q)} f\, d(\cdot,0)^{d-r} \,\mathrm{d}\mu_q^L \,\nu_r(\mathrm{d}L)$$

with a constant c depending on d, q, r. Applications of the special case $d = 3$, $q = 1$ are discussed by Cruz–Orive [190].

4. Vertical Sections. The following special case of Theorem 7.2.4 is of interest in stereology. Let $d = 3$, $q = 1$, $r_1 = 1$, $r_0 = 1$ and let $V \in G(3,1)$ be a fixed line. In some applications, the direction of the 'vertical' line V plays a particular role, and two-dimensional planes parallel to V define 'vertical sections'. Theorem 7.2.4 specializes to

$$\int_{G(3,1)} f \,\mathrm{d}\nu_1 = c \int_{G(V,2)} \int_{G(L,1)} f\,[\cdot, V] \,\mathrm{d}\nu_1^L \,\nu_2^V(\mathrm{d}L).$$

This can be interpreted as saying that an isotropic random line through 0 can be generated by first generating a uniform vertical 2-plane L containing V and then in L a random line through 0 with the distribution defined by the inner integral. Such and more general 'vertical uniform random sampling designs' can be of advantage in practical situations where preferred directions are present. They were suggested by Baddeley [46] and further studied in Baddeley [47, 48], Baddeley, Gundersen and Cruz–Orive [52]; see also Kötzer, Jensen and Baddeley [424] and Beneš and Rataj [90, sect. 4.1.3]. A detailed description is found in Baddeley and Jensen [53, ch. 8].

5. Hug and Reitzner [365] have proved and applied the following formula of Blaschke–Petkantschin type. The initial integration space is $(\mathbb{R}^d)^{d+p}$, where $1 \leq p \leq d$. The pivot determined by $(x_1, \dots, x_{d+p}) \in (\mathbb{R}^d)^{d+p}$ is the pair (H_1, H_2), where $H_1 := \operatorname{aff}\{x_1, \dots, x_d\}$ and $H_2 := \operatorname{aff}\{x_{p+1}, \dots, x_{p+d}\}$. The outer integration space is $A(d, d-1) \times A(d, d-1)$, and for $(H_1, H_2) \in A(d, d-1) \times A(d, d-1)$, the inner integration space is $H_1^p \times (H_1 \cap H_2)^{d-p} \times H_2^p$. The formula reads

$$\begin{aligned}\int_{(\mathbb{R}^d)^{d+p}} f \, \mathrm{d}\lambda^{d+p} = \left(\frac{d!\kappa_d}{2}\right)^2 \int_{A(d,d-1)\times A(d,d-1)} \int_{H_1^p \times (H_1\cap H_2)^{d-p}\times H_2^p} f \\ \times \Delta_d(x_1, \dots, x_d)\Delta_d(x_{p+1}, \dots, x_{p+d})[H_1, H_2]^{p-d} \\ \times \mathrm{d}(\lambda_{H_1}^p \otimes \lambda_{H_1\cap H_2}^{d-p} \otimes \lambda_{H_2}^p)\, \mu_{d-1}^2(\mathrm{d}(H_1, H_2)).\end{aligned}$$

6. There is a spherical counterpart to the affine Blaschke–Petkantschin formula. The initial integration is over $(S^{d-1})^{q+1}$, where $q \in \{1, \dots, d-1\}$. The pivot is the q-flat affinely spanned (almost everywhere) by the integration variable $(x_0, \dots, x_q) \in (S^{d-1})^{q+1}$. The outer integration space is the affine Grassmannian $A(d, q)$, and for $E \in A(d, q)$ hitting S^{d-1}, the inner integration space is $(S^{d-1} \cap E)^{q+1}$. The result appears in Miles [525, Th. 4], with a short sketch of a proof. A detailed proof could be given similarly to Theorem 8.2.3. A typical application is found in Buchta, Müller and Tichy [134].

A related very general transformation formula, involving spheres and linear instead of affine subspaces, appears together with applications in Arbeiter and Zähle [38, Th. 1].

7. The affine Blaschke–Petkantschin formula of Theorem 7.2.7 can be interpreted as a decomposition of the $(q+1)$-fold product of the Lebesgue measure in $\mathbb{R}^d$. Integration with respect to this product measure is decomposed into integration with respect to the $(q+1)$-fold product of Lebesgue measure in a q-dimensional affine subspace, with a suitable Jacobian, followed by an integration over all q-dimensional affine subspaces. A somewhat similar decomposition is possible if the d-dimensional Lebesgue measure is replaced by the k-dimensional Hausdorff measure on a k-surface, $k < d$. In that case, the relative directions of the intersecting affine subspace and the tangent plane of the k-surface at the intersection points enter into the formula and make it complicated. General formulas of this type were proved by Zähle [829] (see Reitzner [628] for a short proof of a useful special case) and Jensen and Kiêu [382] (using an extended coarea formula by Kiêu [410]). A simplified proof was given in Jensen [379]. Stereological applications were presented by Jensen and Gundersen [380], Jensen, Kiêu and Gundersen [383].

7.3 Transformation Formulas Involving Spheres

In this section we prove two formulas of Blaschke–Petkantschin type (in the sense explained in the previous section), where the pivots are spheres. Correspondingly, the outer integration is over the space of all spheres, or equivalently, over the space of all possible centers and all possible radii, with a very simple measure. The inner integrations are conveniently written in terms of the unit sphere instead of variable spheres.

Theorem 7.3.1. *If $f : (\mathbb{R}^d)^{d+1} \to \mathbb{R}$ is a nonnegative measurable function, then*

$$\begin{aligned}\int_{(\mathbb{R}^d)^{d+1}} & f \, \mathrm{d}\lambda^{d+1} \\ &= d! \int_{\mathbb{R}^d} \int_0^\infty \int_{S^{d-1}} \cdots \int_{S^{d-1}} f(z + ru_0, \ldots, z + ru_d) \\ &\quad \times r^{d^2-1} \Delta_d(u_0, \ldots, u_d) \, \sigma(\mathrm{d}u_0) \cdots \sigma(\mathrm{d}u_d) \, \mathrm{d}r \, \lambda(\mathrm{d}z).\end{aligned}$$

Proof. It must be shown that the differentiable mapping

$$T : \mathbb{R}^d \times (0, \infty) \times (S^{d-1})^{d+1} \to (\mathbb{R}^d)^{d+1},$$

which is defined by

$$(z, r, u_0, \ldots, u_d) \mapsto (z + ru_0, \ldots, z + ru_d)$$

and is bijective up to sets of measure zero, has Jacobian given by

$$D(z, r, u_0, \ldots, u_d) = d! r^{d^2-1} \Delta_d(u_0, \ldots, u_d). \tag{7.16}$$

In the proof, we use the block notation for matrices. We write A^t for the transpose of a matrix A; vectors of $\mathbb{R}^d$ are interpreted as columns. We denote by E_k the $k \times k$ unit matrix. In order to prove (7.16) at a given point $(z, r, u_0, \ldots, u_d)$ of $\mathbb{R} \times (0, \infty) \times (S^{d-1})^{d+1}$, we use special local coordinates in a neighborhood of this point. For $i = 0, \ldots, d$ we introduce, in a neighborhood of u_i on S^{d-1}, parameters in such a way that the $d \times d$ matrix $(u_i \, \dot{u}_i)$ becomes orthogonal at the considered point; here $\dot{u}_i$ denotes the $d \times (d-1)$ matrix of the partial derivatives of u_i with respect to the corresponding parameters. This can easily be achieved. If for $u \in S^{d-1}$ the matrix $(u \, \dot{u})$ is orthogonal, then

$$\dot{u}^t u = 0, \qquad \dot{u}^t \dot{u} = E_{d-1}, \qquad E_d - \dot{u}\dot{u}^t = uu^t.$$

For $D = D(z, r, u_0, \ldots, u_d)$ we therefore get

$$D = \begin{vmatrix} E_d & u_0 & r\dot{u}_0 & 0 & \cdots & 0 \\ \cdot & \cdot & 0 & \cdot & \cdots & \cdot \\ \vdots & \vdots & \vdots & \vdots & \ddots & \vdots \\ E_d & u_d & 0 & \cdot & \cdots & r\dot{u}_d \end{vmatrix}.$$

For $\widetilde{D} := r^{1-d^2} D$ we thus obtain

$$
\begin{aligned}
\widetilde{D}^2 &= \left| \begin{array}{cccc} E_d & \cdot & \cdots & E_d \\ u_0^t & \cdot & \cdots & u_d^t \\ \dot{u}_0^t & 0 & \cdots & 0 \\ 0 & \cdot & & \cdot \\ \vdots & \vdots & \ddots & \vdots \\ 0 & \cdot & \cdots & \dot{u}_d^t \end{array} \right| \left| \begin{array}{cccccc} E_d & u_0 & \dot{u}_0 & 0 & \cdots & 0 \\ \cdot & \cdot & 0 & \cdot & \cdots & \cdot \\ \vdots & \vdots & \vdots & \vdots & \ddots & \vdots \\ E_d & u_d & 0 & \cdot & \cdots & \dot{u}_d \end{array} \right| \\
&= \left| \begin{array}{ccccc} (d+1)E_d & \sum u_i & \dot{u}_0 & \cdots & \dot{u}_d \\ \sum u_i^t & d+1 & 0 & \cdots & 0 \\ \dot{u}_0^t & 0 & E_{d-1} & \cdots & 0 \\ \vdots & \vdots & \vdots & \ddots & \vdots \\ \dot{u}_d^t & 0 & 0 & \cdots & E_{d-1} \end{array} \right| \\
&= \left| \begin{array}{cc} (d+1)E_d - \sum \dot{u}_i \dot{u}_i^t & \sum u_i \\ \sum u_i^t & d+1 \end{array} \right| = \left| \begin{array}{cc} \sum u_i u_i^t & \sum u_i \\ \sum u_i^t & d+1 \end{array} \right| \\
&= \left| \begin{pmatrix} u_0 \cdots u_d \\ 1 \cdots 1 \end{pmatrix} \begin{pmatrix} u_0^t & 1 \\ \vdots & \vdots \\ u_d^t & 1 \end{pmatrix} \right| \\
&= (d!)^2 \Delta_d^2(u_0, \dots, u_d),
\end{aligned}
$$

as asserted. □

The previous result was based on the fact that $d+1$ points in general position determine a unique sphere through these points. The following counterpart employs the unique sphere touching $d+1$ hyperplanes in general position and contained in the bounded region determined by the hyperplanes. Let $H_0, \dots, H_d \in A(d, d-1)$ by hyperplanes in general position (that is, they don't have a common point, and any d of their normal vectors are linearly independent). There is a unique simplex S such that $H_0, \dots, H_d$ are the facet hyperplanes of S. We denote by P the set of $(d+1)$-tuples of unit vectors positively spanning $\mathbb{R}^d$, that is, not lying in some closed hemisphere of S^{d-1}.

Theorem 7.3.2. *If $f : A(d, d-1)^{d+1} \to \mathbb{R}$ is a nonnegative measurable function, then*

$$
\begin{aligned}
&\int_{A(d,d-1)^{d+1}} f \, \mathrm{d}\mu_{d-1}^{d+1} \\
&= \frac{d!}{\omega_d^{d+1}} \int_{\mathbb{R}^d} \int_0^\infty \int_{S^{d-1}} \cdots \int_{S^{d-1}} f(H(u_0, \langle z, u_0 \rangle + r), \dots, H(u_d, \langle z, u_d \rangle + r)) \\
&\quad \times \Delta_d(u_0, \dots, u_d) \mathbf{1}_{\mathsf{P}}(u_0, \dots, u_d) \, \sigma(\mathrm{d}u_0) \cdots \sigma(\mathrm{d}u_d) \, \mathrm{d}r \, \lambda(\mathrm{d}z).
\end{aligned}
$$

Proof. Let $A^*(d, d-1)^{d+1}$ denote the set of $(d+1)$-tuples of hyperplanes in general position. Let $H_0, \dots, H_d \in A^*(d, d-1)^{d+1}$, and let Δ be the simplex determined by these hyperplanes. We denote by z the center of the insphere of Δ, by r its radius, and by $z + ru_i$, $i = 0, \dots, d$, the contact points of the insphere with the given hyperplanes. Then $(u_0, \dots, u_d) \in \mathsf{P}$. The mapping

$$(z, r, u_0, \dots, u_d) \mapsto (H(u_0, t_0), \dots, H(u_d, t_d)) \quad \text{with } t_i := \langle z, u_i \rangle + r$$

maps $\mathbb{R}^d \times \mathbb{R}^+ \times \mathsf{P}$ bijectively onto $A^*(d, d-1)^{d+1}$. For fixed $(u_0, \dots, u_d) \in \mathsf{P}$, the mapping $(z, r) \mapsto (t_0, \dots, t_d)$ has Jacobian $d! \Delta_d(u_0, \dots, u_d)$. It follows that

$$\begin{aligned}
&\int_{A(d,d-1)} \cdots \int_{A(d,d-1)} f(H_0, \dots, H_d)\, \mu_{d-1}(\mathrm{d}H_0) \cdots \mu_{d-1}(\mathrm{d}H_d) \\
&= \frac{1}{\omega_d^{d+1}} \int_{S^{d-1}} \cdots \int_{S^{d-1}} \int_{\mathbb{R}} \cdots \int_{\mathbb{R}} f(H(u_0, \tau_0), \dots, H(u_d, \tau_d)) \\
&\quad \times \mathrm{d}\tau_0 \cdots \mathrm{d}\tau_d\, \sigma(\mathrm{d}u_0) \cdots \sigma(\mathrm{d}u_d) \\
&= \frac{d!}{\omega_d^{d+1}} \int_{S^{d-1}} \cdots \int_{S^{d-1}} \int_{\mathbb{R}^d} \int_0^\infty f(H(u_0, \langle z, u_0 \rangle + r), \dots, H(u_d, \langle z, u_d \rangle + r)) \\
&\quad \times \mathrm{d}r\, \lambda(\mathrm{d}z) \mathbf{1}_{\mathsf{P}}(u_0, \dots, u_d) \Delta_d(u_0, \dots, u_d)\, \sigma(\mathrm{d}u_0) \cdots \sigma(\mathrm{d}u_d),
\end{aligned}$$

which gives the assertion, by Fubini's theorem. □

Notes for Section 7.3

1. Theorem 7.3.1 appears, with a sketched proof, in Miles [521], equation (70). It was proved in a different way by Affentranger [9]. The proof given here goes back (for $d = 3$) to Møller [553].

2. Theorem 7.3.2 is taken from Calka [149].

Part III

Selected Topics from Stochastic Geometry

8

Some Geometric Probability Problems

Geometric probability deals with randomly generated objects and configurations of elementary geometry. Whereas stochastic geometry employs more sophisticated models, such as random sets and particle processes, the typical questions of geometric probability involve only finitely many random geometric objects of a simple nature, like points, lines, planes, convex bodies, and simple operations performed with them, for example, taking convex hulls or intersections. The geometric events or random variables under investigation are in general of an elementary nature. Many related questions are easily formulated, but the answers may differ widely in their levels of difficulty.

Geometric probability problems and the early development of integral geometry have been closely related. From the wealth of questions on geometric probabilities that can be considered, we present here a selection, guided by the criteria of intrinsic interest (a personal choice, of course) and of applicability of integral geometry. After a glimpse of the early examples of geometric probability problems in Section 8.1, we devote Section 8.2 to convex hulls of random points, and Section 8.6 to various inequalities for geometric probabilities and expectations of geometric random variables related to convex bodies. In both sections, Blaschke–Petkantschin type transformations are a useful tool. Section 8.3 applies spherical integral geometry to random projections of polytopes, and Section 8.4 treats randomly moving convex bodies and flats by means of Euclidean integral geometry. Section 8.5 develops a theory of randomly touching convex bodies.

8.1 Historical Examples

It seems appropriate to begin our presentation of selected geometric probability problems with a brief look at the classical examples. Geometric probability took its origin, as did probability in general, in considerations about games of chance. In 1733, the naturalist Georges–Louis Leclerc Comte de Buffon presented to the Académie des Sciences his *Mémoire sur le jeu de franc-carreau* (according to [139]). It was published by Buffon [140] in 1777, as Section 23 in a longer article with the remarkable title *Essai d'arithmétique morale*. Buffon's first geometric probability problem reads as follows (in a free translation). 'In a room, parqueted or paved with equal tiles, one tosses a coin in the air; one of the players bets that this coin, after its fall, will lie on a single tile; the second bets that this coin will be placed on two tiles, that is, it will cover one of the joints that separate them; a third player bets that the coin will be placed on two joints; a fourth bets that the coin will be found on three, four or six joints: one asks for the odds of each of these players.' After considering some special cases, the author turns to the question that has become famous as the 'Buffon needle problem': 'I suppose that in a room, in which the parquet is simply divided by parallel joints, one tosses a rod in the air and that one of the players bets that the rod will not cross any of the parquet's parallels while the other bets, on the contrary, that the rod will cross several of the parallels. What are the odds for these two players? *One can play this game on a draughting board with a sewing needle or with a headless pin.*' (Translation from [728]; emphasis by Buffon.) Buffon talked of both, 'baguette' and 'aiguille'; in his calculations he worked, of course, with a line segment. In the case where the distance between two neighboring lines on the floor is D and the needle has length $L < D$, Buffon obtained for the probability p, that the needle hits a line, the value

$$p = \frac{2L}{\pi D}. \tag{8.1}$$

To derive this, he noted that the event of hitting depends only on the distance of the needle from the nearest line and on the angle that the needle forms with the direction of the lines. He tacitly assumed that, as we would say today, both random variables are independent and are uniformly distributed in their respective ranges. Under these assumptions, Buffon found the correct answer (not so, however, for his next question, concerning two orthogonal arrays of equidistant lines – the correct answer in that case was given by Laplace [432, p. 362]).

It was Laplace who pointed out, in his *Théorie analytique des probabilités* of 1812 [432, p. 360], that (8.1) suggests throwing a needle (or rather a very thin cylinder, as he put it) many times, in order to derive from the observed frequency of hits an estimation for the number π. Later, the outcomes of such experiments have repeatedly been reported in the literature. Such reports can only be digested *cum grano salis*. For one reason, it seems arguable whether the assumptions leading to (8.1) are compatible with reality (note that the uniformity assumption for the position of the midpoint only makes sense after going over to a quotient space). Moreover, some of the reported results are too good to be true. The critical discussion of Gridgeman [285] is very illuminating.

'Historically, it would seem that the first question given on local probability, since Buffon, was the remarkable four-point problem of Prof. Sylvester.' This quotation from Crofton's [188] article on *Probability* in the *Encyclopedia*

Britannica of 1885 refers to a question that Sylvester [750] had posed in 1864 in the *Educational Times.* The first part of this question reads: 'Show that the chance of four points forming the apices of a reentrant quadrilateral is 1/4 if they be taken at random in an indefinite plane ... ' Several different answers were received, so that Sylvester came to the conclusion: "This problem does not admit of a determinate solution." A modified version of the question, for which we again quote Crofton, had much more impact: 'Professor Sylvester has remarked that it would be a novel question in the calculus of variations to determine the form of the convex contour which renders the probability a maximum or minimum that four points taken within it shall give a re-entrant quadrilateral.' This asks for the extrema of $p(K)$, taken over all planar convex bodies K, where $p(K)$ is the probability that the convex hull of four independent, uniformly distributed random points in K is a triangle. Some special values of $p(K)$ were readily computed, but the question was only answered in 1917, when Blaschke [102] showed that the minimum of $p(K)$ is attained if and only if K is an ellipse and the maximum if and only if K is a triangle (see Note 1 for Subsection 8.2.3 below). Both the topic – convex hulls of random points – and the tool in part of Blaschke's proof – Steiner symmetrization – later inspired many other geometers.

Relations between geometric probability and an arising integral geometry first appeared implicitly in the work of Barbier [77] in 1860 and explicitly in that of Crofton in 1868. Barbier (following G. Lamé) generalized Buffon's needle problem by replacing the needle by a convex domain K (or even a more general rigid curve) with diameter less than D (the distance between the parallel lines). Under the usual (tacit) assumptions on the distribution of the randomly tossed domain K, he found that the probability that the domain hits a line is given by $L/\pi D$, where L is the boundary length of K. For a convex polygon, he proved this by considering for each of its edges the expectation of the number of hits with a line (as found by Buffon), adding up over all edges, and dividing by two. The obtained formula is equivalent to the planar case of Cauchy's surface area formula (see (6.12)), which Cauchy [166, 167] had already published in 1841, though without relations to geometric probability. That Barbier's formula is in principle an integral geometric result, becomes clearer if one interchanges the roles of the convex domain and the lines. Let C be a circle of diameter D containing the convex domain K, and toss a line randomly on C. The probability that it hits K is given by $L/\pi D$, which is the quotient of the perimeters of K and C. The tacit assumptions on the distribution of the random line are those resulting from the assumptions in Buffon's problem: the direction of the line and its distance from the center of C are independent and uniformly distributed in their ranges. Precisely this distribution is obtained if one derives it from a motion invariant measure (or density) on the space of lines. Such a measure was introduced by Crofton.

The object of Crofton's [186] paper was, as he put it, 'principally, the application of the Theory of Probability to straight lines drawn at random in a plane'. The following considerations led him from the idea of equal probabilities of finitely many cases to a measure on the space of lines. In his own words (though in different order): 'If we take any fixed axes in the plane, a random line may be represented by the equation

$$x \cos\theta + y \sin\theta = p,$$

where p and θ are constants taken at random. Divide the angular space round any point into a number of equal angles $\delta\theta$, and for every direction let the plane be ruled with an infinity of equidistant parallel lines, the common infinitesimal distance being the same for every set of parallels.' In effect, Crofton defined a measure on the lines by the indefinite integral $\int \mathrm{d}p\, \mathrm{d}\theta$. His considerations preceding the definition can be interpreted as indicating that this measure does not depend on the choice of the coordinate system. With this measure at hand, Crofton then determined the measure of the lines (which he called the 'measure of the number of random lines') satisfying certain conditions, for example, hitting a given convex domain, or hitting both of two disjoint convex domains, or separating the latter. He also determined probabilities (strictly speaking, conditional probabilities, derived from the infinite measure, under conditions having finite positive measure). For example, for the probability p that two (independent) random lines hitting a given convex domain K intersect inside that domain he found $p = 2\pi A/L^2$, where A is the area and L is the perimeter of K. He pointed out that $p = 1/2$ for the circle, and $p < 1/2$ for any other convex figure.

A neat modern derivation of the formulas obtained by Crofton, stressing their integral geometric character, was presented by Lebesgue [437].

Twenty-seven years after his four-point problem, Sylvester [751] entered the geometric probability scene again, with *a funicular solution of Buffon's 'problem of the needle' in its most general form.* He generalized the result of Barbier and also results of Crofton, although he did not know of Crofton's paper at the time of writing (see the Postscriptum of [751]). Using Crofton's measure on the space of lines instead of Buffon's approach, one can formulate Sylvester's starting point as follows. Let a finite collection of pairwise disjoint planar convex domains be given, enclosed by some circle C. Determine the probability that a random line hitting C hits a prescribed selection of the given domains and not the others. Sylvester treated the case of three domains in detail and remarked the following about the general case (which explains the title): 'the final result for either probability is a linear homogeneous function of lengths of stretched bands drawn in various ways round the given figures'.

The problematic nature of 'choosing at random' from an infinite collection, as it used to occur in naive geometric probability questions, was illustrated by a striking example that Bertrand gave in his *Calcul des probabilités* [97, pp. 4–5] of 1888. He proposed to draw *at random* a chord of a circle and to ask for the probability that it is longer than an edge of a regular triangle inscribed to the circle. The first suggested reasoning goes like this: by symmetry, one endpoint of the chord can be assumed fixed on the circle; if the other is chosen uniformly on the circle, one obtains 1/3 for the probability in question. Second, by symmetry one may assume that the direction of the chord is fixed; if its distance from the origin is chosen uniformly, one obtains the probability 1/2. Finally, one may choose the midpoint of the chord uniformly in the interior of the circle; this yields the probability 1/4. Bertrand's verdict is that 'the question is ill-posed'. The attribute of *paradox* for this example comes from other authors, for example, Poincaré, in his *Calcul des probabilités* [607, p. 107] of 1896. Poincaré points out that the probability 1/2 is obtained in Bertrand's question if one accepts the usual conventions of the needle problem, in modern words, if one uses the invariant measure on the space of lines. That Crofton's line measure is indeed invariant under rigid motions, was probably felt by Crofton, though not clearly stated. Invariant measures on the lines in the plane and on the planes and lines in three-space were studied systematically by E. Cartan [164]. In the same year, Poincaré in his quoted book introduced the kinematic densities in the plane and on the sphere and showed their invariance; in other words, he established the invariant measure on the motion group of the plane and of the rotation group of three-space. As an application, he considered ([607, p. 118]) a fixed and a moving curve on the sphere and asked for the expected number of their intersection points. He found that it is proportional to the product of the lengths of the two curves. Underlying this result is the assumption that the moving curve undergoes a random rotation with distribution given by the normalized invariant measure. Thus, by the end of the nineteenth century, geometric probability and integral geometry were closely tied together.

Notes for Section 8.1

1. Buffon type problems. Buffon's needle problem and his clean tile problem (*jeu de franc-carreau*) have remained popular until today (as an Internet search will confirm). Variants of the original problems which are still being investigated range from Buffon noodles (Waymire [777]) to various special planar convex bodies that hit at random a given periodic system of lines or figures, with natural uniformity assumptions about the distribution of the randomly placed domains. (Animated demonstrations of Buffon's clean tile problem are found at http://mathworld.wolfram.com/CleanTileProblem.html) We refer the interested reader to the extensive references given in the book by Mathai [456] and mention only a few papers where the randomly placed object is a general convex body in the plane. Ren and Zhang [636], and independently Aleman, Stoka and Zamfirescu [15], consider two families of equidistant lines in the plane and are interested in the angle between them for which the hitting events become independent. Similar questions are treated by Duma and Stoka [212].

A more intricate Buffon type problem is attacked by Bárány [65]. In his work, the lattice to be hit (or not to be hit) is $\mathbb{Z}^d$. It is proved, under the usual distributional assumption for the randomly placed body K, that the probability of non-hitting is of order $1/V_d(K)$ if the volume of K is sufficiently large and its width is sufficiently small.

2. The ideas underlying Sylvester's [751] 'funicular solution' of Buffon's problem admit far-reaching generalizations. They led Ambartzumian [34] to his Combinatorial Integral Geometry. This is not restricted to invariant measures. It has yielded several applications to stochastic geometry and various geometric results, among them a solution of Hilbert's fourth problem in the plane (though not the only solution).

3. The historical article by Seneta, Parshall and Jongmans [728] describes the development of geometric probability in the nineteenth century, paying particular attention to Sylvester, Crofton, Barbier, and Bertrand. The early history is also sketched in the introduction, written by Miles and Serra, to the Proceedings [539] of the Buffon Bicentenary Symposium that was held in Paris in 1977. For the particular history of Sylvester's problem, we refer to Pfiefer [602].

Early collections of problems and solutions of geometric probability problems are Section 6 of Crofton's [188] article in the *Encyclopedia Britannica*, and the books by Czuber [191] and Deltheil [202].

Much information on the later development of geometric probability can be gathered from the book by Kendall and Moran [397] and the survey articles by Moran [558, 559], Little [440], and Baddeley [43]. See also Solomon [731], and in particular the book by Mathai [456].

For introductory surveys to geometrical probability and stochastic geometry, we also refer to the articles by Baddeley [44] and by Weil and Wieacker [806].

8.2 Convex Hulls of Random Points

From the early beginnings of geometric probability until today, convex hulls of independent random points have been a favorite subject of investigation. In the following four subsections and the corresponding notes we hope to give an impression, though necessarily restricted, of the nature and diversity of the questions that have been posed and answered.

8.2.1 A Given Point in a Random Convex Hull

We consider n independent, identically distributed random points $X_1, \dots, X_n$ in $\mathbb{R}^d$ and ask for the probability of the event that 0 is in the convex hull of these points. Evidently, this probability depends heavily on the distribution of the points. It may be surprising, at first sight, that for centrally symmetric distributions (that is, distributions invariant under reflection in the origin) the probability is independent of the distribution. If we denote the probability in question by $q_n^{(d)}$, then $p_n^{(d)} = 1 - q_n^{(d)}$ is the probability that $X_1, \dots, X_n$ lie in some open halfspace with 0 in the boundary. The following is a classical result due to Wendel [810].

Theorem 8.2.1. *If $X_1, \dots, X_n$ are i.i.d. random points in $\mathbb{R}^d$ whose distribution is symmetric with respect to 0 and assigns measure zero to every hyperplane through 0, then*

$$p_n^{(d)} := \mathbb{P}(0 \notin \operatorname{conv}\{X_1, \ldots, X_n\}) = \frac{1}{2^{n-1}} \sum_{k=0}^{d-1} \binom{n-1}{k}.$$

For example, $p_n^{(d)} = 1$ for $n \leq d$, and $p_{d+1}^{(d)} = 1 - 2^{-d}$.

The independence of the distribution indicates that Theorem 8.2.1 is essentially a geometric result. Its proof depends on an observation in combinatorial geometry that goes back to Steiner [736] for $d \leq 3$ and to Schläfli [667, pp. 209 – 212] in general. Let $H_1, \ldots, H_n \subset \mathbb{R}^d$ be hyperplanes through 0 which are in general position, that is, any d or fewer of them have linearly independent normal vectors. Then $\mathbb{R}^d \setminus (H_1 \cup \ldots \cup H_n)$ consists of finitely many connected components (open polyhedral cones), which are called the **cells** induced by $H_1, \ldots, H_n$.

Lemma 8.2.1. *If $C(n,d)$ denotes the number of cells induced by n hyperplanes through 0 in general position in $\mathbb{R}^d$, then*

$$C(n,d) = 2 \sum_{k=0}^{d-1} \binom{n-1}{k}.$$

Proof. For $d = 1$ (where only $n = 1$ is possible), the assertion is true, and also for $d = 2$, since evidently $C(n, 2) = 2n$. The assertion is further true for $n = 1$ and any dimension $d \geq 1$. Suppose that $n \geq 2$ and that the assertion is true for $n-1$ hyperplanes, in any dimension $d \geq 2$. Let $H_1, \ldots, H_n \subset \mathbb{R}^d$ be hyperplanes through 0 in general position. We may assume that $d \geq 3$. The hyperplanes $H_1, \ldots, H_{n-1}$ induce $C(n-1, d)$ cells. Some of these cells, let C' be their number, are cut by H_n into two pieces; the remaining C'' cells are not hit by H_n. It follows that $C(n-1,d) = C' + C''$ and $C(n,d) = 2C' + C''$, hence

$$C(n,d) = C' + C(n-1,d). \tag{8.2}$$

We state that

$$C' = C(n-1, d-1). \tag{8.3}$$

In fact, the intersections $H_j \cap H_n$, $j = 1, \ldots, n-1$, are hyperplanes in the $(d-1)$-dimensional space H_n, and they are in general position. Therefore, $H_n \setminus \bigcup_{j=1}^{n-1} (H_j \cap H_n)$ has $C(n-1, d-1)$ cells of dimension $d-1$, and clearly these are the intersections of the C' cells in $\mathbb{R}^d$ with H_n. This proves (8.3).

From (8.2) and (8.3) we obtain $C(n,d) = C(n-1,d-1) + C(n-1,d)$, which together with the induction hypothesis shows that the assertion is true for n hyperplanes. □

Proof of Theorem 8.2.1. Let ϕ denote the distribution of X_i, and define

$$g(x_1, \ldots, x_n) := \begin{cases} 1, & \text{if } x_1, \ldots, x_n \text{ lie in some open halfspace} \\ & \text{with 0 in the boundary,} \\ 0 & \text{else.} \end{cases}$$

Then

$$p_n^{(d)} = \int_{\mathbb{R}^d} \cdots \int_{\mathbb{R}^d} g(x_1, \ldots, x_n)\, \phi(\mathrm{d}x_1) \cdots \phi(\mathrm{d}x_n)$$
$$= \int_{\mathbb{R}^d} \cdots \int_{\mathbb{R}^d} g(\epsilon_1 x_1, \ldots, \epsilon_n x_n)\, \phi(\mathrm{d}x_1) \cdots \phi(\mathrm{d}x_n)$$

for $\epsilon_i \in \{1, -1\}$ $(i = 1, \ldots, n)$, since ϕ is invariant under reflection in the origin. This gives

$$p_n^{(d)} = \int_{\mathbb{R}^d} \cdots \int_{\mathbb{R}^d} \frac{1}{2^n} \sum_{\epsilon_i \in \{1,-1\}} g(\epsilon_1 x_1, \ldots, \epsilon_n x_n)\, \phi(\mathrm{d}x_1) \cdots \phi(\mathrm{d}x_n).$$

We consider a fixed n-tuple $x_1, \ldots, x_n \in \mathbb{R}^d$. Due to the assumption on ϕ, we can assume that any d or fewer of these vectors are independent (the other n-tuples do not contribute to the integral). An n-tuple $(\epsilon_1, \ldots, \epsilon_n)$ contributes 1 to the sum in the integral if and only if there is a vector y with

$$\langle \epsilon_i x_i, y \rangle > 0 \qquad \text{for } i = 1, \ldots, n,$$

thus if and only if

$$\bigcap_{i=1}^{n} \{z \in \mathbb{R}^d : \langle \epsilon_i x_i, z \rangle > 0\} \neq \emptyset.$$

The nonempty intersections of this kind are precisely the cells that are induced by the hyperplanes through 0 orthogonal to $x_1, \ldots, x_n$, respectively. It follows that

$$\sum_{\epsilon_i \in \{1,-1\}} g(\epsilon_1 x_1, \ldots, \epsilon_n x_n) = C(n, d).$$

Lemma 8.2.1 now completes the proof of the theorem. □

Notes for Subsection 8.2.1

1. Wendel's theorem 8.2.1 is complemented by the (deeper) result saying that the probability in question is extremal for symmetric distributions. The following was proved by Wagner and Welzl [775]. Let μ be a probability measure on $\mathbb{R}^d$ that is absolutely continuous with respect to Lebesgue measure. Let $X_1, \ldots, X_n$ be i.i.d. random points in $\mathbb{R}^d$ with distribution μ. Then the probability $q_n^{(d)}$ of the event $0 \in \operatorname{conv}\{X_1, \ldots, X_n\}$ satisfies

$$q_n^{(d)} \leq 1 - \frac{1}{2^{n-1}} \sum_{k=0}^{d-1} \binom{n-1}{k}.$$

Equality holds if and only if

$$\mu(\{\alpha x : x \in B,\ \alpha \geq 0\} = \mu(\{\alpha x : x \in B,\ \alpha \leq 0\}$$

holds for every Borel set $B \subset \mathbb{R}^d$.

2. Slightly different versions of the proof of Theorem 8.2.1 are found in Mycielski [570] and Bárány [64, Sec. 9].

3. The setting underlying Wendel's theorem, namely the partition of $\mathbb{R}^d$ induced by n hyperplanes through 0 in general position, was studied further by Cover and Efron [179]. Some of their results have the following interpretation in terms of geometric probability. Let $X_1, \dots, X_n$ be independent uniform random points on the sphere S^{d-1} (thus, the distribution of X_i is σ/ω_d; this can be generalized). Under the condition that $X_1, \dots, X_n$ lie in some open hemisphere, their spherical convex hull has expected spherical volume $\binom{n-1}{d-1}/C(n,d)$, and the expectation of its number of k-faces is given by

$$2^{k+1}\binom{n}{k+1}\frac{C(n-k-1,d-k-1)}{C(n,d)},$$

$k = 0, \dots, d-2$. For $n \to \infty$, this converges to $2^{k+1}\binom{d-1}{k+1}$, which happens to be the number of k-faces of the $(d-1)$-dimensional crosspolytope.

For $d = 3$, Miles [524] has determined first and second moments of area, perimeter and vertex number for the spherical random polytopes considered here.

8.2.2 Points in Balls and Spheres

If $1 \le q+1 \le d+1$, then the convex hull of $q+1$ independent uniform random points in the ball B^d is almost surely a q-dimensional simplex. The moments of its q-dimensional volume can be computed explicitly. This was first done by Miles, in fact more generally for independent random points of which some are uniform in B^d and the others are uniform on the boundary S^{d-1} of B^d. We present here only the cases where the points are uniform either in B^d or in S^{d-1}. The computation is a typical application of Blaschke–Petkantschin type formulas. Before these results, which are stated in Theorem 8.2.3 and are needed on various occasions, we consider the q-dimensional volume of the parallelepiped spanned by q independent uniform random vectors in B^d.

Note that the subsequent theorems yield results about moments, after normalization.

Theorem 8.2.2. *For integers $d \ge 1$, $1 \le q \le d$, $k \ge 1$,*

$$\begin{aligned} I(d,q,k) &:= \int_{B^d} \cdots \int_{B^d} \nabla_q(x_1,\dots,x_q)^k \,\lambda(\mathrm{d}x_1)\cdots\lambda(\mathrm{d}x_q) \\ &= \kappa_{d+k}^q \prod_{j=0}^{q-1} \frac{\omega_{d-j}}{\omega_{d+k-j}}. \end{aligned}$$

Proof. An elementary calculation yields

$$I(d,1,k) = \frac{\omega_d}{d+k}$$

and thus the assertion for $q = 1$ (and hence for $d = 1$). Assume now that $d \geq 2$ and $q \geq 2$. Let $L \in G(d, q-1)$. Using Fubini's theorem and (7.9), we obtain

$$\begin{aligned}
\frac{I(d,q,k)}{I(d,q-1,k)} &= \int_{B^d} d(x,L)^k \, \lambda(\mathrm{d}x) \\
&= \int_{L^\perp \cap B^d} \int_{(L+x)\cap B^d} \|x\|^k \, \lambda_{L+x}(\mathrm{d}y) \, \lambda_{L^\perp}(\mathrm{d}x) \\
&= \kappa_{q-1} \int_{L^\perp \cap B^d} \|x\|^k (1-\|x\|^2)^{(q-1)/2} \, \lambda_{L^\perp}(\mathrm{d}x) \\
&= \kappa_{q-1}\omega_{d-q+1} \int_0^1 (1-t^2)^{(q-1)/2} \, t^{d+k-q} \, \mathrm{d}t \\
&= \kappa_{d+k} \frac{\omega_{d-q+1}}{\omega_{d+k-q+1}}.
\end{aligned}$$

Repeated application gives the desired result. □

Theorem 8.2.3. *For integers $d \geq 1$, $1 \leq q \leq d$, $k \geq 0$,*

$$\begin{aligned}
J(d,q,k) &:= \int_{B^d} \cdots \int_{B^d} \Delta_q(x_0,\dots,x_q)^k \, \lambda(\mathrm{d}x_0)\cdots\lambda(\mathrm{d}x_q) \\
&= \frac{1}{(q!)^k} \kappa_{d+k}^{q+1} \frac{\kappa_{q(d+k)+d}}{\kappa_{(q+1)(d+k)}} \frac{b_{dq}}{b_{(d+k)q}}
\end{aligned} \tag{8.4}$$

and

$$\begin{aligned}
S(d,q,k) &:= \int_{S^{d-1}} \cdots \int_{S^{d-1}} \Delta_q(u_0,\dots,u_q)^k \, \sigma(\mathrm{d}u_0)\cdots\sigma(\mathrm{d}u_q) \\
&= \frac{1}{(q!)^k} \omega_{d+k}^{q+1} \frac{\kappa_{q(d+k-2)+d-2}}{\kappa_{(q+1)(d+k-2)}} \frac{b_{dq}}{b_{(d+k)q}}
\end{aligned} \tag{8.5}$$

with b_{dq} given by (7.8).

Proof. An elementary calculation gives

$$J(1,1,k) = \frac{2^{k+3}}{(k+1)(k+2)},$$

and since

$$\frac{\kappa_{k+1}\kappa_{k+2}}{\kappa_{2k+2}} = \frac{2^{k+2}}{k+2},$$

this coincides with the assertion (8.4) for $d = 1$, $q = 1$.

Evidently, $S(1,1,k) = 2^{k+1}$. Since

$$\frac{\kappa_{k+1}\kappa_{k-2}}{\kappa_{2k-2}} = \frac{2^k}{k+1},$$

this coincides with the assertion (8.5) for $d = 1$ and $q = 1$.

Let $d \geq 2$ and assume, first, that $q < d$. For $0 \leq \rho < 1$ we put $B^d_\rho := \{x \in \mathbb{R}^d : \rho \leq \|x\| \leq 1\}$ and

$$J_\rho(d,q,k) := \int_{B^d_\rho} \cdots \int_{B^d_\rho} \Delta_q(x_0,\dots,x_q)^k \, \lambda(\mathrm{d}x_0)\cdots\lambda(\mathrm{d}x_q). \tag{8.6}$$

Using Theorem 7.2.7 and decomposing the measure μ_q according to (13.9), we obtain

$$\begin{aligned} &J_\rho(d,q,k) \\ &= b_{dq}(q!)^{d-q} \int_{G(d,q)} \int_{L^\perp} \int_{(L+y)\cap B^d_\rho} \cdots \int_{(L+y)\cap B^d_\rho} \Delta_q(x_0,\dots,x_q)^{d+k-q} \\ &\quad \times \lambda_{L+y}(\mathrm{d}x_0)\cdots\lambda_{L+y}(\mathrm{d}x_q)\,\lambda_{L^\perp}(\mathrm{d}y)\,\nu_q(\mathrm{d}L). \end{aligned} \tag{8.7}$$

With $\rho = 0$, this gives

$$\begin{aligned} &J(d,q,k) \\ &= b_{dq}(q!)^{d-q} \int_{L^\perp \cap B^d} (1-\|y\|^2)^{q(d+k+1)/2} J(q,q,d+k-q)\,\lambda_{L^\perp}(\mathrm{d}y) \\ &= b_{dq}(q!)^{d-q} \omega_{d-q} J(q,q,d+k-q) \int_0^1 (1-t^2)^{q(d+k+1)/2}\, t^{d-q-1}\,\mathrm{d}t \\ &= \theta(d,q,k) J(q,q,d+k-q), \end{aligned}$$

where

$$\theta(d,q,k) := b_{dq}(q!)^{d-q} \frac{\kappa_{q(d+k)+d}}{\kappa_{q(d+k+1)}}.$$

For $k = 0$ we get

$$\kappa_d^{q+1} = \theta(d,q,0) J(q,q,d-q).$$

Hence, we have

$$\kappa_{d+k}^{q+1} = \theta(d+k,q,0) J(q,q,d+k-q) \tag{8.8}$$

and therefore

$$J(d,q,k) = \kappa_{d+k}^{q+1} \frac{\theta(d,q,k)}{\theta(d+k,q,0)}.$$

This yields the assertion (8.4) for $q < d$. For $k = 0$, (8.4) it is trivially true, and from (8.8) it is obtained for $q = d$, $k \geq 1$, replacing (d,q,k) by $(d+k,d,0)$.

To prove (8.5), we introduce spherical coordinates in (8.6) and employ the mean value theorem of integral calculus. This yields

$$\begin{aligned} J_\rho(d,q,k) = \left(\frac{1-\rho^d}{d}\right)^{q+1} &\int_{S^{d-1}} \cdots \int_{S^{d-1}} \Delta_q(t_0u_0,\dots,t_qu_q)^k \\ &\times \sigma(\mathrm{d}u_0)\cdots\sigma(\mathrm{d}u_q) \end{aligned} \tag{8.9}$$

with suitable numbers $t_i \in [\rho, 1]$ (depending on $u_0, \dots, u_q$, in a measurable way, w.l.o.g.). From (8.9) and (8.7) we get

$$\int_{S^{d-1}} \cdots \int_{S^{d-1}} \Delta_q(t_0 u_0, \dots, t_q u_q)^k \, \sigma(\mathrm{d}u_0) \cdots \sigma(\mathrm{d}u_q)$$
$$= b_{dq}(q!)^{d-q} \int_{G(d,q)} \int_{L^\perp} I_\rho(L, y) \, \lambda_{L^\perp}(\mathrm{d}y) \, \nu_q(\mathrm{d}L)$$

with

$$I_\rho(L, y) := \left(\frac{d}{1-\rho^d}\right)^{q+1} \int_{(L+y)\cap B^d_\rho} \cdots \int_{(L+y)\cap B^d_\rho} \Delta_q(x_0, \dots, x_q)^{d+k-q}$$
$$\times \lambda_{L+y}(\mathrm{d}x_0) \cdots \lambda_{L+y}(\mathrm{d}x_q).$$

For fixed $L \in G(d,q)$ and $y \in L^\perp \cap \operatorname{int} B^d$, we obtain from (8.9), applied in $(L+y) \cap B^d_\rho$, (and denoting by a_+ the positive part of a)

$$I_\rho(L, y)$$
$$= \left(\frac{d}{q}\right)^{q+1} \left(\frac{(1-\|y\|^2)^{q/2} - (\rho^2 - \|y\|^2)_+^{q/2}}{1-\rho^d}\right)^{q+1} (1-\|y\|^2)^{q(d+k-q)/2}$$
$$\times \int_{S^{d-1}\cap L} \cdots \int_{S^{d-1}\cap L} \Delta_q(t_{0,y} u_0, \dots, t_{q,y} u_q)^{d+k-q} \, \sigma_L(\mathrm{d}u_0) \cdots \sigma_L(\mathrm{d}u_q)$$

with suitably chosen intermediate values $t_{i,y}$, satisfying

$$\frac{(\rho^2 - \|y\|^2)_+^{1/2}}{(1-\|y\|^2)^{1/2}} \le t_{i,y} \le 1$$

(and, w.l.o.g., depending in a measurable way on $u_0, \dots, u_q$ and y). Here, we have denoted by σ_L the spherical Lebesgue measure on $S^{d-1} \cap L$. With $\rho \to 1$ and the dominated convergence theorem, we now deduce that

$$S(d, q, k)$$
$$= b_{dq}(q!)^{d-q} \omega_{d-q} S(q, q, d+k-q) \int_0^1 (1-t^2)^{(q(d+k-1)/2)-1} t^{d-q-1} \, \mathrm{d}t$$
$$= \alpha(d, q, k) S(q, q, d+k-q)$$

with

$$\alpha(d, q, k) := b_{dq}(q!)^{d-q} \frac{\kappa_{q(d+k-2)+d-2}}{\kappa_{q(d+k-1)-2}}.$$

(Here $\kappa_{-1} := 1/\pi$, in agreement with the formula $\kappa_p = \pi^{p/2}/\Gamma(1+(p/2))$.) The case $k = 0$ gives

$$\omega_d^{q+1} = \alpha(d, q, 0) S(q, q, d-q),$$

hence

$$\omega_{d+k}^{q+1} = \alpha(d+k, q, 0) S(q, q, d+k-q) \tag{8.10}$$

and, therefore,

$$S(d,q,k) = \omega_{d+k}^{q+1} \frac{\alpha(d,q,k)}{\alpha(d+k,q,0)}.$$

This yields the assertion (8.5) for $q < d$. For $k = 0$, it holds trivially, and for $q = d$ and $k \geq 1$ it follows from (8.10), replacing (d,q,k) by $(d+k,d,0)$. □

Notes for Subsection 8.2.2

1. Theorem 8.2.3 is a special case of a result of Miles [525, (29)]. The value $J(d,d,1)$ was determined earlier by Kingman [412], and for $d = 3$ already by Hostinský [350].

More generally, Miles considered $r+1 \leq d+1$ independent random points in $\mathbb{R}^d$, of which the first p are uniform in the ball B^d, while the last $r+1-p$ are uniform in the sphere S^{d-1}. The convex hull of these points is a.s. an r-dimensional simplex; let $\Delta^{(r,p)}$ denote its r-dimensional volume. Miles [525] determined all moments of $\Delta^{(r,p)}$. Mathai [455] found the probability density of $\Delta^{(r,p)}$ for $r = 2$ and $p = 3$, and Mathai and Tracy [457] did the same for $r = 2$ and $p = 0, 1, 2$. For the general case of arbitrary $r \leq d$ and $p \leq r+1$, Pederzoli [597] obtained explicit series forms for the density function of $\Delta^{(r,p)}$. An extensive study of the set-up introduced by Miles, without the restriction $r \leq d$, was made by Affentranger [3, 4], who extended many of the previous results.

2. A conjecture of Miles [525], that $(2d/3)^{1/2}[r!\Delta^{(r,p)} - (r+1)^{1/2}]$ has asymptotically, for $d \to \infty$, a standard normal distribution, was proved by Ruben [651]; a shorter proof was given by Mathai [455].

3. More general moments of the type $J(d,q,k)$, for rotationally symmetric distributions, were computed by Ruben and Miles [652]. Additional information on volumes of convex hulls of uniform random points in the ball or, more generally, with centrally symmetric distributions, is found in the book by Mathai [456].

8.2.3 Basic Probabilities, Expectations and Moments

For the investigation of convex hulls of random points, we introduce some short notation which will be used throughout this section. Let $\varphi : \mathcal{P}' \to \mathbb{R}^d$ be a measurable function on the set of convex polytopes in $\mathbb{R}^d$. For a given probability distribution μ on $\mathbb{R}^d$, we define the random variable

$$\varphi(\mu, n) := \varphi(\operatorname{conv}\{X_1, \dots, X_n\}), \qquad n \in \mathbb{N},$$

where $X_1, \dots, X_n$ are independent random points, each with distribution μ. If $K \in \mathcal{K}'$ is a convex body with interior points, we write $\varphi(K,n) := \varphi(\mu, n)$, where $\mu := (\lambda \llcorner K)/\lambda(K)$ is the uniform distribution on K. Typical examples of functions φ are V_j, the jth intrinsic volume, and f_k, the number of k-faces. Frequently studied particular cases are the volume V_d and the vertex number

f_0. Also of interest are $D_d(K,n) := V_d(K) - V_d(K,n)$, the **missed volume**, and the indicator function of polytopes with k vertices,

$$\psi_k(P) := \begin{cases} 1, \text{ if } f_0(P) = k, \\ 0, \text{ otherwise.} \end{cases}$$

The classical problem of Sylvester [750], in a precise formulation suggested later, asks for the probability that four independent uniform random points in a two-dimensional convex body K form a triangle, thus it asks for $\mathbb{E}\psi_3(K,4)$. It is easy to see that

$$\mathbb{E}\psi_3(K,4) = 4\frac{\mathbb{E}V_2(K,3)}{V_2(K)}, \tag{8.11}$$

and since $\mathbb{E}f_0(K,4) = 3\mathbb{E}\psi_3(K,4) + 4\mathbb{E}\psi_4(K,4)$, also that

$$\mathbb{E}f_0(K,4) = 4\left(1 - \frac{\mathbb{E}V_2(K,3)}{V_2(K)}\right).$$

The extensions of these relations to dimensions $d \geq 2$ and to arbitrary numbers $n \geq d+1$ of random points read

$$\frac{\mathbb{E}V_d(K,n)}{V_d(K)} = 1 - \frac{\mathbb{E}f_0(K,n+1)}{n+1}, \tag{8.12}$$

$$\mathbb{E}\psi_{d+1}(K,n) = \binom{n}{d+1}\frac{\mathbb{E}V_d(K,d+1)^{n-d-1}}{V_d(K)^{n-d-1}}. \tag{8.13}$$

They are easily obtained by rearranging multiple integrals. The following theorem provides a considerable generalization of (8.12). The proof extends to more general probability distributions, if the volume is replaced by the probability content, but we restrict ourselves here to the uniform distribution in a convex body.

Theorem 8.2.4 (Buchta). *Let $K \in \mathcal{K}'$ be a convex body with interior points, and let $n, k \in \mathbb{N}$. Then*

$$\frac{\mathbb{E}V_d(K,n)^k}{V_d(K)^k} = \mathbb{E}\prod_{i=1}^{k}\left(1 - \frac{f_0(K,n+k)}{n+i}\right) \tag{8.14}$$

and, consequently,

$$\frac{\mathbb{E}D_d(K,n)^k}{V_d(K)^k} = \sum_{j=1}^{k}(-1)^{j-1}\binom{k}{j}\mathbb{E}\left(1 - \prod_{i=1}^{j}\left(1 - \frac{f_0(K,n+j)}{n+i}\right)\right).$$

Proof. Since K is fixed in the following, we write $V_d(n)$ and $f_0(n)$ for $V_d(K,n)$ and $f_0(K,n)$, respectively. Let $X_1, \dots, X_{n+k}$ be independent uniform random points in K. We denote by $P_{n,k}$ the number of k-element subsets of

$\{X_1, \dots, X_{n+k}\}$ which are contained in the convex hull of the remaining n points. For a fixed realization, where $X_1, \dots, X_{n+k}$ are pairwise different, the number $P_{n,k}$ is equal to the number of possibilities of choosing k points from $X_1, \dots, X_{n+k}$ which are not vertices of the convex hull of $X_1, \dots, X_{n+k}$. It follows that

$$P_{n,k} = \binom{n+k-f_0(n+k)}{k} \qquad \text{a.s.}$$

and therefore

$$\mathbb{E}P_{n,k} = \mathbb{E}\binom{n+k-f_0(n+k)}{k}. \tag{8.15}$$

On the other hand, let $p_{n,k}$ denote the probability that $X_1, \dots, X_k$ are contained in the convex hull of $X_{k+1}, \dots, X_{k+n}$. Since $X_1, \dots, X_{n+k}$ are independent and identically distributed, we clearly have

$$\mathbb{E}P_{n,k} = \binom{n+k}{k} p_{n,k}. \tag{8.16}$$

The probability $p_{n,k}$ can be computed as follows. For fixed points $x_1, \dots, x_n$ in K, the probability that a uniform random point X falls within the convex hull of $x_1, \dots, x_n$ is given by $V_d(\operatorname{conv}\{x_1, \dots, x_n\})/V_d(K)$. Hence, the probability that each of the independent points $X_1, \dots, X_k$ falls within the convex hull of $x_1, \dots, x_n$ is the kth power of this value. Since $X_1, \dots, X_{n+k}$ are independent, it follows that

$$p_{n,k} = \frac{\mathbb{E}V_d(n)^k}{V_d(K)^k}. \tag{8.17}$$

From (8.15), (8.16) and (8.17) we get

$$\binom{n+k}{k} \frac{\mathbb{E}V_d(n)^k}{V_d(K)^k} = \mathbb{E}\binom{n+k-f_0(n+k)}{k}. \tag{8.18}$$

Since

$$\binom{n+k}{k}^{-1} \binom{n+k-f_0(n+k)}{k} = \prod_{i=1}^{k} \left(1 - \frac{f_0(n+k)}{n+i}\right),$$

this yields the first assertion of the theorem. The second assertion follows from this, using the identity

$$(1-z)^k = \sum_{j=1}^{k} (-1)^{j-1} \binom{k}{j} (1-z^j).$$

This completes the proof. □

We may write (8.14) in the form

$$\frac{\mathbb{E}V_d(K,n)^k}{V_d(K)^k} = \sum_{r=0}^{k}(-1)^r a_{n,k}^{(r)}\, \mathbb{E}f_0(K,n+k)^r, \tag{8.19}$$

where $a_{n,k}^{(r)}$ denotes the rth elementary symmetric function of the numbers $\frac{1}{n+1},\dots,\frac{1}{n+k}$. Thus, the kth moment of $V_d(K,n)/V_d(K)$ is a linear combination of the first k moments of $f_0(K,n+k)$. For $k=1$, formula (8.19) reduces to (8.12).

Relation (8.19) shows that (if the volume of K is given) all moments $\mathbb{E}V_d(K,n)^k$ are determined by the distribution of $f_0(K,n+k)$. Conversely, the following theorem shows that the distribution of $f_0(K,n)$, for $n \ge d+1$, is determined by the moments $\mathbb{E}V_d(K,j)^{n-j}$, $j=d+1,\dots,n$. The case of $n \le d$ could be included, but is trivial. The case $k=d+1$ of the theorem is formula (8.13).

Theorem 8.2.5. *For $n \ge d+1$ and $k \in \{1,\dots,n\}$,*

$$\mathbb{P}(f_0(K,n)=k) = \binom{n}{k} \sum_{j=d+1}^{k} (-1)^{j+k} \binom{k}{j} \frac{\mathbb{E}V_d(K,j)^{n-j}}{V_d(K)^{n-j}}. \tag{8.20}$$

Proof. Using the notation from the previous proof, together with

$$\mathbb{P}(f_0(K,m)=k) = \mathbb{E}\psi_k(K,m) =: \mathbb{E}\psi_k(m)$$

for $m \ge k$, relation (8.18) can be written in the form

$$\binom{j+k}{k} \frac{\mathbb{E}V_d(j)^k}{V_d(K)^k} = \sum_{i=1}^{j+k} \binom{j+k-i}{k} \mathbb{E}\psi_i(j+k),$$

hence

$$\binom{n}{j} \frac{\mathbb{E}V_d(j)^{n-j}}{V_d(K)^{n-j}} = \sum_{i=1}^{n} \binom{n-i}{n-j} \mathbb{E}\psi_i(n), \qquad j=1,\dots,n.$$

The matrix

$$\left(\binom{n-i}{n-j}\right)_{i,j=1,\dots,n} \qquad \text{has the inverse} \qquad \left((-1)^{j+k}\binom{n-j}{n-k}\right)_{j,k=1,\dots,n}.$$

This is easily checked, noting that $\binom{n-i}{n-j}\binom{n-j}{n-k} \neq 0$ only for $i \le j \le k$ and that

$$\binom{n-i}{n-j}\binom{n-j}{n-k} = \binom{n-i}{n-k}\binom{k-i}{k-j}.$$

Therefore,

$$\mathbb{E}\psi_k(n) = \sum_{j=1}^{n} (-1)^{j+k} \binom{n-j}{n-k} \binom{n}{j} \frac{\mathbb{E}V_d(j)^{n-j}}{V_d(K)^{n-j}}, \qquad k=1,\dots,n.$$

This yields the assertion. ∎

The following identity relates expected volumes of convex hulls for different numbers of i.i.d. random points, with arbitrary distributions.

Theorem 8.2.6 (Buchta). *Let μ be a probability distribution on $\mathbb{R}^d$. For $m \in \mathbb{N}$,*

$$\mathbb{E}V_d(\mu, d+2m) = \frac{1}{2}\sum_{k=1}^{2m-1}(-1)^{k-1}\binom{d+2m}{k}\mathbb{E}V_d(\mu, d+2m-k) \qquad (8.21)$$

and, consequently,

$$\mathbb{E}V_d(\mu, d+2m) = \sum_{k=1}^{m}(2^{2k}-1)\frac{B_{2k}}{k}\binom{d+2m}{2k-1}\mathbb{E}V_d(\mu, d+2m+1-2k), \quad (8.22)$$

where the constants B_{2k} are the Bernoulli numbers.

Proof. The first result is obtained by integrating a geometric identity involving convex hulls. To state this identity, let $x_1, \dots, x_n \in \mathbb{R}^d$ be points (not necessarily distinct) such that their convex hull, denoted by K, is of dimension d. We say that a j-subset $\{i_1, \dots, i_j\}$ from $\{1, \dots, n\}$ *captures* the point $y \in \mathbb{R}^d$ if y is in the convex hull of $x_{i_1}, \dots, x_{i_j}$. Let $c_j(y)$ denote the number of j-subsets from $\{1, \dots, n\}$ that capture the point y ($j \in \mathbb{N}$, with $c_j(y) = 0$ if $j > n$). Let D be the union of all $(d-2)$-flats that are affinely spanned by points from $\{x_1, \dots, x_n\}$. **Cowan's identity** says that

$$c_1(y) - c_2(y) + \ldots + (-1)^{n-1}c_n(y) = (-1)^d \quad \text{if } y \in \operatorname{int} K \setminus D, \quad (8.23)$$

$$c_1(y) - c_2(y) + \ldots + (-1)^{n-1}c_n(y) = 0 \quad \text{if } y \in \operatorname{bd} K. \quad (8.24)$$

For the proof, we first consider the case where $y \in \operatorname{bd} K$, say $y \in K \cap H$, where H is a supporting hyperplane of K. Without loss of generality, let $1, \dots, n-p$ be precisely the indices i for which $x_i \in H$; then $p \in \{1, \dots, n-1\}$. Let $a_j(y)$ denote the number of j-subsets from $\{1, \dots, n-p\}$ that capture the point y. To each such j-subset we can add r indices from $\{n-p+1, \dots, n\}$, where $r \in \{0, \dots, p\}$, to obtain a $(j+r)$-subset from $\{1, \dots, n\}$ capturing y, and all such subsets are obtained in this way. It follows that

$$c_k(y) = \sum_{r=\max\{0,k-n+p\}}^{\min\{p,k-1\}}\binom{p}{r}a_{k-r}(y)$$

for $k = 1, \dots, n$ and hence that

$$\sum_{k=1}^{n}(-1)^{k-1}c_k(y) = \sum_{j=1}^{n-p}(-1)^{j-1}a_j(y)\sum_{r=0}^{p}(-1)^r\binom{p}{r} = 0.$$

This proves (8.24).

By a *spanned hyperplane* we understand a hyperplane that is affinely spanned by points from $x_1, \dots, x_n$. The complement in K of the union of all spanned hyperplanes is the union of finitely many open convex polytopes, which we call *cells*. Let C be a cell, let F be a facet of $\operatorname{cl} C$, and let H be the affine hull of F; then H is a spanned hyperplane. Let H^+ be the closed halfspace bounded by H that does not contain C. Without loss of generality, let $1, \dots, n-p$ be precisely the indices i for which $x_i \in H^+$. Let $y \in C$ and $y' \in \operatorname{relint} F$. By the definition of the cells, no k-flat spanned by points from $x_1, \dots, x_n$ meets C, for $k = 1, \dots, d-1$. It follows that a j-subset from $\{1, \dots, n\}$ that captures y also captures y'. Conversely, a j-subset capturing y' and containing an index from $\{n-p+1, \dots, n\}$ also captures y. Let $b_j(y')$ denote the number of j-subsets from $\{1, \dots, n-p\}$ that capture y'. We have shown that

$$c_j(y') = c_j(y) + b_j(y'). \tag{8.25}$$

We apply (8.25) in two ways. First, we assume that the points $x_1, \dots, x_{n-p}$ not all lie in H. Then their convex hull has dimension d (since H is a spanned hyperplane), hence (8.24), applied to $x_1, \dots, x_{n-p}$, gives

$$\sum_{j=1}^{n} (-1)^{j-1} c_j(y') = \sum_{j=1}^{n} (-1)^{j-1} c_j(y).$$

Since $\operatorname{int} K \setminus D$ is pathwise connected, we conclude that the function

$$\sum_{j=1}^{n} (-1)^{j-1} c_j$$

is constant on $\operatorname{int} K \setminus D$. We show by induction with respect to the dimension that this value is equal to $(-1)^d$. The case $d = 1$ is easy, so we assume that $d \geq 2$ and that the assertion has been proved in dimensions less than d. With the same notations as used before (8.25), we now assume that $x_1, \dots, x_{n-p} \in H$ (that is, we take for F a facet of K and choose C, y and y' appropriately). Using (8.24), (8.25) and the inductive assumption, we get

$$\begin{aligned} 0 = \sum_{j=1}^{n} (-1)^{j-1} c_j(y') &= \sum_{j=1}^{n} (-1)^{j-1} c_j(y) + \sum_{j=1}^{n} (-1)^{j-1} b_j(y') \\ &= \sum_{j=1}^{n} (-1)^{j-1} c_j(y) + (-1)^{d-1} \end{aligned}$$

and hence the assertion (8.23). Thus the identity is proved.

For $y \in \operatorname{int} K \setminus D$, (8.23) can be written as

$$\sum_{j=1}^{n} (-1)^{j-1} \sum_{I \subset \{1, \dots, n\},\, |I| = j} \mathbf{1}_{\operatorname{conv}\{x_i : i \in I\}}(y) = (-1)^d.$$

Integration with respect to Lebesgue measure yields

$$\sum_{j=d+1}^{n} (-1)^{j-1} \sum_{I\subset\{1,\dots,n\},\,|I|=j} V_d(\mathrm{conv}\{x_i : i \in I\}) = (-1)^d V_d(K).$$

Now let $X_1, \dots, X_n$ be independent random points, each with distribution μ. Then we get

$$\sum_{j=d+1}^{n} (-1)^{j-1} \binom{n}{j} \mathbb{E}V_d(\mathrm{conv}\{X_1,\dots,X_j\}) = (-1)^d \mathbb{E}V_d(\mathrm{conv}\{X_1,\dots,X_n\}).$$

Setting $j = n - k$, we obtain

$$\frac{1}{2}\sum_{k=1}^{n-d-1} (-1)^{k-1} \binom{n}{k} \mathbb{E}V_d(\mu, n-k) = \begin{cases} \mathbb{E}V_d(\mu,n), & \text{if } n-d \text{ is even,} \\ 0, & \text{if } n-d \text{ is odd.} \end{cases}$$

With $n - d = 2m$, this gives (8.21). The relation (8.21) can then be used to eliminate the terms with even k from the right side of (8.21); this yields (8.22) (see Badertscher [54]). □

Notes for Subsection 8.2.3

1. Sylvester's problem and related questions. For general two-dimensional convex bodies K, Blaschke [102] (see also [104, §24]) gave a solution to Sylvester's problem (in its later formulation), by determining the range of the probability in question. He proved that

$$\frac{2}{3} \le \mathbb{E}\psi_4(K,4) \le 1 - \frac{35}{12\pi^2}.$$

By (8.11), this is equivalent to

$$\frac{35}{48\pi^2} \le \frac{\mathbb{E}V_2(K,3)}{V_2(K)} \le \frac{1}{12}. \tag{8.26}$$

Equality holds on the left side of (8.26) if and only if K is an ellipse, and on the right side if and only if K is a triangle. Blaschke proved the left side of (8.26) by Steiner symmetrization and the right side by the process of 'shakedown'.

The left side of (8.26) was extended to higher dimensions and higher moments by Groemer [288]. His result, now known as the **Blaschke–Groemer inequality**, says that for a convex body K in $\mathbb{R}^d$ ($d \ge 2$) with given volume $V_d(K) > 0$ and for $n \ge d+1$ and $r \ge 1$, the expectation $\mathbb{E}\, V_d(K,n)^r$ attains its minimum precisely if K is an ellipsoid (see Theorem 8.6.3 for the case $n = d+1$, and see also the references in Note 1 for Section 8.6).

Henze [334] observed that Blaschke's method to prove (8.26) yields more, namely inequalities for distribution functions. Let F_K denote the distribution function of $V_2(K,3)$ for a two-dimensional convex body K with $V_2(K) = 1$, let T be a triangle and E an ellipse, each of area one. Then Henze showed that $F_T(t) \leq F_K(t) \leq F_E(t)$ for $t \in \mathbb{R}$.

Again for $d = 2$, Dalla and Larman [197] showed that $\mathbb{E}\, V_2(K,n) \leq \mathbb{E}\, V_2(T,n)$ for $n \geq 3$ and for a convex body K and a triangle T with the same area, with strict inequality if K is a polygon other than a triangle. Giannopoulos [257] showed for arbitrary K that equality holds only for triangles.

For $d \geq 3$, it has repeatedly been conjectured that $\mathbb{E}\, V_d(K, d+1)/V_d(K)$ is maximal if K is a simplex. This is still one of the major open problems of convex geometry. Its interesting relations to other open problems are discussed in Milman and Pajor [541]. Dalla and Larman [197] proved that on the set of d-polytopes with at most $d+2$ vertices, $\mathbb{E}\, V_d(K, d+1)/V_d(K)$ is maximal precisely for simplices. Strong support for the general conjecture comes from work of Bárány and Buchta [68]. From their more general results they deduced that, for any d-dimensional convex body K and a simplex T^d of the same volume, there is a number n_0 (depending on K) such that $\mathbb{E}\, V_d(K,n) < \mathbb{E}\, V_d(T^d,n)$ for $n \geq n_0$, unless K is a simplex. Restrictions for possible maximizers of the functional $\mathbb{E}\, V_d(K, d+1)/V_d(K)$ were found by Campi, Colesanti and Gronchi [155].

2. Explicit results: volumes. Apart from the formulas for balls mentioned in the preceding subsection, few explicit values of the moments $\mathbb{E} V_d(K,n)^k$ are known. The value of $\mathbb{E} V_d(T^d, d+1)^k$ for a simplex T^d was found by Reed [624]. In the plane, the expectation $\mathbb{E} V_2(K,n)$ was only known if K is an affinely regular polygon and if $n = 3$ (see the survey by Buchta [126] and the references in [125]), until Buchta [125] found a formula by which $\mathbb{E} V_2(K,n)$ can be calculated for any given convex polygon K.

For triangles and parallelograms K, all moments of $V_2(K,3)$ are known (Reed [624]), and for a triangle, Alagar [13] found also the explicit distribution. Henze [334] did the same for parallelograms and circles.

The problem of determining $\mathbb{E} V_d(T^d, d+1)$ for the d-simplex T^d was made popular by Klee [418]. Even the case $d = 3$, in spite of its deceivingly elementary character, remained open for many years. In [135], Buchta and Reitzner announced the formula

$$\mathbb{E} V_3(T^3, 4) = \frac{13}{720} - \frac{\pi^2}{15015} = 0.0173982\ldots, \tag{8.27}$$

as well as a more general formula for $\mathbb{E} V_3(T^3, n)$, and they sketched a proof. Independently, (8.27) was established by Mannion [450], making heavy use of computer algebra. Finally, Buchta and Reitzner [138] published a detailed version of their proof of the formula

$$\mathbb{E} V_3(T^3, n) = p_n - \pi^2 r_n,$$

where p_n and r_n are explicitly given rational numbers.

3. Explicit results: vertex numbers in polygons. Let $X_1, \ldots, X_n$ be independent uniform random points in the triangle with vertices $(0,1)$, $(0,0)$ and $(1,0)$ in $\mathbb{R}^2$. Let N_n be the number of points from $X_1, \ldots, X_n$ that are vertices of the convex hull of $(0,1), X_1, \ldots, X_n, (1,0)$, and let $p_k^{(n)} := \mathbb{P}(N_n = k)$. Bárány, Rote, Steiger and Zhang [74] proved that $p_n^{(n)} = 2^n/[n!(n+1)!]$; they also established a limit shape for random convex chains and a corresponding central limit theorem. Buchta [132] determined the probability $p_k^{(n)}$, for $k = 1, \ldots, n-1$, and thus the distribution of N_n. As an application, he announced the determination of the exact distribution of the vertex number $f_0(P,n)$ for a convex polygon P. For example, if $P = T$ is a triangle, the expectation

$$\mathbb{E} f_0(T,n) = 2\sum_{k=1}^{n-1} \frac{1}{k}$$

is known from Buchta [125], and the new approach yields the variance

$$\operatorname{var} f_0(T,n) = \frac{10}{9}\sum_{k=1}^{n-1} \frac{1}{k} - \frac{4}{3}\sum_{k=1}^{n-1} \frac{1}{k^2}.$$

4. Explicit results for radially symmetric distributions. For independent uniform random points in the ball, Buchta and Müller [133] obtained explicit (though complicated) formulas for $\mathbb{E}\varphi(B^d,n)$, where φ is either the surface area, the mean width, or the number of facets; see also Meilijson [511]. Analogous results for independent uniform random points in the sphere S^{d-1} are due to Buchta, Müller and Tichy [134].

Let μ denote the standard normal distribution in $\mathbb{R}^3$. Bosetto [118] found explicitly the distribution densities of $V_2(\mu,3)$ and $V_3(\mu,4)$.

5. Symmetric convex bodies. If K is a 0-symmetric convex body, one can, together with $\varphi(K,n)$, also consider the random variable

$$\varphi(K,n,\pm) := \varphi(\operatorname{conv}\{\pm X_1,\dots,\pm X_n\},$$

where $X_1,\dots,X_n$ are uniform random points in K. Meckes [507] has calculated the second and fourth moments of $V_d(B_q^d,n,\pm)$, where B_q^d denotes the unit ball of the normed space ℓ_q^d, $1 \le q \le \infty$, and all moments of $V_d(B^d,n,\pm)$. He has also shown in [508] that among planar symmetric convex bodies K the moments of $V_2(K,n)/V_2(K)$ and of $V_2(K,n,\pm)/V_2(K)$ are maximized by parallelograms.

6. Formulas (8.12) and (8.13) appeared in Efron [215]. Theorem 8.2.4 and its proof, as well as the conclusions drawn from it, such as (8.20), are due to Buchta [131]. This paper also discusses the consequences that the results have for the investigation of variances.

7. Points in convex position. The special value $\mathbb{E}\psi_n(K,n) = \mathbb{P}(f_0(K,n) = n)$ gives the probability that n independent uniform random points in K are in convex position, that is, all of them are vertices of their convex hull. Some information on this probability is available for $d = 2$. Valtr [760] determined it explicitly if K is a parallelogram, and in Valtr [761] he did the same for a triangle. For a convex body $K \subset \mathbb{R}^2$ with $V_2(K) = 1$, Bárány [60] (using a result from Bárány [59]) proved the limit relation

$$\lim_{n\to\infty} n^2 (\mathbb{E}\psi_n(K,n))^{1/n} = \frac{1}{4}\mathrm{e}^2 a^3(K),$$

where $a(K)$ denotes the supremum of the affine perimeters of all convex bodies contained in K. He even established the existence of a limit shape and a law of large numbers. As he showed, there exists a unique convex body $\widetilde{K} \subset K$ with affine perimeter $a(K)$. Let K_n denote the convex hull of n independent uniform random points in K. Bárány [60] proved that, for the Hausdorff distance δ and any $\epsilon > 0$,

$$\lim_{n\to\infty} \mathbb{P}(\delta(K_n,\widetilde{K}) > \epsilon \mid f_0(K) = n) = 0.$$

In higher dimensions, Bárány [62] determined the order of $\mathbb{E}\psi_n(K,n)$ as $n \to \infty$.

For balls in increasing dimensions and numbers of points growing with the dimension, Bárány and Füredi [70] (strengthening earlier work of Buchta [127]) established an interesting threshold phenomenon, namely

$$\lim_{d\to\infty} \mathbb{E}\psi_{n(d)}(B^d, n(d)) = \begin{cases} 1, \text{ if } n(d) = 2^{d/2}d^{-\epsilon}, \\ 0, \text{ if } n(d) = 2^{d/2}d^{(3/4)+\epsilon}, \end{cases}$$

for every $\epsilon > 0$.

8. Disjoint convex hulls. Let $X_1, \dots, X_j, Y_1, \dots, Y_k$ be independent, identically distributed random points in $\mathbb{R}^d$, and let p_{jk} denote the probability that the convex hull of $X_1, \dots, X_j$ is disjoint from the convex hull of $Y_1, \dots, Y_k$. This probability was first investigated, for $d = 2$, by Rogers [647]. For example, if the distribution of the points is absolutely continuous, Rogers showed a recursion formula according to which p_{jk} is known for all $j, k \geq 1$ if p_{nn} is known for all $n \geq 2$, or alternatively if p_{n1} is known for all $n \geq 3$. Further work by Buchta [130] was continued by Buchta and Reitzner [136]. For uniform random points in two-dimensional convex bodies K, they connected p_{jk} to equiaffine inner parallel curves of K, found an explicit representation in the case of polygons and proved, among other results, that

$$\lim_{n\to\infty} \frac{p_{nn}}{n^{3/2}4^{-n}} \geq \frac{8\sqrt{\pi}}{3},$$

with equality if K is centrally symmetric. Further results on p_{jk} were obtained in Buchta and Reitzner [137].

9. After Buchta [123] and Affentranger [5, 6] had found special cases of Theorem 8.2.6, the general result was proved by Buchta [129]. The approach presented here, by integrating a pointwise identity, was discovered by Cowan [183]. Badertscher [54] noticed the special form of the coefficients in (8.22).

8.2.4 Convex Hulls: Asymptotic Results

For a given number n and functional φ, the distribution of the random variable $\varphi(K,n)$, and even its expectation, is in general hardly accessible. However, the asymptotic behavior, as $n \to \infty$, has proved to be tractable in several cases. The technical effort to achieve this may be considerable, though. Therefore, we only carry out one example, and give extensive hints in the Notes.

The example that we give concerns the random polytope

$$K_n := \operatorname{conv}\{X_1, \dots, X_n\},$$

where $X_1, \dots, X_n$ are independent uniform random points in the convex body K. However, instead of a real functional applied to K_n, we consider the **selection expectation** of this random set. It can be defined by

$$\mathbb{E}K_n := \{\mathbb{E}Y : Y \text{ is a measurable selection of } K_n\}.$$

Here, a **measurable selection** of the random set K_n is a measurable mapping $Y : \Omega \to \mathbb{R}^d$ satisfying $Y \in K_n$ almost surely. It is known (see, e.g., Molchanov

[548, p. 159]) that $\mathbb{E}K_n$ is a convex body (in this case) and that for the support function h we have

$$h(\mathbb{E}K_n, u) = \mathbb{E}\, h(K_n, u) \qquad \text{for } u \in S^{d-1}.$$

The asymptotic behavior of such expectations, as $n \to \infty$, depends heavily on the boundary structure of K. We consider here only sufficiently smooth convex bodies. The convex body K is said to be of class C^k_+, for $k \geq 2$, if its boundary is a regular, k-times continuously differentiable hypersurface with everywhere positive Gauss–Kronecker curvature (product of the principal curvatures). If K is of class C^2_+ and if $u \in S^{d-1}$, then K has a unique boundary point with outer normal vector u, and we denote by $\kappa(u)$ the Gauss–Kronecker curvature of $\operatorname{bd} K$ at this point.

Theorem 8.2.7. *If K is a convex body of class C^3_+ in $\mathbb{R}^d$ and if $u \in S^{d-1}$, then*

$$h(K,u) - h(\mathbb{E}K_n, u) = c_d \kappa(u)^{1/(d+1)} \left(\frac{n}{V_d(K)}\right)^{-2/(d+1)} + O(n^{-3/(d+1)})$$

as $n \to \infty$, with

$$c_d := \frac{1}{2}\Gamma\left(\frac{d+3}{d+1}\right)\left(\frac{d+1}{\kappa_{d-1}}\right)^{2/(d+1)}.$$

The constant involved in the O-term can be chosen to be independent of u.

Proof. We assume that $0 \in \operatorname{int} K$ and define $v(u,t) := \lambda(\{x \in K : \langle x, u\rangle \geq t\})$ for $u \in S^{d-1}$ and $t \geq 0$. Put $V_d(K) =: V$. There is a number $c_0 > 0$ with $h(K,u) \geq c_0$ for all $u \in S^{d-1}$, and for each $c \in (0, c_0)$ the number

$$\alpha_c := 1 - \min_{u \in S^{d-1}} v(u, h(K,u) - c)/V$$

satisfies $\alpha_c < 1$. Choose $c \in (0, c_0)$. For $A \in \mathcal{K}'$, let

$$g(A,u,t) := \begin{cases} 1, \text{ if } A \cap H(u,t) = \emptyset, \\ 0, \text{ otherwise.} \end{cases}$$

Let $X_1, \dots, X_n$ be independent uniform random points in K. Let $u \in S^{d-1}$. The probability that $\langle X_i, u\rangle \leq h(K,u) - c$ for all i or that $\langle X_i, u\rangle \geq 0$ for all i is less than $2\alpha_c^n$. If we denote by K^n_* the set of all n-tuples $(x_1, \dots, x_n)$ of points in K for which $\langle x_i, u\rangle > h(K,u) - c$ for some i and $\langle x_j, u\rangle < 0$ for some j, then we obtain

$$\begin{aligned} &h(K,u) - \mathbb{E}\, h(K_n, u) \\ &= V^{-n} \int_{K^n_*} \int_{h(K,u)-c}^{h(K,u)} g(\operatorname{conv}\{x_1, \dots, x_n\}, u, t)\, \mathrm{d}t\, \lambda(\mathrm{d}(x_1, \dots, x_n)) + O(\alpha_c^n) \\ &= \int_{h(K,u)-c}^{h(K,u)} V^{-n} \int_{K^n_*} g(\operatorname{conv}\{x_1, \dots, x_n\}, u, t)\, \lambda(\mathrm{d}(x_1, \dots, x_n))\, \mathrm{d}t + O(\alpha_c^n). \end{aligned}$$

The integrand of the inner integral is 1 if and only if $\langle x_k, u\rangle < t$ for all k. It follows that

$$h(K,u) - \mathbb{E}\, h(K_n,u) = \int_{h(K,u)-c}^{h(K,u)} \left(1 - \frac{v(u,t)}{V}\right)^n \mathrm{d}t + O(\alpha_c^n)$$
$$= \int_0^c \left(1 - \frac{w(u,\tau)}{V}\right)^n \mathrm{d}\tau + O(\alpha_c^n),$$

where we have put $w(u,\tau) := v(u, h(K,u) - \tau)$.

Let $u \in S^{d-1}$, and let $y \in \mathrm{bd}\, K$ be the point at which K has the outer normal vector u. Reducing c (independently of u) if necessary, we can assume the following. There is a real C^3 function f defined in a convex neighborhood of 0 in the linear subspace $H(u,0)$ such that every point $x \in H(u, h(K,u) - \tau) \cap \mathrm{bd}\, K$ with $0 \le \tau \le c$ has a representation of the form $x = y + z - f(z)u$ with $z \in H(u,0)$. For given $\tau \in (0,c]$, let $z' \in H(u,0)$ be a point at which the function

$$z \mapsto \left| f(z) - \frac{1}{2} \sum_{i,j=1}^{d-1} f_{ij}(0) z_i z_j \right|$$

attains its maximum, say $b(\tau)$, under the condition $f(z) = \tau$. Here $z_1, \dots, z_{d-1}$ are the coordinates of z with respect to an orthonormal basis of $H(u,0)$, and $f_{ij} := \partial^2 f / \partial z_i \partial z_j$. In the following, $c_1, c_2, \dots$ denote positive constants which can be chosen independently of u. (The necessary arguments to show this independence are carried out in Schneider [684].) It follows from Taylor's theorem that

$$b(\tau) \le c_1 \|z'\|^3. \tag{8.28}$$

The eigenvalues of the matrix $(f_{ij}(0))_{i,j=1,\dots,d-1}$ are the principal curvatures k_i, $i = 1, \dots, d-1$, of $\mathrm{bd}\, K$ at y and hence have a positive lower bound $2c_2$. We have

$$\frac{1}{2} \sum_{i,j=1}^{d-1} f_{ij}(0) z_i' z_j' \ge c_2 \|z'\|^2,$$

hence

$$\tau = f(z') \ge \frac{1}{2} \sum_{i,j=1}^{d-1} f_{ij}(0) z_i' z_j' - b(\tau) \ge c_2 \|z'\|^2 - c_1 \|z'\|^3 \ge c_3 \|z'\|^2, \tag{8.29}$$

provided c has been chosen sufficiently small. Inequalities (8.28) and (8.29) yield

$$b(\tau) \le c_4 \tau^{3/2}.$$

By the definition of $b(\tau)$, every point $z \in H(u,0)$ with $f(z) = \tau$ satisfies

$$\tau - b(\tau) \le \frac{1}{2} \sum_{i,j=1}^{d-1} f_{ij}(0) z_i z_j \le \tau + b(\tau).$$

Defining the $(d-1)$-dimensional ellipsoids

$$E_{\pm} = \left\{ y + z - \tau u : z \in H(u,0) \text{ and } \frac{1}{2} \sum_{i,j=1}^{d-1} f_{ij}(0) z_i z_j \leq \tau \pm b(\tau) \right\},$$

we deduce that $E_- \subset K \cap H(u,\tau) \subset E_+$. Hence, $\lambda_{d-1}(K \cap H(u,\tau))$ lies between the $(d-1)$-volumes of these ellipsoids, which have semi-axes

$$\sqrt{2(\tau \pm b(\tau))/k_i}, \qquad i = 1, \ldots, d-1.$$

It follows that

$$\lambda_{d-1}(K \cap H(u,\tau)) = \kappa_{d-1} \kappa(u)^{-1/2} (2\tau)^{(d-1)/2} (1 + \eta(\tau))$$

with $|\eta(\tau)| \leq c_5 \tau^{1/2}$, and hence

$$w(u,\tau) = \int_0^\tau \lambda_{d-1}(K \cap H(u,\rho))\, \mathrm{d}\rho = \frac{\kappa_{d-1}}{d+1} \kappa(u)^{-1/2} (2\tau)^{(d+1)/2} (1 + \eta_1(\tau))$$

with $|\eta_1(\tau)| \leq c_6 \tau^{1/2}$. This gives

$$\begin{aligned} & h(K,u) - \mathbb{E}\, h(K_n,u) + O(\alpha_c^n) \\ & = J := \int_0^c \left(1 - \frac{2^{(d+1)/2} \kappa_{d-1} \kappa(u)^{-1/2}}{(d+1)V} \tau^{(d+1)/2} (1 + \eta_1(\tau)) \right)^n \mathrm{d}\tau. \end{aligned}$$

We substitute

$$t = na\tau^{(d+1)/2} \qquad \text{with} \qquad a := \frac{2^{(d+1)/2} \kappa_{d-1} \kappa(u)^{-1/2}}{(d+1)V}$$

and obtain

$$J = \frac{2}{d+1} (an)^{-2/(d+1)} \int_0^{\gamma n} \left[1 - \frac{t}{n} - \frac{t}{n} \psi\left(\frac{t}{n}\right) \right]^n t^{\beta - 1}\, \mathrm{d}t$$

with $\gamma := ac^{(d+1)/2}$ and $\beta := 2/(d+1)$; here

$$\left| \psi\left(\frac{t}{n}\right) \right| \leq c_7 \left(\frac{t}{n}\right)^{1/(d+1)}.$$

Without changing c_7, we can decrease c and thus γ, independently of u, such that $t \in [0, \gamma n]$ implies $|\psi(t/n)| \leq 1/2$. We use the estimates

$$0 \leq \mathrm{e}^{-x} - \left(1 - \frac{x}{n}\right)^n \leq \mathrm{e}^{-x} \frac{x^2}{n} \tag{8.30}$$

for $0 \leq x < n$ (see, e.g., Whittaker and Watson [812, p. 242]). For sufficiently large n we get

$$\int_{n^{1/(d+2)}}^{\gamma n} \left[1 - \frac{t}{n} - \frac{t}{n}\psi\left(\frac{t}{n}\right)\right]^n t^{\beta-1}\,\mathrm{d}t$$
$$\leq \int_{n^{1/(d+2)}}^{\gamma n} \mathrm{e}^{-t(1+\psi(t/n))} t^{\beta-1}\,\mathrm{d}t \leq \int_{n^{1/(d+2)}}^{\gamma n} \mathrm{e}^{-t/2} t^{\beta-1}\mathrm{d}t = O(n^{-p})$$

for any $p > 0$. For arbitrary $\rho > 0$ and $q > 0$ we obtain

$$\int_{n^{1/(d+2)}}^{\infty} \mathrm{e}^{-t} t^{\rho-1}\mathrm{d}t < n^{-q/(d+2)} \int_{n^{1/(d+2)}}^{\infty} \mathrm{e}^{-t} t^{\rho-1+q}\mathrm{d}t < n^{-q/(d+2)}\Gamma(\rho+q)$$

and, therefore,

$$\int_0^{n^{1/(d+2)}} \mathrm{e}^{-t} t^{\rho-1}\,\mathrm{d}t = \Gamma(\rho) + O(n^{-k})$$

for arbitrary $k > 0$. In the subsequent integral, we have $0 \leq t \leq n^{1/(d+2)}$ and hence $|t\psi(t/n)| \leq c_7$. We observe that $|\mathrm{e}^x - 1| \leq c_8|x|$ for $|x| \leq c_7$, with a constant c_8 depending only on c_7, and apply this with $x = -t\psi(t/n)$. Together with (8.30), this yields

$$\int_0^{n^{1/(d+2)}} \left[1 - \frac{t}{n} - \frac{t}{n}\psi\left(\frac{t}{n}\right)\right]^n t^{\beta-1}\,\mathrm{d}t$$
$$= \int_0^{n^{1/(d+2)}} \mathrm{e}^{-t-t\psi(t/n)} \left(1 + t^2 O(n^{-1})\right) t^{\beta-1}\,\mathrm{d}t$$
$$= \int_0^{n^{1/(d+2)}} \mathrm{e}^{-t} \left(1 + t^{1+1/(d+1)} O\left(n^{-1/(d+1)}\right)\right) \left(1 + t^2 O(n^{-1})\right) t^{\beta-1}\,\mathrm{d}t$$
$$= \int_0^{n^{1/(d+2)}} \mathrm{e}^{-t} t^{\beta-1}\,\mathrm{d}t + O\left(n^{-1/(d+1)}\right) = \Gamma(\beta) + O\left(n^{-1/(d+1)}\right).$$

Altogether we obtain

$$J = \frac{2}{d+1}(an)^{-2/(d+1)} \left[\Gamma\left(\frac{2}{d+1}\right) + O\left(n^{-1/(d+1)}\right)\right].$$

This completes the proof. □

Corollary 8.2.1. *Under the assumptions of Theorem* 8.2.7,

$$\lim_{n\to\infty} n^{2/(d+1)} \delta(K, \mathbb{E}K_n) = c_d V_d(K)^{2/(d+1)} \kappa_{\max}^{1/(d+1)},$$

where $\kappa_{\max}$ *denotes the maximum of the Gauss–Kronecker curvature of* $\operatorname{bd} K$.

Proof. From Theorem 8.2.7 we have

$$\lim_{n\to\infty} \left(\frac{n}{V_d(K)}\right)^{2/(d+1)} [h(K,u) - h(\mathbb{E}K_n, u)] = c_d \kappa(u)^{1/(d+1)} \tag{8.31}$$

uniformly in u. Let $u_0 \in S^{d-1}$ be a vector with $\kappa(u_0) = \kappa_{\max}$. By (8.31), to every $\epsilon > 0$ there is $n_0 \in \mathbb{N}$ with

$$c_d \kappa_{\max}^{1/(d+1)} - \epsilon \leq \left(\frac{n}{V_d(K)}\right)^{2/(d+1)} [h(K, u_0) - h(\mathbb{E}K_n, u_0)],$$

$$\left(\frac{n}{V_d(K)}\right)^{2/(d+1)} [h(K, u) - h(\mathbb{E}K_n, u)] \leq c_d \kappa_{\max}^{1/(d+1)} + \epsilon$$

for $u \in S^{d-1}$ and $n \geq n_0$. Since $\delta(K, \mathbb{E}K_n) = \max_{u \in S^{d-1}} |h(K, u) - h(\mathbb{E}K_n, u)|$, we obtain

$$c_d \kappa_{\max}^{1/(d+1)} - \epsilon \leq \left(\frac{n}{V_d(K)}\right)^{2/(d+1)} \delta(K, \mathbb{E}K_n) \leq c_d \kappa_{\max}^{1/(d+1)} + \epsilon$$

for $n \geq n_0$. The assertion follows. □

The only further result that we present in this section is a general formula for the expectation $\mathbb{E}\,\varphi(\phi, n)$, where ϕ is a probability distribution on $\mathbb{R}^d$ and φ is a function of the form

$$\varphi(P) = \sum_{F \in \mathcal{F}_{d-1}(P)} \eta(F), \qquad P \in \mathcal{P}',$$

with a measurable function η on $(d-1)$-polytopes. Examples of geometric interest are $\eta = 1$ and $\eta(F) = \lambda_{d-1}(F)$, also $\eta(F) = V_d(\operatorname{conv}(F \cup \{0\}))$ if $0 \in P$. Of the probability distribution ϕ we assume that it has a density g with respect to Lebesgue measure.

Let $X_1, \dots, X_n$, $n \geq d+1$, be independent random points in $\mathbb{R}^d$, each with distribution ϕ, and let P_n be their convex hull. By the assumption on ϕ, every facet of P_n is almost surely a $(d-1)$-simplex. Points $X_1, \dots, X_d$ in general position determine a facet of P_n if and only if the points $X_{d+1}, \dots, X_n$ lie on the same side of the hyperplane spanned by $X_1, \dots, X_d$. For fixed $X_1, \dots, X_d$, this happens with probability

$$\phi(H^+(X_1, \dots, X_d))^{n-d} + (1 - \phi(H^+(X_1, \dots, X_d)))^{n-d},$$

where $H^+(X_1, \dots, X_d)$ is one of the two closed halfspaces bounded by the affine hull of $X_1, \dots, X_d$. Since $X_1, \dots, X_n$ are independent and identically distributed, we deduce that

$$\begin{aligned}\mathbb{E}\,\varphi(\phi, n) = \binom{n}{d} \int_{(\mathbb{R}^d)^d} & [\phi(H^+(x_1, \dots, x_d))^{n-d} + (1 - \phi(H^+(x_1, \dots, x_d)))^{n-d}] \\ & \times \eta(\operatorname{conv}\{x_1, \dots, x_d\}) g(x_1) \cdots g(x_d)\, \lambda^d(\mathrm{d}(x_1, \dots, x_d)).\end{aligned}$$

This can be simplified by introducing as an integration variable the hyperplane spanned (almost everywhere) by $x_1, \dots, x_d$. For a hyperplane H, we denote by H^+ one of the two closed halfspaces bounded by H. The affine Blaschke–Petkantschin formula (7.14) now gives

$$\mathbb{E}\,\varphi(\phi, n) = \frac{d!\kappa_d}{2}\binom{n}{d}\int_{A(d,d-1)}[\phi(H^+)^{n-d} + (1-\phi(H^+))^{n-d}]$$
$$\times \int_{H^d} \eta(\mathrm{conv}\{x_1,\dots,x_d\})\lambda_{d-1}(\mathrm{conv}\{x_1,\dots,x_d\})$$
$$\times\, g(x_1)\cdots g(x_d)\,\lambda_H^d(\mathrm{d}(x_1,\dots,x_d))\,\mu_{d-1}(\mathrm{d}H). \tag{8.32}$$

We have mentioned this formula since it is a typical application of a Blaschke–Petkantschin transformation, and since it and closely related formulas have been the starting point for further asymptotic investigations.

Notes for Subsection 8.2.4

1. Areas, perimeters, and vertex numbers in the plane. The investigation of the asymptotic behavior of convex hulls of independent random points was initiated by Rényi and Sulanke [639, 640]. For a convex r-gon P in the plane they showed in [639] that

$$\mathbb{E} f_0(P,n) = \frac{2r}{3}(\log n + C) + \frac{2}{3}\log\frac{\prod_{i=1}^r A_i}{A(P)^r} + o(1) \tag{8.33}$$

as $n \to \infty$, where C denotes Euler's constant, $A(P)$ is the area of P, and A_i is the area of the triangle spanned by the ith vertex of P and its two neighbors. For a planar convex domain K of class C^2 they obtained

$$\mathbb{E} f_0(K,n) \sim \Gamma\left(\frac{5}{3}\right)\left(\frac{2}{3}\right)^{1/3}\int_{\mathrm{bd}\,K}\kappa^{1/3}\,\mathrm{d}s\left(\frac{n}{A(K)}\right)^{1/3},$$

where κ is the curvature and the integration over the boundary is with respect to arc length. In [640] they proved similar asymptotic results for the area and the perimeter in the case of a sufficiently smooth convex body in the plane. For area and perimeter in the case of polygons, where Rényi and Sulanke had treated only squares, general asymptotic relations were established by Buchta [124]. A considerable strengthening of (8.33), in the form of an asymptotic expansion, was proved by Buchta and Reitzner [136].

It took some time until the first central limit theorems for random variables $\varphi(\mu, n)$ were established. For a convex r-gon P, Groeneboom [293] proved that

$$\frac{f_0(P,n) - \frac{2}{3}r\log n}{\sqrt{\frac{10}{27}r\log n}} \xrightarrow{\mathcal{D}} N(0,1)$$

as $n \to \infty$, where $\xrightarrow{\mathcal{D}}$ denotes convergence in distribution and $N(0,1)$ is the standard normal distribution. Massé [453] deduced from this a law of large numbers,

$$\lim_{n\to\infty}\frac{3f_0(P,n)}{2r\log n} = 1 \qquad \text{in probability.}$$

For the area A and an r-gon P, a result by Cabo and Groeneboom [147], modified by Buchta [131] (see also Groeneboom [294]), says that

$$\frac{n[1 - A(P,n)/A(P)] - \frac{2}{3}r \log n}{\sqrt{\frac{28}{27}r \log n}} \xrightarrow{\mathcal{D}} N(0,1).$$

For the circular disk B^2, a central limit theorem for the vertex number $f_0(B^2, n)$ is due to Groeneboom [293] (see Finch and Hueter [232] for additional information on the variance), and one for the missed area to Hsing [351] (see Buchta [131] for information on the appearing constants). A thorough study of the asymptotic properties of the missed area $A(K) - A(\mu, n)$ and the perimeter difference $L(K) - L(\mu, n)$, for rather general distributions μ in a planar convex body K (either sufficiently smooth and with positive curvature, or a polygon), was made by Bräker and Hsing [120].

2. Random points on convex curves. Let K be a planar convex body of class C^2_+. Let μ be a probability distribution on $\mathrm{bd}\, K$ which has a positive continuous density h with respect to the length measure. Let $X_1, X_2 \dots$ be an independent sequence of random points, each with distribution μ. Schneider [694] proved some laws of large numbers, for example,

$$\lim_{n\to\infty} n^2[A(K) - A(\mathrm{conv}\{X_1, \dots, X_n\})] = \frac{1}{2}\int_{\mathrm{bd}\, K} \kappa h^{-2}\, \mathrm{d}s \qquad \text{a.s.},$$

and a similar relation for the perimeter. Here κ denotes the curvature of $\mathrm{bd}\, K$ and $\mathrm{d}s$ indicates integration with respect to the arc length.

3. The floating body approach. Recall, also for the subsequent notes, that

$$K_n := \mathrm{conv}\,\{X_1, \dots, X_n\}$$

for independent random points $X_1, \dots, X_n$ in K, with uniform distribution unless stated otherwise.

For general convex bodies K, a powerful method for investigating the asymptotic behavior of the polytope K_n was developed by Bárány and Larman [71] (see Bárány [61] for a survey, and Bárány [64] for a highly recommended introduction to the method). For convenience, assume $V_d(K) = 1$. For $x \in K$, let $v(x)$ be the minimum of $V_d(K \cap H)$, taken over all closed halfspaces H with $x \in H$. For each sufficiently small $t > 0$, the **floating body** of K with parameter t is the convex body defined by $K[t] := \{x \in K : v(x) \geq t\}$. What Bárány and Larman discovered is that the asymptotic behavior of K_n, with respect to expectations, can very well be compared to that of $K[1/n]$. Instances of this phenomenon are the following. Bárány and Larman showed that

$$c_1[V_d(K) - V_d(K[1/n])] \leq \mathbb{E}[V_d(K) - V_d(K,n)] \leq c_2[V_d(K) - V_d(K[1/n])]$$

for $n \geq n_0$, with constants c_1, c_2, where c_2 depends on d. Bárány [56] proved

$$c_3 n[V_d(K) - V_d(K[1/n])] \leq \mathbb{E} f_i(K,n) \leq c_4 n[V_d(K) - V_d(K[1/n])]$$

for $i = 0, \dots, d-1$, where c_3, c_4 depend on d. For convex bodies containing a ball of radius r and contained in a ball of radius R, Bárány [57] showed that

$$c_5[V_j(K) - V_j(K[1/n])] \leq \mathbb{E}[V_j(K) - V_j(K,n)] \leq c_6[V_j(K) - V_j(K[1/n])]$$

for $j = 1, \ldots, d$, where the constants c_5, c_6 depend on d, r, R. Bárány and Vitale [75] considered the set-valued expectation (selection expectation) of K_n and showed that, with suitable constants $a, b > 0$,

$$K[a/n] \subset \mathbb{E}K_n \subset K[b/n].$$

The order of $V_d(K) - V_d(K[1/n])$ for $n \to \infty$ can be estimated, and it can be determined for special classes of convex bodies. For example (Bárány and Larman [71]),

$$V_d(P) - V_d(P[1/n]) = \Theta(n^{-1} \log^{d-1} n)$$

for polytopes P and

$$V_d(K) - V_d(K[1/n]) = \Theta(n^{-2/(d+1)})$$

for K of class C^2. (Here and below, $f(K,n) = \Theta(g(K,n))$ means that $c_1 g(K,n) < f(K,n) < c_2 g(K,n)$ for all n, with constants $c_1, c_2 > 0$ independent of n, but possibly depending on d and K.) For applications to $\mathbb{E}\varphi(K,n)$, see the subsequent notes. Another application appears in Bárány [63].

4. Support function and mean width. Theorem 8.2.7 determines the asymptotic behavior of the support function $h(K_n, u)$ at given u, for independent uniform random points in a smooth convex body K. The proof is taken from Schneider and Wieacker [718], where the result is formulated for the mean width, that is, twice the mean value of the support function over all directions. In the final part of the proof, we modified an argument of Ziezold [837], who treated an extension of the result to more general distributions. Under stronger smoothness assumptions on K, an asymptotic expansion for $h(K,u) - \mathbb{E}h(K_n, u)$ was derived by Gruber [296].

Schreiber [723] showed

$$\mathbb{E}[V_1(B^d) - V_1(B^d, n)]^k = (c_d)^k n^{-2k/(d+1)} + o(n^{-2k/(d+1)})$$

for all $k \in \mathbb{N}$ and

$$\lim_{n\to\infty} n^{2/(d+1)} [V_1(B^d) - V_1(B^d, n)] = c_d \qquad \text{in probability,}$$

with an explicitly given constant c_d (note that V_1 is $d\kappa_d/2\kappa_{d-1}$ times the mean width).

5. Hausdorff distance. In contrast to the Hausdorff distance between K and the set-valued expectation of K_n, namely

$$\delta(K, \mathbb{E}K_n) = \max_{u \in S^{d-1}} \mathbb{E}[h(K,u) - h(K_n, u)],$$

to which the previous note refers, the expected Hausdorff distance of K and K_n, given by

$$\mathbb{E}\delta(K, K_n) = \mathbb{E} \max_{u \in S^{d-1}} [h(K,u) - h(K_n, u)],$$

is more delicate. For a planar convex body K, either smooth or a polygon, and under suitable assumptions on the distribution, the asymptotic behavior of $\delta(K, K_n)$ was investigated by Bräker, Hsing and Bingham [121]. For K of class C^2_+ in $\mathbb{R}^d$, Bárány [56] showed that

$$\mathbb{E}\delta(K, K_n) = \Theta\left((n^{-1}\log n)^{2/(d+1)}\right).$$

The following results hold for independent uniform random points on the boundary of a convex body K. We write

$$C_n(\mu) := \operatorname{conv}\{X_1, \dots, X_n\}$$

if $X_1, \dots, X_n$ are independent, identically distributed random points with distribution μ. For a planar convex body K of class C^2_+, Schneider [694] showed, under the same assumptions on μ as in Note 2, that

$$\lim_{n\to\infty} \left(\frac{n}{\log n}\right)^2 \delta(K, C_n(\mu)) = \frac{1}{8} \max \frac{\kappa}{h^2} \qquad \text{a.s.}$$

Dümbgen and Walther [211] proved that $\delta(K, K_n)$ is of order $O((n^{-1}\log n)^{1/(d-1)})$ for general K in $\mathbb{R}^d$, and of order $O((n^{-1}\log n)^{2/(d-1)})$ under a smoothness assumption. For K of class C^3_+ and for a distribution μ on $\operatorname{bd} K$ with a continuous positive density h, Glasauer and Schneider [269] obtained a weak law of large numbers, namely

$$\lim_{n\to\infty} \left(\frac{n}{\log n}\right)^{2/(d-1)} \delta(K, C_n(\mu)) = \frac{1}{2}\left(\frac{1}{\kappa_{d-1}} \max \frac{\sqrt{\kappa}}{h}\right)^{2/(d-1)} \qquad \text{in probability,}$$

where κ is the Gauss–Kronecker curvature.

6. Diameter. Mayer and Molchanov [467] found the limit distribution for $D(B^d, n)$, where D denotes the diameter.

7. For independent uniform random points in a convex body K, Bárány and Dalla [69] have shown that, with high probability, K_n can be obtained by taking the convex hull of $m = o(n)$ points chosen independently and uniformly from a small neighborhood of the boundary of K. A similar investigation for the circular disk was made before by Carnal and Hüsler [163].

8. Intrinsic volumes and face numbers. Early extensions of some of the asymptotic results of Rényi and Sulanke to higher dimensions appear in Efron [215] (formulas of type (8.32)), Raynaud [623], Wieacker [814] (unpublished). Raynaud determined the asymptotic behavior of $\mathbb{E}f_{d-1}(B^d, n)$, and Wieacker did the same (by a different method) for $\mathbb{E}f_{d-1}(K, n)$ if K is smooth, and for $\mathbb{E}V_j(B^d, n)$ for $j = d$ (volume) and $j = d-1$ (surface area). This was later generalized considerably, as follows.

In this note we assume

$$V_d(K) = 1$$

for the considered convex bodies K. Without this assumption, some of the asymptotic formulas have to be modified by inserting suitable powers of $V_d(K)$.

For a d-dimensional convex body K of class C^3_+ and for the intrinsic volume V_j, $j \in \{1, \dots, d\}$, the relation

$$\begin{aligned}&\mathbb{E}[V_j(K) - V_j(K, n)] \\ &= c_{d,j} \int_{\operatorname{bd} K} \kappa^{1/(d+1)} H_{d-j}\, \mathrm{d}S \cdot n^{-2/(d+1)} + O\left(n^{-3/(d+1)} \log^2 n\right) \qquad (8.34)\end{aligned}$$

holds, where $c_{d,j}$ is a constant depending only on d and j, H_{d-j} denotes the $(d-j)$th normalized elementary symmetric function of the principal curvatures of bd K, and dS indicates integration with respect to the surface area. This result is due to Bárány [57] (with the form of the integral given by Reitzner [631]). It was extended by Böröczky, Hoffmann and Hug [115], to convex bodies admitting a freely rolling ball. For $j = d$, (8.34) yields the limit relation

$$\lim_{n \to \infty} n^{2/(d+1)} \mathbb{E}[V_d(K) - V_d(K,n)] = c_{d,d} \int_{\mathrm{bd}\, K} \kappa^{1/(d+1)} \,\mathrm{d}S, \tag{8.35}$$

where the integral on the right side is the **affine surface area** of K, which is denoted by $\Omega(K)$. This relation was extended by Schütt [724] to arbitrary convex bodies of volume one, with the notion of Gauss–Kronecker curvature generalized appropriately. For a convex body K of class C_+^{k+3} and with volume one, Reitzner [631] obtained an asymptotic expansion

$$\mathbb{E}[V_j(K) - V_j(K,n)] = c_{d,j}^{(2)} n^{-2/(d+1)} + \ldots + c_{d,j}^{(k)} n^{-k/(d+1)} + O\left(n^{-(k+1)/(d+1)}\right)$$

as $n \to \infty$, with additional information on the coefficients (for $d = j = 2$, see also Reitzner [627]).

For a d-dimensional polytope P, the relation

$$\mathbb{E}[V_1(P) - V_1(P,n)] \sim c(P) n^{-1/d} \qquad \text{as } n \to \infty$$

was proved by Schneider [692]. The constant $c(P)$ is expressed as an integral that depends only on arbitrarily small neighborhoods of the vertices of P. For a simple polytope P with r vertices, Affentranger and Wieacker [12] deduced from their more general results that

$$\mathbb{E}[V_d(P) - V_d(P,n)] = V_d(P) \frac{rd}{(d+1)^{d-1}} \frac{\log^{d-1} n}{n} + O\left(n^{-1} \log^{d-2} n\right)$$

and derived an asymptotic relation for the vertex number. These results were extended to arbitrary d-polytopes by Bárány and Buchta [68] (announced in [67]). They showed that

$$\mathbb{E} f_0(P,n) = \frac{T(P)}{(d+1)^{d-1}(d-1)!} \log^{d-1} n + O\left(\log^{d-2} n \log\log n\right). \tag{8.36}$$

Here $T(P)$ denotes the number of **towers** of P, that is, of chains $F_0 \subset F_1 \subset \ldots \subset F_{d-1}$, where F_i is an i-dimensional face of P. Bárány and Buchta also mentioned without proof that their method would allow one to extend (8.36) to $\mathbb{E} f_i(P,n)$ for $i = 0, \ldots, d-1$, with the denominator replaced by a constant depending only on d and i. For the intrinsic volumes $V_2, \ldots, V_{d-1}$, only the estimate

$$\mathbb{E}[V_j(P) - V_j(P,n)] = \Theta\left(n^{-1/(d-j+1)}\right),$$

due to Bárány [56], seems to be known.

Reitzner [632] developed a new method for computing face numbers and deduced the following results. For a polytope P, he proved the asymptotic relation

$$\mathbb{E} f_i(P,n) = \bar{c}_{d,i} T(P) \log^{d-1} n + O\left(\log^{d-2} n \log n\right)$$

with $\bar{c}_{d,i} > 0$. For K of class C^2_+, he obtained

$$\mathbb{E} f_i(K,n) = c_{d,i} \Omega(K) n^{(d-1)/(d+1)} + o\left(n^{(d-1)/(d+1)}\right),$$

where $\Omega(K)$ is the affine surface area and the $c_{d,i}$ are positive constants. Both results replace, by precise asymptotic relations, corresponding estimates of the orders that had been obtained earlier by Bárány [57]. Reitzner also obtained a law of large numbers,

$$\lim_{n\to\infty} f_i(K,n) n^{-(d-1)/(d+1)} = c_{d,i} \Omega(K) \qquad \text{in probability.}$$

For convex bodies K more general than those of class C^2, namely admitting a ball of positive radius rolling freely inside K, Böröczky, Fodor, Reitzner and Vígh [113] have extended the asymptotic relation for the expected mean width of K_n from Schneider and Wieacker [718] (see Note 4) and have proved a corresponding strong law of large numbers,

$$\lim_{n\to\infty} [V_1(K) - V_1(K,n)] n^{2/(d+1)} = c_d \int_{\mathrm{bd}\, K} \kappa^{(d+2)/(d+1)} \, \mathrm{d}S \qquad \text{a.s.},$$

with an explicitly given positive constant c_d.

9. Central limit theorems and deviation estimates. Calka and Schreiber [154] proved a large deviation estimate for $f_0(B^d, n)$.

Essential progress towards central limit theorems for $\varphi(K,n)$ in higher dimensions began with two papers of Reitzner. In [630], he used the Efron–Stein jackknife inequality for obtaining a variance estimate,

$$\operatorname{Var} V_d(K,n) \le c_1(K) n^{(d+3)/(d+1)},$$

for convex bodies K of class C^2_+, and he deduced a strong law of large numbers. (For $K = B^d$, the upper estimate for the variance was obtained before by Küfer [429].) In Reitzner [633], under the same assumptions on K, a lower bound for the variance of the same order was proved. Further, instead of K_n, Reitzner considered a stationary Poisson point process Π of intensity $n/V_d(K)$ in $\mathbb{R}^d$, and for the volume of the convex hull of the points of Π falling in K, denoted by K_n^Π, he showed that

$$\left| \mathbb{P}\left(\frac{V_d(K_n^\Pi) - \mathbb{E} V_d(K_n^\Pi)}{\sqrt{\operatorname{Var} V_d(K_n^\Pi)}} \le x \right) - \Phi(x) \right| \le c_2(K) n^{-(d+3)/(d+1)} \log^{2d+4} n$$

for all x. Here Φ is the distribution function of the standard normal distribution. For the proof, he used a central limit theorem of Rinott for weakly dependent random variables, involving their dependency graph. Reitzner's result can be transferred to a central limit theorem for $V_d(K,n)$,

$$\left| \mathbb{P}\left(\frac{V_d(K,n) - \mathbb{E} V_d(K,n)}{\sqrt{\operatorname{Var} V_d(K,n)}} \le x \right) - \Phi(x) \right| \le \epsilon(n)$$

for all x, with $\epsilon(n) \to 0$ for $n \to \infty$. In full generality, this was proved by Vu [774], who used Reitzner's result and his own strong tail estimates for geometric random variables such as $V_d(K) - V_d(K,n)$, which he had obtained in [773]. In that paper,

Vu proved a deviation estimate of the following form. Let a convex body K and a (sufficiently small) number $\epsilon > 0$ be given. There exist numbers $A, B > 0$, depending on ϵ and on the behavior of K near its boundary, and there are positive constants α, c, ϵ_0 such that, for $\alpha n^{-1} \log n < \epsilon \le \epsilon_0$ and $0 < \lambda \le B/4A^2$, one has

$$\mathbb{P}(|V_d(K) - \mathbb{E}V_d(K,n)| \ge \sqrt{B\lambda}) \le 2\mathrm{e}^{-\lambda/4} + \mathrm{e}^{-c\epsilon n}.$$

In the geometric part of the proof, floating bodies play an essential role.

For polytopes P, Bárány and Reitzner [73] showed that $V_d(P) - V_d(P,n)$ satisfies a central limit theorem.

Results corresponding to those for the volume functionals were also obtained for other functionals, so for the vertex number in Reitzner [630], and for the number f_i of i-faces in Reitzner [633], Bárány and Reitzner [73], Vu [774].

10. General convex bodies. If the asymptotic behavior of $\mathbb{E}\varphi(K,n)$, as $n \to \infty$, is of different orders for polytopes and for smooth convex bodies, then it follows from a general principle (see Gruber [295]) that for most convex bodies (in the sense of Baire category) the asymptotic behavior oscillates between these orders and is, thus, highly irregular. (Relation (8.35), which holds for general convex bodies, is consistent with this, since the affine surface area vanishes for most convex bodies.) More precise formulations, in several instances, are found in Schneider [692, p. 305], Bárány and Larman [71, Th. 5], Bárány [56, Cor. 3], [64, Th. 4.7]. Therefore, for general convex bodies, the aim can only be to obtain lower and upper estimates which are asymptotically of optimal order. Bárány [56] showed that

$$c_1(d)(\log n)^{d-1} < \mathbb{E}f_i(K,n) < c_2 n^{(d-1)/(d+1)} \tag{8.37}$$

for $i = 0, \dots, d-1$, with positive constants $c_1(d)$, $c_2(d)$, where the orders are best possible. For the intrinsic volumes, it is only known that, with positive constants $a_i(K)$,

$$a_1(K)n^{-2/(d+1)} < \mathbb{E}[V_1(K) - V_1(K,n)] < a_2(K)n^{-1/d}$$

(Schneider [692]) and

$$a_3(K)n^{-1}\log^{d-1} n < \mathbb{E}[V_d(K) - V_d(K,n)] < a_4(K)n^{-2/(d+1)},$$

as follows from Bárány and Larman [71] or from the case $i = 0$ of (8.37) together with (8.12). Note that polytopes and smooth bodies switch their roles here, which suggests that the intermediate cases may be difficult.

11. Random points on the boundary. Let K be a convex body with interior points and let μ_h be a distribution on its boundary which has a positive, continuous density h with respect to the normalized surface area measure (in particular, μ_1 denotes μ_h for $h \equiv 1$). Some results on $\varphi(\mu_1, n)$ were already mentioned in Notes 2 and 5. For smooth K, Buchta, Müller and Tichy [134] proved an asymptotic relation for $\mathbb{E}[V_1(K) - V_1(\mu_1,n)]$, and they sketched similar results for V_{d-1} and f_{d-1} in the case of B^d. Complete proofs for V_{d-1} and V_d were given by Müller [569], who in [568] also studied $\mathbb{E}[V_1(K) - V_1(\mu_h,n)]$ for smooth K.

For K of class C^2_+ and for $j \in \{1, \dots, d\}$, Reitzner [628] showed that

$$\begin{aligned}&\mathbb{E}[V_j(K) - V_j(\mu_h,n)]\\ &= b_{d,j}\int_{\mathrm{bd}\,K} \kappa^{1/(d-1)} H_{d-j} h^{-2/(d-1)}\,\mathrm{d}S \cdot n^{-2/(d-1)} + o\left(n^{-2/(d-1)}\right)\end{aligned}$$

as $n \to \infty$. Under stronger differentiability assumptions on K and h, he obtained an asymptotic expansion with more terms.

An asymptotic formula in the case of volume, for convex bodies that need not be of class C^2, was established by Schütt and Werner [727] (announced in [726]). With a long and difficult proof they showed that

$$\lim_{n\to\infty} n^{2/(d-1)}[V_d(K) - \mathbb{E}V_d(\mu_h, n)] = b(d) \int_{\mathrm{bd}\, K} \kappa^{1/(d-1)} h^{-2/(d-1)}\, \mathrm{d}S,$$

if the lower and upper curvatures of K are between two fixed positive bounds.

For K of class C^2_+ and the random variable $V_d(\mu_h, n)$, Richardson, Vu and Wu [643] established the precise order of the variance and a concentration result, from which they deduced that the kth moment of $V_d(\mu_h, n)$, $k \geq 2$, is of order $O((n^{-(d+3)/(d-1)})^{k/2})$. Further, they obtained a central limit theorem for $V_d(\operatorname{conv} \Pi_n)$, where Π_n is a Poisson process with intensity measure equal to n times the normalized boundary measure of K.

12. Comparison of random and best approximation. There are several instances where it can be observed, and made precise, that random approximation is of the same order of magnitude as best approximation. Different aspects of this phenomenon are described in Bárány [58], Gruber [297], Reitzner [629], Schütt [725].

13. Giannopoulos and Tsolomitis [259] studied the volume radius, $\varphi(C) := V_d(C)^{1/d}$, and gave lower and upper estimates for $\mathbb{E}\varphi(K, n)$ if n satisfies certain lower and upper estimates depending on the dimension.

14. Normally distributed random points. Let ν denote the standard normal distribution in $\mathbb{R}^d$, and let G_n be the convex hull of n independent random points with distribution ν. Geffroy [254] determined the radius of a ball B_n with center 0 such that a.s. $\lim_{n\to\infty} \delta(G_n, B_n) = 0$. Efron [215] gave integral representations of $\mathbb{E}\varphi(\nu, n)$ for some functions φ in two and three dimensions. Rényi and Sulanke [639] found $\mathbb{E}\, f_{d-1}(\nu, n)$ for $d = 2$ and Raynaud [623] for arbitrary d. Compare also Note 4 of Section 8.3. Groeneboom [293] mentioned that his methods would yield a central limit theorem for $f_0(\nu, n)$ in the plane. Concerning central limit theorems by Hueter [352, 353], see the remark in Bárány and Vu [76].

Hug, Munsonius and Reitzner [364] determined, among other results, the asymptotic behavior of $\mathbb{E}\varphi(\nu, n)$, where $\varphi(P)$ is either the number or the total k-dimensional volume of the k-faces of the polytope P. Analogous results for the intrinsic volumes had been obtained before by Affentranger [10]. Hug and Reitzner [365] proved laws of large numbers for $f_i(\nu, n)$ and $V_j(\nu, n)$, after establishing estimates for the variances by using the Efron–Stein jackknife inequality.

Bárány and Vu [76] succeeded in proving central limit theorems for $V_d(\nu, n)$ and for $f_i(\nu, n)$.

15. Points with rotationally symmetric distributions. Some of the random variables $\varphi(\mu, n)$ have been studied for more general distributions μ with spherical symmetry. See Carnal [162], Kaltenbach [388], Affentranger [10], Aldous, Fristedt, Griffin and Pruitt [14], Devroye [203], Dwyer [213], Chu [176], Hueter [354], Massé [452, 453], and see the book by Mathai [456].

16. Random 0-1-polytopes and generalizations. A random 0-1-polytope with n vertices is the convex hull of n independent randomly chosen vertices (with equal probabilities) of the d-dimensional unit cube. These polytopes pose interesting problems, though somewhat outside the scope of this book. We mention contributions by Füredi [240], Dyer, Füredi and McDiarmid [214], Bárány and Pór [72], Giannopoulos and Hartzoulaki [258], Gatzouras, Giannopoulos and Markoulakis [252, 253], Mendelson, Pajor and Rudelson [512], Gatzouras and Giannopoulos [250, 251]. Some of these papers consider the symmetric (or absolutely) convex hull, or they consider more general random points with independent coordinates; see also Litvak, Pajor, Rudelson and Tomczak–Jaegermann [441].

17. Intersections of random halfspaces. As an alternative to taking the convex hull of random points, a random polytope can be generated by taking the intersection of random closed halfspaces. One instance where such random polytopes appear is the average case analysis of linear programming algorithms. Closer to some of the topics described in the previous notes, and somehow 'dual' (though not in a strict sense), is the question of how well a given convex body is approximated by the intersection of n independent random closed halfspaces containing the body, as n tends to infinity. For the case of the plane, we refer to Carlsson and Grenander [161], Rényi and Sulanke [641], Ziezold [836], Schneider [694]; for higher dimensions, see Kelly and Tolle [394], Buchta [128], Kaltenbach [388], Reitzner [628], Böröczky and Reitzner [116], Böröczky and Schneider [117].

A different version of approximation by circumscribed random convex bodies is studied by Small [730], who considers the intersection of a finite (and increasing) number of cylinders circumscribed about a convex body with independent random directions.

18. Random polytopes in asymptotic convexity. In high-dimensional convex geometry, random constructions often yield examples or counterexamples. However, for the notorious problem of whether the isotropic constant of convex bodies is bounded by a constant independent of dimension, two different approaches using random polytopes have not produced counterexamples. For the convex hull of n independent standard Gaussian random points in $\mathbb{R}^d$, Klartag and Kozma [417] proved that with high probability the isotropic constant is bounded by a universal constant. Dafnis, Giannopoulos and Guédon [192] obtained a similar result for the convex hull and the symmetric convex hull of n independent uniform random points in an isotropic 1-unconditional convex body.

19. The highly recommended survey article by Bárány [66] presents many of the more recent developments about random points in convex bodies, in particular those touched upon in the previous Notes 1, 3, 8, 9, 10. It also treats lattice points in convex bodies, emphasizing the analogies in the asymptotic behavior of convex hulls of independent random points and of lattice points.

8.3 Random Projections of Polytopes

In this section, we introduce a totally different way of generating a d-dimensional random polytope. Every d-dimensional polytope with $N+1 \geq d+1$ vertices is affinely equivalent to an orthogonal projection of an N-dimensional regular simplex (e.g., McMullen and Shephard [475, p. 121]). If one is only interested in affine properties of polytopes, this suggests the following natural way of defining random polytopes. Consider $\mathbb{R}^d$ as a subspace of $\mathbb{R}^N$, $N > d$, and let T^N be a regular simplex in $\mathbb{R}^N$. Let ϑ be a random rotation of $\mathbb{R}^N$, that is, a random element of SO_N with distribution equal to the invariant probability measure on SO_N. Then the orthogonal projection of ϑT^N to $\mathbb{R}^d$ defines a random polytope in $\mathbb{R}^d$. Similarly, every centrally symmetric d-polytope with $2N \geq 2d$ vertices is a projection of an N-dimensional crosspolytope (Grünbaum [299, p. 72]). This suggests replacing the regular simplex T^N by a regular crosspolytope Q^N, to obtain a random centrally symmetric polytope in $\mathbb{R}^d$. Up to (inessential) rigid motions, the same random polytopes are obtained if we orthogonally project the fixed simplex T^N (respectively, the crosspolytope Q^N) to a uniform random d-dimensional subspace of $\mathbb{R}^N$; we prefer here the latter interpretation. This approach to random polytopes is also known as the 'Grassmann approach', since it involves the Grassmannian $G(N,d)$ and its invariant probability measure, which we denote here by $\nu_{N,d}$. We shall derive basic expectation formulas for this model and then describe an interesting application. We restrict ourselves, however, to the geometric background, since the analytic details are outside the scope of this book.

First we consider projections of an arbitrary convex polytope. Let $N > d$, and let P be an N-dimensional convex polytope in $\mathbb{R}^N$. Let L be a d-dimensional linear subspace of $\mathbb{R}^N$. Recall that $P|L$ denotes the image of P under orthogonal projection to L (we use this notation also in $\mathbb{R}^N$). We are interested in $f_k(P|L)$, the number of k-dimensional faces of the d-polytope $P|L$, for $k = 0, \dots, d-1$. We suppose that L is in general position with respect to P, which means that $\dim F|L = k$ for each k-face F of P and each $k \in \{1, \dots, d\}$. It follows from Lemma 13.2.1 that $\nu_{N,d}$-almost all $L \in G(N,d)$ are in general position with respect to P.

For a closed convex cone C and a linear subspace E in $\mathbb{R}^N$ we write

$$\xi(C,E) := \begin{cases} 1, \text{ if } E \cap C = \{0\}, \\ 0, \text{ if } E \cap C \neq \{0\}. \end{cases}$$

Let $k \in \{0, \dots, d-1\}$ and let F be a k-face of P. Choose a point $z \in \operatorname{relint} F$ and let $C(P,F)$ be the cone spanned by $P - z$; this cone does not depend on the choice of z. We have $\dim F|L = k$, and $F|L$ is a face of $P|L$ if and only if the flat $L^\perp + z$ lies in a supporting hyperplane of P. Since $(L^\perp + z) \cap F = \{z\}$ by general position, this is equivalent to $\xi(C(P,F), L^\perp) = 1$. It follows that

$$f_k(P|L) = \sum_{F \in \mathcal{F}_k(P)} \xi(C(P,F), L^\perp).$$

Now let X be a uniform random d-subspace of $\mathbb{R}^N$, that is, a random element of $G(N,d)$ with distribution $\nu_{N,d}$. Then the previous considerations show that

$$\mathbb{E}\, f_k(P|X) = \sum_{F\in\mathcal{F}_k(P)} \gamma^{N-d,N}(P,F)$$

with

$$\gamma^{N-d,N}(P,F) := \int_{G(N,N-d)} \xi(C(P,F),E)\, \nu_{N,N-d}(\mathrm{d}E).$$

We apply some spherical integral geometry from Section 6.5. Introducing the spherical polytope

$$S(P,F) := C(P,F) \cap S^{N-1}$$

and using (6.62), (6.63), (6.50), we obtain

$$\begin{aligned}
\gamma^{N-d,N}(P,F) &= 1 - \int_{G(N,N-d)} \chi(S(P,F)\cap E)\, \nu_{N,N-d}(\mathrm{d}E)\\
&= 1 - 2U_d(S(P,F))\\
&= 1 - 2\sum_{s\geq 0} v_{d+2s}(S(P,F))\\
&= 1 - 2\sum_{s\geq 0}\sum_{G\in\mathcal{F}_{d+1+2s}(C(P,F))} \beta(0,G)\gamma(G,C(P,F))\\
&= 1 - 2\sum_{s\geq 0}\sum_{G\in\mathcal{F}_{d+1+2s}(P)} \beta(F,G)\gamma(G,P).
\end{aligned}$$

Alternatively, we may use (6.55), (6.53) to get

$$1 - 2\sum_{s\geq 0} v_{d+2s}(S(P,F)) = 2\sum_{s\geq 0} v_{d-2-2s}(S(P,F)),$$

which gives

$$\gamma^{N-d,N}(P,F) = 2\sum_{s\geq 0}\sum_{G\in\mathcal{F}_{d-1-2s}(P)} \beta(F,G)\gamma(G,P).$$

Thus, we have obtained the following result.

Theorem 8.3.1. *Let $N > d$, let P be a convex polytope in $\mathbb{R}^N$, and let X be a uniform random d-dimensional subspace of $\mathbb{R}^N$. Then, for $k \in \{0,\dots,d-1\}$,*

$$\begin{aligned}
\mathbb{E} f_k(P|X) &= f_k(P) - 2\sum_{s\geq 0}\sum_{F\in\mathcal{F}_k(P)}\sum_{G\in\mathcal{F}_{d+1+2s}(P)} \beta(F,G)\gamma(G,P)\\
&= 2\sum_{s\geq 0}\sum_{F\in\mathcal{F}_k(P)}\sum_{G\in\mathcal{F}_{d-1-2s}(P)} \beta(F,G)\gamma(G,P). \qquad (8.38)
\end{aligned}$$

The internal and external angles appearing in the preceding formulas are defined as spherical volumes of certain spherical polytopes, hence they can in general not be computed explicitly in concrete cases. An exception is the (regular) cube C^N, for which all internal and external angles are powers of 1/2, and it is not difficult to show that

$$\mathbb{E}f_k(C^N|X) = 2\binom{N}{k}\sum_{s\geq 0}\binom{N-k}{d-1-2s-k}. \tag{8.39}$$

For the regular simplex T^N we obtain, for example,

$$\begin{aligned}&\mathbb{E}f_k(T^N|X)\\ &= 2\sum_{s\geq 0}\binom{N+1}{d-2s}\binom{d-2s}{k+1}\beta(T^k,T^{d-2s-1})\gamma(T^{d-2s-1},T^N).\end{aligned} \tag{8.40}$$

Here $\beta(T^p,T^q)$ is the internal angle and $\gamma(T^p,T^q)$ is the external angle of a regular q-simplex at one of its p-faces ($p<q$).

Random projections of regular polytopes have found applications of a kind that we briefly indicate. Various practical tasks lead to the following reconstruction problem. Let $N,d,k\in\mathbb{N}$ be given numbers with $d,k<N$. Let S be an $(N-d)$-dimensional linear subspace of $\mathbb{R}^N$. Suppose we are given a vector $y'\in\mathbb{R}^N$, and we have to find a vector $y\in S$ that differs from y' in at most k coordinates, or a good approximation of such a vector. A surprisingly good strategy consists in constructing a point y in S that is nearest to y' in the ℓ_1^N-norm $\|\cdot\|_1$. This approach is good since for many subspaces S (in a sense to be made precise) it yields a correct solution. To investigate this phenomenon more closely, we formulate the following property.

Definition 8.3.1. *Let $k\in\{1,\dots,N\}$. The subspace S has* **property $\mathcal{U}_k$** *if the following holds. Whenever $y\in S$ and $y'\in\mathbb{R}^N$ are such that they differ in at most k coordinates, then the optimization problem*

minimize $\|x-y'\|_1$ subject to the condition $x\in S$

has a unique solution and this is equal to y.

The unit ball of the ℓ_1^N-norm in $\mathbb{R}^N$ is the regular crosspolytope $Q^N:=\operatorname{conv}\{\pm e_1,\dots,\pm e_N\}$, where $(e_1,\dots,e_N)$ is the standard orthonormal basis of $\mathbb{R}^N$. The polytope Q^N is centrally N-neighborly. For $k\in\mathbb{N}$, a centrally symmetric polytope P is called **centrally k-neighborly** if every subset of k vertices of P, not containing a pair of opposite vertices, is the set of vertices of a $(k-1)$-face of P (necessarily a $(k-1)$-dimensional simplex). It turns out that property $\mathcal{U}_k$ has to do with neighborliness properties of projections of Q^N.

Lemma 8.3.1. *Let $k \in \{1, \ldots, N\}$. The subspace S has property $\mathcal{U}_k$ if and only if the polytope $Q^N|S^\perp$ satisfies $f_{k-1}(Q^N|S^\perp) = f_{k-1}(Q^N)$, equivalently, it has $2N$ vertices and is centrally k-neighborly.*

Proof. Let F be a $(k-1)$-face of Q^N, and choose $z \in F$. Then z lies in the intersection of Q^N with some k-dimensional coordinate subspace. Therefore, any point $y \in S$ and the point $y' := y + z$ differ in at most k coordinates. Suppose that the linear subspace S has property $\mathcal{U}_k$. Then S touches the crosspolytope $Q^N + y'$ at the unique point y. Equivalently, $S + z$ touches Q^N only at z. Thus, to any point z in a $(k-1)$-face of Q^N, there exists a translate of the subspace S that touches Q^N only at z. It follows that the projection $F|S^\perp$ is a $(k-1)$-face of the centrally symmetric polytope $Q^N|S^\perp$. Since F was an arbitrary $(k-1)$-face of Q^N, we deduce that $f_{k-1}(Q^N|S^\perp) = f_{k-1}(Q^N) = 2^k\binom{N}{k}$. The centrally symmetric polytope $Q^N|S^\perp$ has $2v \le 2N$ vertices and, therefore, at most $2^k\binom{v}{k}$ faces of dimension $k-1$. Hence, $Q^N|S^\perp$ has $2N$ vertices, and any k of these vertices without an antipodal pair determine a $(k-1)$-face of $Q^N|S^\perp$. The arguments can be reversed (replacing Q^N by $\|y' - y\|_1 Q^N$). □

This equivalence will be useful if 'many' subspaces S have property $\mathcal{U}_k$. In particular, we hope that for a random subspace with uniform distribution, the required condition is satisfied with high probability.

Therefore, we consider a d-dimensional uniform random subspace X of $\mathbb{R}^N$. We are interested in those realizations X for which the polytope $Q^N|X$ has the same number of $(k-1)$-faces as Q^N, and hence is centrally k-neighborly. Since

$$
\begin{aligned}
&\mathbb{E}\left(f_{k-1}(Q^N) - f_{k-1}(Q^N|X)\right) \\
&= \int \mathbf{1}\{f_{k-1}(Q^N) > f_{k-1}(Q^N|X)\}\left(f_{k-1}(Q^N) - f_{k-1}(Q^N|X)\right) \mathrm{d}\mathbb{P} \\
&\ge \mathbb{P}\left(f_{k-1}(Q^N) > f_{k-1}(Q^N|X)\right),
\end{aligned}
$$

we have

$$
\mathbb{P}\left(f_{k-1}(Q^N|X) < f_{k-1}(Q^N)\right) \le f_{k-1}(Q^N) - \mathbb{E} f_{k-1}(Q^N|X). \tag{8.41}
$$

This shows that for obtaining an upper estimate for the probability that the random subspace X does not have property $\mathcal{U}_k$, we need information on the expected number $\mathbb{E} f_{k-1}(Q^N|X)$ of $(k-1)$-faces of the random polytope $Q^N|X$. From (8.41) and (8.38), and using elementary information on the faces and face numbers of the regular crosspolytope, we get

$$
\begin{aligned}
&\mathbb{P}\left(f_{k-1}(Q^N|X) < f_{k-1}(Q^N)\right) \\
&\le 2 \sum_{s \ge 0} \sum_{F \in \mathcal{F}_{k-1}(Q^N)} \sum_{G \in \mathcal{F}_{d+1+2s}(Q^N)} \beta(F,G)\gamma(G,Q^N)
\end{aligned}
$$

$$= 2\sum_{s\geq 0} 2^\ell \binom{N}{k}\binom{N-k}{\ell-k+1}\beta(T^{k-1},T^\ell)\gamma(T^\ell,Q^N) \tag{8.42}$$

with with $\ell := d+1+2s$. The expression (8.42) is the starting point for asymptotic investigations; see Note 7 below.

Notes for Section 8.3

1. The Grassmann approach was suggested by J.E. Goodman and R. Pollack in the following general formulation. Every configuration of $N+1$ numbered points in general position in $\mathbb{R}^d$ is affinely equivalent to the orthogonal projection of the set of numbered vertices of a fixed regular simplex T^N in $\mathbb{R}^N$ into a unique d-dimensional linear subspace of $\mathbb{R}^N$. In this way, one obtains a one-to-one correspondence between the orientation-preserving equivalence classes of such configurations and an open dense subset of the Grassmannian $G(N,d)$. The invariant probability measure on $G(N,d)$ can then be transferred to a natural probability distribution on the mentioned set of equivalence classes of $(N+1)$-point configurations.

Independently, versions of the Grassmann approach were proposed by Vershik, see [764], [763]. Vershik and Sporyshev [765, 766] used it for an asymptotic upper estimate for the average number of steps required by a version of the simplex algorithm, with a fixed number of constraints and the number of variables tending to infinity.

2. For the external angles of regular simplices appearing in (8.40), an integral representation for volumes of regular spherical simplices due to Ruben [650] can be used. It yields

$$\gamma(T^{d-1},T^N) = \sqrt{\frac{d}{\pi}}\int_{-\infty}^{\infty} \mathrm{e}^{-dt^2}\left(\frac{1}{\sqrt{\pi}}\int_{-\infty}^{t}\mathrm{e}^{-s^2}\,\mathrm{d}s\right)^{N-d+1}\mathrm{d}t. \tag{8.43}$$

Short proofs are found in Hadwiger [311, sect. 3.3] and Böhm and Hertel [110, p. 283]. Using this formula, Affentranger and Schneider [11] deduced from (8.40) the asymptotic relation

$$\mathbb{E}f_k(T^N|X) \sim \frac{2^d}{\sqrt{d}}\binom{d}{k+1}\beta(T^k,T^{d-1})(\pi\log N)^{(d-1)/2}, \tag{8.44}$$

as $N\to\infty$. It describes the asymptotic behavior of $\mathbb{E}f_k(T^N|X)$ for large N, though with a factor involving the ‘unknown’ constant $\beta(T^k,T^{d-1})$. For this, Böröczky and Henk [114] obtained the asymptotic formula

$$\beta(T^k,T^{d-1}) = \frac{(k+1)^{\frac{d-k-2}{2}}\mathrm{e}^{\frac{d-3k-3}{2}}}{\sqrt{2}^{d-k}\sqrt{\pi}^{d-k-1}d^{\frac{d-k-2}{2}}}\left(1+O\left(\frac{k^2+1}{d}\right)\right).$$

The relation

$$\mathbb{E}f_k(T^N|X^\perp) \sim \binom{N+1}{k+1} = f_k(T^N)$$

as $N\to\infty$, is also proved in Affentranger and Schneider [11].

3. An analog of (8.43) for the external angle of the regular crosspolytope Q^N at a $(d-1)$-face F^{d-1} reads

$$\gamma(F^{d-1}, Q^N) = \sqrt{\frac{d}{\pi}} \int_0^\infty e^{-dt^2} \left(\frac{2}{\sqrt{\pi}} \int_0^t e^{-s^2} \, ds \right)^{N-d} dt;$$

it is due to Betke and Henk [99]. Böröczky and Henk [114] used it to prove a counterpart to (8.44), namely

$$\mathbb{E} f_k(Q^N|X) \sim \frac{2^d}{\sqrt{d}} \binom{d}{k+1} \beta(T^k, T^{d-1}) (\pi \log N)^{(d-1)/2}, \tag{8.45}$$

as $N \to \infty$. Remarkably, the right sides of (8.44) and (8.45) are identical. Equation (8.39) also appears in [114].

4. It was observed in Affentranger and Schneider [11] that the expected facet number $\mathbb{E} f_{d-1}(T^N|X)$, where X is an isotropic random d-subspace, is the same as the expected number of facets of the convex hull of $N+1$ independent random points in $\mathbb{R}^d$ with standard normal distribution; this expectation was determined earlier by Raynaud [623]). This fact found a general explanation in the work of Baryshnikov and Vitale [82], who proved the following result. Let T^N be a regular simplex in $\mathbb{R}^N$ with center of gravity at the origin. Let ϑ be a uniform random rotation of $\mathbb{R}^N$, and project the vertices of ϑT^N into a fixed subspace. Up to an affine transformation, the resulting point set coincides in distribution with a standard Gaussian sample of size $N+1$ in that subspace. As a corollary, an affine-invariant functional of a finite point set follows the same distribution for the Goodman–Pollack model and a standard Gaussian sample. Baryshnikov [81] explained the unique role that is played in this correspondence by the vertex sets of regular simplices. Baryshnikov and Vitale used their correspondence to transcribe to standard Gaussian samples the asymptotic results from [11]. Only later were these, and much stronger, results for convex hulls of standard Gaussian samples obtained in a direct way, see Hug, Munsonius and Reitzner [364], Hug and Reitzner [365].

5. Vershik and Sporyshev [767] studied $\mathbb{E} f_k(T^N|X)$ (where $\dim X = d$) under a linearly coordinated growth of the parameters. For $0 < \gamma < \alpha < 1$, $d = [\alpha N]$, $k = [\gamma N]$, they determined the asymptotic behavior of $\mathbb{E} f_k(T^N|X)$ for $N \to \infty$. Recall that a polytope is j-neighborly if any j of its vertices are the vertices of a $(j-1)$-face of the polytope. Vershik and Sporyshev also studied neighborliness, and a weaker form of it, where the overwhelming majority of j-tuples of vertices has to determine $(j-1)$-faces. They proved the following result. Let $k \in \mathbb{N}_0$ be fixed, let $d = c \log N + o(\log N)$. If $c > 2(k+1)$, then

$$\mathbb{E} f_k(T^N|X) = \binom{N+1}{k+1} (1 + o(1))$$

for $N \to \infty$. If $c < 2(k+1)$, this does not hold.

6. By duality, the results of Böröczky and Henk (see Note 3) can be carried over to random central sections of the cube C^N. Additional results for the numbers of k-faces of such sections, among them lower bounds with consequences for asymptotic results, were obtained by Lonke [442].

7. Formula (8.42) estimates the probability $\mathbb{P}\left(f_{k-1}(Q^N|X) < f_{k-1}(Q^N)\right)$. By analyzing carefully the asymptotic behavior, as $N \to \infty$, of the combinatorial factor, the internal angle and the external angle in (8.42), Donoho [207] obtained upper estimates, in the interesting cases where d and k are proportional to N. He established the existence of a function $\mathsf{P} : (0,1) \to (0,1]$, defined implicitly but computable numerically with sufficient accuracy, such that the following holds. For given $\delta \in (0,1)$, let $0 < \rho < \mathsf{P}(\delta)$ and put $d := \lfloor \delta N \rfloor$ and $k := \lfloor \rho d \rfloor$. Then, for sufficiently small $\epsilon > 0$,

$$\mathbb{P}\left(f_{k-1}(Q^N|X) < f_{k-1}(Q^N)\right) \le N\mathrm{e}^{-N\epsilon}$$

for $N > N_0(\delta, \rho, \epsilon)$.

This estimate is only one example of the author's thorough investigation of applications of ℓ_1^N approximation and its relation to neighborliness properties of randomly projected polytopes. As an approach to sparse solutions of underdetermined linear equations this is explained in Donoho [206]. Donoho and Tanner [208] studied random projections of simplices in a similar way, with applications to sparse nonnegative solutions of underdetermined linear equations. Neighborliness properties of random projections of simplices were investigated by Donoho and Tanner in [209], and in [210] they discussed various applications of (weak) neighborliness of random projections of high-dimensional polytopes.

8.4 Randomly Moving Bodies and Flats

Some of the early interpretations of the kinematic and Crofton formulas of integral geometry were in terms of geometric probabilities for randomly moving convex bodies and flats. In contrast to the applications to particle processes and random sets, which will be explained in Chapter 9, in these elementary geometric probability problems one is dealing with a finite number of randomly moving geometric objects; the randomness concerns the positions of these objects, but not their shapes. In this section, we give a few basic examples of typical questions and results, where formulas from integral geometry provide the answers. We consider probabilities, distributions and expectations that are related to the interaction between randomly moving geometric objects and a fixed one.

8.4.1 Hitting Probabilities

By a **randomly moving** geometric object (for example, a convex body, a q-flat, a cylinder) we understand a random variable with values in a space of congruent geometric objects; here 'congruent' means equivalent by a rigid motion.

From a geometric point of view, it appears natural to impose strong invariance assumptions on the probability distribution of a randomly moving set, and possibly to derive this distribution from an invariant measure. In the cases where such an invariant measure is infinite, as, for example, the Haar measure

μ on the group G_d of rigid motions, only the introduction of additional conditions will allow the definition of probability distributions. A typical example is given by a randomly moving convex body, under the condition that it hits a fixed convex body, that is, has nonempty intersection with it.

Definition 8.4.1. *Let $M, K_0 \in \mathcal{K}'$ be convex bodies, let* $\operatorname{int} K_0 \neq \emptyset$. *A* **random congruent copy of** M **meeting** K_0 *is a random convex body X_{M,K_0} with distribution $\mathbb{Q}$ given by*

$$\mathbb{Q}(A) := \frac{\mu(\{g \in G_d : gM \cap K_0 \neq \emptyset,\ gM \in A\})}{\mu(\{g \in G_d : gM \cap K_0 \neq \emptyset\})}$$

for $A \in \mathcal{B}(\mathcal{K})$.

In other words, if we define the mapping $f_M : G_d \to \mathcal{K}$ by $f_M(g) := gM$, then

$$\mathbb{Q} = \frac{f_M(\mu) \llcorner \mathcal{F}_{K_0}}{f_M(\mu)(\mathcal{F}_{K_0})}.$$

Let K be a convex body contained in K_0. The probability

$$\mathbb{P}(X_{M,K_0} \cap K \neq \emptyset)$$

can, with a slight abuse of language, be considered as the conditional probability that a random congruent copy of M hits K, under the condition that it hits K_0. The value of this probability immediately follows from the principal kinematic formula (Theorem 5.1.3).

Theorem 8.4.1. *If $M, K, K_0 \in \mathcal{K}'$ are convex bodies with $K \subset K_0$ and* $\operatorname{int} K_0 \neq \emptyset$ *and if X_{M,K_0} is a random congruent copy of M meeting K_0, then*

$$\mathbb{P}(X_{M,K_0} \cap K \neq \emptyset) = \frac{\sum_{k=0}^{d} c_{d,0}^{k,d-k} V_k(K) V_{d-k}(M)}{\sum_{k=0}^{d} c_{d,0}^{k,d-k} V_k(K_0) V_{d-k}(M)}.$$

It is evident why we have to assume convexity of the sets K, M, K_0. For sets from the convex ring, for example, the same question can be asked, but the values $\mu(\{g \in G_d : K \cap gM \neq \emptyset\})$ cannot be obtained as special cases of the principal kinematic formula. In the case where M and K_0 are convex and $K \in \mathcal{R}$, one can still not evaluate the intersection probability $\mathbb{P}(X_{M,K_0} \cap K \neq \emptyset)$, but the kinematic formula yields the expectations $\mathbb{E} V_j(X_{M,K_0} \cap K)$, which are of interest, for example, in stereology.

In a similar way, more complicated random events can be treated. For example, let $K_0, M_1, \dots, M_n \in \mathcal{K}'$ be given convex bodies, where $\operatorname{int} K_0 \neq \emptyset$. For $i \in \{1, \dots, n\}$, let X_{M_i,K_0} be a random congruent copy of M_i hitting K_0, and suppose that $X_{M_1,K_0}, \dots, X_{M_n,K_0}$ are independent. We ask for the probability that the random convex bodies $X_{M_1,K_0}, \dots, X_{M_n,K_0}$ have a common intersection point in K_0. Immediately from the definitions we get

$$\mathbb{P}(K_0 \cap X_{M_1,K_0} \cap \ldots \cap X_{M_n,K_0} \neq \emptyset)$$
$$= \frac{\int_{G_d} \cdots \int_{G_d} V_0(K_0 \cap g_1 M_1 \cap \ldots \cap g_n M_n)\, \mu(\mathrm{d}g_1) \cdots \mu(\mathrm{d}g_n)}{\prod_{i=1}^n \int_{G_d} V_0(K_0 \cap g_i M_i)\, \mu(\mathrm{d}g_i)}.$$

While the denominator is obtained directly from the kinematic formula, for the numerator we have to use the iteration given by Theorem 5.1.5. We do not write down the result explicitly.

We turn to an example where the expectation of a random volume defined by moving convex bodies can be calculated by integral geometric means. Let $M, K_0 \in \mathcal{K}'$ be given convex bodies with interior points. We consider n independent random congruent copies of M meeting K_0, and we ask for the expected volume of the set of points in K_0 that are covered exactly r times. To be precise, let $X_1, \ldots, X_n$ be independent, identically distributed random convex bodies, each with the same distribution as X_{M,K_0}. For convex bodies $M_1, \ldots, M_n$ and for $r = 0, \ldots, n$, we define

$$A_r(M_1, \ldots, M_n) := \{x \in K_0 : x \in M_i \text{ for precisely } r \text{ indices } i\}.$$

We want to find $\mathbb{E}V_d(A_r(X_1, \ldots, X_n))$.

Let Y be a uniform random point in K_0, such that $Y, X_1, \ldots, X_n$ are independent. We define random variables N, N_j by

$$N := \operatorname{card}\{i : Y \in X_i\}, \qquad N_j := \mathbf{1}_{X_j}(Y),$$

for $j = 1, \ldots, n$. Then

$$\mathbb{P}(N = r) = \mathbb{E}\mathbf{1}\{Y \in A_r(X_1, \ldots, X_n)\} = \mathbb{E}V_d(A_r(X_1, \ldots, X_n))/V_d(K_0).$$

Since $N_1, \ldots, N_n$ are independent and identically distributed, the random variable $N = N_1 + \ldots + N_n$ has a binomial distribution, namely

$$\mathbb{P}(N = r) = \binom{n}{r} p^r (1-p)^{n-r}$$

with $p := \mathbb{P}(N_j = 1)$. Now,

$$\mathbb{P}(N_j = 1) = \frac{\int_{G_d} \int_{K_0} \mathbf{1}_{gM}(x)\, \lambda(\mathrm{d}x)\, \mu(\mathrm{d}g)}{V_d(K_0) \int_{G_d} V_0(K_0 \cap gM)\, \mu(\mathrm{d}g)}$$
$$= \frac{\int_{G_d} V_d(K_0 \cap gM)\, \mu(\mathrm{d}g)}{V_d(K_0) \int_{G_d} V_0(K_0 \cap gM)\, \mu(\mathrm{d}g)}.$$

Consequently,

$$\mathbb{E}V_d(A_r(X_1, \ldots, X_n)) = V_d(K_0) \binom{n}{r} p^r (1-p)^{n-r}$$

with

$$p = \frac{V_d(M)}{\sum_{k=0}^{d} c_{d,0}^{k,d-k} V_k(K_0) V_{d-k}(M)}.$$

Notes for Subsection 8.4.1

1. For convex bodies $K \subset K_0$, the probability $\mathbb{P}(X_{M,K_0} \cap K \neq \emptyset)$ was interpreted above, somewhat loosely, as the 'conditional probability that a random congruent copy of M hits K, under the condition that it hits K_0'. We remark that Rényi [637] has developed an axiomatic theory of conditional probabilities which, as in this case, are not necessarily derived from a finite measure. He has mentioned integral geometry as one field where such conditional probabilities appear naturally.

2. Results of the types of Theorems 8.4.1 above and 8.4.2 below are classical interpretations of integral geometric formulas in terms of geometric probabilities. A general source for results of the type considered here is the book by Santaló [662]; see also Santaló [656, 659].

8.4.2 Randomly Moving Flats

Now we consider randomly moving flats. We assume that a convex reference body K_0 is given and derive a natural probability distribution for q-flats hitting K_0 from the invariant measure μ_q on the affine Grassmannian $A(d,q)$.

Definition 8.4.2. *Let $q \in \{0,\dots,d-1\}$, and let $K_0 \in \mathcal{K}'$ be a convex body with $V_{d-q}(K_0) > 0$. An* **isotropic random q-flat through** *K_0 is a random q-flat X_{q,K_0} with distribution given by*

$$\mathbb{Q} = \frac{\mu_q \llcorner \mathcal{F}_{K_0}}{\mu_q(\mathcal{F}_{K_0})}.$$

If K is a convex body contained in K_0, then similarly to the previous subsection, we consider the probability $\mathbb{P}(X_{q,K_0} \cap K \neq \emptyset)$ as the conditional probability that an isotropic random q-flat hits K, under the condition that it hits K_0. For this probability, the Crofton formula of Theorem 5.1.1 immediately gives the following result.

Theorem 8.4.2. *Let $K_0, K \in \mathcal{K}'$ be convex bodies with $K \subset K_0$, let $q \in \{0,\dots,d-1\}$ and $V_{d-q}(K_0) > 0$. If X_{q,K_0} is an isotropic random q-flat through K_0, then*

$$\mathbb{P}(X_{q,K_0} \cap K \neq \emptyset) = \frac{V_{d-q}(K)}{V_{d-q}(K_0)}.$$

We take again a convex body $K_0 \in \mathcal{K}'$, and we consider n stochastically independent isotropic q-flats $X_1,\dots,X_n$ through K_0. We assume that $n(d-q) \le d$, then the intersection $X_1 \cap \ldots \cap X_n$ is almost surely of dimension $d - n(d-q) \ge 0$ (as can be shown with the aid of Lemma 13.2.1). We can therefore ask for the probability that this intersection meets a given convex body $K \subset K_0$ (even for $K = K_0$, this is a nontrivial question). From the definition and the Crofton formula of Theorem 5.1.1 we obtain

$$\begin{aligned}
&\mathbb{P}(X_1 \cap \ldots \cap X_n \cap K \neq \emptyset) \\
&= \frac{\int_{A(d,q)} \cdots \int_{A(d,q)} V_0(K \cap E_1 \cap \ldots \cap E_n)\, \mu_q(\mathrm{d}E_n) \cdots \mu_q(\mathrm{d}E_1)}{\left(\int_{A(d,q)} V_0(K_0 \cap E)\, \mu_q(\mathrm{d}E)\right)^n} \\
&= \frac{c_{d,0}^{q,d-q} \int\limits_{A(d,q)} \cdots \int\limits_{A(d,q)} V_{d-q}(K \cap E_1 \cap \ldots \cap E_{n-1})\, \mu_q(\mathrm{d}E_{n-1}) \cdots \mu_q(\mathrm{d}E_1)}{\left(c_{d,0}^{q,d-q} V_{d-q}(K_0)\right)^n} \\
&= \ldots = \frac{c_{d,0}^{q,d-q} c_{d,d-q}^{q,2(d-q)} c_{d,2(d-q)}^{q,3(d-q)} \cdots c_{d,(n-1)(d-q)}^{q,n(d-q)} V_{n(d-q)}(K)}{\left(c_{d,0}^{q,d-q} V_{d-q}(K_0)\right)^n} \\
&= \frac{(n(d-q))!\kappa_{n(d-q)}}{((d-q)!\kappa_{d-q})^n} \frac{V_{n(d-q)}(K)}{V_{d-q}^n(K_0)}.
\end{aligned}$$

Examples. Let $d = 2$, $q = 1$, $n = 2$ and $K = K_0$, so we consider two independent isotropic random lines through K in the plane. The probability that their intersection point lies in K is given by

$$\frac{2\pi}{2^2} \frac{V_2(K)}{V_1(K)^2} = 2\pi \frac{A(K)}{L(K)^2}.$$

By the planar isoperimetric inequality, this probability is at most $1/2$, and it is equal to $1/2$ if and only if K is a circular disk.

For d independent isotropic random hyperplanes through a convex body K in $\mathbb{R}^d$, the probability that their intersection point lies in K, is given by

$$\frac{d!\kappa_d}{2^d} \frac{V_d(K)}{V_1(K)^d} = \frac{(d-1)!\kappa_{d-1}^d}{(d\kappa_d)^{d-1}} \frac{V_d(K)}{b(K)^d},$$

where $b(K)$ is the mean width of K. By (14.31), the maximum of this expression is attained precisely if K is a ball.

For applications and simulations it is important to know how such random q-flats as considered above can be generated, and how their distributions are related to other natural distributions. (Randomly moving convex bodies can be treated along similar lines.)

Let $q \in \{0, \ldots, d-1\}$, and let $K, M \in \mathcal{K}'$ be convex bodies with $V_{d-q}(M) > 0$ and $M \subset K$. For an isotropic random q-flat through K we can consider the conditional distribution, under the condition that it hits M. Immediately from the definitions we obtain, for $A \in \mathcal{B}(A(d,q))$,

$$\begin{aligned}\mathbb{P}(X_{q,K} \in A \mid X_{q,K} \cap M \neq \emptyset) &= \frac{\mu_q(A \cap \mathcal{F}_K \cap \mathcal{F}_M)/\mu_q(\mathcal{F}_K)}{\mu_q(\mathcal{F}_M)/\mu_q(\mathcal{F}_K)} \\ &= \frac{\mu_q(A \cap \mathcal{F}_M)}{\mu_q(\mathcal{F}_M)} = \mathbb{P}(X_{q,M} \in A).\end{aligned}$$

We formulate this as a lemma.

Lemma 8.4.1. *Let $q \in \{0, \dots, d-1\}$. Let $K, M \in \mathcal{K}'$ be convex bodies with $M \subset K$ and $V_{d-q}(M) > 0$. Then the conditional distribution of an isotropic random q-flat through K, under the condition that it hits M, is equal to the distribution of an isotropic random q-flat through M.*

In view of this lemma, the generation of isotropic random q-flats through a given convex body $K \in \mathcal{K}'$ can be achieved as follows. We choose a ball rB^d of radius rB^d containing K and generate independent isotropic q-flats through rB^d until the body K is hit for the first time. The q-flat thus obtained is then a realization of an isotropic q-flat through K. In the case $q = 0$, where $X_{0,K}$ is a uniform random point in K, it is more convenient to replace the ball rB^d by a cube, for instance $C = \alpha[0,1]^d$ with $\alpha > 0$. In fact, a uniform random point x in C is easily generated in the form $x = \alpha(\xi_1, \dots, \xi_d)$, where $\xi_1, \dots, \xi_d$ are d independent uniform random numbers in the interval $[0, 1]$.

In order to generate, for $q \geq 1$, an isotropic q-flat through rB^d, we use the definition of the measure μ_q. Let $L \in G(d,q)$ be fixed and let $rB_q := rB^d \cap L^\perp$. We choose independently a uniform random point $x \in rB_q$ (as described above) and an isotropic random rotation $\vartheta \in SO_d$. Then $E := \vartheta(L + x)$ is an isotropic q-flat through rB^d. In order to generate ϑ, we can choose d independent uniform points $x_1, \dots, x_d$ in B^d, then $x_1/\|x_1\|, \dots, x_d/\|x_d\|$ are unit vectors which are linearly independent with probability one. The mapping ψ described in the proof of Theorem 13.2.9 maps this d-tuple into a rotation ϑ. In this way we have defined a random rotation ϑ with rotation invariant distribution ν, as follows from the proof of Theorem 13.2.9.

We now describe a different, and possibly more natural, way of constructing random q-flats through a given convex body $K \in \mathcal{K}$ with interior points. We choose a uniform random point Y in K and independently an isotropic random subspace $L^{(q)} \in G(d,q)$. Then

$$X_K^{(q)} := L^{(q)} + Y$$

is a random q-flat meeting K. It is called a **q-weighted random q-flat through** K. The corollary to the following theorem describes how its distribution is related to the distribution of an isotropic random q-flat through K.

Theorem 8.4.3. *Let $q \in \{1, \dots, d-1\}$. Let $K \in \mathcal{K}$ be a convex body with interior points. If $X_K^{(q)}$ is a q-weighted random q-flat through K, then*

$$\mathbb{P}(X_K^{(q)} \in A) = \frac{1}{V_d(K)} \int_A V_q(K \cap E)\, \mu_q(\mathrm{d}E)$$

for $A \in \mathcal{B}(A(d,q))$.

Proof. By definition and Fubini's theorem,

$$\begin{aligned}
&V_d(K)\mathbb{P}(X_K^{(q)} \in A) \\
&= \int_{G(d,q)} \int_K \mathbf{1}_A(L+x)\, \lambda(\mathrm{d}x)\, \nu_q(\mathrm{d}L) \\
&= \int_{G(d,q)} \int_{L^\perp} \int_L \mathbf{1}_A(L+y)\mathbf{1}_K(y+z)\, \lambda_L(\mathrm{d}z)\, \lambda_{L^\perp}(\mathrm{d}y)\, \nu_q(\mathrm{d}L) \\
&= \int_{G(d,q)} \int_{L^\perp} \mathbf{1}_A(L+y)V_q(K \cap (L+y))\, \lambda_{L^\perp}(\mathrm{d}y)\, \nu_q(\mathrm{d}L) \\
&= \int_A V_q(K \cap E)\, \mu_q(\mathrm{d}E),
\end{aligned}$$

as asserted. □

Corollary 8.4.1. *Let* $q \in \{1, \dots, d-1\}$. *Suppose that* $X_K^{(q)}$ *is a* q*-weighted random* q*-flat through* K *and* $X_{q,K}$ *is an isotropic random* q*-flat through* K. *Then the distribution of* $X_K^{(q)}$ *is absolutely continuous with respect to the distribution of* $X_{q,K}$, *with density given by*

$$E \mapsto c_{d,0}^{q,d-q} \frac{V_{d-q}(K)}{V_d(K)} V_q(K \cap E).$$

Various random sections of sets play a prominent role in stereology. As an applied discipline, stereology is concerned with estimating geometric parameters of three-dimensional structures from the evaluation of lower-dimensional sections, projections, projected thick sections, etc. We explain briefly, in the setting of d-dimensional space, some ideas underlying the use of sections with random flats. The set in $\mathbb{R}^d$ to be investigated will be denoted by K.

We assume that a random q-flat X_q is given and that $K \cap X_q$ can be observed. We also assume that K is contained in some a priori given, known reference set $K_0 \in \mathcal{K}'$. A plausible choice for X_q would then be an isotropic random q-flat through K_0. In order to apply the Crofton formula, we suppose that K is an element of the convex ring $\mathcal{R}$ (which, from a practical viewpoint, is not a severe restriction) and that the quantity

$$V_j(K \cap X_q), \qquad j \in \{0, \dots, q\},$$

can be measured. We are interested in the expectation of the random variable $V_j(K \cap X_q)$, because the expectation will be a d-dimensional quantity associated with K, for which the q-dimensional quantity $V_j(K \cap X_q)$ is an unbiased estimator. The Crofton formula of Theorem 5.1.1 (extended to the convex ring) immediately gives the following result.

Theorem 8.4.4. *Let $q \in \{0, \dots, d-1\}$ and $j \in \{0, \dots, q\}$. If $K_0 \in \mathcal{K}'$, $K \in \mathcal{R}$ are sets with $K \subset K_0$ and if X_{q,K_0} is an isotropic random q-flat through K_0, then*

$$\mathbb{E}V_j(K \cap X_{q,K_0}) = c_{j,d-q}^{0,d-q+j} \frac{V_{d-q+j}(K)}{V_{d-q}(K_0)}.$$

A situation of practical interest is the case $d = 3$ and $q = 2$. In a section with X_2 one may measure the Euler characteristic $V_0(K \cap X_2)$, the boundary length $2V_1(K \cap X_2)$, or the area $V_2(K \cap X_2)$. If the mean width of K_0 and hence $V_1(K_0)$ is known, one can thus estimate without bias the quantities $V_1(K), V_2(K), V_3(K)$, which are, respectively, proportional to the (additive extension of) the mean width, the surface area, and the volume of K. If measurements at the two-dimensional set $K \cap X_2$ are too involved (for example, in the case of the boundary length $L(K \cap X_2)$), one may in turn apply Theorem 8.4.4, with $d = 2$ and $q = 1$, to the sets $\widetilde{K} := K \cap X_2 \subset K_0 \cap X_2 =: \widetilde{K}_0$. For this, one considers isotropic random lines X_1 through $\widetilde{K}_0$ and determines $V_0(\widetilde{K} \cap X_1)$ or $V_1(\widetilde{K} \cap X_1)$. In this way, $V_1(\widetilde{K})$ and $V_2(\widetilde{K})$ can be estimated without bias, so that in the end one obtains an estimate for surface area and volume of K. Going one step further, we may generate random points X_0 within the set $\widetilde{K}_0$ and then use $V_0(\widetilde{K} \cap X_0)$ to estimate the area of $\widetilde{K}$, and thus finally the volume of K.

Alternatively, we can directly choose random lines X_1 through K_0 or random points X_0 in K_0 and use them for estimating volume and surface area of K, respectively the volume alone. However, it must be pointed out that the two-step method described above, which consists in first choosing an isotropic random plane X_2 through K_0 and subsequently choosing an isotropic random line in X_2 through $K_0 \cap X_2$, will not lead to an isotropic line X_1 through K_0. The distributions of random flats arising by such two-step procedures will be discussed in Subsection 8.4.3 in a general form.

A disadvantage of the formula of Theorem 8.4.4 for practical purposes is to be seen in the fact that one obtains the quotient

$$\frac{V_{d-q+j}(K)}{V_{d-q}(K_0)},$$

whereas the quotient

$$\frac{V_{d-q+j}(K)}{V_d(K_0)}$$

(the specific amount of V_{d-q+j} of K per unit volume) might be of greater interest. In particular, this would be the case if K is obtained from a larger set $\widetilde{K} \in \mathcal{R}$ by intersection with K_0. Now

$$\mathbb{E}V_j(K \cap X_{q,K_0}) = c_{d-q,j}^{d-q+j,0} \frac{V_{d-q+j}(K)}{V_{d-q}(K_0)}$$

and

$$\mathbb{E}V_q(K_0 \cap X_{q,K_0}) = c_{d-q,q}^{d,0} \frac{V_d(K_0)}{V_{d-q}(K_0)},$$

hence the quotient

$$\frac{V_j(K \cap X_{q,K_0})}{V_q(K_0 \cap X_{q,K_0})}$$

may be viewed as a natural estimator for

$$c_{d,j}^{q,d-q+j} \frac{V_{d-q+j}(K)}{V_d(K_0)}.$$

In general, this estimator is biased, that is,

$$\mathbb{E}\frac{V_j(K \cap X_{q,K_0})}{V_q(K_0 \cap X_{q,K_0})} \neq c_{d,j}^{q,d-q+j} \frac{V_{d-q+j}(K)}{V_d(K_0)}.$$

If, however, we do not work with isotropic random q-flats through K_0, but with q-weighted random q-flats, then we obtain an unbiased estimator.

Theorem 8.4.5. *Let $K_0 \in \mathcal{K}'$ and $K \in \mathcal{R}$ be sets with $K \subset K_0$, let $q \in \{0, \dots, d-1\}$ and $j \in \{0, \dots, q\}$, and let $X_{K_0}^{(q)}$ be a q-weighted random q-flat through K_0. Then*

$$\mathbb{E}\frac{V_j(K \cap X_{K_0}^{(q)})}{V_q(K_0 \cap X_{K_0}^{(q)})} = c_{d,j}^{q,d-q+j} \frac{V_{d-q+j}(K)}{V_d(K_0)}.$$

Proof. Let X_{q,K_0} be an isotropic random q-flat through K_0. From Corollary 8.4.1 and Theorem 8.4.4 we get

$$\begin{aligned}\mathbb{E}\frac{V_j(K \cap X_{K_0}^{(q)})}{V_q(K_0 \cap X_{K_0}^{(q)})} &= \mathbb{E}\left(\frac{V_j(K \cap X_{q,K_0})}{V_q(K_0 \cap X_{q,K_0})} c_{d,0}^{q,d-q} \frac{V_{d-q}(K_0)}{V_d(K_0)} V_q(K_0 \cap X_{q,K_0})\right) \\ &= c_{d,0}^{q,d-q} \frac{V_{d-q}(K_0)}{V_d(K_0)} \mathbb{E}V_j(K \cap X_{q,K_0}) \\ &= c_{d,j}^{q,d-q+j} \frac{V_{d-q+j}(K)}{V_d(K_0)},\end{aligned}$$

as asserted. □

Notes for Subsection 8.4.2

1. The systematic investigation of isotropic uniform random flats through a convex body began with an influential paper by Miles [520]. Different ways of generating random lines through a convex body and the relations between their distributions were first investigated by Kingman [411, 412] and Coleman [178]. The importance of q-weighted random q-flats, in particular for stereology, was emphasized by Miles and Davy [536], Davy and Miles [201].

2. Principles of stereology. In stereological notation, for $d = 3$, $q = 2$, the expectation formulas of Theorem 8.4.5 read

$$V_V = A_A,$$
$$S_V = \frac{4}{\pi} L_A,$$
$$M_V = 2\pi\chi_A.$$

Here, V_V, S_V, M_V denote the mean volume, mean surface area and mean integral mean curvature per unit volume, and A_A, L_A and χ_A are the corresponding mean values of area, boundary length and Euler characteristic per unit area (in a planar section). These 'fundamental formulas' of stereology appeared in the non-mathematical literature starting with the volume formula by the geologist A. Delesse in 1847 and subsequently improved and extended to the other functionals over a period of more than a hundred years. The improvements mainly concerned corresponding expectation formulas for $d = 2$, $q = 1$, respectively $d = 1$, $q = 0$, by which the mean values A_A and L_A in planar sections can again be estimated through intersections with linear structures (grids of lines or segments) or even, in the case of A_A, with a grid of points. It was obvious that the validity of these fundamental formulas required some randomness, either of the underlying structure or of the sectioning devices, as well as appropriate invariance assumptions on the corresponding distributions. Only after the foundation of the International Society for Stereology in 1961 was a systematic investigation of the mathematical background and the interrelations with integral geometry started, and this led to the fundamental papers of Miles and Davy mentioned above (see also Davy [199]). Nowadays two different stereological approaches are distinguished, the **design-based** approach and the **model-based** approach. In the former, the set K under consideration is assumed to be fixed and contained in a reference set K_0. The quantities of interest are the quotients $V_V = V(K)/V(K_0)$, $A_V = A(K)/V(K_0)$ and $M_V = M(K)/V(K_0)$. In this case, the quantities on the right side of the fundamental formulas have to be interpreted as expectations with respect to random (two-dimensional) sections of K_0, hence they describe natural estimators. Only through the work of Davy and Miles did it become clear that in order for these estimators to be unbiased, the random sections have to be obtained not from isotropic random planes but from 2-weighted ones. The design-based approach is discussed in detail in the recent book by Baddeley and Jensen [53], which includes many interesting historical remarks and a description of various pitfalls in the statistical analysis of stereological problems.

The model-based approach, in contrast, assumes that the structure $K \subset K_0$ of interest is the realization of a random set Z (intersected with the window K_0). If Z is stationary and isotropic and the sectioning plane E is fixed (or if Z is stationary and E is an isotropic random plane), the fundamental formulas appear as Crofton formulas for specific intrinsic volumes, as discussed at the end of Section 9.4.

3. After Lemma 8.4.1, we have mentioned a method of generating isotropic random rotations. This method may be impractical for actual simulations if d is large. For a different method see, for example, Stewart [737].

4. Let K be a convex body, and let $X_1, \dots, X_n$ be independent uniform random lines through K. For given $r > 0$, these lines are said to be r-close if there exists some point in K from which all the lines have distance at most r. Formulas for the probability of this and similar events (for flats) can be obtained with the help of iterated kinematic formulas, see Hadwiger and Streit [313].

8.4.3 Distributions of Random Flats

In this subsection we determine the distributions of random flats that are derived, by various geometric constructions, from given isotropic random flats. We restrict ourselves to affine subspaces, remarking that linear subspaces can be treated similarly.

Let $K \in \mathcal{K}'$ be a convex body with interior points, and let $0 \le p < q < d$. We imagine a random p-flat X through K, generated by first choosing an isotropic uniform q-flat Y through K and then choosing in Y an isotropic uniform p-flat through $K \cap Y$. We want to determine the distribution of the resulting random p-flat through K.

We denote the distribution of an isotropic random r-flat through the convex body K by $\mathbb{P}_r^{(K)}$. By definition and by the Crofton formula, we have

$$\mathbb{P}_r^{(K)}(A) = \frac{\mu_r(A \cap \{E \in A(d,r) : E \cap K \neq \emptyset\})}{c_{d,0}^{r,d-r} V_{d-r}(K)}$$

for $A \in \mathcal{B}(A(d,r))$. Now let $F \in A(d,q)$ be a q-flat with $\dim(K \cap F) = q$. An isotropic random p-flat through $K \cap F$, taken in F as the surrounding space, can also be considered as a random p-flat in $\mathbb{R}^d$. Denoting its distribution by $\mathbb{P}_p^{(K,F)}$, we have

$$\mathbb{P}_p^{(K,F)}(A) = \frac{\mu_p^F(A \cap \{E \in A(F,p) : E \cap K \neq \emptyset\})}{c_{q,0}^{p,q-p} V_{q-p}(K \cap F)}$$

for $A \in \mathcal{B}(A(d,p))$. It follows from Lemma 13.2.2 that for each $A \in \mathcal{B}(A(d,p))$ the function $F \mapsto \mathbb{P}_p^{(K,F)}(A)$ is measurable on

$$[K]_q := \{F \in A(d,q) : \dim(K \cap F) = q\}.$$

Thus the function $(F, A) \mapsto \mathbb{P}_p^{(K,F)}(A)$ is a transition probability from $([K]_q, \mathcal{B}([K]_q))$ to $(A(d,p), \mathcal{B}(A(d,p)))$. Let

$$\mathbb{Q} := \mathbb{P}_q^{(K)} \otimes \mathbb{P}_p^{(K,\,\cdot\,)}$$

be the probability measure on $\mathcal{B}([K]_q) \otimes \mathcal{B}(A(d,p))$ that is determined by $P_q^{(K)}$ and $\mathbb{P}_p^{(K,\,\cdot\,)}$. The distribution $\mathbb{P}_q^{(K)}$ of an isotropic random q-flat through K is concentrated on $[K]_q$, since the set of q-flats touching K has μ_q-measure zero, as is easy to see.

The generation of the random p-flat X, which was described above in a more heuristic way, can now be modeled more precisely as a two-step random experiment, by requiring that the distribution of X be given by

$$\mathbb{P}_2(A) := \mathbb{Q}([K]_q \times A), \qquad A \in \mathcal{B}(A(d,p)).$$

For this distribution we then obtain

$$\begin{aligned}
&\mathbb{P}_2(A)\\
&= (\mathbb{P}_q^{(K)} \otimes \mathbb{P}_p^{(K,\,\cdot\,)})([K]_q \times A)\\
&= \int_{[K]_q} \int_{A(d,p)} \mathbf{1}_{[K]_q \times A}(F,E)\, \mathbb{P}_p^{(K,F)}(\mathrm{d}E)\, \mathbb{P}_q^{(K)}(\mathrm{d}F)\\
&= \int_{A(d,q)} \int_{A(F,p)} \mathbf{1}_A(E) \frac{V_0(K \cap E)}{c_{q,0}^{p,q-p} V_{q-p}(K \cap F)}\, \mu_p^F(\mathrm{d}E) \frac{V_0(K \cap F)}{c_{d,0}^{q,d-q} V_{d-q}(K)}\, \mu_q(\mathrm{d}F)\\
&= c \int_{A(d,q)} \int_{A(F,p)} \mathbf{1}_A(E) \frac{V_0(K \cap E) V_0(K \cap F)}{V_{q-p}(K \cap F)}\, \mu_p^F(\mathrm{d}E)\, \mu_q(\mathrm{d}F)
\end{aligned}$$

with

$$c = (c_{d,0,0}^{d-q,p,q-p} V_{d-q}(K))^{-1}.$$

The distribution $\mathbb{P}_2$ can be considered as known if one knows its density with respect to the invariant measure μ_p. Using the interchange formula (7.5) and observing that

$$V_0(K \cap E) V_0(K \cap F) = V_0(K \cap E)$$

for $E \subset F$, we obtain

$$\begin{aligned}
\mathbb{P}_2(A) &= c \int_{A(d,p)} \int_{A(E,q)} \mathbf{1}_A(E) \frac{V_0(K \cap E) V_0(K \cap F)}{V_{q-p}(K \cap F)}\, \mu_q^E(\mathrm{d}F)\, \mu_p(\mathrm{d}E)\\
&= c \int_A \left(V_0(K \cap E) \int_{A(E,q)} \frac{1}{V_{q-p}(K \cap F)}\, \mu_q^E(\mathrm{d}F) \right) \mu_p(\mathrm{d}E).
\end{aligned}$$

Thus, the distribution $\mathbb{P}_2$ of the random p-flat X has a density with respect to the invariant measure μ_p on the space $A(d,p)$ which is given by

$$E \mapsto c_{d-q,p,q-p}^{d,0,0} \frac{V_0(K \cap E)}{V_{d-q}(K)} \int_{A(E,q)} \frac{1}{V_{q-p}(K \cap F)}\, \mu_q^E(\mathrm{d}F). \tag{8.46}$$

We state this as a theorem.

Theorem 8.4.6. *Let $0 \le p < q < d$, and let $K \in \mathcal{K}'$ be a convex body with $V_d(K) > 0$. Let Y be an isotropic random q-flat through K, and let X be an isotropic random p-flat through $K \cap Y$ in Y. Then the distribution of X has a density with respect to the invariant measure μ_p which is given by* (8.46).

In a similar way we may, for $0 \le p < q < d$, generate a random q-flat Y through K, by first choosing an isotropic uniform p-flat X through K and then choosing an isotropic q-flat containing X. Let $E \in A(d,p)$ be a p-flat with $\dim(K \cap E) = p$. An isotropic flat in $A(E,q)$ has, by definition, the distribution μ_q^E. Since $\mu_q^{(\cdot)}$ is a transition probability from $([K]_p, \mathcal{B}([K]_p))$ to $(A(d,q), \mathcal{B}(A(d,q)))$, a probability measure

$$\mathbb{Q}' := \mathbb{P}_p^{(K)} \otimes \mu_q^{(\cdot)}$$

on $\mathcal{B}([K]_p) \otimes \mathcal{B}(A(d,q))$ is determined by $\mathbb{P}_p^{(K)}$ and $\mu_q^{(\cdot)}$. Then, similarly to above,

$$\mathbb{P}_2'(A) := \mathbb{Q}'([K]_p \times A), \qquad A \in \mathcal{B}(A(d,q)),$$

can be considered as the distribution of the required random q-flat Y. For this distribution we obtain, as before, for $A \in \mathcal{B}(A(d,q))$,

$$\begin{aligned}
\mathbb{P}_2'(A) &= \int_{[K]_p} \int_{A(E,q)} \mathbf{1}_{[K]_p \times A}(E,F)\, \mu_q^E(\mathrm{d}F)\, \mathbb{P}_p^{(K)}(\mathrm{d}E) \\
&= \int_{A(d,p)} \int_{A(E,q)} \mathbf{1}_A(F)\, \mu_q^E(\mathrm{d}F) \frac{V_0(K \cap E)}{c_{d,0}^{p,d-p} V_{d-p}(K)}\, \mu_p(\mathrm{d}E) \\
&= \frac{c_{p,d-p}^{d,0}}{V_{d-p}(K)} \int_A \int_{A(F,p)} V_0(K \cap E)\, \mu_p^F(\mathrm{d}E)\, \mu_q(\mathrm{d}F) \\
&= \frac{c_{q,d-p}^{d,q-p}}{V_{d-p}(K)} \int_A V_{q-p}(K \cap F)\, \mu_q(\mathrm{d}F),
\end{aligned}$$

where the Crofton formula (5.6) was applied in F. We have obtained the density of the distribution $\mathbb{P}_2'$ of the random q-flat Y with respect to the invariant measure μ_q on the space $A(d,q)$, namely

$$F \mapsto c_{q,d-p}^{d,q-p} \frac{V_{q-p}(K \cap F)}{V_{d-p}(K)}. \tag{8.47}$$

For $p = 0$, this result was already noted in Theorem 8.4.3. We state the general result again as a theorem.

Theorem 8.4.7. *Let $0 \le p < q < d$, and let $K \in \mathcal{K}'$ be a convex body with $V_d(K) > 0$. Let X be an isotropic random p-flat through K, and let Y be an isotropic random q-flat containing X. Then Y is a random q-flat through K, whose distribution has a density with respect to the invariant measure μ_q which is given by* (8.47).

Next, we study random flats spanned by independent uniform random points. Let $K \in \mathcal{K}'$ be a convex body with $V_d(K) > 0$. For $q \in \{1, \dots, d-1\}$ we consider $q+1$ independent uniform random points in K. Almost surely their affine hull is a q-dimensional flat. The distribution of this random flat we denote by $\mathbb{Q}$. For $A \in \mathcal{B}(A(d,q))$, we then have

$$\mathbb{Q}(A) = \frac{1}{V_d(K)^{q+1}} \int_{\mathbb{R}^d} \cdots \int_{\mathbb{R}^d} f(x_0, \dots, x_q)\, \lambda(\mathrm{d}x_0) \cdots \lambda(\mathrm{d}x_q)$$

with

$$f(x_0, \dots, x_q) := \mathbf{1}_A(\operatorname{aff}\{x_0, \dots, x_q\}) \prod_{j=0}^{q} \mathbf{1}_K(x_j).$$

The affine Blaschke–Petkantschin formula of Theorem 7.2.7 immediately yields that the distribution $\mathbb{Q}$ has a density with respect to the invariant measure μ_q on $A(d,q)$, given by

$$E \mapsto \frac{b_{dq}(q!)^{d-q}}{V_d(K)^{q+1}} \int_{E\cap K} \cdots \int_{E\cap K} \Delta_q(x_0, \dots, x_q)^{d-q}\, \lambda_E(\mathrm{d}x_0) \cdots \lambda_E(\mathrm{d}x_q), \tag{8.48}$$

for $E \in A(d,q)$.

Theorem 8.4.8. *Let $q \in \{1, \dots, d-1\}$, and let $K \in \mathcal{K}'$ be a convex body with $V_d(K) > 0$. Let $\mathbb{Q}$ be the distribution of the random q-flat that is spanned (almost surely) by $q+1$ independent uniform random points in K. With respect to the invariant measure μ_q on $A(d,q)$, the distribution $\mathbb{Q}$ has the density given by* (8.48).

Finally, we study the distribution of the intersection of independent isotropic random flats through a convex body. We restrict ourselves to the case of $2 \le q \le d$ independent isotropic random hyperplanes through a convex body $K \in \mathcal{K}'$ with $\dim K \ge 1$. The intersection of these hyperplanes is almost surely a $(d-q)$-flat X. The distribution of this random flat X is, of course, no longer concentrated on the set of $(d-q)$-flats hitting K. Let $\mathbb{Q}$ denote the distribution of X. For $A \in \mathcal{B}(A(d,d-q))$ we have, by the Crofton formula and (14.7),

$$\mathbb{Q}(A) = \frac{1}{b(K)^q} \int_{A(d,d-1)} \cdots \int_{A(d,d-1)} f(H_1, \dots, H_q)\, \mu_{d-1}(\mathrm{d}H_1) \cdots \mu_{d-1}(\mathrm{d}H_q)$$

with

$$f(H_1, \dots, H_q) := \mathbf{1}_A(H_1 \cap \ldots \cap H_q) \prod_{j=1}^{q} \mathbf{1}_{\mathcal{H}_K}(H_j);$$

here we use the notation

$$\mathcal{H}_K := \{H \in A(d,d-1) : H \cap K \neq \emptyset\}.$$

From Theorem 7.2.8, for $s_1 = \ldots = s_q = d-1$, we obtain that the distribution of X has a density with respect to the invariant measure μ_{d-q} which is given by

$$E \mapsto c \int_{\mathcal{H}_K^E} \cdots \int_{\mathcal{H}_K^E} [H_1, \dots, H_q]^{d-q+1} \mu_{d-1}^E(\mathrm{d}H_1) \cdots \mu_{d-1}^E(\mathrm{d}H_q) \tag{8.49}$$

with $\mathcal{H}_K^E := A(E, d-1) \cap \mathcal{H}_K$ and

$$c = b_{dq} \left(\frac{\omega_q}{\omega_d b(K)} \right)^q .$$

We state the result as a theorem.

Theorem 8.4.9. *Let $q \in \{2, \dots, d\}$, and let $K \in \mathcal{K}'$ be a convex body with $\dim K \geq 1$. The intersection of q independent isotropic uniform random hyperplanes through K is almost surely a $(d-q)$-flat. With respect to the invariant measure μ_{d-q} on $A(d, d-q)$, the distribution of this random $(d-q)$-flat has the density given by* (8.49).

In particular, we see from (8.49) that the conditional distribution of the generated $(d-q)$-flat X, under the condition that it meets the body K, is isotropic and uniform; but this is not surprising. Of more interest is the information about the form of the density on the set of $(d-q)$-flats not meeting K. For example, for $d = q = 2$ we obtain the density

$$x \mapsto c \int_{\mathcal{H}_K^{\{x\}}} \int_{\mathcal{H}_K^{\{x\}}} [G, H] \, \mu_1^{\{x\}}(\mathrm{d}G) \, \mu_1^{\{x\}}(\mathrm{d}H).$$

The invariant measure $\mu_1^{\{x\}}$ is in this case induced from the uniform distribution on $[0, 2\pi)$; therefore the right side is proportional to

$$\int_0^{2\pi} \int_0^{2\pi} f(K, x, \alpha) f(K, x, \beta) |\sin(\alpha - \beta)| \, \mathrm{d}\alpha \, \mathrm{d}\beta,$$

where $f(K, x, \alpha)$ denotes the indicator function of the event that the line $G = G(x, \alpha)$ through x and making the angle α with a fixed direction, intersects the body K. Thus, for $x \notin K$ the density is proportional to $\omega - \sin \omega$, where $\omega = \omega(x)$ denotes the angle under which K is seen from x.

Notes for Subsection 8.4.3

1. The distribution of the intersection point of two independent isotropic random lines through a planar convex body was already determined by Crofton [186]; he found the last result mentioned above.

2. Distributions of random lines through convex bodies generated in different ways were investigated by Kingman [411], Coleman [178], Enns and Ehlers [217]. See also Note 2 for Section 8.6.

8.5 Touching Probabilities

In this section, we want to show how integral geometry and curvature measures can be used to treat an unconventional 'dice probability problem'. Imagine that two congruent cubical dice in three-space are thrown randomly in such a way that they touch each other. Under plausible assumptions on the distribution, only the touching positions 'edge against edge' and 'vertex against two-face' will have positive probabilities. If two players bet on these complementary events, who has the better chances? Under suitable natural model assumptions, we shall give an explicit answer to this question.

We consider touching probabilities of a more general kind. Let K and M be nonempty convex bodies in $\mathbb{R}^d$, and let $A \subset \operatorname{bd} K$, $B \subset \operatorname{bd} M$ be two given Borel sets. If we consider the boundary sets A of K and B of M as being colored and if K and M touch at random, what is the probability that the touching occurs at a pair of colored points? Of course, we must first specify a suitable probabilistic model. If not touching, but hitting probabilities are asked, there is a geometrically natural underlying sample space, namely $\{g \in G_d : K \cap gM \neq \emptyset\}$, together with the motion invariant measure μ on G_d, restricted to this space and normalized to a probability measure. The obvious underlying space for the investigation of random touchings is given by the set

$$G_d(K, M) := \{g \in G_d : K \text{ and } gM \text{ touch}\},$$

which was introduced in Section 5.1 (proof of Theorem 5.1.2). This set, however, has μ measure zero, so that a conditional probability

$$\mathbb{P}(A \cap gB \neq \emptyset \mid K \text{ and } gM \text{ touch})$$

cannot be defined in an elementary way. Therefore, the set $G_d(K, M)$ of rigid motions bringing M into a touching position with K has first to be equipped with a positive finite measure, which can then be normalized to a probability measure, and which can be viewed as a natural touching measure. (As usual, this measure will be considered as a measure defined on all of G_d.) The measure can be considered as natural or canonical if it can be deduced in a simple way from the motion invariant measure μ, say in analogy to the deduction of a natural notion of (Minkowski) surface area from the notion of volume. We shall now describe such a construction for given convex bodies $K, M \in \mathcal{K}'$.

We define

$$r(K, M) := \min\{\|x - y\| : x \in K,\ y \in M\}$$

and

$$G_d^*(K, M) := \{g \in G_d : K \cap gM = \emptyset\}.$$

Then

$$G_d^*(K, M) = \bigcup_{r>0} \{g \in G_d : r(K, gM) = r\} = \bigcup_{r>0} G_d(K + rB^d, M),$$

and the latter is a disjoint union. Therefore, the required touching measure $\mu(K, M, \cdot)$ should have the property that

$$\mu \llcorner G_d^*(K, M) = \int_0^\infty \mu(K + rB^d, M, \cdot)\, \mathrm{d}r,$$

and the integrand should be a continuous function of r, with respect to the weak topology. The proof that this is possible in a unique way is preceded by a lemma, which expresses an analogous relationship between the Lebesgue measure and the boundary measure of a convex body.

Throughout this section, we make essential use of the curvature measures $\Phi_0(K, \cdot), \dots, \Phi_{d-1}(K, \cdot)$ of a convex body K (see Section 14.2).

Lemma 8.5.1. *If $K \in \mathcal{K}'$, then*

$$\lambda \llcorner (\mathbb{R}^d \setminus K) = 2 \int_0^\infty \Phi_{d-1}(K + rB^d, \cdot)\, \mathrm{d}r.$$

Proof. It suffices to prove the assertion for the case where K is a polytope; the general case is then obtained by approximation, using the weak convergence of the curvature measures and the dominated convergence theorem.

For a polytope K we have

$$\mathbb{R}^d \setminus K = \bigcup_{j=0}^{d-1} \bigcup_{F \in \mathcal{F}_j(K)} \{x \in \mathbb{R}^d \setminus K : p(K, x) \in \operatorname{relint} F\},$$

and this is a disjoint decomposition, hence

$$\lambda \llcorner (\mathbb{R}^d \setminus K) = \sum_{j=0}^{d-1} \sum_{F \in \mathcal{F}_j(K)} \lambda \llcorner \{x \in \mathbb{R}^d \setminus K : p(K, x) \in \operatorname{relint} F\}. \tag{8.50}$$

Now let $A \in \mathcal{B}(\mathbb{R}^d)$ and suppose that $A \subset \{x \in \mathbb{R}^d \setminus K : p(K, x) \in \operatorname{relint} F\}$ for some face $F \in \mathcal{F}_j(K)$. With $L_{d-j} := (\operatorname{aff} F)^\perp$ we then have

$$\begin{aligned}
\lambda(A) &= (\lambda_F \otimes \lambda_{F^\perp})(A) \\
&= \left(\lambda_F \otimes \int_0^\infty 2\Phi_{d-j-1}\left(rB^d \cap L_{d-j}, \cdot\right)\, \mathrm{d}r\right)(A) \\
&= \int_0^\infty 2\Phi_{d-1}(F + rB^d, A)\, \mathrm{d}r \\
&= 2 \int_0^\infty \Phi_{d-1}(K + rB^d, A)\, \mathrm{d}r.
\end{aligned}$$

Together with (8.50), this yields the assertion. ∎

The construction of the touching measure is based on this lemma and is at the same time a generalization of the procedure leading to it.

Theorem 8.5.1. *There is precisely one mapping that associates with every pair $K, M \in \mathcal{K}'$ a finite measure $\mu(K,M,\cdot)$ on $G_d(K,M)$ that is continuous and satisfies*

$$\mu \llcorner G_d^*(K,M) = \int_0^\infty \mu(K + rB^d, M, \cdot)\, \mathrm{d}r.$$

The measure $\mu(K,M,\cdot)$ is given by

$$\mu(K,M,A) = 2\int_{SO_d} \Phi_{d-1}(K - \vartheta M, T(A,\vartheta))\, \nu(\mathrm{d}\vartheta) \tag{8.51}$$

for $A \in \mathcal{B}(G_d)$, where

$$T(A,\vartheta) := \{x \in \mathbb{R}^d : \gamma(x,\vartheta) \in A\}.$$

(Recall that the map $\gamma : \mathbb{R}^d \times SO_d \to G_d$ is given by $\gamma(x,\vartheta)y := \vartheta y + x$ for $y \in \mathbb{R}^d$.)

Proof. Let $A \subset G_d^*(K,M)$ be a Borel set. We have

$$\begin{aligned}
\mu(A) &= \int_{SO_d}\int_{\mathbb{R}^d} \mathbf{1}_A(\gamma(x,\vartheta))\, \lambda(\mathrm{d}x)\, \nu(\mathrm{d}\vartheta) \\
&= \int_{SO_d}\int_{\mathbb{R}^d \setminus (K-\vartheta M)} \mathbf{1}_A(\gamma(x,\vartheta))\, \lambda(\mathrm{d}x)\, \nu(\mathrm{d}\vartheta),
\end{aligned}$$

since, for given $\vartheta \in SO_d$, the relation $\gamma(x,\vartheta) \in G_d^*(K,M)$ is equivalent to $x \in \mathbb{R}^d \setminus (K - \vartheta M)$. By Lemma 8.5.1, applied to $K - \vartheta M$ instead of K, we have

$$\mu(A) = 2\int_{SO_d}\int_0^\infty \int_{\mathbb{R}^d} \mathbf{1}_A(\gamma(x,\vartheta))\, \Phi_{d-1}(K - \vartheta M + rB^d, \mathrm{d}x)\, \mathrm{d}r\, \nu(\mathrm{d}\vartheta).$$

By Theorem 14.2.2(c), the mapping

$$(\vartheta, r) \mapsto \Phi_{d-1}(K - \vartheta M + rB^d, \cdot)$$

is continuous, hence Lemma 12.1.2 yields the measurability of the mapping

$$(\vartheta, r) \mapsto \int_{\mathbb{R}^d} \mathbf{1}_A(\gamma(x,\vartheta))\, \Phi_{d-1}(K - \vartheta M + rB^d, \mathrm{d}x).$$

An application of Fubini's theorem gives

$$\begin{aligned}
\mu(A) &= 2\int_0^\infty \int_{SO_d} \Phi_{d-1}(K - \vartheta M + rB^d, T(A,\vartheta))\, \nu(\mathrm{d}\vartheta)\, \mathrm{d}r \\
&= \int_0^\infty \mu(K + rB^d, M, A)\, \mathrm{d}r,
\end{aligned}$$

where (8.51) was used as a definition. Since the measure $\Phi_{d-1}(K - \vartheta M, \cdot)$ is concentrated on $\mathrm{bd}\,(K - \vartheta M)$, the measure defined by (8.51) is concentrated on $G_d(K, M)$.

We show that $\mu(K, M, \cdot)$ depends weakly continuously on (K, M). Let K_i, M_i be convex bodies such that $(K_i, M_i) \to (K, M)$ for $i \to \infty$. Let $A \subset G_d$ be open; then for given $\vartheta \in SO_d$, the set $T(A, \vartheta)$ is open, too. Using the weak continuity of Φ_{d-1} and the Lemma of Fatou, we obtain

$$\begin{aligned}
\mu(K, M, A) &= \int_{SO_d} \Phi_{d-1}(K - \vartheta M, T(A, \vartheta))\, \nu(\mathrm{d}\vartheta) \\
&\le \int_{SO_d} \liminf_{i \to \infty} \Phi_{d-1}(K_i - \vartheta M_i, T(A, \vartheta))\, \nu(\mathrm{d}\vartheta) \\
&\le \liminf_{i \to \infty} \int_{SO_d} \Phi_{d-1}(K_i - \vartheta M_i, T(A, \vartheta))\, \nu(\mathrm{d}\vartheta) \\
&= \liminf_{i \to \infty} \mu(K_i, M_i, A).
\end{aligned}$$

Similarly one shows that $\lim_{i \to \infty} \mu(K_i, M_i, G_d) = \mu(K, M, G_d)$, which then yields the assertion.

To prove the uniqueness, we assume that there are two mappings $(K, M) \mapsto \mu^{(k)}(K, M, \cdot)$, $k = 1, 2$, with the required properties; then, in particular,

$$\mu \llcorner G_d^*(K, M) = \int_0^\infty \mu^{(k)}(K + rB^d, M, \cdot)\, \mathrm{d}r.$$

We choose functions f on $[0, \infty)$ and h on G_d, both continuous and with compact support. For $k = 1, 2$ we get

$$\begin{aligned}
&\int_{G_d^*(K,M)} f(r(K, gM)) h(g)\, \mu(\mathrm{d}g) \\
&= \int_0^\infty \int_{G_d} f(r(K, gM)) h(g)\, \mu^{(k)}(K + rB^d, M, \mathrm{d}g)\, \mathrm{d}r \\
&= \int_0^\infty f(r) \int_{G_d} h(g)\, \mu^{(k)}(K + rB^d, M, \mathrm{d}g)\, \mathrm{d}r.
\end{aligned}$$

The inner integral depends continuously on r. Since f was arbitrary, we conclude that

$$\int_{G_d} h(g)\, \mu^{(1)}(K + rB^d, M, \mathrm{d}g) = \int_{G_d} h(g)\, \mu^{(2)}(K + rB^d, M, \mathrm{d}g)$$

for all $r \ge 0$. Since h was arbitrary, we deduce that

$$\mu^{(1)}(K, M, \cdot) = \mu^{(2)}(K, M, \cdot).$$

This completes the proof of the theorem. □

The measure $\mu(K, M, \cdot)$ is called the **touching measure** of the convex bodies K and M.

Remark. The touching measure $\mu(K, M, \cdot)$ is not symmetric in K and M; in the background is the idea that K is fixed and M is moving. However, it is evident that

$$\mu(M, K, A) = \mu(K, M, A^{-1}) \qquad \text{for } A \in \mathcal{B}(G_d).$$

Remark. Let $r_{K,M} : G_n^*(K, M) \to (0, \infty)$ be defined by

$$r_{K,M}(g) := r(K, gM).$$

Then, Theorem 8.5.1 says that the family

$$\{\mu(K + rB^d, M, \cdot)\}_{r>0}$$

is a disintegration of the measure $\mu^* := \mu \llcorner G_d^*(K, M)$ with respect to the function $r_{K,M}$; thus, $\mu(K + rB^d, M, \cdot)$ is the (continuous) regular version of the conditional measure $\mu^*(\cdot \mid r_{K,M} = r)$ (cf. results on regular conditional probabilities in Kallenberg [386, Th. 6.3]).

The touching measure of K and M can be normalized to a probability measure if $\mu(K, M, G_d) > 0$. We first check when the latter condition is satisfied.

Corollary 8.5.1. *We have*

$$\mu(K, M, G_d) = 2 \sum_{k=0}^{d-1} c_{1,d}^{d-k,k+1} V_k(K) V_{d-1-k}(M),$$

where the coefficients are given by (5.5).

Proof. Since $T(G_d, \vartheta) = \mathbb{R}^d$, Theorem 8.5.1 gives

$$\begin{aligned}
\mu(K, M, G_d) &= 2 \int_{SO_d} \Phi_{d-1}(K - \vartheta M, T(G_d, \vartheta))\, \nu(\mathrm{d}\vartheta) \\
&= 2 \int_{SO_d} V_{d-1}(K - \vartheta M)\, \nu(\mathrm{d}\vartheta) \\
&= 2 \sum_{k=0}^{d-1} c_{1,d}^{d-k,k+1} V_k(K) V_{d-1-k}(M),
\end{aligned}$$

where Theorem 6.1.1 was used. $\square$

By the preceding result, in order to have $\mu(K,M,G_d)>0$, the condition $\dim K+\dim M\ge d-1$ is necessary and sufficient. For the following we assume for simplicity that $\dim K\ge d-1$ and $M\ne\emptyset$. By

$$\frac{\mu(K,M,\cdot)}{\mu(K,M,G_d)}$$

we can then define the **touching probability measure** of K and M. We denote it by

$$\mathbb{P}(\cdot\mid g\in G_d(K,M))=\mathbb{P}(\cdot\mid K \text{ and } gM \text{ touch}).$$

Having thus described a probabilistic model for the idea of randomly touching convex bodies, we ask for the probability

$$\mathbb{P}(A\cap gB\ne\emptyset\mid g\in G_d(K,M)),$$

for Borel sets $A\subset \operatorname{bd}K$ and $B\subset\operatorname{bd}M$. In order to apply Theorem 8.5.1, we have to consider, for

$$\widetilde{A}:=\{g\in G_d(K,M): A\cap gB\ne\emptyset\}$$

and $\vartheta\in SO_d$, the set $T(\widetilde{A},\vartheta)$. The relation $x\in T(\widetilde{A},\vartheta)$ is equivalent to

$$\gamma(x,\vartheta)\in G_d(K,M)\quad\text{and}\quad A\cap(\vartheta B+x)\ne\emptyset.$$

Here, the first condition is equivalent to $x\in\operatorname{bd}(K-\vartheta M)$, and the second to $x\in A-\vartheta B$; thus

$$T(\widetilde{A},\vartheta)=(A-\vartheta B)\cap\operatorname{bd}(K-\vartheta M).$$

It follows that

$$\mu(K,M,\widetilde{A})=2\int_{SO_d}\Phi_{d-1}(K-\vartheta M,A-\vartheta B)\,\nu(\mathrm{d}\vartheta).$$

This integral can be computed with the aid of Theorem 6.1.2; we obtain

$$\mu(K,M,\widetilde{A})=2\sum_{k=0}^{d-1}c_{1,d}^{d-k,k+1}\Phi_k(K,A)\Phi_{d-1-k}(M,B).$$

Thus we have proved the following result.

Theorem 8.5.2. *If $K,M\in\mathcal{K}'$ are convex bodies with $\dim K\ge d-1$ and $A,B\in\mathcal{B}(\mathbb{R}^d)$ are Borel sets with $A\subset\operatorname{bd}K$ and $B\subset\operatorname{bd}M$, then*

$$\mathbb{P}(A\cap gB\ne\emptyset\mid K\text{ and } gM\text{ touch})=\frac{\sum_{k=0}^{d-1}c_{1,d}^{d-k,k+1}\Phi_k(K,A)\Phi_{d-1-k}(M,B)}{\sum_{k=0}^{d-1}c_{1,d}^{d-k,k+1}V_k(K)V_{d-1-k}(M)}.$$

Let us consider the special case where K and M are polytopes with interior points and where the specified boundary sets are

$$A^{(i)} := \bigcup_{F \in \mathcal{F}_i(K)} \operatorname{relint} F,$$

$$B^{(j)} := \bigcup_{G \in \mathcal{F}_j(M)} \operatorname{relint} G,$$

with $i, j \in \{0, \ldots, d-1\}$. In this case,

$$\Phi_k(K, A^{(i)}) = \begin{cases} V_i(K) & \text{if } k = i, \\ 0 & \text{if } k \neq i, \end{cases}$$

and similarly for $\Phi_k(M, B^{(j)})$. Therefore, Theorem 8.5.1 yields

$$\mathbb{P}(A^{(i)} \cap gB^{(j)} \neq \emptyset \mid g \in G_d(K, M)) = 0 \quad \text{if } i + j \neq d - 1,$$

and

$$\mathbb{P}\left(A^{(i)} \cap gB^{(d-1-i)} \neq \emptyset \mid g \in G_d(K, M)\right)$$

$$= \frac{c_{1,d}^{d-i,i+1} V_i(K) V_{d-1-i}(M)}{\sum_{k=0}^{d-1} c_{1,d}^{d-k,k+1} V_k(K) V_{d-1-k}(M)}.$$

Example. Let $d = 3$, and let $K = M = C^3$, the unit cube. The total touching measure is given by

$$\mu(C^3, C^3, G_3(C^3, C^3)) = 2 \sum_{k=0}^{2} c_{1,3}^{3-k,k+1} V_k(C^3) V_{2-k}(C^3)$$

$$= 4V_2(C^3) + \frac{\pi}{2} V_1^2(C^3) = 12 + \frac{\pi}{2} 9 = \frac{3}{2}(8 + 3\pi).$$

The only events with positive touching probability are

$$E^{(1)} := \{g \in G_3(C^3, C^3) : A^{(0)} \cap gB^{(2)} \neq \emptyset\}$$

(a vertex of the fixed cube touches a two-face of the moving cube),

$$E^{(2)} := \{g \in G_3(C^3, C^3) : A^{(1)} \cap gB^{(1)} \neq \emptyset\}$$

(edge against edge), and

$$E^{(3)} := \{g \in G_3(C^3, C^3) : A^{(2)} \cap gB^{(0)} \neq \emptyset\}$$

(a two-face of the fixed cube touches a vertex of the moving cube). We have

$$\mathbb{P}(E^{(1)} \mid G_3(C^3, C^3)) = \mathbb{P}(E^{(3)} \mid G_3(C^3, C^3)) = \frac{6}{\frac{3}{2}(8 + 3\pi)} = \frac{4}{8 + 3\pi}$$

and

$$\mathbb{P}(E^{(2)} \mid G_3(C^3, C^3)) = \frac{\frac{9\pi}{2}}{\frac{3}{2}(8 + 3\pi)} = \frac{3\pi}{8 + 3\pi}.$$

Thus, the chances of the bet described at the beginning of this section are given by

$$\mathbb{P}(\text{touching 'vertex against face'} \mid G_3(C^3, C^3)) = \frac{8}{8 + 3\pi} = 0.4591$$

and

$$\mathbb{P}(\text{touching 'edge against edge'} \mid G_3(C^3, C^3)) = \frac{3\pi}{8 + 3\pi} = 0.5409.$$

In an analogous way, we can also treat the touching of convex bodies by q-dimensional planes. We only sketch the corresponding considerations. As for touching convex bodies, we define similarly

$$A(d, q, K) := \{E \in A(d, q) : E \text{ touches } K\}$$

and

$$A(d, q, K)^* := \{E \in A(d, q) : E \cap K = \emptyset\}.$$

Then we have a disjoint decomposition

$$A(d, q, K)^* = \bigcup_{r>0} A(d, q, K + rB^d)$$

and a disintegration

$$\mu_q \llcorner A(d, q, K)^* = \int_0^\infty \mu_q(K + rB^d, \cdot) \, \mathrm{d}r.$$

How the measure $\mu_q(K, \cdot)$ has to be defined for this purpose, can be seen as follows. For fixed $L_q \in G(d, q)$ and for a Borel set $A \subset A(d, q, K)^*$, we have, by the definition of μ_q,

$$\mu_q(A) = \int_{SO_d} \int_{L_q^\perp} \mathbf{1}_A(\gamma_q(x, \vartheta)) \, \lambda_{L_q^\perp}(\mathrm{d}x) \, \nu(\mathrm{d}\vartheta)$$

(recall that $\gamma_q(x, \vartheta) = \vartheta(L_q + x)$). Since $\gamma_q(x, \vartheta) \in A(d, q, K)^*$ is equivalent to $x \notin (\vartheta^{-1}K)|L_q^\perp$, the inner integral extends only over $L_q^\perp \setminus (\vartheta^{-1}K)|L_q^\perp$, and Lemma 8.5.1 yields

$$\mu_q(A) = 2\int_{SO_d}\int_0^\infty\int_{L_q^\perp} \mathbf{1}_A(\vartheta(L_q+x))$$
$$\times\,\Phi_{d-q-1}((\vartheta^{-1}K)|L_q^\perp + r(B^d\cap L_q^\perp), \mathrm{d}x)\,\mathrm{d}r\,\nu(\mathrm{d}\vartheta)$$
$$= 2\int_0^\infty\int_{SO_d} \Phi_{d-q-1}((\vartheta^{-1}(K+rB^d)|L_q^\perp, T_q'(A,\vartheta))\,\nu(\mathrm{d}\vartheta)\,\mathrm{d}r$$
$$= 2\int_0^\infty\int_{SO_d} \Phi_{d-q-1}((K+rB^d)|\vartheta L_q^\perp, T_q(A,\vartheta))\,\nu(d\vartheta)\,\mathrm{d}r,$$

with $T_q'(A,\vartheta) := \{x \in L_q^\perp : \vartheta(L_q+x)\in A\}$ and

$$T_q(A,\vartheta) := \vartheta T_q'(A,\vartheta) = \{x\in L_q^\perp : \vartheta L_q + x \in A\}.$$

Thus, we have to define

$$\mu_q(K,A) := 2\int_{SO_d} \Phi_{d-q-1}(K|\vartheta L_q^\perp, T_q(A,\vartheta))\,\nu(\mathrm{d}\vartheta).$$

The remaining assertions and proofs are then completely analogous to the case of touching convex bodies. We call $\mu_q(K,\cdot)$ the q-**flat touching measure** of K.

Now we prescribe again a Borel set $A\in\mathcal{B}(\mathbb{R}^d)$ with $A\subset \mathrm{bd}\,K$, and we ask for the measure of all q-flats touching K in points of A, that is, for $\mu_q(K,\widetilde{A})$ with

$$\widetilde{A} := \{E\in A(d,q,K) : E\cap A \neq \emptyset\}.$$

Since

$$T_q(\widetilde{A},\vartheta) = (A|\vartheta L_q^\perp)\cap \mathrm{bd}\,(K|\vartheta L_q^\perp),$$

the local projection formula of Theorem 6.2.1 yields the following theorem.

Theorem 8.5.3. *For a convex body $K\in\mathcal{K}'$, a Borel set $A\subset\mathcal{B}(\mathbb{R}^d)$ with $A\subset \mathrm{bd}\,K$, and for $q\in\{0,\dots,d-1\}$, let $\widetilde{A}$ denote the set of all q-flats touching K at A. Then*

$$\mu_q(K,\widetilde{A}) = 2c_{d,1}^{d-q,q+1}\Phi_{d-q-1}(K,A).$$

This shows that the curvature measure $\Phi_j(K,\cdot)$ quantifies the touching of K by $(d-j-1)$-flats. In particular, $V_j(K)$ is proportional to the measure of $(d-j-1)$-flats touching K, $j\in\{0,\dots,d-1\}$.

As before, Theorem 8.5.3 allows the determination of touching probabilities.

Notes for Section 8.5

1. The study of touching probabilities, in the sense described here, was initiated by Firey. This began in [233], with an integral geometric interpretation of the area measures of convex bodies in terms of touching flats. Firey also proposed considering randomly touching three-dimensional dice and asked whether a touching edge-to-edge or a touching vertex-to-face had larger probability. An answer was given by McMullen [468], though without a precise specification of the underlying probability model. Firey [234] investigated general convex bodies touching randomly at given sets of directions. With a different approach and some extensions, this was continued by Schneider [672, 673] (see also [674]). The more intuitive random touching of convex bodies at given sets of boundary points was treated by Schneider [677]; the formula of Theorem 8.5.2 appears there. In all this work, the touching probabilities were defined as certain limits (after first 'thickening' sets of touching positions to obtain sets of positive rigid motion invariant measure). The proper underlying probability measures on the spaces of touching positions, as specified in this section, were constructed by Weil [779, 782]; see also [780]. Further extensions and variants of touching probabilities were treated in Firey [235], Schneider [680], Weil [783], Schneider and Wieacker [719], Molter [557], Weil [789].

2. A similar question, about probabilities of collisions of a randomly moving convex body with a field of convex particles, was investigated by Papaderou–Vogiatzaki and Schneider [595], also with integral geometric methods.

8.6 Extremal Problems for Probabilities and Expectations

Geometric probabilities and expectations of geometric random variables can in general not be computed explicitly. In some cases, sharp estimates for such quantities are known. In particular, some classical inequalities for convex bodies can be interpreted as estimates for moments of geometric random variables defined by convex hulls of independent uniform random points or by intersections of convex bodies with isotropic random flats. In this section, we collect such results, mostly without proofs, for which we refer to the original literature.

First we extend the definitions of the quantities $I(d,q,k)$ (Theorem 8.2.2) and $J(d,q,k)$ (Theorem 8.2.3) from balls to general convex bodies. For a d-dimensional convex body $K \in \mathcal{K}'$ and for $1 \le q \le d$ and $k \ge 0$, we put

$$I(K,q,k) := \int_K \cdots \int_K \nabla_q(x_1,\dots,x_q)^k \,\lambda(\mathrm{d}x_1)\cdots\lambda(\mathrm{d}x_q),$$

$$J(K,q,k) := \int_K \cdots \int_K \Delta_q(x_0,\dots,x_q)^k \,\lambda(\mathrm{d}x_0)\cdots\lambda(\mathrm{d}x_q).$$

For $r > 0$ we have

$$I(rB^d,q,k) = r^{q(d+k)} I(d,q,k),$$

hence

$$I(rB^d,q,k) = I(d,q,k)\left(\frac{V_d(rB^d)}{\kappa_d}\right)^{q(d+k)/d}.$$

The functional $I(\cdot, d, k)$ is invariant under volume-preserving linear transformations, hence for a d-dimensional ellipsoid Q_0 with center 0 we obtain

$$I(Q_0, d, k) = I(d, d, k) \left(\frac{V_d(Q_0)}{\kappa_d} \right)^{d+k}. \tag{8.52}$$

Similarly, using the fact that $J(\cdot, d, k)$ is invariant under volume-preserving affinities, we get

$$J(Q, d, k) = J(d, d, k) \left(\frac{V_d(Q)}{\kappa_d} \right)^{d+k+1}, \tag{8.53}$$

if Q is a d-dimensional ellipsoid.

The special values (8.52) and (8.53) are of interest, since ellipsoids have extremal properties with respect to the functionals we have introduced. We quote the following theorem. The case $k = 1$ is known as the **Busemann random simplex inequality**.

Theorem 8.6.1. *If $K \in \mathcal{K}'$ is a d-dimensional convex body and if $k \geq 1$, then*

$$I(K, d, k) \geq \frac{\kappa_{d+k}^d}{\kappa_d^{d+k}} \left(\prod_{j=1}^{d} \frac{\omega_j}{\omega_{k+j}} \right) V_d(K)^{d+k}.$$

Equality holds if and only if K is an ellipsoid with center 0.

This theorem was proved by Busemann [141] for $k = 1$. His proof, which uses Steiner symmetrization, yields the assertion also for $k \geq 1$; one merely has to use that the function $x \mapsto x^k$, $x \geq 0$, is convex and strictly increasing.

Theorem 8.6.1 admits the following interpretation. We consider d independent uniform random points in the convex body K and denote by $\widetilde{T}(K)$ the random variable given by the volume of the simplex that is the convex hull of these points and the origin. For the kth moment of the random variable $\widetilde{T}(K)/V_d(K)$ we then have

$$\mathbb{E} \left(\frac{\widetilde{T}(K)}{V_d(K)} \right)^k \geq \frac{1}{(d!)^k} \frac{\kappa_{d+k}^d}{\kappa_d^{d+k}} \prod_{j=1}^{d} \frac{\omega_j}{\omega_{k+j}},$$

with equality if and only if K is an ellipsoid with center 0.

Combining Theorem 8.6.1 with the linear Blaschke–Petkantschin formula, we obtain information about volumes of sections of convex bodies by random flats. The case $q = d-1$ of (8.54) is known as the **Busemann intersection inequality**.

Theorem 8.6.2. *If $K \in \mathcal{K}'$ is a d-dimensional convex body and if $1 \leq q \leq d-1$, then*

$$\int_{G(d,q)} V_q(K \cap L)^d \, \nu_q(\mathrm{d}L) \le \frac{\kappa_q^d}{\kappa_d^q} V_d(K)^q. \tag{8.54}$$

Equality holds for $q = 1$ *if and only if* K *is symmetric with respect to* 0, *and for* $q \ge 2$ *if and only if* K *is an ellipsoid with center* 0.

Proof. Let $L \in G(d,q)$. By Theorem 8.6.1, with the triple (K,d,k) replaced by $(K \cap L, q, d-q)$, we have

$$\frac{\kappa_d^q}{\kappa_q^d} V_q(K \cap L)^d \le b_{dq} I(K \cap L, q, d-q).$$

This, together with Theorem 7.2.1, yields

$$\begin{aligned}
&\frac{\kappa_d^q}{\kappa_q^d} \int_{G(d,q)} V_q(K \cap L)^d \, \nu_q(\mathrm{d}L) \\
&\le b_{dq} \int_{G(d,q)} \int_{K \cap L} \cdots \int_{K \cap L} \nabla_q(x_1,\dots,x_q)^{d-q} \, \lambda_L(\mathrm{d}x_1) \cdots \lambda_L(\mathrm{d}x_q) \, \nu_q(\mathrm{d}L) \\
&= \int_K \cdots \int_K \lambda(\mathrm{d}x_1) \cdots \lambda(\mathrm{d}x_q) \\
&= V_d(K)^q.
\end{aligned}$$

Equality holds if and only if, for every subspace $L \in G(d,q)$ satisfying $\dim(K \cap L) = q$, the intersection $K \cap L$ is a q-dimensional ellipsoid (a line segment if $q = 1$) with center 0. The equality condition now follows, due to a theorem of Busemann [142, p. 91]. □

Remark. The fact that (8.54) holds with equality if $q = 1$ and K has symmetry center 0, is just a special case of the usual formula for calculating the volume in spherical coordinates. However, that for $q \ge 2$ an ellipsoid Q_0 with center 0 satisfies the equality

$$\int_{G(d,q)} V_q(Q_0 \cap L)^d \, \nu_q(\mathrm{d}L) = \frac{\kappa_q^d}{\kappa_d^q} V_d(Q_0)^q, \tag{8.55}$$

is not so evident. Equation (8.55) is known as the **Furstenberg–Tzkoni formula**.

Now we turn to analogous considerations for affine instead of linear subspaces. For the proof of the following counterpart to Theorem 8.6.1, which also uses Steiner symmetrization, we refer to Groemer [287] (where, though, the value of $J(Q,d,k)$ for ellipsoids Q and $k > 1$ is not given explicitly). The result is known as the **Blaschke–Groemer inequality**.

Theorem 8.6.3. *If* $K \in \mathcal{K}'$ *is a* d*-dimensional convex body and if* $k \ge 1$, *then*

$$J(K,d,k) \ge \frac{1}{(d!)^k} \frac{\kappa_{d+k}^{d+1}}{\kappa_d^{d+k+1}} \frac{\kappa_{d(d+k)+d}}{\kappa_{(d+1)(d+k)}} \frac{1}{b_{(d+k)d}} V_d(K)^{d+k+1}.$$

Equality holds if and only if K is an ellipsoid.

Thus, if we consider $d+1$ independent uniform random points in the convex body K and if $\widetilde{S}(K)$ is the volume of the simplex which is (a.s.) the convex hull of the points, then

$$\mathbb{E}\left(\frac{\widetilde{S}(K)}{V_d(K)}\right)^k \ge \frac{1}{(d!)^k} \frac{\kappa_{d+k}^{d+1}}{\kappa_d^{d+k+1}} \frac{\kappa_{d(d+k)+d}}{\kappa_{(d+1)(d+k)}} \frac{1}{b_{(d+k)d}},$$

with equality if and only if K is an ellipsoid.

We combine Theorem 8.6.3 with the affine version of the Blaschke–Petkantschin formula and thus obtain the following result about section volumes.

Theorem 8.6.4. *If $K \in \mathcal{K}'$ is a d-dimensional convex body and if $1 \le p < q \le d$, then*

$$\int_{A(d,q)} V_q(K \cap E)^{p+1}\, \mu_q(\mathrm{d}E) \ge \frac{\kappa_q^{p+1}}{\kappa_p^{q+1}} \frac{\kappa_{p(q+1)}}{\kappa_{(p+1)q}} \int_{A(d,p)} V_p(K \cap E)^{q+1}\, \mu_p(\mathrm{d}E).$$

Equality holds always for $p = 1$, and for $p \ge 2$ it holds if and only if K is an ellipsoid.

Proof. We use, in this order, Theorems 7.2.7, 7.1.2, 8.6.3 (from which the values of the constants c_1, c_2 can be seen) and obtain

$$\begin{aligned}
&\int_{A(d,q)} V_q(K \cap F)^{p+1}\, \mu_q(\mathrm{d}F) \\
&= \int_{A(d,q)} \int_{(K\cap F)^{p+1}} \mathrm{d}\lambda_F^{p+1}\, \mu_q(\mathrm{d}F) \\
&= c_1 \int_{A(d,q)} \int_{A(F,p)} \int_{(K\cap F)^{p+1}} \Delta_p^{q-p}\, \mathrm{d}\lambda_E^{p+1}\, \mu_p^F(\mathrm{d}E)\, \mu_q(\mathrm{d}F) \\
&= c_1 \int_{A(d,p)} \int_{(K\cap F)^{p+1}} \Delta_p^{q-p}\, \mathrm{d}\lambda_E^{p+1}\, \mu_p(\mathrm{d}E) \\
&\ge c_1 \int_{A(d,p)} c_2 V_p(K \cap E)^{q+1}\, \mu_p(\mathrm{d}E).
\end{aligned}$$

Equality holds if and only if, for every flat $E \in A(d,p)$ satisfying $\dim(K \cap E) = p$, the intersection $K \cap E$ is a p-dimensional ellipsoid. Again, together with the theorem of Busemann [142, p. 91], this yields the equality conditions. □

We emphasize two special cases of the preceding theorem. For $q = d$, we obtain the inequality

$$\int_{A(d,p)} V_p(K \cap E)^{d+1} \, \mu_p(\mathrm{d}E) \leq \frac{\kappa_p^{d+1}}{\kappa_d^{p+1}} \frac{\kappa_{(p+1)d}}{\kappa_{p(d+1)}} V_d(K)^{p+1}. \tag{8.56}$$

Equality holds always for $p = 1$, and for $p \geq 2$ it holds if and only if K is an ellipsoid.

The case $p = 1$ of Theorem 8.6.4 yields the identity

$$\int_{A(d,q)} V_q(K \cap E)^2 \, \mu_q(\mathrm{d}E) = \frac{\kappa_q}{q+1} \int_{A(d,1)} V_1(K \cap L)^{q+1} \, \mu_1(\mathrm{d}L) \tag{8.57}$$

(if the factor is treated as in the proof of Theorem 8.2.3).

This leads us to the frequently studied random chords of convex bodies. For a convex body $K \in \mathcal{K}'$, the **chord power integrals** are defined by

$$I_k(K) := \frac{\omega_d}{2} \int_{A(d,1)} V_1(K \cap E)^k \, \mu_1(\mathrm{d}E)$$

for $k \geq 1$. (The factor before the integral has only historical reasons and comes from a different normalization of the measure μ_1.) Obviously,

$$I_1(K) = \frac{\omega_d}{2} V_d(K).$$

The case $q = d$ of (8.57) gives

$$I_{d+1}(K) = \frac{d(d+1)}{2} V_d(K)^2. \tag{8.58}$$

For $d = 2$, this is an old result due to Crofton [187].

For the other chord power integrals one can obtain sharp estimates. They are based on the following result, which we quote without proof (see, for example, Pfiefer [603]).

Theorem 8.6.5. *Let $f : (0, \infty) \to \mathbb{R}$ be a strictly decreasing function satisfying $\int_0^a |f(x)| x^{d-1} \, \mathrm{d}x < \infty$ for all $a \in \mathbb{R}$. Then, among all convex bodies K with given volume $V_d(K) > 0$, precisely the balls yield the maximum of the integral*

$$\int_K \int_K f(\|x - y\|) \, \lambda(\mathrm{d}x) \, \lambda(\mathrm{d}y).$$

With this result, the following inequalities for chord power integrals can be obtained.

Theorem 8.6.6. *Every d-dimensional convex body $K \in \mathcal{K}'$ satisfies the inequalities*

$$I_k(K) \le I_k(B^d)\left(\frac{V_d(K)}{\kappa_d}\right)^{(d+k-1)/d}$$

for $1 < k < d+1$, *and*

$$I_k(K) \ge I_k(B^d)\left(\frac{V_d(K)}{\kappa_d}\right)^{(d+k-1)/d}$$

for $k > d+1$. *Equality, for given* k, *holds if and only if* K *is a ball; further,*

$$I_k(B^d) = 2^{k-1}\frac{\kappa_d \kappa_{d+k-1}}{\kappa_k} = \frac{2^{k-1}\pi^{d-\frac{1}{2}}k\Gamma(\frac{1}{2}k)}{\Gamma(\frac{1}{2}d)\Gamma(\frac{1}{2}(d+k+1))}.$$

Proof. Let $j > -d$. Theorem 7.2.7 with $q = 1$ yields

$$\begin{aligned}
&\int_K \int_K ||x_0 - x_1||^j \, \lambda(\mathrm{d}x_0)\, \lambda(\mathrm{d}x_1) \\
&= \frac{\omega_d}{2}\int_{A(d,1)}\int_{K\cap E}\int_{K\cap E} ||x_0 - x_1||^{d+j-1}\, \lambda_E(\mathrm{d}x_0)\, \lambda_E(\mathrm{d}x_1)\, \mu_1(\mathrm{d}E) \\
&= \frac{\omega_d}{2}\int_{A(d,1)} \frac{2}{(d+j)(d+j+1)} V_1(K\cap E)^{d+j+1}\, \mu_1(\mathrm{d}E) \\
&= \frac{2}{(d+j)(d+j+1)} I_{d+j+1}(K).
\end{aligned}$$

To the first double integral, we can apply Theorem 8.6.5, with $f(x) = x^j$ if $-d < j < 0$ and $f(x) = -x^j$ if $j > 0$. Assuming that $V_d(K) = \kappa_d$, we get the inequalities

$$I_{d+j+1}(K) \begin{cases} \le I_{d+j+1}(B^d) & \text{if } -d < j < 0, \\ \ge I_{d+j+1}(B^d) & \text{if } j > 0. \end{cases}$$

For general K with $V_d(K) > 0$, the inequalities of the theorem follow, since I_k is homogeneous of degree $d+k+1$. Moreover, the value

$$\frac{2}{(d+j)(d+j+1)} I_{d+j+1}(B^d) = J(d,1,j)$$

is already known from Theorem 8.2.3, and an easy calculation gives the stated result. □

In order to interpret the obtained results as information about random chords, we consider a d-dimensional convex body $K \in \mathcal{K}'$ and an isotropic random line $\widetilde{G}$ through K, as defined in Section 8.4. Thus, the distribution of $\widetilde{G}$ is given by

$$\frac{\omega_d}{2\kappa_{d-1}}\frac{1}{V_{d-1}(K)}\mu_1 \llcorner \mathcal{E}_K^{(1)} \qquad \text{with } \mathcal{E}_K^{(1)} := \{E \in A(d,1) : K \cap E \neq \emptyset\}.$$

The intersection $\widetilde{G} \cap K$ is a random chord of K. It is called a μ-**random chord** of K, since its definition is based on the invariant measure μ_1. We define the random variable $\widetilde{\sigma}_\mu(K)$ as the length of a μ-random chord of K. For the moments of $\widetilde{\sigma}_\mu(K)$ we then have

$$\mathbb{E}\,\widetilde{\sigma}_\mu(K)^k = \frac{1}{\kappa_{d-1} V_{d-1}(K)} I_k(K).$$

Now Theorem 8.6.6 yields sharp inequalities for these moments involving volume and surface area of K.

Besides isotropic random flats through K, in Section 8.4 we have also considered q-weighted random q-flats through K. Let $\widetilde{G}_\nu$ be a 1-weighted random line through K, and let $\widetilde{\sigma}_\nu(K)$ be the length of the chord $\widetilde{G}_\nu \cap K$. In the literature, $\widetilde{G}_\nu \cap K$ has been called a ν-**random chord** of K. By Corollary 8.4.1, the distribution of $\widetilde{G}_\nu$ has, with respect to the distribution of $\widetilde{G}$, the density

$$E \mapsto \frac{2\kappa_{d-1}}{\omega_d} \frac{V_{d-1}(K)}{V_d(K)} V_1(K \cap E), \qquad E \in A(d,1).$$

Therefore, for the moments of the length $\widetilde{\sigma}_\nu(K)$ of a ν-random chord of K one obtains the representation

$$\mathbb{E}\,\widetilde{\sigma}_\nu(K)^k = \frac{2}{\omega_d} \frac{1}{V_d(K)} I_{k+1}(K). \tag{8.59}$$

Hence, for $0 < k < d$, and given volume $V_d(K)$, the moment $\mathbb{E}\widetilde{\sigma}_\nu(K)^k$ becomes maximal if and only if K is a ball. In particular, the expectation satisfies

$$\mathbb{E}\,\widetilde{\sigma}_\nu(K) \le \frac{4\kappa_{d+1}}{\pi\kappa_d} \left(\frac{V_d(K)}{\kappa_d} \right)^{1/d}, \tag{8.60}$$

with equality precisely if K is a ball. For $d = 2$ and with the normalization $V_2(K) = \kappa_2$, we get

$$\mathbb{E}\,\widetilde{\sigma}_\nu(K) \le \frac{16}{3\pi} = 1.6977,$$

and for $d = 3$ and with the normalization $V_3(K) = \kappa_3$ we have

$$\mathbb{E}\,\widetilde{\sigma}_\nu(K) \le \frac{3}{2} = 1.5.$$

An essentially different result is obtained if random chords are generated in the following way. Let $\widetilde{G}_\lambda$ be the random line that is spanned by two independent uniform random points in K. The intersection $\widetilde{G}_\lambda \cap K$ is called a λ-**random chord** of K. Let $\widetilde{\sigma}_\lambda(K)$ denote the length of this chord. By Theorem 8.4.8, the distribution of $\widetilde{G}_\lambda$ has a density with respect to μ_1 which is given by

$$E \mapsto \frac{\kappa_d}{d+1} \frac{1}{V_d(K)^2} V_1(K \cap E)^{d+1}, \qquad E \in A(d,1).$$

Thus, we have

$$\mathbb{E}\,\widetilde{\sigma}_\lambda(K)^k = \frac{2}{d(d+1)}\frac{1}{V_d(K)^2} I_{d+k+1}(K).$$

From Theorem 8.6.6 we now deduce that, for $k > 0$ and given volume $V_d(K)$, the moment $\mathbb{E}\,\widetilde{\sigma}_\lambda(K)^k$ becomes minimal (not maximal, as above) precisely if K is a ball. In particular, the expectation satisfies

$$\mathbb{E}\,\widetilde{\sigma}_\lambda(K) \geq \frac{2^{d+2}}{d+1}\frac{\kappa_{2d+1}}{\kappa_d \kappa_{d+2}}\left(\frac{V_d(K)}{\kappa_d}\right)^{1/d},$$

with equality if and only if K is a ball. For $d = 2$ and $V_2(K) = \kappa_2$, we get

$$\mathbb{E}\,\widetilde{\sigma}_\lambda(K) \geq \frac{256}{45\pi} = 1.81081,$$

and for $d = 3$ and $V_3(K) = \kappa_3$, we obtain

$$\mathbb{E}\,\widetilde{\sigma}_\lambda(K) \geq \frac{12}{7} = 1.7143.$$

We shall now consider some extremal problems for geometric probabilities related to convex bodies. The proofs combine Blaschke–Petkantschin type formulas with classical inequalities from convex geometry.

Let K be a convex body with interior points. Recall that an isotropic uniform random (isotropic random, for short) q-flat through K is a random q-flat with distribution

$$\frac{\mu_q \llcorner \mathcal{E}_K^{(q)}}{\mu_q(\mathcal{E}_K^{(q)})}, \qquad \mathcal{E}_K^{(q)} := \{E \in A(d,q) : E \cap K \neq \emptyset\},$$

and that

$$\mu_q(\mathcal{E}_K^{(q)}) = c_{0,d}^{q,d-q} V_{d-q}(K),$$

by the Crofton formula (5.6).

Now let r, s be positive integers with $r + s \leq d - 1$, and let E_1, E_2 be independent random flats such that E_1 is an isotropic random r-flat and E_2 is an isotropic random s-flat through K. With probability one, the distance between E_1 and E_2 is realized by a unique pair of points $x_1 \in E_1$ and $x_2 \in E_2$. We denote by $p_{r,s}(K)$ the probability that x_1, x_2 both belong to K.

Theorem 8.6.7. *Let r, s be positive integers with $r + s \leq d - 1$. On the set of d-dimensional convex bodies K, the probability $p_{r,s}(K)$ is maximal if and only if K is a ball.*

Proof. In this proof, $c_1, \dots, c_5$ denote positive constants depending only on d, r, s. By definition,

$$p(K) = \frac{c_1}{V_{d-r}(K)V_{d-s}(K)} I(K)$$

with

$$I(K) := \int_{A(d,s)} \int_{A(d,r)} f(E_1, E_2)\, \mu_r(\mathrm{d}E_1)\, \mu_s(\mathrm{d}E_2),$$

where $f(E_1, E_2) := 1$ if a pair of points realizing the distance of E_1 and E_2 is unique and belongs to K, and 0 otherwise. From Theorem 7.2.9 with $q = 2$ and $p = r + s + 1$ we get

$$I(K) = c_2 \int_{A(d,p)} \int_{A(U,s)} \int_{A(U,r)} f(E_1, E_2) D(E_1, E_2)^{d-p} [E_1, E_2]_{(r,s)}^{d-p} \\ \times \mu_r^U(\mathrm{d}E_1)\, \mu_s^U(\mathrm{d}E_2)\, \mu_p(\mathrm{d}U).$$

To the inner double integral we apply Theorem 7.2.10 (with $\mathbb{R}^d$ replaced by U). The resulting integrals are easily carried out, observing that for $x_1, x_2 \in F \in A(d, 1)$ and subspaces $L_1, L_2 \subset F^\perp$ we have $D(L_1 + x_1, L_2 + x_2) = \|x_1 - x_2\|$. We arrive at

$$I(K) = c_3 \int_{A(d,p)} \int_{A(U,1)} V_1(K \cap F)^{d-r-s+1}\, \mu_1^U(\mathrm{d}F)\, \mu_p(\mathrm{d}U).$$

By Theorem 7.1.2,

$$I(K) = c_4 \int_{A(d,1)} V_1(K \cap F)^{d-r-s+1}\, \mu_1(\mathrm{d}F).$$

Now Theorem 8.6.6 shows that

$$p_{r,s}(K) \le c_5 \frac{V_d(K)^{(d-r)/d}}{V_{d-r}(K)} \frac{V_d(K)^{(d-s)/d}}{V_{d-s}(K)},$$

and inequality (14.31) yields the assertion. □

Let us consider random hyperplanes through a d-dimensional convex body $K \in \mathcal{K}'$ which are not necessarily isotropic, but still uniform, in the following sense. A **uniform random hyperplane through** K is a random hyperplane in $\mathbb{R}^d$ with distribution

$$\frac{\Theta \llcorner \mathcal{H}_K}{\Theta(\mathcal{H}_K)}$$

(recall that $\mathcal{H}_K := \{H \in A(d, d-1) : H \cap K \neq \emptyset\}$), where $\Theta \neq 0$ is a translation invariant, locally finite measure on $A(d, d-1)$. By Theorem 4.4.2 (compare (4.33)), there exist a number $\gamma > 0$ and an even probability measure φ on S^{d-1} such that

$$\int_{A(d,d-1)} f \, \mathrm{d}\Theta = \gamma \int_{S^{d-1}} \int_{-\infty}^{\infty} f(H(u,\tau)) \, \mathrm{d}\tau \, \varphi(\mathrm{d}u) \tag{8.61}$$

for every nonnegative measurable function f on $A(d, d-1)$. We assume that the random hyperplane is nondegenerate, which means that the measure φ is not concentrated on a great subsphere.

Now we consider d independent uniform random hyperplanes through K with the same distribution given above, and we ask for the probability, denoted by $p_d(K, \Theta)$, that their intersection point belongs to K. We shall see that this probability is maximized by a unique homothety class of convex bodies. Similarly to Section 4.6, we define the convex body $B(\Theta)$ with surface area measure

$$S_{d-1}(B(\Theta), \cdot) = \gamma\varphi$$

and symmetry center 0. Existence and uniqueness are guaranteed by Theorem 14.3.1.

Theorem 8.6.8. *Under the above assumptions,*

$$p_d(K, \Theta) \le p_d(B(\Theta), \Theta),$$

and equality holds if and only if K is homothetic to $B(\Theta)$.

Proof. By definition and by (8.61) we have (compare the proof of Theorem 4.4.8)

$$\begin{aligned} & p_d(K, \Theta) \\ & = \frac{1}{\Theta(\mathcal{H}_K)^d} \int_{A(d,d-1)} \cdots \int_{A(d,d-1)} \chi(K \cap H_1 \cap \ldots \cap H_d) \, \Theta(\mathrm{d}H_1) \cdots \Theta(\mathrm{d}H_d) \\ & = \frac{V_d(K)}{\Theta(\mathcal{H}_K)^d} \gamma^d \int_{S^{d-1}} \cdots \int_{S^{d-1}} \nabla_d(u_1, \ldots, u_d) \, \varphi(\mathrm{d}u_1) \cdots \varphi(\mathrm{d}u_d). \end{aligned}$$

From (8.61) and (14.23) we obtain

$$\Theta(\mathcal{H}_K) = \gamma \int_{S^{d-1}} [h(K,u) + h(K,-u)] \, \varphi(\mathrm{d}u) = 2dV(K, B(\Theta), \ldots, B(\Theta)).$$

Minkowski's inequality (14.30) gives

$$\Theta(\mathcal{H}_K)^d \ge (2d)^d V_d(K) V_d(B(\Theta))^{d-1}$$

and hence

$$\frac{V_d(K)}{\Theta(\mathcal{H}_K)^d} \le \frac{1}{(2d)^d} \frac{1}{V_d(B(\Theta))^{d-1}} = \frac{V_d(B(\Theta))}{\Theta(\mathcal{H}_{B(\Theta)})^d}.$$

Together with the equality conditions of Minkowski's inequality, this yields the assertion. □

In a d-dimensional convex body K we now consider a fixed number $m \geq 2$ of independent uniform random points $X_1, \dots, X_m$. We denote by $p_m(K)$ the probability that the circumball of $X_1, \dots, X_m$, that is, the smallest ball containing these points, is contained in K.

Theorem 8.6.9. *The probability $p_m(K)$ is maximal if and only if K is a ball.*

Proof. We denote by $B(x_1, \dots, x_m)$ the circumball of the points $x_1, \dots, x_m$. With probability one, at most $d+1$ and at least two of the points $X_1, \dots, X_m$ lie on the boundary of $B(X_1, \dots, X_m)$, and $B(X_1, \dots, X_m)$ is also the circumball of these points. For $1 \leq q \in \min\{m-1, d\}$, we define

$$f_q(x_1, \dots, x_m) := \begin{cases} 1, \text{ if } B(x_1, \dots, x_m) \subset K, \\ \qquad x_1, \dots, x_{q+1} \in \operatorname{bd} B(x_1, \dots, x_m), \\ \qquad x_{q+2}, \dots, x_m \in \operatorname{int} B(x_1, \dots, x_m), \\ 0, \text{ otherwise.} \end{cases}$$

Then

$$\begin{aligned} p_m(K) &= \frac{1}{V_d(K)^m} \sum_{q=1}^{d} \binom{m}{q+1} \int_K \cdots \int_K f_q(x_1, \dots, x_m)\, \lambda(\mathrm{d}x_1) \cdots \lambda(\mathrm{d}x_m) \\ &= \frac{1}{V_d(K)^m} \sum_{q=1}^{d} \binom{m}{q+1} \kappa_d^{m-q-1} A_q \end{aligned}$$

with

$$A_q := \int_K \cdots \int_K f_q(x_1, \dots, x_{q+1}) r(x_1, \dots, x_{q+1})^{d(m-q-1)}\, \lambda(\mathrm{d}x_1) \cdots \lambda(\mathrm{d}x_{q+1}),$$

where $r(x_1, \dots, x_{q+1})$ is the radius of $B(x_1, \dots, x_{q+1})$. As $f_q(x_1, \dots, x_{q+1}) = 0$ if some $x_i \notin K$, the Blaschke–Petkantschin formula of Theorem 7.2.7 yields

$$\begin{aligned} A_q &= \int_{(\mathbb{R}^d)^{q+1}} f_q r^{d(m-q-1)}\, \mathrm{d}\lambda^{q+1} \\ &= b_{dq} (q!)^{d-q} \int_{A(d,q)} \int_{E^{q+1}} f_q r^{d(m-q-1)} \Delta_q^{d-q}\, \mathrm{d}\lambda_E^{q+1}\, \mu_q(\mathrm{d}E). \end{aligned}$$

To the inner integral over E^{q+1} we apply the transformation formula of Theorem 7.3.1 (with $\mathbb{R}^d$ replaced by E), observing that $f_q(z+su_1, \dots, z+su_{q+1}) = 1$ implies $r(z+su_1, \dots, z+su_{q+1}) = s$ and that

$$\Delta_q(z+su_1, \dots, z+su_{q+1}) = s^q \Delta_q(u_1, \dots, u_{q+1}).$$

Denoting by S_E the intersection of the sphere S^{d-1} with the linear subspace parallel to E and by σ_E its spherical Lebesgue measure, we obtain

$$A_q = b_{dq}(q!)^{d-q+1} \int_{A(d,q)} \int_E \int_0^\infty \int_{S_E} \cdots \int_{S_E} f_q(z + ru_1, \dots, z + ru_{q+1})$$
$$\times\, r^{d(m-1)-1} \Delta_q(u_1, \dots, u_{q+1})^{d-q+1}\, \sigma_E(\mathrm{d}u_1) \cdots \sigma_E(\mathrm{d}u_{q+1})$$
$$\times\, \mathrm{d}r\, \lambda_E(\mathrm{d}z)\, \mu_q(\mathrm{d}E).$$

We have $f_q(z+ru_1, \dots, z+ru_{q+1}) = 1$ if and only if $z + rB^d \subset K$, the points $z + ru_1, \dots, z + ru_{q+1}$ lie on the boundary of $z + rB^d$, and $z + rB^d$ is the circumball of these points. The latter holds if and only if $u_1, \dots, u_{q+1}$ do not all lie in an open hemisphere of S_E. Therefore, defining

$$h_E(u_1, \dots, u_{q+1}) := \begin{cases} 1, & \text{if } u_1, \dots, u_{q+1} \in S_E \text{ are not contained} \\ & \quad \text{in an open hemisphere of } S_E, \\ 0, & \text{otherwise,} \end{cases}$$

and

$$g(z, r) := \begin{cases} 1, & \text{if } z + rB^d \subset K, \\ 0, & \text{otherwise,} \end{cases}$$

we get

$$A_q = b_{dq}(q!)^{d-q+1} \int_{A(d,q)} \int_E \int_0^\infty \int_{S_E} \cdots \int_{S_E} h_E(u_1, \dots, u_{q+1}) g(z, r)$$
$$\times\, r^{d(m-1)-1} \Delta_q(u_1, \dots, u_{q+1})^{d-q+1}\, \sigma_E(\mathrm{d}u_1) \cdots \sigma_E(\mathrm{d}u_{q+1})$$
$$\times\, \mathrm{d}r\, \lambda_E(\mathrm{d}z)\, \mu_q(\mathrm{d}E).$$

We write

$$M(d,k)$$
$$:= \int_{S^{d-1}} \cdots \int_{S^{d-1}} h_{\mathbb{R}^d}(u_1, \dots, u_{d+1}) \Delta_d(u_1, \dots, u_{d+1})^k\, \sigma(\mathrm{d}u_1) \cdots \sigma(\mathrm{d}u_{d+1}),$$

then

$$\int_{S_E} \cdots \int_{S_E} h_E(u_1, \dots, u_{q+1}) \Delta_q(u_1, \dots, u_{q+1})^{d-q+1}\, \sigma_E(\mathrm{d}u_1) \cdots \sigma_E(\mathrm{d}u_{q+1})$$
$$= M(q, d-q+1).$$

Further, we define

$$J_p(K) := \int_K d(z, \operatorname{bd} K)^p\, \lambda(\mathrm{d}z)$$

for $p \in \mathbb{N}$, where $d(z, \operatorname{bd} K) = \max\{r \geq 0 : z + rB^d \subset K\}$ for $z \in K$ is the distance of z from the boundary of K. Then we obtain

$$
\begin{aligned}
A_q &= b_{dq}(q!)^{d-q+1}M(q,d-q+1)\int_{A(d,q)}\int_{E\cap K}\int_0^{d(z,\operatorname{bd}K)} r^{d(m-1)-1} \\
&\quad \times \mathrm{d}r\,\lambda_E(\mathrm{d}z)\,\mu_q(\mathrm{d}E) \\
&= \frac{1}{d(m-1)} b_{dq}(q!)^{d-q+1}M(q,d-q+1)\int_{A(d,q)}\int_{E\cap K} d(z,\operatorname{bd}K)^{d(m-1)} \\
&\quad \times \lambda_E(\mathrm{d}z)\,\mu_q(\mathrm{d}E) \\
&= \frac{1}{d(m-1)} b_{dq}(q!)^{d-q+1}M(q,d-q+1)\int_K d(z,\operatorname{bd}K)^{d(m-1)}\,\lambda(\mathrm{d}z)
\end{aligned}
$$

(the latter, for example, by (13.9)). Thus, we finally arrive at

$$
\begin{aligned}
&p_m(K) \\
&= \frac{1}{d(m-1)}\frac{J_{d(m-1)}(K)}{V_d(K)^m}\sum_{q=1}^{d}\binom{m}{q+1}\kappa_d^{m-q-1} b_{dq}(q!)^{d-q+1}M(q,d-q+1).
\end{aligned}
$$

The inequality

$$
\frac{J_p(K)}{V_d(K)^{(d+p)/d}} \le \frac{1}{\binom{d+p}{d}\kappa_d^{p/d}} \tag{8.62}
$$

was proved for $d=3$ by Bol [112] and was extended in Bauer and Schneider [83]. Equality for some $p\in\mathbb{N}$ holds if and only if K is a ball. This finishes the proof. □

The explicit values of $p_m(B^d)$, which would require the computation of $M(d,k)$, are unknown, with the exception of

$$
p_2(B^d) = 2^d\binom{2d}{d}^{-1} \qquad \text{and} \qquad p_m(B^2) = \frac{m}{2m-1}.
$$

The first follows from the easily established relation $M(1,k)=2^{k+1}$. For the second, we compute the value of

$$
M(2,1) = \int_{S^1}\int_{S^1}\int_{S^1} h_{\mathbb{R}^2}(u_1,u_2,u_3)\Delta_2(u_1,u_2,u_3)\,\sigma(\mathrm{d}u_1)\,\sigma(\mathrm{d}u_2)\,\sigma(\mathrm{d}u_3)
$$

as follows. The integrand does not change if u_1,u_2,u_3 undergo the same rotation. Therefore, we may fix u_1, perform the other two integrations, and multiply the result by 2π, to get $M(2,1)$. If the circle S^1 is oriented, the integration variables u_2,u_3 occur (up to a set of measure zero) either in the order u_1,u_2,u_3 or in the order u_1,u_3,u_2. Each order gives the same contribution to the integral. It follows that

$$
M(2,1) = 4\pi\int_0^{\pi}\int_{\pi-\varphi}^{\pi}\frac{1}{2}[\sin\varphi+\sin\psi-\sin(\varphi+\psi)]\,\mathrm{d}\psi\,\mathrm{d}\varphi = 6\pi^2.
$$

As a variant of the preceding problem, we consider $d+1$ independent uniform random points in a d-dimensional convex body K and ask for the probability, denoted by $p_O(K)$, that the circumsphere of these points is entirely contained in K. The **circumsphere** of $d+1$ points $x_0, \dots, x_d$ in general position is the unique $(d-1)$-sphere through these points (this must be distinguished from the boundary of the circumball).

Theorem 8.6.10. *The probability $p_O(K)$ is maximal if and only if K is a ball.*

Proof. By a similar, though simpler, argument to the foregoing proof, using Theorem 7.3.1, we find that

$$p_O(K) = \frac{J_{d^2}(K)}{V_d(K)^{d+1}} \frac{d!}{d^2} \int_{S^{d-1}} \cdots \int_{S^{d-1}} \Delta_d(u_0, \dots, u_d)\, \sigma(\mathrm{d}u_0) \cdots \sigma(\mathrm{d}u_d).$$

As in the previous proof, the assertion follows from inequality (8.62) and its equality condition. □

The explicit value of $p_O(B^d)$ can be obtained from Theorem 8.2.3.

Notes for Section 8.6

1. The proofs of Theorems 8.6.1, 8.6.3, 8.6.5 rest on Steiner symmetrization. Apparently, Blaschke [102] was the first to apply Steiner symmetrization to solve an extremal problem for a geometric expectation. Following him, Busemann [141] proved Theorem 8.6.1, and Groemer [287] obtained Theorem 8.6.3. Generalizations of the latter result are found in Groemer [288, 292], Schöpf [722], Hartzoulaki and Paouris [322], Campi and Gronchi [156]. Inequality (8.54) for $q = d-1$ is due to Busemann [141]; for the general case, see Busemann and Straus [144] or Grinberg [286]. Equation (8.55) was first observed, in a more general context, by Furstenberg and Tzkoni [241]; a simple proof using the Blaschke–Petkantschin formula was given by Miles [528]. The identity (8.57) (which for $d = 3$ appears in Blaschke and Varga [109]) was pointed out to us by Lothar Heinrich. Inequality (8.56) was proved in Schneider [687] (second part of Theorem 1). It was also shown in [687] that for $q \in \{2, \dots, d-1\}$, $k \in \{q+1, \dots, d\}$, on the set of convex bodies K with given positive volume, the functional

$$\int_{A(d,q)} V_q(K \cap E)^k \, \mu_q(\mathrm{d}E)$$

attains its maximum precisely at the balls.

The integrals appearing in Theorem 8.6.2 are usually renormalized in the form

$$\widetilde{\Phi}_{d-q}(K) := \frac{\kappa_d}{\kappa_q} \left(\int_{G(d,q)} V_q(K \cap L)^d \nu_q(\mathrm{d}L) \right)^{1/d}$$

for $q = 1, \dots, d-1$, supplemented by $\widetilde{\Phi}_0(K) := V_d(K)$ and $\widetilde{\Phi}_d(K) := \kappa_d$. The functional $\widetilde{\Phi}_{d-q}$ is known as the $(d-q)$th **dual affine quermassintegral**. It is invariant under volume-preserving linear transformations (Grinberg [286]). It has been asked whether the inequality

$$\kappa_d^i \widetilde{\Phi}_j(K)^{d-i} \leq \kappa_d^j \widetilde{\Phi}_i(K)^{d-j}$$

holds for $0 \leq i < j \leq d$ (Problem 9.6 in the first edition of [244]). However, Gardner [245] found that this inequality does not hold, even for 0-symmetric convex bodies, if $1 = i < j \leq d-1$.

The functionals appearing in inequality (8.57), renormalized as

$$\overline{\Phi}_{d-q}(K) := \frac{\kappa_d}{\kappa_q} \left(\int_{A(d,q)} V_q(K \cap E)^{d+1} \mu_q(\mathrm{d}E) \right)^{1/(d+1)},$$

$q = 0, \dots, d-1$, are known as **mean dual affine quermassintegrals**. They are invariant under volume-preserving affine transformations of K. This follows from the invariance of the dual quermassintegrals under volume-preserving linear transformations, since

$$\overline{\Phi}_{d-q}(K) = \left(\frac{\kappa_d}{\kappa_q} \int_K \widetilde{\Phi}_{d-q}(K-t)^d \,\lambda(\mathrm{d}t) \right)^{1/(d+1)}. \tag{8.63}$$

In fact, writing $f(E) := V_q(K \cap E)^d$ for $E \in A(d,q)$, we have

$$\begin{aligned}
&\int_K \int_{G(d,q)} V_q((K-t) \cap L)^d \,\nu_q(\mathrm{d}L)\,\lambda(\mathrm{d}t) \\
&= \int_K \int_{G(d,q)} f(L+t)\,\nu_q(\mathrm{d}L)\,\lambda(\mathrm{d}t) \\
&= \int_{G(d,q)} \int_{\mathbb{R}^d} f(L+t) \mathbf{1}_K(t)\,\lambda(\mathrm{d}t)\,\nu_q(\mathrm{d}L) \\
&= \int_{G(d,q)} \int_{L^\perp} \int_L f(L+y) \mathbf{1}_K(y+z)\,\lambda_L(\mathrm{d}z)\,\lambda_{L^\perp}(\mathrm{d}y)\,\nu_q(\mathrm{d}L) \\
&= \int_{G(d,q)} \int_{L^\perp} f(L+y) V_q(K \cap (L+y))\,\lambda_{L^\perp}(\mathrm{d}y)\,\nu_q(\mathrm{d}L) \\
&= \int_{A(d,q)} f(E) V_q(K \cap E)\,\mu_q(\mathrm{d}E),
\end{aligned}$$

from which (8.63) follows.

The $(d-q)$th **affine quermassintegral** of a convex body K is defined by

$$\Phi_{d-q}(K) := \frac{\kappa_d}{\kappa_q} \left(\int_{G(d,q)} V_q(K|L)^{-d} \nu_q(\mathrm{d}L) \right)^{-1/d}$$

for $q = 1, \dots, d-1$, and $\Phi_0(K) := V_d(K)$, $\Phi_d(K) := \kappa_d$. It is invariant under volume-preserving affine transformations (Grinberg [286]). Lutwak [443] has asked whether the inequality

$$\kappa_d^i \Phi_j(K)^{d-i} \geq \kappa_d^j \Phi_i(K)^{d-j}$$

holds for $0 \leq i < j \leq d$. Equality holds if K is a centered ellipsoid.

Some of the inequalities of this section hold for much more general sets than convex bodies. See Pfiefer [603] for an extension of the Blaschke–Groemer inequality to compact sets, and Gardner [245] for generalizations, for example, of Theorem 8.6.2 to bounded Borel sets, complete with equality conditions.

2. For hints to literature on chord power integrals, see Santaló [662, pp. 48, 238] and [664]. Theorem 8.6.5 goes back, in principle, to Blaschke [103] and Carleman [160]; a general version was proved by Pfiefer [603].

The investigation of different types of random secants of a convex body (namely, μ, ν, λ random) goes back to Kingman [411], Coleman [178], Enns and Ehlers [217]. The inequality (8.60), which follows from (8.59) and Theorem 8.6.6, had been conjectured by Enns and Ehlers [217] and was proved independently by Davy [200], Schneider [687], and Santaló [664].

In a series of papers, Enns and Ehlers [218, 219, 220, 221] have studied further types of random chords and rays.

In the plane, Xie and Jiang [821] studied the 'double chord power integrals'

$$\int_{A(2,1)} \int_{A(2,1)} \mathbf{1}\{E \cap F \cap K \neq \emptyset\} V_1(K \cap E)^m V_1(K \cap F)^n \, \mu_1(\mathrm{d}E) \, \mu_1(\mathrm{d}F)$$

for convex bodies K and nonnegative integers m, n.

3. For $d = 3$, formula (8.58) reads

$$\int_{A(3,1)} V_1(K \cap E)^4 \, \mu_1(\mathrm{d}E) = \frac{3}{\pi} V_3(K)^2.$$

Variants of this formula, also for non-convex sets, have applications in stereology; see Cabo and Baddeley [145, 146] and the references given there.

4. In the proof of Theorem 8.6.6, the classical formula

$$\int_{K \times K} \|x_1 - x_2\|^j \, \lambda^2(\mathrm{d}(x_1, x_2)) = \frac{2}{(d+j)(d+j+1)} I_{d+j+1}(K)$$

was used. This formula was generalized by Piefke [605], to pairs of convex bodies and more general functions of the distance.

The distribution of the chord length $\widetilde{\sigma}_\mu(K)$ (of a μ-random chord of K) can be determined from the distribution of the distance between two independent uniform random points in K, and conversely. General formulas to this effect were proved by Piefke [604].

5. Chord length distributions and the covariogram problem. It was an old question of Blaschke (for $d = 2$, see [107, p. 52]) whether the moments of the random chord length $\widetilde{\sigma}_\mu(K)$ determine the convex body K uniquely, up to a rigid motion or reflection. This was disproved by Mallows and Clark [448], who constructed two non-congruent polygons with the same chord length distribution. Gates [247] showed that triangles and quadrangles can be reconstructed from their chord length distributions. Further results on the chord length distribution of planar convex sets are found in Waksman [776] and Gates [249].

A different question arises if one considers, for each direction $u \in S^{d-1}$, the distribution of the length of the uniform random chord of the convex body K with direction u. Determination of a convex body K by these distributions, for all directions, is equivalent to the determination by its **covariogram**, which is the function defined by $C(K,x) := V_d(K \cap (K+x))$, $x \in \mathbb{R}^d$. Matheron [458] introduced this covariogram (more generally, for functions), and in [463] he conjectured that a planar convex body with interior points is uniquely determined (within the class of convex bodies) by its covariogram, up to translation or reflection in a point. An equivalent problem is whether a convex body K is determined, up to translation or reflection, by the distribution of $X-Y$, where X and Y are independent uniform random points in K. In the plane, an affirmative answer to the covariogram problem for convex polygons was given by Nagel [572]. Various partial results were obtained by several authors (see the references in [41]), until Averkov and Bianchi [41] finally settled the problem completely for arbitrary convex bodies in the plane. Examples show that convexity is essential in this characterization. Bianchi [100] found counterexamples to the covariogram conjecture in dimensions $d \geq 4$, and a positive answer for three-dimensional polytopes in [101]. The general three-dimensional case is still open. For $d \geq 3$, most convex bodies, in the sense of Baire category, are determined by their covariogram; this was proved by Goodey, Schneider and Weil [276].

6. Theorem 8.6.7 was proved for $d = 3$ and $r = s = 1$ by Knothe [421], for $d \geq 3$ and $r + s = 1$ by Schneider [687], and in general by Affentranger [7]. An extension to two convex bodies appears in Wu [820].

7. More generally than before Theorem 8.6.8, let $k \in \{2, \dots, d\}$ and consider k independent uniform random hyperplanes through a d-dimensional convex body K with distribution defined by Θ. The hyperplanes are isotropic if Θ is also rotation invariant and thus a multiple of the Haar measure μ_{d-1} on $A(d, d-1)$. Let $p_k(K, \Theta)$ denote the probability that the intersection of the hyperplanes meets the convex body K. Miles [520] has proved that $p_k(K, \mu_{d-1}) \leq p_k(B^d, \mu_{d-1})$, with equality if and only if K is a ball. This was extended by Schneider [687] to intersections of tuples of independent isotropic uniform random flats of different dimensions through K. Miles had also conjectured that $p_k(B^d, \Theta) \leq p_k(B^d, \mu_{d-1})$. This was proved by Schneider [686]. (Essentially equivalent is the proof by which Thomas [756] proved Theorem 4.6.5.) Equality holds if and only if the random hyperplanes are isotropic. Theorem 8.6.8 was also proved in [686]. There it was further shown that for $d = 2$ one has $p_2(K, \Theta) \leq 1/2 = p_2(B^2, \mu_1)$, whereas for $d > 2$ and a d-dimensional cube C, one has $p_d(C, S_{d-1}(C, \cdot)) = d^{-d} d! > p_d(B^d, \mu_{d-1})$. This disproved the conjecture of Miles [520] that $p_k(K, \Theta) \leq p_k(B^d, \mu_{d-1})$ for all K and Θ. In Bauer and Schneider [83], the counterexample was extended to $2 \leq k < d$. The maximum of $p_d(K, \Theta)$ for $d > 2$ is still unknown.

8. Theorem 8.6.9 was proved by Bauer and Schneider [83]. It was motivated by Theorem 8.6.10, which is due to Affentranger [9]. Affentranger has also computed the probability that the $(m-2)$-dimensional circumsphere through $m \in \{3, \dots, d+1\}$ independent uniform random points in the unit ball B^d is contained in the ball; see [8] for $m = 3$ and [9] for the general case. A generalization of the planar case of Theorem 8.6.10, in which the class of circles is replaced by a homothety class of strictly convex closed curves of class C^2, was treated by Bauer and Schneider [83].

9. Let $H_1, \dots, H_n$ be independent, identically distributed uniform random hyperplanes through the unit ball B^d ($n \geq d$). We assume that their directional distribution φ, defined by (8.61), is not concentrated on a great subsphere. Let $P_n := \bigcap_{i=1}^n H_i^-$, where H_i^- is the (almost surely unique) closed halfspace bounded by H_i that contains the origin 0. Let $f_0(P_n)$ be the number of vertices of the random polyhedral set P_n. It was shown by Schneider [685] that

$$\lim_{n\to\infty} \mathbb{E}\, f_0(P_n) = 2^{-d} d! V_d(Z_\varphi) V_d(Z_\varphi^\circ),$$

where Z_φ is the zonoid with support function

$$h(Z_\varphi, u) = \frac{1}{2}\int_{S^{d-1}} |\langle u, v\rangle|\, \varphi(\mathrm{d}v), \qquad u \in \mathbb{R}^d.$$

Now (14.45) yields sharp inequalities for $\lim_{n\to\infty} \mathbb{E}\, f_0(P_n)$; in particular, this limit is maximal if the random hyperplanes are isotropic.

10. The following question seems still to be unanswered. Let K be a convex body with interior points and consider the random triangle which is the convex hull of three independent uniform random points in K. It was asked by G.R. Hall [315] whether the probability that this triangle is acute becomes maximal if K is a ball.

11. Consider n independent isotropic uniform random lines through the unit disk B^2 in the plane. Sulanke [748] and Gates [248] have investigated bounds for the probability that all intersection points of the lines are in B^2.

12. Randomly iterated symmetrizations. Most of the inequalities for geometric expectations mentioned in this section are proved by symmetrization. This gives us an opportunity to mention further relations between symmetrizations of convex bodies and probability.

In $\mathbb{R}^d$, let $L_1, L_2, \dots$ be an i.i.d. sequence of random $(d-1)$-dimensional linear subspaces with uniform distribution (that is, given by the rotation invariant probability measure ν_{d-1} on $G(d, d-1)$). Let K be a convex body with interior points, and define $K_1 := K$, $K_{n+1} := S_n K_n$ for $n \in \mathbb{N}$, where S_n denotes the (orthogonal) Steiner symmetrization with respect to L_n. Then, with probability one, the sequence $K_1, K_2, \dots$ converges to a ball. This was proved by Mani–Levitska [449].

Let $\vartheta_1, \vartheta_2, \dots$ be an i.i.d. sequence of rotations with uniform distribution (that is, given by the invariant probability measure ν on SO_d). Let K be a convex body with more than one point, and define $K_n := \frac{1}{n}(\vartheta_1 K + \dots + \vartheta_n K)$ for $n \in \mathbb{N}$. Then, with probability one, the sequence $K_1, K_2, \dots$ converges to a ball (Vitale [769]).

9

Mean Values for Random Sets

For a stationary random closed set Z in $\mathbb{R}^d$, the volume density or specific volume was defined in Section 2.4 by

$$\overline{V}_d(Z) = \frac{\mathbb{E}\,\lambda(Z \cap B)}{\lambda(B)}, \tag{9.1}$$

where $B \subset \mathbb{R}^d$ can be an arbitrary Borel set with $0 < \lambda(B) < \infty$. This important parameter describes the mean volume of the random set per unit volume of the space. It is obtained by a double averaging, stochastic and spatial. The straightforward definition (9.1) has the advantage that it immediately exhibits $\lambda(Z \cap B)/\lambda(B)$ as an unbiased estimator for the specific volume. The situation becomes less simple if one wants to take other quantitative aspects of point sets into account. For example, in several applications one is interested in the mean surface area (the mean perimeter in the plane) per unit volume. Clearly, one cannot just proceed as in the case of (9.1), since the surface area of $Z \cap B$ is in general not defined. Evidently, we must restrict the realizations of the random set Z as well as the 'observation window' B. For that reason, we shall assume in the following that the realizations of the closed random set Z belong to the extended convex ring $\mathcal{S}$, the sets of which have the property that the intersection with any convex body is a finite union of convex bodies. Moreover, the observation window will be a compact convex set W with positive volume. In that case, $Z \cap W$ has a well-defined surface area. However, part of it generally comes from $Z \cap \operatorname{bd} W$ and not from the boundary of Z. To overcome boundary effects caused by the window W, the definition of densities for functionals other than the volume will require additional devices, for example, limit procedures.

The main purpose of Section 9.2 is the specification of a class of stationary random sets Z (with locally polyconvex realizations) and a class of functionals φ (defined on polyconvex sets) such that the limit

$$\overline{\varphi}(Z) := \lim_{r \to \infty} \frac{\mathbb{E}\,\varphi(Z \cap rW)}{V_d(rW)} \tag{9.2}$$

exists for every convex body W with $V_d(W) > 0$. The parameter $\overline{\varphi}(Z)$ is called the φ-density of Z. Important (but not the only) examples of functionals satisfying the assumptions are the intrinsic volumes (or Minkowski functionals) $V_0, \dots, V_{d-1}$. In this way, the V_j-density, or specific jth intrinsic volume, $\overline{V}_j(Z)$, is defined for a large class of stationary random sets. Included are the specific surface area, $2\overline{V}_{d-1}(Z)$, and the specific Euler characteristic, $\overline{V}_0(Z)$.

For the same class of functionals φ, and for stationary particle processes X with polyconvex grains and satisfying a suitable integrability condition, the φ-density, which was defined in Section 4.1 by

$$\overline{\varphi}(X) = \gamma \int_{\mathcal{C}_0} \varphi \, \mathrm{d}\mathbb{Q},$$

can be represented in the form

$$\overline{\varphi}(X) = \lim_{r \to \infty} \frac{\mathbb{E} \sum_{C \in X} \varphi(C \cap rW)}{V_d(rW)},$$

which is analogous to (9.2).

For Boolean models with convex grains and satisfying suitable invariance assumptions, the existence of the specific intrinsic volumes can be obtained in a more direct way, as a consequence of explicit formulas. These formulas will be derived, together with some other results on Boolean models, in Section 9.1. They show, in particular, how the specific intrinsic volumes of a stationary, isotropic Boolean model with convex grains can be computed from the specific intrinsic volumes of the underlying Poisson particle process, and conversely. Especially, the intensity of the underlying particle process can, in principle, be determined from the specific intrinsic volumes of the union set. This seems surprising at first sight, but is, of course, nothing but another manifestation of the strong independence properties of Poisson processes. In the derivation, the integral geometric results of Chapters 5 and 6 will play an important role.

Instead of (9.2), it may even happen, under suitable assumptions, that the limit

$$\lim_{r \to \infty} \frac{\varphi(Z \cap rW)}{V_d(rW)}$$

exists $\mathbb{P}$-almost surely and is a constant, which is then equal to $\overline{\varphi}(Z)$. This ergodic approach to densities is described in Section 9.3.

As soon as densities of various functionals for stationary random sets are defined, the problem arises to estimate these densities from observations of realizations of the random set within a bounded sampling window, or from observations in a lower-dimensional section. In Section 9.4, results from integral geometry are employed to derive various formulas which are useful in this respect.

Mathematical principles of further estimation procedures are the topic of Section 9.5. This section gives selected examples and is not meant as a systematic exposition of estimation methods.

9.1 Formulas for Boolean Models

In our treatment of germ-grain models in Section 4.3, we have already emphasized the particular role played by the Boolean models. Recall that a Boolean model in $\mathbb{R}^d$ is a random closed set of the form

$$Z = \bigcup_{K \in X} K,$$

where X is a Poisson particle process. The Boolean model Z is stationary (isotropic) if and only if the underlying particle process X is stationary (isotropic).

In this section, we shall show how some characteristic parameters of random closed sets specialize in the case of stationary (and possibly isotropic) Boolean models, in particular those with convex grains, and then appear in rather explicit formulas. We begin with evaluating in closed form the capacity functional and the contact distribution functions H and $H_{[0,u]}$, introduced in Section 2.4.

Let Z be a Boolean model, generated as the union set of the Poisson particle process X with intensity measure Θ. If Z and thus X is stationary, the decomposition of Θ yields the intensity γ and the grain distribution $\mathbb{Q}$ of X. We shall call Θ, γ and $\mathbb{Q}$ also the intensity measure, the intensity, and the grain distribution, respectively, of Z. For stationary Z, we assume that $\gamma > 0$.

According to Theorem 3.6.3, the capacity functional of the Boolean model Z satisfies the equation

$$T_Z(C) = 1 - e^{-\Theta(\mathcal{F}_C)} \tag{9.3}$$

for all $C \in \mathcal{C}$. Now we assume that Z is stationary. As in the proof of Theorem 4.1.2, we then have

$$\Theta(\mathcal{F}_C) = \gamma \int_{\mathcal{C}_0} V_d(K - C)\, \mathbb{Q}(dK). \tag{9.4}$$

In general, this integral cannot be simplified further. If, however, Z is a Boolean model with convex grains and if $C \in \mathcal{K}'$, then the volume $V_d(K - C)$ can, according to (14.20), be expressed in terms of mixed volumes, in the form

$$V_d(K - C) = \sum_{j=0}^{d} \binom{d}{j} V(K[j], -C[d-j]).$$

For $C = rB^d$, $r > 0$, this is the Steiner formula (14.5).

The contact distribution function H_M of a random closed set Z with respect to the structuring element $M \in \mathcal{K}'$ with $0 \in M$ is, according to Section 2.4, given by

$$H_M(r) = 1 - \frac{\mathbb{P}(0 \notin Z - rM)}{\mathbb{P}(0 \notin Z)}$$

for $r \geq 0$, if $\mathbb{P}(0 \notin Z) > 0$. For a stationary Boolean model Z with generating Poisson particle process X we always have

$$\mathbb{P}(0 \notin Z) = 1 - T_Z(\{0\}) = \mathrm{e}^{-\overline{V}_d(X)} > 0, \tag{9.5}$$

by (9.3) and (9.4).

Theorem 9.1.1. *Let Z be a stationary Boolean model in $\mathbb{R}^d$ with intensity γ and grain distribution $\mathbb{Q}$. Then*

$$T_Z(C) = 1 - \exp\left(-\gamma \int_{\mathcal{C}_0} V_d(K - C)\, \mathbb{Q}(\mathrm{d}K)\right), \quad C \in \mathcal{C}.$$

For the structuring element $M \in \mathcal{K}'$ with $0 \in M$, the contact distribution function is given by

$$H_M(r) = 1 - \exp\left(-\gamma \int_{\mathcal{C}_0} [V_d(K - rM) - V_d(K)]\, \mathbb{Q}(\mathrm{d}K)\right), \quad r \geq 0.$$

If Z has convex grains, then, for $M \in \mathcal{K}'$,

$$T_Z(M) = 1 - \exp\left(-\gamma \sum_{k=0}^{d} \binom{d}{k} \int_{\mathcal{K}_0} V(-M[k], K[d-k])\, \mathbb{Q}(\mathrm{d}K)\right)$$

and

$$H_M(r) = 1 - \exp\left(-\gamma \sum_{k=1}^{d} \binom{d}{k} r^k \int_{\mathcal{K}_0} V(-M[k], K[d-k])\, \mathbb{Q}(\mathrm{d}K)\right).$$

In particular, in this case the spherical contact distribution function is given by

$$H(r) = 1 - \exp\left(-\sum_{k=1}^{d} \kappa_k r^k \overline{V}_{d-k}(X)\right), \quad r \geq 0,$$

and for $u \in S^{d-1}$, the linear contact distribution function is given by

$$H_{[0,u]}(r) = 1 - \exp\left(-\gamma r \int_{\mathcal{K}_0} V_{d-1}(K|u^\perp)\, \mathbb{Q}(\mathrm{d}K)\right), \quad r \geq 0.$$

If, moreover, Z is isotropic and $M \in \mathcal{K}'$, then

$$T_Z(M) = 1 - \exp\left(-\sum_{k=0}^{d} c_{0,d}^{k,d-k} V_k(M) \overline{V}_{d-k}(X)\right),$$

where the constants are given by (5.5).

(In the formula for $H_{[0,u]}$, the integrand $V_{d-1}(K|u^\perp)$ is the $(d-1)$-dimensional volume of the orthogonal projection of K onto $u^\perp$.)

Proof. The first two assertions about the capacity functional have already been proved. From these, the formulas for $H_M(r)$ follow because of

$$H_M(r) = 1 - \frac{1 - T_Z(rM)}{1 - T_Z(\{0\})}, \qquad r \geq 0.$$

The special form of $H(r)$ in the case of convex grains is obtained, for $M = B^d$, from (14.20), and the expression for $H_{[0,u]}(r)$ follows from

$$V_d(K + r[0,u]) = V_d(K) + rV_{d-1}(K|u^\perp).$$

Now suppose that Z is also isotropic, so that $\mathbb{Q}$ is rotation invariant. Let $M \in \mathcal{K}'$. In the equation

$$\Theta(\mathcal{F}_M) = \gamma \int_{\mathcal{K}_0} \int_{\mathbb{R}^d} \mathbf{1}_{\mathcal{F}_M}(K + x)\, \lambda(\mathrm{d}x)\, \mathbb{Q}(\mathrm{d}K)$$

we can replace K in the integrand by ϑK with a rotation $\vartheta \in SO_d$; this does not change the integral, since $\mathbb{Q}$ is rotation invariant. Then we integrate over all $\vartheta \in SO_d$ with respect to the invariant measure ν and apply Fubini's theorem and the principal kinematic formula (Theorem 5.1.3). For $M, K' \in \mathcal{K}'$ we have $\mathbf{1}_{\mathcal{F}_M}(K') = V_0(M \cap K')$ (since V_0 is the Euler characteristic). This gives

$$\begin{aligned}
\Theta(\mathcal{F}_M) &= \gamma \int_{SO_d} \int_{\mathcal{K}_0} \int_{\mathbb{R}^d} \mathbf{1}_{\mathcal{F}_M}(\vartheta K + x)\, \lambda(\mathrm{d}x)\, \mathbb{Q}(\mathrm{d}K)\, \nu(\mathrm{d}\vartheta) \\
&= \gamma \int_{\mathcal{K}_0} \int_{SO_d} \int_{\mathbb{R}^d} V_0(M \cap (\vartheta K + x))\, \lambda(\mathrm{d}x)\, \nu(\mathrm{d}\vartheta)\, \mathbb{Q}(\mathrm{d}K) \\
&= \gamma \sum_{k=0}^{d} c_{0,d}^{k,d-k} V_k(M) \int_{\mathcal{K}_0} V_{d-k}(K)\, \mathbb{Q}(\mathrm{d}K) \\
&= \sum_{k=0}^{d} c_{0,d}^{k,d-k} V_k(M) \overline{V}_{d-k}(X),
\end{aligned}$$

which completes the proof. □

The contact distribution function H_M of a stationary random closed set Z can be generalized in various directions. First, one can skip the stationarity and consider the distribution of the M-distance $d_M(x, Z)$ of a point x to Z, provided $x \notin Z$. Then, one can take into account not only distances but also directions, contact points and other local geometric information which can be measured from outside Z. Such generalized contact distributions are discussed in Section 11.2 and in the corresponding Notes. They give us more information about the random set Z; in some cases, they even determine the distribution of Z. An example of that kind is presented in Section 9.5.

As an introduction to the main topic of this section, we consider the (already defined) specific volume

$$\overline{V}_d(Z) = \frac{\mathbb{E}\,\lambda(Z \cap W)}{\lambda(W)}$$

for the case of a stationary Boolean model Z (with general compact grains). Here W may be an arbitrary Borel set with $\lambda(W) > 0$. We can find a connection with the volume density $\overline{V}_d(X)$ of the underlying particle process X. In fact,

$$\begin{aligned}\overline{V}_d(Z) &= \mathbb{P}(0 \in Z) = 1 - \mathbb{P}(0 \notin Z)\\ &= 1 - \mathbb{P}(\mathrm{card}\,(X \cap \mathcal{C}_{\{0\}}) = 0) = 1 - \mathrm{e}^{-\Theta(\mathcal{C}_{\{0\}})}\end{aligned}$$

and

$$\begin{aligned}\Theta(\mathcal{C}_{\{0\}}) &= \gamma \int_{\mathcal{C}_0} \int_{\mathbb{R}^d} \mathbf{1}_{\mathcal{C}_{\{0\}}}(K + x)\, \lambda(\mathrm{d}x)\, \mathbb{Q}(\mathrm{d}K)\\ &= \gamma \int_{\mathcal{C}_0} V_d(K)\, \mathbb{Q}(\mathrm{d}K)\\ &= \overline{V}_d(X).\end{aligned}$$

Thus, we have found that

$$\overline{V}_d(Z) = 1 - \mathrm{e}^{-\overline{V}_d(X)}. \tag{9.6}$$

This equality should have come as a surprise: it says that the volume density $\overline{V}_d(X)$ of the particle process X can be determined from the volume density of the union set. This is surprising, since in a given realization of Z one cannot identify the generating particles, due to overlapping, and some particles may even be covered totally by others. The reason for the existence of the exact relation (9.6) lies in the strong independence properties of Poisson processes. The elegant connection between quantitative properties of a stationary Boolean model and its underlying particle process is not restricted to the volume, as we shall soon see.

Let Z be a (not necessarily stationary) Boolean model, generated by the Poisson process X with intensity measure Θ. For simplicity, we assume that the particles in X are a.s. convex, although the following results hold true for polyconvex particles, under an additional integrability condition (see the remark at the end of this section). Motivated by practical applications (in small dimensions), we assume that a **sampling window**, a convex body W with $V_d(W) > 0$, is given in which $Z \cap W$ can be observed. Our aim is to study random variables of the type

$$\varphi(Z \cap W),$$

with suitable functionals φ replacing the volume. In this way, we want to find out which information on Z and its underlying particle process can be obtained from measuring the realizations of Z within a bounded observation window W. Under appropriate assumptions, this will lead in a natural way to densities of Z and to relations between such densities defined for Z and similar parameters defined for the underlying particle process.

Since we intend to investigate sets arising as unions of convex bodies, we allow measurable functions φ defined on the convex ring $\mathcal{R}$ and having a simple behavior under unions. Therefore, $\varphi : \mathcal{R} \to \mathbb{R}$ is assumed to be **additive**, that is, to satisfy

$$\varphi(K \cup L) = \varphi(K) + \varphi(L) - \varphi(K \cap L) \tag{9.7}$$

for $K, L \in \mathcal{R}$ and $\varphi(\emptyset) = 0$. We further assume that φ is conditionally bounded. Here, we call a function $\varphi : \mathcal{R} \to \mathbb{R}$ **conditionally bounded** if, for each $K \in \mathcal{K}'$, the function φ is bounded on the set $\{L \in \mathcal{K}' : L \subset K\}$. When φ is translation invariant and additive, it is sufficient for this to assume that φ is bounded on the set $\{L \in \mathcal{K}' : L \subset C^d\}$. If φ is given as a functional on $\mathcal{K}'$ and is continuous and additive (the latter means that (9.7) holds whenever $K, L, K \cup L \in \mathcal{K}'$), then Groemer's extension theorem (Theorem 14.4.2) says that the functional φ has an additive extension (which we denote by the same symbol) to the convex ring $\mathcal{R}$. By Theorem 14.4.4, the extension is measurable and, due to the continuity on $\mathcal{K}'$, it is also conditionally bounded. The intrinsic volumes V_j, $j = 0, \dots, d$, are prototypes of measurable, additive and conditionally bounded functionals $\varphi : \mathcal{R} \to \mathbb{R}$; they are also motion invariant.

For a Boolean model Z with convex grains, $Z \cap W$ is a polyconvex set, hence $\varphi(Z \cap W)$ is defined and yields a random variable. We want to investigate how its expectation is related to the characteristics of the underlying particle process, that is, to the intensity measure Θ of X. In applications, such relations may be used to fit a Boolean model to given data, or to estimate densities of functionals for the particle process, in particular its intensity (in the stationary case), from measurements at realizations of the union set.

To begin with the computation of $\mathbb{E}\,\varphi(Z \cap W)$, for an additive, conditionally bounded and measurable function φ, let ν be the random number of particles of X hitting W, and let $M_1, \dots, M_\nu$ be these particles (with any numbering). Then the inclusion–exclusion principle (14.47) gives

$$\begin{aligned}
\varphi(Z \cap W) &= \varphi\left(\bigcup_{K \in X} K \cap W\right) \\
&= \sum_{k=1}^{\nu} (-1)^{k-1} \sum_{1 \le i_1 < \dots < i_k \le \nu} \varphi(W \cap M_{i_1} \cap \dots \cap M_{i_k}) \\
&= \sum_{k=1}^{\nu} \frac{(-1)^{k-1}}{k!} \sum_{(K_1, \dots, K_k) \in X^k_{\neq}} \varphi(W \cap K_1 \cap \dots \cap K_k).
\end{aligned}$$

Here $X^k_{\neq}$ is the set of pairwise distinct k-tuples from X. In the last line, we may extend the first summation to ∞, since $\varphi(\emptyset) = 0$.

Since φ is conditionally bounded, there exists a number c (depending on W) with $|\varphi(L)| \leq c$ for all $L \in \mathcal{K}'$ with $L \subset W$. This gives

$$\begin{aligned} |\varphi(Z \cap W)| &\leq \sum_{k=1}^{\nu} \frac{1}{k!} \left| \sum_{(K_1,\dots,K_k) \in X^k_{\neq}} \varphi(W \cap K_1 \cap \ldots \cap K_k) \right| \\ &\leq \sum_{k=1}^{\nu} \binom{\nu}{k} c \leq 2^{\nu} c = 2^{\operatorname{card}(X \cap \mathcal{K}_W)} c. \end{aligned}$$

Since $\operatorname{card}(X \cap \mathcal{K}_W)$ has a Poisson distribution,

$$\begin{aligned} \mathbb{E}\, 2^{\operatorname{card}(X \cap \mathcal{K}_W)} &= \sum_{k=0}^{\infty} 2^k\, \mathbb{P}(\operatorname{card}(X \cap \mathcal{K}_W) = k) \\ &= \mathrm{e}^{-\Theta(\mathcal{K}_W)} \sum_{k=0}^{\infty} \frac{[2\Theta(\mathcal{K}_W)]^k}{k!} \\ &= \mathrm{e}^{-\Theta(\mathcal{K}_W)} \mathrm{e}^{2\Theta(\mathcal{K}_W)} = \mathrm{e}^{\Theta(\mathcal{K}_W)} < \infty. \end{aligned}$$

It follows that $\varphi(Z \cap W)$ is integrable. By the dominated convergence theorem, we can interchange expectation and summation. Using Theorem 3.1.3 and Corollary 3.2.4, we obtain

$$\begin{aligned} &\mathbb{E}\, \varphi(Z \cap W) \\ &= \sum_{k=1}^{\infty} \frac{(-1)^{k-1}}{k!}\, \mathbb{E} \sum_{(K_1,\dots,K_k) \in X^k_{\neq}} \varphi(W \cap K_1 \cap \ldots \cap K_k) \\ &= \sum_{k=1}^{\infty} \frac{(-1)^{k-1}}{k!} \int_{\mathcal{K}} \cdots \int_{\mathcal{K}} \varphi(W \cap K_1 \cap \ldots \cap K_k)\, \Theta(\mathrm{d}K_1) \cdots \Theta(\mathrm{d}K_k). \end{aligned}$$

So far, we have not used stationarity. But if we now assume that Z is stationary, we can use the decomposition of the intensity measure and get

$$\begin{aligned} &\mathbb{E}\, \varphi(Z \cap W) \\ &= \sum_{k=1}^{\infty} \frac{(-1)^{k-1}}{k!} \gamma^k \int_{\mathcal{K}_0} \cdots \int_{\mathcal{K}_0} \int_{(\mathbb{R}^d)^k} \varphi(W \cap (K_1 + x_1) \cap \ldots \cap (K_k + x_k)) \\ &\quad \times \lambda^k(\mathrm{d}(x_1, \dots, x_k))\, \mathbb{Q}(\mathrm{d}K_1) \cdots \mathbb{Q}(\mathrm{d}K_k). \end{aligned}$$

We summarize the results in the following theorem.

Theorem 9.1.2. *Let Z be a Boolean model in $\mathbb{R}^d$ with convex grains, let $W \in \mathcal{K}'$ and $\varphi : \mathcal{R} \to \mathbb{R}$ be a measurable, additive and conditionally bounded functional. Then we have*

$$\mathbb{E}\,|\varphi(Z \cap W)| < \infty$$

and

$$\begin{aligned} &\mathbb{E}\,\varphi(Z \cap W) \\ &= \sum_{k=1}^{\infty} \frac{(-1)^{k-1}}{k!} \int_{\mathcal{K}} \cdots \int_{\mathcal{K}} \varphi(W \cap K_1 \cap \ldots \cap K_k)\, \Theta(\mathrm{d}K_1) \cdots \Theta(\mathrm{d}K_k). \end{aligned} \tag{9.8}$$

If Z is stationary, then

$$\begin{aligned} &\mathbb{E}\,\varphi(Z \cap W) \\ &= \sum_{k=1}^{\infty} \frac{(-1)^{k-1}}{k!} \gamma^k \int_{\mathcal{K}_0} \cdots \int_{\mathcal{K}_0} \Phi(W, K_1, \ldots, K_k)\, \mathbb{Q}(\mathrm{d}K_1) \cdots \mathbb{Q}(\mathrm{d}K_k) \end{aligned}$$

with

$$\begin{aligned} &\Phi(W, K_1, \ldots, K_k) \\ &:= \int_{(\mathbb{R}^d)^k} \varphi(W \cap (K_1 + x_1) \cap \ldots \cap (K_k + x_k)) \lambda^k(\mathrm{d}(x_1, \ldots, x_k)). \end{aligned}$$

To proceed further, in the stationary case, we need to compute the integrals

$$\int_{\mathcal{K}_0} \cdots \int_{\mathcal{K}_0} \Phi(W, K_1, \ldots, K_k)\, \mathbb{Q}(\mathrm{d}K_1) \cdots \mathbb{Q}(\mathrm{d}K_k).$$

This is possible for special choices of φ, using the translative integral formulas from Section 6.4.

Let us first consider the volume again, $\varphi = V_d$. For convex bodies $K, K_1, \ldots, K_k$, we have

$$\begin{aligned} &\Phi(W, K_1, \ldots, K_k) \\ &= \int_{(\mathbb{R}^d)^k} V_d(W \cap (K_1 + x_1) \cap \ldots \cap (K_k + x_k))\, \lambda^k(\mathrm{d}(x_1, \ldots, x_k)) \\ &= V_d(W) V_d(K_1) \cdots V_d(K_k). \end{aligned}$$

This follows from (6.15), but is also a direct consequence of Fubini's theorem. Thus, we obtain

$$\mathbb{E} V_d(Z \cap W) = \sum_{k=1}^{\infty} \frac{(-1)^{k-1}}{k!} V_d(W) \overline{V}_d(X)^k = V_d(W) \left(1 - \mathrm{e}^{-\overline{V}_d(X)}\right).$$

This is nothing but relation (9.6) again.

Now we consider the intrinsic volume V_{d-1}, which is half the surface area (for convex bodies with interior points). Again from (6.15) or (5.15), we obtain

$$\int_{(\mathbb{R}^d)^k} V_{d-1}(K_0 \cap (K_1 + x_1) \cap \ldots \cap (K_k + x_k))\, \lambda^k(\mathrm{d}(x_1, \ldots, x_k))$$
$$= \sum_{i=0}^{k} V_d(K_0) \cdots V_d(K_{i-1}) V_{d-1}(K_i) V_d(K_{i+1}) \cdots V_d(K_k).$$

Therefore, we get

$$\begin{aligned} &\mathbb{E}V_{d-1}(Z \cap W) \\ &= \sum_{k=1}^{\infty} \frac{(-1)^{k-1}}{k!} \left[V_{d-1}(W)\overline{V}_d(X)^k + kV_d(W)\overline{V}_{d-1}(X)\overline{V}_d(X)^{k-1}\right] \\ &= V_d(W)\overline{V}_{d-1}(X) \sum_{k=1}^{\infty} \frac{[-\overline{V}_d(X)]^{k-1}}{(k-1)!} + V_{d-1}(W)\left(1 - \mathrm{e}^{-\overline{V}_d(X)}\right), \end{aligned}$$

hence

$$\mathbb{E}V_{d-1}(Z \cap W) = V_d(W)\overline{V}_{d-1}(X)\mathrm{e}^{-\overline{V}_d(X)} + V_{d-1}(W)\left(1 - \mathrm{e}^{-\overline{V}_d(X)}\right).$$

In contrast to the case of the volume, the quotient

$$\frac{\mathbb{E}V_{d-1}(Z \cap W)}{V_d(W)} = \overline{V}_{d-1}(X)\mathrm{e}^{-\overline{V}_d(X)} + \frac{V_{d-1}(W)}{V_d(W)}\left(1 - \mathrm{e}^{-\overline{V}_d(X)}\right)$$

still depends on the observation window W. This influence disappears for increasing W. More precisely, we have

$$\lim_{r\to\infty} \frac{\mathbb{E}V_{d-1}(Z \cap rW)}{V_d(rW)} = \overline{V}_{d-1}(X)\mathrm{e}^{-\overline{V}_d(X)}.$$

The limit on the left side is denoted by $\overline{V}_{d-1}(Z)$ and called the **specific surface area** or the **density of the surface area** of Z (not caring about the factor 1/2). Such limits exist under more general assumptions, as we shall study in the next section.

We repeat that so far we have obtained the two relations

$$\begin{aligned} \overline{V}_d(Z) &= 1 - \mathrm{e}^{-\overline{V}_d(X)}, \\ \overline{V}_{d-1}(Z) &= \overline{V}_{d-1}(X)\mathrm{e}^{-\overline{V}_d(X)}, \end{aligned} \tag{9.9}$$

connecting specific intrinsic volumes of the stationary Boolean model Z with corresponding densities of the underlying particle process X.

We return to the case of a general additive functional φ (continuous on $\mathcal{K}'$). An explicit formula can still be obtained if we assume that the particle process X and thus the Boolean model Z is isotropic.

Let Z be a stationary, isotropic Boolean model (always with convex grains). Since the grain distribution $\mathbb{Q}$ of X is rotation invariant, we can insert rotations, integrate over the rotation group, apply Fubini's theorem, and then use the iteration of Hadwiger's general integral geometric theorem (Theorem 5.1.2). In this way, we obtain

$$\int_{\mathcal{K}_0} \cdots \int_{\mathcal{K}_0} \Phi(W, K_1, \ldots, K_k)\, \mathbb{Q}(\mathrm{d}K_1) \cdots \mathbb{Q}(\mathrm{d}K_k)$$

$$= \int_{\mathcal{K}_0} \cdots \int_{\mathcal{K}_0} \int_{SO_d} \int_{\mathbb{R}^d} \cdots \int_{SO_d} \int_{\mathbb{R}^d} \varphi(W \cap (\vartheta_1 K_1 + x_1) \cap \ldots \cap (\vartheta_k K_k + x_k))$$
$$\times\, \lambda(\mathrm{d}x_1)\, \nu(\mathrm{d}\vartheta_1) \cdots \lambda(\mathrm{d}x_k)\, \nu(\mathrm{d}\vartheta_k)\, \mathbb{Q}(\mathrm{d}K_1) \cdots \mathbb{Q}(\mathrm{d}K_k)$$

$$= \int_{\mathcal{K}_0} \cdots \int_{\mathcal{K}_0} \sum_{\substack{r_0, \ldots, r_k = 0 \\ r_0 + \ldots + r_k = kd}}^{d} c^d_{d-r_0}\, \varphi_{r_0}(W) \prod_{i=1}^{k} c_d^{r_i} V_{r_i}(K_i)\, \mathbb{Q}(\mathrm{d}K_1) \cdots \mathbb{Q}(\mathrm{d}K_k).$$

This gives

$$\mathbb{E}\,\varphi(Z \cap W)$$
$$= \sum_{k=1}^{\infty} \frac{(-1)^{k-1}}{k!} \sum_{\substack{r_0, \ldots, r_k = 0 \\ r_0 + \ldots + r_k = kd}}^{d} c^d_{d-r_0}\, \varphi_{r_0}(W) \prod_{i=1}^{k} c_d^{r_i} \overline{V}_{r_i}(X)$$
$$= \sum_{k=1}^{\infty} \frac{(-1)^{k-1}}{k!} \sum_{m=0}^{d} c^d_{d-m}\, \varphi_m(W) \sum_{\substack{m_1, \ldots, m_k = 0 \\ m_1 + \ldots + m_k = kd - m}}^{d} \prod_{i=1}^{k} c_d^{m_i} \overline{V}_{m_i}(X)$$
$$= \varphi(W) \left(1 - \mathrm{e}^{-\overline{V}_d(X)}\right)$$
$$+ \sum_{m=1}^{d} c^d_{d-m}\, \varphi_m(W) \underbrace{\sum_{k=1}^{\infty} \frac{(-1)^{k-1}}{k!} \sum_{\substack{m_1, \ldots, m_k = 0 \\ m_1 + \ldots + m_k = kd - m}}^{d} \prod_{i=1}^{k} c_d^{m_i} \overline{V}_{m_i}(X)}_{S}.$$

We rearrange the last two sums according to the number, say s, of indices among $m_1, \ldots, m_k$ that are smaller than d; here $s \in \{1, \ldots, m\}$. This gives

$$S = \sum_{s=1}^{m} \sum_{r=0}^{\infty} \binom{r+s}{r} \frac{(-1)^{r+s-1}}{(r+s)!} \overline{V}_d(X)^r \sum_{\substack{m_1, \ldots, m_s = 0 \\ m_1 + \ldots + m_s = sd - m}}^{d-1} \prod_{i=1}^{s} c_d^{m_i} \overline{V}_{m_i}(X)$$
$$= -\mathrm{e}^{-\overline{V}_d(X)} \sum_{s=1}^{m} \frac{(-1)^s}{s!} \sum_{\substack{m_1, \ldots, m_s = 0 \\ m_1 + \ldots + m_s = sd - m}}^{d-1} \prod_{i=1}^{s} c_d^{m_i} \overline{V}_{m_i}(X).$$

Thus we have obtained the following result.

Theorem 9.1.3. *Let Z be a Boolean model in $\mathbb{R}^d$, generated by a stationary, isotropic Poisson process X of convex particles. If $\varphi : \mathcal{R} \to \mathbb{R}$ is an additive functional which is continuous on $\mathcal{K}'$, then, for any $W \in \mathcal{K}'$ with $V_d(W) > 0$,*

$$\mathbb{E}\,\varphi(Z \cap W) = \varphi(W)\left(1 - \mathrm{e}^{-\overline{V}_d(X)}\right)$$
$$-\mathrm{e}^{-\overline{V}_d(X)} \sum_{m=1}^{d} c_{d-m}^{d}\,\varphi_m(W) \sum_{s=1}^{m} \frac{(-1)^s}{s!} \sum_{\substack{m_1,\dots,m_s=0\\ m_1+\dots+m_s=sd-m}}^{d-1} \prod_{i=1}^{s} c_d^{m_i}\,\overline{V}_{m_i}(X).$$

A remarkable fact here is that the functional φ and its derived functionals φ_m are applied, on the right side, only to the sampling window W. For given φ and W, the expectation $\mathbb{E}\,\varphi(Z \cap W)$ *depends only on the densities of the intrinsic volumes of the generating particle process X*. Conversely, this means that no information about the stationary isotropic particle process X beyond its specific intrinsic volumes can be obtained from expectations of measurements $\varphi(Z \cap W)$. All the densities $\overline{V}_i(X)$ already occur if we choose for φ the intrinsic volumes $V_0, \dots, V_d$.

For that reason, we now concentrate on $\varphi = V_j$, the jth intrinsic volume. By the Crofton formula (5.6), we have

$$(V_j)_m = c_j^{d-m} c_d^{m+j} V_{m+j},$$

with $V_{m+j} = 0$ if $m+j > d$. Inserting this (and renaming the first summation index), we obtain

$$\mathbb{E}\,V_j(Z \cap W) = V_j(W)\left(1 - \mathrm{e}^{-\overline{V}_d(X)}\right)$$
$$-\mathrm{e}^{-\overline{V}_d(X)} \sum_{m=j+1}^{d} c_j^{m} V_m(W) \sum_{s=1}^{m-j} \frac{(-1)^s}{s!} \sum_{\substack{m_1,\dots,m_s=j\\ m_1+\dots+m_s=sd+j-m}}^{d-1} \prod_{i=1}^{s} c_d^{m_i}\,\overline{V}_{m_i}(X).$$

Here we can replace W by rW with $r > 0$ and then let r tend to infinity. We obtain the following result.

Theorem 9.1.4. *Let Z be a Boolean model in $\mathbb{R}^d$, generated by a stationary, isotropic Poisson process X of convex particles. The limit*

$$\overline{V}_j(Z) := \lim_{r \to \infty} \frac{\mathbb{E}\,V_j(Z \cap rW)}{V_d(rW)}$$

exists and is given by

$$\overline{V}_j(Z) = \mathrm{e}^{-\overline{V}_d(X)} \left[\overline{V}_j(X) - \sum_{s=2}^{d-j} \frac{(-1)^s}{s!} \sum_{\substack{m_1,\dots,m_s=j+1\\ m_1+\dots+m_s=(s-1)d+j}}^{d-1} \prod_{i=1}^{s} c_j^{m_i}\,\overline{V}_{m_i}(X) \right]$$

if $j = 0, \dots, d-1$ *and*

$$\overline{V}_d(Z) = 1 - e^{-\overline{V}_d(X)}.$$

The cases $j = d$ and $j = d-1$ have been obtained earlier without the isotropy assumption.

We call $\overline{V}_j(Z)$ the **density of the** jth **intrinsic volume**, or the **specific** jth **intrinsic volume**, of the Boolean model Z. In the next section, we shall introduce such densities for much more general random sets.

For Boolean models, Theorem 9.1.4 can be used to determine the densities $\overline{V}_i(X)$ of the underlying particle process from the densities $\overline{V}_j(Z)$ of the union set. We demonstrate this only in dimensions two and three. Here we use classical notation:

$$d = 2 \qquad \begin{aligned} &A = V_2, \text{ area} \\ &L = 2V_1, \text{ perimeter} \\ &\chi = V_0, \text{ Euler characteristic} \end{aligned}$$

$$d = 3 \qquad \begin{aligned} &V = V_3, \text{ volume} \\ &S = 2V_2, \text{ surface area} \\ &M = \pi V_1, \text{ integral of mean curvature} \\ &\chi = V_0, \text{ Euler characteristic.} \end{aligned}$$

We obtain the following relations: For $d = 2$,

$$\begin{aligned} \overline{A}(Z) &= 1 - e^{-\overline{A}(X)}, \\ \overline{L}(Z) &= e^{-\overline{A}(X)}\overline{L}(X), \\ \overline{\chi}(Z) &= e^{-\overline{A}(X)}\left(\overline{\chi}(X) - \frac{1}{4\pi}\overline{L}(X)^2\right). \end{aligned}$$

For $d = 3$,

$$\begin{aligned} \overline{V}(Z) &= 1 - e^{-\overline{V}(X)}, \\ \overline{S}(Z) &= e^{-\overline{V}(X)}\overline{S}(X), \\ \overline{M}(Z) &= e^{-\overline{V}(X)}\left(\overline{M}(X) - \frac{\pi^2}{32}\overline{S}(X)^2\right), \\ \overline{\chi}(Z) &= e^{-\overline{V}(X)}\left(\overline{\chi}(X) - \frac{1}{4\pi}\overline{M}(X)\overline{S}(X) + \frac{\pi}{384}\overline{S}(X)^3\right). \end{aligned}$$

In either case, if all the parameters on the left side are known, then all parameters on the right side are known.

In particular, the densities on the left side determine $\overline{\chi}(X)$, which is the intensity γ of X. We point out, however, that the determination of the intensity $\overline{\chi}(X)$ requires the determination of the densities of all the $d+1$ intrinsic volumes of Z.

If we drop the isotropy assumption, hence consider a stationary Boolean model Z with convex grains and $\varphi = V_j$, we can combine Theorem 9.1.2 with the iterated translative formula (6.15). We obtain

$$\begin{aligned}
&\mathbb{E}\, V_j(Z \cap W)\\
&= \sum_{k=1}^{\infty} \frac{(-1)^{k-1}}{k!}\gamma^k \sum_{\substack{m_0,\dots,m_k=j\\ m_0+\dots+m_k=kd+j}}^{d} \int_{\mathcal{K}_0}\cdots\int_{\mathcal{K}_0} V^{(j)}_{m_0,\dots,m_k}(W, K_1,\dots,K_k)\\
&\quad \times \mathbb{Q}(\mathrm{d}K_1)\cdots\mathbb{Q}(\mathrm{d}K_k).
\end{aligned}$$

Again, we replace W by rW, divide by $V_d(rW)$ and let $r \to \infty$. Then, due to the homogeneity properties of the mixed functionals (see Theorem 6.4.1), all summands on the right side with $m_0 < d$ disappear asymptotically. For $m_0 = d$, we can use the decomposability property of the mixed functionals (Theorem 6.4.1) and get, with essentially the same arguments as in the isotropic case,

$$\begin{aligned}
\overline{V}_j(Z) &= \lim_{r\to\infty} \frac{\mathbb{E}\, V_j(Z \cap rW)}{V_d(rW)}\\
&= \sum_{k=1}^{\infty} \frac{(-1)^{k-1}}{k!}\gamma^k \sum_{\substack{m_1,\dots,m_k=j\\ m_1+\dots+m_k=(k-1)d+j}}^{d} \int_{\mathcal{K}_0}\cdots\int_{\mathcal{K}_0} V^{(j)}_{m_1,\dots,m_k}(K_1,\dots,K_k)\\
&\quad \times \mathbb{Q}(\mathrm{d}K_1)\cdots\mathbb{Q}(\mathrm{d}K_k)\\
&= \sum_{s=1}^{d-j}\sum_{r=0}^{\infty} \binom{r+s}{r} \frac{(-1)^{r+s-1}}{(r+s)!} \overline{V}_d(X)^r \gamma^s\\
&\qquad \sum_{\substack{m_1,\dots,m_s=j\\ m_1+\dots+m_s=(s-1)d+j}}^{d-1} \int_{\mathcal{K}_0}\cdots\int_{\mathcal{K}_0} V^{(j)}_{m_1,\dots,m_s}(K_1,\dots,K_s)\,\mathbb{Q}(\mathrm{d}K_1)\cdots\mathbb{Q}(\mathrm{d}K_s)\\
&= -\mathrm{e}^{-\overline{V}_d(X)} \sum_{s=1}^{d-j} \frac{(-1)^s}{s!} \sum_{\substack{m_1,\dots,m_s=j\\ m_1+\dots+m_s=(s-1)d+j}}^{d-1} \overline{V}^{(j)}_{m_1,\dots,m_s}(X,\dots,X)\\
&= \mathrm{e}^{-\overline{V}_d(X)} \left(\overline{V}_j(X) - \sum_{s=2}^{d-j} \frac{(-1)^s}{s!} \sum_{\substack{m_1,\dots,m_s=j+1\\ m_1+\dots+m_s=(s-1)d+j}}^{d-1} \overline{V}^{(j)}_{m_1,\dots,m_s}(X,\dots,X) \right).
\end{aligned}$$

The densities of X appearing here are special cases of the **mixed densities** defined by

$$\overline{V}^{(j)}_{m_1,\dots,m_s}(X,\dots,X,K_{k+1},\dots,K_s)$$

$$:= \gamma^k \int_{\mathcal{K}_0} \cdots \int_{\mathcal{K}_0} V^{(j)}_{m_1,\dots,m_s}(K_1,\dots,K_k,K_{k+1},\dots,K_s)\,\mathbb{Q}(\mathrm{d}K_1)\cdots\mathbb{Q}(\mathrm{d}K_k).$$

Hence, we arrive at the following result.

Theorem 9.1.5. *For a stationary Boolean model Z in $\mathbb{R}^d$ with convex grains, the limit*

$$\overline{V}_j(Z) := \lim_{r\to\infty} \frac{\mathbb{E}\, V_j(Z \cap rW)}{V_d(rW)}$$

exists and satisfies

$$\overline{V}_d(Z) = 1 - \mathrm{e}^{-\overline{V}_d(X)},$$

$$\overline{V}_{d-1}(Z) = \mathrm{e}^{-\overline{V}_d(X)}\overline{V}_{d-1}(X),$$

and

$$\overline{V}_j(Z)$$

$$= \mathrm{e}^{-\overline{V}_d(X)} \left(\overline{V}_j(X) - \sum_{s=2}^{d-j} \frac{(-1)^s}{s!} \sum_{\substack{m_1,\dots,m_s=j+1\\ m_1+\dots+m_s=(s-1)d+j}}^{d-1} \overline{V}^{(j)}_{m_1,\dots,m_s}(X,\dots,X) \right)$$

for $j = 0,\dots,d-2$.

For $d = 2$, only the formula for the Euler characteristic V_0 is different from the isotropic case, and we have

$$\overline{A}(Z) = 1 - \mathrm{e}^{-\overline{A}(X)},$$

$$\overline{L}(Z) = \mathrm{e}^{-\overline{A}(X)}\overline{L}(X),$$

$$\overline{\chi}(Z) = \mathrm{e}^{-\overline{A}(X)}\left(\gamma - \overline{A}(X,-X)\right),$$

where

$$\overline{A}(X,-X) := \gamma^2 \int_{\mathcal{K}_0}\int_{\mathcal{K}_0} A(K,-M)\,\mathbb{Q}(\mathrm{d}K)\,\mathbb{Q}(\mathrm{d}M).$$

Here, we made use of the fact that the mixed functional $V^{(0)}_{1,1}(K,M)$ in the plane equals twice the mixed area $A(K,-M)$ of K and $-M$. It is obvious that the formulas can no longer be used directly for the estimation of γ. Hence, we need more (local) information for the statistical analysis of non-isotropic Boolean models; this will be discussed in Section 9.5.

As an immediate generalization of Theorem 9.1.5, we can replace the intrinsic volumes $V_j(Z \cap W)$ by (additively extended) mixed volumes $V(Z \cap W[j], M[d-j])$, $j = 1,\dots,d-1$, for $M \in \mathcal{K}'$. Applying Theorem 9.1.2 to the functional φ given by

$$\varphi(K) = \binom{d}{j} V(K[j], -M[d-j]) = V^{(0)}_{j,d-j}(K,M)$$

and using (6.15), we obtain the following result. Since the proof is identical to the previous one, we skip it.

Theorem 9.1.6. *Let Z be a stationary Boolean model in $\mathbb{R}^d$ with convex grains, $j \in \{1, \dots, d-1\}$ and $M \in \mathcal{K}'$. Then the limit*

$$\overline{V}^{(0)}_{j,d-j}(Z, M) := \binom{d}{j} \lim_{r\to\infty} \frac{\mathbb{E}\, V(Z \cap rW[j], -M[d-j])}{V_d(rW)}$$

exists, is independent of W and satisfies

$$\overline{V}^{(0)}_{j,d-j}(Z, M) = \mathrm{e}^{-\overline{V}_d(X)} \Bigg(\overline{V}^{(0)}_{j,d-j}(X, M) - \sum_{s=2}^{d-j} \frac{(-1)^s}{s!} \sum_{\substack{m_1,\dots,m_s=j+1 \\ m_1+\dots+m_s=(s-1)d+j}}^{d-1} \overline{V}^{(0)}_{m_1,\dots,m_s,d-j}(X, \dots, X, M) \Bigg).$$

For $j = d-1$, the theorem yields

$$\overline{V}^{(0)}_{d-1,1}(Z, M) = \mathrm{e}^{-\overline{V}_d(X)} \overline{V}^{(0)}_{d-1,1}(X, M).$$

We can transform this into a local formula for area measures, using (14.23). Namely, we can rewrite (14.23) as

$$V^{(0)}_{d-1,1}(K, M) = \int_{S^{d-1}} h^*(M, -u)\, S_{d-1}(K, \mathrm{d}u) \tag{9.10}$$

(where h^* denotes the centered support function, see Section 4.6) and remark that, by additive extension in each variable, (9.10) holds for $K, M \in \mathcal{R}$. Since the vector space generated by the functions $h^*(M, \cdot)$, $M \in \mathcal{K}'$, is dense in the space of centered, continuous functions on S^{d-1} and since area measures have centroid 0, we deduce that the weak limit

$$\overline{S}_{d-1}(Z, \cdot) := \lim_{r\to\infty} \frac{\mathbb{E}\, S_{d-1}(Z \cap rW, \cdot)}{V_d(rW)}. \tag{9.11}$$

exists and satisfies the local density formula

$$\overline{S}_{d-1}(Z, \cdot) = \mathrm{e}^{-\overline{V}_d(X)} \overline{S}_{d-1}(X, \cdot)\,. \tag{9.12}$$

Here, according to (4.43),

$$\overline{S}_{d-1}(X, \cdot) := \gamma \int_{\mathcal{K}_0} S_{d-1}(K, \cdot)\, \mathbb{Q}(\mathrm{d}K).$$

Alternatively, (9.12) and the existence of the limit (9.11) can be obtained directly, with a proof similar to that of the previous results. For this, we use

Theorem 9.1.2 with $\varphi(K) := S_{d-1}(K, A)$, for a fixed Borel set $A \subset S^{d-1}$, together with the translative formula for area measures,

$$\int_{\mathbb{R}^d} S_{d-1}(K \cap (M + x), \cdot)\, \lambda(\mathrm{d}x) = V_d(M) S_{d-1}(K, \cdot) + V_d(K) S_{d-1}(M, \cdot).$$

For $K, M \in \mathcal{K}'$, this formula can either be deduced from the more general results in Section 6.4 or proved directly, using approximation by polytopes.

Note that (9.12) is a local version of (9.9).

We could also use (9.10) to obtain a formula for a local version of $\overline{V}^{(0)}_{j,d-j}(Z, M)$ for $j = 1$. This would involve the limit of the centered, additively extended support function

$$\overline{h}(Z, \cdot) := \lim_{r \to \infty} \frac{\mathbb{E}\, h^*(Z \cap rW, \cdot)}{V_d(rW)}$$

and expresses $\overline{h}(Z, \cdot)$ in terms of (mean values of) iterated versions of the mixed support functions, which appear in Theorem 6.4.6. We mention only the planar case, where the formula is simple. For $d = 2$,

$$\overline{h}(Z, \cdot) = \mathrm{e}^{-\overline{A}(X)} \overline{h}(X, \cdot), \tag{9.13}$$

where

$$\overline{h}(X, \cdot) := \gamma \int_{\mathcal{K}_0} h^*(K, \cdot)\, \mathbb{Q}(\mathrm{d}K).$$

Remark. Starting with Theorem 9.1.2, the results of this section remain true for Boolean models with polyconvex grains, if the additively extended functionals are used and the grain distribution satisfies (9.17).

Notes for Section 9.1 are included in the Notes for Section 9.4.

9.2 Densities of Additive Functionals

In the previous section we have seen that for stationary isotropic Boolean models Z and for arbitrary convex bodies W with positive volume the limit

$$\lim_{r \to \infty} \frac{\mathbb{E}\, V_j(Z \cap rW)}{V_d(rW)} = \overline{V}_j(Z)$$

always exists. In this way, the specific jth intrinsic volume $\overline{V}_j(Z)$ can be defined. The existence of the limit for Boolean models was deduced from explicit formulas. They yielded, at the same time, a representation of this density of the jth intrinsic volume of the Boolean model Z in terms of densities of the underlying particle process X.

Our aim in this section is to show the existence of corresponding densities for more general random closed sets and for rather general functionals. Essentially, the realizations of the random closed sets will locally belong to the convex ring $\mathcal{R}$, consisting of all finite unions of convex bodies in $\mathbb{R}^d$. The functionals to be considered will share with the intrinsic volumes the property of additivity.

The existence proof for the limit will be prepared by two lemmas. We make use of the unit cube $C^d = [0,1]^d$ and the half-open unit cube $C_0^d := [0,1)^d$. The **upper right boundary**

$$\partial^+ C^d := C^d \setminus C_0^d$$

is the union of d facets of C^d and hence belongs to the convex ring $\mathcal{R}$.

For $z \in \mathbb{Z}^d$, we put

$$C_z := C^d + z, \qquad C_{0,z} := C_0^d + z, \qquad \partial^+ C_z := \partial^+ C^d + z.$$

Then

$$\mathbb{R}^d = \bigcup_{z\in\mathbb{Z}^d} C_{0,z}$$

is a disjoint decomposition of $\mathbb{R}^d$.

Let φ be a real function on the convex ring $\mathcal{R}$, and let $K \in \mathcal{R}$. Since $\emptyset \neq K \cap C_{0,z} = K \cap C_{0,y}$ for $z, y \in \mathbb{Z}^d$ implies $z = y$, we can define

$$\varphi(K \cap C_{0,z}) := \varphi(K \cap C_z) - \varphi(K \cap \partial^+ C_z).$$

Lemma 9.2.1. *If $\varphi : \mathcal{R} \to \mathbb{R}$ is an additive function and $K \in \mathcal{R}$ is a polyconvex set, then*

$$\varphi(K) = \sum_{z\in\mathbb{Z}^d} \varphi(K \cap C_{0,z}).$$

Proof. We give two proofs for this crucial lemma. The first one employs the extension theorem 14.4.3. This allows us to work with relatively open polytopes and, therefore, with disjoint decompositions. The second proof does not use the extension theorem and has, therefore, a basic idea which is slightly less obvious.

First proof. Let $K \in \mathcal{R}$. For a polytope $P \in \mathcal{P}$ we define

$$\psi(P) := \varphi(K \cap P).$$

Then ψ is an additive functional on convex polytopes. By Theorem 14.4.3, it has a unique extension to an additive function on $U(\mathcal{P}_{ro})$, the system of finite unions of relatively open polytopes. We denote this extension also by ψ. Since $C_{0,z} = C_z \setminus \partial^+ C_z$ and all sets here belong to $U(\mathcal{P}_{ro})$, the additivity of ψ gives

$$\psi(C_{0,z}) = \psi(C_z) - \psi(\partial^+ C_z).$$

Moreover, $\psi(P) = 0$ for all convex polytopes P with $K \cap P = \emptyset$. We can choose a finite set $S \subset \mathbb{Z}^d$ such that

$$K \subset Q := \bigcup_{z \in S} C_{0,z}$$

and that $\operatorname{cl} Q$ is convex. Then

$$\begin{aligned}
\varphi(K) &= \varphi(K \cap \operatorname{cl} Q) = \psi(\operatorname{cl} Q) = \psi(Q) \\
&= \sum_{z \in \mathbb{Z}^d} \psi(C_{0,z}) \\
&= \sum_{z \in \mathbb{Z}^d} [\psi(C_z) - \psi(\partial^+ C_z)] \\
&= \sum_{z \in \mathbb{Z}^d} [\varphi(K \cap C_z) - \varphi(K \cap \partial^+ C_z)].
\end{aligned}$$

This concludes the first proof.

Second proof. We denote by $<$ the lexicographic order on $\mathbb{Z}^d$, that is,

$$(z^1, \ldots, z^d) < (y^1, \ldots, y^d)$$

if and only if $z^i = y^i$ for $i < k$ and $z_k < y_k$, for some $k \in \{1, \ldots, d\}$. Then

$$\partial^+ C_z = C_z \cap \bigcup_{z < y \in \mathbb{Z}^d} C_y$$

for $z \in \mathbb{Z}^d$. With the inclusion–exclusion principle we get (all sums are finite)

$$\begin{aligned}
\sum_{z \in \mathbb{Z}^d} \varphi(K \cap \partial^+ C_z) &= \sum_{z \in \mathbb{Z}^d} \varphi\left(\bigcup_{z < y \in \mathbb{Z}^d} (K \cap C_z \cap C_y) \right) \\
&= \sum_{z \in \mathbb{Z}^d} \sum_{k=1}^{\infty} (-1)^{k-1} \sum_{z < y_1 < \ldots < y_k} \varphi(K \cap C_z \cap C_{y_1} \cap \ldots \cap C_{y_k}) \\
&= -\sum_{k=2}^{\infty} (-1)^{k-1} \sum_{z_1 < \ldots < z_k} \varphi(K \cap C_{z_1} \cap \ldots \cap C_{z_k}).
\end{aligned}$$

This gives

$$\begin{aligned}
\varphi(K) &= \varphi\left(\bigcup_{z \in \mathbb{Z}^d} (K \cap C_z) \right) \\
&= \sum_{z \in \mathbb{Z}^d} \varphi(K \cap C_z) + \sum_{k=2}^{\infty} (-1)^{k-1} \sum_{z_1 < \ldots < z_k} \varphi(K \cap C_{z_1} \cap \ldots \cap C_{z_k}) \\
&= \sum_{z \in \mathbb{Z}^d} \varphi(K \cap C_z) - \sum_{z \in \mathbb{Z}^d} \varphi(K \cap \partial^+ C_z),
\end{aligned}$$

as asserted. □

We recall that a function $\varphi : \mathcal{R} \to \mathbb{R}$ is conditionally bounded if it is bounded on $\{L \in \mathcal{K}' : L \subset K\}$, for each $K \in \mathcal{K}'$. In particular, if φ is continuous on $\mathcal{K}'$, it is conditionally bounded.

Lemma 9.2.2. *Let the function $\varphi : \mathcal{R} \to \mathbb{R}$ be translation invariant, additive and conditionally bounded. Then*

$$\lim_{r\to\infty} \frac{\varphi(rW)}{V_d(rW)} = \varphi(C_0^d)$$

for every $W \in \mathcal{K}'$ with $V_d(W) > 0$.

Proof. Let $W \in \mathcal{K}'$ and $0 \in \operatorname{int} W$, without loss of generality. For $K \in \mathcal{K}$ and $z \in \mathbb{Z}^d$ we put

$$\varphi(K, z) := \varphi(K \cap C_{0,z}). \tag{9.14}$$

Lemma 9.2.1 shows that

$$\varphi(rW) = \sum_{z\in\mathbb{Z}^d} \varphi(rW, z) \qquad \text{for } r > 0.$$

Define

$$Z_r^1 := \{z \in \mathbb{Z}^d : C_z \cap rW \neq \emptyset,\ C_z \not\subset rW\}$$

and

$$Z_r^2 := \{z \in \mathbb{Z}^d : C_z \subset rW\}.$$

Then

$$\lim_{r\to\infty} \frac{|Z_r^1|}{V_d(rW)} = 0, \qquad \lim_{r\to\infty} \frac{|Z_r^2|}{V_d(rW)} = 1, \tag{9.15}$$

where $|A|$ denotes the number of elements of a set A. The limit relations follow from the fact that one easily shows the existence of numbers $r_0 > s, t > 0$ such that

$$z \in Z_r^1 \Rightarrow C_z \subset (r+s)W \setminus (r-s)W$$

and

$$(r-t)W \subset \bigcup_{z\in Z_r^2} C_z$$

for $r \geq r_0$.

By assumption,

$$|\varphi(rW, z)| = |\varphi(rW - z, 0)| \leq b$$

with some constant b independent of z, W and r. This gives

$$\frac{1}{V_d(rW)} \left| \sum_{z\in Z_r^1} \varphi(rW, z) \right| \leq b \frac{|Z_r^1|}{V_d(rW)} \to 0 \qquad \text{for } r \to \infty.$$

From this we deduce that

$$\begin{aligned}\lim_{r\to\infty}\frac{\varphi(rW)}{V_d(rW)} &= \lim_{r\to\infty}\frac{1}{V_d(rW)}\sum_{z\in\mathbb{Z}^d}\varphi(rW,z)\\ &= \lim_{r\to\infty}\frac{1}{V_d(rW)}\sum_{z\in Z_r^2}\varphi(rW,z)\\ &= \varphi(C_0^d)\lim_{r\to\infty}\frac{|Z_r^2|}{V_d(rW)}\\ &= \varphi(C_0^d).\end{aligned}$$

This proves the lemma. □

Mean Values of Additive Functionals for Random Sets

Now we introduce a suitable class of random closed sets for which the existence of densities for rather general functionals can be shown. Recall that the **extended convex ring** in $\mathbb{R}^d$ is defined by

$$\mathcal{S} := \{F \subset \mathbb{R}^d : F \cap K \in \mathcal{R} \text{ for } K \in \mathcal{K}\}.$$

The elements of $\mathcal{S}$ are called **locally polyconvex sets**. Thus a locally polyconvex set has the property that its intersection with any convex body is a finite union of convex bodies.

If $M \in \mathcal{R}$ is a nonempty polyconvex set, there are a number $m \in \mathbb{N}$ and convex bodies $K_1, \ldots, K_m \in \mathcal{K}'$ such that $M = K_1 \cup \ldots \cup K_m$. The smallest number m with this property is denoted by $N(M)$. We also put $N(\emptyset) = 0$. By Lemma 4.3.1, the function $N : \mathcal{R} \to \mathbb{N}_0$ is measurable. Now we can define the random closed sets which will be admitted in the following.

Definition 9.2.1. *A* **standard random set** *in $\mathbb{R}^d$ is a random closed set Z in $\mathbb{R}^d$ with the following properties:*

(a) *The realizations of Z are a.s. locally polyconvex.*
(b) *Z is stationary.*
(c) *Z satisfies the integrability condition*

$$\mathbb{E}\, 2^{N(Z\cap C^d)} < \infty. \tag{9.16}$$

Important examples of standard random sets are the Boolean models Z with convex grains. As we have seen in Section 9.1, they satisfy (9.16).

We are now in a position to prove the existence of densities of suitable functionals for standard random sets.

Theorem 9.2.1. *Let Z be a standard random set, let the function $\varphi : \mathcal{R} \to \mathbb{R}$ be translation invariant, additive, measurable and conditionally bounded. Let $W \in \mathcal{K}'$ be such that $V_d(W) > 0$. Then the limit*

$$\overline{\varphi}(Z) := \lim_{r \to \infty} \frac{\mathbb{E}\,\varphi(Z \cap rW)}{V_d(rW)}$$

exists and satisfies

$$\overline{\varphi}(Z) = \mathbb{E}\,\varphi(Z \cap C_0^d).$$

Hence, $\overline{\varphi}(Z)$ is independent of W.

Proof. Without loss of generality, we can assume that $W \subset C^d$. For given $\omega \in \Omega$, there is a representation

$$Z(\omega) \cap W = \bigcup_{i=1}^{N_W(\omega)} K_i(\omega) \qquad \text{with } K_i(\omega) \in \mathcal{K}',$$

where $N_W(\omega) := N(Z(\omega) \cap W)$. By the inclusion–exclusion principle,

$$\begin{aligned} &\varphi(Z(\omega) \cap W) \\ &= \sum_{k=1}^{N_W(\omega)} (-1)^{k-1} \sum_{1 \le i_1 < \ldots < i_k \le N_W(\omega)} \varphi(K_{i_1}(\omega) \cap \ldots \cap K_{i_k}(\omega)). \end{aligned}$$

Since φ is conditionally bounded, there is a constant b such that

$$\mathbb{E}\,|\varphi(Z \cap W)| \le b\,\mathbb{E} \sum_{k=1}^{N_W} \binom{N_W}{k} \le b\,\mathbb{E}\,2^{N(Z \cap W)} \le b\,\mathbb{E}\,2^{N(Z \cap C^d)},$$

since $N(Z(\omega) \cap W) \le N(Z(\omega) \cap C^d)$. By assumption, the right side is finite, hence $\varphi(Z \cap W)$ is integrable. For a polyconvex set $M \in \mathcal{R}$, the integrability of $\varphi(Z \cap M)$ then follows from additivity, using the inclusion–exclusion principle again. Therefore, we can define a functional $\phi : \mathcal{R} \to \mathbb{R}$ by

$$\phi(M) := \mathbb{E}\,\varphi(Z \cap M) \qquad \text{for } M \in \mathcal{R}.$$

Then ϕ is additive, translation invariant (as follows from the stationarity of Z) and conditionally bounded (as follows from the last estimate above). Now the assertion of the theorem follows from Lemma 9.2.2. □

With suitable conditions on φ and Z, the preceding result would also hold for general stationary random closed sets with values in $\mathcal{F}$. However, the useful functionals φ on $\mathcal{R}$ that satisfy the assumptions of the theorem have, with the exception of the volume, no reasonable extension to all of $\mathcal{C}$; therefore, the restriction to the convex ring seems appropriate (but see the Notes for Section 9.4, for other set classes).

For $\varphi = V_d$, the density $\overline{V}_d$ was already defined in Section 2.4, for example, by

$$\overline{V}_d(Z) = \frac{\mathbb{E}V_d(Z \cap W)}{V_d(W)},$$

for any Borel set W with $V_d(W) > 0$. Thus, the introduction of the specific volume does not require a limit procedure, and the assertion of Theorem 9.2.1 is trivial, since $V_d(Z \cap \partial^+ C^d) = 0$.

The quantity $\overline{\varphi}(Z)$ in Theorem 9.2.1 is called the φ-**density** of Z. The most important functionals φ are the intrinsic volumes $V_0, \dots, V_{d-1}$. The density $\overline{V}_j(Z)$ is also called the **specific** jth **intrinsic volume** of Z. In particular, $2\overline{V}_{d-1}(Z)$ is the **specific surface area** of Z. In Section 9.4, we shall give an alternative interpretation of $\overline{V}_j(Z)$, as a Radon–Nikodym derivative of the expectation of the curvature measure $\Phi_j(Z, \cdot)$ (which is a stationary random measure on $\mathbb{R}^d$) with respect to the Lebesgue measure λ. This representation will allow us in Section 11.1 to introduce specific intrinsic volumes (as functions on $\mathbb{R}^d$) also for non-stationary random closed sets.

Further functionals φ to which Theorem 9.2.1 can be applied are:

- the mixed volumes, $\varphi(K) := V(K[j], M[d-j])$, for fixed $M \in \mathcal{R}$ and $j \in \{1, \dots, d-1\}$,
- the surface area measure, $\varphi(K) := S_{d-1}(K, A)$, for $A \in \mathcal{B}(S^{d-1})$,
- the centered support function, $\varphi(K) := h^*(K, u)$, for $u \in S^{d-1}$.

Letting $A \in \mathcal{B}(S^{d-1})$ vary, we thus get, under the assumptions of Theorem 9.2.1, a finite Borel measure $\overline{S}_{d-1}(Z, \cdot)$ on S^{d-1}, the **specific surface area measure** or the **mean normal measure** of Z. By (14.15), $\overline{S}_{d-1}(Z, \cdot)$ is always nonnegative, and it is centered. If $\overline{S}_{d-1}(Z, \cdot)$ is not concentrated on a subsphere, it is (by Theorem 14.3.1) the surface area measure of a unique convex body in $\mathcal{K}_0$, which we call the **Blaschke body** $B(Z)$ of Z. Further, letting $u \in S^{d-1}$ vary, we get a centered continuous function $\overline{h}(Z, \cdot)$ on S^{d-1}, the **specific support function** of Z. (The continuity can be shown with the aid of (6.28), cf. Goodey and Weil [280, p. 339].) The function $\overline{h}(Z, \cdot)$ is a support function for $d = 2$, but in general not for $d \geq 3$.

Mixed volumes $V(K[j], M[d-j])$ are only special cases of mixed functionals $V^{(j)}_{m_1,\dots,m_k}(K_1, \dots, K_k)$, as studied in Section 6.4. Since the latter are additive in each component and have an additive extension to $\mathcal{R}$, Theorem 9.2.1 also yields densities of mixed functionals for standard random sets Z. Later, we shall need one series of these mixed densities for Z, but in a local version. Namely, for $j \in \{0, \dots, d\}$, $k \in \{j, \dots, d\}$, $M \in \mathcal{K}'$ and $A \in \mathcal{B}$, the functional

$$\varphi_A : K \mapsto \Phi^{(j)}_{k,d-k+j}(K, M; \mathbb{R}^d \times A)$$

satisfies the assumptions of Theorem 9.2.1. The density $\overline{\varphi_A}(Z)$, as a function of A, is a (signed) measure, which we denote by $\overline{\Phi}^{(j)}_{k,d-k+j}(Z, M; \cdot)$, thus

$$\overline{\Phi}^{(j)}_{k,d-k+j}(Z, M; A) = \lim_{r\to\infty} \frac{\mathbb{E}\,\Phi^{(j)}_{k,d-k+j}(Z \cap rW, M; \mathbb{R}^d \times A)}{V_d(rW)},$$

with a window W as before. We call $\overline{\Phi}^{(j)}_{k,d-k+j}(Z, M; \cdot)$ the **specific** (j, k)th **mixed measure** of Z (for given M). We notice that

$$\overline{\Phi}^{(j)}_{j,d}(Z, M; \cdot) = \overline{V}_j(Z)\, \Phi_d(M, \cdot),$$

$$\overline{\Phi}^{(j)}_{d,j}(Z, M; \cdot) = \overline{V}_d(Z)\, \Phi_j(M, \cdot).$$

We finally remark that we can now give a new interpretation of the intensity of a stationary process of k-flats that was introduced in Theorem 4.4.2. Let Z_X be the union set of a stationary k-flat process X with intensity γ. For $r > 0$ and $W \in \mathcal{K}$ with $V_d(W) > 0$ we have, by the additivity of V_k,

$$\frac{1}{V_d(rW)}\, \mathbb{E} V_k(Z_X \cap rW) = \frac{1}{V_d(rW)}\, \mathbb{E} \sum_{E\in X} \lambda_E(rW) = \gamma,$$

by Theorem 4.4.3. Thus, the left side is independent of r, hence for $r \to \infty$ it converges to $\overline{V}_k(Z_X)$, even without the integrability condition (9.16). Therefore, we have

$$\gamma = \overline{V}_k(Z_X).$$

Mean Values for Particle Processes

For a stationary particle process X, the φ-density $\overline{\varphi}(X)$ was already introduced in Section 4.1, and different representations were established. Further representations in the case of additive functionals can now be obtained in analogy to Theorem 9.2.1. Of the stationary particle process X in $\mathcal{R}$ to be considered we need the integrability condition

$$\int_{\mathcal{R}_0} 2^{N(C)} V_d(C + B^d)\, \mathbb{Q}(\mathrm{d}C) < \infty, \tag{9.17}$$

where $\mathbb{Q}$ is the grain distribution of X. If the particles are convex, then condition (9.17) reduces to the original condition (4.4).

Theorem 9.2.2. *Let X be a stationary particle process in $\mathbb{R}^d$ with particles in $\mathcal{R}$ and with grain distribution $\mathbb{Q}$ satisfying* (9.17). *Let $\varphi : \mathcal{R} \to \mathbb{R}$ be translation invariant, additive, measurable and conditionally bounded. Then φ is $\mathbb{Q}$-integrable, and*

$$\overline{\varphi}(X) = \lim_{r\to\infty} \frac{1}{V_d(rW)}\, \mathbb{E} \sum_{C\in X} \varphi(C \cap rW)$$

holds for all $W \in \mathcal{K}$ with $V_d(W) > 0$. Moreover,

$$\overline{\varphi}(X) = \mathbb{E} \sum_{C\in X} \varphi(C \cap C^d_0).$$

Proof. For given $C \in \mathcal{R}_0$, let

$$Z := \{z \in \mathbb{Z}^d : C \cap C_z \neq \emptyset\}.$$

For $z \in Z$ we have $C_z \subset C + \sqrt{d}B^d$, hence

$$|Z| = \lambda\left(\bigcup_{z \in Z} C_z\right) \leq V_d(C + \sqrt{d}B^d) \leq kV_d(C + B^d), \tag{9.18}$$

if k is chosen such that $\sqrt{d}B^d$ can be covered by k unit balls.

Let $W \in \mathcal{K}'$ and $r > 0$. By

$$\varphi_x(M) := \varphi((M + x) \cap rW), \qquad M \in \mathcal{R},$$

for given $x \in \mathbb{R}^d$, an additive functional φ_x is defined. By Lemma 9.2.1 we get, using the notation of (9.14), that

$$\varphi((C + x) \cap rW) = \varphi_x(C) = \sum_{z \in Z} \varphi_x(C, z). \tag{9.19}$$

As in the proof of Theorem 9.2.1, the additivity and translation invariance of φ lead to an estimate

$$|\varphi_x(C, z)| \leq b2^{N(C)}, \tag{9.20}$$

with

$$b := c(d) \sup_{L \in \mathcal{K},\, L \subset C^d} |\varphi(L)| < \infty$$

and $c(d)$ depending only on d. Together with (9.18), (9.19) and (9.20) this gives

$$|\varphi((C + x) \cap rW)| \leq kb2^{N(C)}V_d(C + B^d).$$

Since the right side is $\mathbb{Q}$-integrable, this yields (with $x = 0$ and $r \to \infty$) the $\mathbb{Q}$-integrability of φ. Further, we obtain

$$\begin{aligned}
\int_{\mathbb{R}^d} |\varphi((C + x) \cap rW)| \, \lambda(\mathrm{d}x) &\leq \sum_{z \in Z} \int_{\mathbb{R}^d} |\varphi_x(C, z)| \, \lambda(\mathrm{d}x) \\
&\leq |Z| b 2^{N(C)} V_d(rW + C^d) \\
&\leq kbV_d(rW + C^d) 2^{N(C)} V_d(C + B^d)
\end{aligned}$$

and hence

$$\int_{\mathcal{R}_0} \int_{\mathbb{R}^d} |\varphi((C + x) \cap rW)| \, \lambda(\mathrm{d}x) \, \mathbb{Q}(\mathrm{d}C) < \infty.$$

Therefore, we can apply the Campbell theorem (Theorem 3.1.2), and with Theorem 4.1.1 we obtain

$$\mathbb{E} \sum_{C \in X} \varphi(C \cap rW) = \gamma \int_{\mathcal{R}_0} \int_{\mathbb{R}^d} \varphi((C + x) \cap rW) \, \lambda(\mathrm{d}x) \, \mathbb{Q}(\mathrm{d}C).$$

Here we can decompose

$$\int_{\mathbb{R}^d} \varphi((C+x)\cap rW)\,\lambda(\mathrm{d}x) = I_1(r) + I_2(r)$$

with

$$I_\nu(r) := \sum_{z\in Z} \int_{A_r^\nu - z} \varphi_x(C,z)\,\lambda(\mathrm{d}x), \quad \nu = 1,2,$$

$$A_r^1 := \{x \in \mathbb{R}^d : (C^d + x)\cap rW \neq \emptyset,\ C^d + x \not\subset rW\},$$
$$A_r^2 := \{x \in \mathbb{R}^d : C^d + x \subset rW\}.$$

We have

$$\lim_{r\to\infty} \frac{\lambda(A_r^1)}{V_d(rW)} = 0, \qquad \lim_{r\to\infty} \frac{\lambda(A_r^2)}{V_d(rW)} = 1.$$

With (9.20) we get

$$|I_1(r)| \leq |Z| b 2^{N(C)} \lambda(A_r^1)$$

and hence

$$\lim_{r\to\infty} \frac{I_1(r)}{V_d(rW)} = 0.$$

Further, we have

$$I_2(r) = \sum_{z\in Z} \varphi(C,z)\lambda(A_r^2) = \varphi(C)\lambda(A_r^2)$$

by Lemma 9.2.1 and thus

$$\lim_{r\to\infty} \frac{I_2(r)}{V_d(rW)} = \varphi(C).$$

This yields

$$\frac{|I_1(r)+I_2(r)|}{V_d(rW)} \leq kb2^{N(C)} V_d(C+B^d) \frac{\lambda(A_r^1)}{V_d(rW)} + |\varphi(C)|.$$

By (9.17) and the $\mathbb{Q}$-integrability of φ we now obtain, using the dominated convergence theorem,

$$\lim_{r\to\infty} \frac{1}{V_d(rW)} \mathbb{E} \sum_{C\in X} \varphi(C\cap rW) = \gamma \int_{\mathcal{R}_0} \varphi(C)\,\mathbb{Q}(\mathrm{d}C) = \overline{\varphi}(X),$$

which is the first assertion of the theorem.

We put

$$\phi(K) := \mathbb{E} \sum_{C\in X} \varphi(C\cap K) \qquad \text{for } K \in \mathcal{R}.$$

For $K \in \mathcal{K}$, the random variable $\sum_{C\in X} \varphi(C\cap K)$ is integrable, as shown, hence by additivity it is also integrable for $K \in \mathcal{R}$. The functional ϕ is additive, translation invariant and conditionally bounded. Now Lemma 9.2.2 yields the second assertion of the theorem. □

As in the case of random closed sets Z, the natural candidates for the functional φ are the intrinsic volumes, the mixed volumes, the surface area measure, and the centered support function. These choices lead to the **specific intrinsic volumes** $\overline{V}_j(X)$ ($j \in \{0,\dots,d\}$) and to the mean values $\overline{V}(X[j], M[d-j])$ ($M \in \mathcal{R}$, $j \in \{1,\dots,d-1\}$), $\overline{S}_{d-1}(X,\cdot)$ and $\overline{h}(X,\cdot)$. Other examples are the densities of mixed measures or mixed functionals. We shall need the **specific** (j,k)th **mixed measure** $\overline{\Phi}^{(j)}_{k,d-k+j}(X,M;\cdot)$ of X (and M), which either arises as an outcome of Theorem 9.2.2 or can be defined directly by

$$\overline{\Phi}^{(j)}_{k,d-k+j}(X,M;\cdot) := \gamma \int_{\mathcal{R}_0} \Phi^{(j)}_{k,d-k+j}(C,M;\mathbb{R}^d \times \cdot)\, \mathbb{Q}(\mathrm{d}C).$$

Again, we have

$$\overline{\Phi}^{(j)}_{j,d}(X,M;\cdot) = \overline{V}_j(X)\, \Phi_d(M,\cdot),$$

$$\overline{\Phi}^{(j)}_{d,j}(X,M;\cdot) = \overline{V}_d(X)\, \Phi_j(M,\cdot).$$

The mixed densities

$$\overline{V}^{(j)}_{m_1,\dots,m_s}(X,\dots,X) \quad \text{and} \quad \overline{V}^{(0)}_{m_1,\dots,m_s,d-j}(X,\dots,X,M)$$

were introduced in Section 9.1, for Poisson particle processes X, by a multiple integral with respect to $(\gamma\mathbb{Q})^s$. Their definition immediately extends to general point processes on $\mathcal{K}'$ (or $\mathcal{R}'$). By an iterated application of Theorem 9.2.2, one gets

$$\overline{V}^{(j)}_{m_1,\dots,m_s}(X,\dots,X) = \lim_{r_1\to\infty} \cdots \lim_{r_s\to\infty} \frac{1}{V_d(r_1W)\cdots V_d(r_sW)}$$

$$\times\, \mathbb{E} \sum_{(C_1,\dots,C_s)\in X_1\times\ldots\times X_s} V^{(j)}_{m_1,\dots,m_s}(C_1\cap r_1W,\dots,C_s\cap r_sW),$$

where $X_1,\dots,X_s$ are independent copies of X. A similar limit relation holds for $\overline{V}^{(0)}_{m_1,\dots,m_s,d-j}(X,\dots,X,M)$. For Poisson processes, we can use Corollary 3.2.4 to obtain

$$\overline{V}^{(j)}_{m_1,\dots,m_s}(X,\dots,X) = \lim_{r_1\to\infty} \cdots \lim_{r_s\to\infty} \frac{1}{V_d(r_1W)\cdots V_d(r_sW)}$$

$$\times\, \mathbb{E} \sum_{(C_1,\dots,C_s)\in X^s_{\neq}} V^{(j)}_{m_1,\dots,m_s}(C_1\cap r_1W,\dots,C_s\cap r_sW).$$

Also for stationary particle processes X, we shall give in Section 9.4 an alternative interpretation of $\overline{V}_j(X)$ as a Radon–Nikodym derivative with respect to the Lebesgue measure λ, namely of the expectation of the stationary random measure

$$\Phi_j(X,\cdot) := \sum_{K\in X} \Phi_j(K,\cdot).$$

This representation will be used in Section 11.1 to introduce specific intrinsic volumes for non-stationary particle processes X, again as functions on $\mathbb{R}^d$.

We remark that the measure $\overline{S}_{d-1}(X,\cdot)$ on the unit sphere S^{d-1} is always nonnegative (by (14.15)) and centered and therefore, if it is not concentrated on a subsphere, is the surface area measure of a unique convex body in $\mathcal{K}_0$, the **Blaschke body** $B(X)$, which was introduced in Section 4.6. There, we assumed that the particles $K \in X$ are convex, but now we see that polyconvex particles can be allowed. For polyconvex particles, $\overline{h}(X,\cdot)$ is a continuous function. If the particles are convex (or if $d = 2$), $\overline{h}(X,\cdot)$ is the support function of a unique convex body in $\mathcal{K}_0$, the **mean body** $M(X)$, which was also introduced and studied in Section 4.6.

Remark. For the introduction of the densities $\overline{\varphi}(Z)$, $\overline{\varphi}(X)$ of translation invariant functionals φ for random sets Z and particle processes X, we have not used the stationarity of Z or X to its full extent. In fact, for a particle process X, for example, we need only the invariance of the expectations

$$\mathbb{E} \sum_{K \in X+t} \varphi(K \cap W)$$

under all translations by $t \in \mathbb{R}^d$, for all windows W. This, in turn, is satisfied if the process X is **weakly stationary**, which means that its intensity measure is translation invariant. The process X is called **weakly isotropic** if its intensity measure is rotation invariant. The analogous terminology is used for processes of flats. For Poisson processes, there is no difference between stationarity and weak stationarity (isotropy and weak isotropy), by Theorem 3.2.1. We remark that most mean value formulas to be proved later for stationary (stationary and isotropic) particle processes X require only that X be weakly stationary (weakly stationary and weakly isotropic). For simplicity, however, we shall stay in the framework of stationarity and isotropy. A similar remark refers to random sets Z, where instead of stationarity it is mostly only needed that the expectations

$$\mathbb{E}\,\varphi((Z+t) \cap W)$$

are invariant under all translations by $t \in \mathbb{R}^d$.

Notes for Section 9.2 are included in the Notes for Section 9.4.

9.3 Ergodic Densities

In the previous section we have seen that for suitable closed random sets Z and functions φ a density $\overline{\varphi}(Z)$ can be defined by

$$\overline{\varphi}(Z) := \lim_{r \to \infty} \frac{\mathbb{E}\,\varphi(Z \cap rW)}{V_d(rW)}.$$

It is a natural question whether a corresponding limit exists also pointwise, that is, without taking the expectation. More precisely, we would like to know under which conditions the limit

$$\overline{\varphi}(Z,\omega) := \lim_{r\to\infty} \frac{\varphi(Z(\omega)\cap rW)}{V_d(rW)}$$

exists for almost all $\omega \in \Omega$. Existence assumed, $\overline{\varphi}(Z,\cdot)$ is a random variable, and we would expect that it satisfies

$$\mathbb{E}\,\overline{\varphi}(Z,\cdot) = \overline{\varphi}(Z).$$

Particularly interesting are the random closed sets Z for which $\overline{\varphi}(Z,\cdot)$ is almost surely equal to a constant, thus satisfying

$$\overline{\varphi}(Z,\cdot) = \overline{\varphi}(Z) \qquad \text{a.s.}$$

For such a random set Z, the φ-density $\overline{\varphi}(Z)$ can be estimated from a single realization $Z(\omega)$, by measuring

$$\frac{\varphi(Z(\omega)\cap W)}{V_d(W)}$$

in a large window W. Results of the type

$$\overline{\varphi}(Z) = \lim_{r\to\infty} \frac{\varphi(Z\cap rW)}{V_d(rW)}, \tag{9.21}$$

where the left side is a constant and the right side is a limit of random variables, are known as ergodic theorems. More precisely, one talks of individual ergodic theorems if the equality holds almost surely, and of statistical ergodic theorems if on the right one has L^p-convergence, for suitable p. We restrict ourselves here to individual ergodic theorems. Such an ergodic result holds for random closed sets Z satisfying certain independence properties, for instance, for ergodic random closed sets, as will be explained below. If the density $\overline{\varphi}(Z)$ can be obtained in the form (9.21), one talks of an ergodic density.

The program thus sketched will now be made precise. We shall, however, not give complete proofs, but for one crucial theorem rely on the literature. In order that the results be applicable not only to random sets, but also to point processes, the following considerations will adopt a more general point of view.

Let $(\Omega, \mathbf{A}, \mathbb{P})$, as always, be the underlying probability space. A bijective map $T : \Omega \to \Omega$ with the property that T and T^{-1} are measurable and leave the probability measure $\mathbb{P}$ invariant (that is, satisfy $\mathbb{P}(TA) = \mathbb{P}(T^{-1}A) = \mathbb{P}(A)$ for all $A \in \mathbf{A}$), is called an **automorphism**. We assume that for $(\Omega, \mathbf{A}, \mathbb{P})$ a set $\mathcal{T} = \{T_x : x \in \mathbb{Z}^d\}$ of automorphisms satisfying $T_xT_y = T_{x+y}$ for $x, y \in \mathbb{Z}^d$ is given; thus the set $\mathcal{T}$ together with the composition is an abelian group. We denote by $\mathbf{T} \subset \mathbf{A}$ the σ-algebra of all events invariant under $\mathcal{T}$, thus

$$\mathbf{T} := \{A \in \mathbf{A} : T_x A = A \text{ for all } x \in \mathbb{Z}^d\}.$$

A family $(\xi_K)_{K\in\mathcal{R}}$ of real random variables on $(\Omega, \mathbf{A})$ is called a **stochastic process with parameter space** $\mathcal{R}$. Since here the parameter $K \in \mathcal{R}$ plays the role of the time (for stochastic processes with continuous time), we also talk of a 'spatial process'. The spatial process $(\xi_K)_{K\in\mathcal{R}}$ is called **additive** if for $K, K' \in \mathcal{R}$ almost surely

$$\xi_{K\cup K'} + \xi_{K\cap K'} = \xi_K + \xi_{K'}$$

holds and, in addition, $\xi_\emptyset = 0$. It is called $\mathcal{T}$**-covariant** if for all $K \in \mathcal{R}$ and all $x \in \mathbb{Z}^d$ the equation

$$\xi_{K+x}(T_x\omega) = \xi_K(\omega)$$

holds for almost all $\omega \in \Omega$. Further, $(\xi_K)_{K\in\mathcal{R}}$ is called **bounded** if there exists an integrable random variable $\eta \geq 0$ with

$$|\xi_K| \leq \eta \qquad \text{a.s. for all } K \in \mathcal{K} \text{ with } K \subset C^d. \tag{9.22}$$

In the following theorem, $\mathbb{E}(\cdot \mid \mathbf{T})$ denotes the conditional expectation with respect to the σ-algebra $\mathbf{T}$ of $\mathcal{T}$-invariant events.

Theorem 9.3.1. *Let $(\xi_W)_{W\in\mathcal{R}}$ be an additive, $\mathcal{T}$-covariant, bounded stochastic process with parameter space $\mathcal{R}$. Then, for $W \in \mathcal{K}$ with $0 \in \operatorname{int} W$, the relation*

$$\lim_{r\to\infty} \frac{\xi_{rW}}{V_d(rW)} = \mathbb{E}\left(\xi_{C^d} - \xi_{\partial^+ C^d} \mid \mathbf{T}\right)$$

holds a.s.

Proof. First we proceed as in the proof of Lemma 9.2.2 and also use the same notation. Let W be as above and assume, without loss of generality, that $W \subset C^d$. For $z \in \mathbb{Z}^d$ and $K \in \mathcal{K}$ we put

$$\xi_{K,z} := \xi_{K\cap C_z} - \xi_{K\cap\partial^+ C_z}.$$

Then, by Lemma 9.2.1, for $r > 0$ we have

$$\begin{aligned}
\xi_{rW}(\omega) &= \sum_{z\in\mathbb{Z}^d} \xi_{rW,z}(\omega) \\
&= \sum_{z\in Z_r^1} \xi_{rW,z}(\omega) + \sum_{z\in Z_r^2} \xi_{rW,z}(\omega) \\
&= \sum_{z\in Z_r^1} \xi_{rW,z}(\omega) + \sum_{z\in Z_r^2} \left[\xi_{C^d}(T_{-z}\omega) - \xi_{\partial^+ C^d}(T_{-z}(\omega))\right].
\end{aligned}$$

If (9.22) holds, we obtain as in the proof of Theorem 9.2.1 that

$$|\xi_{rW,z}| \leq |\xi_{rW\cap C_z}| + |\xi_{rW\cap\partial^+ C_z}| \leq c_d\, \eta \circ T_{-z}$$

with a constant c_d, hence

$$\left| \sum_{z \in Z_r^1} \xi_{rW,z}(\omega) \right| \le c_d \sum_{z \in Z_r^1} \eta(T_{-z}\omega).$$

Now we apply a version of the individual ergodic theorem, for which we refer to Tempel'man [755, Th. 6.1]. If ζ is an integrable random variable on Ω and $(Z_k)_{k \in \mathbb{N}}$ is an increasing sequence of sets $Z_k \subset \mathbb{Z}^d$, satisfying certain assumptions, then

$$\lim_{k \to \infty} \frac{1}{|Z_k|} \sum_{z \in Z_k} \zeta(T_{-z}\omega) = \mathbb{E}(\zeta \mid \mathbf{T})(\omega)$$

holds for almost all $\omega \in \Omega$.

We apply this theorem, first, to $\zeta = \eta$ and $Z_k = Z_{r_k}^1 \cup Z_{r_k}^2$, respectively $Z_k = Z_{r_k}^2$, where $(r_k)_{k \in \mathbb{N}}$ is an increasing real sequence with $r_k \to \infty$. The required assumptions on the sequence $(Z_z)_{k \in \mathbb{N}}$ are satisfied in either case. Observing (9.15), we obtain

$$\lim_{k \to \infty} \frac{1}{V_d(r_k W)} \sum_{z \in Z_{r_k}^1} \eta(T_{-z}\omega) = 0$$

for almost all ω, hence also

$$\lim_{k \to \infty} \frac{1}{V_d(r_k W)} \sum_{z \in Z_{r_k}^1} \xi_{r_k W,z}(\omega) = 0.$$

Second, with $\zeta = \xi_{C^d} - \xi_{\partial^+ C^d}$ and $Z_k = Z_{r_k}^2$ and with the result just obtained, we get

$$\begin{aligned} \lim_{k \to \infty} \frac{\xi_{r_k W}(\omega)}{V_d(r_k W)} &= \lim_{k \to \infty} \frac{1}{V_d(r_k W)} \sum_{z \in Z_{r_k}^2} \left[\xi_{C^d}(T_{-z}\omega) - \xi_{\partial^+ C^d}(T_{-z}\omega)\right] \\ &= \mathbb{E}\left(\xi_{C^d} - \xi_{\partial^+ C^d} \mid \mathbf{T}\right)(\omega) \end{aligned}$$

for almost all ω. This yields the assertion, also for the limit $r \to \infty$ (cf. Tempel'man, *loc.cit.* §8). □

The quadruple $(\Omega, \mathbf{A}, \mathbb{P}, \mathcal{T})$ underlying our considerations is often called a **dynamical system**. This system is called **ergodic** if $\mathbb{P}(A) \in \{0, 1\}$ for all $A \in \mathbf{T}$. In the ergodic case we have

$$\mathbb{E}\left(\xi_{C^d} - \xi_{\partial^+ C^d} \mid \mathbf{T}\right) = \mathbb{E}\left(\xi_{C^d} - \xi_{\partial^+ C^d}\right) \qquad \text{a.s.},$$

thus the limit in Theorem 9.3.1 is almost surely constant. The system $(\Omega, \mathbf{A}, \mathbb{P}, \mathcal{T})$ is called **mixing** if the automorphisms $T_x \in \mathcal{T}$ have the asymptotic independence property

$$\lim_{||x||\to\infty} \mathbb{P}(A \cap T_x B) = \mathbb{P}(A)\mathbb{P}(B) \tag{9.23}$$

for all $A, B \in \mathbf{A}$. Every mixing system is ergodic, since (9.23) with $A \in \mathbf{T}$ and $B = A$ implies $\mathbb{P}(A) = \mathbb{P}(A)^2$. The next lemma shows that it is sufficient to check (9.23) for a restricted class of sets.

Lemma 9.3.1. *The dynamical system $(\Omega, \mathbf{A}, \mathbb{P}, \mathcal{T})$ is mixing if there is a semialgebra $\mathbf{A}_0 \subset \mathbf{A}$ generating $\mathbf{A}$ and satisfying*

$$\lim_{||x||\to\infty} \mathbb{P}(A \cap T_x B) = \mathbb{P}(A)\mathbb{P}(B) \tag{9.24}$$

for all $A, B \in \mathbf{A}_0$.

Proof. Suppose such a semialgebra $\mathbf{A}_0$ exists. The algebra $\mathbf{A}_1$ generated by $\mathbf{A}_0$ consists of all finite disjoint unions of sets from $\mathbf{A}_0$. Therefore, (9.24) holds also for $A, B \in \mathbf{A}_1$. Now let $A, B \in \mathbf{A}$. For given $\epsilon > 0$ there are elements $A', B' \in A_1$ with $\mathbb{P}(A \triangle A') \le \epsilon$ and $\mathbb{P}(B \triangle B') \le \epsilon$ (see, for example, Chow and Teicher [175, p. 23]). From $\mathbb{P}((A \cap B) \triangle (A' \cap B')) \le \mathbb{P}(A \triangle A') + \mathbb{P}(B \triangle B')$ and the $\mathcal{T}$-invariance of $\mathbb{P}$ we obtain

$$\mathbb{P}((A \cap T_x B) \triangle (A' \cap T_x B')) \le 2\epsilon$$

for all $x \in \mathbb{Z}^d$. This gives

$$|\mathbb{P}(A \cap T_x B) - \mathbb{P}(A)\mathbb{P}(B)| \le |\mathbb{P}(A' \cap T_x B') - \mathbb{P}(A')\mathbb{P}(B')| + 4\epsilon,$$

which yields the assertion. □

The preceding general considerations will now be applied to more concrete situations. Let Z be a stationary random closed set in $\mathbb{R}^d$. We can choose $(\mathcal{F}, \mathcal{B}(\mathcal{F}), \mathbb{P}_Z)$ as the underlying probability space $(\Omega, \mathbf{A}, \mathbb{P})$ and $\mathcal{T}$ as the group of the ordinary lattice translations of $\mathbb{R}^d$. Here, $T_x F := F + x$ for $F \in \mathcal{F}$ and $T_x \in \mathcal{T}$. Since Z is stationary, the probability measure $\mathbb{P}_Z$ is invariant under all translations $T_x \in \mathcal{T}$. We call the random closed set Z **mixing**, respectively **ergodic**, if the dynamical system $(\mathcal{F}, \mathcal{B}(\mathcal{F}), \mathbb{P}_Z, \mathcal{T})$ has this property. The following theorem expresses the mixing property of Z in terms of the capacity functional T_Z.

Theorem 9.3.2. *The stationary random closed Z in $\mathbb{R}^d$ is mixing if and only if*

$$\lim_{||x||\to\infty} (1 - T_Z(C_1 \cup T_x C_2)) = (1 - T_Z(C_1))(1 - T_Z(C_2)) \tag{9.25}$$

holds for all $C_1, C_2 \in \mathcal{C}$.

Proof. By Lemma 2.2.2, the system

$$\mathbf{A}_0 := \{\mathcal{F}^{C_0}_{C_1,\dots,C_k} : C_0, \dots, C_k \in \mathcal{C},\ k \in \mathbb{N}_0\}$$

is a semialgebra, which by Lemma 2.1.1 generates the σ-algebra $\mathcal{B}(\mathcal{F})$.

Let $A, B \in \mathbf{A}_0$, say

$$A = \mathcal{F}^{C_0}_{C_1,\dots,C_p}, \quad B = \mathcal{F}^{D_0}_{D_1,\dots,D_q}.$$

First we assume that $p, q \geq 1$. Using (2.2) and (2.3), we obtain

$$\begin{aligned}
&\mathbb{P}_Z(A \cap T_x B) \\
&= \mathbb{P}_Z\left(\mathcal{F}^{C_0 \cup T_x D_0}_{C_1,\dots,C_p,T_x D_1,\dots,T_x D_q}\right) \\
&= \sum_{r=0}^{p}\sum_{s=0}^{q}(-1)^{r+s-1} \sum_{\substack{0=i_0<i_1<\dots<i_r\leq p\\ 0=j_0<j_1<\dots<j_s\leq q}} T_Z\left(\bigcup_{\nu=0}^{r} C_{i_\nu} \cup \bigcup_{\mu=0}^{s} T_x D_{j_\mu}\right) \\
&= \sum_{r=0}^{p}\sum_{s=0}^{q}(-1)^{r+s} \sum_{\substack{0=i_0<i_1<\dots<i_r\leq p\\ 0=j_0<j_1<\dots<j_s\leq q}} \left(1 - T_Z\left(\bigcup_{\nu=0}^{r} C_{i_\nu} \cup T_x \bigcup_{\mu=0}^{s} D_{j_\mu}\right)\right).
\end{aligned}$$

This shows that (9.25) implies

$$\begin{aligned}
&\lim_{||x||\to\infty} \mathbb{P}_Z(A \cap T_x B) \\
&= \sum_{r=0}^{p}\sum_{s=0}^{q}(-1)^{r+s} \sum_{\substack{0=i_0<i_1<\dots<i_r\leq p\\ 0=j_0<j_1<\dots<j_s\leq q}} \left(1 - T_Z\left(\bigcup_{\nu=0}^{r} C_{i_\nu}\right)\right)\left(1 - T_Z\left(\bigcup_{\mu=0}^{s} D_{j_\mu}\right)\right) \\
&= \mathbb{P}_Z(A)\mathbb{P}_Z(B).
\end{aligned}$$

The argument is similar if $p = 0$ or $q = 0$, where, for example, $\mathbb{P}(\mathcal{F}^{C_0}) = 1 - T_Z(C_0)$ has to be used. From Lemma 9.3.1 it now follows that Z is mixing. The converse direction is clear. □

We remark that Theorem 9.3.2 and its proof verbally carry over to the case where $\mathbb{R}^d$ is replaced by $E = \mathcal{F}'(\mathbb{R}^d)$ as base space and the operation of $\mathcal{T}$ on $\mathcal{F}(E)$ is defined by $T_x F := F + x$ (with $F + x := \{A + x : A \in F\}$).

Now we apply Theorem 9.3.1 to the situation described in Theorem 9.2.1. Let Z be a stationary random closed set with values in $\mathcal{S}$. If $\varphi : \mathcal{R} \to \mathbb{R}$ is translation invariant, additive, measurable and conditionally bounded and if Z satisfies the integrability condition of Theorem 9.2.1, then

$$\varphi_K(Z) := \varphi(Z \cap K), \qquad K \in \mathcal{R},$$

defines an additive, $\mathcal{T}$-covariant and bounded stochastic process with parameter space $\mathcal{R}$. The $\mathcal{T}$-covariance of $(\varphi_K)_{K\in\mathcal{R}}$ follows from the translation invariance of φ, and the boundedness is a consequence of the integrability condition on Z, since φ is conditionally bounded. Recall that we work with the canonical probability space $(\Omega, \mathbf{A}, \mathbb{P}) = (\mathcal{F}, \mathcal{B}(\mathcal{F}), \mathbb{P}_Z)$, so that

$$\mathbf{T} = \{A \in \mathcal{B}(\mathcal{F}) : A + x = A \text{ for all } x \in \mathbb{Z}^d\},$$

where $A+x := \{F+x : F \in A\}$. In order to stay within the general framework of this section, we nevertheless continue to use notations such as $Z(\omega)$. From Theorem 9.3.1 we obtain the following result.

Theorem 9.3.3. *Let Z be a standard random set and let $\varphi : \mathcal{R} \to \mathbb{R}$ be additive, translation invariant, measurable and conditionally bounded. Then, for $W \in \mathcal{K}'$ with $0 \in \operatorname{int} W$, the limit*

$$\overline{\varphi}(Z,\omega) := \lim_{r\to\infty} \frac{\varphi(Z(\omega) \cap rW)}{V_d(rW)}$$

exists, for almost all $\omega \in \Omega$, and this limit is independent of W. Further,

$$\overline{\varphi}(Z,\cdot) = \mathbb{E}(\widetilde{\varphi}(Z) \mid \mathbf{T}) \qquad a.s.,$$

where $\widetilde{\varphi}(S) := \varphi(S \cap C^d) - \varphi(S \cap \partial^+ C^d)$ for $S \in \mathcal{S}$.

If Z is ergodic, then

$$\overline{\varphi}(Z,\cdot) = \overline{\varphi}(Z) \qquad a.s.$$

The last assertion follows from the fact that in the ergodic case we have

$$\overline{\varphi}(Z,\cdot) = \mathbb{E}(\widetilde{\varphi}(Z)) = \mathbb{E}\left[\varphi(Z \cap C^d) - \varphi(Z \cap \partial^+ C^d)\right] \qquad \text{a.s.}$$

and that this is equal to $\overline{\varphi}(Z)$, by Theorem 9.2.1.

Theorem 9.3.1 can also be applied to the situation of Theorem 9.2.2. Let X be a stationary particle process in $\mathbb{R}^d$. We consider the dynamical system $(\mathsf{N}, \mathcal{N}, \mathbb{P}_X, \mathcal{T})$ with $\mathsf{N} = \mathsf{N}(\mathcal{F}'(\mathbb{R}^d))$, $\mathcal{N} = \mathcal{N}(\mathcal{F}'(\mathbb{R}^d))$, where $\mathbb{P}_X$ is the distribution of X and $\mathcal{T} = \{T_x : x \in \mathbb{Z}^d\}$ is defined by $(T_x\eta)(B) := \eta(B - x)$ for $B \in \mathcal{B}(\mathcal{F}')$ and $\eta \in \mathsf{N}$. Because of the stationarity of X, the probability measure $\mathbb{P}_X$ is invariant under the mappings $T_x \in \mathcal{T}$. The invariant σ-algebra $\mathbf{T}$ is given by

$$\mathbf{T} = \{A \in \mathcal{N} : T_x A = A \text{ for all } x \in \mathbb{Z}^d\},$$

where $T_x A := \{T_x \eta : \eta \in A\}$. The particle process X is called **mixing**, respectively **ergodic**, if the dynamical system $(\mathsf{N}, \mathcal{N}, \mathbb{P}_X, \mathcal{T})$ has this property.

Now suppose that X and the functional φ satisfy the assumptions of Theorem 9.2.2. Then by

$$\varphi_K(X) := \sum_{C \in X} \varphi(C \cap K), \qquad K \in \mathcal{R},$$

we define an additive, $\mathcal{T}$-covariant and bounded stochastic process with parameter space $\mathcal{R}$. This is verified similarly to above, as well as the following result.

Theorem 9.3.4. *Let X be a stationary particle process in $\mathbb{R}^d$ with particles in $\mathcal{R}$ and with grain distribution $\mathbb{Q}$ satisfying* (9.17). *Let $\varphi : \mathcal{R} \to \mathbb{R}$ be additive, translation invariant, measurable and conditionally bounded. Then for $W \in \mathcal{K}'$ with $0 \in \operatorname{int} W$ the limit*

$$\overline{\varphi}(X, \omega) := \lim_{r\to\infty} \frac{1}{V_d(rW)} \sum_{C\in X(\omega)} \varphi(C \cap rW)$$

exists for almost all $\omega \in \Omega$, and this limit is independent of W. Further,

$$\overline{\varphi}(X, \cdot) = \mathbb{E}(\widetilde{\varphi}(X) \mid \mathbf{T}) \qquad a.s.,$$

where the function $\widetilde{\varphi}$ is defined by

$$\widetilde{\varphi}(\eta) := \sum_{C\in\operatorname{supp}\eta} [\varphi(C \cap C^d) - \varphi(C \cap \partial^+ C^d)], \quad \eta \in \mathsf{N}.$$

If X is ergodic, then

$$\overline{\varphi}(X, \cdot) = \overline{\varphi}(X) \qquad a.s.$$

At least for the most important examples of stationary random closed sets, respectively particle processes, we want to show that they are mixing and thus ergodic.

Theorem 9.3.5. *Stationary Boolean models are mixing.*

Proof. For the stationary Boolean model Z with intensity γ and grain distribution $\mathbb{Q}$, the capacity functional is, according to Theorem 9.1.1, given by

$$1 - T_Z(C) = \mathrm{e}^{-\gamma \int_{\mathcal{C}_0} V_d(C-K)\,\mathbb{Q}(\mathrm{d}K)}, \quad C \in \mathcal{C}.$$

For $C_1, C_2 \in \mathcal{C}$ we have

$$V_d((C_1 \cup T_x C_2) - K) = V_d((C_1 - K) \cup (C_2 - K + x)).$$

For given $K \in \mathcal{C}_0$ and sufficiently large $||x||$ we get $(C_1 - K) \cap (C_2 - K + x) = \emptyset$, hence

$$\lim_{||x||\to\infty} V_d((C_1 \cup T_x C_2) - K) = V_d(C_1 - K) + V_d(C_2 - K).$$

Further,

$$V_d((C_1 \cup T_x C_2) - K) \leq V_d(C_1 - K) + V_d(C_2 - K).$$

The dominated convergence theorem yields (9.25) and thus the assertion. □

The preceding theorem allows us, in particular, to interpret the intrinsic volume densities of a stationary Boolean model Z with grains in $\mathcal{R}$ as ergodic densities. In the case of convex grains, the integrability condition (9.16) is satisfied automatically, since for the Poisson particle process X that generates Z we have, for $K \in \mathcal{K}$,

$$\mathbb{E}\, 2^{N(Z\cap K)} \le \mathbb{E}\, 2^{X(\mathcal{F}_K)} = \sum_{k=0}^{\infty} 2^k \mathrm{e}^{-\Theta(\mathcal{F}_K)} \frac{\Theta(\mathcal{F}_K)^k}{k!} = \mathrm{e}^{\Theta(\mathcal{F}_K)} < \infty.$$

Hence, for any convex body $W \in \mathcal{K}$ with $V_d(W) > 0$ and for $j = 0, \dots, d$ we conclude from Theorems 9.3.3 and 9.3.5 that

$$\overline{V}_j(Z) = \lim_{r\to\infty} \frac{V_j(Z \cap rW)}{V_d(rW)} \qquad \text{a.s.}$$

A counterpart to Theorem 9.3.5 is true for particle processes.

Theorem 9.3.6. *Stationary Poisson particle processes in $\mathbb{R}^d$ are mixing.*

Proof. Given the stationary Poisson particle process X, we consider, as before Theorem 9.3.4, the dynamical system $(\mathsf{N}(E), \mathcal{N}(E), \mathbb{P}_X, \mathcal{T})$ for the base space $E = \mathcal{F}'(\mathbb{R}^d)$. By Lemma 3.1.4, $Z := \operatorname{supp} X$ defines a locally finite random closed set in E. For $T_x \in \mathcal{T}$ and $F \in \mathcal{F}(E)$ we let $T_x F := F + x$ (with $F + x := \{A + x : A \in F\}$). We show that the dynamical system $(\mathcal{F}(E), \mathcal{B}(\mathcal{F}(E)), \mathbb{P}_Z, \mathcal{T})$ is mixing. Since the operations of $\mathcal{T}$ on $\mathsf{N}(E)$ respectively $\mathcal{F}(E)$ commute with the mapping $i : \eta \mapsto \operatorname{supp} \eta$ of Lemma 3.1.4, we can then deduce that also $(\mathsf{N}(E), \mathcal{N}(E), \mathbb{P}_X, \mathcal{T})$ is mixing, which is the assertion.

As remarked after the proof of Theorem 9.3.2, that theorem holds also for $\mathcal{F}(\mathbb{R}^d)$ instead of $\mathbb{R}^d$. In this form it will be used in the following.

The capacity functional of the random closed set Z is given by

$$T_Z(C) = \mathbb{P}(C \cap \operatorname{supp} X \ne \emptyset) = \mathbb{P}(X(C) \ne 0)$$

for $C \in \mathcal{C}(E)$. If Θ denotes the intensity measure of the Poisson process X, then

$$1 - T_Z(C) = \mathrm{e}^{-\Theta(C)}. \tag{9.26}$$

In order to show (9.25) (in its generalized form), let $C_1, C_2 \in \mathcal{C}(E)$, thus these are compact subsets of $\mathcal{F}'(\mathbb{R}^d)$. There are (according to the proof of Lemma 2.3.1) compact subsets K_1, K_2 of $\mathbb{R}^d$ with $C_i \subset \mathcal{F}_{K_i}$, $i = 1, 2$. For $x \in \mathbb{Z}^d$ we have $T_x C_2 \subset \mathcal{F}_{K_2+x}$, hence

$$\begin{aligned} \Theta(C_1 \cap T_x C_2) &\le \Theta(\mathcal{F}_{K_1, K_2+x}) \\ &= \gamma \int_{\mathcal{C}_0} \int_{\mathbb{R}^d} \mathbf{1}_{\mathcal{F}_{K_1,K_2+x}}(C + t)\, \lambda(\mathrm{d}t)\, \mathbb{Q}(\mathrm{d}C), \end{aligned}$$

by Theorem 4.1.1. For given $C \in \mathcal{C}_0$ and sufficiently large $\|x\|$, there is no t satisfying $(C + t) \cap K_1 \ne \emptyset$ and $(C + t) \cap (K_2 + x) \ne \emptyset$; therefore,

$$\lim_{||x|| \to \infty} \int_{\mathbb{R}^d} \mathbf{1}_{\mathcal{F}_{K_1, K_2+x}}(C+t)\, \lambda(\mathrm{d}t) = 0.$$

Moreover,

$$\int_{\mathbb{R}^d} \mathbf{1}_{\mathcal{F}_{K_1, K_2+x}}(C+t)\, \lambda(\mathrm{d}t) \le V_d(K_1 - C),$$

and the function $C \mapsto V_d(K_1 - C)$ is $\mathbb{Q}$-integrable, by (4.4). The dominated convergence theorem yields

$$\lim_{||x|| \to \infty} \Theta(C_1 \cap T_x C_2) = 0$$

and thus

$$\lim_{||x|| \to \infty} \mathrm{e}^{-\Theta(C_1 \cup T_x C_2)} = \mathrm{e}^{-\Theta(C_1)} \mathrm{e}^{-\Theta(C_2)}.$$

Now (9.26) and Theorem 9.3.2 yield the assertion. □

If X is, in particular, a stationary Poisson particle process in $\mathcal{R}$ that satisfies (9.17), then for $W \in \mathcal{K}$ with $V_d(W) > 0$ we obtain

$$\overline{V}_j(X) = \lim_{r \to \infty} \frac{1}{V_d(rW)} \sum_{C \in X} V_j(C \cap rW) \qquad \text{a.s.}$$

Note for Section 9.3

First uses of ergodic theorems in stochastic geometry were made by Miles [517]; see also [521, 523]. In special situations, he has proved a number of convergence results for 'increasing observation windows'. A unified and general treatment of such convergence theorems was given by Nguyen and Zessin [584], building on work of Tempel'man [755]. In Section 9.3, we followed their approach. In their application to Boolean models, however, Nguyen and Zessin did not mention that the conditional expectation obtained as a limit function is almost surely constant, as a consequence of the mixing property of stationary Poisson processes. The importance of mixing properties in stochastic geometry was pointed out by Cowan [180, 181]. A simple proof of the mixing property of stationary Boolean models was given by Wieacker [815]. In using the capacity functional in establishing mixing properties, we here followed Heinrich [324]; there one also finds further information on germ-grain models. For ergodic theory in general, we refer to Krengel [428].

9.4 Intersection Formulas and Unbiased Estimators

The following is a typical question arising from practical applications of random sets. Suppose that the realizations $Z(\omega)$ of a standard random set Z can be observed in a window, say, a compact convex set W with $V_d(W) > 0$. By 'observation' we mean that, in principle, values such as $V_j(Z(\omega) \cap W)$ can be measured. We want to use the random variables $V_j(Z \cap W)/V_d(W)$ to estimate the densities $\overline{V}_j(Z)$. In general, however, $V_j(Z \cap W)/V_d(W)$ will depend on W and will not be an unbiased estimator for $\overline{V}_j(Z)$. To estimate the bias, we have to determine the expectation of $V_j(Z \cap W)$. Under suitable assumptions on the random set Z, this can be achieved by means of integral geometry. From the obtained set of expectations, one can then also derive unbiased estimators for the densities of the intrinsic volumes.

Analogous situations arise for stationary processes X of polyconvex particles or k-dimensional flats. In both cases, the total jth intrinsic volumes of the visible parts in a sampling window W,

$$\sum_{F \in X(\omega)} V_j(F \cap W),$$

are observable for certain realizations $X(\omega)$ of X and we need the corresponding expectation to derive unbiased estimators for $\overline{V}_j(X)$.

Finally, a problem, also motivated by practical applications, consists in the estimation of densities of a stationary random set Z or a stationary process X of particles or k-flats from measurements in lower-dimensional sections. Here, stochastic versions of the Crofton formulas yield an answer.

Intersection Formulas for Random Sets

We begin this program with an extension of the local translative formula (5.17) to standard random sets.

Theorem 9.4.1. *Let Z be a standard random set in $\mathbb{R}^d$, let $W \in \mathcal{K}'$ and $j \in \{0, \dots, d\}$. Then*

$$\mathbb{E}\,\Phi_j(Z \cap W, \cdot) = \sum_{k=j}^{d} \overline{\Phi}^{(j)}_{k,d-k+j}(Z, W; \cdot), \tag{9.27}$$

If Z is isotropic, then

$$\mathbb{E}\,\Phi_j(Z \cap W, \cdot) = \sum_{k=j}^{d} c_{j,d}^{k,d-k+j}\,\overline{V}_k(Z)\Phi_{d-k+j}(W, \cdot), \tag{9.28}$$

where the constants are given by (5.5).

Proof. Let $B \in \mathcal{B}(\mathbb{R}^d)$ be bounded. The function

$$\begin{aligned} &\mathbb{R}^d \times \Omega \to \mathbb{R} \\ &(x, \omega) \ \mapsto \Phi_j(Z(\omega) \cap W \cap (B^d + x), B) \end{aligned}$$

is measurable, by Theorems 14.2.2 and 14.4.4. It is also integrable with respect to the product measure $\lambda \otimes \mathbb{P}$. This follows as in the proof of Theorem 9.2.1

(using $\Phi_j(K,B) \le V_j(K)$ for convex bodies) if we additionally assume that $W \subset C^d$. This assumption is not a restriction of generality, since in the arguments the cube C^d can clearly be replaced by a larger cube.

For $x \in \mathbb{R}^d$ and $r > 0$, we deduce from the translation covariance of Φ_j and the stationarity of Z that

$$\begin{aligned}\mathbb{E}\,\Phi_j(Z \cap W \cap (rB^d + x), B) &= \mathbb{E}\,\Phi_j((Z - x) \cap (W - x) \cap rB^d, B - x)\\ &= \mathbb{E}\,\Phi_j(Z \cap (W - x) \cap rB^d, B - x).\end{aligned}$$

Using Fubini's theorem and the invariance properties of λ, we get

$$\begin{aligned}&\mathbb{E}\int_{\mathbb{R}^d} \Phi_j(Z \cap W \cap (rB^d + x), B)\,\lambda(\mathrm{d}x)\\ &= \mathbb{E}\int_{\mathbb{R}^d} \Phi_j(Z \cap (W + x) \cap rB^d, B + x)\,\lambda(\mathrm{d}x).\end{aligned}$$

We apply the local translative formula (5.17) (for polyconvex sets, see Theorem 5.2.4) to either side (with one of the sets A, B in the quoted formula equal to $\mathbb{R}^d$) and obtain

$$\sum_{k=j}^{d} \mathbb{E}\,\Phi^{(j)}_{k,d-k+j}(Z \cap W, rB^d; B \times \mathbb{R}^d) = \sum_{k=j}^{d} \mathbb{E}\,\Phi^{(j)}_{k,d-k+j}(Z \cap rB^d, W; \mathbb{R}^d \times B).$$

Now we divide both sides by $V_d(rB^d)$ and let r tend to infinity. Because of

$$\Phi^{(j)}_{k,d-k+j}(Z \cap W, rB^d; B \times \mathbb{R}^d) = r^{d-k+j}\Phi^{(j)}_{k,d-k+j}(Z \cap W, B^d; B \times \mathbb{R}^d)$$

and the decomposability property (Theorem 6.4.1), the left side tends to $\mathbb{E}\,\Phi_j(Z \cap W, B)$ and, by Theorem 9.2.1, the right side tends to

$$\sum_{k=j}^{d} \overline{\Phi}^{(j)}_{k,d-k+j}(Z, W; B).$$

If Z is isotropic,

$$\begin{aligned}&\overline{\Phi}^{(j)}_{k,d-k+j}(Z, W; B)\\ &= \lim_{r\to\infty} \frac{1}{V_d(rB^d)} \mathbb{E}\,\Phi^{(j)}_{k,d-k+j}(Z \cap rB^d, W; \mathbb{R}^d \times B)\\ &= \lim_{r\to\infty} \frac{1}{V_d(rB^d)} \mathbb{E}\int_{SO_d} \Phi^{(j)}_{k,d-k+j}(\vartheta Z \cap rB^d, W; \mathbb{R}^d \times B)\,\nu(\mathrm{d}\vartheta)\\ &= \lim_{r\to\infty} \frac{1}{V_d(rB^d)} c^{k,d-k+j}_{j,d}\,\mathbb{E}\,V_k(Z \cap rB^d)\Phi_{d-k+j}(W, B).\end{aligned}$$

Here we have used Fubini's theorem and Theorem 6.4.2. This completes the proof. □

The special case $j = d$ of formula (9.27) reduces to (2.20); it holds for arbitrary stationary random closed sets.

We note two consequences of Theorem 9.4.1. Let Z be a standard random set in $\mathbb{R}^d$ and $B \subset \mathbb{R}^d$ a bounded Borel set. We choose a convex body $W \in \mathcal{K}'$ with $B \subset \operatorname{int} W$. Then $\Phi_j(Z \cap W, B) = \Phi_j(Z, B)$. Since $\Phi^{(j)}_{k,d-k+j}(C, W; \mathbb{R}^d \times B) = 0$ for $k > j$ and $\Phi^{(j)}_{j,d}(C, W; \mathbb{R}^d \times B) = V_j(C)\lambda(B)$, for arbitrary $C \in \mathcal{R}$, we obtain

$$\overline{\Phi}^{(j)}_{k,d-k+j}(Z, W; B) = 0, \ k > j, \quad \text{and} \quad \overline{\Phi}^{(j)}_{j,d}(Z, W; B) = \overline{V}_j(Z)\lambda(B).$$

Therefore, Theorem 9.4.1 implies the following result, which was already announced in Section 9.2.

Corollary 9.4.1. *If Z is a standard random set in $\mathbb{R}^d$ and $B \subset \mathbb{R}^d$ is a bounded Borel set, then*

$$\mathbb{E}\, \Phi_j(Z, B) = \overline{V}_j(Z)\lambda(B)$$

for $j \in \{0, \dots, d\}$.

For $j = d - 1$, Theorem 9.4.1 yields

$$\mathbb{E}\, \Phi_{d-1}(Z \cap W, B) = \overline{V}_{d-1}(Z)\lambda(W \cap B) + \overline{V}_d(Z)\Phi_{d-1}(W, B),$$

and Corollary 9.4.1 reads

$$\mathbb{E}\, \Phi_{d-1}(Z, B) = \overline{V}_{d-1}(Z)\lambda(B).$$

Since $\Phi_{d-1}(C, \cdot) \geq 0$ for $C \in \mathcal{R}$, by (14.15), both results and the limit relation

$$\overline{V}_{d-1}(Z) = \lim_{r \to \infty} \frac{\mathbb{E} V_{d-1}(Z \cap rW)}{V_d(rW)}$$

hold for stationary random closed sets Z with values in $\mathcal{S}$, even without the integrability condition (9.16), but the corresponding expressions may be infinite.

As a further consequence of Theorem 9.4.1, we note the global case of (9.28), which is the formula

$$\mathbb{E} V_j(Z \cap W) = \sum_{k=j}^{d} c^{k,d-k+j}_{j,d} \overline{V}_k(Z) V_{d-k+j}(W).$$

It holds for isotropic Z, but also in the non-isotropic case, if W is a ball or if W is replaced by a randomly rotated version ϑW, with ϑ uniform and independent of Z, and if in addition the expectation over ϑ is taken on the left side. In that case, we can even consider functionals without motion invariance.

Theorem 9.4.2. *Let Z be a standard random set in $\mathbb{R}^d$, let $\varphi : \mathcal{R} \to \mathbb{R}$ be additive, translation invariant and continuous on $\mathcal{K}'$, let $W \in \mathcal{K}'$. Further, let θ be a random rotation with distribution ν and independent of Z. Then*

$$\mathbb{E}_\nu \, \mathbb{E} \, \varphi(Z \cap \theta W) = \sum_{k=0}^{d} V_k(W) \overline{\varphi_{d-k}}(Z).$$

Proof. Similar to the proof of Theorem 9.4.1, one shows that

$$(x, \vartheta, \omega) \mapsto \varphi(Z(\omega) \cap \vartheta W \cap (B^d + x))$$

is $\lambda \otimes \nu \otimes \mathbb{P}$-integrable. The translation invariance of φ and the stationarity of Z show that

$$\mathbb{E}_\nu \, \mathbb{E} \, \varphi(Z \cap \theta W \cap (rB^d + x)) = \mathbb{E}_\nu \, \mathbb{E} \, \varphi(Z \cap (\theta W - x) \cap rB^d).$$

Integration over $\mathbb{R}^d$ and Fubini's theorem give

$$\begin{aligned} &\mathbb{E}_\nu \, \mathbb{E} \int_{\mathbb{R}^d} \varphi(Z \cap \theta W \cap (rB^d + x)) \, \lambda(\mathrm{d}x) \\ &= \mathbb{E} \int_{SO_d} \int_{\mathbb{R}^d} \varphi(Z \cap (\vartheta W - x) \cap rB^d) \, \lambda(\mathrm{d}x) \, \nu(\mathrm{d}\vartheta). \end{aligned}$$

In the first integral, we can replace B^d by ρB^d with $\rho \in SO_d$ and then integrate over all $\rho \in SO_d$ with respect to ν. This gives

$$\mathbb{E}_\nu \, \mathbb{E} \int_{G_d} \varphi(Z \cap \theta W \cap grB^d) \, \mu(\mathrm{d}g) = \mathbb{E} \int_{G_d} \varphi(Z \cap gW \cap rB^d) \, \mu(\mathrm{d}g).$$

Now Theorem 5.1.2 (Hadwiger's general integral geometric theorem) yields

$$\sum_{k=0}^{d} \mathbb{E}_\nu \, \mathbb{E} \, \varphi_{d-k}(Z \cap \theta W) V_k(rB^d) = \sum_{k=0}^{d} \mathbb{E} \, \varphi_{d-k}(Z \cap rB^d) V_k(W). \tag{9.29}$$

Recall that

$$\varphi_{d-k}(K) = \int_{A(d,k)} \varphi(K \cap E) \, \mu_k(\mathrm{d}E)$$

for $K \in \mathcal{K}$. This definition can also be used for $K \in \mathcal{R}$ and then provides the additive extension of φ_{d-k} to the convex ring $\mathcal{R}$. Since φ is continuous on $\mathcal{K}'$ and therefore conditionally bounded, also φ_{d-k} is conditionally bounded. Hence, Theorem 9.2.1 applies, and the density

$$\overline{\varphi_{d-k}}(Z) = \lim_{r \to \infty} \frac{\mathbb{E} \, \varphi_{d-k}(Z \cap rB^d)}{V_d(rB^d)}$$

exists. Therefore, dividing the obtained equation (9.29) by $V_d(rB^d)$ and letting r tend to infinity, we obtain the assertion. □

The special choice $\varphi = V_j$, where

$$\varphi_{d-k} = (V_j)_{d-k} = c_{j,d}^{k,d-k+j} V_{d-k+j},$$

gives

$$\mathbb{E}_\nu \mathbb{E}\, V_j(Z \cap \vartheta W) = \sum_{k=j}^{d} c_{j,d}^{k,d-k+j} V_k(W) \overline{V}_{d-k+j}(Z),$$

for $j = 0, \dots, d$.

Intersection Formulas for Particle Processes

Now we consider similar intersection formulas for particle processes. For simplicity, we restrict ourselves to convex particles, although under suitable integrability conditions the results are also valid for point processes in the convex ring $\mathcal{R}$. For convex particles, the only integrability condition needed is (4.1), which by Theorem 4.1.2 is equivalent to the integrability of the intrinsic volumes with respect to the grain distribution.

Theorem 9.4.3. *Let X be a stationary process of convex particles in $\mathbb{R}^d$, let $j \in \{0, \dots, d\}$ and $W \in \mathcal{K}'$. Then*

$$\mathbb{E} \sum_{K \in X} \Phi_j(K \cap W, \cdot) = \sum_{k=j}^{d} \overline{\Phi}_{k,d-k+j}^{(j)}(X, W; \cdot). \tag{9.30}$$

If X is isotropic, then

$$\mathbb{E} \sum_{K \in X} \Phi_j(K \cap W, \cdot) = \sum_{k=j}^{d} c_{j,d}^{k,d-k+j} \overline{V}_k(X) \Phi_{d-k+j}(W, \cdot). \tag{9.31}$$

Proof. Let $B \in \mathcal{B}(\mathbb{R}^d)$. With Campbell's theorem (Theorem 3.1.2) and the decomposition of Theorem 4.1.1, we obtain

$$\mathbb{E} \sum_{K \in X} \Phi_j(K \cap W, B) = \gamma \int_{\mathcal{K}_0} \int_{\mathbb{R}^d} \Phi_j((K + x) \cap W, B)\, \lambda(\mathrm{d}x)\, \mathbb{Q}(\mathrm{d}K),$$

where γ and $\mathbb{Q}$ are, respectively, the intensity and the grain distribution of X. Now the translative formula (5.17) immediately yields (9.30).

For isotropic X, we can in addition integrate over rotations of the particles K and then either apply the kinematic formula for curvature measures (Theorem 5.3.2) or the rotation formula for mixed measures (Theorem 6.4.2) to get (9.31). □

The consequences of this result are similar to those of Theorem 9.4.1. Namely, for $j = d$ and $j = d - 1$, (9.30) reduces to the simple relations

$$\mathbb{E} \sum_{K \in X} \Phi_d(K \cap W, \cdot) = \overline{V}_d(X) \Phi_d(W, \cdot),$$

$$\mathbb{E} \sum_{K \in X} \Phi_{d-1}(K \cap W, \cdot) = \overline{V}_{d-1}(X) \Phi_d(W, \cdot) + \overline{V}_d(X) \Phi_{d-1}(W, \cdot).$$

If B is a bounded Borel set and W is large enough such that $B \subset \operatorname{int} W$, then (9.30) implies

$$\mathbb{E} \sum_{K \in X} \Phi_j(K, B) = \overline{V}_j(X) \lambda(B).$$

Since both sides are nonnegative, they define locally finite measures and the equality holds for arbitrary Borel sets B.

Corollary 9.4.2. *Let X be a stationary process of convex particles in $\mathbb{R}^d$ and let $B \in \mathcal{B}(\mathbb{R}^d)$ be a Borel set. Then*

$$\mathbb{E} \sum_{K \in X} \Phi_j(K, B) = \overline{V}_j(X) \lambda(B).$$

for $j \in \{0, \dots, d\}$.

The global case of (9.31), which can be written as

$$\mathbb{E} \sum_{K \in X} V_j(K \cap W) = \sum_{k=j}^{d} c_{j,d}^{k,d-k+j} V_k(W) \overline{V}_{d-k+j}(X), \tag{9.32}$$

holds for isotropic X or for general X if either W is a ball or if we average over random rotations of W. The following result is the analog of Theorem 9.4.2.

Theorem 9.4.4. *Let X be a stationary process of convex particles in $\mathbb{R}^d$, let $\varphi : \mathcal{R} \to \mathbb{R}$ be additive, translation invariant and continuous on $\mathcal{K}'$, let $W \in \mathcal{K}'$. If θ is a random rotation with distribution ν and independent of Z, then*

$$\mathbb{E}_\nu \mathbb{E} \sum_{K \in X} \varphi(K \cap \theta W) = \sum_{k=0}^{d} V_k(W) \overline{\varphi_{d-k}}(X).$$

Proof. In complete analogy to the proof of Theorem 9.4.2, we obtain

$$\sum_{k=0}^{d} \mathbb{E}_\nu \mathbb{E} \sum_{K \in X} \varphi_{d-k}(K \cap \theta W) V_k(rB^d) = \sum_{k=0}^{d} \mathbb{E} \sum_{K \in X} \varphi_{d-k}(K \cap rB^d) V_k(W).$$

Dividing by $V_d(rB^d)$, letting $r \to \infty$, and using Theorem 9.2.2, we complete the proof. □

Processes of Flats

Instead of particle processes, we now consider k-flat processes in $\mathbb{R}^d$. For $k \in \{0, \ldots, d-1\}$, let X be a stationary k-flat process with intensity γ and directional distribution $\mathbb{Q}$, let $j \in \{0, \ldots, k\}$ and $W \in \mathcal{K}'$. In analogy to the corresponding notion for particle processes, we define the **specific** (j, k)th **mixed measure** $\overline{\Phi}^{(j)}_{k,d-k+j}(X, W; \cdot)$ of X (for given W) by

$$\overline{\Phi}^{(j)}_{k,d-k+j}(X, W; \cdot) := \gamma \int_{G(d,k)} \Phi^{(j)}_{k,d-k+j}(L, W; B_L \times \cdot)\, \mathbb{Q}(\mathrm{d}L),$$

where $B_L \subset L$ is the ball with center 0 and $\lambda_k(B_L) = 1$.

Theorem 9.4.5. *Let X be a stationary k-flat process in $\mathbb{R}^d$, $k \in \{0, \ldots, d-1\}$, and let $j \in \{0, \ldots, k\}$, $W \in \mathcal{K}'$ and $B \subset \mathbb{R}^d$ a Borel set. Then*

$$\mathbb{E} \sum_{E \in X} \Phi_j(E \cap W, B) = \overline{\Phi}^{(j)}_{k,d-k+j}(X, W; B).$$

If X is isotropic, then

$$\mathbb{E} \sum_{E \in X} \Phi_j(E \cap W, B) = \gamma c^{k,d-k+j}_{j,d} \Phi_{d-k+j}(W, B). \tag{9.33}$$

Proof. Using Campbell's theorem, the decomposition of Theorem 4.4.2, the local determination of curvature measures and the translative Crofton formula from Theorem 6.4.3, we obtain

$$\begin{aligned}
&\mathbb{E} \sum_{E \in X} \Phi_j(E \cap W, B) \\
&= \gamma \int_{G(d,k)} \int_{L^\perp} \Phi_j(W \cap (L + x), B \cap (L + x))\, \lambda_{L^\perp}(\mathrm{d}x)\, \mathbb{Q}(\mathrm{d}L) \\
&= \gamma \int_{G(d,k)} \Phi^{(j)}_{k,d-k+j}(L, W; B_L \times B)\, \mathbb{Q}(\mathrm{d}L) \\
&= \overline{\Phi}^{(j)}_{k,d-k+j}(X, W; B).
\end{aligned}$$

In the isotropic case, we can perform an additional integration over all rotations of L and then use either the Crofton formula for curvature measures or the rotation formula for mixed measures. □

The relations (9.33) provide $k + 1$ interpretations of the intensity γ, including those given by (4.27) and Theorem 4.4.3.

As before, we get a simpler result if we apply an independent uniform random rotation to the sampling window. We state it only in the global version.

Theorem 9.4.6. *Let X be a stationary k-flat process of intensity γ in $\mathbb{R}^d$, let $k \in \{1, \ldots, d-1\}$, $j \in \{0, \ldots, k\}$ and $W \in \mathcal{K}'$. If θ is a random rotation with distribution ν and independent of X, then*

$$\mathbb{E}_\nu \mathbb{E} \sum_{E \in X} V_j(E \cap \theta W) = \gamma c_{j,d}^{k,d-k+j} V_{d-k+j}(W).$$

If X is isotropic or if W is a ball, the result holds without the expectation $\mathbb{E}_\nu$.

Crofton Formulas

Theorems 9.4.1 and 9.4.3 also immediately yield Crofton formulas for random sets and particle processes. If we talk of a standard random set Z or a stationary particle process X in some affine subspace E, the stationarity (and possibly isotropy) of Z and X refers to E, and densities of intrinsic volumes have to be computed in E. For the following results, we denote by

$$\overline{V}^{(j)}_{d-k+j,k}(Y, K) := \overline{\Phi}^{(j)}_{d-k+j,k}(Y, K; \mathbb{R}^d)$$

the **specific** $(j, d-k+j)$th **mixed functional** of the random set or particle process Y and $K \in \mathcal{K}'$.

Theorem 9.4.7. *Let Z be a standard random set in $\mathbb{R}^d$, let $E \in A(d, k)$ be a k-dimensional flat, where $k \in \{1, \ldots, d-1\}$, $B_E \subset E$ a ball with $\lambda_k(B_E) = 1$, and let $j \in \{0, \ldots, k\}$. Then $Z \cap E$ is a standard random set in E, and*

$$\overline{V}_j(Z \cap E) = \overline{V}^{(j)}_{d-k+j,k}(Z, B_E).$$

If Z is isotropic, then $Z \cap E$ is isotropic and

$$\overline{V}_j(Z \cap E) = c_{j,d}^{k,d-k+j}\, \overline{V}_{d-k+j}(Z).$$

Proof. We omit the (not difficult) proof that $Z \cap E$ is, with respect to E, again a standard random set (and isotropic, if Z is isotropic). For that reason, the density $\overline{V}_j(Z \cap E)$ exists. Theorem 9.4.1 yields

$$\mathbb{E}V_j(Z \cap B_E) = \sum_{m=d-k+j}^{d} \overline{V}^{(j)}_{m,d-m+j}(Z, B_E) \tag{9.34}$$

where only terms with $m \geq d-k+j$ appear since $\overline{V}^{(j)}_{m,d-m+j}(Z, B_E) = 0$ for $m < d-k+j$. Since Z is stationary, we can assume that $0 \in E$ and hence $rB_E \subset E$ for $r > 0$. In (9.34), we replace B_E by rB_E and divide the equation by $V_k(rB_E)$. For $r \to \infty$, the left side tends to $\overline{V}_j(Z \cap E)$, since $V_j(Z \cap rB_E) = V_j(Z \cap E \cap rB_E)$ (and the intrinsic volumes do not depend on the dimension of the surrounding space). Since $\overline{V}^{(j)}_{m,d-m+j}(Z, rB_E)$ is homogeneous of degree $d-m+j$ in r, the right side tends to $\overline{V}^{(j)}_{d-k+j,k}(Z, B_E)$.

As in earlier proofs, the result for isotropic Z follows from the rotation formula in Theorem 6.4.2. □

In analogy to Theorem 9.4.7, the following Crofton formula for particle processes can be stated. For simplicity, we assume that the resulting intersection processes $X \cap E$ are simple, though it is not difficult to extend the results to the general case.

Theorem 9.4.8. *Let X be a stationary process of convex particles in $\mathbb{R}^d$, let $E \in A(d,k)$ be a k-dimensional flat, where $k \in \{1,\dots,d-1\}$, $B_E \subset E$ a ball with $\lambda_k(B_E) = 1$, and let $j \in \{0,\dots,k\}$. Then the intersection process $X \cap E$ is a stationary process of convex particles with respect to E, and*

$$\overline{V}_j(X \cap E) = \overline{V}^{(j)}_{d-k+j,k}(X, B_E).$$

If X is isotropic, then $X \cap E$ is isotropic and

$$\overline{V}_j(X \cap E) = c_{j,d}^{k,d-k+j}\overline{V}_{d-k+j}(X).$$

Proof. It is clear that $X \cap E$ is a stationary process of convex particles in E (and isotropic if X is isotropic). In view of the stationarity of X, we may assume that $0 \in B_E$. From Theorem 9.4.3, and since $\overline{V}^{(j)}_{m,d-m+j}(X, B_E) = 0$ for $m < d-k+j$, we get

$$\begin{aligned}\mathbb{E} \sum_{K' \in X \cap E} V_j(K' \cap B_E) &= \mathbb{E} \sum_{K \in X} V_j(K \cap B_E)\\ &= \sum_{m=d-k+j}^{d} \overline{V}^{(j)}_{m,d-m+j}(X, B_E).\end{aligned}$$

We replace B_E by rB_E with $r > 0$ and divide by $V_k(rB_E)$. For $r \to \infty$, by Theorem 9.2.2, applied in E, the left side converges to $\overline{V}_j(X \cap E)$, and the right side converges to $\overline{V}^{(j)}_{d-k+j,k}(X, B_E)$.

For the result in the isotropic case, we use Theorem 6.4.2 again. □

As an example, let X be a stationary and isotropic process of line segments in $\mathbb{R}^d$. For a hyperplane $E \in A(d, d-1)$ we obtain from the preceding theorem

$$\overline{\chi}(X \cap E) = c_{0,d}^{d-1,1}\overline{V}_1(X).$$

For $v \in S^{d-1}$, we let $E := v^{\perp}$ and put $\gamma(v) := \overline{\chi}(X \cap E)$. Observing that $c_{0,d}^{d-1,1} = 2\kappa_{d-1}/d\kappa_d$, we get

$$\gamma(v) = \frac{2\kappa_{d-1}}{d\kappa_d}\overline{V}_1(X).$$

This is also obtained from (4.40), since in the isotropic case the spherical directional distribution φ is given by

$$\varphi = \frac{1}{\sigma(S^{d-1})}\sigma = \frac{1}{d\kappa_d}\sigma$$

(with the spherical Lebesgue measure σ), and

$$\int_{S^{d-1}} |\langle u, v\rangle| \, \sigma(\mathrm{d}u) = 2\kappa_{d-1}.$$

Unbiased Estimators

The intersection formulas proved so far can be used, in an obvious way, to provide estimators for the specific intrinsic volumes, which are unbiased or asymptotically unbiased.

We first discuss the situation for a standard random set Z. Let $j \in \{0, \dots, d\}$. Since the estimation of the specific volume $\overline{V}_d(Z)$ is of a special and simple nature (and was discussed earlier), we concentrate on the cases $j \le d-1$. An unbiased estimator $\widehat{V}_j$ for $\overline{V}_j(Z)$, based on the observation of Z in a sampling window W with $W \in \mathcal{K}'$ and $V_d(W) > 0$, is immediately given by Corollary 9.4.1, namely

$$\widehat{V}_j := \frac{\Phi_j(Z \cap W, \operatorname{int} W)}{V_d(W)}.$$

For example, for $j = d-1$, this estimator requires us to evaluate the total surface area of the boundary parts of $Z(\omega)$ inside the window W.

Since the evaluation of curvature measures Φ_j with $j < d-1$ is more complicated, it seems natural to use the intrinsic volume $V_j(Z \cap W)$ (normalized by $V_d(W)$) as an estimator. This estimator is, in general, not unbiased. In fact, the bias is given by (9.27), namely through the right side of

$$\mathbb{E} V_j(Z \cap W) = \sum_{k=j}^{d} \overline{V}^{(j)}_{k,d-k+j}(Z, W). \tag{9.35}$$

Writing (9.35), for the sampling window rW, $r > 0$, in the form

$$\frac{\mathbb{E}\, V_j(Z \cap rW)}{V_d(rW)} = \overline{V}_j(Z) + \frac{1}{V_d(W)} \sum_{k=j+1}^{d} r^{j-k} \overline{V}^{(j)}_{k,d-k+j}(Z, W),$$

we see how the mean error tends to 0 for increasing windows W.

In the isotropic case, one can also obtain an unbiased estimator from (9.35). Recall that, for isotropic Z, (9.35) transforms into

$$\mathbb{E} V_j(Z \cap W) = \sum_{k=j}^{d} c_{j,d}^{k,d-k+j} V_k(W) \overline{V}_{d-k+j}(Z), \qquad j = 0, \dots, d.$$

This system of equations can be solved for $\overline{V}_0(Z), \dots, \overline{V}_d(Z)$, since the coefficient matrix is triangular. The resulting formulas are of the form

$$\overline{V}_i(Z) = \mathbb{E}\left(\sum_{m=i}^{d} \beta_{dim}(W) V_m(Z \cap W)\right), \qquad i = 0, \ldots, d,$$

hence

$$\widehat{V}_j := \sum_{m=j}^{d} \beta_{djm}(W) V_m(Z \cap W)$$

is an unbiased estimator for $\overline{V}_j(Z)$. As an example, we write down the two-dimensional case, using the notations A, L, χ for area, perimeter and Euler characteristic, respectively:

$$\begin{aligned}
\overline{A}(Z) &= \mathbb{E}\,\frac{A(Z \cap W)}{A(W)}, \\
\overline{L}(Z) &= \mathbb{E}\left[\frac{L(Z \cap W)}{A(W)} - \frac{L(W)A(Z \cap W)}{A(W)^2}\right], \\
\overline{\chi}(Z) &= \mathbb{E}\left[\frac{\chi(Z \cap W)}{A(W)} - \frac{1}{2\pi}\frac{L(W)L(Z \cap W)}{A(W)^2}\right. \\
&\qquad \left. + \left(\frac{1}{2\pi}\frac{L(W)^2}{A(W)^3} - \frac{1}{A(W)^2}\right) A(Z \cap W)\right].
\end{aligned}$$

The method just described requires the evaluation of all the intrinsic volumes $V_0(Z \cap W), \ldots, V_d(Z \cap W)$. A similar method is based on the evaluation of one functional, the Euler characteristic $V_0(Z \cap W)$, but in different sampling windows $r_0 W, \ldots, r_d W$. The system of equations then reads

$$\mathbb{E} V_0(Z \cap r_j W) = \sum_{k=0}^{d} c_{0,d}^{k,d-k} r_j^k V_k(W) \overline{V}_{d-k}(Z), \qquad j = 0, \ldots, d.$$

If the parameters $r_0, \ldots, r_d$ are chosen such that the square matrix with entries $c_{0,d}^{k,d-k} r_j^k V_k(W)$ is regular, this system of equations can again be solved for $\overline{V}_0(Z), \ldots, \overline{V}_d(Z)$ and yields unbiased estimators

$$\widehat{V}_j := \sum_{m=0}^{d} \alpha_{djm}(W) V_0(Z \cap r_m W)$$

for $\overline{V}_j(Z)$.

Returning to random sets Z without the isotropy condition, there is also an unbiased estimator for $\overline{V}_j(Z)$ coming from Theorem 9.2.1, namely

$$\widehat{V}_j := V_j(Z \cap C^d) - V_j(Z \cap \partial^+ C^d).$$

This estimator has been described in the stereological literature.

If Z is a stationary (or stationary and isotropic) Boolean model, the estimators described so far in a sampling window W are strongly consistent, for increasing W, due to Theorem 9.3.3. For example,

$$\widehat{V}_j := \frac{V_j(Z \cap rW)}{V_d(rW)} \to \overline{V}_j(Z) \qquad \text{a.s.}$$

as $r \to \infty$.

With respect to stationary particle processes X on $\mathcal{K}'$ (or $\mathcal{R}'$), the situation is completely analogous. We therefore skip the details. The basic result here is Theorem 9.4.3. It provides unbiased estimators

$$\widehat{V}_j := \frac{\sum_{K \in X} \Phi_j(K \cap W, \operatorname{int} W)}{V_d(W)}$$

for $\overline{V}_j(X)$, whereas the estimator

$$\widehat{V}_j := \frac{\sum_{K \in X} V_j(K \cap W)}{V_d(W)}$$

is asymptotically unbiased. A different unbiased estimator is given by

$$\widehat{V}_j := \sum_{K \in X} \left(V_j(K \cap C^d) - V_j(K \cap \partial^+ C^d) \right).$$

Of course, the different representations of φ-densities in Theorem 4.1.3 yield further unbiased or asymptotically unbiased estimators.

If the particles $K \in X$ are polyconvex and uniformly bounded and the window W is large enough, such that $V_d(W \ominus K) > 0$, for $\mathbb{Q}$-almost all K, another simple estimator is given by

$$\widehat{V}_j := \sum_{K \in X,\, K \subset W} \frac{V_j(K)}{V_d(W \ominus K)}.$$

In fact, since

$$\begin{aligned}
&\mathbb{E} \sum_{K \in X,\, K \subset W} \frac{V_j(K)}{V_d(W \ominus K)} \\
&= \gamma \int_{\mathcal{R}_0} \int_{\mathbb{R}^d} \frac{V_j(K + x)}{V_d(W \ominus (K + x))} \mathbf{1}\{K + x \subset W\}\, \lambda(\mathrm{d}x)\, \mathbb{Q}(\mathrm{d}K) \\
&= \gamma \int_{\mathcal{R}_0} \frac{V_j(K)}{V_d(W \ominus K)} \int_{\mathbb{R}^d} \mathbf{1}\{K + x \subset W\}\, \lambda(\mathrm{d}x)\, \mathbb{Q}(\mathrm{d}K) \\
&= \gamma \int_{\mathcal{R}_0} V_j(K)\, \mathbb{Q}(\mathrm{d}K) \\
&= \overline{V}_j(X),
\end{aligned}$$

this estimator is unbiased.

For isotropic X, the linear equation method yields unbiased estimators

$$\widehat{V}_j := \sum_{m=j}^{d} \beta_{djm}(W) \sum_{K \in X} V_m(K \cap W),$$

respectively

$$\widehat{V}_j := \sum_{m=j}^{d} \alpha_{djm}(W) \sum_{K \in X} V_0(K \cap r_m W).$$

Notice that the coefficients $\beta_{djm}(W)$ and $\alpha_{djm}(W)$ are the same as in the case of random sets. Therefore, also the given explicit formulas in the planar case transfer immediately to particle processes.

For Poisson processes, Theorem 9.3.5 implies that the estimators are strongly consistent.

Let us now come to applications of the Crofton formulas. We concentrate on stationary and isotropic random sets Z. Then we can work with a fixed plane E. Analogous estimators for non-isotropic sets Z follow, if a random plane E with isotropic distribution (and independent of Z) is chosen. Also, the formulas for particle processes X are totally analogous.

We have seen how the densities $\overline{V}_j(Z)$ of an isotropic standard random set admit asymptotically unbiased or even unbiased estimators. If Z is observed in a k-dimensional section $Z \cap E$, then we can obtain estimators for $\overline{V}_j(Z \cap E)$. Theorem 9.4.7 tells us that these are at the same time (asymptotically) unbiased estimators for the densities $c_{j,d}^{k,d-k+j}\overline{V}_{d-k+j}(Z)$.

As an example, we consider the practically relevant case where $d = 3$ and $k = 2$. We deal with the three-dimensional densities $\overline{V}$ (volume), $\overline{S}$ (surface area), $\overline{M}$ (integral of mean curvature) and with the two-dimensional densities $\overline{A}$ (area), $\overline{L}$ (boundary length), $\overline{\chi}$ (Euler characteristic). The equations of Theorem 9.4.7 now read

$$\overline{V}(Z) = \overline{A}(Z \cap E), \tag{9.36}$$

$$\overline{S}(Z) = \frac{4}{\pi}\overline{L}(Z \cap E), \tag{9.37}$$

$$\overline{M}(Z) = 2\pi\overline{\chi}(Z \cap E). \tag{9.38}$$

These equations provide an exact theoretical foundation for the 'fundamental equations of stereology', which are traditionally written in the form

$$V_V = A_A,$$

$$S_V = \frac{4}{\pi} L_A,$$

$$M_V = 2\pi\chi_A.$$

In this way, formula (9.35) and Theorem 9.4.7 provide theoretical justifications for some practical procedures of stereology, at least in those cases where it is reasonable to model probes of real materials by realizations of isotropic standard random sets. From the practical point of view, the consideration of locally polyconvex sets only does not seem to be very restrictive. Of the invariance properties, stationarity is always unrealistic, requiring unbounded sets, but it may well be satisfied approximately at close range. The most critical assumption is that of isotropy. For that reason, the applicability of motion invariant stereology is limited, and the employment of translative integral geometry is appropriate.

Notes for Sections 9.1, 9.2, 9.4

1. The introduction of densities of functionals for random $\mathcal{S}$-sets, intersection formulas as in Section 9.4, and formulas for Boolean models as in Section 9.1, go back to various sources, where they can be found in varying degrees of generality, in part under special assumptions or treated heuristically. We mention the following references, roughly in chronological order: Matheron [462], Davy [198, 199], Miles [530], Miles and Davy [536], Stoyan [739], A.M. Kellerer [390, 391], H.G. Kellerer [392], Weil [787], Wieacker [815], Weil and Wieacker [804], Zähle [826].

The starting point for much of the presentation in Sections 9.2, 9.4 and 9.1 was the work of Weil and Wieacker [804]. We gratefully acknowledge simplifications suggested orally by Markus Kiderlen (proof of Theorem 9.4.1) and Lars Diening (second proof of Lemma 9.2.1).

2. Theorem 9.4.1 and its counterpart for particle processes, Theorem 9.4.3, which provide unbiased estimators for the specific intrinsic volumes without isotropy assumptions, were proved by Weil [787, 788].

3. Special cases of the intersection formulas of Section 9.4 first came up in stereology (see also Note 2 of Subsection 8.4.2). We have treated them here rigorously and generally, for suitable stationary random closed sets or particle processes as the employed models. An alternative approach of stochastic geometry to section stereology consists in working with deterministic (and bounded) structures and investigating them with the aid of random sections. Different distributions of intersection planes that are relevant in this context are discussed in Section 8.4. A presentation of stereological problems and formulas from a geometric point of view is found in Weil [785]. A reader interested in the practical side of stereology is referred to the two volumes of Weibel [778]. More recent developments in the stereology of non-stationary structures are presented in the book by Jensen [379]. For a modern view on stereology in general, we refer to the volume *Stereology for Statisticians* by Baddeley and Jensen [53].

We have restricted ourselves here to standard random sets. Other classes of random closed sets can be treated according to the availability of suitable integral geometric formulas. For example, a counterpart to the second formula of Theorem 9.4.7, for stationary, isotropic random closed sets which are rectifiable manifolds, appears in Mecke [479]. A very general investigation of intersection formulas for random processes of Hausdorff rectifiable closed sets is due to Zähle [822].

4. Applications of Boolean models to various questions of statistical physics (percolation, complex fluids, structure of the universe) were suggested and investigated by K. Mecke [505, 506]; see also Beisbart *et al.* [89], Beisbart *et al.* [88]. Here specific intrinsic volumes (under the name of means of Minkowski functionals) are used as morphological parameters for the description of spatial structures.

5. Concerning the estimation of the specific intrinsic volumes of standard random sets Z, Schmidt and Spodarev [669] proposed a further method, based on the additively extended Steiner formula (14.70). In global form, with $\rho_\epsilon(K) := \rho_\epsilon(K, \mathbb{R}^d)$, the latter says that

$$\rho_\epsilon(K) = \sum_{j=0}^{d} \epsilon^{d-j} \kappa_{d-j} V_j(K)$$

for $\epsilon \geq 0$ and $K \in \mathcal{R}$. Since $K \mapsto \rho_\epsilon(K)$ is additive, translation invariant and locally bounded (it is even continuous on $\mathcal{K}'$), the density

$$\overline{\rho}_\epsilon(Z) := \lim_{r\to\infty} \frac{\mathbb{E}\rho_\epsilon(Z \cap rW)}{V_d(rW)}$$

exists and satisfies

$$\overline{\rho}_\epsilon(Z) = \sum_{j=0}^{d} \epsilon^{d-j} \kappa_{d-j} \overline{V}_j(Z).$$

Choosing pairwise different values $\epsilon_0, \dots, \epsilon_d$ and inverting the system of linear equations yields

$$\overline{V}_j(Z) = \sum_{m=0}^{d} \gamma_{djm} \overline{\rho}_{\epsilon_m}(Z), \qquad j = 0, \dots, d.$$

As estimators of $\overline{\rho}_{\epsilon_m}(Z)$, again the values $\rho_{\epsilon_m}(Z \cap W)/V_d(W)$ in a window W can be used; then

$$\widehat{V}_j := \frac{1}{V_d(rW)} \sum_{m=0}^{d} \gamma_{djm} \rho_{\epsilon_m}(Z \cap rW)$$

is an asymptotically unbiased estimator, as $r \to \infty$. The evaluation of $\rho_{\epsilon_m}(Z \cap W)$ is based on the integral of the index function over $W + \epsilon_m B^d$ (see Note 3 of Section 14.4). A variant, which is also studied in Schmidt and Spodarev [669], is to integrate the index function only over $W \ominus \epsilon_m B^d$ (this method is sometimes called 'minus sampling'), then the corresponding estimator is unbiased. Under additional assumptions, the authors also give a consistent estimation of the asymptotic covariance matrix of these estimators and show that, for germ-grain models satisfying some mixing conditions, a central limit theorem holds.

An algorithmic version of the estimation procedure, for digitized images of random sets, is developed in Klenk, Schmidt and Spodarev [420], and further in Guderlei, Klenk, Mayer, Schmidt and Spodarev [301].

6. Limit theorems. For ergodic standard random sets Z, Theorem 9.3.3 provides an a.s. limit theorem for additive functionals φ of $Z \cap W$, as the sampling window W increases to the whole space. This raises the natural question of more refined results for the corresponding estimators (asymptotic normality, large deviations, etc.). We mention here some of the more recent advances and refer to Molchanov [546] for further and, in particular, earlier results.

Heinrich and Molchanov [329] show a central limit theorem for quite general random measures associated with stationary Boolean models. Their results include the positive extensions of the curvature measures (intrinsic volumes) and also generalize to germ-grain models with suitable ergodicity or mixing conditions.

Pantle, Schmidt and Spodarev [594] study the asymptotic normality of estimators for additive functionals (valuations) for stationary Boolean models. In particular, this includes the (additively extended) intrinsic volumes. Also here, the results extend to more general germ-grain models satisfying a mixing condition.

Heinrich [326] proved a large deviations result for the empirical volume fraction of a stationary Boolean model.

9.5 Further Estimation Problems

In the previous section, we have discussed several methods of estimating important characteristics of stationary random closed sets Z or particle processes X, the specific intrinsic volumes. For stationary and isotropic Boolean models Z, we have also seen in Section 9.1 how measurements on Z can be used to estimate the specific intrinsic volumes of the underlying Poisson process X of particles. These mean values give first quantitative information about Z or X. However, even for a Poisson process X of random balls, which are distributed according to a radius distribution function G on $(0, \infty)$, the specific intrinsic volumes of X, though yielding certain moments of G, in general do not determine the whole distribution.

In the following, we continue these considerations and discuss three particular estimation problems in more detail. The first problem concerns the determination of the intensity γ for a stationary Boolean model Z. We shall describe different estimation methods which work under various assumptions, in particular one which is based on the formulas in Theorem 9.1.5. The second problem is to estimate the radius distribution of a stationary process X of balls in $\mathbb{R}^d$ from measurements of the section process $X \cap L$ in a k-dimensional section plane L, $k \in \{1, \dots, d-1\}$. For $d = 3$, $k = 2$, this is the classical Wicksell problem. In the third problem, we consider a stationary Boolean model with spherical grains and show how the radius distribution can be estimated using generalized contact distributions.

Intensity Estimation for Boolean Models

We have seen in Theorem 9.1.4 that, for a stationary and isotropic Boolean model Z with convex grains, the $(d + 1)$-tuple of specific intrinsic volumes $\overline{V}_0(Z), \dots, \overline{V}_d(Z)$ determines the corresponding $(d+1)$-tuple $\overline{V}_0(X), \dots, \overline{V}_d(X)$ of the underlying Poisson particle process X uniquely (we also emphasized the cases $d = 2$ and $d = 3$). Since the formulas in Theorem 9.1.4 follow a triangular array, although not linear, they can easily be solved for $\overline{V}_0(X), \dots, \overline{V}_d(X)$ yielding equations

$$\overline{V}_j(X) = f_{dj}(\overline{V}_0(Z), \dots, \overline{V}_d(Z)), \qquad j = 0, \dots, d,$$

with rational functions f_{dj}. Using the estimators for $\overline{V}_i(Z)$ from the previous section, we thus obtain estimators for $\overline{V}_0(X), \dots, \overline{V}_d(X)$ (which are no longer unbiased). Since

$$\overline{V}_0(X) = \overline{\chi}(X) = \gamma$$

due to the convexity of the grains, this includes an estimator of the intensity γ. As we have mentioned, Theorem 9.1.4 remains valid for polyconvex grains (under appropriate integrability conditions). If the grains have Euler characteristic one (which in the plane is the case if they are simply connected), then $\overline{\chi}(X) = \gamma$, so we still obtain an estimator for γ. However, if we drop the isotropy of Z, the situation becomes more complicated.

We first consider a stationary Boolean model Z with convex grains. According to Theorem 9.1.1, the spherical contact distribution function H of Z is given by

$$H(r) = 1 - \exp\left(-\sum_{k=1}^{d} \kappa_k r^k \overline{V}_{d-k}(X)\right), \qquad r \geq 0.$$

Hence,

$$f(r) := -\ln(1 - H(r)) = \sum_{k=1}^{d} c_k r^k$$

is a polynomial in r with coefficients $c_k := \kappa_k \overline{V}_{d-k}(X)$ (and without constant term). Since $H(r)$ can be expressed in terms of the volume fractions p of Z and $p(r)$ of $Z + rB^d$, simple estimators for $f(r)$ exist (for example, in the planar case by counting pixels in a digitized image of $Z(\omega) \cap W$). If $\widehat{f}_1, \dots, \widehat{f}_m$ are corresponding estimated values of $f(r)$, for different values $r_1, \dots, r_m$, then fitting a polynomial $\widehat{f}$ of degree d (and with $\widehat{f}(0) = 0$) to these values yields estimators for c_k, $k = 1, \dots, d$. Here, $c_d = \gamma$.

Another method, which also requires convex grains, is based on the lower tangent point $\widetilde{z}(C)$, $C \in \mathcal{C}'$, which we have introduced in Section 4.2. Since there we concentrated on the planar case, we shall do this again, although the method can be extended to higher dimensions. Hence, we consider a stationary Boolean model Z in $\mathbb{R}^2$ with convex grains. Let X be the underlying Poisson particle process and γ the intensity. Since $\widetilde{z}$ is a center function, the points $\widetilde{z}(K)$, $K \in X$, constitute a stationary Poisson process $\widetilde{X}$ in $\mathbb{R}^2$ which also has intensity γ (by Theorems 4.2.1 and 4.2.2). Since $\widetilde{z}(K)$ is a boundary point of K, some points of $\widetilde{X}$ lie on the boundary of Z and the others in the interior. Let X' be the thinning of $\widetilde{X}$ consisting of all points $x \in \widetilde{X}$ which lie in the boundary of Z, hence they are observable from Z. These points are the lower tangent points of particles from X which are not covered by any other particle (the case that the lower tangent point x of one particle is also in the boundary of another particle from X has probability 0). Since $X' = \widetilde{X} \cap \mathrm{cl}\, Z^c$, the (simple) point process X' is stationary. Using the common notation from stereology, we denote the intensity of X' by $\overline{\chi}^+(Z)$.

Theorem 9.5.1. *Let Z be a stationary Boolean model in $\mathbb{R}^2$ with convex grains and X the underlying Poisson particle process with intensity γ. Then*

$$\overline{\chi}^+(Z) = \gamma \mathrm{e}^{-\overline{A}(X)}. \tag{9.39}$$

Proof. In the following proof, we make the identifications explained before Theorems 3.3.5 and 3.5.9. For $K \in \mathcal{K}'$, $\eta \in \mathsf{N}_s(\mathcal{K}')$ and $x \in \mathbb{R}^2$, let

$$Z(K,\eta) := \bigcup_{C \in \eta \setminus \{K\}} C$$

and

$$f(x,K,\eta) := \frac{1}{\pi} \mathbf{1}_{B^2 \cap Z(K,\eta)^c}(x).$$

We apply Theorem 4.2.4 to the particle process X and the center function $\widetilde{z}$. Let $\widetilde{\mathcal{K}} := \{K \in \mathcal{K}' : \widetilde{z}(K) = 0\}$ be the corresponding mark space, $\widetilde{\mathbb{Q}}$ the mark distribution and $(\mathbb{P}^{0,K})_{K \in \widetilde{\mathcal{K}}}$ the regular family occurring in the theorem. We then use Slivnyak's theorem (Theorem 3.5.9) for the stationary marked Poisson process $X_{\widetilde{z}}$. Since we view $\mathbb{P}^{0,K}$ as a measure on $\mathcal{B}(\mathcal{F}(\mathcal{K}'))$, as described in the proof of Theorem 4.2.4, Slivnyak's theorem gives

$$\mathbb{P}^{0,K}(A) = \mathbb{P}(X \cup \{K\} \in A)$$

for $A \in \mathcal{B}(\mathcal{F}(\mathcal{K}'))$ and $\widetilde{\mathbb{Q}}$-almost all $K \in \widetilde{\mathcal{K}}$, hence

$$\int_{\mathsf{N}_s(\mathcal{K}')} g(\eta)\, \mathbb{P}^{0,K}(\mathrm{d}\eta) = \int_{\mathsf{N}_s(\mathcal{K}')} g(\eta \cup \{K\})\, \mathbb{P}_X(\mathrm{d}\eta)$$

for all measurable functions $g \geq 0$. By the definition of $\overline{\chi}^+(Z)$, we thus obtain

$$\begin{aligned}
\overline{\chi}^+(Z) &= \mathbb{E} \sum_{K \in X} f(\widetilde{z}(K), K, X) \\
&= \gamma \int_{\mathbb{R}^2} \int_{\widetilde{\mathcal{K}}} \int_{\mathsf{N}_s(\mathcal{K}')} f(x, K+x, \eta+x)\, \mathbb{P}^{0,K}(\mathrm{d}\eta)\, \widetilde{\mathbb{Q}}(\mathrm{d}K)\, \lambda(\mathrm{d}x) \\
&= \frac{\gamma}{\pi} \int_{\mathbb{R}^2} \int_{\widetilde{\mathcal{K}}} \int_{\mathsf{N}_s(\mathcal{K}')} \mathbf{1}_{B^2}(x) \mathbf{1}_{Z(K,\eta)^c}(0)\, \mathbb{P}^{0,K}(\mathrm{d}\eta)\, \widetilde{\mathbb{Q}}(\mathrm{d}K)\, \lambda(\mathrm{d}x) \\
&= \gamma \int_{\widetilde{\mathcal{K}}} \int_{\mathsf{N}_s(\mathcal{K}')} \mathbf{1}_{Z(K,\eta)^c}(0)\, \mathbb{P}_X(\mathrm{d}\eta)\, \widetilde{\mathbb{Q}}(\mathrm{d}K) \\
&= \gamma \int_{\widetilde{\mathcal{K}}} \int_{\mathsf{N}_s(\mathcal{K}')} \mathbf{1}_{Z_\eta^c}(0)\, \mathbb{P}_X(\mathrm{d}\eta)\, \widetilde{\mathbb{Q}}(\mathrm{d}K) \\
&= \gamma \mathbb{P}(0 \notin Z) \\
&= \gamma \mathrm{e}^{-\overline{A}(X)},
\end{aligned}$$

where $Z_\eta := \bigcup_{K\in\eta} K$ and where we have used that, for fixed K, the relation $K \notin \eta$ holds for $\mathbb{P}_X$-almost all η. □

Combining (9.39) with (9.6) we obtain a simple estimator for the intensity, namely by counting the number $\chi^+(Z\cap W)$ of lower tangent points of Z in the window W and dividing by the area of the uncovered part,

$$\widehat{\gamma} := \frac{\chi^+(Z\cap W)}{A(Z^c \cap W)}.$$

This estimator is ratio-unbiased and strongly consistent, but depends very much on the convexity of the grains.

We next describe a method for stationary Boolean models Z in the plane, which may have arbitrarily shaped compact grains, but they should be connected and their circumradius should be bounded from above by some constant r_0. We make use of the formula

$$\mathbb{P}(Z\cap C = \emptyset) = 1 - T_Z(C) = \mathrm{e}^{-\Theta(\mathcal{F}_C)}$$

for $C \in \mathcal{C}$.

For $\epsilon > 0$, we put

$$\begin{aligned}
C_1 &:= [0, 2r_0+\epsilon] \times [0,\epsilon],\\
C_2 &:= [0,\epsilon] \times [0, 2r_0+\epsilon],\\
C_0 &:= ([0, 2r_0+\epsilon] \times \{0\}) \cup (\{0\} \times [0, 2r_0+\epsilon]).
\end{aligned}$$

Then

$$\begin{aligned}
&\ln \frac{\mathbb{P}(Z\cap(C_0\cup C_1\cup C_2)=\emptyset)\mathbb{P}(Z\cap C_0=\emptyset)}{\mathbb{P}(Z\cap(C_0\cup C_1)=\emptyset)\mathbb{P}(Z\cap(C_0\cup C_2)=\emptyset)}\\
&= \Theta(\mathcal{F}_{C_0\cup C_1}) + \Theta(\mathcal{F}_{C_0\cup C_2}) - \Theta(\mathcal{F}_{C_0\cup C_1\cup C_2}) - \Theta(\mathcal{F}_{C_0})\\
&= \Theta(\mathcal{F}^{C_0}_{C_1,C_2}). \qquad (9.40)
\end{aligned}$$

In order to calculate $\Theta(\mathcal{F}^{C_0}_{C_1,C_2})$, we use Theorem 4.2.1 with the lower left corner z' as center function. Let $\mathbb{Q}'$ be the corresponding mark distribution. Due to our assumptions, we have $r(C) \le r_0$ for $\mathbb{Q}'$-almost all $C \in \mathcal{C}_{z',0} := \{D \in \mathcal{C}' : z'(D) = 0\}$. Therefore, for these C and for $x \in \mathbb{R}^2$, the condition $C + x \in \mathcal{F}^{C_0}_{C_1,C_2}$ is equivalent to $x \in (0,\epsilon]^2$ (here we need the assumption that C is connected). From this, we obtain

$$\Theta(\mathcal{F}^{C_0}_{C_1,C_2}) = \gamma \int_{\mathcal{C}_{z',0}} \int_{\mathbb{R}^2} \mathbf{1}_{(0,\epsilon]^2}(x)\,\lambda(\mathrm{d}x)\,\mathbb{Q}'(\mathrm{d}C) = \gamma\epsilon^2.$$

Since ϵ is known, this can be used for the estimation of γ. Because of

$$\mathbb{P}(Z \cap C = \emptyset) = 1 - \mathbb{P}(0 \in Z - C),$$

one would have to estimate the area densities of $Z - C_0, Z - (C_0 \cup C_1), Z - (C_0 \cup C_2)$ and $Z - (C_1 \cup C_2)$. The resulting estimator only makes sense if the observed area fractions of these four outer parallel sets of Z are smaller than one or even bounded away from one, since otherwise the logarithm of the quotient above is not defined or rather unstable. This implies that the estimation method requires that both the intensity has to be small and the particles need to be small, in comparison to the observation window W.

Now we return to restricted shapes and describe an estimation method based on the formulas of Theorem 9.1.5. We assume a stationary Boolean model Z in $\mathbb{R}^2$ and, since Theorem 9.1.5 was formulated for convex grains, we make the same assumption, although the method also works for simply connected polyconvex grains. We recall the density formulas for this case:

$$\overline{A}(Z) = 1 - \mathrm{e}^{-\overline{A}(X)},$$

$$\overline{L}(Z) = \mathrm{e}^{-\overline{A}(X)}\overline{L}(X),$$

$$\overline{\chi}(Z) = \mathrm{e}^{-\overline{A}(X)}\left(\gamma - \overline{A}(X,-X)\right).$$

If the densities on the left side are estimated, we obtain estimators for $\overline{A}(X), \overline{L}(X)$ and $\gamma - \overline{A}(X,-X)$. However, $\overline{A}(X,-X)$ cannot be expressed in terms of $\overline{L}(X)$ or $\overline{A}(X)$. We therefore replace the second equation above by its local counterparts (9.12),

$$\overline{S}_1(Z,\cdot) = \mathrm{e}^{-\overline{A}(X)}\overline{S}_1(X,\cdot),$$

and (9.13),

$$\overline{h}(Z,\cdot) = \mathrm{e}^{-\overline{A}(X)}\overline{h}(X,\cdot).$$

The connection with $\overline{A}(X,-X)$ is given by

$$\begin{aligned}\overline{A}(X,-X) &= \frac{1}{2}\gamma^2 \int_{\mathcal{K}_0}\int_{\mathcal{K}_0}\int_{S^1} h(K_1,u)\,S_1(-K_2,\mathrm{d}u)\,\mathbb{Q}(\mathrm{d}K_1)\,\mathbb{Q}(\mathrm{d}K_2)\\ &= \frac{1}{2}\int_{S^1}\overline{h}(X,u)\,\overline{S}_1(-X,\mathrm{d}u).\end{aligned}$$

Hence, we can estimate $\overline{A}(X,-X)$, and therefore also γ, if we can estimate $\overline{h}(X,\cdot)$ and $\overline{S}_1(X,\cdot)$. Fortunately, it is sufficient to estimate only one of these quantities. Namely, the Blaschke body $B(X)$ of X, satisfying $S_1(B(X),\cdot) = \overline{S}_1(X,\cdot)$, is identical with the mean body of X, since in the plane, Blaschke addition coincides with Minkowski addition. It follows that $\overline{h}(X,\cdot) = h(B(X),\cdot)$. Therefore, $\overline{S}_1(X,\cdot)$ determines $B(X)$ and thus $\overline{h}(X,\cdot)$, and conversely.

It is obvious that a corresponding analysis of higher-dimensional Boolean models becomes more and more complicated. We refer to the Notes of this section, for a corresponding analysis of the three-dimensional case.

The Wicksell Problem

Let X be a stationary process of balls in $\mathbb{R}^d$, that is, a stationary particle process with intensity measure concentrated on the set of balls with positive radius. The **radius distribution** $\mathbb{G}$ of X can be defined by

$$\mathbb{G}(A) := \frac{1}{\gamma}\mathbb{E}\sum_{K\in X} \mathbf{1}_B(c(K))\mathbf{1}_A(r(K))$$

for $B \in \mathcal{B}$ with $\lambda(B) = 1$ and $A \in \mathcal{B}(\mathbb{R}^+)$, where $c(K)$ is the center and $r(K)$ is the radius of the ball K, and where $\gamma > 0$ denotes the intensity of X. We assume that $\mathbb{G}(\{0\}) = 0$. Of course, $\mathbb{G}$ is also the image of the grain distribution $\mathbb{Q}$ under the mapping $K \mapsto r(K)$. If we represent X as the marked point process

$$\widetilde{X} := \sum_{K\in X} \delta_{(c(K),r(K))}$$

(with mark space $\mathbb{R}^+$), the radius distribution of X is just the mark distribution of $\widetilde{X}$. For a k-dimensional linear subspace $L \in G(d,k)$, $k \in \{1,\dots,d-1\}$, the section process $X \cap L$ is a stationary process of (k-dimensional) balls; we denote its radius distribution by $\mathbb{G}_L$. We shall now establish a connection between $\mathbb{G}$ and $\mathbb{G}_L$.

For $x \in \mathbb{R}^d$ we use the orthogonal decomposition $x = x_L + x^L$ with $x_L \in L$ and $x^L \in L^\perp$. The Euclidean norm in $\mathbb{R}^d$ is denoted by $\|\cdot\|$. For a ball $K \subset \mathbb{R}^d$, the intersection $K \cap L$ is a ball in L with radius $\sqrt{r(K)^2 - \|c(K)^L\|^2}$, if $\|c(K)^L\| \le r(K)$ (otherwise $K \cap L = \emptyset$). With the section process $X \cap L$ we therefore associate the marked point process

$$\widetilde{X}_L := \sum_{K\in X,\, \|c(K)^L\|\le r(K)} \delta_{\left(c(K)_L, \sqrt{r(K)^2 - \|c(K)^L\|^2}\right)}$$

in L with mark space $\mathbb{R}^+$ (assuming here that it is simple); it is stationary in L. The radius distribution $\mathbb{G}_L$ of the section process $X \cap L$ is the mark distribution of $\widetilde{X}_L$. The intensity $\gamma_{X\cap L}$ of $X \cap L$ is also the intensity of $\widetilde{X}_L$. By Theorem 3.5.1, the intensity measure of $\widetilde{X}$ is given by $\gamma\lambda \otimes \mathbb{G}$, and the intensity measure of $\widetilde{X}_L$ is given by $\gamma_{X\cap L}\lambda_L \otimes \mathbb{G}_L$. Therefore, for $B \in \mathcal{B}(L)$ and $A \in \mathcal{B}(\mathbb{R}^+)$ we obtain

$$\begin{aligned}
&\gamma_{X\cap L}\lambda_L(B)\mathbb{G}_L(A)\\
&= \mathbb{E}\sum_{(x,a)\in\widetilde{X}_L} \mathbf{1}_{B\times A}(x,a)\\
&= \mathbb{E}\sum_{(x,a)\in\widetilde{X}} \mathbf{1}_{B\times A}\left(x_L, \sqrt{\max\{0, a^2 - \|x^L\|^2\}}\right)\\
&= \gamma\int_{\mathbb{R}^+}\int_{\mathbb{R}^d} \mathbf{1}_B(x_L)\mathbf{1}_A\left(\sqrt{\max\{0, a^2 - \|x^L\|^2\}}\right)\lambda(\mathrm{d}x)\,\mathbb{G}(\mathrm{d}a)
\end{aligned}$$

$$= \gamma \lambda_L(B) \int_{\mathbb{R}^+} \int_{L^\perp} \mathbf{1}_{[0,a]}(\|z\|) \mathbf{1}_A \left(\sqrt{a^2 - \|z\|^2}\right) \lambda_{L^\perp}(\mathrm{d}z)\, \mathbb{G}(\mathrm{d}a)$$

$$= \gamma \lambda_L(B) \int_{L^\perp} \int_{\mathbb{R}^+} \mathbf{1}_{(\|z\|,\infty)}(a) \mathbf{1}_A \left(\sqrt{a^2 - \|z\|^2}\right) \mathbb{G}(\mathrm{d}a)\, \lambda_{L^\perp}(\mathrm{d}z)$$

$$= \gamma \lambda_L(B)(d-k)\kappa_{d-k} \int_0^\infty \int_t^\infty \mathbf{1}_A \left(\sqrt{a^2 - t^2}\right) \mathbb{G}(\mathrm{d}a) t^{d-k-1}\, \mathrm{d}t.$$

In particular, for $A = [x, \infty)$ with $x > 0$, we get

$$\gamma_{X \cap L} \mathbb{G}_L([x, \infty)) = \gamma (d-k) \kappa_{d-k} \int_0^\infty \mathbb{G}\left(\left[\sqrt{x^2 + t^2}, \infty\right)\right) t^{d-k-1}\, \mathrm{d}t. \quad (9.41)$$

With $x \to 0$ we obtain

$$\gamma_{X \cap L} = \kappa_{d-k} \gamma M_{d-k},$$

where M_{d-k} is the $(d-k)$th moment of the radius distribution $\mathbb{G}$.

The **Wicksell corpuscle problem** of stereology is the task to determine, in the case $d = 3$, $k = 2$, the distribution $\mathbb{G}$ from the distribution $\mathbb{G}_L$. If we denote (as is common in stereology) by D_V and D_A the distribution function of $\mathbb{G}$ and $\mathbb{G}_L$, respectively, and if d_V denotes the first moment of $\mathbb{G}$, then (9.41), for $d = 3$, $k = 2$, is equivalent to

$$D_A(r) = 1 - \frac{1}{d_V} \int_0^\infty \left(1 - D_V\left(\sqrt{r^2 + x^2}\right)\right) \mathrm{d}x \qquad \text{for } r > 0.$$

Thus, to determine D_V from D_A, one has (besides the determination of d_V) to solve an Abel type integral equation. An inversion formula exists, but is numerically unstable. In practice, where D_A can only be estimated, this inverse problem presents considerable difficulties.

Boolean Models with Spherical Grains

As a second situation, where an estimation of the radius distribution is possible, we consider a stationary Boolean model Z where the primary grain Z_0 is a random ball with radius distribution $\mathbb{G}$ (again, we assume $\mathbb{G}(\{0\}) = 0$). We recall the contact distribution function H_B (with structuring element B) which we have discussed earlier (see Sections 2.4 and 9.1). For a Boolean model of balls and $t > 0$,

$$\begin{aligned} H_B(t) &= \mathbb{P}(d_B(0, Z) \le t \mid 0 \notin Z) \\ &= 1 - \exp\left(-\gamma \sum_{j=1}^d \kappa_{d-j} V_j(B) t^j \int_0^\infty r^{d-j}\, \mathbb{G}(\mathrm{d}r)\right). \end{aligned} \quad (9.42)$$

We shall consider a variant of H_B which includes the radii of the observable boundaries of the balls.

Namely, if $0 \notin Z$, the set $d_B(0,Z)B$ touches Z almost surely at a boundary point of precisely one grain. We state this fact, for later use, in a more general form.

Lemma 9.5.1. *Let* $X = \{(\xi_1, Z_1), (\xi_2, Z_2), \ldots\}$ *be an independently marked Poisson process on* $\mathbb{R}^d$ *with mark space* $\mathcal{K}_0$ *and intensity measure*

$$\Theta = \left(\int h \, \mathrm{d}\lambda\right) \otimes \mathbb{Q}.$$

Let $x \in \mathbb{R}^d$. *Then*

$$\mathbb{P}(0 < d_B(x, \xi_m + Z_m) = d_B(x, \xi_n + Z_n) < \infty) = 0, \quad m \neq n.$$

Proof. Using Campbell's theorem and then Corollary 3.2.4, we obtain

$$\begin{aligned}
&\mathbb{P}\left(\bigcup_{m \neq n} \{0 < d_B(x, \xi_m + Z_m) = d_B(x, \xi_n + Z_n) < \infty\}\right) \\
&\leq \mathbb{E}\left(\frac{1}{2} \sum_{m \neq n} \mathbf{1}\{0 < d_B(x, \xi_m + Z_m) = d_B(x, \xi_n + Z_n) < \infty\}\right) \\
&= \frac{1}{2} \int_{(\mathbb{R}^d \times \mathcal{K}_0)^2} \mathbf{1}\{y \in \mathrm{bd}(x - K + d_B(x, z + M)B)\} \\
&\quad \times \mathbf{1}\{0 < d_B(x, z + M) < \infty\} \, \Lambda^{(2)}(\mathrm{d}((y, K), (z, M))) \\
&= \frac{1}{2} \int_{\mathcal{K}_0} \int_{\mathcal{K}_0} \int_{\mathbb{R}^d} \int_{\mathbb{R}^d} \mathbf{1}\{y \in \mathrm{bd}(x - K + d_B(x, z + M)B)\} \, h(y) h(z) \\
&\quad \times \mathbf{1}\{0 < d_B(x, z + M) < \infty\} \, \lambda(\mathrm{d}y) \, \lambda(\mathrm{d}z) \, \mathbb{Q}(\mathrm{d}K) \, \mathbb{Q}(\mathrm{d}M).
\end{aligned}$$

The last expression vanishes, since the boundary of a convex body has Lebesgue measure zero. □

We return to our stationary Boolean model Z with spherical grains and assume that $x \notin Z$. Applying the lemma to the underlying (stationary) Poisson process X of balls, we almost surely obtain a unique grain $\widetilde{Z}$ in X with $d_B(x, Z)B \cap \widetilde{Z} \neq \emptyset$. We define $r_B(x, Z)$ as the radius $r(\widetilde{Z})$ of $\widetilde{Z}$. Then the following result holds.

Theorem 9.5.2. *Let* Z *be a stationary Boolean model in* $\mathbb{R}^d$ *with spherical grains and with intensity* γ *and radius distribution* $\mathbb{G}$. *Let* $g \geq 0$ *be a measurable function on* $\mathbb{R}^+ \times \mathbb{R}^+$. *Then we have*

$$\begin{aligned}
&\mathbb{E}(g(d_B(0, Z), r_B(0, Z)) \,|\, 0 \notin Z) \\
&= \gamma \sum_{j=0}^{d-1} (j+1) \kappa_{d-1-j} V_{j+1}(B) \int_0^\infty \int_0^\infty r^{d-1-j} t^j (1 - H_B(t)) g(t, r) \, \mathrm{d}t \, \mathbb{G}(\mathrm{d}r).
\end{aligned}$$

Proof. We use

$$Z = U(X),$$

where $X = \{Z_1, Z_2, \ldots\}$ is a measurable enumeration of the stationary Poisson process of balls underlying Z, with intensity γ and radius distribution $\mathbb{G}$, and where $U(Y)$, for a particle process Y, denotes the union set,

$$U(Y) := \bigcup_{K \in Y} K.$$

For $n \in \mathbb{N}$, we define the events

$$A_n := \{0 < d_B(0, Z_n) < \infty\}$$

and

$$B_n := \{d_B(0, U(X \setminus \{Z_n\})) > d_B(0, Z_n)\}.$$

Then

$$(d_B(0, Z), r_B(0, Z)) = (d_B(0, Z_n), r(Z_n))$$

on $A_n \cap B_n$ and

$$\{0 < d_B(0, Z) < \infty\} = \bigcup_{n=1}^{\infty} (A_n \cap B_n) \qquad \text{a.s.}$$

Using this and Theorem 3.2.5, we obtain

$$\begin{aligned}
&\mathbb{E}\left(\mathbf{1}\{0 < d_B(0, Z) < \infty\} g(d_B(0, Z), r_B(0, Z))\right) \\
&= \mathbb{E} \sum_{n=1}^{\infty} \mathbf{1}_{A_n \cap B_n} g(d_B(0, Z_n), r(Z_n)) \\
&= \mathbb{E}\Bigg(\sum_{K \in X} \mathbf{1}\{0 < d_B(0, K) < \infty\} g(d_B(0, K), r(K)) \\
&\qquad \times \mathbf{1}\{d_B(0, U(X \setminus \{K\})) > d_B(0, K)\} \Bigg) \\
&= \int_{\mathcal{K}'} \mathbf{1}\{0 < d_B(0, K) < \infty\} g(d_B(0, K), r(K)) \\
&\qquad \times \mathbb{P}\left(d_B(0, U(X)) > d_B(0, K)\right) \Theta(\mathrm{d}K) \\
&= \mathbb{P}(0 \notin Z) \gamma \int_0^{\infty} \int_{\mathbb{R}^d} \mathbf{1}\{0 < d_B(0, z + rB^d) < \infty\} g(d_B(0, z + rB^d), r) \\
&\qquad \times (1 - H_B(d_B(0, z + rB^d)))\, \lambda(\mathrm{d}z)\, \mathbb{G}(\mathrm{d}r).
\end{aligned}$$

To the inner integral, we can apply formulas (14.27) (with $K = rB^d$) and (14.25). Using $d_B(0, z + rB^d) = d_B(-z, rB^d)$ and the reflection invariance of λ, we obtain

$$\int_{\mathbb{R}^d} \mathbf{1}\{0 < d_B(0, z + rB^d) < \infty\} g(d_B(0, z + rB^d), r)$$
$$\times (1 - H_B(d_B(0, z + rB^d)))\, \lambda(\mathrm{d}z)$$
$$= \int_{\mathbb{R}^d} \mathbf{1}\{0 < d_B(z, rB^d) < \infty\} g(d_B(z, rB^d), r)$$
$$\times (1 - H_B(d_B(z, rB^d)))\, \lambda(\mathrm{d}z)$$
$$= \sum_{m=0}^{d-1} (d-m)\kappa_{d-m} \int_0^\infty \int_{\mathbb{R}^d \times \mathbb{R}^d} g(t, r)(1 - H_B(t)) t^{d-1-m}$$
$$\times\, \Xi_m(rB^d; B; \mathrm{d}(y, b))\, \mathrm{d}t$$
$$= \sum_{j=0}^{d-1} \binom{d-1}{j} dV(B^d[d-1-j], B[j+1]) r^{d-1-j} \int_0^\infty (1 - H_B(t)) t^j g(t, r) \mathrm{d}t.$$

Since (9.42) implies $\mathbb{P}(d_B(0, Z) < \infty) = 1$, division by $\mathbb{P}(0 \notin Z)$ and formula (14.18) yield

$$\mathbb{E}(g(d_B(0, Z), r_B(0, Z)) \,|\, 0 \notin Z)$$
$$= \gamma \sum_{j=0}^{d-1} (j+1)\kappa_{d-1-j} V_{j+1}(B) \int_0^\infty \int_0^\infty r^{d-1-j} t^j (1 - H_B(t)) g(t, r)\, \mathrm{d}t\, \mathbb{G}(\mathrm{d}r).$$

This proves the theorem. □

For $g(t, r) := \mathbf{1}\{t \leq s\}$, $s \geq 0$, the theorem yields

$$H_B(s) = \int_0^s h_B(t)(1 - H_B(t))\, \mathrm{d}t$$

with

$$h_B(t) := \gamma \sum_{j=0}^{d-1} (j+1)\kappa_{d-1-j} V_{j+1}(B) t^j \int_0^\infty r^{d-1-j}\, \mathbb{G}(\mathrm{d}r).$$

Equation (9.42) shows that $H_B(s) < 1$ and that H_B is a continuous function satisfying $H_B(0) = 0$. Using the monotonicity of H_B, we obtain that

$$\int_0^s h_B(t)\, \mathrm{d}t \leq \frac{H_B(s)}{1 - H_B(s)} < \infty$$

for all $s \geq 0$. Hence, the exponential formula of Lebesgue–Stieltjes calculus (see, for example, Last and Brandt [434, Theorem A4.12]) shows that

$$H_B(s) = 1 - \exp\left\{ - \int_0^s h_B(t)\, \mathrm{d}t \right\}.$$

Consequently, formula (9.42) is contained in Theorem 9.5.2 as a special case.

We now exploit the result for other suitable functions g. Let $W \in \mathcal{K}$ be a sampling window with $\lambda(W) > 0$. Choosing

$$g(t,r) := \frac{f(t)}{h(t,r)} \mathbf{1}_C(r)$$

for a Borel set $C \subset \mathbb{R}^+$, a measurable function $f \geq 0$ and

$$h(t,r) := \sum_{j=0}^{d-1} (j+1)\kappa_{d-1-j} V_{j+1}(B) r^{d-1-j} t^j,$$

we see that

$$\widehat{\mathbb{G}}(C) := \frac{\int_{W\setminus Z} \mathbf{1}_C(r_B(x,Z)) f(d_B(x,Z)) h(d_B(x,Z), r_B(x,Z))^{-1} \lambda(\mathrm{d}x)}{\int_{W\setminus Z} f(d_B(x,Z)) h(d_B(x,Z), r_B(x,Z))^{-1} \lambda(\mathrm{d}x)}$$

is a ratio-unbiased estimator of $\mathbb{G}(C)$. In fact,

$$\begin{aligned} &\mathbb{E} \int_{W\setminus Z} \mathbf{1}_C(r_B(x,Z)) \frac{f(d_B(x,Z))}{h(d_B(x,Z), r_B(x,Z))} \lambda(\mathrm{d}x) \\ &= \gamma\lambda(W)\mathbb{P}(0 \notin Z) \int_0^\infty (1 - H_B(t)) f(t)\, \mathrm{d}t \cdot \mathbb{G}(C) \end{aligned}$$

and (putting $C = \mathbb{R}^+$)

$$\mathbb{E} \int_{W\setminus Z} \frac{f(d_B(x,Z))}{h(d_B(x,Z), r_B(x,Z))} \lambda(\mathrm{d}x) = \gamma\lambda(W)\mathbb{P}(0 \notin Z) \int_0^\infty (1 - H_B(t)) f(t)\, \mathrm{d}t.$$

It should be emphasized that this estimator uses information outside the sampling window W. Namely, for each $x \in W \setminus Z$, the B-distance $d_B(x,Z)$ and the radius $r_B(x,Z)$ of the grain determined by the corresponding contact point have to be observed and the latter may lie outside W.

However, the above considerations show that the generalized contact distributions which we considered give sufficient information to determine the radius distribution $\mathbb{G}$.

We discuss a special case, where the estimator $\widehat{\mathbb{G}}$ has a simpler form. Namely, we consider a planar Boolean model, choose a square as sampling window and $B = [0, u]$, where the unit vector u is parallel to one side of W. Then $h(t,r) = 2r$, hence

$$\widehat{\mathbb{G}} = \frac{1}{\sum_{i=1}^n w_i} \sum_{i=1}^n w_i \delta_{r_i}. \tag{9.43}$$

Here, $r_1, \dots, r_n$ are the radii of the arcs $C_1, \dots, C_n$ in $\operatorname{bd} Z$ which appear as projections from points in W in direction u. If A_i is the region that 'projects' onto C_i, namely the union of all segments $\overline{xy}$, $x \in W \setminus Z$, $y = x + d_{[0,u]}(x, Z)u \in C_i$, the weights w_i are given by

$$w_i = \frac{1}{2r_i} \int_{A_i} f(d_{[0,u]}(z, C_i))\, \lambda(\mathrm{d}z), \qquad i = 1, \ldots, n.$$

In the simplest case, $f = 1$, the weights are proportional to the area of A_i. On the other hand, if $f(t) = \frac{1}{\epsilon}\mathbf{1}\{t \le \epsilon\}$ with $\epsilon \to 0$, we get in the limit an estimator of the form (9.43) where the weights are proportional to the lengths of the arcs. This estimator is studied in the book by Hall [317].

Notes for Section 9.5

1. The estimation of the parameters of a stationary Boolean model (with or without isotropy) is discussed in the books of Serra [729], Cressie [185], Stoyan, Kendall and Mecke [743] and, in particular, in Molchanov [544, 546]. The method of fitting a polynomial to the logarithm of the (empirical) spherical contact distribution function is known as **minimum contrast method**. As a variant, one can investigate the contact distribution function $H_M(r)$, for a fixed value $r = 1$, say, but for different structuring elements M. For example, in the planar case, M can be chosen to be 0-dimensional (point), 1-dimensional (segment) and 2-dimensional (square). The resulting equations can then be solved for γ. In this way, an estimation method for γ was constructed in Hall [316], which is based on counting the number of cells, edges and vertices of a square lattice which are intersected by the given planar Boolean model.

2. Formula (9.39) for the intensity of the uncovered lower tangent points in Theorem 9.5.1 seems to occur first in Serra [729]. The uncovered lower tangent points are dependent, so they no longer form a Poisson process. However, the following result holds. Consider the stationary Boolean model Z with convex grains and with intensity γ in the half plane

$$\mathbb{R}^2_+ := \{(x^1, x^2) \in \mathbb{R}^2 : x^1 \ge 0\}.$$

The **Laslett transform** $L : \mathbb{R}^2_+ \to \mathbb{R}^2_+$ (depending on Z) shifts the points of $\mathbb{R}^2_+$ to the 'left' as far as possible, treating Z as 'empty space' and its complement as solid. More precisely, $L(x_1, x_2) := (\widehat{x}^1, x^2)$ with

$$\widehat{x}^1 := \lambda_1(([0, x^1] \times \{x^2\}) \cap Z^c).$$

The images of the uncovered lower tangent points of Z under this transformation form the restriction to $\mathbb{R}^2_+$ of a stationary Poisson process with intensity γ. This was first explained in Cressie [185]; a short and elegant proof based on a martingale argument was given by Barbour and Schmidt [78]. They mention that the approach also holds in the d-dimensional setting. A further proof in $\mathbb{R}^d$ was given by Černý [168].

For the estimator $\widehat{\gamma}$ based on the uncovered lower tangent points, asymptotic normality was shown by Molchanov and Stoyan [549].

3. The estimation method for γ based on formula (9.40) is due to Schmitt [670]; it has been extended to non-stationary Boolean models as well (Schmitt [671]).

4. The use of Theorem 9.1.4 for the estimation in stationary and isotropic Boolean models Z is classical (see Molchanov [546]). The procedure is sometimes called the **method of moments** since it yields estimators for all the specific intrinsic volumes $\overline{V}_0(X), \dots, \overline{V}_d(X)$. Because of $\overline{V}_0(X) = \gamma$ (in the case of convex grains), this determines the mean values

$$\int_{\mathcal{K}_0} V_j(K)\, \mathbb{Q}(\mathrm{d}K), \qquad j = 1, \dots, d.$$

If the grains are balls, we thus obtain the first d moments of the distribution of the radii. The extension of the method of moments to non-isotropic Boolean models in the plane is due to Weil [794], based on earlier results in [793]. As we mentioned already, a corresponding analysis in $\mathbb{R}^3$ is still possible. We sketch the corresponding approach from Weil [798].

For $d = 3$, we consider the density equations

$$\begin{aligned}
\overline{V}(Z) &= 1 - \mathrm{e}^{-\overline{V}(X)}, \\
\overline{S}_2(Z, \cdot) &= \mathrm{e}^{-\overline{V}(X)} \overline{S}_2(X, \cdot), \\
\overline{h}(Z, \cdot) &= \mathrm{e}^{-\overline{V}(X)} \left(\overline{h}(X, \cdot) - \overline{h}_2(X, X, \cdot) \right), \\
\overline{\chi}(Z) &= \mathrm{e}^{-\overline{V}(X)} \left(\gamma - \overline{V}^{(0)}_{1,2}(X, X) + \overline{V}^{(0)}_{2,2,2}(X, X, X) \right).
\end{aligned}$$

Here, the first equation is the usual one, and the fourth results from Theorem 9.1.5. The second equation is (9.12), and the third is the three-dimensional analog of (9.13). It involves the **specific mixed support function**

$$\overline{h}_2(X, X, \cdot) := \gamma^2 \int_{\mathcal{K}_0} \int_{\mathcal{K}_0} h_2^*(K, M; \cdot)\, \mathbb{Q}(\mathrm{d}M)\, \mathbb{Q}(\mathrm{d}K)$$

(see Theorem 6.4.6). The first equation serves to remove the exponential expression, so we can assume that the quantities $\overline{S}_2(X, \cdot)$, $\overline{h}(X, \cdot) - \overline{h}_2(X, X, \cdot)$ and $\gamma - \overline{V}^{(0)}_{1,2}(X, X) + \overline{V}^{(0)}_{2,2,2}(X, X, X)$ are determined by the left sides. Using the representation (6.30) of $h_2^*(K, M; \cdot)$ for polytopes K, M, we obtain

$$h_2^*(K, M; u) = \int_{S^2} \int_{S^2} f(-u, v, w)\, S_2(K, \mathrm{d}v)\, S_2(M, \mathrm{d}w), \qquad u \in S^2,$$

with a function f, given explicitly by (6.30). By approximation, this representation extends to all convex bodies K, M, therefore we get

$$\overline{h}_2(X, X; u) = \int_{S^2} \int_{S^2} f(-u, v, w)\, \overline{S}_2(X, \mathrm{d}v)\, \overline{S}_2(X, \mathrm{d}w), \qquad u \in S^2.$$

It follows that $\overline{h}_{2,2}(X, X, \cdot)$ is determined and thus also $\overline{h}(X, \cdot)$. It remains to show that $\overline{V}^{(0)}_{1,2}(X, X)$ and $\overline{V}^{(0)}_{2,2,2}(X, X, X)$ can be expressed in terms of $\overline{h}(X, \cdot)$, $\overline{h}_2(X, X, \cdot)$ and $\overline{S}_2(X, \cdot)$, since then we obtain γ. For the first density, this is easy since (9.10) immediately yields

$$\overline{V}^{(0)}_{1,2}(X, X) = \int_{S^2} \overline{h}(X, u)\, \overline{S}_2(X, \mathrm{d}u).$$

For the second density, it turns out that similarly

$$\overline{V}^{(0)}_{2,2,2}(X,X,X) = \int_{S^2} \overline{h}_2(X,X,-u)\,\overline{S}_2(X,\mathrm{d}u)$$
$$= \int_{S^2}\int_{S^2}\int_{S^2} f(u,v,w)\,\overline{S}_2(X,\mathrm{d}v)\,\overline{S}_2(X,\mathrm{d}w)\,\overline{S}_2(X,\mathrm{d}u)$$

holds.

In Weil [801], these estimation problems for $d = 2$ and $d = 3$ were reviewed from the point of densities of mixed volumes (compare Theorem 9.1.6). It was shown that, for $d = 2$, the densities $\overline{V}_0(Z)$, $\overline{V}(Z[1], M[1])$ for all $M \in \mathcal{K}'$, and $\overline{V}_2(Z)$, determine γ uniquely, whereas in dimension $d = 3$, the densities $\overline{V}_0(Z)$, $\overline{V}(Z[1], M[2])$ and $\overline{V}(Z[2], M[1])$, for all $M \in \mathcal{K}'$, as well as $\overline{V}_3(Z)$ are needed. In [801], also the four-dimensional situation was discussed and it was claimed that the intensity γ is determined by the densities of mixed volumes of Z. The proof, however, is incomplete, since a summand $\overline{V}^{(0)}_{2,2}(X,X)$ is missing in the formula for the specific Euler characteristic (see the remarks in Goodey and Weil [280]). Therefore, the four-dimensional case is still open, as are all the higher-dimensional situations.

The approach with local densities (of surface area measures and support functions) or specific mixed volumes can be applied also to non-stationary Boolean models. The specific intrinsic volumes and their local counterparts, the specific surface area measure and the specific support function, then also depend on the location in space. For their definition and further details, see Section 11.1 and the corresponding Note 2.

5. The estimation procedure described before the Wicksell problem requires in practice an estimation of the densities $\overline{h}(Z,\cdot)$ and/or $\overline{S}_1(Z,\cdot)$ from measurements in an observation window. Methods to achieve this are described, for $\overline{h}(Z,\cdot)$ in Weil [794, p. 112 ff], and for $\overline{S}_1(Z,\cdot)$ in Rataj [612], Kiderlen and Jensen [407].

6. The Wicksell corpuscle problem is a classic of stochastic geometry, since Wicksell [813] first treated it and gave an explicit solution of the corresponding Abel type integral equation. The use of marked point processes for the derivation of (9.41) goes back to Mecke and Stoyan [502]. For more details on the Wicksell problem, we refer to Stoyan, Kendall and Mecke [743, sect. 11.4]; see also Ripley [644, sect. 9.4]. Limit distributions of stereological estimators in Wicksell's problem were studied by Heinrich [327]. Zähle [830] treated Wicksell's corpuscle problem in spherical space.

7. Theorem 9.5.2 is a special case of more general results in Hug, Last and Weil [358]. We shall present some of them in Section 11.2. In [358] also various situations are discussed where (generalized) contact distributions of a Boolean model can be used to obtain information on the underlying grain distribution $\mathbb{Q}$. The particular case of spherical grains in the plane, which we presented here, was explained in Weil [802] and is based on work in progress by Hug, Last and Weil.

8. Estimating the intensity of stationary flat processes. Let X be a stationary process of k-flats in $\mathbb{R}^d$ ($k \in \{1,\dots,d-1\}$) with intensity γ . Let $W \in \mathcal{K}$ be a convex sampling window with $V_{d-k}(K) > 0$. The 'weighted estimator'

$$\widehat{\gamma} := \sum_{E \in X \cap \mathcal{F}_W} \frac{1}{V_{d-k}(W|E^\perp)}$$

is an unbiased estimator for the intensity γ, as follows immediately from the Campbell theorem and (4.25). On the other hand, if X has a known directional distribution

$\mathbb{Q}$, then

$$\widehat{\gamma} := \frac{1}{V_{\mathbb{Q}}} \sum_{E \in X \cap \mathcal{F}_W} 1 \qquad \text{with} \qquad V_{\mathbb{Q}} := \int_{G(d,k)} V_{d-k}(W|L^{\perp})\, \mathbb{Q}(\mathrm{d}L)$$

can be used as an unbiased estimator. Schladitz [666] has interpolated between these two extreme cases (of no knowledge and of complete knowledge about the directional distribution), defining an unbiased estimator for the intensity, the 'R-estimator', in the case where the directional distribution of X is known to belong to a given family R of probability measures on $G(d,k)$. She gave sufficient conditions for the R-estimator to be the uniformly best unbiased estimator for the intensity of stationary Poisson k-flat processes with directional distribution in R. For stationary ergodic flat processes, the R-estimator is still uniformly better than the 'naive' one based on Theorem 4.4.3, that is (for $V_d(W) > 0$),

$$\widehat{\gamma} := \frac{1}{V_d(W)} \sum_{E \in X} V_k(E \cap W).$$

9. Estimating the Euler characteristic. The system of formulas (9.36)–(9.38) (as well as the corresponding system in other dimensions) does not include an intersection formula for the specific Euler characteristic $\overline{\chi}(Z)$. In fact, this density, as well as the mean particle number for processes of convex particles, cannot be estimated from the information provided by lower-dimensional sections. To overcome this difficulty, estimators have been suggested that use the information coming jointly from two close parallel hyperplane sections, or from the slab between them. Unbiasedness of these estimators is only guaranteed if the sets under investigation satisfy additional assumptions. We refer to the papers by Ohser and Nagel [588] and by Rataj [616] and to the literature quoted there.

10. Estimating the directional distribution of fiber processes. If X is a stationary fiber process in $\mathbb{R}^d$, with specific length $\overline{V}_1(X)$ and spherical directional distribution φ, then (4.40) says that

$$\overline{V}_0(X \cap v^{\perp}) = \overline{V}_1(X) \int_{S^{d-1}} |\langle u, v \rangle|\, \varphi(\mathrm{d}u)$$

for $v \in S^{d-1}$. If the specific length has already been estimated, this can be used to estimate the directional distribution by means of intersection point counts in hyperplanes $v^{\perp}$. (The function $v \mapsto \overline{V}_0(X \cap v^{\perp})$ is known as the 'rose of intersections', and the even probability measure φ as the 'rose of directions'.) Although the measure φ is uniquely determined by the rose of intersections, there are practical difficulties, since the inversion of the cosine transform is unstable, and only finitely many values of the rose of intersections will be available. Different methods of nonparametric estimation to overcome these difficulties have been described by Kiderlen [403] and by Kiderlen and Pfrang [408].

11. Estimating mean normal measures. Similar problems to those described in the previous note arise if one wants to use (4.41) for the estimation of the directional distribution of a stationary hypersurface process. A related notion is the mean normal measure of a stationary process X of convex particles or of a stationary standard random set Z, denoted by $\overline{S}_{d-1}(X, \cdot)$ and $\overline{S}_{d-1}(Z, \cdot)$, respectively. Since outer normal vectors are used in their definitions, these measures are also called **oriented mean normal measures**, to distinguish them from their even parts, which are called **unoriented mean normal measures** and correspond to the directional distributions of the boundaries. Various estimation procedures for both oriented and unoriented mean normal measures, by means of lower-dimensional sections, have been investigated; we refer to Schneider [704], Kiderlen [404, 406] and the literature quoted there.

12. Estimating particle orientation. If the Blaschke body $B(X)$ of a stationary process X of convex particles is distinctly non-spherical, it reveals anisotropy of X. It may, therefore, be of interest to estimate the Blaschke body by stereological means. Weil [796] has shown that $B(X)$ is uniquely determined by the statistical properties of two-dimensional sections of X, but in practice an estimation based on this fact may be difficult. If only mean particle orientation is of interest, one can replace the Blaschke body by a suitable ellipsoid (equivalently, a positive definite symmetric matrix), which is more accessible to stereological estimation. For a convex body $K \in \mathcal{K}$ with interior points, the **area moment tensor** $T(K)$ is the symmetric tensor of rank two with cartesian coordinates $T_{ij}(K)$ given by

$$T_{ij}(K) := \int_{S^{d-1}} u_i u_j \, S_{d-1}(K, \mathrm{d}u).$$

The eigenvalues and eigendirections of the matrix $(T_{ij}(K))_{i,j=1}^{d}$ can be used to describe the anisotropy of K. The **specific area moment tensor** of the stationary particle process X (with convex particles, intensity γ and grain distribution $\mathbb{Q}$) is defined by

$$\overline{T}(X) := \gamma \int_{\mathcal{K}_0} T(K) \, \mathbb{Q}(\mathrm{d}K) = \lim_{r \to \infty} \frac{1}{V_d(rW)} \mathbb{E} \sum_{K \in X} T(K \cap rW),$$

for arbitrary $W \in \mathcal{K}$ with $V_d(W) > 0$. It turns out that $\overline{T}(X) = T(B(X))$. A method for estimating $\overline{T}(X)$ from sections with hyperplanes was described by Schneider and Schuster [714].

13. Estimation from digitized images. Practical estimation in two and three dimensions may meet the additional difficulty that only digitized images are available, or the sets under investigation are accessible only via their intersections with sufficiently fine scaled grids. Estimation can then be based, for example, on pixel or voxel configuration counts. Methods for the estimation from digitized images have been developed in several investigations, for the Euler characteristic by Nagel, Ohser and Pischang [573], Ohser, Nagel and Schladitz [589, 590], Kiderlen [405], for specific intrinsic volumes by Lang, Ohser and Hilfer [431], and for directional distributions and oriented mean normal measures by Jensen and Kiderlen [381], Kiderlen and Jensen [407], Gutkowski, Jensen and Kiderlen [302], Ziegel and Kiderlen [835].

10

Random Mosaics

By a mosaic we understand a system of convex polytopes in $\mathbb{R}^d$ that cover the whole space and have pairwise no common interior points. A random mosaic can alternatively be described as a special random closed set (formed by the boundaries of the cells of the mosaic), or as a special point process of convex polytopes. The k-dimensional faces of these polytopes themselves generate point processes of k-dimensional sets. Thus, a random mosaic is in a natural way associated with $d+1$ particle processes (the processes of vertices, edges, ..., cells). This special structure and the copious relations between the intensities and specific intrinsic volumes of the various face processes make random mosaics a rich topic for mathematical studies. The planar case has been investigated most thoroughly, but also for three-dimensional random mosaics there are many particular results. In this chapter, we maintain the general frame as before and study random mosaics in $\mathbb{R}^d$, though on some occasions we restrict ourselves to the two- or three-dimensional case, when general results are not known or would be too complicated.

After a general treatment of random mosaics in Section 10.1, we study two special types of random mosaics in more detail, namely the Voronoi mosaics (together with the Delaunay mosaics arising by duality) and the hyperplane mosaics.

A random Voronoi mosaic X is generated by an ordinary point process $\widetilde{X}$ in $\mathbb{R}^d$, by associating with each point $x \in \widetilde{X}$ its Voronoi cell with respect to the Euclidean metric, that is,

$$C(x, \widetilde{X}) := \{z \in \mathbb{R}^d : \|z - x\| \leq \|z - y\| \text{ for all } y \in \widetilde{X}\}.$$

Under suitable assumptions on $\widetilde{X}$ (for instance, stationarity), each cell $C(x, \widetilde{X})$ is bounded and hence a convex polytope. If $\widetilde{X}$ is, in particular, a stationary Poisson process, then all the mean values of X are determined by the intensity of $\widetilde{X}$. Some formulas of this type are proved in Section 10.2. Corresponding results for Delaunay mosaics are obtained by duality.

The second special class of mosaics, to be considered in Section 10.3, are the hyperplane mosaics. Such a mosaic consists of the cells that are induced by a hyperplane process $\widehat{X}$ in $\mathbb{R}^d$, where suitable assumptions on $\widehat{X}$ have to guarantee that the cells are bounded. Again, the Poisson hyperplane mosaics play a prominent role.

In Section 10.4, we collect various results about zero cells and typical cells of general and special random mosaics. Section 10.5 is devoted to mixing properties of mosaics.

10.1 Mosaics as Particle Processes

Generally, one may speak of a mosaic in $\mathbb{R}^d$ whenever the space is covered by d-dimensional closed sets (the cells of the mosaic) which have pairwise no common interior points. In principle, one could allow unbounded, non-convex and even multiply connected cells and still develop certain parts of the theory. Here, however, we restrict ourselves to mosaics which are defined by locally finite systems of compact convex cells.

Definition 10.1.1. *A* **mosaic** *in* $\mathbb{R}^d$ *is a countable system* m *of subsets satisfying the following conditions:*

(a) $\mathsf{m} \in \mathcal{F}_{\ell f}(\mathcal{F}')$, *that is,* m *is a locally finite system of nonempty closed sets.*
(b) *The sets* $K \in \mathsf{m}$ *are compact, convex and have interior points.*
(c) *The sets of* m *cover the space,*

$$\bigcup_{K \in \mathsf{m}} K = \mathbb{R}^d.$$

(d) *If* $K, K' \in \mathsf{m}$ *and* $K \neq K'$, *then* $\operatorname{int} K \cap \operatorname{int} K' = \emptyset$.

According to this definition, a mosaic is a special element of the set

$$\mathcal{F}_{\ell fc}(\mathcal{F}') := \mathcal{F}_{\ell f}(\mathcal{F}') \cap \mathcal{F}(\mathcal{K}').$$

On the subset $\mathcal{K}' \subset \mathcal{F}$ of nonempty convex bodies in $\mathbb{R}^d$ occurring here, the topology induced by $\mathcal{F}$ coincides with the topology derived from the Hausdorff metric (cf. Theorem 12.3.4). The space $\mathcal{K}'$ is locally compact. The elements of $\mathcal{F}_{\ell fc}(\mathcal{F}')$ are sets of convex bodies and at the same time locally finite subsets of $\mathcal{F}'$. Hence, in every such set of convex bodies only finitely many of them meet a given compact subset of $\mathbb{R}^d$.

The elements of a mosaic m are often called the **cells** of m. They are necessarily of a special type.

Lemma 10.1.1. *The cells of a mosaic are convex polytopes.*

Proof. Let m be a mosaic and $K \in \mathsf{m}$. Due to the local finiteness of m, there are only finitely many cells $K_1, \dots, K_m \in \mathsf{m} \setminus \{K\}$ such that $K_i \cap K \neq \emptyset$. From $\mathbb{R}^d = \bigcup_{K' \in \mathsf{m}} K'$ it follows that

$$\operatorname{bd} K = \bigcup_{i=1}^{m} (K_i \cap K).$$

Let $i \in \{1, \dots, m\}$. Since $\operatorname{int} K \cap \operatorname{int} K_i = \emptyset$, the convex bodies K and K_i can be separated by a hyperplane H_i, that is, the closed halfspaces H_i^+ and H_i^- bounded by H_i satisfy, say, $K \subset H_i^+$ and $K_i \subset H_i^-$ $(i = 1, \dots, m)$. We assert that this implies

$$K = \bigcap_{i=1}^{m} H_i^+. \tag{10.1}$$

The inclusion $K \subset \bigcap_{i=1}^m H_i^+$ is trivial. Let $x \in \bigcap_{i=1}^m H_i^+$, and suppose that $x \notin K$. There is a point $y \in \operatorname{int} K \subset \operatorname{int} \bigcap_{i=1}^m H_i^+$. The line segment with end points y and x contains a boundary point x' of K. Since $x' \neq x$, we have $x' \in \operatorname{int} \bigcap_{i=1}^m H_i^+$. On the other hand, $x' \in K_j$ for some $j \in \{1, \dots, m\}$, a contradiction. Hence, (10.1) holds, and K, being compact and an intersection of finitely many closed halfspaces, is a polytope. □

The (proper) **faces** of a convex polytope P are the intersections of P with its supporting hyperplanes. A face of dimension k is called a k-**face**, $k \in \{0, \dots, d-1\}$. The 0-faces are the **vertices** (here we identify $\{x\}$ with x), the 1-faces are the **edges**, and the $(d-1)$-faces of a d-dimensional polytope are its **facets**. If the polytope P is represented, as in (10.1), by the intersection of finitely many closed halfspaces H_i^+, then each k-face of P is the intersection of P with suitable $d-k$ of the corresponding hyperplanes H_i. Every boundary point of P is a relatively interior point of a uniquely determined face of P. Sometimes it is convenient to consider the d-dimensional polytope P as the d-face of P. We write $\mathcal{F}_k(P)$ for the set of all k-faces of P, $k = 0, \dots, d$, and put $\mathcal{F}_\bullet(P) := \bigcup_{k=0}^d \mathcal{F}_k(P)$.

A mosaic can have the property that faces of different cells overlap; for example, a vertex of some cell P can be a relatively interior point of a facet of a neighboring cell. In the following, we exclude this phenomenon and consider only mosaics m which induce cell complexes, that is, satisfy

$$P \cap P' \in (\mathcal{F}_\bullet(P) \cap \mathcal{F}_\bullet(P')) \cup \{\emptyset\} \qquad \text{for all } P, P' \in \mathsf{m}. \tag{10.2}$$

A mosaic satisfying condition (10.2) is called **face-to-face**. For such a mosaic, we write

$$\mathcal{F}_k(\mathsf{m}) := \bigcup_{P \in \mathsf{m}} \mathcal{F}_k(P) \qquad \text{for } k = 0, \dots, d$$

and $\mathcal{F}_\bullet(\mathsf{m}) := \bigcup_{k=0}^d \mathcal{F}_k(\mathsf{m})$.

We denote the set of all mosaics in $\mathbb{R}^d$ by $\mathbb{M}$ and the subset of all face-to-face mosaics by $\mathbb{M}^*$. It is for reasons of simplicity that in the treatment of random mosaics we restrict ourselves to face-to-face mosaics; for the particular mosaics considered in Sections 10.2 and 10.3, this condition is satisfied automatically.

A face-to-face mosaic m is called **normal** if every k-face of m is contained in the boundary of precisely $d-k+1$ cells, $k=0,\dots,d-1$. For $k=d-1$, this condition is always satisfied; every facet of a mosaic belongs to two neighboring cells. The further conditions are, for example, not satisfied by a hyperplane mosaic. In the planar case ($d=2$), normality means that every vertex of m is contained in exactly three cells, and hence in exactly three edges. Generally, for a normal mosaic every j-face belongs to $\binom{d-j+1}{k-j}$ k-faces, $0 \le j \le k \le d$.

The definition and treatment of random mosaics require some measurability assertions.

Lemma 10.1.2. *The set* $\mathbb{M}$ *of all mosaics and the set* $\mathbb{M}^*$ *of face-to-face mosaics in* $\mathbb{R}^d$ *are Borel sets in* $\mathcal{F}(\mathcal{F}')$. *The map*

$$\begin{aligned} \varphi_k : \mathbb{M}^* &\to \mathcal{F}(\mathcal{F}') \\ \mathsf{m} &\mapsto \mathcal{F}_k(\mathsf{m}) \end{aligned}$$

is measurable, $k = 0,\dots,d-1$.

Proof. By Lemma 3.1.4, the set $\mathcal{F}_{\ell fc}(\mathcal{F}')$ is a Borel set in $\mathcal{F}(\mathcal{F}')$. For $r \in \mathbb{N}$, the set $\mathcal{K}_r^{(d-1)} := \{K \in \mathcal{K}' : \dim K \le d-1,\ K \subset rB^d\}$ is closed in $\mathcal{F}'$. We have

$$\mathbb{M} = \left\{ \mathsf{m} \in \mathcal{F}_{\ell fc}(\mathcal{F}') : \bigcup_{K \in \mathsf{m}} K = \mathbb{R}^d \right\} \cap$$

$$\bigcap_{r=1}^{\infty} \left\{ \mathsf{m} \in \mathcal{F}_{\ell fc}(\mathcal{F}') : \mathsf{m} \cap \mathcal{K}_r^{(d-1)} = \emptyset,\ \sum_{K \in \mathsf{m}} V_d(K \cap rB^d) = V_d(rB^d) \right\}.$$

The mapping $\mathsf{m} \mapsto \bigcup_{K \in \mathsf{m}} K$ from $\mathcal{F}_{\ell fc}(\mathcal{F}')$ into $\mathcal{F}'$ is measurable, as follows from the proof of Theorem 3.6.2. The mapping $\mathsf{m} \mapsto \mathsf{m} \cap \mathcal{K}_r^{(d-1)}$ is measurable by Theorem 12.2.6, and the measurability of the mapping $\mathsf{m} \mapsto \sum_{K \in \mathsf{m}} V_d(K \cap rB^d)$ follows as in the proof of Theorem 3.1.2. Hence, $\mathbb{M}$ is a Borel set in $\mathcal{F}(\mathcal{F}')$.

The set $\mathcal{P}'$ of all nonempty convex polytopes in $\mathbb{R}^d$ is a Borel set in $\mathcal{F}'$. We assert that the map

$$\begin{aligned} \psi_k : \mathcal{P}' &\to \mathcal{F}(\mathcal{F}') \\ P &\mapsto \mathcal{F}_k(P) \end{aligned}$$

is measurable. For the proof we consider, for $r,s,t \in \mathbb{N}$, the set $\mathcal{P}_{r,s,t} \subset \mathcal{P}'$ of polytopes P with the following properties: $P \subset rB^d$, P has exactly s vertices, for $q \in \{1,\dots,d\}$ and any $q+1$ vertices $x_1,\dots,x_{q+1}$ of P we have either $V_q(\operatorname{conv}\{x_1,\dots,x_{q+1}\}) = 0$ or $V_q(\operatorname{conv}\{x_1,\dots,x_{q+1}\}) \ge 1/t$. It is easy to see that $\mathcal{P}_{r,s,t}$ is closed and that the restriction of ψ_k to $\mathcal{P}_{r,s,t}$ is continuous. From $\bigcup_{r,s,t\in\mathbb{N}} \mathcal{P}_{r,s,t} = \mathcal{P}'$ we now see that ψ_k is measurable.

In particular, the map $\psi : P \mapsto \mathcal{F}_\bullet(P)$ is measurable. Therefore, the set

$$\begin{aligned} A &:= \{(P,Q) \in \mathcal{P}' \times \mathcal{P}' : P \cap Q = \emptyset \text{ or } P \cap Q \in \mathcal{F}_\bullet(P) \cap \mathcal{F}_\bullet(Q)\} \\ &= \{(P,Q) \in \mathcal{P}' \times \mathcal{P}' : P \cap Q = \emptyset \text{ or } \psi(P) \cap \psi(Q) = \psi(P \cap Q)\} \end{aligned}$$

is a Borel set. For $\mathsf{m} \in \mathbb{M}$, let η_{m} denote the simple counting measure on $\mathcal{F}'$ with support m. By Lemma 3.1.4, the mapping $\mathsf{m} \mapsto \eta_{\mathsf{m}}$ is measurable. Now from

$$\mathbb{M}^* = \{\mathsf{m} \in \mathbb{M} : \eta_{\mathsf{m}} \otimes \eta_{\mathsf{m}}(\mathcal{F}' \times \mathcal{F}' \setminus A) = 0\}$$

the measurability of $\mathbb{M}^*$ follows.

Let $k \in \{0, \dots, d-1\}$. For a polytope P, let $\eta_{P,k}$ be the simple counting measure on $\mathcal{F}'$ with support $\psi_k(P) = \mathcal{F}_k(P)$. By Lemma 3.1.4, the mapping $P \mapsto \eta_{P,k}$ is measurable. For $\mathsf{m} \in \mathbb{M}^*$, the (non-simple) counting measure

$$\nu_{\mathsf{m},k} := \sum_{P \in \mathsf{m}} \eta_{P,k}$$

has support $\mathcal{F}_k(\mathsf{m})$. The mapping $\mathsf{m} \mapsto \nu_{\mathsf{m},k}$ from $\mathbb{M}^*$ into $\mathsf{N}(\mathcal{F}')$ is measurable, by the Campbell theorem (observe that the measurability of $\mathsf{m} \mapsto \nu_{\mathsf{m},k}$ is equivalent to the measurability of $\mathsf{m} \mapsto \nu_{\mathsf{m},k}(A)$ for all Borel sets $A \in \mathcal{B}(\mathcal{F}')$). Now Lemma 3.1.4 yields the measurability of the mapping φ_k. □

Definition 10.1.2. *By a* **random mosaic** *in* $\mathbb{R}^d$ *we understand a particle process* X *in* $\mathbb{R}^d$ *satisfying* $X \in \mathbb{M}^*$ *almost surely.*

Thus, a random mosaic in our terminology is a point process of convex polytopes, pairwise not overlapping, covering the whole space, and satisfying condition (10.2). The random mosaic X is called **normal** if $\mathbb{P}$-almost all realizations of X are normal.

For a random mosaic X,

$$X^{(k)} := \mathcal{F}_k(X), \qquad k \in \{0, \dots, d\},$$

(with $X^{(d)} := X$) defines a particle process, more precisely, a point process of k-dimensional polytopes. The measurability follows from Lemma 10.1.2. The local finiteness of $X^{(k)}(\omega)$ is a consequence of the corresponding property of $X(\omega)$. We denote the intensity measure of $X^{(k)}$ by $\Theta^{(k)}$. It must be stressed that the local finiteness of $\Theta^{(d)}$ does not imply the local finiteness of the intensity measures $\Theta^{(k)}$ for $k < d$ (cf. Note 2 of this section).

Convention. For the random mosaics X treated in this section we make the general assumption that all their face processes $X^{(k)}$, $k = 0, \dots, d$, have locally finite intensity measures.

From now on, we consider only stationary random mosaics X. Then also the face processes $X^{(k)}$, $k = 0, \dots, d$, are stationary. We denote by $\gamma^{(k)}$ the intensity and by

$$d_j^{(k)} := \overline{V}_j(X^{(k)}), \qquad j = 0, \dots, k,$$

the density of the jth intrinsic volume of $X^{(k)}$ (thus, $\gamma^{(k)} = d_0^{(k)}$). Further, $\mathbb{Q}^{(k)}$ denotes the grain distribution of $X^{(k)}$. In particular, intensity and grain distribution of the particle process X itself are now denoted by $\gamma^{(d)}$ and $\mathbb{Q}^{(d)}$, for reasons of greater clarity, and not by γ and $\mathbb{Q}$, as formerly. For $X^{(d)}$, however, we continue to write X, partly because the mosaic X, independently of its definition as a cell process, is meant to comprise the whole collection $X^{(d)}, \dots, X^{(0)}$.

A random polytope $Z^{(k)}$ with distribution $\mathbb{Q}^{(k)}$ is called the **typical k-face** (for $k = d$, the **typical cell**) of X. With this terminology, the expectation

$$\mathbb{E} V_j(Z^{(k)}) = \frac{d_j^{(k)}}{\gamma^{(k)}} = \int_{\mathcal{K}_0} V_j(K)\, \mathbb{Q}^{(k)}(\mathrm{d}K) \tag{10.3}$$

represents the mean jth intrinsic volume of the typical k-face $Z^{(k)}$ of X. The typical cell $Z^{(d)}$ is often denoted by Z.

The volume density $d_d^{(d)}$ seems to be of little interest, since Theorem 9.2.2 yields $\overline{V}_d(X) = 1$. However, from (4.37) we obtain the equation

$$\mathbb{E} V_d(Z) = \frac{1}{\gamma^{(d)}}, \tag{10.4}$$

and hence the information that the mean volume of the typical cell of a stationary random mosaic X is the reciprocal intensity of X.

To motivate the subsequent considerations, we begin with a few elementary geometric observations, which will then be generalized. For further investigations of the specific intrinsic volumes $d_j^{(k)}$ we employ the representation (14.14),

$$V_j(P) = \sum_{S \in \mathcal{F}_j(P)} \gamma(S, P) V_j(S),$$

for the intrinsic volume $V_j(P)$ of a polytope P ($j \in \{0, \dots, d\}$). Here $\gamma(S, P)$ denotes the external angle of the polytope P at its face S. By Theorem 9.2.2,

$$d_j^{(k)} = \lim_{r \to \infty} \frac{1}{r^d} \mathbb{E} \sum_{K \in X^{(k)}} V_j(K \cap rC^d). \tag{10.5}$$

For every polytope K we have

$$V_j(K \cap rC^d) \geq \sum_{S \in \mathcal{F}_j(K)} \gamma(S, K) V_j(S \cap rC^d).$$

The monotonicity of V_j on $\mathcal{K}$ implies

$$V_j(K \cap rC^d) - \sum_{S \in \mathcal{F}_j(K)} \gamma(S,K) V_j(S \cap rC^d) \le V_j(K \cap rC^d) \le r^j V_j(C^d).$$

For $j \le k$ this gives

$$\begin{aligned} d_j^{(k)} &= \lim_{r\to\infty} \frac{1}{r^d} \mathbb{E} \sum_{K \in X^{(k)}} \sum_{S \in \mathcal{F}_j(K)} \gamma(S,K) V_j(S \cap rC^d) \\ &= \lim_{r\to\infty} \frac{1}{r^d} \mathbb{E} \sum_{S \in X^{(j)}} V_j(S \cap rC^d) \sum_{K \in X^{(k)}} \gamma(S,K). \end{aligned}$$

The measurability that was used here can be shown as in the proof of Lemma 10.1.2. Further, we have made use of the fact that the set of all j-faces of $X^{(k)}$ is equal to $X^{(j)}$.

In the case $k = d$, $j = d-1$ we have $\gamma(S,K) = \frac{1}{2}$, hence $\sum_{K \in X} \gamma(S,K) = 1$. This gives

$$d_{d-1}^{(d)} = \lim_{r\to\infty} \frac{1}{r^d} \mathbb{E} \sum_{S \in X^{(d-1)}} V_{d-1}(S \cap rC^d) = d_{d-1}^{(d-1)}, \tag{10.6}$$

by (10.5), a relation which is intuitively plausible.

In the case $k = 1$, $j = 0$ we also have $\gamma(S,K) = \frac{1}{2}$, hence $2\sum_{K \in X^{(1)}} \gamma(S,K)$ is the number of edges of X emanating from the vertex S, which we denote by $N_1(S,X)$. We obtain

$$\gamma^{(1)} = \lim_{r\to\infty} \frac{1}{2r^d} \mathbb{E} \sum_{S \in X^{(0)}} V_0(S \cap rC^d) N_1(S,X). \tag{10.7}$$

If the mosaic X is normal, then $N_1(S,X) = d+1$, hence in this case we get

$$\gamma^{(1)} = \frac{d+1}{2} \gamma^{(0)}. \tag{10.8}$$

Such observations are a motivation for considering, besides the specific intrinsic volumes of the face processes, further quantities of random mosaics that depend on the 'combinatorial neighborhood' of the faces of a given dimension. A simple example is the mean number, n_{01}, of edges emanating from a (typical) vertex. Another example is the mean total edge length of a typical cell. More generally, we consider for a mosaic the totality of the k-faces that are incident with a given j-face, and functions on pairs of incident j-faces and k-faces.

It will be convenient to use marked particle processes, where the particles are the j-faces of the mosaic and some aspects of the combinatorial neighborhood of the j-faces are considered as marks. We prepare this by some definitions.

We denote by $\mathcal{F}_f(\mathcal{K}')$ the system of all finite sets of nonempty convex bodies in $\mathbb{R}^d$. In the following, the letter $\mathcal{S}$ is used as a variable. There is no danger of confusion with the extended convex ring, since the latter is not used in the present chapter. Recall that we use c as center function, where $c(C)$ is the center of the circumball of the compact set C.

Definition 10.1.3. *For $j, k \in \{0, \dots, d\}$, a (j,k)-**face star** is a pair $(T, \mathcal{S}) \in \mathcal{K}' \times \mathcal{F}_f(\mathcal{K}')$, where T is a j-dimensional polytope and $\mathcal{S}$ is a finite set of k-dimensional polytopes satisfying the following conditions:*

$$\begin{aligned} &\mathcal{S} = \mathcal{F}_k(T - c(T)) && \text{if } j \geq k, \\ &T - c(T) \in \mathcal{F}_j(S) \quad \text{for all } S \in \mathcal{S} && \text{if } j < k. \end{aligned}$$

If $\mathsf{m} \in \mathbb{M}^*$ is a mosaic and T is a j-face of m, then

$$\begin{aligned} &\mathcal{F}_k(T, \mathsf{m}) := \mathcal{F}_k(T - c(T)) && \text{for } j \geq k, \\ &\mathcal{F}_k(T, \mathsf{m}) := \{S - c(T) : S \in \mathcal{F}_k(\mathsf{m}),\, T \subset S\} && \text{for } j < k. \end{aligned}$$

Thus, $(T, \mathcal{F}_k(T, \mathsf{m}))$ is a (j,k)-face star. We call it a (j,k)-**face star of the mosaic** m, and we denote by

$$\mathcal{T}_{jk}(\mathsf{m}) := \{(T, \mathcal{F}_k(T, \mathsf{m})) : T \in \mathcal{F}_j(\mathsf{m})\}$$

the set of all (j,k)-face stars of m. Again, we prove a measurability statement.

Lemma 10.1.3. *The mapping*

$$\begin{aligned} \varphi_{jk} : \mathbb{M}^* &\to \mathcal{F}(\mathcal{K}' \times \mathcal{F}(\mathcal{K}')) \\ \mathsf{m} &\mapsto \mathcal{T}_{jk}(\mathsf{m}) \end{aligned}$$

is measurable, $j, k = 0, \dots, d$.

Proof. Let $f : \mathcal{K}' \times \mathcal{F}(\mathcal{K}') \to \mathbb{R} \cup \{\infty\}$ be nonnegative and measurable. Then the function defined by

$$\mathsf{m} \mapsto \sum_{T \in \mathcal{F}_j(\mathsf{m})} f(T, \mathsf{m}), \qquad \mathsf{m} \in \mathbb{M}^*,$$

is measurable. For a proof, we may restrict ourselves to functions of the form $f = \mathbf{1}_{A_1 \times A_2}$ with $A_1 \in \mathcal{B}(\mathcal{K}')$, $A_2 \in \mathcal{B}(\mathcal{F}(\mathcal{K}'))$ and then apply Lemma 10.1.2 and Lemma 3.1.4.

Suppose, first, that $j \geq k$. For $T \in \mathcal{K}'$, the set

$$\mathcal{K}_{\subset}(T) := \{K \in \mathcal{K}' : K \subset T\}$$

is closed. The map $T \mapsto \mathcal{K}_{\subset}(T)$ from $\mathcal{K}'$ into $\mathcal{F}(\mathcal{K}')$ is continuous and, hence, measurable.

For $A \in \mathcal{B}(\mathcal{K}' \times \mathcal{F}(\mathcal{K}'))$ we define

$$\rho_{jk}(\mathsf{m}, A) := \sum_{T \in \mathcal{F}_j(\mathsf{m})} \mathbf{1}_A(T, [\mathcal{F}_k(\mathsf{m}) \cap \mathcal{K}_\subset(T)] - c(T)) \qquad \text{for } \mathsf{m} \in \mathbb{M}^*.$$

Then $\rho_{jk}(\mathsf{m}, \{(K, \mathcal{S})\}) > 0$ holds if and only if $(K, \mathcal{S}) \in \mathcal{T}_{jk}(\mathsf{m})$. By the preceding observations and by Lemma 10.1.2, the mapping $\rho_{jk}(\cdot, A)$ is measurable. For every fixed $\mathsf{m} \in \mathbb{M}^*$, the function $\rho_{jk}(\mathsf{m}, \cdot)$ is a counting measure on $\mathcal{K}' \times \mathcal{F}(\mathcal{K}')$. The mapping $\mathbb{M}^* \to N(\mathcal{K}' \times \mathcal{F}(\mathcal{K}'))$, defined by $\mathsf{m} \mapsto \rho_{jk}(\mathsf{m}, \cdot)$, is measurable. From

$$\mathcal{T}_{jk}(\mathsf{m}) = \operatorname{supp} \rho_{jk}(\mathsf{m}, \cdot)$$

and Lemma 3.1.4, we now deduce the measurability of φ_{jk}.

The case $j < k$ can be treated analogously, using the set

$$\mathcal{K}_\supset(T) := \{K \in \mathcal{K}' : K \supset T\},$$

which is closed in $\mathcal{K}'$. Also the map $T \mapsto \mathcal{K}_\supset(T)$ from $\mathcal{K}'$ into $\mathcal{F}(\mathcal{K}')$ is continuous. □

Let X be a stationary random mosaic. For $j, k \in \{0, \dots, d\}$, the set

$$\mathcal{X}^{(j,k)} := \mathcal{T}_{jk}(X)$$

of all (j, k)-face stars of the mosaic defines, by Lemma 10.1.3, a point process in the space $\mathcal{K}' \times \mathcal{F}(\mathcal{K}')$; it is concentrated on $\mathcal{K}' \times \mathcal{F}_f(\mathcal{K}')$. Here we may identify $\mathcal{X}^{(j,j)}$ with $X^{(j)}$. The process $\mathcal{X}^{(j,k)}$ is a stationary marked particle process in $\mathcal{K}'$ with mark space $\mathcal{F}_f(\mathcal{K}')$. The intensity of $\mathcal{X}^{(j,k)}$ is equal to $\gamma^{(j)}$. The grain-mark distribution $\mathbb{Q}^{(j,k)}$ can be considered as a probability measure on $\mathcal{K}_0 \times \mathcal{F}_f(\mathcal{K}')$. It is concentrated on the (j, k)-face stars in $\mathcal{K}_0 \times \mathcal{F}_f(\mathcal{K}')$, more precisely, on those appearing in face-to-face mosaics.

The marginal distribution of the grain-mark distribution $\mathbb{Q}^{(j,k)}$ with respect to projection to the first factor of $\mathcal{K}_0 \times \mathcal{F}_f(\mathcal{K}')$ is nothing but the grain distribution $\mathbb{Q}^{(j)}$ of the face process $X^{(j)}$. Hence, for every nonnegative measurable function f on $\mathcal{K}_0$ we have

$$\int_{\mathcal{K}_0 \times \mathcal{F}_f(\mathcal{K}')} f(T)\, \mathbb{Q}^{(j,k)}(\mathrm{d}(T, \mathcal{S})) = \int_{\mathcal{K}_0} f \, \mathrm{d}\mathbb{Q}^{(j)}. \tag{10.9}$$

In view of $\mathbb{E}\, \mathcal{X}^{(j,k)}(\cdot \times \mathcal{F}_f(\mathcal{K}')) = \mathbb{E}\, X^{(j)}$, this follows from the definition of the distributions $\mathbb{Q}^{(j,k)}$ and $\mathbb{Q}^{(j)}$. In particular, for every nonnegative measurable function $f : \mathcal{K} \times \mathcal{K}_0 \to \mathbb{R}$ and for $0 \le k \le j \le d$, the equation

$$\begin{aligned} &\int_{\mathcal{K}_0 \times \mathcal{F}_f(\mathcal{K}')} \sum_{S \in \mathcal{S}} f(S, T)\, \mathbb{Q}^{(j,k)}(\mathrm{d}(T, \mathcal{S})) \\ &= \int_{\mathcal{K}_0} \sum_{S \in \mathcal{F}_k(T)} f(S, T)\, \mathbb{Q}^{(j)}(\mathrm{d}T) \end{aligned} \tag{10.10}$$

holds.

For the process $\mathcal{X}^{(j,k)}$ of the (j,k)-face stars of a stationary random mosaic X we define quantities analogous to the specific intrinsic volumes. For $i = 0,\dots,d$ and for $\mathcal{S} \in \mathcal{F}_f(\mathcal{K}')$, we put

$$V_i(\mathcal{S}) := \sum_{S\in\mathcal{S}} V_i(S)$$

and define

$$v_i^{(j,k)} := \int_{\mathcal{K}_0\times\mathcal{F}_f(\mathcal{K}')} V_i(\mathcal{S})\,\mathbb{Q}^{(j,k)}(\mathrm{d}(T,\mathcal{S})) \tag{10.11}$$

and

$$d_i^{(j,k)} := \gamma^{(j)} v_i^{(j,k)}.$$

From the assumptions made so far one cannot conclude that the quantities $v_i^{(j,k)}$ are finite, as simple examples show (see Note 2, second paragraph, of this section). Although some of the relations to be proved below hold also for infinite values of these parameters, for simplicity we make the following assumption.

Convention. For the considered stationary random mosaics, the parameters $v_i^{(j,k)}$ are assumed to be finite.

For some of these quantities, we use a special notation. Namely, we write $N_k(T,\mathcal{S}) := V_0(\mathcal{S})$ for the number of k-faces in a (j,k)-face star $(T,\mathcal{S})$, and we put $n_{jk} := v_0^{(j,k)}$. Thus, n_{jk} is the expectation of N_k with respect to $\mathbb{Q}^{(j,k)}$, in other words, the mean number of k-faces of the typical (j,k)-face star ($n_{jj} = 1$). For $j > k$, the number n_{jk} is, therefore, the mean number of k-faces of the typical j-face of the random mosaic X.

The various numbers thus defined are connected by many relations. Some of these can be obtained immediately.

First we note that equation (10.7) can be written in the form

$$\gamma^{(0)} n_{01} = 2\gamma^{(1)}, \tag{10.12}$$

by applying (4.10) to the marked particle process $\mathcal{X}^{(0,1)}$. Intuitively, equation (10.12) is obvious. It is obtained by 'counting' the incident pairs (vertex, edge) in two different ways, summing first either over the vertices or over the edges. Below, this principle will be elaborated. For the number

$$\ell_{01} := v_1^{(0,1)},$$

the mean length of the typical edge star (that is, (0,1)-face star), the relation

$$\gamma^{(0)} \ell_{01} = 2d_1^{(1)} \tag{10.13}$$

is plausible for similar reasons; it can be proved as above.

Now we apply (10.10) with $f \equiv 1$, $j \in \{0, \dots, d\}$ and $k \in \{0, \dots, j\}$. For a (j,k)-face star $(T, \mathcal{S})$, the sum $\sum_{S \in \mathcal{S}} f(S,T) = N_k(T, \mathcal{S}) = f_k(T)$ is the number of k-faces of the j-polytope T. Using Euler's relation (14.63), we therefore obtain

$$\begin{aligned}\sum_{k=0}^{j} (-1)^k n_{jk} &= \sum_{k=0}^{j} (-1)^k \int_{\mathcal{K}_0 \times \mathcal{F}_f(\mathcal{K}')} \sum_{S \in \mathcal{S}} f(S,T) \, \mathbb{Q}^{(j,k)}(\mathrm{d}(T, \mathcal{S})) \\ &= \sum_{k=0}^{j} (-1)^k \int_{\mathcal{K}_0} \sum_{S \in \mathcal{F}_k(T)} f(S,T) \, \mathbb{Q}^{(j)}(\mathrm{d}T) \\ &= \int_{\mathcal{K}_0} \sum_{k=0}^{j} (-1)^k f_k(T) \, \mathbb{Q}^{(j)}(\mathrm{d}T) = 1,\end{aligned}$$

hence

$$\sum_{k=0}^{j} (-1)^k n_{jk} = 1 \tag{10.14}$$

for $j = 0, \dots, d$. The special cases $j = 1$ and $j = 2$ are

$$n_{10} = 2, \tag{10.15}$$

$$n_{20} = n_{21}. \tag{10.16}$$

A counterpart to (10.14) is the relation

$$\sum_{k=j}^{d} (-1)^{d-k} n_{jk} = 1 \tag{10.17}$$

for $j = 0, \dots, d$, with the special cases

$$n_{d-1,d} = 2, \tag{10.18}$$

$$n_{d-2,d} = n_{d-2,d-1}. \tag{10.19}$$

The proof of relation (10.17) uses the limit representation (4.11) for densities and relation (14.65):

$$\begin{aligned}&\gamma^{(j)} \sum_{k=j}^{d} (-1)^{d-k} n_{jk} \\ &= \sum_{k=j}^{d} (-1)^{d-k} \lim_{r \to \infty} \frac{1}{V_d(rB^d)} \mathbb{E} \sum_{(T, \mathcal{S}) \in \mathcal{X}^{(j,k)},\, T \subset rB^d} V_0(\mathcal{S}) \\ &= \lim_{r \to \infty} \frac{1}{V_d(rB^d)} \mathbb{E} \sum_{k=j}^{d} (-1)^{d-k} \sum_{T \in X^{(j)},\, T \subset rB^d} N_k(T, \mathcal{F}_k(T, X))\end{aligned}$$

$$= \lim_{r\to\infty} \frac{1}{V_d(rB^d)} \mathbb{E} \sum_{T\in X^{(j)},\, T\subset rB^d} 1$$

$$= \gamma^{(j)}.$$

Now we turn to a more systematic study of relations such as (10.12) and (10.13).

Theorem 10.1.1. *Let X be a stationary random mosaic in $\mathbb{R}^d$, and let the function $f : \mathcal{K} \times \mathcal{K} \to \mathbb{R}$ be nonnegative, measurable and jointly translation invariant (that is, satisfying $f(K + x, L + x) = f(K, L)$ for $x \in \mathbb{R}^d$). Then, for $j, k \in \{0, \dots, d\}$,*

$$\gamma^{(j)} \int_{\mathcal{K}_0\times\mathcal{F}_f(\mathcal{K}')} \sum_{S\in\mathcal{S}} f(S,T)\, \mathbb{Q}^{(j,k)}(\mathrm{d}(T,\mathcal{S}))$$

$$= \gamma^{(k)} \int_{\mathcal{K}_0\times\mathcal{F}_f(\mathcal{K}')} \sum_{T\in\mathcal{T}} f(S,T)\, \mathbb{Q}^{(k,j)}(\mathrm{d}(S,\mathcal{T})).$$

(Observe that one side of the equation can be simplified by means of (10.10).)

Proof. For $s > 0$, let $f_s(S, T) := f(S, T)$ if the diameters of S and T are not larger than s, and put $f_s(S, T) := 0$ otherwise. The sum $\sum_{S\in\mathcal{S}} f_s(S, T)$ converges increasingly to $\sum_{S\in\mathcal{S}} f(S, T)$ as $s \to \infty$ (and similarly $\sum_{T\in\mathcal{T}} f_s(S, T)$ to $\sum_{T\in\mathcal{T}} f(S, T)$), hence it suffices to prove the assertion for the function f_s.

We make use of Theorem 4.1.5. First, we apply the limit relation (4.11) to the marked particle process $\mathcal{X}^{(j,k)}$ and the function defined by

$$\varphi(T, \mathcal{S}) := \sum_{S\in\mathcal{S}} f_s(S + c(T), T),$$

which is translation invariant in its first variable. Then we use the relation

$$\{(S + c(T), T) : T \in \mathcal{F}_j(\mathsf{m}),\ S \in \mathcal{F}_k(T, \mathsf{m})\}$$

$$= \{(S, T + c(S)) : S \in \mathcal{F}_k(\mathsf{m}),\ T \in \mathcal{F}_j(S, \mathsf{m})\},$$

valid for $\mathsf{m} \in \mathbb{M}^*$, and apply (4.11) to the marked particle process $\mathcal{X}^{(k,j)}$. In this way we get, observing that $c(T) = 0$ if $(T, \mathcal{S})$ is in the support of $\mathbb{Q}^{(j,k)}$,

$$\gamma^{(j)} \int_{\mathcal{K}_0\times\mathcal{F}_f(\mathcal{K}')} \sum_{S\in\mathcal{S}} f_s(S,T)\, \mathbb{Q}^{(j,k)}(\mathrm{d}(T,\mathcal{S}))$$

$$= \lim_{r\to\infty} \frac{1}{V_d(rB^d)} \mathbb{E} \sum_{T\in X^{(j)},\, T\subset rB^d} \ \sum_{S\in\mathcal{F}_k(T,X)} f_s(S + c(T), T)$$

$$\le \lim_{r\to\infty} \frac{1}{V_d(rB^d)} \mathbb{E} \sum_{S\in X^{(k)},\, S\subset(r+s)B^d} \ \sum_{T\in\mathcal{F}_j(S,X)} f_s(S, T + c(S))$$

$$= \gamma^{(k)} \int_{\mathcal{K}_0\times\mathcal{F}_f(\mathcal{K}')} \sum_{T\in\mathcal{T}} f_s(S,T)\, \mathbb{Q}^{(k,j)}(\mathrm{d}(S,\mathcal{T})).$$

In a similar way, we obtain the inequality

$$\gamma^{(k)} \int_{\mathcal{K}_0 \times \mathcal{F}_f(\mathcal{K}')} \sum_{T \in \mathcal{T}} f_s(S,T)\, \mathbb{Q}^{(k,j)}(\mathrm{d}(S,\mathcal{T}))$$

$$\leq \gamma^{(j)} \int_{\mathcal{K}_0 \times \mathcal{F}_f(\mathcal{K}')} \sum_{S \in \mathcal{S}} f_s(S,T)\, \mathbb{Q}^{(j,k)}(\mathrm{d}(T,\mathcal{S})),$$

and thus the assertion. □

We mention an alternative approach to Theorem 10.1.1, which is based on Neveu's exchange formula (Theorem 3.4.5). For this, we choose the canonical setting $(\Omega, \mathbf{A}, \mathbb{P}) = (\mathsf{N}(\mathcal{K}'), \mathcal{N}(\mathcal{K}'), \mathbb{P}_X)$. For a particle process Y, let $c(Y)$ be the point process of center points $c(K)$, $K \in Y$. We consider the point processes $Y_j := c(X^{(j)})$ and $Y_k := c(X^{(k)})$. Both are stationary and have intensities $\gamma^{(j)} > 0$ and $\gamma^{(k)} > 0$, respectively. If we denote the corresponding Palm distributions by $\mathbb{P}^{0,j}$ and $\mathbb{P}^{0,k}$, Theorem 3.4.5 yields

$$\gamma^{(j)} \int_{\Omega} \sum_{y \in Y_j} g(y, X-y)\, \mathrm{d}\mathbb{P}^{0,j} = \gamma^{(k)} \int_{\Omega} \sum_{x \in Y_k} g(-x, X)\, \mathrm{d}\mathbb{P}^{0,k}, \quad (10.20)$$

for any nonnegative measurable function g on $\mathbb{R}^d \times \Omega$. We consider $\mathcal{X}^{(j,k)}$ as a marked point process on $\mathbb{R}^d$; then the underlying unmarked point process is Y_j and the marks are face stars $(T, \mathcal{S})$ with $c(T) = 0$. For $x \in \mathbb{R}^d$, the mapping $\zeta_x : \widetilde{\eta} \mapsto \zeta_x(\widetilde{\eta})$, introduced after Theorem 3.5.5, maps $\mathcal{X}^{(j,k)}$ to the face star $\zeta_x(\mathcal{X}^{(j,k)}) = (T, \mathcal{S})$ such that $(x, (T, \mathcal{S})) \in \mathcal{X}^{(j,k)}$, if $x \in Y_j$ (and $\zeta_x(\mathcal{X}^{(j,k)}) = (T_0, \mathcal{S}_0)$, for a fixed face star $(T_0, \mathcal{S}_0)$, otherwise). The image measure of $\mathbb{P}^{0,j}$ under ζ_0 is $\mathbb{Q}^{(j,k)}$ (see again the remark after Theorem 3.5.5). If

$$g(0, X) = \sum_{(T,S) \in \zeta_0(\mathcal{X}^{(j,k)})} f(S,T)$$

and $g(x, X) = 0$, for $x \neq 0$, and if f is the nonnegative, measurable and translation invariant function f on $\mathcal{K} \times \mathcal{K}$ given in Theorem 10.1.1, then on the left side of (10.20) we obtain

$$\int_{\Omega} \sum_{y \in Y_j} g(y, X-y)\, \mathrm{d}\mathbb{P}^{0,j} = \int_{\mathcal{K}_0 \times \mathcal{F}_f(\mathcal{K}')} \sum_{S \in \mathcal{S}} f(S,T)\, \mathbb{Q}^{(j,k)}(\mathrm{d}(T,\mathcal{S})).$$

Similarly, on the right side we get

$$\int_{\Omega} \sum_{x \in Y_k} g(-x, X)\, \mathrm{d}\mathbb{P}^{0,k} = \int_{\mathcal{K}_0 \times \mathcal{F}_f(\mathcal{K}')} \sum_{T \in \mathcal{T}} f(S,T)\, \mathbb{Q}^{(k,j)}(\mathrm{d}(S,\mathcal{T})).$$

Thus, (10.20) implies the assertion of Theorem 10.1.1.

Relation (10.12) is the special case $j = 0$, $k = 1$, $f = 1$ of Theorem 10.1.1, and (10.13) is obtained if we put $f(S,T) := V_1(S)$. The choice

$f(S,T) := V_i(S)V_l(T)$ yields relations for expected values of products of intrinsic volumes. We note only the special case $l = 0$, where $f(S,T) := V_i(S)$; this gives the following result.

Theorem 10.1.2. *For a stationary random mosaic X in $\mathbb{R}^d$ and for $i, j, k \in \{0, \dots, d\}$,*

$$d_i^{(j,k)} = \gamma^{(k)} \int_{\mathcal{K}_0 \times \mathcal{F}_f(\mathcal{K}')} V_i(S) N_j(S, \mathcal{T})\, \mathbb{Q}^{(k,j)}(\mathrm{d}(S, \mathcal{T}))$$

and especially, for $i = 0$,

$$\gamma^{(j)} n_{jk} = \gamma^{(k)} n_{kj}.$$

For another consequence of Theorem 10.1.1, we use the internal angles $\beta(F, P)$ of a polytope P at a face F. The **internal angle** of the d-dimensional polytope P at its face F is defined by

$$\beta(F, P) := \lambda(S(P, F) \cap B^d)/\kappa_d,$$

where $S(P, F)$ is the cone spanned by P at an arbitrary relatively interior point z of F, thus $S(P, F) := \{\alpha(x - z) : x \in P,\ \alpha \geq 0\}$. **Gram's relation** (see, for example, Grünbaum [299]) says that

$$\sum_{i=0}^{d} (-1)^i \sum_{F \in \mathcal{F}_i(P)} \beta(F, P) = 0.$$

Theorem 10.1.3. *Let X be a stationary random mosaic in $\mathbb{R}^d$, let the function $g : \mathcal{K}' \to \mathbb{R}$ be translation invariant, nonnegative, and measurable, and let $j \in \{0, \dots, d\}$. Then*

$$\gamma^{(d)} \int_{\mathcal{K}_0} \sum_{S \in \mathcal{F}_j(P)} \beta(S, P) g(S)\, \mathbb{Q}^{(d)}(\mathrm{d}P) = \gamma^{(j)} \int_{\mathcal{K}_0} g \,\mathrm{d}\mathbb{Q}^{(j)},$$

in particular (case $g = 1$),

$$\gamma^{(j)} = \gamma^{(d)} \int_{\mathcal{K}_0} \sum_{S \in \mathcal{F}_j(P)} \beta(S, P)\, \mathbb{Q}^{(d)}(\mathrm{d}P).$$

Moreover,

$$\sum_{i=0}^{d} (-1)^i \gamma^{(i)} = 0. \tag{10.21}$$

Proof. In Theorem 10.1.1, we choose $k = d$ and $f(S, T) := \beta(T, S) g(T)$ (with $\beta(T, S) := 0$ unless S is a d-polytope and T is a face of S). The measurability of the function f can be proved with the methods employed in the proof of Lemma 10.1.2. The (j, d)-face stars $(T, \mathcal{S})$ on which the distribution $\mathbb{Q}^{(j,d)}$ is concentrated belong to face-to-face mosaics and hence satisfy $\sum_{S \in \mathcal{S}} \beta(T, S) = 1$. Therefore, we obtain

$$\begin{aligned} &\gamma^{(j)} \int_{\mathcal{K}_0 \times \mathcal{F}_f(\mathcal{K}')} g(T)\, \mathbb{Q}^{(j,d)}(\mathrm{d}(T, \mathcal{S})) \\ &= \gamma^{(d)} \int_{\mathcal{K}_0 \times \mathcal{F}_f(\mathcal{K}')} \sum_{T \in \mathcal{F}_j(S)} \beta(T, S) g(T)\, \mathbb{Q}^{(d,j)}(\mathrm{d}(S, \mathcal{T})) \end{aligned}$$

and thus, in view of (10.9) and (10.10), the first equation of the theorem. Then the second equation for $j = 0, \dots, d$ and Gram's relation yield the third equation. □

Equation (10.21) is the special case $j = 0$ of a more general relation for the densities of the intrinsic volumes.

Theorem 10.1.4. *If X is a stationary random mosaic in $\mathbb{R}^d$ and if $j \in \{0, \dots, d-1\}$, then*

$$\sum_{i=j}^{d} (-1)^i d_j^{(i)} = 0. \tag{10.22}$$

Proof. First we consider a fixed mosaic m. Let $S_1, \dots, S_p$ be the different cells of m that meet the ball B^d. Let $j \in \{0, \dots, d-1\}$. Since the intrinsic volume V_j is additive on the convex ring, the inclusion–exclusion principle gives

$$\begin{aligned} V_j(B^d) &= V_j\left(B^d \cap \bigcup_{i=1}^{p} S_i\right) \\ &= \sum_{r=1}^{p} (-1)^{r-1} \sum_{i_1 < \dots < i_r} V_j(S_{i_1} \cap \dots \cap S_{i_r} \cap B^d) \\ &= \sum_{i=j}^{d} \sum_{F \in \mathcal{F}_i(\mathsf{m})} V_j(F \cap B^d) \sum_{r=1}^{p} (-1)^{r-1} \nu(F, r). \end{aligned}$$

Here $\nu(F, r)$ denotes the number of r-tuples $(S_{i_1}, \dots, S_{i_r})$ with $S_{i_1} \cap \dots \cap S_{i_r} = F$. We have used the fact that every nonempty intersection $S_{i_1} \cap \dots \cap S_{i_r}$ is a face of m and that every face of m meeting B^d is of this type. Moreover, $V_j(F) = 0$ for $\dim F < j$ was used. By (14.66),

$$\sum_{r=1}^{p} (-1)^{r-1} \nu(F, r) = (-1)^{d - \dim F}.$$

Thus we obtain

$$V_j(B^d) = \sum_{i=j}^{d} (-1)^{d-i} \sum_{F \in \mathcal{F}_i(\mathsf{m})} V_j(F \cap B^d).$$

Now let X be a stationary random mosaic. Then

$$\sum_{i=j}^{d}(-1)^{d-i}\sum_{F\in\mathcal{F}_i(X)} V_j(F\cap B^d) = V_j(B^d) \qquad \text{a.s.} \tag{10.23}$$

Because of $V_j(F\cap B^d)\le V_j(B^d)\chi(F\cap B^d)$ and the generally assumed local finiteness of the intensity measure of $X^{(i)}$, we have $\mathbb{E}\sum_{F\in\mathcal{F}_i(X)} V_j(F\cap B^d) < \infty$. Here and in (10.23), we may replace B^d by rB^d with $r>0$. Since $j<d$, we obtain

$$\begin{aligned} 0 &= \lim_{r\to\infty}\frac{V_j(rB^d)}{V_d(rB^d)} \\ &= \lim_{r\to\infty}\frac{1}{V_d(rB^d)}\mathbb{E}\sum_{i=j}^{d}(-1)^{d-i}\sum_{F\in X^{(i)}} V_j(F\cap rB^d) \\ &= \sum_{i=j}^{d}(-1)^{d-i}\overline{V}_j(X^{(i)}) \\ &= \sum_{i=j}^{d}(-1)^{d-i}d_j^{(i)}, \end{aligned}$$

by Theorem 9.2.2. The integrability condition required by that theorem is satisfied, since the process $X^{(i)}$ has convex particles and locally finite intensity measure. □

For normal mosaics, there are further relations between the intensities.

Theorem 10.1.5. *Let X be a stationary normal random mosaic in $\mathbb{R}^d$, and let $k\in\{1,\dots,d\}$. Then*

$$(1-(-1)^k)\gamma^{(k)} = \sum_{j=0}^{k-1}(-1)^j\binom{d+1-j}{k-j}\gamma^{(j)}.$$

Proof. For $j\le k$, every j-face of a normal mosaic is contained in exactly $\binom{d+1-j}{k-j}$ k-faces. Hence, for X we have

$$n_{jk} = \binom{d+1-j}{k-j}.$$

Theorem 10.1.2 therefore yields

$$\sum_{j=0}^{k}(-1)^j\binom{d+1-j}{k-j}\gamma^{(j)} = \sum_{j=0}^{k}(-1)^j\gamma^{(j)}n_{jk}$$

$$= \gamma^{(k)} \sum_{j=0}^{k} (-1)^j n_{kj}$$

$$= \gamma^{(k)} \int_{\mathcal{K}_0} \sum_{j=0}^{k} (-1)^j f_j(Q)\, \mathbb{Q}^{(k)}(\mathrm{d}Q)$$

$$= \gamma^{(k)},$$

where the Euler relation (14.63) for k-dimensional polytopes was applied. □

For $k = 1$, we obtain equation (10.8) again.

For dimensions 2 and 3, we shall now collect the obtained results, with some reformulations and supplements. We use some special notation, similar to earlier terminology. First let $d = 2$. Then we put

$$d_2^{(2)} = \overline{V}_2(X) =: \gamma^{(2)} a, \quad d_1^{(2)} = \overline{V}_1(X) =: \frac{1}{2}\gamma^{(2)} p,$$
$$d_1^{(1)} = \overline{V}_1(X^{(1)}) =: \gamma^{(1)} l_1$$

and collect the interpretations of the introduced parameters in the following list:

$\gamma^{(2)}, \gamma^{(1)}, \gamma^{(0)}$	intensities (of cells, edges, vertices, respectively),
a, p	mean area, mean perimeter of the typical cell,
l_1, l_{01}	mean length of the typical edge or the typical edge star,
$n_{20} = n_{21}$	mean number of vertices (= mean number of edges) of the typical cell,
$n_{01} = n_{02}$	mean number of edges of the typical edge star (= mean number of cells of the typical $(0, 2)$-star).

Parameters n_{ij} that are not listed can be obtained from (10.15), (10.16), (10.18), (10.19).

The following theorem summarizes the main relations between these parameters. It shows, in particular, that in the planar case all the considered parameters of the random mosaic X can be expressed in terms of the intensities $\gamma^{(0)}$ and $\gamma^{(2)}$ and the mean edge length l_1.

Theorem 10.1.6. *The parameters of a stationary random mosaic X in the plane satisfy*

(a)
$$\gamma^{(1)} = \gamma^{(0)} + \gamma^{(2)},$$

(b)
$$n_{02} = 2 + 2\frac{\gamma^{(2)}}{\gamma^{(0)}}, \quad n_{20} = 2 + 2\frac{\gamma^{(0)}}{\gamma^{(2)}},$$

(c)
$$l_{01} = 2\frac{\gamma^{(1)}}{\gamma^{(0)}} l_1,$$

(d) $$a = \frac{1}{\gamma^{(2)}}, \quad p = 2\frac{\gamma^{(1)}}{\gamma^{(2)}} l_1,$$

(e) $$3 \leq n_{02}, n_{20} \leq 6.$$

If X is normal, then $n_{02} = 3$ and $n_{20} = 6$.

Proof. (a) is equation (10.21) for $n = 2$. The equations (b) follow from (a) and from the equations

$$n_{02} = n_{01} = 2\frac{\gamma^{(1)}}{\gamma^{(0)}}, \quad n_{20} = 2\frac{\gamma^{(1)}}{\gamma^{(2)}},$$

which result from Theorem 10.1.2 together with (10.15), (10.16), (10.18), (10.19). Equation (c) is just a reformulation of (10.13). The first equality in (d) is (10.4) for $n = 2$, and the second follows from (10.6) or (10.22).

Trivially, every vertex of a planar mosaic is contained in at least three cells, and every cell has at least three vertices, hence the corresponding mean values are at least three. From (b) it follows that

$$\frac{1}{n_{02}} + \frac{1}{n_{20}} = \frac{1}{2},$$

which gives the second inequality of (e). For a normal mosaic, $n_{02} = 3$ holds by definition, and hence $n_{20} = 6$ follows from the last equation. □

We remark that (b) and (e) together yield the inequalities

$$\frac{1}{2}\gamma^{(2)} \leq \gamma^{(0)} \leq 2\gamma^{(2)}. \tag{10.24}$$

Equality holds on the right side for normal mosaics and on the left side for triangle mosaics.

Now let $d = 3$. We restrict ourselves to a selection of the many possible parameters, and put

$$d_3^{(3)} = \overline{V}_3(X) =: \gamma^{(3)} v, \quad d_2^{(3)} = \overline{V}_2(X) =: \frac{1}{2}\gamma^{(3)} s, \quad d_1^{(3)} = \overline{V}_1(X) =: \frac{1}{\pi}\gamma^{(3)} m,$$

$$d_2^{(2)} = \overline{V}_2(X^{(2)}) =: \gamma^{(2)} a_2, \quad d_1^{(2)} = \overline{V}_1(X^{(2)}) =: \frac{1}{2}\gamma^{(2)} p_2,$$

$$d_1^{(1)} = \overline{V}_1(X^{(1)}) =: \gamma^{(1)} l_1,$$

$$v_1^{(3,1)} =: l_{31}, \quad v_1^{(0,1)} =: l_{01}, \quad v_2^{(0,2)} =: a_{02}, \quad v_2^{(1,2)} =: a_{12}.$$

The interpretations are again collected in a list:

$\gamma^{(3)}, \gamma^{(2)}, \gamma^{(1)}, \gamma^{(0)}$	intensities of cells, facets, edges, vertices,
v, s, m, l_{31}	mean volume, mean surface area, mean integral of mean curvature, mean edge length sum of the typical cell,
a_2, p_2	mean area, mean perimeter of the typical facet,
l_1, l_{01}	mean length of the typical edge, of the typical edge star,
a_{02}, a_{12}	mean area of the typical $(0,2)$-star, of the typical $(1,2)$-star,
n_{32}, n_{02}, n_{12}	mean facet number of the typical cell, of the typical (0,2)-star, of the typical (1,2)-star
n_{31}, n_{21}, n_{01}	mean edge number of the typical cell, of the typical facet, of the typical edge star,
n_{30}, n_{03}	mean vertex number of the typical cell, mean cell number of the typical (0,3)-star.

Parameters n_{ij} that are not listed can again be obtained from (10.15), (10.16), (10.18), (10.19).

Further, we consider the weighted mean values

$$w_{1i} := \int_{\mathcal{K}_0 \times \mathcal{F}_f(\mathcal{K}')} V_1(S) N_i(S,T)\, \mathbb{Q}^{(1,i)}(\mathrm{d}(S,T)), \qquad i = 2,3,$$

$$w_{2i} := \int_{\mathcal{K}_0 \times \mathcal{F}_f(\mathcal{K}')} V_2(S) N_i(S,T)\, \mathbb{Q}^{(2,i)}(\mathrm{d}(S,T)), \qquad i = 0,1.$$

We have $w_{12} = w_{13}$ and $w_{20} = w_{21}$, as is easy to see from the limit relations (4.10)–(4.12); these yield, for example,

$$\gamma^{(1)} w_{1i} = \lim_{r\to\infty} \frac{1}{V(rB^3)} \mathbb{E} \sum_{(S,T)\in\mathcal{X}^{(1,i)},\, S\subset rB^3} V_1(S) N_i(S,T).$$

The subsequent theorem collects the most important relations between the listed parameters.

Theorem 10.1.7. *The parameters of a stationary random mosaic X in $\mathbb{R}^3$ satisfy*

(a)
$$\gamma^{(1)} + \gamma^{(3)} = \gamma^{(0)} + \gamma^{(2)},$$

(b)
$$\gamma^{(0)} n_{01} = 2\gamma^{(1)}, \quad \gamma^{(3)} n_{32} = 2\gamma^{(2)}, \quad \gamma^{(0)} n_{03} = \gamma^{(3)} n_{30},$$
$$\gamma^{(0)} n_{02} = \gamma^{(1)} n_{12} = \gamma^{(2)} n_{21} = \gamma^{(3)} n_{31},$$

(c)
$$n_{01} - n_{02} + n_{03} = 2, \quad n_{30} - n_{31} + n_{32} = 2,$$
$$2n_{02} = n_{01} n_{12}, \quad 2n_{31} = n_{21} n_{32}, \quad n_{02} n_{30} = n_{03} n_{31},$$

(d)
$$\gamma^{(0)} l_{01} = 2\gamma^{(1)} l_1,$$
$$\gamma^{(2)} p_2 = \gamma^{(1)} w_{13} = \gamma^{(3)} l_{31}, \qquad \gamma^{(0)} a_{02} = \gamma^{(2)} w_{21} = \gamma^{(1)} a_{12},$$

(e)
$$v = \frac{1}{\gamma^{(3)}}, \quad s = 2\frac{\gamma^{(2)}}{\gamma^{(3)}} a_2, \quad m = \frac{\pi}{2}\frac{\gamma^{(2)}}{\gamma^{(3)}} p_2 - \pi \frac{\gamma^{(1)}}{\gamma^{(3)}} l_1.$$

If X is normal, then $n_{01} = n_{03} = 4$, $n_{12} = 3$, and $n_{02} = 6$.

Proof. Equation (a) is nothing but (10.21) for $n = 3$. The equations (b) follow from Theorem 10.1.2, in connection with (10.15), (10.16), (10.18), (10.19). The first two equations of (c) are special cases of (10.14) and (10.17). The remaining equations in (c) are obtained by eliminating the intensities in suitable pairs of equations in (b).

The first equation in (d) is merely a reformulation of (10.13).

In Theorem 10.1.1, we choose $d = 3$, $j = 1$, $k = 2$, $f(S,T) := V_1(T)$. Then we obtain
$$\gamma^{(2)} u_2 = \gamma^{(1)} w_{12} = \gamma^{(1)} w_{13} = \gamma^{(3)} v_1^{(3,1)} = \gamma^{(3)} l_{31}$$
by Theorem 10.1.2 and thus the second equation in (d).

The third equation in (d) results from Theorem 10.1.1 with $d = 3$, $j = 1$, $k = 2$, $f(S,T) := V_2(S)$. This gives
$$\gamma^{(1)} a_{12} = \gamma^{(2)} w_{21} = \gamma^{(2)} w_{20} = \gamma^{(0)} v_2^{(0,2)} = \gamma^{(0)} a_{02},$$
where Theorem 10.1.2 was used again.

The first equation in (e) is just (10.4) for $= 3$; the other two equations result by reformulating (10.22).

The assertion about normal mosaics is a simple consequence of the definition. □

For a stationary random mosaic, our main interest so far has been in the processes of faces and their combinatorial neighborhoods. Now we consider the unions of the faces of a fixed dimension, thus obtaining a collection of random closed sets associated with the random mosaic.

Definition 10.1.4. *For a mosaic* m *and for* $k \in \{0, \dots, d-1\}$, *the set*
$$\mathrm{skel}_k \mathsf{m} := \bigcup_{S \in \mathcal{F}_k(\mathsf{m})} S$$
is the k-**skeleton** *of* m.

Let X be a random mosaic, and let Z_k be its k-skeleton, $k \in \{0, \dots, d\}$. By Lemma 10.1.2 and Theorem 3.6.2, Z_k is a random closed set. If X is stationary, then Z_k is stationary. The realizations of Z_k are a.s. locally polyconvex. We cannot guarantee, however, that Z_k is a standard random set, since the integrability condition (9.16) need not be satisfied. Therefore, we cannot use Theorem 9.2.1 to obtain the existence of densities for additive functionals, without additional assumptions. However, the existence of the specific Euler characteristic (the density of the intrinsic volume V_0) of the k-skeleton can be shown in a direct way, and it can be represented in terms of the intensities $\gamma^{(j)}$ of the j-face processes, $j = 0, \dots, k$.

Theorem 10.1.8. *Let X be a stationary random mosaic in $\mathbb{R}^d$, and let Z_k be its k-skeleton, $k \in \{0, \dots, d\}$. The specific Euler characteristic*

$$\overline{\chi}(Z_k) := \lim_{r\to\infty} \frac{1}{r^d} \mathbb{E}\chi(Z_k \cap rC^d) \tag{10.25}$$

exists and is given by

$$\overline{\chi}(Z_k) = \mathbb{E}\chi(Z_k \cap C_0^d); \tag{10.26}$$

it satisfies

$$\overline{\chi}(Z_k) = \sum_{j=0}^{k} (-1)^j \gamma^{(j)}. \tag{10.27}$$

Proof. By Theorem 14.4.3 (or Theorem 14.4.5), the Euler characteristic has an additive extension, also denoted by χ, to the system $U(\mathcal{P}_{ro})$ of finite unions of relatively open polytopes. If Q is a relatively open polytope of dimension j, then $\chi(Q) = (-1)^j$. In the following proof, we denote the relative interior of a polytope S by S^0. The representation

$$Z_k \cap rC^d = \bigcup_{j=0}^{k} \bigcup_{S\in X^{(j)}} (S^0 \cap rC^d)$$

is a disjoint union, hence

$$\begin{aligned}\chi(Z_k \cap rC^d) &= \sum_{j=0}^{k} \sum_{S\in X^{(j)}} \chi(S^0 \cap rC^d)\\ &= \sum_{j=0}^{k} (-1)^j \sum_{S\in X^{(j)},\, S\subset rC^d} 1 + \sum_{j=0}^{k} \sum_{S\in X^{(j)},\, S\cap \mathrm{bd}\, rC^d\neq\emptyset} \chi(S^0 \cap rC^d).\end{aligned}$$

By Theorem 4.1.3(b),

$$\lim_{r\to\infty} \frac{1}{r^d}\, \mathbb{E} \sum_{S\in X^{(j)},\, S\subset rC^d} 1 = \overline{\chi}(X^{(j)}) = \gamma^{(j)}.$$

From

$$S^0 \cap rC^d = \bigcup_{F\in\mathcal{F}_\bullet(rC^d)} (S^0 \cap F^0)$$

and $\chi(S^0 \cap F^0) = \pm 1$ it follows that $|\chi(S^0 \cap rC^d)| \le c(d)$, where $c(d)$ depends only on d. Now the relations (b) and (c) of Theorem 4.1.3 together yield

$$\lim_{r\to\infty} \frac{1}{r^d} \mathbb{E} \sum_{S\in X^{(j)},\, S\cap \mathrm{bd}\, rC^d \neq \emptyset} \chi(S^0 \cap rC^d) = 0$$

(the integrability condition for (c) is satisfied, since $X^{(j)}$ has a locally finite intensity measure). The obtained limit relations show that the limit (10.25) exists and is equal to (10.27).

Let $n \in \mathbb{N}$, and let $\Lambda \subset \mathbb{Z}^d$ be the set of integer vectors for which

$$nC_0^d = \bigcup_{z\in\Lambda} (C_0^d + z).$$

Then

$$\begin{aligned} \mathbb{E} \sum_{S\in X^{(j)}} \chi(S^0 \cap nC_0^d) &= \sum_{z\in\Lambda} \mathbb{E} \sum_{S\in X^{(j)}} \chi(S^0 \cap (C_0^d + z)) \\ &= n^d\, \mathbb{E} \sum_{S\in X^{(j)}} \chi(S^0 \cap C_0^d), \end{aligned}$$

by stationarity. This gives

$$\begin{aligned} &\mathbb{E}\chi(Z_k \cap C_0^d) \\ &= \sum_{j=0}^{k} \mathbb{E} \sum_{S\in X^{(j)}} \chi(S^0 \cap C_0^d) \\ &= \sum_{j=0}^{k} \frac{1}{n^d}\, \mathbb{E} \sum_{S\in X^{(j)}} \chi(S^0 \cap nC_0^d) \\ &= \sum_{j=0}^{k} \frac{1}{n^d}\, \mathbb{E} \sum_{S\in X^{(j)},\, S\subset nC_0^d} (-1)^j + \sum_{j=0}^{k} \frac{1}{n^d}\, \mathbb{E} \sum_{S\in X^{(j)},\, S\cap \mathrm{bd}\, nC^d\neq\emptyset} \chi(S^0 \cap nC_0^d). \end{aligned}$$

For $n \to \infty$, this proves (10.26). □

For $k = d$, equation (10.27) yields (10.21) again.

We conclude this section with a brief comment on the sections of random mosaics by fixed planes. Let X be a stationary and isotropic random mosaic in $\mathbb{R}^d$, and let $E \in A(d,s)$ be a fixed s-dimensional plane, $s \in \{1, \dots, d-1\}$. Then $X \cap E$ is a random mosaic in E, which is stationary and isotropic with respect to E. For $k \in \{d-s, \dots, d\}$, we can apply Theorem 9.4.8 to the process $X^{(k)}$ of k-faces of X and obtain

$$\overline{V}_j(X^{(k)} \cap E) = c_{j,d}^{s,d-s+j}\, \overline{V}_{d-s+j}(X^{(k)}) \tag{10.28}$$

for $j \in \{0, \dots, k+s-d\}$. The nonempty sections of E with $X^{(k)}$ are almost surely precisely the $(k+s-d)$-faces of the mosaic $X \cap E$. Hence, if $Z^{(k)}$ denotes the typical k-face of X, $Z_E^{(m)}$ is the typical m-face of $X \cap E$, and if $\gamma_E^{(m)}$ is the intensity of the process of m-faces of $X \cap E$, then (10.28) can be written in the form

$$\gamma_E^{(k+s-d)} \mathbb{E} V_j(Z_E^{(k+s-d)}) = c_{j,d}^{s,d-s+j} \gamma^{(k)} \mathbb{E} V_{d-s+j}(Z^{(k)}).$$

In particular, we have

$$\gamma_E^{(k+s-d)} = c_{0,d}^{s,d-s} \gamma^{(k)} \mathbb{E} V_{d-s}(Z^{(k)})$$

and

$$\mathbb{E} V_j(Z_E^{(k+s-d)}) = c_{j,d-s}^{0,d-s+j} \frac{\mathbb{E} V_{d+j-s}(Z^{(k)})}{\mathbb{E} V_{d-s}(Z^{(k)})}.$$

Notes for Section 10.1

1. The systematic study of random mosaics began, after a few sporadic papers on planar mosaics, essentially with the important investigations of Miles [517, 521, 527] and Matheron [460], [462, ch. 6] on stationary Poisson hyperplane mosaics. Early papers on more general random mosaics and especially on Voronoi mosaics are due to Meijering [510], Ambartzumian [32, 33], Miles [522, 526], Cowan [180, 181]. Whereas these papers follow an ergodic approach, Palm methods were predominant later, beginning with Mecke [478]. A general, comprehensive presentation of random mosaics in d-dimensional space was given by Møller [551]; we have benefited from this work at several places in the present chapter. Further general information on random mosaics and their applications is found in the books by Ambartzumian, Mecke and Stoyan [36], Stoyan, Kendall and Mecke [743], Mecke, Schneider, Stoyan and Weil [500]. For a first glimpse, the encyclopedia article by Miles [534] is recommended.

The reader should be warned that in stochastic geometry and in discrete geometry the terminology connected with mosaics is not uniform. Mosaics are also called **tessellations**. Instead of 'face-to-face', some authors on random mosaics use 'regular', which has a different meaning in discrete geometry. Also 'normal' occurs with different meanings.

We have introduced a random mosaic as a special point process in the space of convex polytopes and not, as some authors do, as a random closed set (describing the union of the boundaries of the cells). We feel that this approach facilitates the proof of some measurability assertions (often neglected in the literature), for example for the face processes.

2. By a general assumption, we have assured that the k-face processes have locally finite intensity measures $\Theta^{(k)}$. As mentioned, for $k < d$ this local finiteness cannot be deduced from the local finiteness of $\Theta^{(d)}$. We demonstrate this by an example which is due to Ulrich Brehm. It suffices to sketch the construction in $\mathbb{R}^3$. For sufficiently large $k \in \mathbb{N}$, we decompose the unit cube C^3 into $2k$ polytopes, in the following way. The first k polytopes are 'horizontal plates' with a k-fold rotational symmetry around the vertical axis of the cube, with diameters decreasing towards the center of the cube; the last k polytopes are 'vertical conical plates'. This can be done in such a way that all polytopes of the decomposition are convex and that each horizontal plate intersects each vertical plate in a 2-face. By periodic continuation we obtain, if the construction is suitably done, a mosaic m_k. Inside C^3, it contains $2k$ cells and at least k^2 two-dimensional faces. Next, we can construct a random mosaic X such that each realization of X is a mosaic $\mathbf{m}_k$, where $k \in \mathbb{N}$, and m_k is attained with probability p_k, where the numbers p_k are chosen such that $\sum p_k = 1$, $\sum kp_k < \infty$, and $\sum k^2 p_k = \infty$. Translating X by a random vector which is uniformly distributed in C^3, we obtain a stationary random mosaic the intensity measures $\Theta^{(k)}$ of which have the property that $\Theta^{(3)}$ is locally finite, but $\Theta^{(2)}$ is not locally finite.

By another example, we show that the quantities $v_i^{(j,k)}$ defined by (10.11) need not be finite, even if the face processes have locally finite intensity measures. Let m_k be the planar mosaic arising from the standard tessellation of the plane into squares of side length k ($k \in \mathbb{N}$). Let $(p_k)_{k\in\mathbb{N}}$ be a sequence of positive numbers with $\sum p_k = 1$ and $\sum kp_k = \infty$, and let $\xi_1, \xi_2, \dots$ be a sequence of independent random vectors such that ξ_k is uniformly distributed in kC^2. We construct a stationary random mosaic X, putting $X = \mathsf{m}_k + \xi_k$ with probability p_k. The face processes of X have locally finite intensity measures, but $v_1^{(0,1)} = \infty$.

3. The results collected in Theorems 10.1.6 and 10.1.7 are due, in different degrees of generality, to several authors, among them Matschinski [464], Ambartzumian [33], Cowan [180, 181], Mecke [478, 481], Radecke [611], Møller [551]. Essentially, we have followed Møller [551]. This author also applies an exchange formula for Palm distributions, an approach which we have mentioned after Theorem 10.1.1. In contrast to this, we have preferred to give a proof of Theorem 10.1.1 which is direct and only uses the limit relations for densities. We emphasize, however, that more general versions are possible; see, for example, Proposition 2.2 in Baumstark and Last [85] which, as the authors point out, holds for general stationary face-to-face tessellations.

Under stronger assumptions, the considered densities also have ergodic interpretations, as follows, in particular, from sections 9.3 and 10.5. For these, we also refer to Miles [517, 522, 523], Cowan [180, 181], Zähle [823].

4. A counterpart to Theorem 10.1.5 holds for a stationary random mosaic in $\mathbb{R}^d$ which is **simplicial** (all faces are simplices). In this case, for $k \in \{0, \dots, d-1\}$,

$$(1 - (-1)^{d-k})\gamma^{(k)} = \sum_{j=k+1}^{d} (-1)^{d-j} \binom{j+1}{k+1} \gamma^{(j)}.$$

This can be proved in a similar way to Theorem 10.1.5.

5. That for a planar mosaic the considered parameters can be expressed in terms of the three quantities $\gamma^{(0)}, \gamma^{(1)}, l_1$, as in Theorem 10.1.6, goes back to Mecke [481]. The exact range of these three parameters is given by the inequalities (10.24) together with the trivial inequalities $\gamma^{(0)}, \gamma^{(2)}, l_1 > 0$. More precisely, every parameter triple satisfying these inequalities can be realized by a random mosaic which is stationary, isotropic and ergodic. This was shown by Kendall and Mecke [399].

Nagel and Weiss [574] have extended the system of the three parameters $\gamma^{(0)}, \gamma^{(2)}, l_1$ by two distributional 'parameters': the distribution of the direction of the typical edge and the 'rose of directions' of the segment process of the edges;

the latter can be interpreted as the distribution of the direction of the typical edge weighted by its length. The authors found necessary and sufficient conditions on these five parameters to correspond to a planar stationary tessellation.

In Theorems 10.1.6 and 10.1.7, we have restricted ourselves to selected relations and to dimensions two and three. The comprehensive paper of Møller [551] elaborates on random mosaics in $\mathbb{R}^d$ and contains further relations. Extensions to mosaics with not necessarily convex cells are found in Weiss and Zähle [809], Zähle [828], Leistritz and Zähle [438].

The special case $k = 1$ of Theorem 10.1.8 was treated and discussed by Mecke and Stoyan [503]. The general case was mentioned there without proof; a proof was indicated by Mecke [493].

Intersections of a stationary random mosaic with a fixed plane have been treated by Miles [533] and Møller [551].

Inequalities of isoperimetric type for planar stationary random mosaics have been investigated by Mecke [482, 485, 488].

For random mosaics on the sphere, see Miles [524], Arbeiter and Zähle [39].

6. Superpositions. Mecke [481] and Santaló [663] have studied superpositions of planar random mosaics. Nagel and Weiss [575] have shown that (suitably rescaled) superpositions of i.i.d. stationary random tessellations in the plane converge weakly to Poisson line tessellations.

7. Iterations, or nesting. An interesting class of stationary random tessellations, which are not face-to-face, is obtained by the process of **iteration**, also called **nesting**. The idea is to subdivide the cells of a given stationary tessellation independently by a sequence of other stationary tessellations, and to repeat this process, combined with rescalings, to keep the density $d_{d-1}^{(d-1)}$ constant. Fundamental results were obtained by Nagel and Weiss [577]. Let Y_0 be stationary tessellation of $\mathbb{R}^d$, and for each $k \in \mathbb{N}$ let $\mathcal{Y}_k$ be a sequence of tessellations, such that all involved tessellations, including Y_0, are i.i.d. Assume that the cells of Y_0 are numbered, and replace the ith cell of Y_0 by its d-dimensional intersections with the cells of $\mathcal{Y}_i$. The result is a new stationary tessellation, denoted by $I(Y_0, \mathcal{Y}_1)$. Define

$$\begin{aligned} I_2(Y_0) &:= I(2Y_0, 2\mathcal{Y}_1), \\ I_m(Y_0) &:= I(mY_0, m\mathcal{Y}_1, \dots, m\mathcal{Y}_{m-1}) \\ &:= I(I(mY_0, m\mathcal{Y}_1, \dots, m\mathcal{Y}_{m-2}), m\mathcal{Y}_{m-1}), \qquad m = 3, 4, \dots, \end{aligned}$$

where the factors denote dilatation. A stationary random tessellation Y is called **stable with respect to iteration** (**STIT**) if $Y \stackrel{\mathcal{D}}{=} I_m(Y)$ for $m = 2, 3, \dots$ Nagel and Weiss [577] proved the existence of STIT tessellations, depending on a given translation invariant, locally finite, nondegenerate measure on the space of hyperplanes. They are called **crack STIT** tessellations, because the construction proceeds by successively 'cracking' (that is, dividing by means of random hyperplanes) the cells within a given polytopal window W, and then showing consistency for increasing windows and using Theorem 2.3.1. If Y is STIT, then trivially $I_n(Y) \Rightarrow Y$ as $n \to \infty$, where $\Rightarrow$ denotes weak convergence of the distributions (equivalently, convergence of the capacity functionals) of the $(d-1)$-skeletons. Nagel and Weiss had already proved in [575] that $I_n(Y_0) \Rightarrow Y$, for stationary random tessellations Y_0 and Y, implies that Y is STIT. In [577] they showed that any nondegenerate stationary tessellation Y (with $0 < d_{d-1}^{(d-1)} < \infty$) satisfies $I_n(Y) \Rightarrow Y(1)$, where $Y(1)$ is a crack STIT tessellation. Moreover, for given $Y(1)$, all tessellations Y with this property were determined, and the following uniqueness theorem was obtained. A stationary random tessellation Y in $\mathbb{R}^d$ with $0 < d_{d-1}^{(d-1)} < \infty$ and given directional distribution is STIT if and only if $Y \stackrel{\mathcal{D}}{=} Y(1)$, where $Y(1)$ is the crack STIT tessellation corresponding to the given data.

Further investigations of STIT tessellations, including global constructions, structural and distributional properties, are found in Nagel and Weiss [576, 578, 579], Mecke, Nagel and Weiss [496, 497, 498].

Nagel and Weiss pointed out in [575] that the typical cell of a planar STIT tessellation has the same distribution as the typical cell of a suitable stationary Poisson line tessellation. Other stationary tessellations in the plane that share this property, and also the property that all nodes are T-shaped, but which are not STIT, were constructed before by Miles and Mackisack [537].

At the same time, a general theory of iterated tessellations, comprising superpositions and nesting, was developed by Maier and Schmidt [446]. They derived sufficient conditions for iterated tessellations to be stationary and isotropic and obtained formulas for the intensities of the facet processes of stationary and isotropic iterated tessellations, in terms of specific intrinsic volumes of the component tessellations. Also obtained were formulas for the expected intrinsic volumes of typical facets, and many other results. Maier, Mayer and Schmidt [447] investigated distributional properties of the typical cell of stationary iterated tessellations in the plane, which are generated by Poisson–Voronoi or Poisson line tessellations.

8. Mackisack and Miles [445] investigated a class of homogeneous rectangular tessellations in the plane, which are not face-to-face.

9. Limit theorems. For stationary and ergodic random tessellations with independently generated inner cell structures, Heinrich, H. Schmidt and V. Schmidt [331] have proved a strong law of large numbers and a multivariate central limit theorem, for measurements in convex sampling windows that increase unboundedly. In [333], they obtained similar results for tessellations where the random cell structures are generated in the facets of the cells.

10.2 Voronoi and Delaunay Mosaics

Let $A \neq \emptyset$ be a locally finite set in $\mathbb{R}^d$. To each $x \in A$, there corresponds the set

$$C(x, A) := \{z \in \mathbb{R}^d : \|z - x\| \leq \|z - a\| \text{ for all } a \in A\}$$

of all points for which x is the nearest point in A. For $x \neq y$, let $H_y^+(x)$ denote the closed halfspace containing x that is bounded by the mid-hyperplane of x and y, that is, the hyperplane through $(x + y)/2$ and orthogonal to $y - x$; explicitly

$$H_y^+(x) = \{z \in \mathbb{R}^d : \langle z, y - x \rangle \leq \frac{1}{2}(\|y\|^2 - \|x\|^2)\}.$$

Then the set $C(x, A)$ can also be written in the form

$$C(x,A) = \bigcap_{y\in A,\, y\neq x} H_y^+(x).$$

From this, it is evident that $C(x,A)$ is a closed convex set with interior points. It is called the **Voronoi cell** of x (with respect to A), and x is called the **nucleus** of $C(x,A)$. The collection $\mathsf{m} := \{C(x,A) : x \in A\}$ is again locally finite. In fact, suppose that $C(x_i,A) \cap rB^d \neq \emptyset$ for $i \in I$ (an index set). Then, to each $i \in I$ there exists $y_i \in C(x_i,A) \cap rB^d$, and it follows that $\|x_i\| \leq \|x_i - y_i\| + \|y_i\| \leq \|x_1 - y_i\| + \|y_i\| \leq \|x_1\| + 2\|y_i\| \leq \|x_1\| + 2r$, hence $x_i \in (\|x_1\| + 2r)B^d$; thus I is finite.

Therefore, the system m has the properties (a), (c) and (d) of Definition 10.1.1 that are required of a mosaic. Condition (b), however, is in general not satisfied, since the cells $C(x,A)$ need not be bounded. Sufficient (though not necessary) for the boundedness of the Voronoi cells is the assumption that $\operatorname{conv} A = \mathbb{R}^d$. Suppose that this is satisfied, and assume that some Voronoi cell $C(x,A)$ is unbounded. By the convexity of $C(x,A)$, there is then a direction $u \in S^{d-1}$ so that the ray $S := \{x + \alpha u : \alpha \geq 0\}$ is contained in $C(x,A)$. For every $\alpha > 0$, the ball with center $x + \alpha u$ and radius α contains x in its boundary and does not contain a point of A in its interior. Letting $\alpha \to \infty$, we deduce the existence of an open halfspace not containing a point of A. This is in contradiction to the assumption that $\operatorname{conv} A = \mathbb{R}^d$.

Theorem 10.2.1. *Let $A \subset \mathbb{R}^d$ be locally finite, nonempty and such that the corresponding Voronoi cells $C(x,A)$, $x \in A$, are bounded. Then $\mathsf{m} := \{C(x,A) : x \in A\}$ is a face-to-face mosaic.*

Proof. It only remains to show that m is face-to-face. Assume this were false. Then there exist two cells $C_1 := C(x_1,A)$ and $C_2 := C(x_2,A)$ such that $S := C_1 \cap C_2 \neq \emptyset$, but S is not a face of C_1. Therefore, the affine hull of S contains a point $z \in C_1$ such that $z \notin S$. This affine hull and the set S both lie in the mid-hyperplane $H_{x_2}^+(x_1) \cap H_{x_1}^+(x_2)$ of x_1 and x_2, hence

$$\langle z, x_2 - x_1\rangle = \frac{1}{2}(\|x_2\|^2 - \|x_1\|^2). \tag{10.29}$$

Since $z \notin C_2$, there is $y \in A$ with $z \notin H_y^+(x_2)$, thus

$$\langle z, y - x_2\rangle > \frac{1}{2}(\|y\|^2 - \|x_2\|^2).$$

Since $z \in C_1$,

$$\langle z, y - x_1\rangle \leq \frac{1}{2}(\|y\|^2 - \|x_1\|^2).$$

The latter two inequalities together contradict (10.29). It follows that the assumption was false, hence m is face-to-face. □

General assumption. All stationary ordinary point processes $\widetilde{X}$ appearing in this section are assumed to satisfy $\widetilde{X} \neq \emptyset$ a.s.

Theorem 10.2.2. *Let $\widetilde{X}$ be a stationary point process in $\mathbb{R}^d$, and let $X := \{C(x, \widetilde{X}) : x \in \widetilde{X}\}$ be the collection of the corresponding Voronoi cells. Then X is a stationary (face-to-face) random mosaic, provided that X has locally finite intensity measure.*

Proof. By Theorem 2.4.4 we have $\operatorname{conv} \widetilde{X} = \mathbb{R}^d$ almost surely, hence the sets $C(x, \widetilde{X}(\omega))$, $x \in \widetilde{X}(\omega)$, are bounded for almost all realizations $\widetilde{X}(\omega)$. Now Theorem 10.2.1 implies that X is a (face-to-face) random mosaic. The stationarity is evident. □

The mosaic X defined by Theorem 10.2.2 is called a **Voronoi mosaic** (with generating point process $\widetilde{X}$).

Remark. The local finiteness of the intensity measure of the process X in Theorem 10.2.2 must be assumed and does not follow automatically from the fact that $\widetilde{X}$ has locally finite intensity measure, even if $\widetilde{X}$ is stationary and isotropic (see the counterexample in Note 1). Further, the mosaic X is not necessarily normal. For example, if $\widetilde{X}(\omega) = \mathbb{Z}^d$, the cells of the corresponding Voronoi mosaic are cubes, and every vertex of the mosaic is contained in 2^d of the cubes. However, we shall see in a moment that X has locally finite intensity measure and is almost surely normal if the generating point process $\widetilde{X}$ is Poisson. In that case, X is called a **Poisson–Voronoi mosaic.**

In the following, it will be essential that for a stationary Poisson process the points are almost surely in general position, that is, any $k+1$ of the points are not contained in a $(k-1)$-dimensional plane, $k = 1, \dots, d$. If $m \in \{1, \dots, d\}$ and the $m+1$ points $x_0, \dots, x_m \in \mathbb{R}^d$ are in general position, there is a unique m-dimensional ball, $B^m(x_0, \dots, x_m)$, containing these points in its relative boundary. Let $z(x_0, \dots, x_m)$ denote the center of this ball, and let $F(x_0, \dots, x_m)$ be the $(d-m)$-dimensional affine subspace through $z(x_0, \dots, x_m)$ and orthogonal to $B^m(x_0, \dots, x_m)$. If $A \subset \mathbb{R}^d$ and $x_0, \dots, x_m \in A$ are points in general position, we define

$$S(x_0, \dots, x_m; A) := \{y \in F(x_0, \dots, x_m) : B_0(y, \|y - x_0\|) \cap A = \emptyset\},$$

where $B_0(y, r)$ denotes the open ball with center y and radius r. If $x_0, \dots, x_m$ are not in general position, we put $z(x_0, \dots, x_m) := 0$ and $S(x_0, \dots, x_m; A) := \emptyset$.

If now X is the Poisson–Voronoi mosaic generated by the point process $\widetilde{X}$ and if $(x_0, \dots, x_m) \in \widetilde{X}^{m+1}_{\neq}$, then the affine subspace $F(x_0, \dots, x_m)$ is the affine hull of some $(d-m)$-face S of X if and only if $S(x_0, \dots, x_m; \widetilde{X}) \neq \emptyset$. If this is the case, we have $S = S(x_0, \dots, x_m; \widetilde{X})$, and every $(d-m)$-face S of X is obtained in this way.

Theorem 10.2.3. *Every Poisson–Voronoi mosaic in $\mathbb{R}^d$ is normal.*

Proof. Let S be a k-face of X, $k \in \{0, \dots, d-2\}$. Let S be in the boundary of precisely the cells $C(x_0, \widetilde{X}), \dots, C(x_m, \widetilde{X})$ of X. Then the affine hull of S, aff S, is the set of all points $y \in \mathbb{R}^d$ having the same distance from $x_0, \dots, x_m$. Since almost surely no $d+2$ points of $\widetilde{X}$ are on the same sphere, we must have $m \le d$. Now it follows that aff $S = F(x_0, \dots, x_m)$, hence $k = d - m$. □

A stationary Poisson process $\widetilde{X}$ is uniquely determined, up to stochastic equivalence, by its intensity γ. Hence, all distributional parameters of the generated Poisson–Voronoi mosaic depend on γ alone. We shall provide explicit formulas for the densities $d_k^{(j,k)}$ of the (j,k)-face stars of X with $k \le j$. The finiteness of these quantitites, and with it the local finiteness of the intensity measures of all face processes (including the mosaic itself), follows from the finiteness of γ, as the proof shows. Thus, the finiteness conventions made earlier are now unnecessary.

Theorem 10.2.4. *Let X be a Poisson–Voronoi mosaic of (cell) intensity γ in $\mathbb{R}^d$. Then, for $k \in \{0, \dots, d\}$,*

$$d_k^{(k)} = \frac{2^{d-k+1}\pi^{\frac{d-k}{2}}}{d(d-k+1)!} \frac{\Gamma\left(\frac{d^2-kd+k+1}{2}\right)\Gamma\left(1+\frac{d}{2}\right)^{d-k+\frac{k}{d}}\Gamma\left(d-k+\frac{k}{d}\right)}{\Gamma\left(\frac{d^2-kd+k}{2}\right)\Gamma\left(\frac{d+1}{2}\right)^{d-k}\Gamma\left(\frac{k+1}{2}\right)}\gamma^{\frac{d-k}{d}}$$

and

$$d_k^{(j,k)} = \binom{d-k+1}{j-k} d_k^{(k)} \tag{10.30}$$

for $j \in \{k, \dots, d\}$.

Proof. By Theorem 10.2.3, X is normal, hence for $(S, \mathcal{T}) \in \mathcal{X}^{(k,j)}$ we have $N_j(S, \mathcal{T}) = \binom{d-k+1}{j-k}$. Therefore, Theorem 10.1.2 yields

$$\begin{aligned}
d_k^{(j,k)} &= \gamma^{(k)} \int_{\mathcal{K}_0 \times \mathcal{F}_f(\mathcal{K}')} V_k(S) N_j(S, \mathcal{T})\, \mathbb{Q}^{(k,j)}(\mathrm{d}(S, \mathcal{T})) \\
&= \binom{d-k+1}{j-k} \gamma^{(k)} \int_{\mathcal{K}_0 \times \mathcal{F}_f(\mathcal{K}')} V_k(S)\, \mathbb{Q}^{(k,j)}(\mathrm{d}(S, \mathcal{T})) \\
&= \binom{d-k+1}{j-k} \gamma^{(k)} \int_{\mathcal{K}_0} V_k(K)\, \mathbb{Q}^{(k)}(\mathrm{d}K) \\
&= \binom{d-k+1}{j-k} d_k^{(k)}
\end{aligned}$$

by (10.3). Thus (10.30) holds.

Let $\widetilde{X}$ be the Poisson point process generating X. Before Theorem 10.2.3 we have explained how the k-faces of X correspond to $(d-k+1)$-tuples of points from $\widetilde{X}$ in general position. This gives

$$d_k^{(k)} = \frac{1}{\kappa_d (d-k+1)!} \mathbb{E} \sum_{(x_0,\dots,x_{d-k}) \in \widetilde{X}^{d-k+1}} V_k(S(x_0,\dots,x_{d-k};\widetilde{X}) \cap B^d).$$

Now we apply the Slivnyak–Mecke formula (Corollary 3.2.3) and get

$$\begin{aligned} &\mathbb{E} \sum_{(x_0,\dots,x_{d-k}) \in \widetilde{X}^{d-k+1}} V_k(S(x_0,\dots,x_{d-k};\widetilde{X}) \cap B^d) \\ &= \gamma^{d-k+1} \int_{\mathbb{R}^d} \cdots \int_{\mathbb{R}^d} \mathbb{E} V_k(S(x_0,\dots,x_{d-k};\widetilde{X} \cup \{x_0,\dots,x_{d-k}\}) \cap B^d) \\ &\quad \times \lambda(\mathrm{d}x_0) \cdots \lambda(\mathrm{d}x_{d-k}). \end{aligned}$$

For points $x_0,\dots,x_{d-k} \in \mathbb{R}^d$ in general position, we have

$$\begin{aligned} &\mathbb{E} V_k(S(x_0,\dots,x_{d-k};\widetilde{X} \cup \{x_0,\dots,x_{d-k}\}) \cap B^d) \\ &= \int_{F(x_0,\dots,x_{d-k}) \cap B^d} \mathbb{P}(\widetilde{X} \cap B_0(y,\|y-x_0\|) = \emptyset)\, \lambda_k(\mathrm{d}y) \\ &= \int_{F(x_0,\dots,x_{d-k}) \cap B^d} \mathrm{e}^{-\gamma\kappa_d \|y-x_0\|^d}\, \lambda_k(\mathrm{d}y), \end{aligned}$$

hence altogether we get

$$\begin{aligned} d_k^{(k)} = \frac{1}{\kappa_d (d-k+1)!} \gamma^{d-k+1} \int_{\mathbb{R}^d} \cdots \int_{\mathbb{R}^d} \int_{F(x_0,\dots,x_{d-k}) \cap B^d} \mathrm{e}^{-\gamma\kappa_d \|y-x_0\|^d} \\ \times \lambda_k(\mathrm{d}y)\, \lambda(\mathrm{d}x_0) \cdots \lambda(\mathrm{d}x_{d-k}). \end{aligned}$$

The outer $(d-k+1)$-fold integral can be transformed by means of the affine Blaschke–Petkantschin formula (Theorem 7.2.7). Thus we obtain

$$\begin{aligned} d_k^{(k)} &= \frac{1}{\kappa_d (d-k+1)!} \gamma^{d-k+1} [(d-k)!]^k b_{d(d-k)} \int_{A(d,d-k)} \int_E \cdots \int_E \\ &\int_{(z(x_0,\dots,x_{d-k})+E^\perp) \cap B^d} \mathrm{e}^{-\gamma\kappa_d \|y-x_0\|^d} \Delta_{d-k}(x_0,\dots,x_{d-k})^k \\ &\times \lambda_k(\mathrm{d}y)\, \lambda_E(\mathrm{d}x_0) \cdots \lambda_E(\mathrm{d}x_{d-k})\, \mu_{d-k}(\mathrm{d}E), \end{aligned}$$

with the constant $b_{d(d-k)}$ as given in Theorem 7.2.7.

Let y_0 be the projection of the origin to E. We apply Theorem 7.3.1 to the $(d-k+1)$-fold integration over E, replacing d by $d-k$ and interpreting E as $\mathbb{R}^{d-k}$, with y_0 as origin. Let S_E^{d-k-1} be the unit sphere in the linear subspace parallel to E, and let σ_E be the spherical Lebesgue measure on S_E^{d-k-1}. Then we obtain

$$\int_E \cdots \int_E \int_{(z(x_0,\dots,x_{d-k})+E^\perp) \cap B^d} \mathrm{e}^{-\gamma\kappa_d \|y-x_0\|^d} \Delta_{d-k}(x_0,\dots,x_{d-k})^k\, \lambda_k(\mathrm{d}y)$$

$$\times \lambda_E(\mathrm{d}x_0)\cdots\lambda_E(\mathrm{d}x_{d-k})$$
$$= (d-k)! \int_E \int_0^\infty \int_{S_E^{d-k-1}} \cdots \int_{S_E^{d-k-1}} \int_{(z+E^\perp)\cap B^d}$$
$$\mathrm{e}^{-\gamma\kappa_d \|y-(z+ru_0)\|^d} r^{d(d-k)-1} \Delta_{d-k}(u_0,\dots,u_{d-k})^{k+1}$$
$$\times \lambda_k(\mathrm{d}y)\,\sigma_E(\mathrm{d}u_0)\cdots\sigma_E(\mathrm{d}u_{d-k})\,\mathrm{d}r\,\lambda_{d-k}(\mathrm{d}z).$$

Using the abbreviations

$$A := \frac{[(d-k)!]^{k+1} b_{d(d-k)}}{(d-k+1)!\kappa_d}\,\gamma^{d-k+1}$$

and

$$J(E,u_0) := \int_E \int_0^\infty \int_{(z+E^\perp)\cap B^d} \mathrm{e}^{-\gamma\kappa_d \|y-(z+ru_0)\|^d} r^{d(d-k)-1}$$
$$\times \lambda_k(\mathrm{d}y)\,\mathrm{d}r\,\lambda_{d-k}(\mathrm{d}z)$$

for $E \in A(d,d-k)$ and $u_0 \in S_E^{d-k-1}$, we get

$$d_k^{(k)} = A \int_{A(d,d-k)} \int_{S_E^{d-k-1}} \cdots \int_{S_E^{d-k-1}} J(E,u_0)\Delta_{d-k}(u_0,\dots,u_{d-k})^{k+1}$$
$$\times \sigma_E(\mathrm{d}u_0)\cdots\sigma_E(\mathrm{d}u_{d-k})\,\mu_{d-k}(\mathrm{d}E).$$

The integral $J(E,u_0)$ has the same value for all $u_0 \in S_E^{d-k-1}$. The $(d-k+1)$-fold integral over S_E^{d-k-1} can be evaluated by means of Theorem 8.2.3. Thus, denoting by u_E an arbitrary unit vector parallel to E, we get

$$d_k^{(k)} = A \int_{A(d,d-k)} J(E,u_E) S(d-k,d-k,k+1)\,\mu_{d-k}(\mathrm{d}E),$$

where the constant $S(d-k,d-k,k+1)$ is given by Theorem 8.2.3. The function $E \mapsto J(E,u_E)$ is invariant under rotations, hence with a fixed subspace $L \in G(d,d-k)$ we get

$$I := \int_{A(d,d-k)} J(E,u_E)\,\mu_{d-k}(\mathrm{d}E)$$
$$= \int_{L^\perp} J(L+y_0,u_L)\,\lambda_k(\mathrm{d}y_0)$$
$$= \int_{L^\perp} \int_L \int_0^\infty \int_{(z+L^\perp)\cap B^d} \mathrm{e}^{-\gamma\kappa_d \|y-(y_0+z+ru_L)\|^d} r^{d(d-k)-1}\,\lambda_k(\mathrm{d}y)$$
$$\times \mathrm{d}r\,\lambda_{d-k}(\mathrm{d}z)\,\lambda_k(\mathrm{d}y_0)$$
$$= \int_{\mathbb{R}^d} \int_0^\infty \int_{(x+L^\perp)\cap B^d} \mathrm{e}^{-\gamma\kappa_d \|y-(x+ru_L)\|^d} r^{d(d-k)-1} \lambda_k(\mathrm{d}y)\,\mathrm{d}r\,\lambda(\mathrm{d}x)$$
$$= \int_{\mathbb{R}^d} \int_0^\infty \int_{L^\perp\cap(B^d-x)} \mathrm{e}^{-\gamma\kappa_d \|y-ru_L\|^d} r^{d(d-k)-1} \lambda_k(\mathrm{d}y)\,\mathrm{d}r\,\lambda(\mathrm{d}x).$$

The condition $y \in B^d - x$ is equivalent to $x \in B^d - y$, hence we obtain

$$I = \kappa_d \int_0^\infty \int_{L^\perp} \mathrm{e}^{-\gamma\kappa_d(\|y\|^2+r^2)^{\frac{d}{2}}} r^{d(d-k)-1} \lambda_k(\mathrm{d}y)\,\mathrm{d}r$$
$$= \kappa_d\omega_k \int_0^\infty \int_0^\infty \mathrm{e}^{-\gamma\kappa_d(r^2+s^2)^{\frac{d}{2}}} r^{d(d-k)-1} s^{k-1}\,\mathrm{d}s\,\mathrm{d}r.$$

Here we substitute

$$s = u\sqrt{1-t}, \qquad r = u\sqrt{t}$$

with $t \in [0,1]$, $u \in [0,\infty)$; the Jacobian is equal to $u/2\sqrt{t(1-t)}$. Thus we get

$$I = \frac{\kappa_d\omega_k}{2}\left(\int_0^1 t^{\frac{d(d-k)}{2}-1}(1-t)^{\frac{k}{2}-1}\mathrm{d}t\right)\left(\int_0^\infty \mathrm{e}^{-\gamma\kappa_d u^d} u^{d^2-dk+k-1}\mathrm{d}u\right).$$

The integrals are Euler's Beta integral and, after a substitution, Euler's Gamma integral; explicitly,

$$\int_0^1 t^{\frac{d(d-k)}{2}-1}(1-t)^{\frac{k}{2}-1}\mathrm{d}t = \frac{\Gamma(\frac{d(d-k)}{2})\Gamma(\frac{k}{2})}{\Gamma(\frac{d(d-k)+k}{2})}$$

and

$$\int_0^\infty \mathrm{e}^{-\gamma\kappa_d u^d} u^{d^2-dk+k-1}\mathrm{d}u = \frac{\Gamma(\frac{d^2-dk+k}{d})}{d(\gamma\kappa_d)^{\frac{d^2-dk+k}{d}}}.$$

Collecting the results and inserting the values of the constants $b_{d(d-k)}$ and $S(d-k,d-k,k+1)$, we obtain the assertion. □

The case $k = 0$ of Theorem 10.2.4 gives

$$\gamma^{(0)} = \frac{2^{d+1}\pi^{\frac{d-1}{2}}}{d^2(d+1)} \frac{\Gamma\left(\frac{d^2+1}{2}\right)}{\Gamma\left(\frac{d^2}{2}\right)} \left[\frac{\Gamma\left(1+\frac{d}{2}\right)}{\Gamma\left(\frac{d+1}{2}\right)}\right]^d \gamma \tag{10.31}$$

and (as can also be deduced from Theorem 10.1.2 and the normality)

$$\gamma^{(j)} n_{j0} = \binom{d+1}{j}\gamma^{(0)}$$

for $j = 1,\dots,d$. In particular, we obtain (10.3) again, that is,

$$2\gamma^{(1)} = (d+1)\gamma^{(0)}.$$

Further, we have

$$\gamma^{(d)} = \gamma. \tag{10.32}$$

In fact, (4.8) gives $\gamma^{(d)} \le \gamma$, and (4.9) gives $\gamma \le \gamma^{(d)}$. The equality (10.32) also follows directly from Note 1 of Section 4.2, since the nucleus is a generalized center function.

For $d = 3$, we obtain the value of $\gamma^{(2)}$ from Theorem 10.1.5 or from (10.21). Now, for planar or spatial Poisson–Voronoi mosaics, all the mean values appearing in Theorems 10.1.6 and 10.1.7 can be expressed in terms of the intensity γ, by applying Theorem 10.2.4 for $k > 0$ and then Theorems 10.1.6 and 10.1.7, and using the normality. In the following theorem, we collect only a selection of these mean values.

Theorem 10.2.5. *Let X be a Poisson–Voronoi mosaic of intensity γ in $\mathbb{R}^d$. Then, for $d = 2$,*

$$\gamma^{(0)} = 2\gamma, \quad \gamma^{(1)} = 3\gamma, \quad \gamma^{(2)} = \gamma,$$

$$n_{02} = 3, \quad n_{20} = 6,$$

$$l_1 = \frac{2}{3\sqrt{\gamma}}, \quad a = \frac{1}{\gamma}, \quad p = \frac{4}{\sqrt{\gamma}},$$

and for $d = 3$,

$$\gamma^{(0)} = \frac{24\pi^2}{35}\gamma, \quad \gamma^{(1)} = \frac{48\pi^2}{35}\gamma, \quad \gamma^{(2)} = \left(\frac{24\pi^2}{35} + 1\right)\gamma, \quad \gamma^{(3)} = \gamma,$$

$$n_{01} = n_{03} = 4, \quad n_{02} = 6, \quad n_{12} = 3,$$

$$n_{21} = \frac{144\pi^2}{24\pi^2 + 35} \simeq 5.23, \quad n_{30} = \frac{96\pi^2}{35} \simeq 27.07,$$

$$n_{31} = \frac{144\pi^2}{35} \simeq 40.61, \quad n_{32} = \frac{48\pi^2}{35} + 2 \simeq 15.54,$$

$$l_1 = \frac{7\Gamma\left(\frac{1}{3}\right)}{9(36\pi\gamma)^{\frac{1}{3}}}, \quad a_2 = \frac{35 \cdot 2^{\frac{8}{3}}\Gamma\left(\frac{2}{3}\right)\pi^{\frac{1}{3}}}{(24\pi^2 + 35)(9\gamma)^{\frac{2}{3}}}, \quad p_2 = \frac{7 \cdot 2^{\frac{10}{3}}\Gamma\left(\frac{1}{3}\right)\pi^{\frac{5}{3}}}{(24\pi^2 + 35)(9\gamma)^{\frac{1}{3}}},$$

$$v = \frac{1}{\gamma}, \quad s = \frac{2^{\frac{11}{3}}\Gamma\left(\frac{2}{3}\right)\pi^{\frac{1}{3}}}{(9\gamma)^{\frac{2}{3}}}, \quad m = \frac{2^3\Gamma\left(\frac{1}{3}\right)\pi^{\frac{8}{3}}}{15(36\gamma)^{\frac{1}{3}}}.$$

Now we turn to Delaunay mosaics. They are, in a sense, dual to suitable Voronoi mosaics. We start with a locally finite set $A \subset \mathbb{R}^d$ for which $\operatorname{conv} A = \mathbb{R}^d$. It induces a Voronoi mosaic $\mathsf{m} = \{C(x, A) : x \in A\}$. For $e \in \mathcal{F}_0(\mathsf{m})$ we put

$$D(e, A) := \operatorname{conv}\{x \in A : e \in \mathcal{F}_0(C(x, A))\}.$$

Theorem 10.2.6. *Let $A \subset \mathbb{R}^d$ be a locally finite set with* $\operatorname{conv} A = \mathbb{R}^d$, *and let m be the corresponding Voronoi mosaic. Then*

$$\mathsf{d} := \{D(e, A) : e \in \mathcal{F}_0(\mathsf{m})\}$$

is a face-to-face mosaic.

Proof. Since $\{x \in A : e \in \mathcal{F}_0(C(x,A))\}$ is finite, the sets $D(e,A)$, $e \in \mathcal{F}_0(\mathsf{m})$, are compact; further, they have interior points. Since $\mathcal{F}_0(\mathsf{m})$ is locally finite, also d is locally finite. The points $x \in A$ for which e is a vertex of $C(x,A)$ have the same distance from e, hence they lie in the boundary of a ball $K(e)$ with center e and circumscribed to $D(e,A)$; they are vertices of $D(e,A)$. There are no points of A in the interior of $K(e)$, and each point of $A \cap \operatorname{bd} K(e)$ is a vertex of $D(e,A)$.

For $e, e' \in \mathcal{F}_0(\mathsf{m})$, $e \neq e'$, we consider the intersection $D(e,A) \cap D(e',A)$ and assume that it is not empty. Then all vertices of $D(e,A)$ lie in $K(e) \setminus \operatorname{int} K(e')$, and all vertices of $D(e',A)$ lie in $K(e') \setminus \operatorname{int} K(e)$. Let z be the center of $K(e) \cap K(e')$ and let E be the hyperplane through z orthogonal to $e - e'$. Then each point of $D(e,A) \cap D(e',A)$ lies in E, and it follows that

$$D(e,A) \cap D(e',A) = \operatorname{conv}\{x \in A : x \in \operatorname{bd} K(e) \cap \operatorname{bd} K(e')\}.$$

From this, we see that $D(e,A) \cap D(e',A)$ has no interior points and that d is face-to-face.

Let $y \in \mathbb{R}^d$. Since $y \in \operatorname{conv} A$, there are affinely independent points $x_1, \dots, x_{d+1} \in A$ with $y \in \operatorname{conv}\{x_1, \dots, x_{d+1}\}$. We suppose that these points are chosen so that the ball K with $x_1, \dots, x_{d+1} \in \operatorname{bd} K$ has minimal radius. Assume there is a point $x \in A \cap \operatorname{int} K$. Then one of the points $x_1, \dots, x_{d+1}$, say x_1, can be replaced by x in such a way that still $y \in \operatorname{conv}\{x, x_2, \dots, x_{d+1}\}$ and $x, x_2, \dots, x_{d+1}$ are affinely independent. The ball containing $x, x_2, \dots, x_d$ in its boundary has radius smaller than that of K, a contradiction. Thus, the interior of K does not contain a point of A. Therefore, the center e of K satisfies $e \in C(x_i, A)$, $i = 1, \dots, d+1$, hence $e \in \mathcal{F}_0(\mathsf{m})$ and $y \in D(e,A)$. This proves that

$$\bigcup_{e \in \mathcal{F}_0(\mathsf{m})} D(e,A) = \mathbb{R}^d,$$

which completes the proof. □

We call d the **Delaunay mosaic** corresponding to the set A. If the points of A are in general position and if any $d+2$ of them do not lie on a sphere, then the corresponding Voronoi mosaic m is normal. In this case, every cell of d has $d+1$ vertices and hence is a d-simplex. Then all faces of d are simplices. A mosaic with this property is called **simplicial**.

In order to emphasize the duality between m and d, we now restrict ourselves to the case where m is normal, and hence d is simplicial. For the Delaunay mosaics coming from stationary Poisson processes, which will be considered below, this condition is satisfied almost surely, by Theorem 10.2.3.

Now let $m \in \{0, \dots, d\}$, and let $F \in \mathcal{F}_m(\mathsf{m})$ be an m-face of the Voronoi mosaic m. With a notation introduced earlier in this section, we have $F = S(x_0, \dots, x_{d-m}; A)$ for suitable pairwise different points $x_0, \dots, x_{d-m} \in A$, which are uniquely determined up to order. We define

$$\Sigma(F) := \operatorname{conv}\{x_0, \dots, x_{d-m}\}.$$

Lemma 10.2.1. *If the Voronoi mosaic* m *corresponding to the set* A *is normal, then* Σ *is an antitone (inclusion reversing) bijection from* $\mathcal{F}_\bullet(\mathsf{m})$ *to* $\mathcal{F}_\bullet(\mathsf{d})$.

Proof. First, let $F = \{e\}$ be a vertex of m. For $\{e\} = S(x_0, \dots, x_d; A)$ we have $e \in \mathcal{F}_0(C(x_i, A))$, $i = 0, \dots, d$. Due to the normality of m this implies

$$D(e, A) = \operatorname{conv}\{x \in A : e \in \mathcal{F}_0(C(x, A))\} = \operatorname{conv}\{x_0, \dots, x_d\},$$

hence $\Sigma(\{e\}) \in \mathcal{F}_d(\mathsf{d})$, and the map $\Sigma : \mathcal{F}_0(\mathsf{m}) \to \mathcal{F}_d(\mathsf{d})$ is bijective.

Now let $m > 0$, and let $F = S(x_0, \dots, x_{d-m}; A)$ be an m-face of m. We choose a vertex e of F. Then $\{e\} = S(x_0, \dots, x_d; A)$ holds with suitable $x_{d-m+1}, \dots, x_d \in A$. Thus $\Sigma(F) = \operatorname{conv}\{x_0, \dots, x_{d-m}\}$ is a $(d-m)$-face of $D(e, A) = \operatorname{conv}\{x_0, \dots, x_d\}$. It is evident that $\Sigma : \mathcal{F}_m(\mathsf{m}) \to \mathcal{F}_{d-m}(\mathsf{d})$ is bijective, hence the map $\Sigma : \mathcal{F}_\bullet(\mathsf{m}) \to \mathcal{F}_\bullet(\mathsf{d})$ is bijective, and it is obviously antitone. □

Now we consider random Delaunay mosaics. From Theorems 10.2.6 and 10.2.3, we immediately obtain the following assertion.

Theorem 10.2.7. *Let* $\widetilde{X}$ *be a stationary point process in* $\mathbb{R}^d$, *let* X *be the corresponding Voronoi mosaic and*

$$Y := \Sigma(X^{(0)}) = \{D(e, \widetilde{X}) : e \in \mathcal{F}_0(X)\}.$$

Then Y *is a stationary (face-to-face) random mosaic. If* $\widetilde{X}$ *is a Poisson process, then* Y *is simplicial.*

We restrict ourselves again to stationary Poisson processes $\widetilde{X}$ (of intensity $\gamma > 0$) in $\mathbb{R}^d$. The corresponding Delaunay mosaic Y is called the **Poisson–Delaunay mosaic** of intensity γ. As before, $X^{(j)}$, $j = 0, \dots, d$, are the face processes of X, and $\gamma^{(j)}$ are the corresponding intensities. The face processes of Y and their intensities are denoted by $Y^{(j)}$ and $\beta^{(j)}$, respectively.

Theorem 10.2.8. *Let* $\gamma^{(j)}$, $j = 0, \dots, d$, *be the face intensities of the Poisson–Voronoi mosaic corresponding to the stationary Poisson process* $\widetilde{X}$ *in* $\mathbb{R}^d$, *and let* $\beta^{(j)}$, $j = 0, \dots, d$, *be the face intensities of the associated Poisson–Delaunay mosaic* Y. *Then*

$$\beta^{(j)} = \gamma^{(d-j)}$$

for $j = 0, \dots, d$.

Proof. For $j \in \{0, \dots, d\}$ we consider the (stationary) marked point process

$$\widehat{X}^{(d-j)} := \{(z(x_0, \dots, x_j), S(x_0, \dots, x_j; \widetilde{X}) - z(x_0, \dots, x_j)) : \\ (x_0, \dots, x_j) \in \widetilde{X}^{j+1},\ S(x_0, \dots, x_j; \widetilde{X}) \neq \emptyset\}.$$

The corresponding particle process is $X^{(d-j)}$, hence $\widehat{X}^{(d-j)}$ has intensity $\gamma^{(d-j)}$. This is also the intensity of the unmarked, ordinary point process

$$\widetilde{Y}^{(j)} := \{z(x_0,\ldots,x_j) : (x_0,\ldots,x_j) \in \widetilde{X}^{j+1},\ S(x_0,\ldots,x_j;\widetilde{X}) \neq \emptyset\}.$$

In evaluating the intensity of $Y^{(j)}$, we may choose as a center function the mapping that associates with every j-dimensional simplex the center of the $(j-1)$-dimensional sphere through its vertices. Now it is clear that the intensity of $Y^{(j)}$ is the same as that of $\widetilde{Y}^{(j)}$. □

The next theorem collects the most important quantities of the planar Poisson–Delaunay mosaic.

Theorem 10.2.9. *Let Y be a stationary planar Poisson–Delaunay mosaic, corresponding to the intensity γ, and let $\beta^{(0)}, \beta^{(1)}, \beta^{(2)}$ be its face intensities. Then*

$$\beta^{(0)} = \gamma, \quad \beta^{(1)} = 3\gamma, \quad \beta^{(2)} = 2\gamma,$$

$$n_{02} = 6, \quad n_{20} = 3,$$

$$l_1 = \frac{32}{9\pi\sqrt{\gamma}}, \quad a = \frac{1}{2\gamma}, \quad p = \frac{32}{3\pi\sqrt{\gamma}}.$$

Proof. The equations for the intensities follow from Theorems 10.2.8 and 10.2.5. From these values and the formula for the mean edge length, the remaining values are obtained according to Theorem 10.1.6. The mean edge length is given by

$$l_1 = \frac{1}{3}\mathbb{E}\, L(Z)$$

(where L is the perimeter). Its computation can be reduced to Theorem 10.4.4, which is proved in Section 10.4. From that theorem, one obtains

$$\begin{aligned} l_1 &= \frac{d_2}{3}\gamma^2 \int_0^\infty \int_{S^1}\int_{S^1}\int_{S^1} L(\mathrm{conv}\,\{u_0,u_1,u_2\})\mathrm{e}^{-\gamma\pi r^2} r^4 \Delta_2(u_0,u_1,u_2) \\ &\quad \times \sigma(\mathrm{d}u_0)\,\sigma(\mathrm{d}u_1)\,\sigma(\mathrm{d}u_2)\,\mathrm{d}r \\ &= \frac{1}{48\pi^2\sqrt{\gamma}} \int_{S^1}\int_{S^1}\int_{S^1} L(\mathrm{conv}\,\{u_0,u_1,u_2\}) A(\mathrm{conv}\,\{u_0,u_1,u_2\}) \\ &\quad \times \sigma(\mathrm{d}u_0)\,\sigma(\mathrm{d}u_1)\,\sigma(\mathrm{d}u_2). \end{aligned}$$

For reasons of symmetry, the integral

$$I := \int_{S^1}\int_{S^1}\int_{S^1} L(\mathrm{conv}\,\{u_0,u_1,u_2\}) A(\mathrm{conv}\,\{u_0,u_1,u_2\})\,\sigma(\mathrm{d}u_0)\sigma(\mathrm{d}u_1)\sigma(\mathrm{d}u_2)$$

can be written in the form

$$I = 6\pi \int_{S^1}\int_{S^1} \|u_1 - u_2\| A(\mathrm{conv}\,\{e_1,u_1,u_2\})\,\sigma(\mathrm{d}u_1)\,\sigma(\mathrm{d}u_2),$$

with a fixed unit vector e_1. Let (e_1, e_2) be an orthonormal basis of $\mathbb{R}^2$. We put

$$u_i = e_1 \cos\varphi_i + e_2 \sin\varphi_i, \quad 0 \le \varphi_i \le 2\pi, \quad i = 1, 2.$$

Then

$$\|u_1 - u_2\| = 2\left|\sin\frac{\varphi_1 - \varphi_2}{2}\right|$$

and

$$A(\mathrm{conv}\,\{e_1, u_1, u_2\}) = \frac{1}{2}|\det(u_1 - e_1, u_2 - e_1)|$$

with

$$\begin{aligned}
&\det(u_1 - e_1, u_2 - e_1)\\
&= \det\begin{pmatrix} \cos\varphi_1 - 1 & \sin\varphi_1 \\ \cos\varphi_2 - 1 & \sin\varphi_2 \end{pmatrix}\\
&= -\sin(\varphi_1 - \varphi_2) + \sin\varphi_1 - \sin\varphi_2\\
&= -2\sin\frac{\varphi_1 - \varphi_2}{2}\cos\frac{\varphi_1 - \varphi_2}{2} + 2\sin\frac{\varphi_1 - \varphi_2}{2}\cos\frac{\varphi_1 + \varphi_2}{2}.
\end{aligned}$$

Since

$$\cos\frac{\varphi_1 - \varphi_2}{2} - \cos\frac{\varphi_1 + \varphi_2}{2} = 2\sin\frac{\varphi_1}{2}\sin\frac{\varphi_2}{2} \ge 0,$$

we get

$$A(\mathrm{conv}\,\{e_1, u_1, u_2\}) = \left|\sin\frac{\varphi_1 - \varphi_2}{2}\right|\left(\cos\frac{\varphi_1 - \varphi_2}{2} - \cos\frac{\varphi_1 + \varphi_2}{2}\right)$$

and thus

$$I = 12\pi\int_0^{2\pi}\int_0^{2\pi}\sin^2\frac{\varphi_1 - \varphi_2}{2}\left(\cos\frac{\varphi_1 - \varphi_2}{2} - \cos\frac{\varphi_1 + \varphi_2}{2}\right)\mathrm{d}\varphi_1\,\mathrm{d}\varphi_2 = \frac{2^9\pi}{3}.$$

This yields the value of l_1 as stated. □

Notes for Section 10.2

1. As announced after Theorem 10.2.2, we give an example of a stationary, isotropic point process $\widetilde{X}$ for which the induced process of Voronoi cells does not have locally finite intensity measure. For $k \in \mathbb{N}$, consider the lattice $L_k := \{(km, n/k\} \in \mathbb{R}^2 : m, n \in \mathbb{Z}\}$, let ξ_k be a uniform random vector in the rectangle $[0, k] \times [0, 1/k]$, and let ϑ_k be a uniform random rotation, independent of ξ_k. Let $X_k := \vartheta_k L_k + \xi_k$. Let $p_1, p_2, \dots$ be a sequence of positive numbers with $\sum_{k=1}^\infty p_k = 1$ and $\sum_{k=1}^\infty kp_k = \infty$. There exists a point process $\widetilde{X}$ in $\mathbb{R}^2$ such that $\widetilde{X} = X_k$ with probability p_k. The process $\widetilde{X}$ is stationary and isotropic and has intensity one. Let X be the system of Voronoi cells induced by $\widetilde{X}$. With probability p_k, the ball B^2 is hit by at least $2k$

cells of X. Therefore, the number of cells of X hitting B^2 has infinite expectation, thus the intensity measure of X is not locally finite.

2. For Poisson–Voronoi mosaics, already Meijering [510] has calculated some of the mean values listed in Theorem 10.2.5. The variance of the cell volume and other distributional parameters were considered by Gilbert [261]. More systematically, two- and three-dimensional Poisson–Voronoi mosaics were studied by Miles [522, 526], and extensions to higher dimensions by Møller [551, 553]. This work we have used here.

3. Theorem 10.2.4 can be found already, without proof, in Miles [521]; proofs were later provided by Miles [533] and Møller [551]. Much material about Voronoi mosaics is found in the book by Okabe, Boots, Sugihara and Chiu [591], which contains also a chapter on Poisson–Voronoi mosaics. For random Voronoi mosaics, see also the survey of Møller [554].

4. Planar sections of Voronoi mosaics were studied by Miles [526, 533], Møller [551], and others. Chiu, van de Weygaert and Stoyan [172] have proved that the intersection of a d-dimensional stationary Poisson–Voronoi mosaic with a fixed k-plane, $k \in \{2, \dots, d-1\}$, is almost surely not a k-dimensional Voronoi mosaic of any point process. (That the section cannot be a Poisson–Voronoi mosaic, can be seen more easily, for example, from Theorem 10.2.4 in connection with the Crofton type formula of Theorem 9.4.8; compare also Mecke [481].) Indeed a stronger result is proved in [172]: in a sectional Poisson–Voronoi tessellation, almost surely each cell is a non-Voronoi cell.

5. For several characteristic quantities of stationary Poisson–Voronoi and –Delaunay mosaics, more precise information on distributional properties is available. Without going into details, we mention papers by Muche and Stoyan [566], Rathie [622], Zuyev [838], Muche [561, 562, 563, 564, 565], Møller [553], Mecke and Muche [494], Schlather [668], Heinrich and Muche [330]. Very general distributional results were obtained by Baumstark and Last [85]; see Note 7 of Section 10.4.

Heinrich [325] investigated contact distributions for more general stationary Voronoi mosaics.

6. Limit theorems. Central limit theorems for Poisson–Voronoi tessellations in the plane were established by Avram and Bertsimas [42], and in arbitrary dimensions by Heinrich and Muche [330]. A general approach to central limit theorems in geometry, yielding also results for a quite large class of functionals related to Poisson–Voronoi tessellations, is presented by Penrose and Yukich [600].

7. Approximation of Borel sets. For a Poisson process X in $\mathbb{R}^d$ and a Borel set $A \subset \mathbb{R}^d$, let $v_X(A)$ be the union of all Voronoi cells $C(x, X)$ with $x \in A$. This set can be considered as an approximation of A. If X has intensity measure $\Theta = n \int f \, d\lambda$, where the density f is supported by C^d and is positive in the interior $(0,1)^d$, let $v_n(A) := v_X(A) \cap C^d$, for $n \in \mathbb{N}$ and $A \subset C^d$. Khmaladze conjectured a strong law of large numbers, in the sense that $\lambda(v_n(A) \triangle A) \to 0$ a.s., as $n \to \infty$ (here $\triangle$ denotes the symmetric difference). For $d = 1$, this was shown in Khmaladze and Toronjadze [402]. Penrose [599] proved the result for arbitrary dimension d under the additional assumption that f is bounded from below by some constant $c > 0$. The general case is apparently still open.

Reitzner and Heveling [634] considered stationary X and convex bodies A with interior points. They gave formulas for the expectations and variances of $\lambda(v_X(A))$ and $\lambda(v_X(A)\triangle A)$ and estimates for large deviations.

8. The construction of random Voronoi mosaics can be generalized in different directions. The Voronoi mosaic of order $k \in \mathbb{N}$ corresponding to a locally finite set $A \subset \mathbb{R}^d$ is obtained if the set of points of $\mathbb{R}^d$ having the same k nearest neighbors in A is defined as a cell. For stationary Poisson point processes, such mosaics have been studied by Miles [522], Miles and Maillardet [538]. Voronoi mosaics with an anisotropic growth were investigated by Scheike [665]. The underlying idea of the **Johnson–Mehl model** is cell growth starting at kernels which are randomly generated at different times. A comprehensive investigation of random Johnson–Mehl mosaics (the cells of which are in general not convex) is provided by Møller [552]; see also the fifth chapter, written by Møller, in the book edited by Barndorff–Nielsen, Kendall and van Lieshout [80].

9. Random Laguerre mosaics. A further generalization of the Voronoi mosaics are the Laguerre mosaics, where the generating points carry weights. With a pair $(x,w) \in \mathbb{R}^d \times \mathbb{R}^+$ one can associate the sphere $S(x,\sqrt{w})$ with center x and radius $\sqrt{w}$, and then the **power** of a point $y \in \mathbb{R}^d$ with respect to this sphere is defined by

$$\mathrm{pow}(y,(x,w)) := \|y-x\|^2 - w.$$

If $S \subset \mathbb{R}^d \times \mathbb{R}^+$ is a countable set such that $\min_{(x,w)\in S} \mathrm{pow}(y,(x,w))$ exists for each $y \in \mathbb{R}^d$, the **Laguerre cell** of $(x,w) \in S$ is defined by

$$C((x,w),S) := \{y \in \mathbb{R}^d : \mathrm{pow}(y,(x,w)) \le \mathrm{pow}(y,(x',w')) \text{ for all } (x',w') \in S\}.$$

The **Laguerre mosaic** induced by S is the set of all nonempty Laguerre cells arising in this way. If $w = 0$ for all $(x,w) \in S$, a Voronoi mosaic is obtained. In contrast to this case, the Laguerre cell of some (x,w) can be empty, and if it is not, it need not necessarily contain the 'nucleus' x. Laguerre mosaics are very general: every normal mosaic in $\mathbb{R}^d$ for $d \ge 3$ is a Laguerre mosaic.

Let X be a stationary marked Poisson process in $\mathbb{R}^d$ with mark space $\mathbb{R}^+$ and mark distribution $\mathbb{F}$, satisfying $\int_0^\infty r^d\, \mathbb{F}(\mathrm{d}r) < \infty$. Then the Laguerre mosaic induced by X almost surely exists; it is called a **Poisson–Laguerre mosaic**. A thorough investigation of these random tessellations is due to Lautensack [436]. Many results from the Voronoi case can be extended, though often in a less explicit form.

10. Let $\widetilde{X}$ be a stationary Poisson process in $\mathbb{R}^d$. The Stienen ball associated with a point x of $\widetilde{X}$ is the ball with center x and radius equal to half the distance from x to the nearest point in $\widetilde{X}$. The random set defined by the union of all Stienen balls is known as the **Stienen model**. Olsbo [592] has established an integral formula for the correlation between the volume of the typical cell of the Poisson–Voronoi tessellation and the volume of the Stienen ball in the typical cell. He proved that this correlation is positive (not surprisingly, but the proof is not easy).

11. Poisson–Voronoi tessellations in three-dimensional hyperbolic space were investigated by Isokawa [377]. He derived formulas for the expected area, perimeter, and vertex number of the typical cell.

10.3 Hyperplane Mosaics

Let $\mathcal{H}$ be a locally finite system of hyperplanes in $\mathbb{R}^d$. The connected components of the complement of the union $\bigcup_{H\in\mathcal{H}} H$ are open polyhedral sets. Their closures are the **cells induced by** $\mathcal{H}$. A mosaic m (in the sense of Section 10.1) in $\mathbb{R}^d$ is called a **hyperplane mosaic** if its cells are induced by a system of hyperplanes. It is said to be in **general position** if the system $\mathcal{H}$ is in general position, and this means that every k-dimensional plane of $\mathbb{R}^d$ is contained in at most $d-k$ hyperplanes of the system, $k = 0, \dots, d-1$.

In the present section, we study random hyperplane mosaics. These are random mosaics induced by hyperplane processes. If the random mosaic X is induced by the hyperplane process $\widehat{X}$, we call $\widehat{X}$ the hyperplane process **generating** X, and X is said to be in **general position** if $\widehat{X}$ is almost surely in general position. The mosaic X is stationary if and only if this holds for $\widehat{X}$. For stationary random hyperplane mosaics the parameters introduced in Section 10.1 satisfy additional relations. After deriving some general relations of this type, we turn to mosaics induced by stationary Poisson hyperplane processes. These mosaics, and in particular their zero cells, have interesting properties, and they permit the treatment of some extremal problems (in Section 10.4).

We begin with a result holding for general distributions.

Theorem 10.3.1. *Let X be a stationary random hyperplane mosaic in general position in $\mathbb{R}^d$ and assume that the generating hyperplane process has intersection processes with finite intensities. For $0 \le j \le k \le d$,*

$$d_j^{(k)} = \binom{d-j}{d-k} d_j^{(j)}, \tag{10.33}$$

in particular

$$\gamma^{(k)} = \binom{d}{k} \gamma^{(0)} \tag{10.34}$$

and

$$n_{kj} = 2^{k-j}\binom{k}{j}. \tag{10.35}$$

Proof. For the stationary hyperplane process $\widehat{X}$ generating X, let $\widehat{X}_{d-k}$ be the intersection process of order $d-k$ ($\widehat{X}_1 = \widehat{X}$); this is a stationary k-flat process, $k = 0, \dots, d-1$.

We show first that the face process $X^{(k)}$ of X has locally finite intensity measure, $k = 0, \dots, d$. For given $r > 0$, let ν_k be the number of k-dimensional faces of X that meet the interior of the ball rB^d. From (14.67) we have

$$\nu_k = \sum_{j=d-k}^{d} \binom{j}{d-k} \alpha_j,$$

where α_j is the number of j-tuples from $\widehat{X}$ whose intersection meets the interior of rB^d (with $\alpha_0 := 1$). Since $\widehat{X}$ is in general position, the number α_j, for $j > 0$, is the number of $(d-j)$-flats from $\widehat{X}_j$ meeting the interior of rB^d. By assumption, this number has a finite expectation. Hence $\mathbb{E}\,\nu_k < \infty$, which implies that the intensity measure of $X^{(k)}$ is locally finite.

Now let $k \in \{1, \dots, d-1\}$ (the case $k = 0$ is trivial). Let $j \in \{0, \dots, k-1\}$ and $r > 0$. In every k-plane $E \in \widehat{X}_{d-k}$, a mosaic is induced by $\widehat{X} \cap E$. From (10.23), applied in E instead of $\mathbb{R}^d$, we get that almost surely

$$\sum_{E \in \widehat{X}_{d-k}} V_j(E \cap rB^d) = \sum_{i=j}^{k} (-1)^{k-i} \sum_{E \in \widehat{X}_{d-k}} \sum_{F \in X^{(i)},\, F \subset E} V_j(F \cap rB^d)$$
$$= \sum_{i=j}^{k} (-1)^{k-i} \binom{d-i}{d-k} \sum_{F \in X^{(i)}} V_j(F \cap rB^d). \quad (10.36)$$

Here we have used the fact that every i-face of X is almost surely contained in exactly $\binom{d-i}{d-k}$ planes of the intersection process $\widehat{X}_{d-k}$, since X is in general position.

In (10.36) we take the expectation and divide by $V_d(rB^d)$. On the left side we get

$$\frac{1}{V_d(rB^d)} \mathbb{E} \sum_{E \in \widehat{X}_{d-k}} V_j(E \cap rB^d) \leq \frac{1}{V_d(rB^d)} \mathbb{E} \sum_{E \in \widehat{X}_{d-k}} r^j V_j(B^k) \chi(E \cap rB^d)$$
$$= \frac{r^j V_j(B^k)}{r^d \kappa_d} \mathbb{E}\, \widehat{X}_{d-k}(\mathcal{F}_{rB^d})$$
$$= \frac{r^j V_j(B^k)}{r^d \kappa_d} \kappa_{d-k} r^{d-k} \widehat{\gamma}_{d-k},$$

by (4.27), for rB^d instead of B^d. Here, $\widehat{\gamma}_{d-k}$ is the intensity of $\widehat{X}_{d-k}$. Since $j < k$, the latter expression converges to 0 as $r \to \infty$. Hence, (10.36) together with Theorem 9.2.2 gives

$$\sum_{i=j}^{k} (-1)^{k-i} \binom{d-i}{d-k} d_j^{(i)} = \sum_{i=j}^{k} (-1)^{k-i} \binom{d-i}{d-k} \overline{V}_j(X^{(i)}) = 0.$$

This is true for $0 \leq j < k \leq d$, since for $k = d$ it holds by Theorem 10.1.4.

For $j \in \{0, \dots, d-1\}$, the obtained system of linear equations,

$$\sum_{i=j}^{k} (-1)^i \binom{d-i}{d-k} d_j^{(i)} = 0, \qquad k = j+1, \dots, d,$$

has the solution

$$d_j^{(i)} = \binom{d-j}{d-i} d_j^{(j)}, \qquad i = j, \ldots, d.$$

These are the relations (10.33). For $j = 0$, we get (10.34).

Theorem 10.1.2 gives $\gamma^{(k)} n_{kj} = \gamma^{(j)} n_{jk}$. For a hyperplane mosaic in general position, every j-face with $j \le k$ lies in precisely $2^{k-j}\binom{d-j}{d-k}$ k-faces. This leads to

$$\gamma^{(k)} n_{kj} = 2^{k-j} \binom{d-j}{d-k} \gamma^{(j)}.$$

Together with (10.34), this yields (10.35). □

We point out that Theorem 10.3.1 does not require special assumptions on the distribution of X, except the stationarity and the finiteness of the intensities of the intersection processes (the latter is satisfied in the case of a Poisson process, by Theorem 4.4.8). In contrast to this, we now restrict the distributions considerably, namely, we consider hyperplane mosaics generated by stationary Poisson hyperplane processes.

Let $\widehat{X}$ be a stationary Poisson hyperplane process in $\mathbb{R}^d$ with intensity $\widehat{\gamma} > 0$ and spherical direction distribution $\widehat{\varphi}$. We assume that it is **nondegenerate**, which means that $\widehat{\varphi}$ is not concentrated on a great subsphere. This is equivalent to the assumption that there does not exist a line to which almost surely all the hyperplanes of $\widehat{X}$ are parallel. Every realization of $\widehat{X}$ is a.s. a locally finite system of hyperplanes and induces a decomposition of $\mathbb{R}^d$ into d-dimensional cells, as explained at the beginning of this section.

Theorem 10.3.2. *If $\widehat{X}$ is a nondegenerate stationary Poisson hyperplane process in $\mathbb{R}^d$, then the system X of its induced cells is a random mosaic in general position.*

Proof. If $\widehat{X}$ is as described, then the origin 0 lies almost surely in no hyperplane of the process; it is, therefore, contained in a uniquely determined induced cell Z_0, which is again called the **zero cell**. We show first that it is almost surely bounded. Let $U \subset S^{d-1}$ denote the support of the spherical directional distribution $\widehat{\varphi}$ of $\widehat{X}$. The even measure $\widehat{\varphi}$ is not concentrated on a great subsphere, hence $0 \in \operatorname{int} \operatorname{conv} U$. By the theorem of Steinitz (see, for example, Schneider [695, p. 15]), there exist $2d$ (not necessarily distinct) points $u_1, \ldots, u_{2d} \in U$ such that $0 \in \operatorname{int} \operatorname{conv} \{u_1, \ldots, u_{2d}\}$. To each $i \in \{1, \ldots, 2d\}$, we can choose a neighborhood $U_i \subset S^{d-1}$ of u_i such that

$$0 \in \operatorname{int} \operatorname{conv} \{v_1, \ldots, v_{2d}\} \text{ for all } (v_1, \ldots, v_{2d}) \in U_1 \times \ldots \times U_{2d}. \qquad (10.37)$$

Since U is the support of $\widehat{\varphi}$, we have $\widehat{\varphi}(U_i) > 0$ for $i = 1, \ldots, 2d$. Let A_i be the set of hyperplanes $H \in A(d, d-1)$ with $0 \notin H$ for which the unit normal vector pointing away from 0 belongs to U_i, $i = 1, \ldots, 2d$. For the intensity measure $\widehat{\Theta}$ of $\widehat{X}$ we have, by (4.25),

$$\widehat{\Theta}(A_i) = \widehat{\gamma} \int_{S^{d-1}} \int_{-\infty}^{\infty} \mathbf{1}_{A_i}(u^\perp + \tau u) \, \mathrm{d}\tau \, \widehat{\varphi}(\mathrm{d}u) = \infty.$$

Since $\widehat{X}$ is a Poisson process, this implies $\mathbb{P}(\widehat{X}(A_i) = \infty) = 1$ for $i = 1, \dots, 2d$ and thus

$$\mathbb{P}(\widehat{X}(A_i) > 0 \text{ for } i = 1, \dots, 2d) = 1. \tag{10.38}$$

If $(H_1, \dots, H_{2d}) \in A_1 \times \dots \times A_{2d}$, then the zero cell induced by $\{H_1, \dots, H_{2d}\}$ is bounded, by (10.37). Hence, (10.38) implies that the zero cell Z_0 of $\widehat{X}$ is bounded almost surely. By definition, it is closed.

Next we show that Z_0 is a random closed set. Since $\mathcal{B}(\mathcal{F})$ is generated by the system $\{\mathcal{F}_G : G \in \mathcal{G}\}$ (Lemma 2.2.1), for the proof of the measurability it is sufficient to show that for every open set $G \subset \mathbb{R}^d$ the set $A := \{\omega \in \Omega : Z_0(\omega) \cap G \neq \emptyset\}$ is measurable. Let $G \in \mathcal{G}$, let $(x_i)_{i \in \mathbb{N}}$ be a dense sequence in G, and let

$$A_i := \{\omega \in \Omega : H \cap [0, x_i) = \emptyset \text{ for } H \in \widehat{X}(\omega)\}.$$

Then $A = \bigcup_{i \in \mathbb{N}} A_i$. With $\mathcal{E}_i := \{H \in A(d, d-1) : H \cap [0, x_i) \neq \emptyset\}$ we have $\mathcal{E}_i \in \mathcal{B}(A(d, d-1))$ and $A_i = \{\omega \in \Omega : \widehat{X}(\omega) \cap \mathcal{E}_i = \emptyset\}$. Thus A_i and, therefore, A are measurable. This proves that Z_0 is measurable.

Now we choose a dense sequence $(z_i)_{i \in \mathbb{N}}$ in $\mathbb{R}^d$. By the same argument as applied above to the origin, almost surely to each $i \in \mathbb{N}$ there is a uniquely determined cell Z_i containing z_i, and it is bounded. Hence, almost surely all cells Z_i, $i \in \mathbb{N}$, are bounded. Each map $Z_i : \Omega \to \mathcal{K}'$ is measurable.

The set $X(\omega) := \{Z_i(\omega) : i \in \mathbb{N}\}$ belongs to $\mathcal{F}_{lfc}(\mathcal{F}') = \mathcal{F}_{lf}(\mathcal{F}') \cap \mathcal{F}(\mathcal{K}')$. The map $X : \Omega \to \mathcal{F}_{lfc}(\mathcal{F}')$ defined in this way is measurable, since the measurability of Z_i implies that for every compact set $C \in \mathcal{C}(\mathcal{F})$ the set

$$\{\omega \in \Omega : X(\omega) \cap C = \emptyset\} = \bigcap_{i \in \mathbb{N}} \{\omega \in \Omega : Z_i(\omega) \notin C\}$$

is measurable.

It follows that X is a random mosaic in the sense of Section 10.1, since the defining properties of mosaics, including the face-to-face property, are satisfied. As $\widehat{X}$ is a stationary Poisson process, one also sees (with similar arguments to those in the proof of Theorem 4.4.5) that X is in general position.

The local finiteness of the intensity measures of the face processes $X^{(k)}$, $k \in \{0, \dots, d\}$, has already been obtained in the proof of Theorem 10.3.1. □

We remark that without an additional assumption such as the Poisson property, the boundedness of the cells in Theorem 10.3.2 cannot be proved. As an example, we consider a stationary line process in the plane with the property that with probability $\frac{1}{2}$ a realization contains only horizontal lines, and with probability $\frac{1}{2}$ it contains only vertical lines. Such a process is non-degenerate, but it does not generate a mosaic.

A random mosaic generated by a stationary Poisson hyperplane process, as in Theorem 10.3.2, is called a (stationary) **Poisson hyperplane mosaic**. The parameters of such a mosaic depend only on the intensity and the spherical directional distribution of the generating Poisson hyperplane process. We shall now derive corresponding representations of the most important quantities. For this, we use the associated zonoids introduced in Section 4.6.

To the nondegenerate stationary Poisson hyperplane process $\widehat{X}$ in $\mathbb{R}^d$ with intensity $\widehat{\gamma}$ and spherical directional distribution $\widehat{\varphi}$, the associated zonoid $\Pi_{\widehat{X}}$ is, according to (4.59), defined by its support function

$$h(\Pi_{\widehat{X}}, \cdot) = \frac{\widehat{\gamma}}{2} \int_{S^{d-1}} |\langle \cdot, v \rangle| \, \widehat{\varphi}(\mathrm{d}v). \tag{10.39}$$

The random mosaic X generated by $\widehat{X}$ is a stationary process of convex particles and, therefore, has an associated zonoid, too; this is denoted by Π_X. By (4.47), its support function is given by

$$h(\Pi_X, \cdot) = \frac{1}{2} \int_{S^{d-1}} |\langle \cdot, v \rangle| \, \overline{S}_{d-1}(X, \mathrm{d}v), \tag{10.40}$$

where

$$\overline{S}_{d-1}(X, \cdot) = \gamma^{(d)} \int_{\mathcal{K}_0} S_{d-1}(K, \cdot) \, \mathbb{Q}^{(d)}(\mathrm{d}K).$$

Here, as in Section 10.1, $\gamma^{(d)}$ is the intensity and $\mathbb{Q}^{(d)}$ is the grain distribution of $X = X^{(d)}$. We shall establish a connection between the two associated zonoids.

For a hyperplane H, we denote by $\pm u_H$ its two unit normal vectors. Let $A \in \mathcal{B}(S^{d-1})$ and $r > 0$. By Campbell's theorem,

$$\begin{aligned} &\mathbb{E} \sum_{H \in \widehat{X}} \lambda_H(rB^d) \frac{1}{2} [\mathbf{1}_A(u_H) + \mathbf{1}_A(-u_H)] \\ &= \widehat{\gamma} \int_{S^{d-1}} \int_{-\infty}^{\infty} \lambda_{u^\perp + \tau u}(rB^d) \frac{1}{2} [\mathbf{1}_A(u) + \mathbf{1}_A(-u)] \, \mathrm{d}\tau \, \widehat{\varphi}(\mathrm{d}u) \\ &= \widehat{\gamma} \lambda(rB^d) \widehat{\varphi}(A). \end{aligned} \tag{10.41}$$

For the mean normal measure $\overline{S}_{d-1}(X, \cdot)$ of the particle process X, we use the representation

$$\overline{S}_{d-1}(X, \cdot) = \lim_{r \to \infty} \frac{1}{\lambda(rB^d)} \mathbb{E} \sum_{K \in X} S_{d-1}(K \cap rB^d, \cdot), \tag{10.42}$$

which follows from Theorem 9.2.2. This yields

$$\begin{aligned} &\overline{S}_{d-1}(X, A) \\ &= \lim_{r \to \infty} \frac{1}{\lambda(rB^d)} \mathbb{E} \left(\sum_{H \in \widehat{X}} \lambda_H(rB^d) [\mathbf{1}_A(u_H) + \mathbf{1}_A(-u_H)] + O(r^{d-1}) \right), \end{aligned}$$

where the term $O(r^{d-1})$ comprises the contributions coming from the curved boundary parts of the bodies $K \cap rB^d$, $K \in X$. Together with (10.41), this gives

$$\overline{S}_{d-1}(X, \cdot) = 2\widehat{\gamma}\widehat{\varphi},$$

and this together with (10.39) and (10.40) yields

$$\Pi_X = 2\Pi_{\widehat{X}}.$$

Now, some parameters of the mosaic X can conveniently be expressed in terms of functionals of the associated zonoid $\Pi_{\widehat{X}}$.

Theorem 10.3.3. *Let X be a stationary Poisson hyperplane mosaic in $\mathbb{R}^d$, let $\widehat{\gamma}$ be the intensity and let $\Pi_{\widehat{X}}$ be the associated zonoid of the generating Poisson hyperplane process $\widehat{X}$. Then the intensities $\gamma^{(k)}$ and specific intrinsic volumes $d_j^{(k)}$ of the face processes $X^{(k)}$ of X are given by*

$$d_j^{(k)} = \binom{d-j}{d-k} V_{d-j}(\Pi_{\widehat{X}}) \tag{10.43}$$

for $0 \le j \le k \le d$, in particular for $j = 0$,

$$\gamma^{(k)} = \binom{d}{k} V_d(\Pi_{\widehat{X}}). \tag{10.44}$$

If X is isotropic, then

$$d_j^{(k)} = \binom{d-j}{d-k}\binom{d}{j} \frac{\kappa_{d-1}^{d-j}}{d^{d-j}\kappa_d^{d-j-1}\kappa_j} \widehat{\gamma}^{d-j} \tag{10.45}$$

and especially

$$\gamma^{(k)} = \binom{d}{k} \frac{\kappa_{d-1}^d}{d^d \kappa_d^{d-1}} \widehat{\gamma}^d. \tag{10.46}$$

Proof. Let $j \in \{0, \dots, d-1\}$. Let $\widehat{X}_{d-j}$ be the intersection process of order $d-j$ corresponding to the stationary Poisson hyperplane process $\widehat{X}$. Thus, $\widehat{X}_{d-j}$ is a stationary process of j-flats, and by (4.63) its intensity $\widehat{\gamma}_{d-j}$ is given by

$$\widehat{\gamma}_{d-j} = V_{d-j}(\Pi_{\widehat{X}}).$$

By Theorem 4.4.3,

$$\mathbb{E} \sum_{E \in \widehat{X}_{d-j}} \lambda_E = \widehat{\gamma}_{d-j}\lambda.$$

On the other hand,

$$\lim_{r\to\infty} \frac{1}{V_d(rB^d)} \mathbb{E} \sum_{E \in \widehat{X}_{d-j}} \lambda_E(rB^d) = \lim_{r\to\infty} \frac{1}{V_d(rB^d)} \mathbb{E} \sum_{K \in X^{(j)}} V_j(K \cap rB^d)$$

$$= d_j^{(j)}.$$

This gives $d_j^{(j)} = V_{d-j}(\Pi_{\widehat{X}})$ for $j \in \{0, \dots, d\}$ (trivially for $j = d$). Now (10.33) yields (10.43).

If X is isotropic, then $\Pi_{\widehat{X}}$ is a ball, thus $\Pi_{\widehat{X}} = rB^d$, and the radius r is determined by

$$\widehat{\gamma} = \widehat{\gamma}_1 = V_1(\Pi_{\widehat{X}}) = \frac{d\kappa_d}{\kappa_{d-1}} r,$$

by (14.8). From (10.43) and (10.44) we now get (10.45) and (10.46). □

In the isotropic case, a slight reformulation will allow easier comparison with other literature. If $Z^{(k)}$ denotes again the typical k-face of the stationary and isotropic Poisson hyperplane mosaic X, then (10.3), (10.45) and (10.46) lead to

$$\mathbb{E}V_j(Z^{(k)}) = \binom{k}{j} \left(\frac{d\kappa_d}{\kappa_{d-1}}\right)^j \frac{1}{\kappa_j \widehat{\gamma}^j}.$$

Similarly to Section 4.6, we now want to consider a hyperplane mosaic inside a convex observation window $K \in \mathcal{K}'$ with interior points. If $\mathcal{H}$ is a locally finite system of hyperplanes in $\mathbb{R}^d$, we denote by $\nu_k(\mathcal{H}, K)$ the number of k-dimensional cells that this system induces in K; more precisely, this is the number of k-faces of the mosaic generated by $\mathcal{H}$ in $\mathbb{R}^d$ that meet the interior of K. The following theorem collects some results about the expectation of the random variable

$$\nu_k(\widehat{X}, K) := \mathrm{card}\{F \in \mathcal{F}_k(X) : F \cap \mathrm{int}\, K \neq \emptyset\},$$

where $\widehat{X}$ is a stationary Poisson hyperplane process in $\mathbb{R}^d$ and X is the induced mosaic.

Theorem 10.3.4. *Let $\widehat{X}$ be a nondegenerate stationary Poisson hyperplane process in $\mathbb{R}^d$, let $K \in \mathcal{K}'$ be a convex body with interior points, and let $\nu_k(\widehat{X}, K)$ be the number of k-dimensional faces of the induced hyperplane mosaic X meeting the interior of K. Then, for $k \in \{0, \dots, d\}$ and $r > 0$,*

$$\mathbb{E}\,\nu_k(\widehat{X}, rB^d) = \sum_{j=d-k}^{d} \binom{j}{d-k} \kappa_j r^j V_j(\Pi_{\widehat{X}}).$$

For given intensity of $\widehat{X}$, this expectation is maximal if and only if $\widehat{X}$ is isotropic.

If $\widehat{X}$ is isotropic, then

$$\mathbb{E}\,\nu_k(\widehat{X}, K) = \sum_{j=d-k}^{d} \binom{j}{d-k} \left(\frac{\kappa_{d-1}}{d\kappa_d}\right)^j \kappa_j \widehat{\gamma}^j V_j(K).$$

If the intensity of $\widehat{X}$ is given, then for given mean width of K this expectation is maximal if and only if K is a ball, and for $k > 0$ and given volume of K, the expectation is minimal if and only if K is a ball.

Proof. Let $\widehat{\Theta}$ be the intensity measure of $\widehat{X}$, and let $k \in \{0, \dots, d\}$. By Theorem 3.2.3, we have (taking $A = \mathcal{F}_K \cap A(d, d-1)$)

$$\begin{aligned}&\mathbb{E}\,\nu_k(\widehat{X}, K)\\&= \mathrm{e}^{-\widehat{\Theta}(\mathcal{F}_K)} \sum_{m=1}^{\infty} \frac{1}{m!} \int_{\mathcal{F}_K} \cdots \int_{\mathcal{F}_K} \nu_k(\{H_1, \dots, H_m\}, K)\, \widehat{\Theta}(\mathrm{d}H_1) \cdots \widehat{\Theta}(\mathrm{d}H_m).\end{aligned}$$

If $H_1, \dots, H_m$ are hyperplanes in general position meeting K, then (14.67) says that

$$\nu_k(\{H_1, \dots, H_m\}, K) = \sum_{j=d-k}^{d} \binom{j}{d-k} \alpha_j(H_1, \dots, H_m).$$

Here, $\alpha_j(H_1, \dots, H_m)$ denotes the number of j-tuples of hyperplanes from $H_1, \dots, H_m$ with the property that their intersection meets $\operatorname{int} K$ (with $\alpha_0 := 1$). For $j \in \{1, \dots, d\}$, let $p_j(K, \widehat{\Theta})$ denote the probability that j independent, identically distributed random hyperplanes with distribution $\widehat{\Theta} \llcorner \mathcal{F}_K / \widehat{\Theta}(\mathcal{F}_K)$ have a common point in K, thus

$$\widehat{\Theta}(\mathcal{F}_K)^j p_j(K, \widehat{\Theta}) = \int_{\mathcal{F}_K} \cdots \int_{\mathcal{F}_K} \chi(K \cap H_1 \cap \ldots \cap H_j)\, \widehat{\Theta}(\mathrm{d}H_1) \cdots \widehat{\Theta}(\mathrm{d}H_j).$$

Then

$$\int_{\mathcal{F}_K} \cdots \int_{\mathcal{F}_K} \alpha_j(H_1, \dots, H_m)\, \widehat{\Theta}(\mathrm{d}H_1) \cdots \widehat{\Theta}(\mathrm{d}H_m) = \binom{m}{j} \widehat{\Theta}(\mathcal{F}_K)^m p_j(K, \widehat{\Theta})$$

and hence

$$\mathbb{E}\,\nu_k(\widehat{X}, K) = \sum_{j=d-k}^{d} \binom{j}{d-k} \frac{1}{j!} \widehat{\Theta}(\mathcal{F}_K)^j p_j(K, \widehat{\Theta}).$$

Let $\widehat{\gamma}$ be the intensity and $\widehat{\varphi}$ the spherical directional distribution of $\widehat{X}$. As in the proof of Theorem 4.4.8 we get

$$\begin{aligned}&\widehat{\Theta}(\mathcal{F}_K)^j p_j(K, \widehat{\Theta})\\&= \widehat{\gamma}^j \int_{S^{d-1}} \cdots \int_{S^{d-1}} V_j(K|\operatorname{lin}\{u_1, \dots, u_j\}) \nabla_j(u_1, \dots, u_j)\, \widehat{\varphi}(\mathrm{d}u_1) \cdots \widehat{\varphi}(\mathrm{d}u_j).\end{aligned}$$

In particular, we have

$$\widehat{\Theta}(\mathcal{F}_{rB^d})^j p_j(rB^d, \widehat{\Theta}) = \kappa_j r^j j! V_j(\Pi_{\widehat{X}})$$

by (14.35), hence

$$\mathbb{E}\,\nu_k(\widehat{X}, rB^d) = \sum_{j=d-k}^{d} \binom{j}{d-k} \kappa_j r^j V_j(\Pi_{\widehat{X}}).$$

If $\widehat{X}$ is isotropic, the directional distribution $\widehat{\varphi}$ is the normalized spherical Lebesgue measure. Therefore, we can use (14.39) and Theorem 6.2.2 and get

$$\widehat{\Theta}(\mathcal{F}_K)^j p_j(K, \widehat{\Theta}) = \left(\frac{\kappa_{d-1}}{d\kappa_d}\right)^j \kappa_j j! \widehat{\gamma}^j V_j(K).$$

Thus we obtain the stated identities.

The extremal properties follow from (14.31) where, in the first case, $V_1(\Pi_{\widehat{X}}) = \widehat{\gamma}$ has to be observed. □

Notes for Section 10.3

1. The main sources for results about stationary Poisson hyperplane mosaics are the thesis of Miles [517], his papers [521, 523], and the sixth chapter of the book by Matheron [462]. Poisson line mosaics in the plane were investigated by Goudsmit [283], Miles [518, 519, 527, 535], Richards [642], Solomon [731, ch. 3], Tanner [752, 753]. These papers contain many further results, which have not been treated here.

2. Equations (10.34) and (10.35) are due to Mecke [483]. The fact that they do not require special distributional assumptions points to a purely geometric kernel. In fact, there are counterparts for deterministic hyperplane systems, see Schneider [691], where also the assumption of general position is weakened. The equations (10.33) (as well as Theorem 10.1.4) go back to Weiss [807], with a different proof.

Equations (10.42) and similar representations of the measure $\overline{S}_{d-1}(X, \cdot)$ can be found in Weil [796].

Theorem 10.3.3 determines the expected values $\mathbb{E}V_j(Z^{(k)}) = d_j^{(k)}/\gamma^{(k)}$ for the typical k-face $Z^{(k)}$ of a stationary Poisson hyperplane process. For the typical cell $Z = Z^{(d)}$ (with an ergodic interpretation), a number of related expectations have been determined. Miles [517] has considered the following quantities (with different normalization). For a convex d-polytope P in $\mathbb{R}^d$ and for $0 \le j \le k \le d$, let

$$Y_{j,k}(P) := \sum_{F \in \mathcal{F}_k(P)} V_j(F).$$

In particular, $Y_{j,j}(P) = L_j(P)$ is the total j-dimensional volume of the j-faces, $Y_{j,d}(P) = V_j(P)$ is the jth intrinsic volume, and $Y_{0,k}(P) = f_k(P)$ is the number of k-faces of P. Miles has determined, also in the non-isotropic case, the expectation $\mathbb{E}Y_{j,k}(Z)$. Special cases are the expectations $\mathbb{E}V_j(Z)$, also found in Matheron [462]; further, $\mathbb{E}L_j(Z)$ and $\mathbb{E}N_k(Z)$, for which, by (10.35), no Poisson assumption is necessary. Moreover, Miles has obtained an equation equivalent to (10.49) and thus (by Theorem 10.4.1) the expectation $\mathbb{E}V_d^2(Z)$. In the isotropic case he has determined $\mathbb{E}L_r(Z_0)$ (for the zero cell Z_0), which yields $\mathbb{E}[V_d(Z)L_r(Z)]$, and finally the expectation $\mathbb{E}[L_j(Z)L_k(Z)]$ for $0 \le j,\ k \le d$. For further moments of second and higher order in dimensions two and three, we refer to the list and the references in Santaló [662, pp. 57–58, 297]; cf. also Weiss [808], Favis and Weiss [226].

Theorem 10.3.4 transfers a result of Schneider [686] from finitely many random hyperplanes to Poisson hyperplane processes.

3. Limit theorems. For stationary Poisson hyperplane processes, central limit theorems for measurements in increasing convex sampling windows were obtained by Paroux [596] (in the plane), by Heinrich, H. Schmidt and V. Schmidt [332], and by Heinrich [328].

10.4 Zero Cells and Typical Cells

In this section, we study properties of zero cells and typical cells, for general and for special stationary random mosaics, and under various aspects. Recall that the **typical cell** Z of a stationary random mosaic X is a random polytope with distribution $\mathbb{Q}^{(d)}$, the grain distribution of the particle process X. The **zero cell** Z_0 of X is defined by

$$Z_0 := \bigcup_{K \in X} \mathbf{1}_{\operatorname{int} K}(0) K$$

(observe that, due to the stationarity of X, there is almost surely a cell $K \in X$ with $0 \in \operatorname{int} K$). The zero cell, too, is a.s. a random polytope with interior points.

First we investigate relations between the typical cell and the zero cell of a stationary random mosaic in $\mathbb{R}^d$.

Theorem 10.4.1. *Let X be a stationary random mosaic in $\mathbb{R}^d$ with typical cell Z and zero cell Z_0. If $f : \mathcal{K}' \to \mathbb{R}$ is a translation invariant, nonnegative, measurable function, then*

$$\mathbb{E} f(Z_0) = \gamma^{(d)} \mathbb{E}[f(Z) V_d(Z)].$$

This can be expressed by saying that the distribution of the zero cell is, up to translations, the volume-weighted distribution of the typical cell. More precisely: the distribution of $Z_0 - c(Z_0)$ has, with respect to the distribution of Z, a density which is given by $\gamma^{(d)} V_d = V_d / \mathbb{E} V_d(Z)$.

Proof. An application of Campbell's theorem gives

$$\begin{aligned} \mathbb{E} f(Z_0) &= \mathbb{E} \sum_{K \in X} f(K) \mathbf{1}_{\operatorname{int} K}(0) \\ &= \gamma^{(d)} \int_{\mathcal{K}_0} \int_{\mathbb{R}^d} f(K + x) \mathbf{1}_{\operatorname{int}(K+x)}(0)\, \lambda(\mathrm{d}x)\, \mathbb{Q}^{(d)}(\mathrm{d}K) \\ &= \gamma^{(d)} \int_{\mathcal{K}_0} f(K) V_d(K)\, \mathbb{Q}^{(d)}(\mathrm{d}K) \\ &= \gamma^{(d)} \mathbb{E}[f(Z) V_d(Z)], \end{aligned}$$

as stated. ◻

A consequence of Theorem 10.4.1 is, for example, the fact that the zero cell Z_0 has stochastically larger volume than the typical cell Z. A more precise formulation involves the distribution functions of $V_d(Z)$ and $V_d(Z_0)$.

Theorem 10.4.2. *Let X be a stationary random mosaic in $\mathbb{R}^d$ with typical cell Z and zero cell Z_0, let F be the distribution function of $V_d(Z)$ and F_0 the distribution function of $V_d(Z_0)$. Then*

$$F_0(x) \le F(x) \quad \text{for} \quad 0 \le x < \infty.$$

Proof. For $x \ge 0$ we have

$$\begin{aligned}
\int_0^x (F(x) - F(t))\,\mathrm{d}t &= \int_0^x [\mathbb{P}(V_d(Z) \le x) - \mathbb{P}(V_d(Z) \le t)]\,\mathrm{d}t \\
&= \int_0^x \mathbb{E}[\mathbf{1}_{[0,x]}(V_d(Z)) - \mathbf{1}_{[0,t]}(V_d(Z))]\,\mathrm{d}t \\
&= \mathbb{E}\int_0^x [\mathbf{1}_{[0,x]}(V_d(Z)) - \mathbf{1}_{[0,t]}(V_d(Z))]\,\mathrm{d}t \\
&= \mathbb{E}[V_d(Z)\mathbf{1}_{[0,x]}(V_d(Z))] \\
&= (\gamma^{(d)})^{-1}\mathbb{E}\mathbf{1}_{[0,x]}(V_d(Z_0)) \\
&= F_0(x)\mathbb{E}V_d(Z),
\end{aligned}$$

by Theorem 10.4.1 and (10.4). Now

$$\mathbb{E}V_d(Z) = \int_0^\infty (1 - F(x))\,\mathrm{d}x,$$

which gives

$$\begin{aligned}
&F_0(x)\mathbb{E}V_d(Z) \\
&= \int_0^x (F(x) - F(t))\,\mathrm{d}t \\
&= F(x)\int_0^x (1 - F(t))\,\mathrm{d}t - (1 - F(x))\int_0^x F(t)\,\mathrm{d}t \\
&= F(x)\mathbb{E}V_d(Z) - F(x)\int_x^\infty (1 - F(t))\,\mathrm{d}t - (1 - F(x))\int_0^x F(t)\,\mathrm{d}t.
\end{aligned}$$

We conclude that

$$\begin{aligned}
&(F(x) - F_0(x))\mathbb{E}V_d(Z) \\
&= F(x)\int_x^\infty (1 - F(t))\,\mathrm{d}t + (1 - F(x))\int_0^x F(t)\,\mathrm{d}t \\
&\ge 0,
\end{aligned}$$

which completes the proof. □

Theorem 10.4.3. *Let X be a stationary random mosaic in $\mathbb{R}^d$ with typical cell Z and zero cell Z_0. Then, for $k \in \mathbb{N}$,*

$$\mathbb{E}V_d^k(Z_0) \geq \mathbb{E}V_d^k(Z).$$

Proof. This follows from Theorem 10.4.2 and the identities

$$\mathbb{E}V_d^k(Z_0) = k\int_0^\infty x^{k-1}(1 - F_0(x))\,\mathrm{d}x,$$
$$\mathbb{E}V_d^k(Z) = k\int_0^\infty x^{k-1}(1 - F(x))\,\mathrm{d}x$$

for kth moments of nonnegative real random variables. □

The Typical Cell of a Poisson–Delaunay Tessellation

For a stationary Poisson–Delaunay mosaic Y, it is possible to give an explicit integral expression for the distribution of the typical cell, with respect to a suitable center function. For this, we consider Y as a particle process (with all particles being d-simplices) and choose a particular center function z for d-simplices, namely the center of the sphere through the vertices. More precisely, let $\Delta^{(d)}$ be the set of d-dimensional simplices in $\mathbb{R}^d$. For $K \in \Delta^{(d)}$, let $x_0 = x_0(K), \dots, x_d = x_d(K)$ be the vertices of K, say in lexicographic order. With the notation introduced before Theorem 10.2.3, let $B^d(x_0, \dots, x_d)$ be the unique ball having $x_0, \dots, x_d$ in its boundary, and let $z(K) := z(x_0, \dots, x_d)$ be its center. The map $K \mapsto (x_0(K), \dots, x_d(K))$ is measurable. Let $\mathbb{Q}_0$ denote the grain distribution of Y with respect to the center function z. Thus, $\mathbb{Q}_0$ is a probability measure on $\Delta_0^{(d)} := \{K \in \Delta^{(d)} : z(K) = 0\}$. The following theorem describes $\mathbb{Q}_0$ according to the dependence of the intensity γ of $\widetilde{X}$, the underlying point process.

Theorem 10.4.4. *Let Y be a stationary Poisson–Delaunay mosaic in $\mathbb{R}^d$ corresponding to the intensity γ, and let $A \subset \Delta_0^{(d)}$ be a Borel set. Then*

$$\mathbb{Q}_0(A) = a_d\gamma^d \int_0^\infty \int_{S^{d-1}} \cdots \int_{S^{d-1}} \mathbf{1}_A(\operatorname{conv}\{ru_0, \dots, ru_d\})\mathrm{e}^{-\gamma\kappa_d r^d} r^{d^2-1}$$
$$\times \Delta_d(u_0, \dots, u_d)\,\sigma(\mathrm{d}u_0)\cdots\sigma(\mathrm{d}u_d)\,\mathrm{d}r$$

with

$$a_d := \frac{d^2}{2^{d+1}\pi^{\frac{d-1}{2}}} \frac{\Gamma\left(\frac{d^2}{2}\right)}{\Gamma\left(\frac{d^2+1}{2}\right)} \left[\frac{\Gamma\left(\frac{d+1}{2}\right)}{\Gamma\left(1+\frac{d}{2}\right)}\right]^d.$$

Proof. From the definition of $\mathbb{Q}_0$ and from Campbell's theorem we get

$$\begin{aligned}
&\beta^{(d)}\mathbb{Q}_0(A)\\
&= \mathbb{E}\sum_{K\in Y} \mathbf{1}_A(K - z(K))\mathbf{1}_{C^d}(z(K))\\
&= \frac{1}{(d+1)!}\,\mathbb{E}\sum_{(x_0,\dots,x_d)\in \widetilde{X}^{d+1}_{\neq}} \mathbf{1}_A(\operatorname{conv}\{x_0,\dots,x_d\} - z(x_0,\dots,x_d))\\
&\qquad \times \mathbf{1}_{C^d}(z(x_0,\dots,x_d))\mathbf{1}\{\widetilde{X}\cap \operatorname{int} B^d(x_0,\dots,x_d)=\emptyset\}.
\end{aligned}$$

As in the proof of Theorem 10.2.4, we apply the Slivnyak–Mecke formula (Corollary 3.2.3) to the Poisson process $\widetilde{X}$. Using the transformation theorem 7.3.1, we obtain

$$\begin{aligned}
&\beta^{(d)}\mathbb{Q}_0(A)\\
&= \frac{\gamma^{d+1}}{(d+1)!}\int_{\mathbb{R}^d}\cdots\int_{\mathbb{R}^d} \mathbf{1}_A(\operatorname{conv}\{x_0,\dots,x_d\} - z(x_0,\dots,x_d))\\
&\quad \times \mathbf{1}_{C^d}(z(x_0,\dots,x_d))\mathbb{P}(\widetilde{X}\cap B^d(x_0,\dots,x_d)=\emptyset)\,\lambda(\mathrm{d}x_0)\cdots\lambda(\mathrm{d}x_d)\\
&= \frac{\gamma^{d+1}}{(d+1)!}d!\int_{\mathbb{R}^d}\int_0^\infty\int_{S^{d-1}}\cdots\int_{S^{d-1}} \mathbf{1}_A(\operatorname{conv}\{ru_0,\dots,ru_d\})\mathbf{1}_{C^d}(z)\\
&\quad \times \mathrm{e}^{-\gamma\kappa_d r^d} r^{d^2-1}\Delta_d(u_0,\dots,u_d)\,\sigma(\mathrm{d}u_0)\cdots\sigma(\mathrm{d}u_d)\,\mathrm{d}r\,\lambda(\mathrm{d}z).
\end{aligned}$$

Here $\beta^{(d)} = \gamma^{(0)}$ by Theorem 10.2.8, and this value is given by (10.31). This completes the proof. □

A random closed set Z with distribution $\mathbb{Q}_0$ is called a **Poisson–Delaunay simplex** (corresponding to the intensity γ). By Theorem 10.4.4, the distributions of all geometric quantities of Z are determined, in principle. Their explicit calculation, however, will only be possible in special cases. As an example, we consider the volume of a Poisson–Delaunay simplex. For this, all the moments can be computed.

Theorem 10.4.5. *Let Z be the Poisson–Delaunay simplex in $\mathbb{R}^d$ corresponding to the intensity γ. Then*

$$\mathbb{E}V_d(Z)^k = a_d S(d,d,k+1)\frac{(d+k-1)!}{d\kappa_d^{d+k}}\frac{1}{\gamma^k}$$

for $k = 1, 2, \dots$, where $S(d,d,k+1)$ is given by Theorem 8.2.3.

Proof. By Theorem 10.4.4,

$$\begin{aligned}
\mathbb{E}V_d(Z)^k &= a_d\gamma^d\int_0^\infty\int_{S^{d-1}}\cdots\int_{S^{d-1}} \mathrm{e}^{-\gamma\kappa_d r^d} r^{d^2+dk-1}\Delta_d(u_0,\dots,u_d)^{k+1}\\
&\qquad \times \sigma(\mathrm{d}u_0)\cdots\sigma(\mathrm{d}u_d)\,\mathrm{d}r\\
&= a_d\gamma^d S(d,d,k+1)\int_0^\infty \mathrm{e}^{-\gamma\kappa_d r^d} r^{d^2+dk-1}\,\mathrm{d}r,
\end{aligned}$$

from which the assertion follows. □

In the rest of this section, we restrict ourselves to a stationary Poisson hyperplane mosaic X. The zero cell Z_0 of X is known as the **Poisson zero polytope** or **Crofton polytope**. In contrast to this, the typical cell of X is called the **Poisson polytope**. We want to determine some distributional properties or parameters of the Poisson polytope or the Poisson zero polytope. We assume again that the mosaic X is generated by the nondegenerate stationary Poisson hyperplane process $\widehat{X}$.

Typical Cells and Zero Cells of Poisson Hyperplane Tessellations

We want to represent the distribution of the typical cell of a stationary Poisson hyperplane mosaic in a similar manner to that for Poisson–Delaunay mosaics in Theorem 10.4.4. There can, however, only be a partial analogy, since the Poisson polytope may have arbitrarily many facets, therefore the process $\widehat{X}$ cannot be eliminated from the representation. The analogy consists in the fact that the center of the inball is used as a center function, and the inradius enters the representation. At first, we restrict ourselves to the isotropic case; a possible extension is explained after the proof. First we collect some notation for the subsequent proof.

For $u \in S^{d-1}$ and $\tau \in \mathbb{R}$, we write

$$H^-(u,\tau) := \{x \in \mathbb{R}^d : \langle x, u\rangle \le \tau\}$$

for the closed halfspace with outer normal vector u that is bounded by the hyperplane $H(u,\tau)$. If H is a hyperplane and $z \in \mathbb{R}^d \setminus H$, then H_z^+ denotes the closed halfspace bounded by H that contains z.

Let $H_0, \dots, H_d$ be hyperplanes in $\mathbb{R}^d$ in general position (that is, with linearly independent normal vectors and not all passing through one point). Let $\Delta(H_0,\dots,H_d)$ be the unique simplex for which $H_0,\dots,H_d$ are the facet hyperplanes. We denote by $z(H_0,\dots,H_d)$ the center and by $r(H_0,\dots,H_d)$ the radius of its inball. We denote by $B^0(z,r)$ the open ball with center z and radius r. The set $\mathsf{P} \subset (S^{d-1})^{d+1}$ is the set of all $(d+1)$-tuples of unit vectors not lying in a closed hemisphere.

Theorem 10.4.6. *Let $\widehat{X}$ be a stationary, isotropic Poisson hyperplane process in $\mathbb{R}^d$ of intensity $\widehat{\gamma}$, and let $\mathbb{Q}_0$ be the distribution of the typical cell of the induced hyperplane mosaic X with respect to the inball center as center function. Then, for Borel sets $A \in \mathcal{B}(\mathcal{K})$,*

$$\mathbb{Q}_0(A) = \frac{1}{(d+1)\gamma^{(d)}}\left(\frac{\widehat{\gamma}}{\omega_d}\right)^{d+1} \int_0^\infty \int_{S^{d-1}} \cdots \int_{S^{d-1}} \mathrm{e}^{-2\widehat{\gamma} r}$$
$$\times\ \mathbb{P}\left(\bigcap_{H \in \widehat{X} \cap \mathcal{F}_{B^0(0,r)}} H_0^+ \cap \bigcap_{j=0}^{d} H^-(u_j, r) \in A\right)$$

$$\times\ \Delta_d(u_0,\dots,u_d)\mathbf{1}_{\mathsf{P}}(u_0,\dots,u_d)\,\sigma(\mathrm{d}u_0)\cdots\sigma(\mathrm{d}u_d)\,\mathrm{d}r,$$

where the cell intensity $\gamma^{(d)}$ *is given by* (10.46).

Proof. As in the proof of Theorem 4.4.5 one shows that almost surely any $d+1$ hyperplanes of $\widehat{X}$ are in general position. Therefore, the inballs of the cells of X are almost surely unique. It is also easy to show that almost surely every inball of a cell is touched by precisely $d+1$ hyperplanes of $\widehat{X}$. In the proof, we denote the inball center by ζ. In the subsequent formulas, the arguments $H_0,\dots,H_d$ of the functions Δ, z, r are omitted, but have to be kept in mind. For $A \in \mathcal{B}(\mathcal{K})$ we obtain, using the Slivnyak–Mecke formula (Corollary 3.2.3),

$$\begin{aligned}\gamma^{(d)}\mathbb{Q}_0(A) &= \mathbb{E}\sum_{K\in X}\mathbf{1}_A(K-\zeta(K))\mathbf{1}_{C^d}(\zeta(K))\\
&= \frac{1}{(d+1)!}\,\mathbb{E}\sum_{(H_0,\dots,H_d)\in\widehat{X}^{d+1}_{\neq}}\mathbf{1}\{\widehat{X}\cap\mathcal{F}_{B^0(z,r)}=\emptyset\}\mathbf{1}_{C^d}(z)\\
&\quad\times\ \mathbf{1}\left\{\bigcap_{H\in\widehat{X}\cap\mathcal{F}^{B^0(z,r)}}H_z^+ - z\in A\right\}\\
&= \frac{1}{(d+1)!}\int_{A(d,d-1)}\cdots\int_{A(d,d-1)}\mathbb{E}\,\mathbf{1}\{\widehat{X}\cap\mathcal{F}_{B^0(z,r)}=\emptyset\}\mathbf{1}_{C^d}(z)\\
&\quad\times\ \mathbf{1}\left\{\bigcap_{H\in\widehat{X}\cap\mathcal{F}^{B^0(z,r)}}(\Delta\cap H_z^+) - z\in A\right\}\widehat{\Theta}(\mathrm{d}H_0)\cdots\widehat{\Theta}(\mathrm{d}H_d),\end{aligned}$$

where $\widehat{\Theta}$ is the intensity measure of $\widehat{X}$. Since $\widehat{X}\cap\mathcal{F}_{B^0(z,r)}$ and $\widehat{X}\cap\mathcal{F}^{B^0(z,r)}$ are independent, and

$$\mathbb{P}(\widehat{X}\cap\mathcal{F}_{B^0(z,r)}=\emptyset)=\exp\{-\widehat{\Theta}(\mathcal{F}_{B^0(z,r)})\}=\mathrm{e}^{-2\widehat{\gamma}r},$$

we obtain

$$\begin{aligned}&\gamma^{(d)}\mathbb{Q}_0(A)\\
&= \frac{1}{(d+1)!}\int_{A(d,d-1)}\cdots\int_{A(d,d-1)}\mathrm{e}^{-2\widehat{\gamma}r}\mathbf{1}_{C^d}(z)\\
&\quad\times\ \mathbb{P}\left(\bigcap_{H\in\widehat{X}\cap\mathcal{F}^{B^0(z,r)}}(\Delta\cap H_z^+) - z\in A\right)\widehat{\Theta}(\mathrm{d}H_0)\cdots\widehat{\Theta}(\mathrm{d}H_d).\end{aligned}$$

Recall that Δ, z, r all have the arguments $H_0,\dots,H_d$; further, $\widehat{\Theta}=\widehat{\gamma}\mu_{d-1}$. Now we apply the transformation of Theorem 7.3.2 (r, z then becoming independent variables). Making use of the fact that, due to the stationarity of $\widehat{X}$, the events

$$\left(\bigcap_{H\in\widehat{X}\cap\mathcal{F}_{B^0(z,r)}} H_z^+ \cap \bigcap_{j=0}^{d} H^-(u_j,\langle z,u_j\rangle + r)\right) - z \in A$$

and

$$\bigcap_{H\in\widehat{X}\cap\mathcal{F}_{B^0(0,r)}} H_0^+ \cap \bigcap_{j=0}^{d} H^-(u_j,r) \in A$$

are stochastically equivalent, we obtain the stated result. □

In the first step of the proof, essential use was made of the fact that almost surely every cell of X is in one-to-one correpondence with a $(d+1)$-tuple of hyperplanes of X touching the inball of the cell. This is not the case for general stationary Poisson hyperplane mosaics. However, the above theorem and its proof extend easily to the non-isotropic case if it is assumed that the directional distribution of the generating hyperplane process is absolutely continuous with respect to the invariant measure μ_{d-1}.

We derive now a second representation of the typical cell of a stationary Poisson hyperplane mosaic X, using as center function the highest vertex with respect to a given height function. Let $\xi \in S^{d-1}$ be fixed. For a polyhedral set P, the **highest vertex** $t(P)$ in direction ξ is the vertex of P at which $\langle \xi, \cdot \rangle$ attains its maximum on P, if this vertex exists and is unique. To discuss the uniqueness, we say that the vector ξ is **admissible** for a system $\mathcal{H}$ of hyperplanes if $\mathcal{H}$ is in general position and any $d-1$ different hyperplanes from $\mathcal{H}$ have normal vectors which together with ξ are linearly independent. Let ξ be admissible for the hyperplanes $H_1,\dots,H_d$. We denote by $s = s(H_1,\dots,H_d)$ the intersection point of $H_1,\dots,H_d$. The hyperplanes $H_1,\dots,H_d$ decompose $\mathbb{R}^d$ into 2^d simplicial cones with apex s. For precisely one of these cones, the point s is the highest vertex; let $T_\xi(H_1,\dots,H_d)$ be this cone. In fact, there is a unique choice of unit normal vectors $u_1,\dots,u_d$ of $H_1,\dots,H_d$ such that $\xi \in \mathrm{pos}\{u_1,\dots,u_d\}$, then

$$T_\xi(H_1,\dots,H_d) = \bigcap_{i=1}^{d} H^-(u_i,\langle s,u_i\rangle).$$

For linearly independent unit vectors $u_1,\dots,u_{d-1}$, let $w(u_1,\dots,u_{d-1})$ be the unit vector orthogonal to $\mathrm{lin}\{u_1,\dots,u_{d-1}\}$ and such that the d-tuple $(u_1,\dots,u_{d-1},w(u_1,\dots,u_{d-1}))$ is positively oriented. We denote by $w(\widehat{\varphi}^{d-1})$ the image measure of the product measure $\widehat{\varphi}^{d-1}$, restricted to linearly independent $(d-1)$-tuples, under the mapping w. We say that ξ is **admissible** for $\widehat{\varphi}$ if $w(\widehat{\varphi}^{d-1})(\xi^\perp \cap S^{d-1}) = 0$. If ξ is admissible for $\widehat{\varphi}$, then a.s. ξ is admissible for $\widehat{X}$ (this can be proved by similar arguments to those in the proof of Theorem 4.4.5), hence all cells of the tessellation X have a unique highest vertex. Clearly, there are vectors that are admissible for $\widehat{\varphi}$.

We define a probability measure on the space of d-tuples of hyperplanes through 0. For $A \in \mathcal{B}(G(d,d-1)^d)$, let

$$\phi_d(A) := \frac{\widehat{\gamma}^d}{d!\widehat{\gamma}_d} \int_{(S^{d-1})^d} \mathbf{1}_A(u_1^\perp, \dots, u_d^\perp) \nabla_d(u_1, \dots, u_d)\, \widehat{\varphi}^d(\mathrm{d}(u_1, \dots, u_d)).$$

Here $\widehat{\gamma}_d$ is the intensity of the intersection process of order d of $\widehat{X}$. It follows from Theorem 4.4.8 that ϕ_d is indeed a probability measure.

Theorem 10.4.7. *Let $\widehat{X}$ be a nondegenerate stationary Poisson hyperplane process in $\mathbb{R}^d$ with intensity $\widehat{\gamma}$ and spherical directional distribution $\widehat{\varphi}$. Let $\xi \in S^{d-1}$ be a vector which is admissible for $\widehat{\varphi}$, and let $\mathbb{Q}_0$ be the distribution of the typical cell of the induced hyperplane mosaic X with respect to the highest vertex in direction ξ as center function. Then, for Borel sets $A \in \mathcal{B}(\mathcal{K})$,*

$$\mathbb{Q}_0(A) = \int_{G(d,d-1)^d} \mathbb{P}\left(Z_0 \cap T_\xi(H_1, \dots, H_d) \in A\right) \phi_d(\mathrm{d}(H_1, \dots, H_d)).$$

Proof. Let P be a cell of X. Its highest vertex $t(P)$ is the intersection of d hyperplanes $H_1, \dots, H_d$ of $\widehat{X}$ with linearly independent normal vectors, thus $t(P) = s(H_1, \dots, H_d)$ and

$$P = \bigcap_{H \in \widehat{X} \setminus \{H_1, \dots, H_d\}} H^+_{s(H_1,\dots,H_d)} \cap T_\xi(H_1, \dots, H_d). \tag{10.47}$$

Conversely, almost surely for every choice of different hyperplanes $H_1, \dots, H_d$ from $\widehat{X}$, the right side of (10.47) is a cell of X, and $s(H_1, \dots, H_d)$ is its highest vertex.

We define $\mathbf{1}_{C^d}(s(H_1, \dots, H_d)) := 0$ if the normal vectors of $H_1, \dots, H_d$ are not linearly independent. In the subsequent formulas, the arguments $H_1, \dots, H_d$ of s and T_ξ are omitted, but have to be kept in mind. For Borel sets $A \in \mathcal{B}(\mathcal{K})$ we obtain, again using the Slivnyak–Mecke formula,

$$\begin{aligned}
&\gamma^{(d)}\mathbb{Q}_0(A) \\
&= \mathbb{E} \sum_{P \in X} \mathbf{1}_A(P - t(P))\mathbf{1}_{C^d}(t(P)) \\
&= \frac{1}{d!}\mathbb{E} \sum_{(H_1,\dots,H_d) \in \widehat{X}^d_{\neq}} \mathbf{1}_A\left(\bigcap_{H \in \widehat{X} \setminus \{H_1,\dots,H_d\}} (H_s^+ \cap T_\xi) - s \right) \mathbf{1}_{C^d}(s) \\
&= \frac{1}{d!} \int_{A(d,d-1)} \cdots \int_{A(d,d-1)} \mathbb{E}\mathbf{1}_A\left(\bigcap_{H \in \widehat{X}} (H_s^+ \cap T_\xi) - s \right) \mathbf{1}_{C^d}(s) \\
&\quad \times \widehat{\Theta}(\mathrm{d}H_1) \cdots \widehat{\Theta}(\mathrm{d}H_d) \\
&= \frac{1}{d!} \int_{A(d,d-1)} \cdots \int_{A(d,d-1)} \mathbb{P}(Z_0 \cap (T_\xi - s) \in A)\mathbf{1}_{C^d}(s)\, \widehat{\Theta}(\mathrm{d}H_1) \cdots \widehat{\Theta}(\mathrm{d}H_d),
\end{aligned}$$

where we have used the stationarity of $\widehat{X}$ and the fact that $\bigcap_{H\in\widehat{X}} H_0^+$ is the zero cell Z_0 of X. We insert the representation of the intensity measure given by (4.33). Further, we observe that $\gamma^{(d)} = \gamma^{(0)}$ by Theorem 10.3.1 and $\gamma^{(0)} = \widehat{\gamma}_d$. Thus, we obtain

$$\begin{aligned}\mathbb{Q}_0(A) = \frac{\widehat{\gamma}^d}{d!\widehat{\gamma}_d} \int_{(S^{d-1})^d} \int_{-\infty}^{\infty} \cdots \int_{-\infty}^{\infty} \mathbb{P}\left(Z_0 \cap T_\xi(u_1^\perp, \dots, u_d^\perp) \in A\right) \\ \times \mathbf{1}_{C^d}(s(H(u_1,t_1),\dots,H(u_d,t_d)))\mathbf{1}\{\nabla_d(u_1,\dots,u_d) \neq 0\} \\ \times \, \mathrm{d}t_1 \cdots \mathrm{d}t_d \, \widehat{\varphi}^d(\mathrm{d}(u_1,\dots,u_d)).\end{aligned}$$

For fixed linearly independent unit vectors $u_1, \dots, u_d$, let

$$F(t_1,\dots,t_d) := s(H(u_1,t_1),\dots,H(u_d,t_d)).$$

This defines a bijective mapping F from $\mathbb{R}^d$ to $\mathbb{R}^d$. Its inverse has Jacobian $\nabla_d(u_1,\dots,u_d)$. Therefore, we obtain

$$\begin{aligned}\mathbb{Q}_0(A) &= \frac{\widehat{\gamma}^d}{d!\widehat{\gamma}_d} \int_{(S^{d-1})^d} \mathbb{P}\left(Z_0 \cap T_\xi(u_1^\perp, \dots, u_d^\perp) \in A\right) \nabla_d(u_1,\dots,u_d) \\ &\quad \times \widehat{\varphi}^d(\mathrm{d}(u_1,\dots,u_d)) \\ &= \int_{G(d,d-1)^d} \mathbb{P}\left(Z_0 \cap T_\xi(H_1 \dots, H_d) \in A\right) \phi_d(\mathrm{d}(H_1,\dots,H_d)),\end{aligned}$$

which completes the proof. □

Corollary 10.4.1. *Under the assumptions of Theorem* 10.4.7, *there is a random polytope Z' stochastically equivalent to the typical cell Z of X such that $Z' \subset Z_0$ a.s.*

Proof. Let $(H_1,\dots,H_d)$ be a random element of $G(d,d-1)^d$ with distribution ϕ_d such that $(H_1,\dots,H_d)$ and $\widehat{X}$ are stochastically independent. Then it follows from Theorem 10.4.7 that the random polytope $Z' := Z_0 \cap T_\xi(H_1,\dots,H_d)$ ist stochastically equivalent to Z. □

From Theorem 10.4.6, we can obtain the distribution of the inradius of the typical cell, but only under the conditions on the directional distribution under which that theorem holds. The following alternative approach works for general nondegenerate stationary Poisson hyperplane processes. Recall that the inradius $r(K)$ of a convex body K is the maximal radius of the balls contained in K. In general, a maximal ball contained in K need neither be unique nor be touched by $d+1$ supporting hyperplanes of K.

Theorem 10.4.8. *Let $\widehat{X}$ be a nondegenerate stationary Poisson hyperplane process in $\mathbb{R}^d$ of intensity $\widehat{\gamma}$, and let Z be the typical cell of the induced hyperplane mosaic X. Then*

$$\mathbb{P}(r(Z) \leq a) = 1 - \mathrm{e}^{-2\widehat{\gamma}a} \qquad \textit{for } a \geq 0.$$

Proof. Let $a \geq 0$. First we prove a particular reproduction property of the Poisson hyperplane process $\widehat{X}$. As always, $(\Omega, \mathbf{A}, \mathbb{P})$ denotes the underlying probability space. We put

$$\Omega_a := \{\omega \in \Omega : \widehat{X}(\omega)(\mathcal{F}_{aB^d}) = 0\},$$

then $\mathbb{P}(\Omega_a) = \mathrm{e}^{-2\widehat{\gamma}a}$. For $\omega \in \Omega_a$, every hyperplane of $\widehat{X}(\omega)$ has a distance from the origin 0 that is larger than a, hence it can be represented in the form $H(u,\tau) = \{x \in \mathbb{R}^d : \langle x, u\rangle = \tau\}$ with $u \in S^{d-1}$ and $\tau > a$. For such hyperplanes, we put $T_a H(u,\tau) := H(u, \tau - a)$. On the probability space $(\Omega_a, \mathbf{A}_a, \mathbb{P}_a)$, where $\mathbf{A}_a$ is the trace σ-algebra of $\mathbf{A}$ on Ω_a and $\mathbb{P}_a := \mathrm{e}^{2\widehat{\gamma}a}\mathbb{P} \llcorner \Omega_a$, we define a simple hyperplane process $\widehat{X}_a$ by

$$\widehat{X}_a(\omega) := \{T_a H : H \in X(\omega)\}$$

(where, as usual, simple counting measures are identified with their supports). We shall show that $\widehat{X}_a$ and $\widehat{X}$ are stochastically equivalent. For the proof, let $A \in \mathcal{B}(A(d,d-1))$ and suppose, without loss of generality, that $0 \notin H$ for all $H \in A$. For $k \in \mathbb{N}_0$ we have, due to the independence properties of Poisson processes (Theorem 3.2.2(a)), that

$$\begin{aligned}
\mathbb{P}_a(\widehat{X}_a(A) = k) &= \mathrm{e}^{2\widehat{\gamma}a}\mathbb{P}\left(\widehat{X}\left(T_a^{-1}(A)\right) = k,\ \widehat{X}(\mathcal{F}_{aB^d}) = 0\right)\\
&= \mathbb{P}\left(\widehat{X}\left(T_a^{-1}(A)\right) = k\right)\\
&= \mathrm{e}^{-\widehat{\Theta}(T_a^{-1}(A))}\frac{\widehat{\Theta}\left(T_a^{-1}(A)\right)^k}{k!}.
\end{aligned}$$

Let $\widehat{\varphi}$ be the spherical directional distribution of $\widehat{X}$. From

$$\begin{aligned}
&\widehat{\Theta}\left(T_a^{-1}(A)\right)\\
&= 2\widehat{\gamma}\int_{S^{d-1}}\int_a^\infty \mathbf{1}_{T_a^{-1}(A)}(H(u,\tau))\,\mathrm{d}\tau\,\widehat{\varphi}(\mathrm{d}u)\\
&= 2\widehat{\gamma}\int_{S^{d-1}}\int_a^\infty \mathbf{1}_A(H(u,\tau-a))\,\mathrm{d}\tau\,\widehat{\varphi}(\mathrm{d}u)\\
&= 2\widehat{\gamma}\int_{S^{d-1}}\int_0^\infty \mathbf{1}_A(H(u,\tau))\,\mathrm{d}\tau\,\widehat{\varphi}(\mathrm{d}u)\\
&= \widehat{\Theta}(A)
\end{aligned}$$

we get $\mathbb{P}_a(\widehat{X}_a(A) = k) = \mathbb{P}(\widehat{X}(A) = k)$ and hence $\widehat{X}_a \stackrel{\mathcal{D}}{=} \widehat{X}$, as stated.

Now we replace every hyperplane $H \in \widehat{X}$ by the strip $H_a := H + aB^d$. The connected components of the complement of $\bigcup_{H\in\widehat{X}} H_a$ are open polyhedral sets; their closures are called the **cells induced by** $\widehat{X}$ **and** a. The system

X_a of the cells induced by $\widehat{X}$ and a is a stationary particle process (but not a mosaic, if $a > 0$). Let $\gamma_a^{(d)}$ denote the intensity and $\mathbb{Q}_a^{(d)}$ the grain distribution of X_a. The **typical cell** $Z^{(a)}$ of X_a is defined as the random polytope with distribution $\mathbb{Q}_a^{(d)}$ (thus $X_0 = X$, $\gamma_0^{(d)} = \gamma^{(d)}$, $Z^{(0)} = Z$). Conditionally upon $0 \notin \bigcup_{H \in \widehat{X}(\omega)} H_a$ (which is equivalent to $\omega \in \Omega_a$), the **zero cell** $Z_0^{(a)}$ of X_a is defined as the uniquely determined polytope $P \in X_a$ with $0 \in P$.

Let f be a translation invariant, nonnegative, measurable function on $\mathcal{K}'$. By Campbell's theorem,

$$\begin{aligned}\int_{\Omega_a} f(Z_0^{(a)})\,\mathrm{d}\mathbb{P} &= \mathbb{E} \sum_{K \in X_a} f(K)\mathbf{1}_K(0)\\ &= \gamma_a^{(d)} \int_{\mathcal{K}_0} \int_{\mathbb{R}^d} f(K+x)\mathbf{1}_{K+x}(0)\,\lambda(\mathrm{d}x)\,\mathbb{Q}_a^{(d)}(\mathrm{d}K)\\ &= \gamma_a^{(d)} \int_{\mathcal{K}_0} f(K)V_d(K)\,\mathbb{Q}_a^{(d)}(\mathrm{d}K).\end{aligned}$$

Choosing $f = 1/V_d$, we obtain

$$\gamma_a^{(d)} = \int_{\Omega_a} V_d^{-1}(Z_0^{(a)})\,\mathrm{d}\mathbb{P} = \mathrm{e}^{-2\widehat{\gamma}a} \int_{\Omega_a} V_d^{-1}(Z_0^{(a)})\,\mathrm{d}\mathbb{P}_a.$$

The case $a = 0$ reads

$$\gamma^{(d)} = \int_{\Omega} V_d^{-1}(Z_0)\,\mathrm{d}\mathbb{P}.$$

Due to the stochastic equivalence of $\widehat{X}_a$ and $\widehat{X}$ shown above, the random polytopes $Z_0^{(a)}$ (defined on $(\Omega_a, \mathbf{A} \cap \Omega_a, \mathbb{P}_a)$) and Z_0 have the same distribution, hence

$$\int_{\Omega_a} V_d^{-1}(Z_0^{(a)})\,\mathrm{d}\mathbb{P}_a = \int_{\Omega} V_d^{-1}(Z_0)\,\mathrm{d}\mathbb{P}.$$

This gives

$$\gamma_a^{(d)} = \mathrm{e}^{-2\widehat{\gamma}a}\gamma^{(d)}. \tag{10.48}$$

In the following, we use as a center function the incenter. For a convex body $K \in \mathcal{K}'$, the **incenter** $z(K)$ is the center of the inball of K, if this is unique. If the inball is not uniquely determined, we define $z(K)$ as the circumcenter of the set of centers of all inballs of K; the latter set is a convex body and hence contains its circumcenter.

Now let $B \in \mathcal{B}(\mathbb{R}^d)$ and $\lambda(B) = 1$. We have

$$\begin{aligned}\mathbb{P}(r(Z) > a) &= \int_{\mathcal{K}_0} \mathbf{1}_{(a,\infty)}(r(K))\,\mathbb{Q}_0^{(d)}(\mathrm{d}K)\\ &= \frac{1}{\gamma^{(d)}}\,\mathbb{E} \sum_{K \in X,\, z(K) \in B} \mathbf{1}_{(a,\infty)}(r(K))\end{aligned}$$

by Theorem 4.1.3. By Theorem 4.2.1, the replacement of the circumcenter by the incenter is legitimate, since the inradius is translation invariant. Now, the cells $K \in X$ with $r(K) > a$ are in bijective correspondence with the cells $K_a \in X_a$, so that $z(K_a) = z(K)$ (and $r(K_a) = r(K) - a$). It follows that

$$\mathbb{P}(r(Z) > a) = \frac{1}{\gamma^{(d)}} \sum_{K_a \in X_a,\, z(K_a) \in B} 1 = \frac{1}{\gamma^{(d)}} \gamma_a^{(d)} = \mathrm{e}^{-2\widehat{\gamma} a},$$

by (10.48). □

Parameters of the Poisson Zero Polytope

For the Poisson zero polytope, some parameters can be computed explicitly. The expected volume of the zero cell can be determined in the same way as the mean visible volume of a stationary Boolean model was computed (Theorem 4.6.1). For the radius function $\rho(Z_0, \cdot)$ of Z_0 we have, for $u \in S^{d-1}$ and $r > 0$,

$$\begin{aligned}\mathbb{P}(\rho(Z_0, u) \le r) &= \mathbb{P}(\widehat{X}(\mathcal{F}_{[0,ru]}) > 0)\\ &= 1 - \exp\left(-\mathbb{E}\widehat{X}(\mathcal{F}_{[0,ru]})\right)\\ &= 1 - \exp\left(-2rh(\Pi_{\widehat{X}}, u)\right),\end{aligned}$$

by (4.60). Thus, $\rho(Z_0, u)$ has an exponential distribution with parameter $2h(\Pi_{\widehat{X}}, u)$, and as in the proof of Theorem 4.6.1 we get

$$\mathbb{E}V_d(Z_0) = 2^{-d} d! V_d(\Pi^o_{\widehat{X}}). \tag{10.49}$$

Further quantities for which the expected values can be determined, are the total k-dimensional volumes of the k-faces of the zero cell. For a polytope P and for $k \in \{0, \dots, d\}$, let

$$\mathrm{skel}_k P := \bigcup_{F \in \mathcal{F}_k(P)} F$$

be the k-**skeleton** of P, and let $\mathcal{H}^k$ denote the k-dimensional Hausdorff measure. We put

$$L_k(P) := \mathcal{H}^k(\mathrm{skel}_k P) = \sum_{F \in \mathcal{F}_k(P)} V_k(F).$$

To determine the expected value of $L_{d-k}(Z_0)$, we note that almost surely each $(d-k)$-face of Z_0 is the intersection of Z_0 with precisely k hyperplanes of $\widehat{X}$. We use the Slivnyak–Mecke formula (Corollary 3.2.3) and the decomposition (4.33) of the intensity measure $\widehat{\Theta}$ of $\widehat{X}$ (which has intensity $\widehat{\gamma}$ and spherical directional distribution $\widehat{\varphi}$). Thus we obtain

$$\begin{aligned}
&\mathbb{E}L_{d-k}(Z_0)\\
&= \frac{1}{k!}\,\mathbb{E} \sum_{(H_1,\dots,H_k)\in\widehat{X}^k_{\neq}} \mathcal{H}^{d-k}(Z_0\cap H_1\cap\ldots\cap H_k)\\
&= \frac{1}{k!}\int_{A(d,d-1)}\cdots\int_{A(d,d-1)} \mathbb{E}\,\mathcal{H}^{d-k}(Z_0\cap H_1\cap\ldots\cap H_k)\,\widehat{\Theta}(\mathrm{d}H_1)\cdots\widehat{\Theta}(\mathrm{d}H_k)\\
&= \frac{\widehat{\gamma}^k}{k!}\int_{S^{d-1}}\cdots\int_{S^{d-1}}\int_{-\infty}^{\infty}\cdots\int_{-\infty}^{\infty} \mathbb{E}\,\mathcal{H}^{d-k}(Z_0\cap H(u_1,\tau_1)\cap\ldots\cap H(u_k,\tau_k))\\
&\quad\times \mathrm{d}\tau_1\cdots\mathrm{d}\tau_k\,\widehat{\varphi}(\mathrm{d}u_1)\cdots\widehat{\varphi}(\mathrm{d}u_k)\\
&= \frac{\widehat{\gamma}^k}{k!}\,\mathbb{E}V_d(Z_0)\int_{S^{d-1}}\cdots\int_{S^{d-1}} \nabla_k(u_1,\dots,u_k)\,\widehat{\varphi}(\mathrm{d}u_1)\cdots\widehat{\varphi}(\mathrm{d}u_k)\\
&= d^{(d-k)}_{d-k}\,\mathbb{E}V_d(Z_0),
\end{aligned}$$

where we have used (14.35) and (10.43).

We collect the obtained results, together with corresponding inequalities.

Theorem 10.4.9. *Let $\widehat{X}$ be a nondegenerate stationary Poisson hyperplane process in $\mathbb{R}^d$ of intensity $\widehat{\gamma}$, and let Z_0 be the zero cell of the induced hyperplane mosaic. Then*

$$\mathbb{E}V_d(Z_0) = 2^{-d}d!V_d(\Pi^o_{\widehat{X}}) \geq d!\kappa_d\left(\frac{2\kappa_{d-1}}{d\kappa_d}\,\widehat{\gamma}\right)^{-d}, \tag{10.50}$$

with equality if and only if $\widehat{X}$ is isotropic.

For $k=0,\dots,d-1$,

$$\mathbb{E}\,L_k(Z_0) = d^{(k)}_k\,\mathbb{E}V_d(Z_0) = 2^{-d}d!V_{d-k}(\Pi_{\widehat{X}})V_d(\Pi^o_{\widehat{X}}). \tag{10.51}$$

In particular, the vertex number $f_0(Z_0)$ of the zero cell satisfies

$$2^d \leq \mathbb{E}f_0(Z_0) \leq 2^{-d}d!\kappa_d^2. \tag{10.52}$$

Equality holds on the left side if and only if the hyperplanes of $\widehat{X}$ are almost surely parallel to d fixed hyperplanes. Equality on the right side holds if and only if there is a nondegenerate linear transformation α of $\mathbb{R}^d$ such that the hyperplane process $\alpha\widehat{X}$ is isotropic.

Proof. The inequality in (10.50) is analogous to (4.52) and is in the same way a consequence of (14.43). If equality holds, then $\Pi_{\widehat{X}}$ is a ball. From this, we can deduce that $\widehat{X}$ is isotropic, since $\Pi_{\widehat{X}}$ determines the intensity and the directional distribution and thus the intensity measure of $\widehat{X}$; for a Poisson process, the latter determines the distribution. Conversely, if $\widehat{X}$ is isotropic, then $\Pi_{\widehat{X}}$ is a ball, and equality holds in (10.50).

The inequalities (10.52) are analogous to (4.56) and are obtained in a similar way from (14.45). Equality on the left side holds if and only if $\Pi_{\widehat{X}}$ is a parallelepiped. This is the case if and only if the spherical directional distribution of $\widehat{X}$ is concentrated in d pairs of antipodal unit vectors, and these are a.s. the normal vectors of the hyperplanes occurring in $\widehat{X}$. Equality on the right side of (10.52) holds if and only if $\Pi_{\widehat{X}}$ is an ellipsoid. This is equivalent to the existence of a nondegenerate linear transformation α of $\mathbb{R}^d$ such that $\alpha^{-t}\Pi_{\widehat{X}}$ is a ball. From (4.60) we get, for $u \in \mathbb{R}^d$,

$$\begin{aligned} 2h(\Pi_{\alpha\widehat{X}}, u) &= \mathbb{E}(\alpha\widehat{X})(\mathcal{F}_{[0,u]})) = \mathbb{E}\widehat{X}(\alpha^{-1}\mathcal{F}_{[0,u]})) = \mathbb{E}\widehat{X}(\mathcal{F}_{[0,\alpha^{-1}u]}) \\ &= 2h(\Pi_{\widehat{X}}, \alpha^{-1}u) = 2h(\alpha^{-t}\Pi_{\widehat{X}}, u), \end{aligned}$$

hence $\Pi_{\alpha\widehat{X}} = \alpha^{-t}\Pi_{\widehat{X}}$. This yields the remaining assertion. □

By the theorem, the expected vertex number of the zero cell of a stationary Poisson hyperplane mosaic attains its minimal value 2^d precisely for parallel mosaics (where the generating hyperplanes belong to d translation classes). In contrast to this, for the typical cell Z of a stationary hyperplane mosaic, the expected vertex number is always equal to 2^d, independently of the distribution, by Theorem 10.3.1.

The inequality for the expectation of the volume of the Poisson zero polytope given by (10.50) can be extended to higher moments of this volume. For a nondegenerate Poisson hyperplane process $\widehat{X}$ with intensity $\widehat{\gamma}$ and spherical directional distribution $\widehat{\varphi}$, we write

$$M_k(\widehat{\gamma}, \widehat{\varphi}) := \mathbb{E}V_d^k(Z_0) \qquad \text{for } k \in \mathbb{N}_0.$$

We obtain

$$\begin{aligned} M_k(\widehat{\gamma}, \widehat{\varphi}) &= \int_{\mathbb{R}^d} \cdots \int_{\mathbb{R}^d} \mathbb{E}[\mathbf{1}_{Z_0}(x_1) \cdots \mathbf{1}_{Z_0}(x_k)]\, \lambda(\mathrm{d}x_1) \cdots \lambda(\mathrm{d}x_k) \\ &= \int_{\mathbb{R}^d} \cdots \int_{\mathbb{R}^d} \exp\left[-\widehat{\Theta}(\mathcal{F}_{\mathrm{conv}\,\{0,x_1,\dots,x_k\}})\right] \lambda(\mathrm{d}x_1) \cdots \lambda(\mathrm{d}x_k) \\ &= \int_{\mathbb{R}^d} \cdots \int_{\mathbb{R}^d} \exp\left[-\widehat{\gamma} \int_{S^{d-1}} b(K_x, u)\, \widehat{\varphi}(\mathrm{d}u)\right] \lambda(\mathrm{d}x_1) \cdots \lambda(\mathrm{d}x_k), \end{aligned}$$

where we have put $K_x := \mathrm{conv}\,\{0, x_1, \dots, x_k\}$ and where $b(K_x, u) = h(K_x, u) + h(K_x, -u)$ is the width of the convex body K_x in direction u.

For every rotation $\vartheta \in SO_d$, we have

$$b(\mathrm{conv}\,\{0, \vartheta^{-1}x_1, \dots, \vartheta^{-1}x_k\}, u) = b(\mathrm{conv}\,\{0, x_1, \dots, x_k\}, \vartheta u).$$

The rotation invariance of the Lebesgue measure gives

$$M_k(\widehat{\gamma}, \widehat{\varphi}) = \int_{\mathbb{R}^d} \cdots \int_{\mathbb{R}^d} \exp\left[-\widehat{\gamma} \int_{S^{d-1}} b(K_x, \vartheta u)\, \widehat{\varphi}(\mathrm{d}u)\right] \lambda(\mathrm{d}x_1) \cdots \lambda(\mathrm{d}x_k).$$

We integrate this over all $\vartheta \in SO_d$ with respect to the invariant probability measure ν on SO_d. Using Fubini's theorem and Jensen's integral inequality, which can be applied due to the convexity of the exponential function, we obtain

$$
\begin{aligned}
&M_k(\widehat{\gamma}, \widehat{\varphi}) \\
&= \int_{\mathbb{R}^d} \cdots \int_{\mathbb{R}^d} \int_{SO_d} \exp\left[-\widehat{\gamma} \int_{S^{d-1}} b(K_x, \vartheta u)\, \widehat{\varphi}(\mathrm{d}u)\right] \nu(\mathrm{d}\vartheta)\, \lambda(\mathrm{d}x_1) \cdots \lambda(\mathrm{d}x_k) \\
&\geq \int_{\mathbb{R}^d} \cdots \int_{\mathbb{R}^d} \exp\left[-\widehat{\gamma} \int_{SO_d} \int_{S^{d-1}} b(K_x, \vartheta u)\, \widehat{\varphi}(\mathrm{d}u)\, \nu(\mathrm{d}\vartheta)\right] \lambda(\mathrm{d}x_1) \cdots \lambda(\mathrm{d}x_k).
\end{aligned}
$$

Here the equality sign holds if $\widehat{\varphi}$ is rotation invariant and thus coincides with the normalized spherical Lebesgue measure $\overline{\sigma} := \sigma/\sigma(S^{d-1})$. Since

$$
\int_{SO_d} \int_{S^{d-1}} b(K_x, \vartheta u)\, \widehat{\varphi}(\mathrm{d}u)\, \nu(\mathrm{d}\vartheta) = \int_{S^{d-1}} b(K_x, u)\, \overline{\sigma}(\mathrm{d}u),
$$

we have proved the inequality

$$
M_k(\widehat{\gamma}, \widehat{\varphi}) \geq M_k(\widehat{\gamma}, \overline{\sigma}). \tag{10.53}
$$

A similar inequality holds if the Poisson zero polytope is replaced by the Poisson polytope. For $k \in \mathbb{N}$ let

$$
m_k(\widehat{\gamma}, \widehat{\varphi}) := \mathbb{E} V_d^k(Z),
$$

where Z denotes the typical cell of the mosaic induced by $\widehat{X}$. From Theorem 10.4.1, we have

$$
M_{k-1}(\widehat{\gamma}, \widehat{\varphi}) = \gamma^{(d)} m_k(\widehat{\gamma}, \widehat{\varphi}).
$$

By (10.34), $\gamma^{(d)} = \gamma^{(0)}$, and this is the dth intersection density $\widehat{\gamma}_d$ of $\widehat{X}$. By Theorem 4.6.5 it becomes maximal, given the intensity, if $\widehat{X}$ is isotropic; thus

$$
m_k(\widehat{\gamma}, \widehat{\varphi}) \geq m_k(\widehat{\gamma}, \overline{\sigma}).
$$

Suppose that equality holds in (10.53) for a number $k \in \mathbb{N}$. Then in the inequality

$$
\begin{aligned}
&\int_{SO_d} \exp\left[-\widehat{\gamma} \int_{S^{d-1}} b(K_x, \vartheta u)\, \widehat{\varphi}(\mathrm{d}u)\right] \nu(\mathrm{d}\vartheta) \\
&\geq \exp\left[-\widehat{\gamma} \int_{SO_d} \int_{S^{d-1}} b(K_x, \vartheta u)\, \widehat{\varphi}(\mathrm{d}u)\, \nu(\mathrm{d}\vartheta)\right],
\end{aligned}
$$

equality holds for almost all $(x_1, \dots, x_k) \in (\mathbb{R}^d)^k$, and by continuity for all $(x_1, \dots, x_k)$. Since the exponential function is strictly convex, equality in the latter inequality implies that the integral

$$\int_{S^{d-1}} b(K_x, \vartheta u)\, \widehat{\varphi}(\mathrm{d}u)$$

is independent of ϑ. Let $v \in S^{d-1}$, and choose $x_1, \dots, x_k = v$, then $b(K_x, \vartheta u) = |\langle v, \vartheta u \rangle|$. For given $\vartheta \in SO_d$, let $\vartheta\widehat{\varphi}$ be the image measure of $\widehat{\varphi}$ under ϑ. Then

$$\int_{S^{d-1}} |\langle v, u \rangle| (\vartheta\widehat{\varphi} - \widehat{\varphi})(\mathrm{d}u) = 0.$$

Since this holds for all $v \in S^{d-1}$, we deduce that $\vartheta\widehat{\varphi} - \widehat{\varphi} = 0$, by Theorem 14.3.4. Since $\vartheta \in SO_d$ was arbitrary, we conclude that $\widehat{\varphi} = \overline{\sigma}$.

We collect these observations in the following theorem.

Theorem 10.4.10. *Let $\widehat{X}$ be a nondegenerate stationary Poisson hyperplane process in $\mathbb{R}^d$ with intensity $\widehat{\gamma}$ and spherical directional distribution $\widehat{\varphi}$. Let $M_k(\widehat{\gamma}, \widehat{\varphi})$ be the kth moment of the volume of the zero cell, and $m_k(\widehat{\gamma}, \widehat{\varphi})$ the kth moment of the volume of the typical cell of the induced hyperplane mosaic. Then*

$$M_k(\widehat{\gamma}, \widehat{\varphi}) \ge M_k(\widehat{\gamma}, \overline{\sigma}),$$
$$m_k(\widehat{\gamma}, \widehat{\varphi}) \ge m_k(\widehat{\gamma}, \overline{\sigma})$$

for $k \in \mathbb{N}$. Equality for some $k \in \mathbb{N}$ in one of these inequalities holds if and only if the process is isotropic.

Finally we prove, in analogy to Theorem 4.6.9, the following extremal assertion.

Theorem 10.4.11. *Let $\widehat{X}$ be a nondegenerate stationary Poisson hyperplane process in $\mathbb{R}^d$, and let Z_0 be the zero cell of the induced hyperplane mosaic. Among all convex bodies $K \in \mathcal{K}$ with $0 \in \mathcal{K}$ and given volume $V_d(K) > 0$, precisely the homothets of the Blaschke body $B(\widehat{X})$ of $\widehat{X}$ yield the maximal value for the probability $\mathbb{P}(K \subset Z_0)$.*

Proof. Because of $0 \in K$, the inclusion $K \subset Z_0$ holds if and only if $H \cap \operatorname{int} K = \emptyset$ for all $H \in \widehat{X}$. With probability one, every hyperplane from $\widehat{X}$ meeting K also meets $\operatorname{int} K$, hence

$$\mathbb{P}(K \subset Z_0) = \mathbb{P}(\widehat{X}(\mathcal{F}_K) = 0) = \mathrm{e}^{-\mathbb{E}\widehat{X}(\mathcal{F}_K)},$$

where

$$\mathbb{E}\widehat{X}(\mathcal{F}_K) = \mathbb{E} \sum_{H \in \widehat{X}} V_0(K \cap H) = 2dV(K, B(\widehat{X}), \dots, B(\widehat{X}))$$

by (4.69). Now the assertion follows from (14.30). □

Notes for Section 10.4

1. The equation of Theorem 10.4.1 can be found (for hyperplane mosaics) in Miles [517, 521] and Matheron [462]. Theorems 10.4.2 and 10.4.3 are due to Mecke [492].

A very general version of Theorem 10.4.1 for stationary partitions appears in Last [433].

2. Theorems 10.4.4 and 10.4.5 on the typical cell of a Poisson–Delaunay mosaic are due to Miles [521]. Theorem 10.4.4 is now a very special case of distribution results of Baumstark and Last [85].

Direct constructions of random polytopes equivalent to the typical cell of a stationary, isotropic Poisson hyperplane mosaic have been described repeatedly in the literature. We refer to Miles [517, sect. 8.4], [521, p. 220], Ambartzumian [35, sect. 9.3], Mecke [492], Calka [148, 149]. Theorems 10.4.6 and 10.4.7 collect and extend these results; the proof of 10.4.7 is essentially taken from Hug and Schneider [373].

3. Theorem 10.4.8 goes back to Miles [517], who has proved the corresponding result for ergodic distributions. The proof given here can be considered as an elaboration of the hints given in Miles [519]. The result has also been deduced, at least in the isotropic case, from the 'Complementary Theorem' of Miles; see Miles [523, sect. 5.3], and Møller and Zuyev [555] for a version of this theorem using Palm distributions. This general theorem yields a number of conditional distributions for Poisson hyperplane mosaics. Its application to the distribution of the inradius requires that the inball of the typical cell touches a.s. $d+1$ facets of the typical cell; this is not satisfied, for example, if the spherical directional distribution of the generating hyperplane process is concentrated in d pairs of antipodal points.

4. Theorem 10.4.9 is essentially due to Wieacker [817]. He computed $\mathbb{E}\, L_k(Z_0)$ in a different way; the present approach is taken from Favis and Weiss [226].

The double-sided inequality (10.52) has quite an interesting history. In a different but equivalent form, the right side was first proved by Schneider [685], by applying the Blaschke–Santaló inequality to an auxiliary zonoid. The proof gave Reisner [625] an idea for proving his reverse Blaschke–Santaló inequality for zonoids, which could then again be applied in stochastic geometry (see Schneider [689]) and yields, for instance, the left side of (10.52).

Theorem 10.4.11 is Theorem 7.2 of Weil and Wieacker [806].

Theorem 10.4.10 and its proof are due to Mecke [490] (except the equality condition). There it is shown, more generally, that the moments $M_k(\widehat{\gamma}, \widehat{\varphi})$ and $m_k(\widehat{\gamma}, \widehat{\varphi})$ are not increased if one passes from the (spherical) directional distribution $\widehat{\varphi}$ to a directional distribution which is a mixture of rotational images of $\widehat{\varphi}$. A further generalization to certain Cox processes of hyperplanes is found in Mecke [491]. This raises the question whether Mecke's [490] results on moments of the volume have counterparts for the distributions (with respect to a suitable order) of the volume of the typical or the zero cell, or for the other intrinsic volumes. For orthogonal parallel mosaics, corresponding assertions have been proved by Favis [224, 225].

5. Corollary 10.4.1 was emphasized by Mecke [492]; see also the references given there.

6. Typical Poisson–Voronoi cells in the plane. Let Z_{PV} denote the typical cell (with respect to the nucleus) of a planar Poisson–Voronoi mosaic, generated by a stationary Poisson process of intensity γ. Calka [151, 152] has derived integral formulas for the distributions of the vertex numbers and the area of Z_{PV}. Hilhorst [343, 344] has given an asymptotic expansion for the probability that Z_{PV} has n sides and has deduced that, as $n \to \infty$, the n-sided typical cell behaves asymptotically like a circle of radius $(n/4\pi\gamma)^{1/2}$.

Let R_m (R_M) denote the largest (smallest) radius of a disk centered at 0 contained in (containing) Z_{PV}. Calka [150] has obtained an explicit formula for the joint distribution of the pair (R_m, R_M), and has drawn several conclusions. For R_m tending to infinity, Calka and Schreiber [153] have proved a law of large numbers for the vertex number and for the area of Z_{PV} outside the indisk. For the latter, they have also established a central limit theorem.

7. Distributional properties connected with typical faces. In the generalization of the typical cell and the zero cell, the **typical k-face** $Z^{(k)}$ and the **typical k-weighted k-face** $Z_0^{(k)}$ of a stationary random mosaic X can be considered, for $k \in \{0, \dots, d\}$. The typical k-face $Z^{(k)}$ of a stationary mosaic X is a random (k-dimensional) polytope with distribution $\mathbb{Q}^{(k)}$. Alternatively, $Z^{(k)}$ can be described as the (a.s. unique) k-face of X containing the origin, under the Palm distribution $\mathbb{P}^0_{N_k}$ of X with respect to the stationary point process N_k of centers of $X^{(k)}$. The distribution $\mathbb{Q}_0^{(k)}$ of the typical k-weighted k-face arises from the Palm distribution $\mathbb{P}^0_{M_k}$ of X with respect to the stationary random measure that is defined by the restriction of the k-dimensional Hausdorff measure M_k to the union of all k-faces of X. Under $\mathbb{P}^0_{M_k}$, there is almost surely a unique k-face of X containing 0, and this is $Z_0^{(k)}$. In analogy to Theorem 10.4.1 one gets that, for a translation invariant, nonnegative, measurable function $f : \mathcal{K}' \to \mathbb{R}$,

$$d_k^{(k)} \mathbb{E} f(Z_0^{(k)}) = \gamma^{(k)} \mathbb{E}[f(Z^{(k)}) V_k(Z^{(k)})].$$

For typical faces of Poisson–Voronoi mosaics, a rather general investigation was undertaken by Baumstark and Last [85, 86]. In the context of typical (k-weighted) k-faces, it is convenient to use a generalized center function (instead of the circumcenter); compare Note 1 for Section 4.2. Using the terminology introduced before Theorem 10.2.3, for a k-face S of the Voronoi mosaic generated by the set A, this generalized center of S is $z(x_0, \dots, x_{d-k})$ where $S = S(x_0, \dots, x_{d-k}; A)$. The points $x_0, \dots, x_{d-k} \in A$ which determine the k-face S are also called the **neighbors** of S. Assume that $0 \in S$, then $\|x_i\| = R$, $i = 0, \dots, d-k$. Let R' be the radius of the k-dimensional ball $B^k(x_0, \dots, x_{d-k})$, $z = z(x_0, \dots, x_{d-k})$ the midpoint and $U_i := (x_i - z)/R'$, $i = 0, \dots, d-k$, the corresponding directions. Finally, $R'' := (R^2 - R'^2)^{1/2}$ is the distance from 0 to z and (if $R'' > 0$) $V := z/R''$ is the corresponding direction.

For the Poisson–Voronoi mosaic X, under the Palm distribution $\mathbb{P}^0_{M_k}$, the neighbors $x_0, \dots, x_{d-k}$ of $Z_0^{(k)}$ are random points and so U_i, z, V, R, R', R'' are random, too. We have $0 \in S(x_0, \dots, x_{d-k}; \widetilde{X})$ and $z(x_0, \dots, x_{d-k}) \neq 0$ a.s., hence $R'' > 0$. Under the Palm distribution $\mathbb{P}^0_{N_k}$, the x_i, U_i, R are again random variables (and $z = 0$, $R' = R$, $R'' = 0$).

For the typical k-weighted k-face, the following distributional results are shown in Baumstark and Last [85].

Theorem.

(a) *The random variables* $(\widetilde{X} \llcorner (\mathbb{R}^d \setminus RB^d), R)$, R'^2/R^2 *and* $(U_0, \ldots, U_{d-k}, V)$ *are independent.*
(b) R^d *is gamma distributed with shape parameter* $d - k + k/d$ *and scale parameter* $\gamma\kappa_d$.
(c) *The conditional distribution of* $\widetilde{X} \llcorner (\mathbb{R}^d \setminus RB^d)$, *given* $R = r$, *is the distribution of a Poisson process restricted to* $\mathbb{R}^d \setminus rB^d$.
(d) *For* $k \in \{1, \ldots, d-1\}$, R'^2/R^2 *has a beta distribution with parameters* $d(d-k)/2$ *and* $k/2$.
(e) *The joint distribution of* $U_0, \ldots, U_{d-k}, V$ *is given by*

$$c \int_{SO_d} \int_{S^{k-1}_{L^\perp}} \int_{S^{d-k-1}_L} \cdots \int_{S^{d-k-1}_L} \mathbf{1}\{(\vartheta u_0, \ldots, \vartheta u_{d-k}, \vartheta v) \in \cdot\}$$
$$\times \Delta_{d-k}(u_0, \ldots, u_{d-k})^{k+1} \sigma_L(\mathrm{d}u_0) \cdots \sigma_L(\mathrm{d}u_{d-k}) \, \sigma_{L^\perp}(\mathrm{d}v) \, \nu(\mathrm{d}\vartheta),$$

where L *is a fixed* $(d-k)$*-space and* c *is an explicitly given constant.*

The proof of these results uses similar arguments to those employed in the proof of Theorem 10.2.4. Utilizing the connection between the distributions of $Z_0^{(k)}$ and $Z^{(k)}$, one obtains similar distributional results for the typical k-face. We mention the case of the typical vertex ($k = 0$), where $Z_0^{(0)} = Z^{(0)}$. Here, the above results show that:

(a) R^d *is gamma distributed with shape parameter* d *and scale parameter* $\gamma\kappa_d$.
(b) *The joint distribution of* $U_0, \ldots, U_d$ *has density* $c\Delta_d$ *with respect to* σ^{d+1}, *where* c *is an explicitly given constant.*

For the typical edge (the case $k = 1$), more specific results are obtained.

For a k-face $S = S(x_0, \ldots, x_{d-k}; A)$ of a Voronoi mosaic generated by the set A, the **fundamental region** or **Voronoi flower** $F(S)$ of S is the union of all balls centered at the vertices of S and containing the neighbors $x_0, \ldots, x_{d-k}$ in the boundary. $F(S)$ is an example of a **stopping set** associated with A. From more general results about stopping sets, Baumstark and Last [86] (see also Baumstark [84]) deduce that the volume $\lambda(F(Z_0^{(k)}))$ of the fundamental region of the typical k-weighted k-face of a stationary Poisson–Voronoi mosaic X has a gamma distribution. There are also results for the typical k-face, for functionals other than the volume and for stopping sets associated with Poisson processes of k-flats ($k \geq 1$).

8. Scaled vacancies. The Poisson zero polytope Z_0 also appears in a limit theorem for scaled vacancies of Boolean models. The following is a special case of a more general result in Molchanov [545]. Let Z be a stationary Boolean model in $\mathbb{R}^d$ with convex grains and let ν be the expected surface area measure of the typical grain. Let Y denote the connected component of the complement of Z which contains the origin. Then, as the intensity γ of Z tends to infinity, the open random set γY converges in distribution to the interior of the zero polytope Z_0, generated by the Poisson hyperplane process with direction measure ν.

Michel and Paroux [514] prove a corresponding result for a Boolean model of shells (parallel sets of boundaries of convex bodies) and the zero polytope of a thickened Poisson hyperplane process.

9. D.G. Kendall's problem on the asymptotic shape of large cells. In his foreword to the first edition (1987) of the book by Stoyan, (W.S.) Kendall and Mecke [743], D.G. Kendall recalled a conjecture that he had made decades ago and which has later become widely known as Kendall's conjecture. He considered a stationary, isotropic Poisson line process in the plane and its zero cell Z_0, and said in equivalent words: 'One would have preferred to be able to say something about ... my conjecture that the conditional law for the shape of Z_0, given the area $A(Z_0)$ of Z_0, converges weakly, as $A(Z_0) \to \infty$, to the degenerate law concentrated at the circular shape. Unfortunately nothing substantial is known about either of these questions even today ... ' Contributions to the conjecture, though not leading to a complete proof, are due to Miles [535] and Goldman [270]. Miles also suggested measuring the size of the zero cell not only by the area, but also by other functionals, such as the perimeter or the width in a given direction, and pointed out in the latter case that asymptotic shapes other than the circular one were to be expected. A proof of Kendall's original conjecture was given by Kovalenko [425], and a simpler proof in [427]. Kovalenko [426] also obtained a similar result for the typical cell of a stationary Poisson–Voronoi tessellation in the plane. In higher dimensions, Mecke and Osburg [499] treated the shape of large Crofton parallelotopes (generated by a stationary Poisson hyperplane process with a directional distribution concentrated on pairwise orthogonal directions). In a series of papers, Hug, Reitzner and Schneider [366, 367] and Hug and Schneider [370, 371, 372, 373] treated very general higher-dimensional versions, variants and analogs of Kendall's problem. The following is a brief description of their main results.

In Hug and Schneider [372], the following situation is considered. A Poisson hyperplane process X in $\mathbb{R}^d$ is given, with an intensity measure of the form

$$\Theta(A) = 2\gamma \int_{S^{d-1}} \int_0^\infty \mathbf{1}_A(H(u,t))\, t^{r-1}\, \mathrm{d}t\, \varphi(\mathrm{d}u) \tag{10.54}$$

for Borel sets $A \subset A(d, d-1)$ (with $H(u,t)$ as in (4.32)). Here $\gamma > 0$ is a constant (the intensity of X in the case where X is stationary), $r \geq 1$ is a given number, and the directional distribution φ is a probability measure on S^{d-1}, not concentrated on a closed hemisphere. The object of study is the random polytope

$$Z_0 := \bigcap_{H \in X} H^-,$$

where H^- denotes the (a.s. unique) closed halfspace bounded by $H \in X$ that contains the origin 0. The general form of the intensity measure includes two special cases of particular geometric interest: for $r = 1$ and even φ, the random polytope Z_0 is the zero cell of a (not necessarily isotropic) stationary Poisson hyperplane tessellation, and for $r = d$ and rotation invariant φ, the distribution of Z_0 is that of the typical cell of a stationary Poisson–Voronoi tessellation.

The size of Z_0 can be measured by any functional $\Sigma : \mathcal{K}' \to \mathbb{R}$ satisfying the following axioms: Σ is increasing under set inclusion, homogeneous of some degree $k > 0$, continuous, and not identically zero. Any such Σ is called a **size functional**. Examples are volume, surface area, mean width, diameter, thickness, inradius, circumradius, width in a given direction, and many others.

The Poisson hyperplane process X determines a second functional $\Phi : \mathcal{K}' \to \mathbb{R}$, by

$$\Phi(K) := \frac{1}{2\gamma}\mathbb{E}\,\mathrm{card}\,\{H \in X : H \cap K \neq \emptyset\}.$$

For obvious reasons, Φ is called the **hitting functional**. Due to the assumed form (10.54) of the intensity measure Θ of X, the hitting functional is given by

$$\Phi(K) = \frac{1}{r}\int_{S^{d-1}} h(K,u)^r \,\varphi(\mathrm{d}u).$$

For reasons of homogeneity and continuity, the size functional and the hitting functional satisfy a sharp inequality of isoperimetric type, namely

$$\Phi(K) \geq \tau \Sigma(K)^{r/k}. \tag{10.55}$$

Here 'sharp' means that there are extremal bodies, that is, bodies which contain more than one point and for which equality holds (this equality determines the number τ). The extremal bodies have a probabilistic characterization: among all convex bodies K containing 0 and of size $\Sigma(K) = 1$, precisely the extremal bodies maximize the probability $\mathbb{P}(K \subset Z_0)$. A main observation is now that the extremal bodies of (10.55) determine the asymptotic shape of the random polytope Z_0 under the condition that it has large Σ-size. We make this more precise in some typical cases. Let G denote either the group of similarities, or of homotheties, or of positive dilatations of $\mathbb{R}^d$. The G-shape $s_{\mathsf{G}}(K)$ of a convex body K is defined as its G-equivalence class $\{gK : g \in \mathsf{G}\}$. The space of all G-shapes, with the quotient topology, is denoted by $\mathcal{S}_{\mathsf{G}}$. The **conditional law** of the G-shape of Z_0, given the lower bound $a > 0$ for the size Σ, is the probability measure μ_a on $\mathcal{S}_{\mathsf{G}}$ defined by

$$\mu_a(A) := \mathbb{P}(s_{\mathsf{G}}(Z_0) \in A \mid \Sigma(Z_0) \geq a)$$

for $A \in \mathcal{B}(\mathcal{S}_{\mathsf{G}})$.

Theorem. *Suppose that (for some group G as above) the extremal bodies of* (10.55) *belong to a unique G-shape $s_{\mathsf{G}}(B)$. Then $s_{\mathsf{G}}(B)$ is the limit shape of Z_0 for increasing Σ, in the sense that*

$$\lim_{a\to\infty} \mu_a = \delta_{s_{\mathsf{G}}(B)} \quad \textit{weakly},$$

where $\delta_{s_{\mathsf{G}}(B)}$ denotes the Dirac measure concentrated at $s_{\mathsf{G}}(B)$.

This follows from a stronger result, estimating the probability of large deviations of Z_0 from the class of extremal bodies. To measure the deviation, we denote by $\mathcal{K}_0$ the set of convex bodies in $\mathbb{R}^d$ containing 0 and consider a function $\vartheta : \{K \in \mathcal{K}_0 : \Sigma(K) > 0\} \to \mathbb{R}$ which is continuous, nonnegative, homogeneous of degree zero, and satisfies $\vartheta(K) = 0$ for $K \in \mathcal{K}_0$ if and only if K is an extremal body of (10.55). Such functions always exist and are called **deviation functionals** (for given Σ and Φ). Further, there exists a continuous function $f : \mathbb{R}^+ \to \mathbb{R}^+$ with $f(\epsilon) > 0$ for $\epsilon > 0$ and $f(0) = 0$ such that

$$\vartheta(K) \geq \epsilon \;\Rightarrow\; \Phi(K) \geq (1 + f(\epsilon))\tau\Sigma(K)^{r/k} \tag{10.56}$$

for $K \in \mathcal{K}_0$. In the following, we assume that X (and thus Φ and γ), Σ, ϑ and f are given.

Theorem. *For $\epsilon > 0$ and $a > 0$,*

$$\mathbb{P}(\vartheta(Z_0) \geq \epsilon \mid \Sigma(Z_0) \geq a) \leq c \exp\left(-c_0 f(\epsilon) a^{r/k}\gamma\right) \tag{10.57}$$

with positive constants c (depending on X, Σ, f, ϵ) and c_0 (depending only on τ).

This estimate shows clearly the crucial role of the extremal bodies of the isoperimetric inequality (10.55): the probability that the random polytope Z_0 deviates by some given amount from the class of extremal bodies, under the condition that the size of Z_0 is at least a, becomes exponentially small if a tends to infinity.

As a byproduct of the proof, one can obtain information on the asymptotic behavior of the distribution function of the size of the zero cell, namely

$$\lim_{a\to\infty} a^{-r/k} \ln \mathbb{P}(\Sigma(Z_0) \geq a) = -2\tau\gamma.$$

A special case ($d = 2$, X stationary and isotropic, Σ two-dimensional volume) goes back to Goldman [270]

Kendall's original question involved 'the conditional law for the shape of Z_0, given the area $A(Z_0)$ of Z_0' (and not a lower bound for the area). The methods leading to the preceding result (10.57) are strong enough to yield also such results. The random variable Z_0 takes its values in $\mathcal{K}'$, which is a Polish space. Hence, the regular conditional probability distribution of Z_0 with respect to $\Sigma(Z_0)$ exists. In Hug and Schneider [372], also the estimate

$$\mathbb{P}(\vartheta(Z_0) \geq \epsilon \mid \Sigma(Z_0) = a) \leq c \exp\left(-c_0 f(\epsilon) a^{r/k} \gamma\right)$$

was obtained.

The applicability of the preceding general results to concrete cases, that is, special hyperplane processes X and size functionals Σ, depends on the nature of the isoperimetric inequality (10.55), for which the extremal bodies have to be determined and an explicit stability improvement (10.56) has to be established. In Hug, Reitzner and Schneider [366], the zero cell of a stationary (not necessarily isotropic) Poisson hyperplane process X was considered, with its size measured by the volume; this is the extension of Kendall's original problem to higher dimensions and non-isotropic processes. In that case, one uses Minkowski's existence theorem from the theory of convex bodies to obtain a 0-symmetric convex body B for which the spherical directional distribution of X is the surface area measure. With this body, the hitting functional can be expressed as a mixed volume, namely $\Phi(K) = dV(K, B, \dots, B)$, and the crucial isoperimetric inequality becomes Minkowski's inequality

$$V(K, B, \dots, B) \geq V_d(B)^{1-1/d} V_d(K)^{1/d}.$$

It is known that here the extremal bodies are precisely the homothets of B, and also a suitable stability estimate is known. Thus, the homothety class of B is the limit shape for zero cells of large volume, and the deviation estimate holds (with a simple explicit deviation functional ϑ and with $f(\epsilon) = \epsilon^{d+1}$). If X is isotropic, then B is a ball. In Hug, Reitzner and Schneider [367], the typical cell of a stationary Poisson–Voronoi tessellation was treated, with its size measured by the kth intrinsic volume. The limit shape is that of a ball with center 0. The paper by Hug and Schneider [372] lists several more specific examples, including some where the limit shapes are classes of segments. For typical cells (in contrast to zero cells) of stationary Poisson hyerplane processes and for selected size functionals, similar results were obtained in Hug and Schneider [373].

Large typical cells of Poisson–Delaunay tessellations were investigated by Hug and Schneider [370, 371].

10.5 Mixing Properties

For the most important of the previously considered special mosaics, namely those induced by stationary Poisson processes, we shall now show that they are mixing, and hence ergodic. For this, we use Theorem 9.3.2; therefore, we prove relation (9.25) for the considered stationary mosaics X. For these, the crucial relation can be written in the form

$$\lim_{\|x\|\to\infty} \mathbb{P}(X \cap C_1 = \emptyset,\, X \cap (C_2 + x) = \emptyset)$$
$$= \mathbb{P}(X \cap C_1 = \emptyset)\mathbb{P}(X \cap C_2 = \emptyset) \tag{10.58}$$

for all $C_1, C_2 \in \mathcal{C}(\mathcal{F}'(\mathbb{R}^d))$. If this is proved, the mixing property is obtained, in each of the considered cases, by arguments similar to those in the proof of Theorem 9.3.6.

Theorem 10.5.1. *Stationary Poisson–Voronoi mosaics are mixing.*

Proof. Let X be the Voronoi mosaic corresponding to the stationary Poisson process $\widetilde{X}$ in $\mathbb{R}^d$. Every cell $P \in X$ contains a uniquely determined point of $\widetilde{X}$, which we denote by $k(P)$. Let $K_1, \dots, K_m$ be a covering of the ball $5B^d$ by balls of radius 1.

For $r > 0$, we consider the event

$$E_r := \{\text{there exists } P \in X \text{ with } P \cap rB^d \neq \emptyset \text{ and } P \not\subset 5rB^d\}.$$

If $P \in X$ has the property required by E_r, then there are points $x \in P \cap rB^d$ and $y \in P \cap \operatorname{bd} 5rB^d$. The distance between x and y is at least $4r$, hence at least one of the two points has distance at least $2r$ from $k(P)$. If this holds for x, then the interior of rB^d does not contain a point of X. The probability of the latter event is $\mathrm{e}^{-\gamma\kappa_d r^d}$. If, on the other hand, $\|y - k(P)\| \geq 2r$, then y is contained in one of the balls $rK_1, \dots, rK_m$, and the interior of that ball does not contain a point of X. The probability of the latter event can be estimated from above by $m\mathrm{e}^{-\gamma\kappa_d r^d}$. Altogether we get

$$\mathbb{P}(E_r) \leq (1 + m)\mathrm{e}^{-\gamma\kappa_d r^d}. \tag{10.59}$$

In order to prove (10.58), we start with given $C_1, C_2 \in \mathcal{C}(\mathcal{F}'(\mathbb{R}^d))$ and $\epsilon > 0$ and choose $r > 0$ sufficiently large so that

$$C_i \subset \mathcal{F}_{rB^d} \qquad \text{for } i = 1, 2$$

and

$$\mathbb{P}(E_r) < \epsilon;$$

this is possible by (10.59). Let $\omega \in \Omega \setminus E_r$ and $P \in X(\omega) \cap C_1$. Then $P \cap rB^d \neq \emptyset$ and $P \subset 5rB^d$. The Voronoi cell P is of the form

$$P = \bigcap_{y \in \widetilde{X}(\omega),\, y \neq x} H_y^+(x)$$

with $x = k(P) \in P \subset 5rB^d$. This implies $y \in 15rB^d$ for all points $y \in \widetilde{X}(\omega)$ that are required for the determination of P, that is, for which the boundary of $H_y^+(x)$ contains a facet of P. Therefore, the Voronoi cells meeting rB^d that belong to the point sets $\widetilde{X}(\omega)$ (respectively $\widetilde{X}(\omega) \cap 15rB^d$), are identical. We denote by V the system of the Voronoi cells induced by the point process $\widetilde{X} \cap 15rB^d$; then we have shown that

$$X(\omega) \cap C_1 = V(\omega) \cap C_1 \qquad \text{for } \omega \in \Omega \setminus E_r.$$

Now let $x \in \mathbb{R}^d$ and $\|x\| > 30r$. For the event

$$E_r^x := \{\text{there exists } P \in X(\omega) \text{ with } P \cap (rB^d + x) \neq \emptyset \text{ and } P \not\subset 5rB^d + x\}$$

we also have $\mathbb{P}(E_r^x) < \epsilon$. Let V_x denote the system of Voronoi cells induced by $\widetilde{X} \cap (15rB^d + x)$. As above, we have

$$X(\omega) \cap (C_2 + x) = V_x(\omega) \cap (C_2 + x) \qquad \text{for } \omega \in \Omega \setminus E_r^x.$$

Since $15rB^d \cap (15rB^d + x) = \emptyset$, the point processes $\widetilde{X} \cap 15rB^d$ and $\widetilde{X} \cap (15rB^d + x)$ are independent, by Theorem 3.2.2. This gives

$$\mathbb{P}(V \cap C_1 = \emptyset, V_x \cap (C_2 + x) = \emptyset) = \mathbb{P}(V \cap C_1 = \emptyset)\mathbb{P}(V_x \cap (C_2 + x) = \emptyset).$$

The events $A := \{X \cap C_1 = \emptyset\}$, $B := \{X \cap (C_2 + x) = \emptyset\}$, $\overline{A} := \{V \cap C_1 = \emptyset\}$, $\overline{B} := \{V_x \cap (C_2 + x) = \emptyset\}$, $E := (E_r \cup E_r^x)^c$ satisfy $A \cap E = \overline{A} \cap E$, $B \cap E = \overline{B} \cap E$ and hence $|\mathbb{P}(A) - P(\overline{A})| \leq \mathbb{P}(E^c) < 2\epsilon$, $|\mathbb{P}(B) - \mathbb{P}(\overline{B})| < 2\epsilon$, $|\mathbb{P}(A \cap B) - \mathbb{P}(\overline{A} \cap \overline{B})| < 2\epsilon$. We deduce that

$$\begin{aligned}
&|\mathbb{P}(A \cap B) - \mathbb{P}(A)\mathbb{P}(B)| \\
&\leq |\mathbb{P}(A \cap B) - \mathbb{P}(\overline{A} \cap \overline{B})| + |\mathbb{P}(\overline{A})\mathbb{P}(\overline{B}) - \mathbb{P}(A)\mathbb{P}(B)| \\
&\leq |\mathbb{P}(A \cap B) - \mathbb{P}(\overline{A} \cap \overline{B})| + |\mathbb{P}(A) - \mathbb{P}(\overline{A})| + |\mathbb{P}(B) - \mathbb{P}(\overline{B})| \\
&< 6\epsilon.
\end{aligned}$$

The assertion (10.58) now follows from the stationarity of X. □

Theorem 10.5.2. *Stationary Poisson–Delaunay mosaics are mixing.*

Proof. Let X be the Delaunay mosaic corresponding to the stationary Poisson process $\widetilde{X}$. Almost surely all its faces are simplices. For a d-simplex P, we denote by $B(P)$ the ball whose boundary contains the vertices of P. Then, for $P \in X$ we have $\widetilde{X} \cap \operatorname{int} B(P) = \emptyset$ (cf. the proof of Theorem 10.2.6). Let $K_1, \dots, K_m$ be a covering of $\operatorname{bd} 2B^d$ by balls of radius $1/2$.

For $r > 0$, we consider the event

$$E_r := \{\text{there exists } P \in X \text{ with } P \cap rB^d \neq \emptyset \text{ and } B(P) \not\subset 3rB^d\}.$$

Let $\omega \in E_r$, and let $P \in X(\omega)$ be a simplex with $P \cap rB^d \neq \emptyset$ and $B(P) \not\subset 3rB^d$. Since $B(P) \cap rB^d \neq \emptyset$ and $B(P) \cap (\mathbb{R}^d \setminus 3rB^d) \neq \emptyset$, there exists $z \in \operatorname{bd} 2rB^d$ with $B(z,r) \subset B(P)$, hence with $\widetilde{X}(\omega) \cap \operatorname{int} B(z,r) = \emptyset$. There exists a number $i \in \{1,\dots,m\}$ with $rK_i \subset B(z,r)$ and thus with $\widetilde{X}(\omega) \cap \operatorname{int} rK_i = \emptyset$. This gives

$$\mathbb{P}(E_r) \leq me^{-\gamma\kappa_d(\frac{r}{2})^d}.$$

Now let $\widetilde{X}_{3r} := \widetilde{X} \cap 3rB^d$. By a Delaunay cell of $\widetilde{X}_{3r}$ we understand a d-simplex P with vertices in $\widetilde{X}_{3r}$ and the property that $\widetilde{X}_{3r} \cap \operatorname{int} B(P) = \emptyset$. If $\omega \in \Omega \setminus E_r$ and $P \in X(\omega)$ is a cell with $P \cap rB^d \neq \emptyset$ and thus $P \subset 3rB^d$, then P is a Delaunay cell of $\widetilde{X}_{3r}(\omega)$. Conversely, every Delaunay cell of $\widetilde{X}_{3r}(\omega)$ that meets rB^d belongs to $X(\omega)$, because the cells of $X(\omega)$ meeting rB^d cover an open neighborhood of rB^d.

Now the proof can be completed in a similar way to that of Theorem 10.5.1, using the fact that the point processes $\widetilde{X}_{3r}$ and $\widetilde{X}_{3r} + x$ are stochastically independent if $\|x\| > 6r$. □

Theorem 10.5.3. *Let $\widehat{X}$ be a stationary Poisson hyperplane process of intensity $\widehat{\gamma} > 0$ in $\mathbb{R}^d$, and suppose that its spherical directional distribution is zero on every great subsphere of S^{d-1}. Then the hyperplane mosaic X induced by $\widehat{X}$ is mixing.*

Proof. Let U be the support of the spherical directional distribution $\widehat{\varphi}$ of $\widehat{X}$. As in the proof of Theorem 10.3.2, we choose points $u_1,\dots,u_{2d} \in U$ with $0 \in \operatorname{int conv} \{u_1,\dots,u_{2d}\}$. We can find a number $s > 0$ and neighborhoods $U_i \subset S^{d-1}$ of u_i, $i = 1,\dots,2d$, so that $s^{-1}B^d \subset \operatorname{conv}\{v_1,\dots,v_{2d}\}$ for all $(v_1,\dots,v_{2d}) \in U_1 \times \ldots \times U_{2d}$. For $r > 0$, let

$$\mathcal{E}_{i,r} := \{H(u,\tau) : u \in U_i,\ r < \tau < 2r\}$$

(with $H(u,\tau) := \{x \in \mathbb{R}^d : \langle x,u \rangle = \tau\}$); then

$$\widehat{\Theta}(\mathcal{E}_{i,r}) = \widehat{\gamma} \int_{S^{d-1}} \int_r^{2r} \mathbf{1}_{\mathcal{E}_{i,r}}(u^\perp + \tau u)\, \mathrm{d}\tau\, \widehat{\varphi}(\mathrm{d}u) = \widehat{\gamma} r \widehat{\varphi}(U_i) > 0.$$

For the event

$$E_r := \{\widehat{X}(\mathcal{E}_{i,r}) = 0 \text{ for some } i \in \{1,\dots,2d\}\}$$

this yields

$$\mathbb{P}(E_r) \leq \sum_{i=1}^{2d} \mathbb{P}(\widehat{X}(\mathcal{E}_{i,r}) = 0) = \sum_{i=1}^{2d} \mathrm{e}^{-\widehat{\gamma} r \widehat{\varphi}(U_i)}. \tag{10.60}$$

For the hyperplane $H = H(u,\tau)$ with $\tau > 0$, we put $H^- = H^-(u,\tau) := \{x \in \mathbb{R}^d : \langle x,u \rangle \leq \tau\}$. Define

$$Q := \bigcap \{H^- : H \in \widehat{X},\ H \cap rB^d = \emptyset,\ H \cap 2rB^d \neq \emptyset\}.$$

Let $\omega \in \Omega \setminus E_r$. Then $\widehat{X}(\omega)$ contains, to each $i \in \{1, \dots, 2d\}$, a hyperplane $H(v_i, \tau_i)$ with $v_i \in U_i$ and $r < r_i < 2r$. This implies

$$Q(\omega) \subset 2r \bigcap_{i=1}^{2d} H^-(v_i, 1).$$

Here, $\bigcap_{i=1}^{2d} H^-(v_i, 1)$ is the polytope polar to $\operatorname{conv}\{v_1, \dots, v_{2d}\}$. Because of $s^{-1}B^d \subset \operatorname{conv}\{v_1, \dots, v_{2d}\}$, it is contained in sB^d, thus $Q(\omega) \subset 2rsB^d$.

For $v \in S^{d-1}$ and $\alpha \geq 0$, let

$$S(v, \alpha) := \{u \in S^{d-1} : |\langle u, v\rangle| \leq \alpha\}.$$

For the spherical directional distribution $\widehat{\varphi}$ of the hyperplane process $\widehat{X}$ we have assumed that $\widehat{\varphi}(S(v, 0)) = 0$ for every great sphere $S(v, 0)$. This implies that for every $\epsilon > 0$ there exists $\alpha > 0$ such that $\widehat{\varphi}(S(v, \alpha)) < \epsilon$ for all $v \in S^{d-1}$.

For $r > 0$, $z \in S^{d-1}$ and $a > 2r$, let

$$\mathcal{B}_{r,z,a} := \{H \in A(d, d-1) : H \cap rB^d \neq \emptyset,\ H \cap (rB^d + az) \neq \emptyset\}.$$

For $H(u, \tau) \in \mathcal{B}_{r,z,a}$ we have $|\langle u, z\rangle| \leq 2r/a$, hence

$$\widehat{\Theta}(\mathcal{B}_{r,z,a}) \leq \widehat{\gamma} \cdot 2r\widehat{\varphi}(S(z, 2r/a)$$

and therefore

$$\mathbb{P}(\widehat{X}(\mathcal{B}_{r,z,a}) > 0) \leq 1 - e^{-2r\widehat{\gamma}\widehat{\varphi}(S(z,2r/a))}. \tag{10.61}$$

In order to prove (10.58), let $C_1, C_2 \in \mathcal{C}(\mathcal{F}'(\mathbb{R}^d))$ and $\epsilon > 0$ be given. By (10.60) we can choose $r > 0$ so large that

$$C_i \subset \mathcal{F}_{rB^d}(\mathbb{R}^d) \qquad \text{for } i = 1, 2$$

and

$$\mathbb{P}(E_r) < \epsilon.$$

The numbers $s > 0$, $\epsilon > 0$, $r > 0$ being given, we can choose the number a so large that $a > 4rs$ and

$$\mathbb{P}(\widehat{X}(\mathcal{B}_{2rs,z,a}) > 0) < \epsilon$$

for all $z \in S^{d-1}$; this is possible by (10.61).

Let $x \in \mathbb{R}^d$ be a vector with $\|x\| > a$, and let $z := x/\|x\|$. The event

$$E_r^x := \{\widehat{X}(\mathcal{E}_{i,r} + x) = 0 \text{ for some } i \in \{1, \dots, 2d\}\}$$

also satisfies $\mathbb{P}(E_r^x) < \epsilon$. For

$$E := E_r \cup E_r^x \cup \{\omega \in \Omega : \widehat{X}(\mathcal{B}_{2rs,z,a}) > 0\}$$

we get $\mathbb{P}(E) < 3\epsilon$.

Now let $\omega \in \Omega \setminus E$. Every cell of $X(\omega) \cap C_1$ is contained in the ball $2rsB^d$. Hence, for its determination only the hyperplanes of $\widehat{X}(\omega)$ in the set

$$\mathcal{E} := \{H \in \mathcal{E}_{d-1}^d : H \cap 2rsB^d \neq \emptyset\}$$

are needed. Analogously, the determination of a cell from $X(\omega) \cap (C_2 + x)$ requires only the hyperplanes of $\widehat{X}(\omega)$ in the set

$$\mathcal{E}_x := \{H \in \mathcal{E}_{d-1}^d : H \cap (2rsB^d + x) \neq \emptyset\}.$$

Because of $\widehat{X}(\omega)(\mathcal{B}_{2rs,z,a}) = 0$, no hyperplane of $\widehat{X}(\omega)$ belongs to $\mathcal{E} \cap \mathcal{E}_x$, hence every cell of $X(\omega) \cap (C_2 + x)$ depends only on the hyperplanes of $\widehat{X}(\omega)$ belonging to $\mathcal{E}_x \setminus \mathcal{E}$. Since the processes $\widehat{X} \cap \mathcal{E}$ and $\widehat{X} \cap (\mathcal{E}_x \setminus \mathcal{E})$ are stochastically independent, a similar argument to that in the proof of Theorem 10.5.1 yields the estimate

$$|\mathbb{P}(X \cap C_1) = \emptyset, X \cap (C_2 + x) = \emptyset) - \mathbb{P}(X \cap C_1 = \emptyset)\mathbb{P}(X \cap C_2 = \emptyset)| < 9\epsilon.$$

This yields (10.58) and thus the assertion. □

Finally, we remark that the assumption on the directional distribution made in Theorem 10.5.3 is necessary for the mixing property. To see this, let $\widehat{X}$ be a nondegenerate stationary Poisson hyperplane process in $\mathbb{R}^d$, the spherical directional distribution $\widehat{\varphi}$ of which satisfies $\widehat{\varphi}(S(z,0)) > 0$ for some vector $z \in S^{d-1}$. Let X be the hyperplane mosaic generated by $\widehat{X}$. By definition, X is mixing if and only if the dynamical system $(\mathsf{N}, \mathcal{N}, \mathbb{P}_X, \mathcal{T})$ defined after Theorem 9.3.3 is mixing, which means that

$$\lim_{\|x\| \to \infty} \mathbb{P}_X(A \cap T_x B) = \mathbb{P}_X(A)\mathbb{P}_X(B)$$

for $A, B \in \mathcal{N}$. We consider the set $A := \{\eta \in \mathsf{N} : \operatorname{supp} \eta \in \mathbb{M}^*$, $\operatorname{supp} \eta$ contains a cell Z having a facet F with normal vector in $S(z,0)$, and $F \cap \operatorname{int} B^d \neq \emptyset\}$. With

$$\mathcal{E}_z := \{H(u,\tau) : u \in S(z,0),\ |\tau| < 1\},$$

the condition $X \in A$ is equivalent to $\widehat{X}(\mathcal{E}_z) > 0$, hence

$$p := \mathbb{P}_X(A) = \mathbb{P}(\widehat{X}(\mathcal{E}_z) > 0) = 1 - \mathrm{e}^{-2\widehat{\gamma}\widehat{\varphi}(S(z,0))}$$

and $0 < p < 1$. For all vectors $x \in \mathbb{R}^d$ which are multiples of z we have $\mathcal{E}_z + x = \mathcal{E}_z$ and hence

$$\mathbb{P}_X(A \cap T_x A) = \mathbb{P}(X \in A, X - x \in A) = \mathbb{P}(\widehat{X}(\mathcal{E}_z) > 0,\ \widehat{X}(\mathcal{E}_z + x) > 0) = p.$$

This gives

$$\lim_{t \to \infty} \mathbb{P}_X(A \cap T_{tz} A) = p \neq \mathbb{P}_X(A)^2.$$

Hence, X is not mixing.

11

Non-stationary Models

Although the main theme of this book is random geometric structures with invariance properties, such as stationarity or isotropy, we conclude with an outlook to some of the extensions that are possible without such assumptions. The invariance properties in previous chapters allowed us to employ integral geometric formulas for obtaining results on geometric mean values. Our set-up followed also the historical development of the field, where from the beginning stationarity and isotropy seemed to be natural and convenient conditions to get simple and applicable formulas. Their counterparts for non-isotropic random sets and particle processes are necessarily more complicated, as we have seen in some of the previous sections. However, once the step from isotropic to non-isotropic structures is made, the question arises whether a similar generalization from stationary to non-stationary structures is possible. Although random sets and point processes without any invariance properties have been studied by many authors under different aspects, one might get the impression that, for example, the mean value formulas for Boolean models, which are at the heart of stochastic geometry, rely on the invariance of the model. Surprisingly, this is not the case. As the dissertation of Fallert [222] showed (see also [223]), specific intrinsic volumes for Boolean models with convex or polyconvex grains can be introduced without any invariance requirements, and the formulas obtained in Section 9.1 transfer to this situation in a suitably generalized form. Even more astonishing is the fact that these local mean value formulas for non-stationary Boolean models (and Poisson particle processes) make heavy use of the iterated formulas of translative integral geometry, as we have discussed in Section 6.4. Thus, although we do not require that the distributions of our random structures are invariant with respect to the translation group, the corresponding integral geometric setting still plays an essential role.

Fallert's dissertation, which contained results on several non-stationary models (particle processes, Boolean models, processes of flats, random mosaics), initiated various further publications in which counterparts to formulas in the stationary case were established without the assumption of stationarity.

In this chapter, we present some of these generalizations, mostly concentrating on results which are in analogy to the ones discussed in previous sections.

11.1 Particle Processes and Boolean Models

We consider a particle process X on $\mathcal{C}'$ in $\mathbb{R}^d$. Although we do not assume any invariance of the distribution of X, we require some regularity of the intensity measure Θ (which is assumed to be locally finite, as always). Namely, we assume that a decomposition

$$\Theta(A) = \int_{\mathcal{C}_0} \int_{\mathbb{R}^d} \mathbf{1}_A(C + x)\eta(C, x)\, \lambda(\mathrm{d}x)\, \mathbb{Q}(\mathrm{d}C), \qquad A \in \mathcal{B}(\mathcal{C}'), \tag{11.1}$$

exists, with a probability measure $\mathbb{Q}$ on $\mathcal{C}_0$ and a measurable function $\eta \geq 0$ on $\mathcal{C}_0 \times \mathbb{R}^d$. How restrictive is this assumption? Due to the topological properties of $\mathcal{C}'$, respectively those of $\mathcal{C}_0 \times \mathbb{R}^d$, a locally finite measure Θ on $\mathcal{C}'$ always has a decomposition

$$\Theta(A) = \int_{\mathcal{C}_0} \int_{\mathbb{R}^d} \mathbf{1}_A(C + x)\, \rho(C, \mathrm{d}x)\, \mathbb{Q}(\mathrm{d}C), \qquad A \in \mathcal{B}(\mathcal{C}'), \tag{11.2}$$

with a probability measure $\mathbb{Q}$ on $\mathcal{C}_0$ and a kernel $\rho : \mathcal{C}_0 \times \mathcal{B} \to \mathbb{R}^+$, that is, a function that is measurable in the first variable and is a locally finite measure in the second variable. This follows from the disintegration result for probability measures (see, e.g., Kallenberg [386, Th. 6.3]) by a simple extension argument (compare Kallenberg [387, Lemma 3.1]). Our additional assumption is that $\rho(C, \cdot)$ be absolutely continuous with respect to λ, for each C. In fact, if we assume this and denote the density by $\eta(C, \cdot)$, then the decomposition (11.2) transforms into (11.1).

We say that a locally finite measure Θ on $\mathcal{C}'$ admitting a decomposition (11.1) is **translation regular**. This name is chosen since Θ is translation regular if and only if it is absolutely continuous with respect to some translation invariant, locally finite measure $\widetilde{\Theta}$. In fact, for a given translation regular measure Θ with decomposition (11.1), one can choose $\widetilde{\Theta}$ as

$$\widetilde{\Theta}(A) = \int_{\mathcal{C}_0} \int_{\mathbb{R}^d} \mathbf{1}_A(C + x)\, \lambda(\mathrm{d}x)\, \mathbb{Q}(\mathrm{d}C), \qquad A \in \mathcal{B}(\mathcal{C}').$$

The other direction follows from Theorem 4.1.1. One should be aware of the fact that the decomposition (11.1) is not unique, in general. In fact, if $f > 0$ is a measurable function on $\mathcal{C}_0$ with $\int f \mathrm{d}\mathbb{Q} = 1$, then we can replace η by η/f and $\mathbb{Q}$ by $A \mapsto \int f \mathbf{1}_A \, \mathrm{d}\mathbb{Q}$, and (11.1) remains valid. We therefore say that the translation regular measure Θ is **represented by the pair** $(\eta, \mathbb{Q})$ if (11.1) holds.

It is sometimes convenient to modify this set-up slightly by imposing additional conditions. For example, we may require that η is continuous or that η depends only on the location, so that we have

$$\Theta(A) = \int_{\mathcal{C}_0} \int_{\mathbb{R}^d} \mathbf{1}_A(C + x)\eta(x)\,\lambda(\mathrm{d}x)\,\mathbb{Q}(\mathrm{d}C), \qquad A \in \mathcal{B}(\mathcal{C}'). \tag{11.3}$$

The former condition is sometimes helpful, since it implies that densities of geometric functionals exist at every point and not only almost everywhere. The latter condition has the advantage that it ensures that η and $\mathbb{Q}$ are uniquely determined. Namely, if we interpret X as a marked point process $\widehat{X}$ on $\mathbb{R}^d$ with mark space $\mathcal{C}_0$ (such that X is the image of $\widehat{X}$ under $(x, C) \mapsto C + x$), then $A \mapsto \int \eta \mathbf{1}_A \,\mathrm{d}\lambda$ is the intensity measure of the underlying unmarked point process in $\mathbb{R}^d$, and $\mathbb{Q}$ is the mark distribution. It is therefore natural to call the measure $\mathbb{Q}$ in (11.3) the **distribution of the typical grain** and η the **(spatial) intensity function** of X. If X is stationary, $\eta = \gamma$ is a constant.

If X is a Poisson process, (11.3) implies that $\widehat{X}$ is independently marked, whereas (11.1) allows dependencies between the marks (or between the marks and the points).

Up to here, we did not impose additional conditions on the shape of the particles and, in fact, some of the following results hold in this generality, for compact particles. This is particularly the case for the results on contact distributions of Boolean models, and we shall comment on these in the Notes. But since we now aim at defining specific intrinsic volumes, the restriction to convex particles seems natural. Some of the results can be generalized easily to particles in the convex ring, using the additivity of the functionals involved. This would require additional integrability conditions, therefore we leave such generalizations to the reader and assume convex grains, from now on. If the intensity measure Θ of a process X of convex particles has a representation (11.1), then its local finiteness is equivalent to

$$\int_{\mathcal{K}_0} \int_{\mathbb{R}^d} \mathbf{1}\{(K + x) \cap C \neq \emptyset\}\eta(K, x)\,\lambda(\mathrm{d}x)\,\mathbb{Q}(\mathrm{d}K) < \infty \quad \text{for } C \in \mathcal{C}. \tag{11.4}$$

If X is stationary, (11.4) is equivalent to (4.4).

General assumption. We assume throughout Sections 11.1 and 11.2 that the occurring particle processes satisfy (11.1) with locally finite Θ, and thus also (11.4).

In analogy to Sections 4.1 and 9.2, we now want to define densities of translation invariant and measurable functionals φ for the particle process X. Since these densities will depend on the location in space, they will be functions and not constants. Therefore, we need an appropriate local concept. As in Section 9.2, we start with a translation invariant, additive, and measurable functional $\varphi : \mathcal{R} \to \mathbb{R}$. In addition, we require that the restriction of φ to $\mathcal{K}$ is continuous and nonnegative. For simplicity, in this chapter, we call φ a **standard functional**. We say that φ has a **local extension** Φ if $\Phi : \mathcal{R} \times \mathcal{B} \to \mathbb{R}$ is a kernel, in the sense that $\Phi(\cdot, A)$ is a measurable function on $\mathcal{R}$ for each $A \in \mathcal{B}$ and $\Phi(K, \cdot)$ is a finite signed Borel measure on $\mathbb{R}^d$ for each $K \in \mathcal{R}$, and if Φ has the following properties:

- $\varphi(K) = \Phi(K, \mathbb{R}^d)$ for all $K \in \mathcal{R}$,
- $\Phi(K, \cdot) \geq 0$ for $K \in \mathcal{K}$,
- $\Phi(K, \cdot)$ is additive in K, for $K \in \mathcal{K}$,
- Φ is translation covariant, that is, satisfies $\Phi(K + x, A + x) = \Phi(K, A)$ for $K \in \mathcal{R}$, $A \in \mathcal{B}$, $x \in \mathbb{R}^d$,
- Φ is locally determined, that is, $\Phi(K, A) = \Phi(M, A)$ for $K, M \in \mathcal{R}$, $A \in \mathcal{B}$, if there is an open set $U \subset \mathbb{R}^d$ with $K \cap U = M \cap U$ and $A \subset U$,
- $K \mapsto \Phi(K, \cdot)$ is weakly continuous on $\mathcal{K}'$.

Typical examples of standard functionals having a local extension are, of course, the intrinsic volumes, but there are many others.

For a standard functional φ with local extension Φ, we define the φ-**density** $\overline{\varphi}(X, \cdot)$, as a function on $\mathbb{R}^d$, by

$$\overline{\varphi}(X, z) := \int_{\mathcal{K}_0} \int_{\mathbb{R}^d} \eta(K, z - x)\, \Phi(K, \mathrm{d}x)\, \mathbb{Q}(\mathrm{d}K).$$

If X is stationary, then

$$\overline{\varphi}(X, z) = \gamma \int_{\mathcal{K}_0} \varphi(K)\, \mathbb{Q}(\mathrm{d}K) = \overline{\varphi}(X)$$

is the φ-density defined in Section 9.2.

Theorem 11.1.1. *Let X be a process of convex particles in $\mathbb{R}^d$, and let φ be a standard functional with local extension Φ. Then*

$$\mathbb{E} \sum_{K \in X} \Phi(K, \cdot)$$

is a locally finite measure on $\mathbb{R}^d$ which is absolutely continuous with respect to λ, and $\overline{\varphi}(X, \cdot)$ is a corresponding density.

Moreover, we have

$$\overline{\varphi}(X, z) = \lim_{r \to 0} \frac{1}{V_d(rW)}\, \mathbb{E} \sum_{K \in X} \Phi(K, z + rW) \tag{11.5}$$

for λ-almost all $z \in \mathbb{R}^d$ and all $W \in \mathcal{K}$ with $V_d(W) > 0$.

Proof. In order to show the local finiteness, let $B \in \mathcal{B}$ be a bounded Borel set. Choose $r > 0$ with $B \subset \operatorname{int} rB^d$. Then, using Campbell's theorem, the facts that Φ is locally determined and that φ is continuous on $\mathcal{K}'$, we obtain

$$\begin{aligned} &\mathbb{E} \sum_{K \in X} \Phi(K, B) \\ &= \int_{\mathcal{K}_0} \int_{\mathbb{R}^d} \Phi(K + y, B) \eta(K, y)\, \lambda(\mathrm{d}y)\, \mathbb{Q}(\mathrm{d}K) \end{aligned}$$

$$\leq \int_{\mathcal{K}_0} \int_{\mathbb{R}^d} \varphi((K+y) \cap rB^d) \mathbf{1}\{(K+y) \cap rB^d \neq \emptyset\} \eta(K,y) \, \lambda(\mathrm{d}y) \, \mathbb{Q}(\mathrm{d}K)$$

$$\leq c(rB^d) \int_{\mathcal{K}_0} \int_{\mathbb{R}^d} \mathbf{1}\{(K+y) \cap rB^d \neq \emptyset\} \eta(K,y) \, \lambda(\mathrm{d}y) \, \mathbb{Q}(\mathrm{d}K)$$

$$< \infty,$$

by (11.4).

In a similar manner, we get

$$\begin{aligned} \mathbb{E} \sum_{K \in X} \Phi(K,B) &= \int_{\mathcal{K}_0} \int_{\mathbb{R}^d} \Phi(K+y,B) \eta(K,y) \, \lambda(\mathrm{d}y) \, \mathbb{Q}(\mathrm{d}K) \\ &= \int_{\mathcal{K}_0} \int_{\mathbb{R}^d} \int_{\mathbb{R}^d} \mathbf{1}_{B-y}(x) \eta(K,y) \, \Phi(K,\mathrm{d}x) \, \lambda(\mathrm{d}y) \, \mathbb{Q}(\mathrm{d}K) \\ &= \int_{\mathcal{K}_0} \int_{\mathbb{R}^d} \int_{\mathbb{R}^d} \mathbf{1}_{B}(z) \eta(K,z-x) \, \Phi(K,\mathrm{d}x) \, \lambda(\mathrm{d}z) \, \mathbb{Q}(\mathrm{d}K) \\ &= \int_B \left(\int_{\mathcal{K}_0} \int_{\mathbb{R}^d} \eta(K,z-x) \, \Phi(K,\mathrm{d}x) \, \mathbb{Q}(\mathrm{d}K) \right) \lambda(\mathrm{d}z) \end{aligned}$$

which proves the absolute continuity and the stated form of the density.

The limit relation follows from Lebesgue's differentiation theorem (see, e.g., Rudin [654, Th. 8.8] or Wheeden and Zygmund [811, Th. 7.2]). □

If $\eta(K,\cdot)$ is continuous, uniformly in K, then the function $\overline{\varphi}(X,\cdot)$ is continuous and, therefore, the limit relation (11.5) holds for all z.

As a first example of the application of Theorem 11.1.1, we choose $\varphi = V_j$, the jth intrinsic volume. The local extension of V_j is given by the curvature measure Φ_j. Thus, we obtain the following generalization of Corollary 9.4.2.

Corollary 11.1.1. *Let X be a process of convex particles in $\mathbb{R}^d$ and let $j \in \{0,\dots,d\}$. Then*

$$\mathbb{E} \sum_{K \in X} \Phi_j(K,\cdot)$$

is a locally finite measure on $\mathbb{R}^d$ which is absolutely continuous with respect to λ, and a density is given by

$$\begin{aligned} \overline{V}_j(X,z) &:= \int_{\mathcal{K}_0} \int_{\mathbb{R}^d} \eta(K,z-x) \, \Phi_j(K,\mathrm{d}x) \, \mathbb{Q}(\mathrm{d}K) \\ &= \lim_{r \to 0} \frac{1}{V_d(rW)} \, \mathbb{E} \sum_{K \in X} \Phi_j(K, z+rW) \end{aligned} \tag{11.6}$$

for λ-almost all $z \in \mathbb{R}^d$ and all $W \in \mathcal{K}$ with $V_d(W) > 0$.

One could have expected that a locally defined intrinsic volume $\overline{V}_j(X,z)$ should satisfy

$$\overline{V}_j(X,z) = \lim_{r\to 0} \frac{1}{V_d(rW)} \mathbb{E} \sum_{K\in X} V_j(K\cap(z+rW)),$$

but, for $j \in \{0,\dots,d-1\}$, this does not even make sense for stationary and isotropic X, since the limit on the right side does not exist in general, as one can see from (9.32).

For the second example, we choose $\varphi(K) = \binom{d}{j} V(K[j], -M[d-j])$, $j \in \{1,\dots,d-1\}$, with fixed $M \in \mathcal{K}'$. According to (6.25), the local extension is given by the mixed measure $\Phi^{(0)}_{j,d-j}(K,M;\cdot\times\mathbb{R}^d)$.

Corollary 11.1.2. *Let X be a process of convex particles in $\mathbb{R}^d$, let $M \in \mathcal{K}'$ and $j \in \{0,\dots,d\}$. Then*

$$\mathbb{E} \sum_{K\in X} \Phi^{(0)}_{j,d-j}(K,M;\cdot\times\mathbb{R}^d)$$

is a locally finite measure on $\mathbb{R}^d$ which is absolutely continuous with respect to λ, and a density is given by

$$\begin{aligned}
&\binom{d}{j}\overline{V}(X[j],-M[d-j];z)\\
&:= \int_{\mathcal{K}_0}\int_{\mathbb{R}^d} \eta(K,z-x)\,\Phi^{(0)}_{j,d-j}(K,M;\mathrm{d}x\times\mathbb{R}^d)\,\mathbb{Q}(\mathrm{d}K)\\
&= \lim_{r\to 0}\frac{1}{V_d(rW)}\,\mathbb{E}\sum_{K\in X}\Phi^{(0)}_{j,d-j}(K,M;(z+rW)\times\mathbb{R}^d)
\end{aligned}$$

for λ-almost all $z \in \mathbb{R}^d$ and all $W \in \mathcal{K}$ with $V_d(W) > 0$.

For $M = B^d$, Corollary 11.1.2 reduces to Corollary 11.1.1.

However, we may also let M vary and apply Theorem 11.1.1 a second time. Since this would involve independent copies of X, we state the corresponding result only for Poisson processes, to which Corollary 3.2.4 applies. Then we get a density for mixed volumes of the particle process $X^2_{\neq}$. For simplicity, we also omit the corresponding local limit relations in the following results.

Corollary 11.1.3. *Let X be a Poisson process of convex particles in $\mathbb{R}^d$ and let $j \in \{0,\dots,d\}$. Then*

$$\mathbb{E} \sum_{(K,M)\in X^2_{\neq}} \Phi^{(0)}_{j,d-j}(K,M;\cdot)$$

is a locally finite measure on $(\mathbb{R}^d)^2$ which is absolutely continuous with respect to λ^2, and a density is given by

$$\binom{d}{j}\overline{V}(X[j],-X[d-j];z_1,z_2)$$
$$:= \int_{\mathcal{K}_0}\int_{\mathcal{K}_0}\int_{(\mathbb{R}^d)^2} \eta(K_1,z_1-x_1)\eta(K_2,z_2-x_2)\,\Phi^{(0)}_{j,d-j}(K_1,K_2;\mathrm{d}(x_1,x_2))$$
$$\times\,\mathbb{Q}(\mathrm{d}K_1)\,\mathbb{Q}(\mathrm{d}K_2)$$

for λ^2-almost all $(z_1,z_2)\in(\mathbb{R}^d)^2$.

As a further generalization of all three corollaries, we may consider the mixed functional $V^{(j)}_{m_1,\dots,m_k}(K_1,\dots,K_k)$ introduced in Section 6.4. We can keep some of the K_i fixed and let the others vary in X. Repeated application of Theorem 11.1.1 to the Poisson process X, where the local extension at each step uses the mixed measures $\Phi^{(j)}_{m_1,\dots,m_k}(K_1,\dots,K_k;\cdot)$ in a suitable way, yields the existence of the density $\overline{V}^{(j)}_{m_1,\dots,m_n,m_{n+1},\dots,m_k}(X,\dots,X,K_{n+1},\dots,K_k;\cdot)$ as a function on $(\mathbb{R}^d)^n$. We formulate this result only for the case $n=k$.

Corollary 11.1.4. *Let X be a Poisson process of convex particles in $\mathbb{R}^d$, let $k\in\mathbb{N}$, $j\in\{0,\dots,d\}$ and $m_1,\dots,m_k\in\{j,\dots,d\}$ with*

$$\sum_{i=1}^k m_i=(k-1)d+j.$$

Then,

$$\mathbb{E}\sum_{(K_1,\dots,K_k)\in X^k_{\neq}}\Phi^{(j)}_{m_1,\dots,m_k}(K_1,\dots,K_k;\cdot)$$

is a locally finite measure on $(\mathbb{R}^d)^k$ which is absolutely continuous with respect to λ^k, and a density is given by

$$\overline{V}^{(j)}_{m_1,\dots,m_k}(X,\dots,X;z_1,\dots,z_k)$$
$$:= \int_{\mathcal{K}_0}\cdots\int_{\mathcal{K}_0}\int_{(\mathbb{R}^d)^k}\eta(K_1,z_1-x_1)\cdots\eta(K_k,z_k-x_k)$$
$$\times\,\Phi^{(j)}_{m_1,\dots,m_k}(K_1,\dots,K_k;\mathrm{d}(x_1,\dots,x_k))\,\mathbb{Q}(\mathrm{d}K_1)\cdots\,\mathbb{Q}(\mathrm{d}K_k)$$

for λ^k-almost all $(z_1,\dots,z_k)\in(\mathbb{R}^d)^k$.

We remark that the densities $\overline{V}^{(j)}_{m_1,\dots,m_k}(X,\dots,X;\cdot,\dots,\cdot)$ inherit the important properties of the mixed functionals and mixed measures, namely they are symmetric with respect to a permutation of the indices $m_1,\dots,m_k$ (and the corresponding variables), and they obey a decomposition property: if $m_1=d$, then

$$\overline{V}^{(j)}_{m_1,\dots,m_k}(X,\dots,X;z_1,\dots,z_k)$$
$$=\overline{V}_d(X,z_1)\overline{V}^{(j)}_{m_2,\dots,m_k}(X,\dots,X;z_2,\dots,z_k). \qquad (11.7)$$

We now turn to Boolean models $Z = Z_X$, where X is a Poisson process on $\mathcal{K}'$ (we still assume that X satisfies (11.1) and (11.4)). Let Z be a Boolean model with convex grains in $\mathbb{R}^d$, let $K \in \mathcal{K}'$ and $\varphi : \mathcal{R} \to \mathbb{R}$ be a measurable, additive and conditionally bounded functional. Then we have

$$\mathbb{E}\,|\varphi(Z \cap K)| < \infty \tag{11.8}$$

and (recall that $K^x := K + x$)

$$\begin{aligned} &\mathbb{E}\,\varphi(Z \cap K) \\ &= \sum_{k=1}^{\infty} \frac{(-1)^{k-1}}{k!} \int_{\mathcal{K}_0} \cdots \int_{\mathcal{K}_0} \int_{(\mathbb{R}^d)^k} \varphi(K \cap K_1^{x_1} \cap \ldots \cap K_k^{x_k}) \\ &\quad \times \eta(K_1, x_1) \cdots \eta(K_k, x_k)\, \lambda^k(\mathrm{d}(x_1, \ldots, x_k))\, \mathbb{Q}(\mathrm{d}K_1) \cdots \mathbb{Q}(\mathrm{d}K_k). \end{aligned} \tag{11.9}$$

This follows from (9.8), together with the special form of the intensity measure.

In addition, we now assume that φ is a standard functional (hence translation invariant) with local extension Φ. As in Section 6.4, we can infer that there are uniquely determined kernels $\Phi_{(k)} : \mathcal{K}^k \times \mathcal{B}((\mathbb{R}^d)^k) \to \mathbb{R}^+$, for $k = 1, 2, \ldots$, such that

$$\begin{aligned} &\int_{(\mathbb{R}^d)^{k-1}} \Phi(K_1 \cap K_2^{x_2} \cap \ldots \cap K_k^{x_k}, A_1 \cap A_2^{x_2} \cap \ldots \cap A_k^{x_k})\, \lambda^{k-1}(\mathrm{d}(x_2, \ldots, x_k)) \\ &= \Phi_{(k)}(K_1, \ldots, K_k; A_1 \times \ldots \times A_k) \end{aligned} \tag{11.10}$$

holds for all $k \in \mathbb{N}$, $K_1, \ldots, K_k \in \mathcal{K}$, $A_1, \ldots, A_k \in \mathcal{B}$. Namely, (11.10) for all Borel sets $A_1, \ldots, A_k$ is equivalent to

$$\begin{aligned} &\int_{(\mathbb{R}^d)^{k-1}} \int_{\mathbb{R}^d} g(x_1, x_1 - x_2, \ldots, x_1 - x_k)\, \Phi(K_1 \cap K_2^{x_2} \cap \ldots \cap K_k^{x_k}, \mathrm{d}x_1) \\ &\times \lambda^{k-1}(\mathrm{d}(x_2, \ldots, x_k)) \\ &= \int_{(\mathbb{R}^d)^k} g(x_1, \ldots, x_k)\, \Phi_{(k)}(K_1, \ldots, K_k; \mathrm{d}(x_1, \ldots, x_k)) \end{aligned} \tag{11.11}$$

for all continuous functions g on $(\mathbb{R}^d)^k$, provided that the measure on the right side exists. Due to the properties of Φ, the mapping

$$g \mapsto \int_{\mathbb{R}^d} g(x_1, x_1 - x_2, \ldots, x_1 - x_k)\, \Phi(K_1 \cap K_2^{x_2} \cap \ldots \cap K_k^{x_k}, \mathrm{d}x_1)$$

is continuous on $\mathbf{C}_c((\mathbb{R}^d)^k)$, for λ^{k-1}-almost all $(x_2, \ldots, x_k)$. Therefore, the left side of (11.11) defines a positive linear functional T on $\mathbf{C}_c((\mathbb{R}^d)^k)$ through

$$\begin{aligned} T(g) := &\int_{(\mathbb{R}^d)^{k-1}} \int_{\mathbb{R}^d} g(x_1, x_1 - x_2, \ldots, x_1 - x_k)\, \Phi(K_1 \cap K_2^{x_2} \cap \ldots \cap K_k^{x_k}, \mathrm{d}x_1) \\ &\times \lambda^{k-1}(\mathrm{d}(x_2, \ldots, x_k)). \end{aligned}$$

The existence and uniqueness of the measure $\Phi_{(k)}(K_1, \ldots, K_k; \cdot)$ now follows from the Riesz representation theorem. Since

$$(K_1, \ldots, K_k) \mapsto \Phi(K_1 \cap K_2^{x_2} \cap \ldots \cap K_k^{x_k}, \cdot)$$

is continuous on $(\mathcal{K}')^k$, for λ^{k-1}-almost all $(x_2, \ldots, x_k) \in (\mathbb{R}^d)^{k-1}$ (and by our assumptions on Φ), we obtain the continuity (and hence measurability) of

$$(K_1, \ldots, K_k) \mapsto \Phi_{(k)}(K_1, \ldots, K_k; \cdot).$$

Finally, $\Phi_{(k)}(K_1, \ldots, K_k; A_1 \times \ldots \times A_k)$ is invariant under simultaneous permutations of the bodies K_i and the sets A_i.

We call $\Phi_{(1)}, \Phi_{(2)}, \ldots$ the **associated kernels** of Φ. We remark that, since Φ is locally determined, the same is true for the kernel $\Phi_{(k)}$. Therefore, we can replace the convex body K_i by an unbounded convex set, as long as the corresponding Borel set A_i is bounded. Also, the translation covariance of Φ implies that $\Phi_{(k)}$ is translation covariant in each variable K_i (with associated Borel set A_i).

Since φ and the local extension Φ are defined for sets $K \in \mathcal{R}$ and since Φ is locally determined, the value $\Phi(Z, A)$ exists for bounded Borel sets $A \in \mathcal{B}$ and yields a (random) signed Radon measure $\Phi(Z, \cdot)$. We now show that $\mathbb{E}\,\Phi(Z, \cdot)$ is absolutely continuous and prove a representation of the density.

Theorem 11.1.2. *Let Z be a Boolean model in $\mathbb{R}^d$ with convex grains and φ a standard functional with local extension Φ and associated kernels $\Phi_{(k)}$, $k \in \mathbb{N}$. Then*

$$\mathbb{E}\,\Phi(Z, \cdot)$$

is a signed Radon measure on $\mathbb{R}^d$ which is absolutely continuous with respect to λ. For λ-almost all $z \in \mathbb{R}^d$, its density $\overline{\varphi}(Z, \cdot)$ satisfies

$$\overline{\varphi}(Z, z) = \sum_{k=1}^{\infty} \frac{(-1)^{k-1}}{k!} \int_{\mathcal{K}_0} \cdots \int_{\mathcal{K}_0} \int_{(\mathbb{R}^d)^k} \eta(K_1, z - x_1) \cdots \eta(K_k, z - x_k)$$
$$\times\, \Phi_{(k)}(K_1, \ldots, K_k; \mathrm{d}(x_1, \ldots, x_k))\, \mathbb{Q}(\mathrm{d}K_1) \cdots \mathbb{Q}(\mathrm{d}K_k).$$

Proof. For the local finiteness, let $B \in \mathcal{B}$ be a bounded Borel set with $B \subset \operatorname{int} rB^d$, for some $r > 0$. Applying (11.8) with $\varphi = \Phi(\cdot, B)$ and $K = rB^d$, we obtain

$$\mathbb{E}\,|\Phi(Z, B)| < \infty.$$

Moreover, from (11.9) and (11.10), it follows that

$$\mathbb{E}\,\Phi(Z, B)$$
$$= \sum_{k=1}^{\infty} \frac{(-1)^{k-1}}{k!} \int_{\mathcal{K}_0} \cdots \int_{\mathcal{K}_0} \int_{(\mathbb{R}^d)^k} \int_{\mathbb{R}^d} \mathbf{1}_B(x_0)\, \Phi(rB^d \cap K_1^{x_1} \cap \ldots \cap K_k^{x_k}, \mathrm{d}x_0)$$
$$\times\, \eta(K_1, x_1) \cdots \eta(K_k, x_k)\, \lambda^k(\mathrm{d}(x_1, \ldots, x_k))\, \mathbb{Q}(\mathrm{d}K_1) \cdots \mathbb{Q}(\mathrm{d}K_k)$$

$$= \sum_{k=1}^{\infty} \frac{(-1)^{k-1}}{k!} \int_{\mathcal{K}_0} \cdots \int_{\mathcal{K}_0} \int_{(\mathbb{R}^d)^{k+1}} \mathbf{1}_B(x_0)\eta(K_1, x_0 - x_1) \cdots \eta(K_k, x_0 - x_k)$$

$$\times \Phi_{(k+1)}(rB^d, K_1, \ldots, K_k; \mathrm{d}(x_0, \ldots, x_k))\, \mathbb{Q}(\mathrm{d}K_1) \cdots \mathbb{Q}(\mathrm{d}K_k)$$

$$= \sum_{k=1}^{\infty} \frac{(-1)^{k-1}}{k!} \int_{\mathcal{K}_0} \cdots \int_{\mathcal{K}_0} \int_{(\mathbb{R}^d)^{k+1}} \mathbf{1}_B(x_0)\eta(K_1, x_0 - x_1) \cdots \eta(K_k, x_0 - x_k)$$

$$\times \Phi_{(k+1)}(\mathbb{R}^d, K_1, \ldots, K_k; \mathrm{d}(x_0, \ldots, x_k))\, \mathbb{Q}(\mathrm{d}K_1) \cdots \mathbb{Q}(\mathrm{d}K_k).$$

Here we have used that $\Phi_{(k+1)}$ is locally determined.

The translation covariance of $\Phi_{(k+1)}$ in the first variable shows that

$$\Phi_{(k+1)}(\mathbb{R}^d, K_1, \ldots, K_k; \cdot) = \lambda \otimes \Phi_{(k)}(K_1, \ldots, K_k; \cdot).$$

In fact, for bounded $A \in \mathcal{B}$,

$$\Phi_{(k+1)}(\mathbb{R}^d, K_1, \ldots, K_k; A \times A_1 \times \ldots \times A_k)$$

$$= \int_{(\mathbb{R}^d)^k} \Phi(K_1^{x_1} \cap \ldots \cap K_k^{x_k}, A \cap A_1^{x_1} \cap \ldots \cap A_k^{x_k})\, \lambda^k(\mathrm{d}(x_1, \ldots, x_k))$$

$$= \int_{(\mathbb{R}^d)^{k-1}} \int_{\mathbb{R}^d} \Phi(K_1 \cap K_2^{x_2} \cap \ldots \cap K_k^{x_k}, (A - x) \cap A_1 \cap A_2^{x_2} \ldots \cap A_k^{x_k})$$

$$\times \lambda(\mathrm{d}x)\, \lambda^{k-1}(\mathrm{d}(x_2, \ldots, x_k))$$

$$= \lambda(A) \int_{(\mathbb{R}^d)^{k-1}} \Phi(K_1 \cap K_2^{x_2} \cap \ldots \cap K_k^{x_k}, A_1 \cap A_2^{x_2} \ldots \cap A_k^{x_k})$$

$$\times \lambda^{k-1}(\mathrm{d}(x_2, \ldots, x_k)),$$

by Theorem 5.2.1.

Hence, we conclude from Fubini's theorem that

$$\mathbb{E}\,\Phi(Z, B)$$

$$= \sum_{k=1}^{\infty} \frac{(-1)^{k-1}}{k!} \int_{\mathcal{K}_0} \cdots \int_{\mathcal{K}_0} \int_{(\mathbb{R}^d)^k}$$

$$\times \int_{\mathbb{R}^d} \mathbf{1}_B(x_0)\eta(K_1, x_0 - x_1) \cdots \eta(K_k, x_0 - x_k)\, \lambda(\mathrm{d}x_0)$$

$$\times \Phi_{(k)}(K_1, \ldots, K_k; \mathrm{d}(x_1, \ldots, x_k))\, \mathbb{Q}(\mathrm{d}K_1) \cdots \mathbb{Q}(\mathrm{d}K_k)$$

$$= \int_B \left(\sum_{k=1}^{\infty} \frac{(-1)^{k-1}}{k!} \int_{\mathcal{K}_0} \cdots \int_{\mathcal{K}_0} \int_{(\mathbb{R}^d)^k} \eta(K_1, z - x_1) \cdots \eta(K_k, z - x_k) \right.$$

$$\left. \times \Phi_{(k)}(K_1, \ldots, K_k; \mathrm{d}(x_1, \ldots, x_k))\, \mathbb{Q}(\mathrm{d}K_1) \cdots \mathbb{Q}(\mathrm{d}K_k) \right) \lambda(\mathrm{d}z).$$

This confirms the result. □

We apply Theorem 11.1.2 with $\varphi = V_j$. The local extension of $V_j(K)$ is the jth curvature measure $\Phi_j(K,\cdot)$. For the associated kernel $(\Phi_j)_{(k)}$, Theorem 6.4.1 yields

$$(\Phi_j)_{(k)}(K_1,\dots,K_k;\cdot) = \sum_{\substack{m_1,\dots,m_k=j \\ m_1+\dots+m_k=(k-1)d+j}}^{d} \Phi^{(j)}_{m_1,\dots,m_k}(K_1,\dots,K_k;\cdot).$$

Hence, $\mathbb{E}\,\Phi_j(Z,\cdot)$ is absolutely continuous with density a.e. given by

$$\begin{aligned}
&\overline{V}_j(Z,z) \\
&= \sum_{k=1}^{\infty} \frac{(-1)^{k-1}}{k!} \sum_{\substack{m_1,\dots,m_k=j \\ m_1+\dots+m_k=(k-1)d+j}}^{d} \int_{\mathcal{K}_0} \cdots \int_{\mathcal{K}_0} \int_{(\mathbb{R}^d)^k} \eta(K_1, z-x_1)\cdots \\
&\quad \times \eta(K_k, z-x_k)\, \Phi^{(j)}_{m_1,\dots,m_k}(K_1,\dots,K_k; \mathrm{d}(x_1,\dots,x_k))\, \mathbb{Q}(\mathrm{d}K_1)\cdots \mathbb{Q}(\mathrm{d}K_k).
\end{aligned}$$

From Corollary 11.1.4 we obtain

$$\begin{aligned}
&\overline{V}_j(Z,z) \\
&= \sum_{k=1}^{\infty} \frac{(-1)^{k-1}}{k!} \sum_{\substack{m_1,\dots,m_k=j \\ m_1+\dots+m_k=(k-1)d+j}}^{d} \overline{V}^{(j)}_{m_1,\dots,m_k}(X,\dots,X; z,\dots,z).
\end{aligned}$$

We use the decomposition property (11.7) and get, with arguments similar to those in the deduction of Theorem 9.1.3,

$$\begin{aligned}
&\overline{V}_j(Z,z) \\
&= \sum_{s=1}^{d-j} \sum_{r=0}^{\infty} \binom{r+s}{r} \frac{(-1)^{r+s-1}}{(r+s)!} \overline{V}_d(X,z)^r \\
&\quad \times \sum_{\substack{m_1,\dots,m_s=j \\ m_1+\dots+m_s=(s-1)d+j}}^{d-1} \overline{V}^{(j)}_{m_1,\dots,m_s}(X,\dots,X; z,\dots,z) \\
&= -\mathrm{e}^{-\overline{V}_d(X,z)} \sum_{s=1}^{d-j} \frac{(-1)^s}{s!} \sum_{\substack{m_1,\dots,m_s=j \\ m_1+\dots+m_s=(s-1)d+j}}^{d-1} \overline{V}^{(j)}_{m_1,\dots,m_s}(X,\dots,X; z,\dots,z) \\
&= \mathrm{e}^{-\overline{V}_d(X,z)} \Bigg(\overline{V}_j(X,z) - \sum_{s=2}^{d-j} \frac{(-1)^s}{s!} \\
&\quad \times \sum_{\substack{m_1,\dots,m_s=j+1 \\ m_1+\dots+m_s=(s-1)d+j}}^{d-1} \overline{V}^{(j)}_{m_1,\dots,m_s}(X,\dots,X; z,\dots,z) \Bigg).
\end{aligned}$$

Hence, we arrive at the following result.

Theorem 11.1.3. *Let Z be a Boolean model in $\mathbb{R}^d$ with convex grains. Then, for λ-almost all z,*

$$\overline{V}_d(Z, z) = 1 - \mathrm{e}^{-\overline{V}_d(X,z)}, \tag{11.12}$$

$$\overline{V}_{d-1}(Z, z) = \mathrm{e}^{-\overline{V}_d(X,z)}\overline{V}_{d-1}(X, z),$$

and

$$\overline{V}_j(Z, z) = \mathrm{e}^{-\overline{V}_d(X,z)} \left(\overline{V}_j(X, z) - \sum_{s=2}^{d-j} \frac{(-1)^s}{s!} \right. \\ \left. \times \sum_{\substack{m_1,\dots,m_s=j+1 \\ m_1+\dots+m_s=(s-1)d+j}}^{d-1} \overline{V}^{(j)}_{m_1,\dots,m_s}(X, \dots, X; z, \dots, z) \right),$$

for $j = 0, \dots, d-2$.

If Z is stationary, this reduces to Theorem 9.1.5, and if Z is also isotropic, we get Theorem 9.1.3.

Notes for Section 11.1

1. As we have already mentioned, specific intrinsic volumes for non-stationary (Poisson) particle processes and Boolean models were introduced by Fallert [222, 223]. There, one also finds Corollaries 11.1.1, 11.1.4 and Theorem 11.1.3. Corollaries 11.1.2, 11.1.3 are special cases of more general results in Weil [801]. Theorem 6 in [801] gives formulas for the density of mixed volumes,

$$\overline{V}(Z[j], M[d-j], z)$$

for a Boolean model Z with polyconvex grains and a fixed body $M \in \mathcal{K}'$, which are in analogy to Theorem 11.1.3. The proof of Theorem 6 contains some misprints ($z_1, \dots, z_k$ have to be replaced by $z, \dots, z$ and $\lambda(\mathrm{d}z_1) \cdots \lambda(\mathrm{d}z_k)$ by $\lambda(\mathrm{d}z)$). The paper [801] also presents more explicit formulas for the densities $\overline{V}_i(Z, \cdot)$, $i = 0, 1, 2$, for a planar Boolean model with circular grains.

Formulas for densities of some of the intrinsic volumes (volume density, surface area density) for non-stationary Boolean models of (deterministic or random) balls have also been obtained by Hahn, Micheletti, Pohlink and Stoyan [314], K. Mecke [505, 506], Micheletti and Stoyan [516], Quintanilla and Torquato [609, 610].

2. In Note 4 to Section 9.1 we have remarked that, for a stationary Boolean model Z with convex grains in dimensions 2 and 3, densities for mixed volumes of Z determine the intensity γ uniquely. For non-stationary Boolean models in $\mathbb{R}^2$, a corresponding result was obtained by Weil [799]. It was shown that the values $\overline{V}_0(Z, z)$, $\overline{V}(Z[1], M[1]; z)$, for all $M \in \mathcal{K}'$ and $\overline{V}_2(Z, z)$, determine the specific Euler characteristic $\overline{V}_0(X, z)$ at z uniquely. The corresponding three-dimensional case was settled in Goodey and Weil [280] under a symmetry condition. Without this, a uniqueness result for $\overline{V}_0(X, z)$ was shown, if instead of the local mean mixed volumes $\overline{V}(Z[1], M[2]; z)$ and $\overline{V}(Z[2], M[1]; z)$ the densities of support functions and surface area measures for Z at z are given (see also the following note).

3. We have applied formula (11.9) mainly to real functionals φ. It can also be applied to measure- or function-valued functionals. In particular, this yields the existence of the density $\overline{h}(Z,u;z)$ of the centered support function $h^*(K,u)$, $u \in \mathbb{R}^d$, and a formula expressing it in terms of densities of mixed centered support functions of the particles in X. In view of (6.28), the necessary local extension is given by the **support kernel** $\rho(K,u;\cdot)$, defined as

$$\rho(K,u;B) = \Phi^{(0)}_{1,d-1}(K,u^+;B \times A_{u^\perp}),$$

for $u \in S^{d-1}$ and $B \in \mathcal{B}$. This notion was first studied by Goodey and Weil [281]. Similarly, the existence of the density $\overline{S}(Z,B;z)$ of the surface area measure $S_{d-1}(K,B)$, $B \in \mathcal{B}(S^{d-1})$, follows. Its relation to the corresponding notion for X is given by

$$\overline{S}(Z,\cdot;z) = \mathrm{e}^{-\overline{V}_d(Z,z)}\,\overline{S}(X,\cdot;z).$$

The local extension is given here by a suitable support measure. These results were obtained in Goodey and Weil [280]. In contrast to the stationary case and due to the occurrence of the intensity function η, the function $\overline{h}(Z,\cdot;z)$ need no longer be centered (and for $d \geq 3$ also not convex), and the measure $\overline{S}(Z,\cdot;z)$ need no longer be a surface area measure. This indicates some of the difficulties arising in the non-stationary setting.

4. For a non-stationary particle process X and a functional φ, the densities $\overline{\varphi}(X,z)$ were introduced as functions depending on the location $z \in \mathbb{R}^d$, whereas, for stationary X, they do not depend on z. Conversely, one can ask whether invariance properties of X can be inferred from invariance properties of $\overline{\varphi}(X,z)$, for suitable functionals φ. Results of this type were obtained by Hoffmann [345, 347]. Assume that the intensity measure of the particle process X is of the form (11.3) with a continuous function η. Hoffmann defined the **generalized local mean normal measure** of X at $z \in \mathbb{R}^d$ by

$$\mu_z(A,B) := \int_{\mathcal{K}_0} \mathbf{1}_B(K) \int_{\mathbb{R}^d} \eta(z-x)\, \Xi_{d-1}(K, \mathrm{d}x \times A)\, \mathbb{Q}(\mathrm{d}K)$$

for $A \in \mathcal{B}(S^{d-1})$, $B \in \mathcal{B}(\mathcal{K}_0)$. An intuitive interpretation is obtained from

$$\mathbb{E} \sum_{K \in X} \mathbf{1}_B(K - c(K))\mathcal{H}^{d-1}(C \cap \tau(K,A)) = \int_C \mu_z(A,B)\, \lambda(\mathrm{d}z)$$

for $C \in \mathcal{B}$, where $\tau(K,A)$ denotes the set of boundary points of K for which an outer normal vector belongs to A. Under the assumption that $\dim K \geq d-1$ for $\mathbb{Q}$-almost all $K \in \mathcal{K}_0$ and that the support of $\mathbb{Q}$ contains some strictly convex body, Hoffmann proved that X is weakly stationary and weakly isotropic if and only if μ_z is rotation invariant, which means that $\mu_z(\vartheta A, \vartheta B) = \mu_z(A,B)$ for all $z \in \mathbb{R}^d$, $A \in \mathcal{B}(S^{d-1})$, $B \in \mathcal{B}(\mathcal{K}_0)$ and $\vartheta \in SO_d$. Hoffmann also showed a corresponding result for processes of convex cylinders. This comprises Theorem 1 of Schneider [707] (see Theorem 11.3.2 below), which was the motivation for Hoffmann's investigation.

5. Theorem 11.1.3 has been extended to Boolean models of cylinders by Hoffmann [345, 348]. Due to the local nature of the mixed measures, such an extension seems natural; the main effort went into finding the special form of the mixed measures for cylinders. Special cylinder processes were also studied by Spiess and Spodarev [732].

6. For a stationary Poisson process X on $\mathcal{K}'$ and the corresponding Boolean model Z the intersection density $\gamma_d(X)$ and the mean visible volume $\overline{V}_s(Z)$ were introduced and studied in Section 4.6, and sharp lower and upper estimates for the product $\gamma_d(X)\overline{V}_s(Z)$ were given in Theorem 4.6.3. Hoffmann [345] has studied intersection densities and mean visible volumes for non-stationary Poisson processes and Boolean models and has obtained some generalizations of Theorem 4.6.3. He has also considered intersection densities of a different kind, where the Hausdorff measure is replaced by a curvature measure.

11.2 Contact Distributions

We continue the investigation of general Boolean models Z with convex grains and consider generalized contact distributions. As an immediate generalization of the function introduced in Section 2.4, in the stationary case, we define the **contact distribution function** $H_B(x,\cdot)$ of a random closed set $Z \subset \mathbb{R}^d$ as the distribution function of the B-distance $d_B(x,Z)$ from a point $x \notin Z$ to Z, hence, for $r \geq 0$,

$$\begin{aligned} H_B(x,r) &:= \mathbb{P}((x+rB)\cap Z \neq \emptyset \,|\, x \notin Z) \\ &= \mathbb{P}(d_B(x,Z) \leq r \,|\, x \notin Z). \end{aligned}$$

Here the **gauge body** (or structuring element) B is a convex body containing 0, and we assume that the local volume fraction $\mathbb{P}(x \in Z) = \overline{V}_d(Z,x)$ is less than one (so that $\mathbb{P}(x \notin Z) > 0$).

For the Boolean model $Z = Z_X$, we use the same notations as in the previous section. Since

$$H_B(x,r) = \frac{\overline{V}_d(Z - rB, x) - \overline{V}_d(Z,x)}{1 - \overline{V}_d(Z,x)},$$

we obtain from (11.12) and (11.6)

$$\begin{aligned} H_B(x,r) &= \frac{\mathrm{e}^{-\overline{V}_d(X,x)} - \mathrm{e}^{-\overline{V}_d(X-rB,x)}}{\mathrm{e}^{-\overline{V}_d(X,x)}} \\ &= 1 - \exp\left(-\int_{\mathcal{K}_0}\int_{(K-rB)\setminus K} \eta(K,x-y)\,\lambda(\mathrm{d}y)\,\mathbb{Q}(\mathrm{d}K)\right). \quad (11.13) \end{aligned}$$

To the inner integral, we apply formula (14.26) involving the relative support measures $\Xi_j(K;B;\cdot)$ and get

$$\begin{aligned} &\int_{(K-rB)\setminus K} \eta(K,x-y)\,\lambda(\mathrm{d}y) \quad (11.14) \\ &= \sum_{j=0}^{d-1}(d-j)\kappa_{d-j}\int_0^r\int_{(\mathbb{R}^d)^2} t^{d-1-j}\eta(K,x-y-tb)\,\Xi_j(K;B;\mathrm{d}(y,b))\,\mathrm{d}t. \end{aligned}$$

The definition of the relative support measures requires that K and B have independent support sets (see Section 14.3). This is satisfied, for example, if one of the bodies K, B is strictly convex.

Inserting (11.14) into (11.13), we obtain the following theorem.

Theorem 11.2.1. *Let Z be a Boolean model in $\mathbb{R}^d$ with convex grains and let B be a gauge body. Assume that K and B have independent support sets, for $\mathbb{Q}$-almost all K. Then*

$$H_B(x,r) = 1 - \exp\left(-\int_0^r h_B(x,t)\,\mathrm{d}t\right)$$

for $r \geq 0$, with

$$h_B(x,t) := \sum_{j=0}^{d-1} (j+1)\kappa_{j+1} t^j \int_{\mathcal{K}_0} \int_{(\mathbb{R}^d)^2} \eta(K, x-y-tb) \\ \times \Xi_{d-1-j}(K; B; \mathrm{d}(y,b))\,\mathbb{Q}(\mathrm{d}K).$$

If $B = B^d$, the measure $\Xi_j(K;B;\cdot)$ is the (ordinary) support measure $\Xi_j(K,\cdot)$ of K. Hence, we obtain a formula for the **spherical contact distribution function** $H(x,\cdot)$ of Z.

Corollary 11.2.1. *For a Boolean model Z with convex grains, we have*

$$H(x,r) = 1 - \exp\left(-\int_0^r h(x,t)\,\mathrm{d}t\right)$$

for $r \geq 0$, with

$$h(x,t) := \sum_{j=0}^{d-1} (j+1)\kappa_{j+1} t^j \int_{\mathcal{K}_0} \int_{(\mathbb{R}^d)^2} \eta(K, x-y-tb)\, \Xi_{d-1-j}(K, \mathrm{d}(y,b))\,\mathbb{Q}(\mathrm{d}K).$$

Theorem 11.2.1 shows that $H_B(x,\cdot)$ is differentiable. In particular, if for $\mathbb{Q}$-almost all K the function $\eta(K,\cdot)$ is continuous, we get

$$\begin{aligned} \frac{\partial}{\partial r} H(x,r)\Big|_{r=0} &= h(x,0) \\ &= 2 \int_{\mathcal{K}_0} \int_{\mathbb{R}^d} \eta(K, x-y)\, \Phi_{d-1}(K, \mathrm{d}y)\,\mathbb{Q}(\mathrm{d}K) \\ &= 2\overline{V}_{d-1}(X, x) \end{aligned}$$

and thus

$$2\overline{V}_{d-1}(Z,x) = (1 - \overline{V}_d(Z,x)) \frac{\partial}{\partial r} H(x,r)\Big|_{r=0}.$$

Now we consider generalized contact distribution functions, involving directions and local geometric information in the contact points. As we have shown in Lemma 9.5.1, the distance $d_B(x, Z)$ is almost surely attained at a single particle Z_i of the underlying Poisson process X, thus $d_B(x, Z) = d_B(x, Z_i)$. This implies $x \in \operatorname{bd}(Z_i - rB)$ with $r := d_B(x, Z)$. If Z_i and B have almost surely independent support sets (as we shall assume below), the decomposition $x = z + rb$, $z \in \operatorname{bd} Z_i$, $b \in \operatorname{bd}(-B)$, is unique. With the notation introduced before Theorem 14.3.2, we have $z =: p_B(Z, x)$ and $b =: u_B(Z, x)$, thus

$$p_B(Z, x) = x - d_B(x, Z) u_B(Z, x).$$

We call $p_B(Z, x)$ the B-**contact point** in Z and $-u_B(Z, x)$ the B-**direction** from x to Z. For simplicity, we just speak of the contact point and the direction.

It is possible to exploit additional local information at the contact point. For this, we assume that a mapping $\rho : \mathcal{S} \times \mathbb{R}^d \to \mathbb{R}$ (where $\mathcal{S}$ is the extended convex ring) is given which is measurable and translation covariant, that is, satisfies $\rho(F + y, x + y) = \rho(F, x)$ for $F \in \mathcal{S}$ and $x, y \in \mathbb{R}^d$. Moreover, we assume that $\rho(F, x) = 0$ if $x \notin \operatorname{bd} F$ and that ρ is 'local' in the sense that, for any $x \in \mathbb{R}^d$ and any neighborhood U of x, we have $\rho(F, x) = \rho(F \cap U, x)$. For example, $\rho(F, x)$ could be the value of a curvature function of $\operatorname{bd} F$ at x, and 0 if this is not defined. In the following, we write

$$l_B(Z, x) := \rho(Z, p_B(Z, x)).$$

As a generalization of Theorem 11.2.1 (and also of Theorem 9.5.2) we show the following result.

Theorem 11.2.2. *Let Z be a Boolean model in $\mathbb{R}^d$ with convex grains and let B be a gauge body. Assume that K and B have independent support sets, for $\mathbb{Q}$-almost all K. Let $g \geq 0$ be a measurable function on $\mathbb{R}^+ \times \mathbb{R}^d \times \mathbb{R}$. Then, for $x \in \mathbb{R}^d$ with $\mathbb{P}(x \notin Z) > 0$, we have*

$$\begin{aligned}
&\mathbb{E}\left(\mathbf{1}\{d_B(x, Z) < \infty\} g(d_B(x, Z), u_B(Z, x), l_B(Z, x)) \,|\, x \notin Z\right) \\
&= \sum_{j=0}^{d-1} (j+1)\kappa_{j+1} \int_0^\infty t^j (1 - H_B(x, t)) \int_{\mathcal{K}_0} \int_{(\mathbb{R}^d)^2} g(t, b, \rho(K, y)) \\
&\quad \times \eta(K, x - y - tb)\, \Xi_{d-1-j}(K; B; \mathrm{d}(y, b))\, \mathbb{Q}(\mathrm{d}K)\, \mathrm{d}t.
\end{aligned}$$

Proof. We fix x with $\mathbb{P}(x \notin Z) > 0$. The following arguments are quite similar to those employed in the proof of Theorem 9.5.2; we even use some of the notation introduced there. Namely, for an enumeration $X = \{Z_1, Z_2, \ldots\}$ we define the events

$$A_n := \{0 < d_B(x, Z_n) < \infty\},$$

$$B_n := \{d_B(x, U(X \setminus \{Z_n\})) > d_B(x, Z_n)\}$$

and

$$C_n := \{(B, Z_n) \in \mathcal{K}^2_{ind}\},$$

where $\mathcal{K}^2_{ind}$ denotes the set of pairs $(K, M) \in (\mathcal{K}')^2$ of convex bodies with independent support sets. Then

$$(d_B(x,Z), u_B(Z,x), l_B(Z,x)) = (d_B(x,Z_n), u_B(Z_n,x), l_B(Z_n,x))$$

on $A_n \cap B_n \cap C_n$ and almost surely

$$\{0 < d_B(x,Z) < \infty\} = \bigcup_{n=1}^{\infty} (A_n \cap B_n \cap C_n).$$

We abbreviate

$$\widetilde{g}(K) := g(d_B(x,K), u_B(K,x), l_B(K,x))$$

for $K \in \mathcal{K}'$. Using Theorem 3.2.5 and formula (14.27), we obtain

$$\begin{aligned}
&\mathbb{E}\left(\mathbf{1}\{0 < d_B(x,Z) < \infty\} g(d_B(x,Z), u_B(Z,x), l_B(Z,x))\right) \\
&= \mathbb{E} \sum_{n=1}^{\infty} \mathbf{1}_{A_n \cap B_n \cap C_n} \widetilde{g}(Z_n) \\
&= \mathbb{E}\left(\sum_{K \in X} \mathbf{1}\{0 < d_B(x,K) < \infty\} \mathbf{1}\{(B,K) \in \mathcal{K}^2_{ind}\} \widetilde{g}(K) \right. \\
&\qquad \left. \times \mathbf{1}\{d_B(x, U(X \setminus \{K\})) > d_B(x,K)\} \right) \\
&= \int_{\mathcal{K}'} \mathbf{1}\{0 < d_B(x,K) < \infty\} \mathbf{1}\{(B,K) \in \mathcal{K}^2_{ind}\} \widetilde{g}(K) \\
&\qquad \times \mathbb{P}(d_B(x, U(X)) > d_B(x,K))\, \Theta(\mathrm{d}K) \\
&= \mathbb{P}(x \notin Z) \int_{\mathcal{K}_0} \int_{\mathbb{R}^d} \mathbf{1}\{0 < d_B(x, z+K) < \infty\} \widetilde{g}(z+K) \\
&\qquad \times (1 - H_B(x, d_B(x, z+K))) \eta(K,z)\, \lambda(\mathrm{d}z)\, \mathbb{Q}(\mathrm{d}K) \\
&= \mathbb{P}(x \notin Z) \sum_{j=0}^{d-1} (j+1)\kappa_{j+1} \int_0^{\infty} \int_{\mathcal{K}_0} \int_{(\mathbb{R}^d)^2} g(t, b, \rho(K,y)) \eta(K, x-y-tb) \\
&\qquad \times (1 - H_B(x,t)) t^j\, \Xi_{d-1-j}(K; B; \mathrm{d}(y,b))\, \mathbb{Q}(\mathrm{d}K)\, \mathrm{d}t.
\end{aligned}$$

Here we have used that $(y,b) \in \operatorname{supp} \Xi_{d-1-j}(K;B;\cdot)$ implies $\widetilde{g}(x-y-tb+K) = g(t,b,\rho(K,y))$. Division by $\mathbb{P}(x \notin Z)$ yields the assertion. □

We mention some special cases of this result. First, if g depends only on the distance $d_B(x,Z)$, Theorem 11.2.2 reduces to Theorem 11.2.1. This follows from the exponential formula of Lebesgue–Stieltjes calculus (see the corresponding more detailed argument given in Section 9.5). Next, for $B = B^d$, we obtain a result for the spherical contact distribution function, as a generalization of Corollary 11.2.1. Note that, for the spherical contact distribution, the condition $d(x,Z) < \infty$ is satisfied almost surely.

Corollary 11.2.2. *For a Boolean model Z with convex grains, a point $x \in \mathbb{R}^d$ with $\mathbb{P}(x \notin Z) > 0$, and a measurable function $g \geq 0$ on $\mathbb{R}^+ \times \mathbb{R}^d \times \mathbb{R}$, we have*

$$\begin{aligned} &\mathbb{E}\left(g(d(x,Z), u(Z,x), l(Z,x)) \,|\, x \notin Z\right) \\ &= \sum_{j=0}^{d-1} (j+1)\kappa_{j+1} \int_0^\infty t^j (1 - H(x,t)) \int_{\mathcal{K}_0} \int_{\mathbb{R}^d \times \mathbb{R}^d} g(t,u,\rho(K,y)) \\ &\quad \times \eta(K, x-y-tu)\, \Xi_{d-1-j}(K, \mathrm{d}(y,u))\, \mathbb{Q}(\mathrm{d}K)\, \mathrm{d}t. \end{aligned}$$

If Z is stationary, the formulas in Theorem 11.2.2 and Corollary 11.2.2 simplify only slightly, in that x can be replaced by 0 and the function η by the constant γ. Theorem 11.2.1 and Corollary 11.2.1 then reduce to the corresponding results in Theorem 9.1.1. A further simplification of Corollary 11.2.2 is possible if, for stationary Z (and $x = 0$), the function g depends only on $d(0,Z)$ and $u(Z,0)$. Then, the support measure $\Xi_{d-1-j}(K,\cdot)$ can be replaced by its image under $(y,u) \mapsto u$, the area measure $\Psi_{d-1-j}(K,\cdot)$.

Corollary 11.2.3. *For a stationary Boolean model Z with convex grains and a measurable function $g \geq 0$ on $\mathbb{R}^+ \times S^{d-1}$, we have*

$$\begin{aligned} &\mathbb{E}\left(g(d(0,Z), u(Z,0)) \,|\, 0 \notin Z\right) \\ &= \gamma \sum_{j=0}^{d-1} (j+1)\kappa_{j+1} \int_0^\infty t^j (1 - H(t)) \int_{\mathcal{K}_0} \int_{S^{d-1}} g(t,u) \\ &\quad \times \Psi_{d-1-j}(K, \mathrm{d}u)\, \mathbb{Q}(\mathrm{d}K)\, \mathrm{d}t. \end{aligned}$$

Notes for Section 11.2

1. The results on contact distributions for non-stationary Boolean models and their generalized versions have been obtained in Hug [356], Hug and Last [357], and Hug, Last and Weil [358]; a survey with additional results is Hug, Last and Weil [359] (see also shorter presentations in Weil [802, 803]). In Hug, Last and Weil [358], an even more general version of Theorem 11.2.1 and the subsequent results is obtained, where the point x is replaced by a test body $L \in \mathcal{K}'$. The considered function g can then also depend on the contact point $p_B(L,Z)$ in which the B-distance of L from Z is realized (provided this point is unique). The formula is proved in exactly the same way, but uses a more general Steiner-type result which involves mixed relative support measures depending on three convex bodies.

2. As was shown in Hug and Last [357] and Hug, Last and Weil [358], [359], the results on generalized contact distributions, which we proved here for Boolean models, hold true for random closed sets Z which are the union of a point process X on $\mathcal{K}'$, where the intensity measure Θ of X satisfies (11.4) and the second factorial moment measure $\Lambda^{(2)}$ of X has a certain smoothness property. The resulting formulas for Z are then formulated and proved with the Palm distribution of X. In this general framework, (generalized) contact distributions of Gibbs processes, Cox processes, Poisson cluster processes and more general cluster models (grain models where the underlying ordinary point process is a Poisson cluster process) can be subsumed; the results are surveyed in Hug, Last and Weil [359].

3. In Section 9.5, we have already stated and proved a special case of Theorem 11.2.1, namely Theorem 9.5.2. The latter result concerned a stationary Boolean model Z, where the grains are balls with random radius. In the discussion, we remarked that the intensity γ and the radius distribution $\mathbb{G}$ are determined by the generalized contact distributions of Z. One may expect that this result holds for more general Boolean models Z. The question, which information on the intensity function η and the distribution of the typical grain $\mathbb{Q}$ can be inferred from the generalized contact distributions of Z, is discussed in detail in Hug, Last and Weil [358, §4], and several uniqueness results are given.

4. Boolean models with compact grains. For the spherical contact distribution function and its variants, a far-reaching generalization was obtained by Hug, Last and Weil [361]. They proved a Steiner formula for arbitrary closed sets $F \subset \mathbb{R}^d$, by which support measures $\Xi_j(F,\cdot)$, $j = 0,\dots,d-1$, of F are defined. The latter are signed Radon-type measures on the normal bundle $\operatorname{Nor} F$ of F; they are defined on Borel sets $A \subset \operatorname{Nor} F$, for which the reach function $\delta(F,\cdot)$ is bounded away from 0 and ∞. Here, the **reach** $\delta(F,x,u)$, $(x,u) \in \operatorname{Nor} F$, is the largest $r \geq 0$ such that $x+ru$ has a unique projection point x in F. With the help of this Steiner formula, Boolean models with arbitrary compact grains can be considered (satisfying a condition analogous to (11.1)). The following is one of the results obtained in Hug, Last and Weil [361] (d_{dj} are given constants).

Let Z be a stationary Boolean model with compact grains. Then

$$H(r) = 1 - \exp\left(-\int_0^r h(t)\,\mathrm{d}t\right), \qquad r \geq 0,$$

with

$$h(t) := \sum_{j=0}^{d-1} d_{dj} t^j \gamma \int_{\mathcal{C}_0} \int_{\mathbb{R}^d \times S^{d-1}} \mathbf{1}\{t < \delta(C,x,u)\}\, \Xi_{d-1-j}(C, \mathrm{d}(x,u))\, \mathbb{Q}(\mathrm{d}C).$$

5. The general Steiner formula also yields results for (generalized) contact distributions of arbitrary stationary random closed sets $Z \subset \mathbb{R}^d$. Namely, let

$$H(t,A) := \mathbb{P}(d(0,Z) \leq t, u(0,Z) \in A \,|\, 0 \notin Z), \qquad t \geq 0,\ A \in \mathcal{B}(S^{d-1}).$$

Then,

$$(1-p)H(t,A) = \sum_{j=0}^{d-1} c_{dj} \int_0^t s^j \Gamma_{d-1-j}(A \times (s,\infty])\,\mathrm{d}s$$

with constants c_{di} and

$$\Gamma_i(\cdot) := \mathbb{E}\int_{C^d\times S^{d-1}} \mathbf{1}\{(u,\delta(Z,x,u))\in\cdot\}\,\Xi_i(Z,\mathrm{d}(x,u)).$$

Thus, $H(\cdot,A)$ is absolutely continuous and we have an explicit formula for the density.

Moreover, $H(\cdot,A)$ is differentiable with the exception of at most countably many points, but it need not be differentiable at 0. If

$$\mathbb{E}|\Xi_i|(Z,B\times S^{d-1})<\infty \qquad \text{for some } B\in\mathcal{B}(\mathbb{R}^d),\ \lambda(B)>0,\ i=0,\dots,d-1,$$

(which excludes fractal behavior, for example), then

$$\lim_{t\to 0+} t^{-1}(1-p)H(t,A)=\overline{S}_{d-1}(Z,A),$$

for $A\in\mathcal{B}(S^{d-1})$, where

$$\overline{S}_{d-1}(Z,A) := 2\mathbb{E}\,\Xi_{d-1}(Z;C^d\times A)\ <\infty.$$

In particular, for such random closed sets we have

$$(1-p)H'(0)=2\overline{V}_{d-1}(Z):=\overline{S}_{d-1}(Z,S^{d-1}).$$

Hence, for a stationary random set Z fulfilling the expectation condition above, the surface area density is defined and, even more, a mean surface area measure $\overline{S}_{d-1}(Z,\cdot)$ exists. The normalized measure

$$R(Z,\cdot)=\frac{\overline{S}_{d-1}(Z,\cdot)}{\overline{S}_{d-1}(Z,S^{d-1})}$$

is called the **rose of directions** of Z. It is the distribution of the (outer) normal in a typical point of $\mathrm{bd}\,Z$.

6. Characterization of convex grains. For a stationary Boolean model Z with convex grains and a gauge body B, Theorem 9.1.1 shows that

$$-\ln(1-H_B(r))$$

is a polynomial in $r\geq 0$ (of degree d). As we have mentioned earlier, this can be used, for $B=B^d$ or $B=[0,u]$, to obtain simple estimators for the intensity γ, the specific intrinsic volumes $\overline{V}_j(X)$ and other mean functionals, such as $\overline{S}_{d-1}(X,\cdot)$. For non-convex grains, the occurrence of the reach function in the formula, explained in Note 4 above, shows that we can no longer expect a polynomial behavior of contact distributions. This was made more precise by Heveling, Hug and Last [338] and Hug, Last and Weil [362] and yields a possibility to check the convexity of the grains.

Namely, let us consider a stationary Boolean model Z with compact and regular grains (the latter means that $C=\mathrm{cl\,int}\,C$ holds for $\mathbb{Q}$-almost all C). We assume that

$$\int_{\mathcal{C}_0} V_d(\mathrm{conv}\,K+B^d)\,\mathbb{Q}(\mathrm{d}K)<\infty \tag{11.15}$$

and define the **ALLC-function** (average logarithmic linear contact distribution function) L of Z by

$$L(r) := -\int_{S^{d-1}} \ln(1 - H_{[0,u]}(r))\,\sigma_{d-1}(\mathrm{d}u), \qquad r \geq 0.$$

The following result was presented in Hug, Last and Weil [362].

Let Z be a stationary Boolean model in $\mathbb{R}^d$ with regular compact grains satisfying (11.15). Then the ALLC-function L of Z is linear if and only if the grains are almost surely convex.

We sketch the proof of the non-obvious direction. Thus, we assume that L is linear. Then

$$f := \overline{V}_d(X) + L$$

is a polynomial and

$$f(r) = \gamma \int_{S^{d-1}} \int_{\mathcal{C}_0} \lambda(K + r[0,u])\,\mathbb{Q}(\mathrm{d}K)\,\sigma_{d-1}(\mathrm{d}u).$$

We have

$$\begin{aligned} f(r) \leq \widetilde{f}(r) &:= \gamma \int_{S^{d-1}} \int_{\mathcal{C}_0} \lambda(K_u + r[0,u])\,\mathbb{Q}(\mathrm{d}K)\,\sigma_{d-1}(\mathrm{d}u) \\ &= \gamma \int_{S^{d-1}} \int_{\mathcal{C}_0} \Big(\lambda(K_u) + \lambda_{d-1}(K \,|\, u^{\perp})r\Big)\,\mathbb{Q}(\mathrm{d}K)\,\sigma_{d-1}(\mathrm{d}u) \\ &= \widetilde{a} + \widetilde{b}r, \end{aligned}$$

where K_u is the convexification of K in direction u (for each line l in direction u, we replace $K \cap l$ by its convex hull). Hence,

$$f(r) = a + br \qquad \text{with } a \leq \widetilde{a},\, b \leq \widetilde{b}.$$

For sufficiently large r, we have $K + r[0,u] = K_u + r[0,u]$, uniformly in u. This implies $f = \widetilde{f}$, for large r and therefore for all $r \geq 0$. But then $K = K_u$, for all u, which implies convexity. (Notice that in some of these arguments the regularity of the grains is used.)

There is a corresponding result in Hug, Last and Weil [362] for the two-dimensional unit disk B^2, which concerns the **ALDC-function** (average logarithmic disk contact distribution function) D of Z,

$$D(r) := -\int_{SO_d} \ln(1 - H_{\vartheta B^2}(r))\,\nu(\mathrm{d}\vartheta), \qquad r \geq 0.$$

Instead of (11.15), we need the stronger assumption of uniformly bounded grains.

Let Z be a stationary Boolean model in $\mathbb{R}^d$ with regular, uniformly bounded compact grains. Then the ALDC-function D of Z is a polynomial if and only if the grains are almost surely convex.

The proof is more complicated and is based on the following steps.

First, by Fubini's theorem, it is sufficient to show the result for $d = 2$ and a fixed (regular) grain $K \in \mathcal{C}'$. More precisely, it is sufficient to show that $K \subset \mathbb{R}^2$ is convex if $\lambda_2(K + rB^2)$ is a polynomial in $r \geq 0$.

As in the linear case, we compare $\lambda_2(K(r)) = \sum_{i=0}^m a_i r^i$, $K(r) := K + rB^2$, with the volume of the convex hull $\bar{K}(r) := \bar{K} + rB^2$, $\bar{K} := \text{conv}K$,

$$\lambda_2(\bar{K}(r)) = V_2(\bar{K}) + 2rV(\bar{K}, B^2) + r^2 V_2(B^2).$$

Since $\lambda_2(rB^2) \le \lambda_2(K(r)) \le \lambda_2(\bar{K}(r))$, we obtain

$$\lambda_2(K(r)) = a_0 + a_1 r + a_2 r^2, \qquad r \ge 0,$$

with $a_2 = V_2(B^2)$ and $a_0 + a_1 r \le V_2(\bar{K}) + 2rV(\bar{K}, B^2)$.

For λ_1-almost all $r \ge 0$, we have

$$\frac{\mathrm{d}}{\mathrm{d}r} V_2(K(r)) = \int_{\text{bd}\, K(r)} h(B^2, u_{K(r)}(x))\, \mathcal{H}^1(\mathrm{d}x),$$

where $u_{K(r)}$ is the ($\mathcal{H}^1$-almost everywhere existing) outer unit normal vector in x and $h(M, \cdot)$ is the support function of the convex body M. Moreover, $K(r)$ is star-shaped, for sufficiently large r. Hence,

$$\int_{\text{bd}\, K(r)} h(B^2, u_{K(r)}(x))\, \mathcal{H}^1(\mathrm{d}x) \le \int_{\text{bd}\, \bar{K}(r)} h(B^2, u_{\bar{K}(r)}(x))\, \mathcal{H}^1(\mathrm{d}x),$$

which, under the spherical image map, transforms to

$$\int_{S^1} h(B^2, u)\, S_1(K(r), \mathrm{d}u) \le \int_{S^1} h(B^2, u)\, S_1(\bar{K}(r), \mathrm{d}u). \tag{11.16}$$

Here, the image measure $S_1(\bar{K}(r), \cdot)$ is the surface area measure of $\bar{K}(r)$ and $S_1(K(r), \cdot)$ is, by Minkowski's theorem, also the surface area measure of some convex body $\widetilde{K}(r)$, the **convexification** of $K(r)$.

As one can show, $K(r) \subset \widetilde{K}(r)$, after a suitable translation, and therefore $\bar{K}(r) \subset \widetilde{K}(r)$. This implies $h(\bar{K}(r), \cdot) \le h(\widetilde{K}(r), \cdot)$. On the other hand, (11.16) and the symmetry of planar mixed volumes yield

$$\begin{aligned}
\int_{S^1} h(\bar{K}(r), u)\, \sigma_1(\mathrm{d}u) &= \int_{S^1} h(B^2, u)\, S_1(\bar{K}(r), \mathrm{d}u) \\
&\ge \int_{S^1} h(B^2, u)\, S_1(K(r), \mathrm{d}u) \\
&= \int_{S^1} h(B^2, u)\, S_1(\widetilde{K}(r), \mathrm{d}u) \\
&= \int_{S^1} h(\widetilde{K}(r), u)\, \sigma_1(\mathrm{d}u).
\end{aligned}$$

Therefore $h(\bar{K}(r), \cdot) = h(\widetilde{K}(r), \cdot)$, hence $\bar{K}(r) = \widetilde{K}(r)$, which implies that $K(r)$ and the convex hull $\bar{K}(r)$ have the same boundary length, $\mathcal{H}^1(K(r)) = \mathcal{H}^1(\bar{K}(r))$. For a planar star-shaped set this implies $K(r) = \bar{K}(r)$. Consequently, $K = \bar{K} = \text{conv}\, K$.

The result holds in a more general version, with the unit disk B^2 replaced by a smooth planar body B, and the proof is essentially the same.

However, as was shown in Heveling, Hug and Last [338], a corresponding result for three-dimensional gauge bodies B is wrong. In particular, for $d = 3$ and $B = B^3$, $\ln(1 - H)$ can be a polynomial without Z having convex grains. An example is given by a Boolean model Z, the primary grain of which is the union of two touching balls of equal radius.

11.3 Processes of Flats

Our aim in this section is to see how some of the results on flat processes obtained in Sections 4.4 and 4.6 carry over to the non-stationary case. For simplicity, we assume that all k-flat processes occurring in the following are simple. If we omit the stationarity assumption, some regularity property of the intensity measure will be necessary, similarly to Section 11.1, to get smooth results. We say that a measure on the space $A(d,k)$ of k-flats in $\mathbb{R}^d$ is **translation regular** if it is absolutely continuous with respect to some translation invariant, locally finite measure on $A(d,k)$.

Let $k \in \{1,\dots,d-1\}$, and let X be a k-flat process in $\mathbb{R}^d$ with a translation regular intensity measure $\Theta \neq 0$ (assumed to be locally finite, as always). By assumption, there exist a locally finite, translation invariant measure $\widetilde{\Theta}$ on $A(d,k)$ and a nonnegative, locally $\widetilde{\Theta}$-integrable function η on $A(d,k)$ such that

$$\Theta(A) = \int_A \eta \, \mathrm{d}\widetilde{\Theta}$$

for $A \in \mathcal{B}(A(d,k))$. The density η is only determined $\widetilde{\Theta}$-almost everywhere. If $\widetilde{\Theta}$ and η can be chosen such that η is continuous on $A(d,k)$, then we say that Θ is translation regular **with continuous density**.

By Theorem 4.4.1, the measure $\widetilde{\Theta}$ has a decomposition

$$\widetilde{\Theta}(A) = \int_{G(d,k)} \int_{L^\perp} \mathbf{1}_A(L+x) \, \lambda_{L^\perp}(\mathrm{d}x) \, \mathbb{Q}(\mathrm{d}L)$$

for $A \in \mathcal{B}(A(d,k))$, with a finite measure $\mathbb{Q}$ on $G(d,k)$, without loss of generality a probability measure (since $\mathbb{Q} \neq 0$, and $\widetilde{\Theta}$ and η can be changed by constant factors). For the intensity measure of X, this yields the representation

$$\Theta(A) = \int_{G(d,k)} \int_{L^\perp} \mathbf{1}_A(L+x)\eta(L+x) \, \lambda_{L^\perp}(\mathrm{d}x) \, \mathbb{Q}(\mathrm{d}L). \tag{11.17}$$

As in Section 11.1 we say that Θ is **represented by the pair** $(\eta, \mathbb{Q})$ (which, *nota bene*, is not uniquely determined).

For a stationary k-flat process X, the intensity γ, given by

$$\mathbb{E} \sum_{E \in X} \lambda_E = \gamma \lambda$$

(Theorem 4.4.3), and the directional distribution determine the intensity measure, by (4.25). For a k-flat process with a translation regular intensity measure there are corresponding quantities, but depending on the location. They are obtained from the following result. For $E \in A(d,k)$ we denote here by $E_0 \in G(d,k)$ the translate of E through 0.

Theorem 11.3.1. *Let X be a k-flat process in $\mathbb{R}^d$ with a translation regular intensity measure represented by $(\eta, \mathbb{Q})$. Let $A \in \mathcal{B}(G(d,k))$. Then*

$$\mathbb{E} \sum_{E \in X} \mathbf{1}_A(E_0)\lambda_E = \int_{(\cdot)} \int_{G(d,k)} \mathbf{1}_A(L)\eta(L+z)\,\mathbb{Q}(\mathrm{d}L)\,\lambda(\mathrm{d}z).$$

In particular, the measure

$$\mathbb{E} \sum_{E \in X} \lambda_E$$

has a density with respect to Lebesgue measure, given by

$$\gamma(z) := \int_{G(d,k)} \eta(L+z)\,\mathbb{Q}(\mathrm{d}L) \tag{11.18}$$

for almost all $z \in \mathbb{R}^d$.

Proof. Let $B \in \mathcal{B}(\mathbb{R}^d)$. Using Campbell's theorem and (11.17), we obtain

$$\begin{aligned}
&\mathbb{E} \sum_{E \in X} \mathbf{1}_A(E_0)\lambda_E(B) \\
&= \int_{A(d,k)} \mathbf{1}_A(E_0)\lambda_E(B)\,\Theta(\mathrm{d}E) \\
&= \int_{G(d,k)} \int_{L^\perp} \mathbf{1}_A(L)\lambda_{L+x}(B)\eta(L+x)\,\lambda_{L^\perp}(\mathrm{d}x)\,\mathbb{Q}(\mathrm{d}L) \\
&= \int_{G(d,k)} \int_{L^\perp} \int_L \mathbf{1}_B(y+x)\mathbf{1}_A(L)\eta(L+x)\,\lambda_L(\mathrm{d}y)\,\lambda_{L^\perp}(\mathrm{d}x)\,\mathbb{Q}(\mathrm{d}L) \\
&= \int_{G(d,k)} \int_B \mathbf{1}_A(L)\eta(L+z)\,\lambda(\mathrm{d}z)\,\mathbb{Q}(\mathrm{d}L) \\
&= \int_B \int_{G(d,k)} \mathbf{1}_A(L)\eta(L+z)\,\mathbb{Q}(\mathrm{d}L)\,\lambda(\mathrm{d}z),
\end{aligned}$$

which completes the proof. □

The function γ is called the **intensity function** of the k-flat process X. It replaces the constant intensity appearing in the stationary case.

If the density η is continuous on $A(d,k)$, then it follows from the compactness of $G(d,k)$ and the finiteness of $\mathbb{Q}$ that there is a uniquely determined continuous version of the intensity function on $\mathbb{R}^d$. In this case an intuitive interpretation is easily obtained as follows. If K is a convex body with interior points, then, for $z \in \mathbb{R}^d$ and $r > 0$,

$$\mathbb{E} \sum_{E \in X} \lambda_E(rK+z) = \int_{A(d,k)} \lambda_E(rK+z)\,\Theta(\mathrm{d}E)$$

$$= \int_{G(d,k)} \int_{L^\perp} \lambda_{L+x}(rK + z)\eta(L + x)\, \lambda_{L^\perp}(\mathrm{d}x)\, \mathbb{Q}(\mathrm{d}L)$$

$$= \int_{G(d,k)} \int_{L^\perp} \int_L \mathbf{1}_{rK+z}(x + y)\eta(L + x)\, \lambda_L(\mathrm{d}y)\, \lambda_{L^\perp}(\mathrm{d}x)\, \mathbb{Q}(\mathrm{d}L)$$

$$= \int_{G(d,k)} \int_{\mathbb{R}^d} \mathbf{1}_{rK+z}(t)\eta(L + t)\, \lambda(\mathrm{d}t)\, \mathbb{Q}(\mathrm{d}L).$$

Since η is continuous, the inner integral is equal to $\eta(L+z_{r,L})V_d(rK)$ for some point $z_{r,L} \in rK + z$. The continuity of η now gives

$$\gamma(z) = \lim_{r\to 0} \frac{1}{V_d(rK)} \mathbb{E} \sum_{E\in X} \lambda_E(rK + z).$$

It is also easy to see that

$$\gamma(z) = \lim_{r\to 0} \frac{1}{\kappa_{d-k} r^{d-k}} \mathbb{E}X(\mathcal{F}_{rB^d+z})$$

holds for all $z \in \mathbb{R}^d$. This extends (4.27).

As a counterpart to the directional distribution in the stationary case, we define a measure $\varphi(z, \cdot)$ on $G(d, k)$ by

$$\varphi(z, \cdot) := \int_{(\cdot)} \eta(L + z)\, \mathbb{Q}(\mathrm{d}L)$$

for $z \in \mathbb{R}^d$, which is finite almost everywhere. From Theorem 11.3.1 it follows that

$$\mathbb{E} \sum_{E\in X} \mathbf{1}_A(E_0)\lambda_E(B) = \int_B \varphi(z, A)\, \lambda(\mathrm{d}z)$$

for $A \in \mathcal{B}(G(d, k))$ and $B \in \mathcal{B}(\mathbb{R}^d)$. This relation, together with the fact that the σ-algebra $\mathcal{B}(G(d, k))$ is countably generated, shows that for λ-almost all z the measure $\varphi(z, \cdot)$ is uniquely determined and hence depends only on the process X and not on the choice of $\widetilde{\Theta}$ and η. The measure $\varphi(z, \cdot)$ is called the **directional measure** of X at z. At the points z with $0 < \gamma(z) < \infty$, the **directional distribution** $\varphi(z, \cdot)/\gamma(z)$ can be defined.

If X is stationary, then the directional measure $\varphi(z, \cdot)$ does not depend on z. If X is also isotropic, then $\varphi(z, \cdot)$ is rotation invariant. We prove a certain converse statement. Here we use the terminology introduced in the remark at the end of Section 9.2.

Theorem 11.3.2. *Let X be a k-flat process in $\mathbb{R}^d$ whose intensity measure is translation regular with a continuous density. If the directional measure $\varphi(z, \cdot)$ is rotation invariant for all $z \in \mathbb{R}^d$, then X is weakly stationary and weakly isotropic.*

Proof. Under the assumptions, the intensity function γ is continuous, hence the set

$$M := \{z \in \mathbb{R}^d : \gamma(z) > 0\}$$

is open (and not empty). Let $z \in M$. The finite, rotation invariant measure $\varphi(z, \cdot)$ is a multiple of ν_k. Since $\varphi(z, G(d,k)) = \gamma(z)$, it follows that

$$\varphi(z, \cdot) = \gamma(z)\nu_k,$$

hence from (11.18) we get

$$\nu_k(A) = \int_A \frac{\eta(L+z)}{\gamma(z)}\, \mathbb{Q}(\mathrm{d}L) \tag{11.19}$$

for $A \in \mathcal{B}(G(d,k))$. Let also $y \in M$, then

$$\int_A \frac{\eta(L+z)}{\gamma(z)}\, \mathbb{Q}(\mathrm{d}L) = \int_A \frac{\eta(L+y)}{\gamma(y)}\, \mathbb{Q}(\mathrm{d}L)$$

for $A \in \mathcal{B}(G(d,k))$, hence

$$\frac{\eta(L+z)}{\gamma(z)} = \frac{\eta(L+y)}{\gamma(y)} \tag{11.20}$$

for all $L \in \operatorname{supp} \mathbb{Q}$, by the continuity of η. By (11.19), $\operatorname{supp} \mathbb{Q} = G(d,k)$, since $\nu_k(A) > 0$ for every nonempty open set $A \subset G(d,k)$.

The set

$$N_z := \{L \in G(d,k) : \eta(L+z) = 0\}$$

satisfies $\nu_k(N_z) = 0$, by (11.19). Let $U \subset M$ be a neighborhood of y. The directions E_0 of the k-flats E through z and a point of U fill a nonempty open set in $G(d,k)$. Therefore, $y_1 \in U$ can be chosen such that there exists a subspace $L \in G(d,k)$ with $L + z = L + y_1$ and $\eta(L+z) > 0$. Relation (11.20) implies that $\gamma(y_1) = \gamma(z)$, and since y_1 can be chosen arbitrarily close to y, we deduce that $\gamma(y) = \gamma(z)$, by the continuity of γ.

We have proved that the continuous intensity function γ is constant on the set M where it is positive. Hence, γ is constant on $\mathbb{R}^d$. From (11.17), (11.19) and (11.20) we conclude that

$$\Theta(A) = \gamma \int_{G(d,k)} \int_{L^\perp} \mathbf{1}_A(L+x)\, \lambda_{L^\perp}(\mathrm{d}x)\, \nu_k(\mathrm{d}L)$$

for $A \in \mathcal{B}(G(d,k))$. This shows that Θ is invariant under translations and rotations and thus completes the proof. □

The rest of this section is devoted to Poisson hyperplane processes. We want to extend Theorem 4.6.5 on intersection densities to the non-stationary, translation regular case. This requires the introduction of associated zonoids depending on the location.

For hyperplanes, we use the representation (4.32), but we consider only hyperplanes not passing through 0. Every such hyperplane has a unique representation

$$H(u,\tau) := \{x \in \mathbb{R}^d : \langle x, u\rangle = \tau\}$$

with $u \in S^{d-1}$ and $\tau > 0$.

Let X be a hyperplane process in $\mathbb{R}^d$ with a translation regular intensity measure represented by the pair $(\eta, \mathbb{Q})$. It is convenient to use the function $g : S^{d-1} \times (0,\infty) \to [0,\infty)$ defined by

$$g(u,\tau) := \eta(H(u,\tau))$$

for $u \in S^{d-1}$ and $\tau > 0$ and by $g(u,\tau) := 0$ for $\tau \leq 0$ or if $\eta(H(u,\tau))$ is not defined. Instead of $\mathbb{Q}$, we use the measure ϕ on the sphere S^{d-1} with the property

$$\phi(A) = \frac{1}{2}\mathbb{Q}(\{H(u,0) : u \in A\}$$

for $A \in \mathcal{B}(S^{d-1})$ without antipodal points. The intensity measure of X is then given by

$$\Theta(A) = 2\int_{S^{d-1}}\int_0^\infty \mathbf{1}_A(H(u,\tau))g(u,\tau)\,\mathrm{d}\tau\,\phi(\mathrm{d}u) \tag{11.21}$$

for $A \in \mathcal{B}(A(d,d-1))$.

We assume now in addition that X is a Poisson process. For $k \in \{1,\dots,d\}$, let X_k be the intersection process of order k of the process X. Modifying the method of proof for Theorem 4.4.5, one can show that X_k is a.s. simple. Let Θ_k be its intensity measure. As in the proof of Theorem 4.4.8 (where the stationarity assumption is not needed in the beginning), one obtains

$$\Theta_k(A) = \frac{1}{k!}\int_{A(d,d-1)^{k*}} \mathbf{1}_A(H_1 \cap \ldots \cap H_k)\,\Theta^k(\mathrm{d}(H_1,\dots,H_k)) \tag{11.22}$$

for $A \in \mathcal{B}(A(d,d-k))$, where $A(d,d-1)^{k*}$ denotes the set of k-tuples of hyerplanes with linearly independent normal vectors. Thus, $k!\Theta_k$ is the image measure of $\Theta^k \llcorner A(d,d-1)^{k*}$ under the intersection mapping $(H_1,\dots,H_k) \mapsto H_1 \cap \ldots \cap H_k$. It follows that Θ_k is locally finite. Since Θ is absolutely continuous with respect to a translation invariant measure $\widetilde{\Theta}$ on $A(d,d-1)$, the measure Θ_k is absolutely continuous with respect to the image measure of $\widetilde{\Theta}^k \llcorner A(d,d-1)^{k*}$ under the same intersection mapping. Hence, the intersection process X_k has a translation regular intensity measure, too. We compute its intensity function. Let $B \in \mathcal{B}(\mathbb{R}^d)$ and $\lambda(B) < \infty$. Then, using (11.22) and (11.21),

$$\begin{aligned} &k!\mathbb{E}\sum_{E\in X_k} \lambda_E(B) \\ &= k!\int_{A(d,d-k)} \lambda_{d-k}(B\cap E)\,\Theta_k(\mathrm{d}E) \end{aligned}$$

$$= \int_{A(d,d-1)^{k*}} \lambda_{d-k}(B \cap H_1 \cap \ldots \cap H_k)\, \Theta^k(\mathrm{d}(H_1,\ldots,H_k))$$

$$= 2^k \int_{(S^{d-1})^{k*}} \int_{(0,\infty)^k} \lambda_{d-k}(B \cap H(u_1,\tau_1) \cap \ldots \cap H(u_k,\tau_k))$$

$$\times\, g(u_1,\tau_1)\cdots g(u_k,\tau_k)\, \mathrm{d}(\tau_1,\ldots,\tau_k)\, \phi^k(\mathrm{d}(u_1,\ldots,u_k)),$$

where $(S^{d-1})^{k*}$ denotes the set of k-tuples of linearly independent unit vectors. We use the same transformation as at the end of the proof of Theorem 4.4.8. If $u_1,\ldots,u_k \in S^{d-1}$ are linearly independent and $H(u_1,\tau_1)\cap\ldots\cap H(u_k,\tau_k) =: L$, then

$$\int_{(0,\infty)^k} \lambda_{d-k}(B \cap H(u_1,\tau_1) \cap \ldots \cap H(u_k,\tau_k)) g(u_1,\tau_1)\cdots g(u_k,\tau_k)$$

$$\times\, \mathrm{d}(\tau_1,\ldots,\tau_k)$$

$$= \int_{L^\perp} \lambda_{d-k}(B \cap (L+x)) g(u_1,\langle u_1,x\rangle)\cdots g(u_k,\langle u_k,x\rangle)\, \lambda_{L^\perp}(\mathrm{d}x)$$

$$\times\, \nabla_k(u_1,\ldots,u_k)$$

$$= \int_B g(u_1,\langle u_1,z\rangle)\cdots g(u_k,\langle u_k,z\rangle)\, \lambda(\mathrm{d}z)\cdot \nabla_k(u_1,\ldots,u_k).$$

Since $\nabla_k(u_1,\ldots,u_k) = 0$ if $u_1,\ldots,u_k$ are linearly dependent, we conclude that

$$\mathbb{E} \sum_{E\in X_k} \lambda_E(B) = \frac{2^k}{k!} \int_{(S^{d-1})^k} \int_B g(u_1,\langle u_1,z\rangle)\cdots g(u_k,\langle u_k,z\rangle)\, \nabla_k(u_1,\ldots,u_k)$$

$$\times\, \lambda(\mathrm{d}z)\, \phi^k(\mathrm{d}(u_1,\ldots,u_k))$$

$$= \int_B \gamma_k(z)\, \lambda(\mathrm{d}z)$$

with

$$\gamma_k(z) = \frac{2^k}{k!} \int_{(S^{d-1})^k} \nabla_k(u_1,\ldots,u_k)$$

$$\times\, g(u_1,\langle u_1,z\rangle)\cdots g(u_k,\langle u_k,z\rangle)\, \phi^k(\mathrm{d}(u_1,\ldots,u_k)). \quad (11.23)$$

This is the intensity function of the intersection process X_k. We rewrite it, using the measure $\widetilde{\varphi}_z$ on S^{d-1} defined by

$$\widetilde{\varphi}_z(A) := 2 \int_A g(u,\langle u,z\rangle)\, \phi(\mathrm{d}u)$$

for $A \in \mathcal{B}(S^{d-1})$. Then

$$\gamma_k(z) = \frac{1}{k!} \int_{(S^{d-1})^k} \nabla_k(u_1, \dots, u_k)\, \widetilde{\varphi}_z^k(\mathrm{d}(u_1, \dots, u_k)). \tag{11.24}$$

Now we define the **local associated zonoid** $\Pi_X(z)$ of X at z as the convex body with support function given by

$$h(\Pi_X(z), u) := \frac{1}{2} \int_{S^{d-1}} |\langle u, v \rangle|\, \widetilde{\varphi}_z(\mathrm{d}v)$$

for $u \in \mathbb{R}^d$.

From (11.24) and (14.35) we obtain the generalization of formula (4.63) to Poisson hyperplane processes with translation regular intensity measure, namely

$$\gamma_k(z) = V_k(\Pi_X(z)) \tag{11.25}$$

for $z \in \mathbb{R}^d$. We can state the following result.

Theorem 11.3.3. *Let X be a Poisson hyperplane process in $\mathbb{R}^d$ with a translation regular intensity measure. Let $k \in \{2, \dots, d\}$, and let X_k be the intersection process of X of order k. Let γ be the intensity function of X and γ_k the intensity function of X_k. Then*

$$\gamma_k(z) \le \frac{\binom{d}{k} \kappa_{d-1}^k}{d^k \kappa_{d-k} \kappa_d^{k-1}} \gamma(z)^k \tag{11.26}$$

for almost all $z \in \mathbb{R}^d$.

If the intensity measure of X is translation regular with a continuous density and if equality holds in (11.26) *for all z, then the process X is stationary and isotropic.*

Proof. The inequality (11.26) follows from (11.25) in the same way as in the stationary case (see Theorem 4.6.5). Assume that X is translation regular with a continuous density and that equality holds in (11.26) for all $z \in \mathbb{R}^d$. Then, for each z, the local associated zonoid $\Pi_X(z)$ is a ball (possibly one-pointed). Hence, the even part of the measure $\widetilde{\varphi}_z$ is proportional to the spherical Lebesgue measure. Since

$$\varphi(z, A) = \widetilde{\varphi}_z(\{u \in S^{d-1} : H(u, 0) \in A\} \qquad \text{for } A \in \mathcal{B}(G(d, d-1)),$$

it follows that $\varphi(z, \cdot)$ is rotation invariant. Now Theorem 11.3.2 shows that X is weakly stationary and weakly isotropic. Since the intensity measure of a Poisson process determines its distribution, X is stationary and isotropic. $\square$

A remarkable aspect of Theorem 11.3.3 can be seen in the fact that together with isotropy also the stationarity, and thus invariance under all rigid motions, is characterized by an extremal property.

Notes for Section 11.3

1. The results of this section are taken from Schneider [707]. Fallert [222] has studied k-flat processes which instead of (11.17) satisfy

$$\Theta(A) = \int_{G(d,k)} \int_{L^\perp} \mathbf{1}_A(L+x) f(x)\, \lambda_{L^\perp}(\mathrm{d}x)\, \mathbb{Q}(\mathrm{d}L)$$

with a locally integrable nonnegative function f. For $0 < k < d-1$, this assumption is more restrictive than the assumption (11.17) with a locally integrable nonnegative function η.

2. Intersections with fixed flats. Let X be a translation regular k-flat process, let $j \in \{0, \dots, k-1\}$, and let $S \in A(d-k+j)$. The intersection process $X \cap S$, defined by

$$X \cap S := \sum_{E_i \cap S \neq \emptyset} \delta_{E_i \cap S} \quad \text{if} \quad X = \sum \delta_{E_i},$$

is a translation regular j-flat process in S. Its intensity function at $z \in S$ is given by

$$\gamma_{X \cap S}(z) = \int_{G(d,k)} [L, S]\, \varphi(z, \mathrm{d}L),$$

where $\varphi(z, \cdot)$ is the direction measure of X at z. See Schneider [707], also for some results on the determination of translation regular flat processes from information on section processes.

3. Hoffmann [346] has given a common generalization of results of Schneider [707] and of Wieacker [817], by investigating intersection densities and local associated zonoids for non-stationary Poisson processes of hypersurfaces, which are cylinders with $(\mathcal{H}^k, k)$-rectifiable bases.

4. Hug, Last and Weil [360] study generalized contact distribution functions, in the sense of Section 11.2, for **Poisson networks** Z. The latter are union sets of Poisson k-flat processes X with translation regular intensity measures. They prove an analog of Theorem 11.2.2. As a consequence, it is shown that, for processes X with continuous density η and $z \in \mathbb{R}^d$, the distribution of $(d(z, Z), u(Z, z))$ determines the Radon transform $R_{k,d-1}\varphi(z, \cdot)$. Hence for line or hyperplane networks, the directional measure $\varphi(z, \cdot)$ is determined by measuring the distance and direction from the point z to Z. Various generalizations are also treated in [360]. For example, the point z is replaced by a flat $F \in A(d, j)$, with $k + j < d$, and the intensity measure Θ^F of the process of midpoints $m(E, F)$, $E \in X$, is assumed to be given. As another generalization, z is replaced by a flag of (linear) subspaces with increasing dimension and a uniqueness result for stationary Poisson networks is proved.

11.4 Tessellations

The purpose of this section is to extend from the stationary to the non-stationary case a basic result on random hyperplane mosaics, namely the relations of Theorem 10.3.1 between the specific intrinsic volumes of the face processes. However, for reasons explained later, we do this only under Poisson assumptions.

Let $\widehat{X}$ be a Poisson hyperplane process in $\mathbb{R}^d$ with a translation regular intensity measure. It generates a random tessellation X of $\mathbb{R}^d$, and our first aim will be to formulate a condition which ensures that the cells of the tessellation are a.s. bounded. A given point x is a.s. contained in a unique cell, denoted by Z_x, of the mosaic; this follows from the translation regularity.

Definition 11.4.1. *The hyperplane process $\widehat{X}$ is* **nondegenerate** *if the following holds.*

(a) *With positive probability, the zero cell Z_0 is bounded.*

(b) *If $U \subset S^{d-1}$ is a measurable set and if $\widehat{X}$ contains with positive probability a hyperplane with normal vector in U, then $\widehat{X}$ contains with positive probability infinitely many such hyperplanes.*

This is an appropriate geometric condition for obtaining a random mosaic, as the following theorem shows.

Theorem 11.4.1. *Let $\widehat{X}$ be a nondegenerate Poisson hyperplane process in $\mathbb{R}^d$ with a translation regular intensity measure. The system X of the induced cells is a random mosaic in general position. The process $X^{(k)}$ of the k-faces of X has a translation regular intensity measure, for $k = 1, \dots, d$.*

Proof. The intensity measure $\widehat{\Theta}(A)$ of $\widehat{X}$ has the representation (11.21),

$$\widehat{\Theta}(A) = 2 \int_{S^{d-1}} \int_0^\infty \mathbf{1}_A(H(u,\tau)) g(u,\tau)\, \mathrm{d}\tau\, \phi(\mathrm{d}u)$$

for $A \in \mathcal{B}(A(d, d-1))$, with a locally integrable, nonnegative function g and a finite measure ϕ. We put

$$P := \left\{ u \in S^{d-1} : \int_0^\infty g(u,\tau)\, \mathrm{d}\tau > 0 \right\}$$

and assume without loss of generality that the measure ϕ is **reduced**, in the sense that

$$\phi(S^{d-1} \setminus P) = 0.$$

Let $U := \operatorname{supp} \phi$ and suppose that $0 \notin \operatorname{int} \operatorname{conv} U$. Then $\operatorname{conv} U$ and 0 can be separated weakly by a hyperplane, hence there is a vector $v \in S^{d-1}$ with $\langle u, v \rangle \le 0$ for $u \in U$. We denote by M the set of hyperplanes $H(u,\tau)$, $\tau > 0$, meeting the ray $R := \{\lambda v : \lambda > 0\}$. If $H(u,\tau) \cap R \neq \emptyset$ and (w.l.o.g.) $R \not\subset H(u,\tau)$, then $\langle u, v \rangle > 0$, hence

$$\widehat{\Theta}(M) = 2 \int_{S^{d-1}} \int_0^\infty \mathbf{1}_M(H(u,\tau)) g(u,\tau)\, \mathrm{d}\tau\, \phi(\mathrm{d}u) = 0.$$

But then $R \subset Z_0$ a.s., which contradicts the assumption that Z_0 is bounded with positive probability. It follows that $0 \in \operatorname{int} \operatorname{conv} U$. As in the proof of

Theorem 10.3.2, this implies the existence of vectors $u_1, \dots, u_{2d} \in U$ and neighborhoods U_i of u_i in S^{d-1} for $i = 1, \dots, 2d$ such that

$$0 \in \operatorname{int} \operatorname{conv}\{v_1, \dots, v_{2d}\} \qquad \text{for all } (v_1, \dots, v_{2d}) \in U_1 \times \dots \times U_{2d}.$$

Let $x \in \mathbb{R}^d$, and let $A_i(x)$ be the set of all hyperplanes $H(u, \tau)$ with $u \in U_i$ and $\tau > \max\{\langle x, u_i \rangle, 0\}$ for $i = 1, \dots, 2d$. Then

$$\widehat{\Theta}(A_i(x)) = 2 \int_{U_i} \int_{\langle x, u_i \rangle}^{\infty} g(u, \tau)\, \mathrm{d}\tau\, \phi(\mathrm{d}u).$$

We have

$$\widehat{\Theta}(A_i(0)) = 2 \int_{U_i} \int_0^{\infty} g(u, \tau)\, \mathrm{d}\tau\, \phi(\mathrm{d}u) > 0,$$

since $\phi(U_i) > 0$, which follows from $u_i \in \operatorname{supp} \phi$ and the assumption that ϕ is reduced. Since $\widehat{X}$ is nondegenerate, this implies $\widehat{\Theta}(A(0)) = \infty$, thus

$$\widehat{\Theta}(A_i(x)) + 2 \int_{U_i} \int_0^{\langle x, u_i \rangle} g(u, \tau)\, \mathrm{d}\tau\, \mathrm{d}\phi(u) = \infty.$$

Here the second summand is finite since $\widehat{\Theta}$ is finite on compact sets. We conclude that $\widehat{\Theta}(A_i(x)) = \infty$.

Now we can continue as in the proof of Theorem 10.3.2 and deduce that the cell Z_x is almost surely bounded. The rest of that proof also carries over, showing that the system of induced cells is a random mosaic in general position.

Let $X^{(k)}$ be the system of k-faces of X, $k \in \{0, \dots, d\}$. As in Section 10.1 one sees that $X^{(k)}$ is a particle process. Since the intersection processes of $\widehat{X}$ have locally finite intensity measures, the proof of Theorem 10.3.1 shows that $X^{(k)}$ has locally finite intensity measure. It remains to show that this measure is translation regular.

Let $\widehat{X}_s$ be the stationary Poisson hyperplane process with spherical directional distribution ϕ and with intensity 1. It exists by Theorem 4.4.4 and has intensity measure

$$\widehat{\Theta}_s(A) = 2 \int_{S^{d-1}} \int_0^{\infty} \mathbf{1}_A(H(u, \tau))\, \mathrm{d}\tau\, \phi(\mathrm{d}u),$$

for $A \in \mathcal{B}(A(d, d-1))$. The random hyperplane mosaic generated by $\widehat{X}_s$ is denoted by X_s, and the particle process of its k-faces by $X_s^{(k)}$. Let $\Theta_s^{(k)}$ be the intensity measure of $X_s^{(k)}$. We will show that $\Theta^{(k)}$ is absolutely continuous with respect to $\Theta_s^{(k)}$.

Let $A \in \mathcal{B}(\mathcal{K}')$ be a set with

$$\Theta_s^{(k)}(A) = 0. \tag{11.27}$$

In order to show that

$$\Theta^{(k)}(A) = 0, \tag{11.28}$$

it is sufficient to show that $\Theta^{(k)}(A_r) = 0$ for each $r \in \mathbb{N}$, where $A_r := \{K \in A : K \subset rB^d\}$. Let $\mathcal{H}_r$ be the set of hyperplanes meeting rB^d. To prove (11.28), it is sufficient to prove for $r, m \in \mathbb{N}$ that

$$\mathbb{E}(X^{(k)}(A_r) \mid \widehat{X}(\mathcal{H}_r) = m) = 0.$$

We choose r so large that $\widehat{\Theta}(\mathcal{H}_r) \neq 0$, which is possible since $\widehat{\Theta} \neq 0$. The process $\widehat{X}$ restricted to $\mathcal{H}_r$ and under the condition $\widehat{X}(\mathcal{H}_r) = m$ is stochastically equivalent to the process defined by m independent, identically distributed hyperplanes with distribution $\widehat{\Theta} \llcorner \mathcal{H}_r / \widehat{\Theta}(\mathcal{H}_r)$ (Theorem 3.2.2). We denote by $f(u_1, \dots, u_m, \tau_1, \dots, \tau_m)$ the number of k dimensional polytopes in the set A_r that are faces of the tessellation of $\mathbb{R}^d$ generated by the hyperplanes $H(u_1, \tau_1), \dots, H(u_m, \tau_m)$. Then

$$\begin{aligned} &\mathbb{E}(X^{(k)}(A_r) \mid \widehat{X}(\mathcal{H}_r) = m) \\ &= \widehat{\Theta}(\mathcal{H}_r)^{-m} 2^m \int_{(S^{d-1})^m} \int_{(0,\infty)^m} f(u_1, \dots, u_m, \tau_1, \dots, \tau_m) \\ &\quad \times g(u_1, \tau_1) \cdots g(u_m, \tau_m) \, \mathrm{d}(\tau_1, \dots, \tau_m) \, \phi^m(\mathrm{d}(u_1, \dots, u_m)). \end{aligned}$$

Similarly, for the stationary Poisson hyperplane process $\widehat{X}_s$ we get

$$\begin{aligned} &\mathbb{E}(X_s^{(k)}(A_r) \mid \widehat{X}_s(\mathcal{H}_r) = m) \\ &= \widehat{\Theta}_s(\mathcal{H}_r)^{-m} 2^m \int_{(S^{d-1})^m} \int_{(0,\infty)^m} f(u_1, \dots, u_m, \tau_1, \dots, \tau_m) \\ &\quad \times \mathrm{d}(\tau_1, \dots, \tau_m) \, \phi^m(\mathrm{d}(u_1, \dots, u_m)). \end{aligned}$$

Let M be the set of all m-tuples $(u_1, \dots, u_m) \in (S^{d-1})^m$ for which

$$\int_{(0,\infty)^m} f(u_1, \dots, u_m, \tau_1, \dots, \tau_m) g(u_1, \tau_1) \cdots g(u_m, \tau_m) \, \mathrm{d}(\tau_1, \dots, \tau_m) > 0.$$

For $(u_1, \dots, u_m) \in M$ we also have

$$\int_{(0,\infty)^m} f(u_1, \dots, u_m, \tau_1, \dots, \tau_m) \, \mathrm{d}(\tau_1, \dots, \tau_m) > 0.$$

Since $\mathbb{E}(X_s^{(k)}(A_r) \mid \widehat{X}_s(\mathcal{H}_r) = m) = 0$ by (11.27), it follows that $\phi^m(M) = 0$ and, therefore, that $\mathbb{E}(X^{(k)}(A_r) \mid \widehat{X}(\mathcal{H}_r) = m) = 0$. This proves (11.28). Thus $\Theta^{(k)}$ is absolutely continuous with respect to the translation invariant measure $\Theta_s^{(k)}$. This shows that the face process $X^{(k)}$ has a translation regular intensity measure. $\square$

Let $\widehat{X}$ and $\widehat{\Theta}$ be as in the previous theorem. Since the induced hyperplane mosaic $X = X^{(d)}$ and its processes $X^{(k)}$ of k-faces ($k = 0, \ldots, d-1$) have (locally finite) translation regular intensity measures, they admit specific intrinsic volumes $\overline{V}_j(X^{(k)}, z) =: d_j^{(k)}(z)$, $j = 0, \ldots, k$, satisfying

$$d_j^{(k)}(z) = \overline{V}_j(X^{(k)}, z) = \lim_{r \to 0} \frac{1}{V_d(rW)} \mathbb{E} \sum_{K \in X^{(k)}} \Phi_j(K, rW + z) \tag{11.29}$$

for λ-almost all z, where $W \in \mathcal{K}$ with $V_d(W) > 0$; see Corollary 11.1.1. We write

$$\gamma^{(k)}(z) := d_j^{(0)}(z)$$

and call $\gamma^{(k)}$ the **intensity function** of the k-face process $X^{(k)}$.

The following result extends Theorem 10.3.1, for Poisson processes. This restriction was made since it allows us to deduce the translation regularity of the intensity measures of the face processes, which otherwise would have to be an additional assumption.

Theorem 11.4.2. *Let $\widehat{X}$ be a nondegenerate Poisson hyperplane process in $\mathbb{R}^d$ with a translation regular intensity measure, let X be the induced hyperplane mosaic, and let $X^{(k)}$ be its k-face process, for $k = 0, \ldots, d$. For $0 \le j \le k \le d$, the relation*

$$d_j^{(k)} = \binom{d-j}{d-k} d_j^{(j)}$$

holds λ-almost everywhere, in particular

$$\gamma^{(k)} = \binom{d}{k} \gamma^{(0)}.$$

Proof. Let $j \in \{0, \ldots, d-1\}$, $z \in \mathbb{R}^d$, and $r > 0$. Given a realization of $\widehat{X}$ inducing a mosaic X in general position (without loss of generality), we can choose finitely many cells $S_1, \ldots, S_p$ of X such that $P := \bigcup_{i=1}^p S_i$ is a convex polytope with $rB^d + z \subset \operatorname{int} P$. Then $\Phi_j(P, rB^d + z) = 0$ since $j < d$. Since the curvature measure Φ_j is additive on the convex ring $\mathcal{R}$, the inclusion–exclusion principle gives

$$\begin{aligned} 0 &= \Phi_j(P, rB^d + z) \\ &= \Phi_j\left(\bigcup_{i=1}^p S_i, rB^d + z\right) \\ &= \sum_{m=1}^p (-1)^{m-1} \sum_{i_1 < \ldots < i_m} \Phi_j(S_{i_1} \cap \ldots \cap S_{i_m}, rB^d + z). \end{aligned}$$

Each intersection $S_{i_1} \cap \ldots \cap S_{i_m}$ is either empty or an i-face of the mosaic X and thus an element of $X^{(i)}$ for some $i \in \{1, \ldots, d\}$. Conversely, each element of

$X^{(i)}$ meeting the ball $rB^d + z$ is obtained in this way. For a face F, let $\nu(F, m)$ denote the number of m-tuples $(S_{i_1}, \ldots, S_{i_m})$ with $S_{i_1} \cap \ldots \cap S_{i_m} = F$. Taking into account the fact that $\Phi_j(F, \cdot) = 0$ if $\dim F < j$, we deduce that

$$\begin{aligned} 0 &= \sum_{i=j}^{d} \sum_{F \in X^{(i)}} \Phi_j(F, rB^d + z) \sum_{m=1}^{p} (-1)^{m-1} \nu(F, m) \\ &= \sum_{i=j}^{d} (-1)^{d-i} \sum_{F \in X^{(i)}} \Phi_j(F, rB^d + z) \end{aligned} \tag{11.30}$$

(compare the proof of Theorem 10.1.4). Taking the expectation, dividing by $V_d(rB^d)$ and letting r tend to 0, we obtain from (11.29) the relation

$$\sum_{i=j}^{d} (-1)^{d-i} d_j^{(i)}(z) = 0$$

for almost all $z \in \mathbb{R}^d$ and for $j = 0, \ldots, d-1$.

Let $k \in \{1, \ldots, d-1\}$, $j \in \{0, \ldots, k-1\}$, and $r > 0$. Let E be a k-plane of the intersection process $\widehat{X}_{d-k}$. We apply (11.30) to the mosaic induced in the k-plane E. This gives

$$0 = \sum_{i=j}^{k} (-1)^{k-i} \sum_{F \in X^{(i)},\, F \subset E} \Phi_j(F, rB^d + z).$$

We sum over all k-planes $E \in \widehat{X}_{n-k}$ and observe that X is almost surely in general position. Hence, each i-face of X is contained in precisely $\binom{d-i}{d-k}$ k-planes of $\widehat{X}_{d-k}$. This gives

$$0 = \sum_{i=j}^{k} (-1)^{k-i} \binom{d-i}{d-k} \sum_{F \in X^{(i)}} \Phi_j(F, rB^d + z).$$

As above, (11.29) yields

$$\sum_{i=j}^{k} (-1)^{i} \binom{d-i}{d-k} d_j^{(i)}(z) = 0$$

for almost all $z \in \mathbb{R}^d$. The remaining part of the proof is identical to that of Theorem 10.3.1. □

Finally, we observe that the specific j-volume $d_j^{(j)}$ of the j-faces can be expressed in a different way. Let $\widehat{X}_{d-j}$ be the intersection process of order $d-j$ of the hyperplane process $\widehat{X}$, and let $\widehat{\gamma}_{d-j}$ be the intensity function of

$\widehat{X}_{d-j}$. It is the Radon–Nikodym derivative of the measure $\mathbb{E}\sum_{E\in\widehat{X}_{d-j}}\lambda_E$ and hence can be obtained by differentiation, in particular

$$\lim_{r\to 0}\frac{1}{V_d(rB^d}\,\mathbb{E}\sum_{E\in\widehat{X}_{d-j}}\lambda_E(rB^d+z)=\widehat{\gamma}_{d-j}(z)$$

for almost all z. Since

$$\sum_{E\in\widehat{X}_{d-j}}\lambda_E(rB^d+z)=\sum_{K\in X^{(j)}}\Phi_j(K,rB^d+z),$$

we deduce that $d^{(j)}=\widehat{\gamma}_{d-j}$ almost everywhere. Together with (11.26), this yields the inequality

$$d_j^{(k)}(z)\le\binom{d-j}{d-k}\binom{d}{j}\frac{\kappa_{d-1}^{d-j}}{d^{d-j}\kappa_j\kappa_d^{d-j-1}}\widehat{\gamma}(z)^{d-j}.$$

Equality holds if the hyperplane process $\widehat{X}$ is stationary and isotropic.

Notes for Section 11.4

1. The results of this section are taken from Schneider [707]. The proof of Theorem 11.4.2 uses ideas from Weiss [807] and Weiss and Zähle [809].

2. Fallert [222, chap. 6] has investigated Voronoi and Delaunay mosaics induced by Poisson processes in $\mathbb{R}^d$ with translation regular intensity measures.

Part IV

Appendix

12

Facts from General Topology

Throughout this book, we have to use various facts from general topology, the theory of invariant measures, and the geometry of convex sets. In order not to delay the access to stochastic geometry by lengthy preparations, we have collected them here in the Appendix, so that we can refer to the results when they are needed.

The present chapter deals with topological notions and results, including some basic facts on Borel measures. Invariant measures are the topic of Chapter 13, and the necessary material from convex geometry is presented in Chapter 14.

12.1 General Topology and Borel Measures

Most of the topological spaces used in this book are locally compact and have a countable base. By convention, the notions **locally compact** and **compact** are understood to include the Hausdorff separation property. Every locally compact and second countable space is metrizable.

Let E be a topological space. We use the abbreviations $\mathrm{cl}\, A$, $\mathrm{bd}\, A$, $\mathrm{int}\, A$ and A^c, in this order, for the closure, the boundary, the interior and the complement of a set $A \subset E$. We denote by $\mathcal{F}(E)$, $\mathcal{C}(E)$, $\mathcal{G}(E)$, respectively, the system of closed, compact, or open subsets of E (always including the empty set $\emptyset$). When there is no danger of ambiguity, we simply write $\mathcal{F}$, $\mathcal{C}$, $\mathcal{G}$ for these set systems. We denote by $\mathcal{F}'$, $\mathcal{C}'$, $\mathcal{G}'$ the corresponding systems of nonempty sets.

The subsequent theorem collects a number of topological facts which are used frequently.

Theorem 12.1.1. *The following assertions hold for every locally compact space E with a countable base.*

(a) *The topology of* E *has a countable base* $\mathcal{D}$ *consisting of open, relatively compact sets such that every open set* $G \subset E$ *is the union of the sets* $D \in \mathcal{D}$ *satisfying* $\operatorname{cl} D \subset G$.

(b) *There is a sequence* $(G_i)_{i\in\mathbb{N}}$ *of open, relatively compact sets in* E *satisfying* $\operatorname{cl} G_i \subset G_{i+1}$ *for* $i \in \mathbb{N}$ *and* $\bigcup_{i\in\mathbb{N}} G_i = E$.

(c) *For every compact set* $C \subset E$ *there exists a decreasing sequence* $(G_i)_{i\in\mathbb{N}}$ *of open, relatively compact neighborhoods of* C *in* E *such that to every open set* $G \subset E$ *with* $C \subset G$ *there is an* $i \in \mathbb{N}$ *with* $G_i \subset G$.
Further, there is a decreasing sequence $(H_i)_{i\in\mathbb{N}}$ *of open, relatively compact sets with* $\operatorname{cl} H_{i+1} \subset H_i$ *and* $\bigcap_{i\in\mathbb{N}} H_i = C$.

(d) *If* $C \subset E$ *is compact and* $G_1, G_2 \subset E$ *are open sets with* $C \subset G_1 \cup G_2$, *then there are compact sets* $C_1 \subset G_1$ *and* $C_2 \subset G_2$ *with* $C = C_1 \cup C_2$.

Proof. (a) Let $\mathcal{D}'$ be a countable base of the topology of E and let $\mathcal{D} \subset \mathcal{D}'$ be the subsystem of the relatively compact sets in $\mathcal{D}'$. Let $G \subset E$ be open. For $x \in G$ there is an open neighborhood U such that $\operatorname{cl} U$ is compact. The locally compact space E is regular, hence there is an open neighborhood V of x with $\operatorname{cl} V \subset U \cap G$. A suitable base set $D \in \mathcal{D}'$ satisfies $x \in D \subset V$. From $\operatorname{cl} D \subset \operatorname{cl} V$ it follows that $\operatorname{cl} D \subset G$ and $\operatorname{cl} D \subset \operatorname{cl} U$, thus D is relatively compact and, therefore, an element of $\mathcal{D}$. This proves (a).

We fix the system $\mathcal{D}$ for the following.

(b) Let $\mathcal{D} = \{D_i : i \in \mathbb{N}\}$. Put $G_1 := D_1$. If the open, relatively compact set G_m has been defined, choose a number $k > m$ with $\operatorname{cl} G_m \subset \bigcup_{j=1}^{k} D_j$; this is possible due to the compactness of $\operatorname{cl} G_m$. Then put $\bigcup_{j=1}^{k} D_j =: G_{m+1}$. The sequence $(G_m)_{m\in\mathbb{N}}$ thus defined has the required properties.

(c) Let $C \subset E$ be compact. Let $(U_k)_{k\in\mathbb{N}}$ be the sequence (in any order) of all finite unions of sets from $\mathcal{D}$ that cover C. Put $G_m := \bigcap_{k=1}^{m} U_k$. Then $(G_m)_{m\in\mathbb{N}}$ is a decreasing sequence of open neighborhoods of C. Let $G \in \mathcal{G}(E)$ be an open set with $C \subset G$. To each $x \in C$ there exists $D_x \in \mathcal{D}$ with $x \in D_x \subset G$. Finitely many of these D_x cover C, their union is a set U_k, and $G_k \subset U_k \subset G$.

To construct $(H_i)_{i\in\mathbb{N}}$, let $H_1 := G_1$ and suppose that H_m has been defined. To each $x \in C$ there exists $D_x \in \mathcal{D}$ with $x \in D_x \subset \operatorname{cl} D_x \subset H_m$. Finitely many of these D_x cover C, their union is a set U_k, put $H_{m+1} := G_k \cap G_m$. Then $\operatorname{cl} H_{m+1} \subset \operatorname{cl} U_k \subset H_m$.

(d) Let $K_1, K_2 \subset E$ be disjoint compact sets, w.l.o.g. nonempty. Let $x \in K_1$. To each $y \in K_2$ there are disjoint open neighborhoods U_y of x and V_y of y. Since K_2 is compact, there are finitely many points $y_1, \ldots, y_n \in K_2$ with $K_2 \subset \bigcup_{i=1}^{n} V_{y_i} =: V_x$. The sets V_x and $U_x := \bigcap_{i=1}^{n} U_{y_i}$ are open and disjoint. Since K_1 is compact, there are finitely many points $x_1, \ldots, x_k \in K_1$ with $K_1 \subset \bigcup_{i=1}^{k} U_{x_i} =: U_1$. Put $U_2 := \bigcap_{i=1}^{k} V_{x_i}$.

If now C, G_1, G_2 are sets as described in (d), then put $K_i := C \setminus G_i$ $(i = 1, 2)$. The sets K_1 and K_2 are disjoint and compact. For these sets, choose U_1, U_2 as above, and let $C_i := C \setminus U_i$. Then C_i is compact, $C_i \subset G_i$ $(i = 1, 2)$, and $C = C_1 \cup C_2$ (since $U_1 \cap U_2 = \emptyset$). □

Products of topological spaces always carry the product topology. For a topological space E we denote by $\mathbf{C}(E)$ the vector space of continuous real functions on E, and if E is locally compact, $\mathbf{C}_c(E)$ is the subspace of $\mathbf{C}(E)$ of functions with compact support.

By a **Borel measure** ρ on the topological space E we understand a measure (nonnegative, σ-additive set function) on the σ-algebra $\mathcal{B}(E)$ of Borel sets of E (the smallest σ-algebra containing the open sets of E). It is called **locally finite** if it satisfies $\rho(K) < \infty$ for every compact set $K \subset E$. Instead of 'Borel measure', we often say 'measure', for short. The notion 'measurable', without extra specification, means 'Borel measurable'.

In a locally compact space E with a countable base, the values of a locally finite Borel measure ρ are determined by the integrals of continuous functions with compact support, in the following way. For open sets $A \subset E$,

$$\rho(A) = \sup\left\{\int_E f \, \mathrm{d}\rho : f \in \mathbf{C}_c(E),\ 0 \le f \le \mathbf{1}_A\right\}, \tag{12.1}$$

and for arbitrary Borel sets $A \in \mathcal{B}(E)$,

$$\rho(A) = \inf\{\rho(U) : A \subset U,\ U \text{ open}\} \tag{12.2}$$

(see, e.g., Cohn [177, sect. 7.2]).

Concerning the application of Fubini's theorem for Borel measures, we remark the following. Whenever this theorem is used in this book, the topological spaces occurring are locally compact and second countable, thus they are σ-compact. Moreover, all the measures that occur are locally finite. Therefore, all measure spaces under consideration are σ-finite, so that Fubini's theorem can be applied in its usual form (for example, [177, p. 159]). If E and Y are topological spaces with countable bases, then $\mathcal{B}(E \times Y) = \mathcal{B}(E) \otimes \mathcal{B}(Y)$ (see, e.g., [177, p. 242]). This has to be observed occasionally in measurability proofs. For example, in the proof of Theorem 13.1.2, the measurability of the function $(x, y) \mapsto f(y^{-1}x)$ with respect to $\mathcal{B}(G) \otimes \mathcal{B}(G)$ follows from the continuity of the map $(x, y) \mapsto y^{-1}x$ and the measurability of the function f. If in applications of Fubini's theorem the measurability of the integrand with respect to the product σ-algebra is not mentioned explicitly, it can be verified in a similar manner.

The following auxiliary results are used occasionally in order to verify the measurability of some maps.

Lemma 12.1.1. *Let E be a locally compact space with a countable base, let $(T, \mathcal{T})$ be a measurable space and*

$$\psi : T \times \mathcal{B}(E) \to \overline{\mathbb{R}}$$

a map such that $\psi(t, \cdot)$ is a Borel measure for every $t \in T$. Suppose that for every function $f \in \mathbf{C}_c(E)$ the map

$$t \mapsto \int_E f(x)\,\psi(t, \mathrm{d}x) \tag{12.3}$$

is $\mathcal{T}$-measurable. Then, for every nonnegative measurable function f on E the map (12.3) *is $\mathcal{T}$-measurable; in particular, $\psi(\cdot, B)$ is a $\mathcal{T}$-measurable function for every $B \in \mathcal{B}(E)$.*

Proof. First let $B \subset E$ be compact. Since E is locally compact and second countable, the indicator function $\mathbf{1}_B$ is the limit of a decreasing sequence $(f_i)_{i\in\mathbb{N}}$ in $\mathbf{C}_c(E)$. Thus we have

$$\psi(t, B) = \lim_{i\to\infty} \int_E f_i(x)\,\psi(t, \mathrm{d}x)$$

for all $t \in T$, which implies the $\mathcal{T}$-measurability of $\psi(\cdot, B)$.

Let $\mathcal{D}$ be the system of all sets $A \in \mathcal{B}(E)$ for which $\psi(\cdot, A)$ is a $\mathcal{T}$-measurable function. Since $\psi(t, \cdot)$ is a Borel measure and E is a σ-compact space, it is easy to see that $\mathcal{D}$ is a Dynkin system. It contains the intersection stable system of compact sets and therefore the σ-algebra generated by these sets, which is $\mathcal{B}(E)$. Thus the function $\psi(\cdot, A)$ is $\mathcal{T}$-measurable for all $A \in \mathcal{B}(E)$. The $\mathcal{T}$-measurability of the map (12.3) for nonnegative measurable functions f on E is now obtained by a standard argument. □

In typical applications of Lemma 12.1.1, also T is a topological space (with its Borel σ-algebra), and the map (12.3) is continuous if $f \in \mathbf{C}_c(E)$. The function f may also depend on t.

Lemma 12.1.2. *If E and T are locally compact spaces with countable bases and if ψ satisfies the assumptions of* Lemma 12.1.1 *with $\mathcal{T} = \mathcal{B}(T)$, then for every nonnegative measurable function f on $T \times E$ the mapping*

$$t \mapsto \int_E f(t, x)\,\psi(t, \mathrm{d}x)$$

is measurable.

Proof. By Lemma 12.1.1, for $B \in \mathcal{B}(E)$ the function $\psi(\cdot, B)$ is measurable. For $g(t, x) = \mathbf{1}_A(t)\mathbf{1}_B(x)$ with $A \in \mathcal{B}(T)$ and $B \in \mathcal{B}(E)$ we deduce the measurability of the map

$$t \mapsto \mathbf{1}_A(t)\psi(t, B) = \int_E g(t, x)\,\psi(t, \mathrm{d}x).$$

Since $\mathcal{B}(T \times E) = \mathcal{B}(T) \otimes \mathcal{B}(E)$ (Cohn [177, p. 242]), the assertion is obtained by a standard argument. □

12.2 The Space of Closed Sets

The standard treatment of random closed sets makes use of a suitable topology on the system of closed subsets of a given locally compact topological space. We collect here the basic notions and results on this topology which are used in the present book.

Throughout this section, E is a locally compact space with a countable base.

The following subsystems of the system $\mathcal{F} = \mathcal{F}(E)$ of closed subsets of E play an important role. For $A, A_1, \dots, A_k \subset E$ one defines

$$\mathcal{F}^A := \{F \in \mathcal{F} : F \cap A = \emptyset\},$$
$$\mathcal{F}_A := \{F \in \mathcal{F} : F \cap A \neq \emptyset\}$$

and

$$\mathcal{F}^A_{A_1,\dots,A_k} := \mathcal{F}^A \cap \mathcal{F}_{A_1} \cap \dots \cap \mathcal{F}_{A_k}, \qquad k \in \mathbb{N}_0$$

(with $\mathcal{F}^A_{A_1,\dots,A_k} := \mathcal{F}^A$ if $k = 0$). In particular, $\mathcal{F}_A = \mathcal{F}^{\emptyset}_A$ and $(\mathcal{F}_A)^c = \mathcal{F}^A$. Notice that always $\emptyset \in \mathcal{F}^A$, but $\emptyset \notin \mathcal{F}^A_{A_1,\dots,A_k}$ for $k \geq 1$. Occasionally in proofs, this fact requires the consideration of different cases.

On $\mathcal{F}$ we introduce the topology generated by the set system

$$\{\mathcal{F}^C : C \in \mathcal{C}\} \cup \{\mathcal{F}_G : G \in \mathcal{G}\}. \tag{12.4}$$

Because of the relation

$$\mathcal{F}^C_{G_1,\dots,G_k} \cap \mathcal{F}^{C'}_{G'_1,\dots,G'_m} = \mathcal{F}^{C \cup C'}_{G_1,\dots,G_k,G'_1,\dots,G'_m},$$

the system

$$\tau := \{\mathcal{F}^C_{G_1,\dots,G_k} : C \in \mathcal{C},\ G_1,\dots,G_k \in \mathcal{G},\ k \in \mathbb{N}_0\}$$

is $\cap$-stable (stable under intersections), and $\mathcal{F} = \mathcal{F}^{\emptyset} \in \tau$, hence τ is a base of the topology generated by (12.4). Thus, the open sets of this topology are the unions of the sets of τ. This topology is called the **topology of closed convergence** or the **Fell topology**. In this book, the set $\mathcal{F}$ is always equipped with this topology, unless stated otherwise.

Theorem 12.2.1. *$\mathcal{F}$ is a compact space with a countable base.*

Proof. Let $\mathcal{D}$ be a countable base of the topology of E with the properties listed in Theorem 12.1.1(a).

To prove the Hausdorff separation property, let F, F' be distinct elements of $\mathcal{F}$. Without loss of generality, there exists a point $x \in F \setminus F'$. There is a set $D \in \mathcal{D}$ with $x \in D$ and $F' \cap \mathrm{cl}\, D = \emptyset$. Thus $\mathcal{F}_D$ is a neighborhood of F, while $\mathcal{F}^{\mathrm{cl}\, D}$ is a neighborhood of F', and $\mathcal{F}_D \cap \mathcal{F}^{\mathrm{cl}\, D} = \emptyset$. Hence, $\mathcal{F}$ is a Hausdorff space.

For the proof of the compactness it suffices by Alexander's theorem (see, for example, Kelley [393, p. 139]) to prove that every covering of $\mathcal{F}$ by sets of the subbasis (12.4) contains a finite covering of $\mathcal{F}$. Suppose, therefore, that

$$\bigcup_{i\in I} \mathcal{F}^{C_i} \cup \bigcup_{j\in J} \mathcal{F}_{G_j} = \mathcal{F}$$

with a pair of families

$$(C_i)_{i\in I},\ C_i \in \mathcal{C}, \quad \text{and} \quad (G_j)_{j\in J},\ G_j \in \mathcal{G}.$$

Taking complements, we get

$$\bigcap_{i\in I} \mathcal{F}_{C_i} \cap \bigcap_{j\in J} \mathcal{F}^{G_j} = \emptyset.$$

Writing $G := \bigcup_{j\in J} G_j$, we have $\bigcap_{j\in J} \mathcal{F}^{G_j} = \mathcal{F}^G$, hence our assumption gives

$$\bigcap_{i\in I} \mathcal{F}^G_{C_i} = \emptyset.$$

There is an index $i_0 \in I$ with $C_{i_0} \subset G$, since otherwise $G^c \cap C_i \neq \emptyset$ for all $i \in I$, which would imply

$$G^c \in \bigcap_{i\in I} \mathcal{F}^G_{C_i},$$

a contradiction. For the compact set $C_{i_0} \subset G = \bigcup_{j\in J} G_j$, there is a finite subcovering and hence a finite subset $J_0 \subset J$ with $C_{i_0} \subset \bigcup_{j\in J_0} G_j$. This implies

$$\bigcap_{j\in J_0} \mathcal{F}^{G_j}_{C_{i_0}} = \emptyset$$

and thus

$$\mathcal{F}^{C_{i_0}} \cup \bigcup_{j\in J_0} \mathcal{F}_{G_j} = \mathcal{F}.$$

This proves the compactness of $\mathcal{F}$.

To show the existence of a countable base, we consider the countable system

$$\tau' := \left\{ \mathcal{F}^{\mathrm{cl}\, D'_1 \cup \ldots \cup \mathrm{cl}\, D'_m}_{D_1,\ldots,D_k} : D_i, D'_j \in \mathcal{D},\ k \in \mathbb{N}_0,\ m \in \mathbb{N} \right\}.$$

It satisfies $\tau' \subset \tau$. Now let $F \in \mathcal{F}$ and

$$F \in \mathcal{F}^C_{G_1,\ldots,G_k} \in \tau.$$

It suffices to show the existence of a set $\mathcal{A} \in \tau'$ with

$$F \in \mathcal{A} \subset \mathcal{F}^C_{G_1,\ldots,G_k}.$$

Since $F \cap C = \emptyset$, to every $x \in C$ there exists a set $D(x) \in \mathcal{D}$ with $x \in D(x)$ and $F \cap \operatorname{cl} D(x) = \emptyset$. The family $(D(x))_{x \in C}$ is an open covering of the compact set C, hence there exist $D'_1, \dots, D'_m \in \mathcal{D}$ with $C \subset \operatorname{cl} D'_1 \cup \dots \cup \operatorname{cl} D'_m$ and $F \cap \operatorname{cl} D'_i = \emptyset$. If $k = 0$, then

$$F \in \mathcal{F}^{\operatorname{cl} D'_1 \cup \dots \cup \operatorname{cl} D'_m} \subset \mathcal{F}^C.$$

If $k \geq 1$, then for each $i \in \{1, \dots, k\}$ there exist $x_i \in F \cap G_i$ and $D_i \in \mathcal{D}$ with $x_i \in D_i \subset G_i$. This gives

$$F \in \mathcal{F}_{D_1,\dots,D_k}^{\operatorname{cl} D'_1 \cup \dots \cup \operatorname{cl} D'_m} \subset \mathcal{F}_{G_1,\dots,G_k}^C.$$

Hence, also the system τ' is a base of the topology of $\mathcal{F}$. □

Remarks. (a) Due to Theorem 12.2.1, in convergence or continuity arguments it is sufficient to work with sequences instead of nets or filters. By Urysohn's theorem, $\mathcal{F}$ is metrizable.

(b) The space $\mathcal{F}' := \mathcal{F} \setminus \{\emptyset\}$ is locally compact, but for non-compact E it is not compact. This is seen from the fact that

$$\bigcup_{D \in \mathcal{D}} \mathcal{F}_D = \mathcal{F}_E = \mathcal{F} \setminus \{\emptyset\} = \mathcal{F}',$$

but no finite subfamily of $\{\mathcal{F}_D : D \in \mathcal{D}\}$ covers $\mathcal{F}'$.

(c) The point $\emptyset \in \mathcal{F}$ has the system $\{\mathcal{F}^C : C \in \mathcal{C}\}$ as a base of neighborhoods. The space $\mathcal{F}$ is the one-point (Aleksandrov) compactification of $\mathcal{F}'$.

Now we consider convergence in $\mathcal{F}$ and give, in the next theorem, useful equivalent descriptions. As usual, 'almost all $j \in \mathbb{N}$' means: all $j \in \mathbb{N}$ with at most finitely many exceptions.

Theorem 12.2.2. *Let $(F_j)_{j \in \mathbb{N}}$ be a sequence in $\mathcal{F}$, and let $F \in \mathcal{F}$. Then the following assertions* (a), (b), (c) *are equivalent:*

(a) *$F_j \to F$ as $j \to \infty$.*

(b) *Conditions* (b_1) *and* (b_2) *hold:*
  - (b_1) *If $G \in \mathcal{G}$ and $G \cap F \neq \emptyset$, then $G \cap F_j \neq \emptyset$ for almost all j.*
  - (b_2) *If $C \in \mathcal{C}$ and $C \cap F = \emptyset$, then $C \cap F_j = \emptyset$ for almost all j.*

(c) *Conditions* (c_1) *and* (c_2) *hold:*
  - (c_1) *If $x \in F$, then for almost all j there is $x_j \in F_j$ so that $x_j \to x$ as $j \to \infty$.*
  - (c_2) *If $(F_{j_k})_{k \in \mathbb{N}}$ is a subsequence and the points $x_{j_k} \in F_{j_k}$ are such that $x := \lim_{k \to \infty} x_{j_k}$ exists, then $x \in F$.*

Proof. By definition, $F_j \to F$ holds if and only if every neighborhood from the system τ contains almost all F_j. Therefore, the definition of τ implies the equivalence of (a) and (b). We prove now the equivalence of (b_1) with (c_1) and that of (b_2) with (c_2).

(b_1) $\Rightarrow$ (c_1): Let $x \in F$. Let $G_1 \supset G_2 \supset \ldots$ be a base of open neighborhoods of x. Then $G_i \cap F \neq \emptyset$, hence ($b_1$) implies the existence of $k_i \in \mathbb{N}$ with $G_i \cap F_k \neq \emptyset$ for $k \geq k_i$ ($i \in \mathbb{N}$). W.l.o.g., we may assume that $k_1 < k_2 < \ldots$ Hence, there exists a sequence $(x_p)_{p \geq k_1}$ such that

$$x_p \in G_i \cap F_p \quad \text{for } p = k_i, \ldots, k_{i+1} - 1, \quad i \in \mathbb{N}.$$

Thus $x_p \to x$.

(c_1) $\Rightarrow$ (b_1): Let $G \in \mathcal{G}$ and $G \cap F \neq \emptyset$. There exists $x \in G \cap F$. By (c_1), for almost all j there is $x_j \in F_j$ so that $x_j \to x$. Then $x_j \in G$ and thus $G \cap F_j \neq \emptyset$ for almost all j.

(b_2) $\Rightarrow$ (c_2): Let $(F_{j_k})_{k \in \mathbb{N}}$ be a subsequence, $x_{j_k} \in F_{j_k}$ for $k \in \mathbb{N}$, and $\lim_{k \to \infty} x_{j_k} = x$. If $x \notin F$, then there exists a compact neighborhood C of x with $C \cap F = \emptyset$, and (b_2) yields $C \cap F_j = \emptyset$ for almost all j, a contradiction.

(c_2) $\Rightarrow$ (b_2): Suppose that (b_2) is false. Then there is $C \in \mathcal{C}$ with $C \cap F = \emptyset$ and $C \cap F_j \neq \emptyset$ for infinitely many j. Hence, there is a subsequence $(F_{j_k})_{k \in \mathbb{N}}$ with corresponding points $x_{j_k} \in C \cap F_{j_k}$. A suitable subsequence of $(x_{j_k})_{k \in \mathbb{N}}$ converges to some $x \in C$. Then $x \notin F$, which contradicts (c_2). □

In the treatment of random closed sets, the required measurability of set-theoretic operations often follows from their continuity or semicontinuity. For that reason, we now consider assertions of the latter type.

Theorem 12.2.3. *The union map*

$$\begin{aligned} \mathcal{F} \times \mathcal{F} &\to \quad \mathcal{F} \\ (F, F') &\mapsto F \cup F' \end{aligned}$$

is continuous.

Proof. Let $(F_i)_{i \in \mathbb{N}}$ and $(F_i')_{i \in \mathbb{N}}$ be convergent sequences in $\mathcal{F}$ with $F_i \to F$ and $F_i' \to F'$. We have to show that $F_i \cup F_i' \to F \cup F'$. Corresponding to Theorem 12.2.2, we proceed in two steps.

(α) Let $x \in F \cup F'$, w.l.o.g. $x \in F$. For almost all i there exists $x_i \in F_i \subset F_i \cup F_i'$ with $x_i \to x$.

(β) Suppose (w.l.o.g.) that $x_i \to x$ with $x_i \in F_i \cup F_i'$. Then there is a subsequence $(x_{j_k})_{k \in \mathbb{N}}$ satisfying $x_{j_k} \in F_{j_k}$, say. This implies $x \in F$, hence $x \in F \cup F'$.

It follows that $F_i \cup F_i' \to F \cup F'$. □

Other set operations, such as intersection or forming the closure of the complement, are not continuous. If, for example, $(F_i)_{i\in\mathbb{N}}$ is a sequence of singletons $F_i = \{x_i\}$ with $F_i \to F = \{x\}$ (which is equivalent to $x_i \to x$) and $F_i \cap F = \emptyset$, then it is not true that $F_i \cap F \to F \cap F$. If $(F_i)_{i\in\mathbb{N}}$ is a convergent sequence of finite sets with $F_i \to E$, then $\operatorname{cl} F_i^c = E$, hence it is not true that $\operatorname{cl} F_i^c \to \operatorname{cl} E^c$. Also the boundary operation is not continuous. Semicontinuity properties, however, can be proved.

Let $\varphi : T \to \mathcal{F}$ be a map from some topological space T into $\mathcal{F}$. The map φ is called **upper semicontinuous** if $\varphi^{-1}(\mathcal{F}^C)$ is open (in T) for all $C \in \mathcal{C}$; it is called **lower semicontinuous** if $\varphi^{-1}(\mathcal{F}_G)$ is open for all $G \in \mathcal{G}$. The following notions are helpful in the treatment of semicontinuity. For a sequence $(F_i)_{i\in\mathbb{N}}$ in $\mathcal{F}$, one denotes by $\limsup F_i$ the union of all accumulation points (in $\mathcal{F}$) of the sequence $(F_i)_{i\in\mathbb{N}}$, and by $\liminf F_i$ the intersection of all these accumulation points (recall that a 'point' in $\mathcal{F}$ is a set in E). These two sets can be characterized as follows.

Theorem 12.2.4. *Let $(F_i)_{i\in\mathbb{N}}$ be a sequence in $\mathcal{F}$. Then*

$$\begin{aligned} &\limsup F_i \\ &= \{x \in E : \text{every neighborhood of } x \text{ meets infinitely many } F_i\}, \qquad (12.5) \\ &\liminf F_i \\ &= \{x \in E : \text{every neighborhood of } x \text{ meets almost all } F_i\}. \qquad (12.6) \end{aligned}$$

Both sets are closed.

Proof. We denote the right side of (12.5) by A and that of (12.6) by B.

Let $x \in A$. Then there are a subsequence $(F_{i_j})_{j\in\mathbb{N}}$ and points $x_{i_j} \in F_{i_j}$ with $x_{i_j} \to x$ as $j \to \infty$. Since $\mathcal{F}$ is compact and second countable, the sequence $(F_{i_j})_{j\in\mathbb{N}}$ has a subsequence converging to some $F \in \mathcal{F}$. By Theorem 12.2.2, $x \in F \subset \limsup F_i$. Conversely, suppose that $x \in \limsup F_i$. Then there is a convergent subsequence $(F_{i_j})_{j\in\mathbb{N}}$ with limit F such that $x \in F$. Theorem 12.2.2 implies that $x = \lim x_{i_j}$ for suitable $x_{i_j} \in F_{i_j}$ and $j \geq j_0$, say. It follows that $x \in A$. Thus $A = \limsup F_i$. This also shows that $\limsup F_i$ is closed.

Let $x \in B$. Then there are points $x_i \in F_i$ $(i \geq i_0)$ with $\lim x_i = x$. If F is the limit of a convergent subsequence of $(F_i)_{i\in\mathbb{N}}$, then it follows that $x \in F$. Thus $x \in \liminf F_i$. Conversely, let $y \in E$ be a point such that $y \notin B$. Then there is a neighborhood U of y and a subsequence $(F_{i_j})_{j\in\mathbb{N}}$ with $U \cap F_{i_j} = \emptyset$ for $j \in \mathbb{N}$. This sequence has a subsequence converging to some $F \in \mathcal{F}$. Theorem 12.2.2 gives $y \notin F$, hence $y \notin \liminf F_i$. This completes the proof of $B = \liminf F_i$. Clearly $\liminf F_i$ is closed. □

Remark. For a sequence $(F_i)_{i\in\mathbb{N}}$ in $\mathcal{F}$ we obviously have $\lim F_i = F$ if and only if

$$\limsup F_i = \liminf F_i = F. \qquad (12.7)$$

More generally, one can define $\limsup F_i$ and $\liminf F_i$ by (12.5) and (12.6) for sequences $(F_i)_{i\in\mathbb{N}}$ of not necessarily closed subsets of E, and one can then define the convergence of $(F_i)_{i\in\mathbb{N}}$ to F by (12.7). Since $\limsup F_i$ and $\liminf F_i$ are always closed, one calls F the **closed limit** of the sequence $(F_i)_{i\in\mathbb{N}}$. This explains the name '**topology of closed convergence**' for the topology introduced on $\mathcal{F}$.

The following is a useful criterion for semicontinuity.

Theorem 12.2.5. *Let T be a topological space with a countable base and $\varphi : T \to \mathcal{F}$ a mapping.*

(a) *φ is upper semicontinuous if and only if*

$$\limsup \varphi(t_i) \subset \varphi(t) \qquad \text{for all } t, t_i \in T \text{ with } t_i \to t.$$

(b) *φ is lower semicontinuous if and only if*

$$\liminf \varphi(t_i) \supset \varphi(t) \qquad \text{for all } t, t_i \in T \text{ with } t_i \to t.$$

Proof. (a) The set $\varphi^{-1}(\mathcal{F}^C)$ is open for all $C \in \mathcal{C}$ if and only if $\varphi^{-1}(\mathcal{F}_C)$ is closed for all $C \in \mathcal{C}$. This is equivalent to the following condition:

(a_1) If $C \in \mathcal{C}$, $t_i \to t$ in T and $\varphi(t_i) \cap C \neq \emptyset$ for all i, then $\varphi(t) \cap C \neq \emptyset$.

On the other hand, (a_1) is equivalent to (a_2):

(a_2) If $t_i \to t$ in T, then $\limsup \varphi(t_i) \subset \varphi(t)$.

To prove this latter equivalence, assume that (a_1) holds. If $t_i \to t$ in T and $x \in \limsup \varphi(t_i)$, then every neighborhood of x, in particular every compact neighborhood C, meets infinitely many $\varphi(t_i)$. By (a_1) (applied to a subsequence) this implies $\varphi(t) \cap C \neq \emptyset$ and hence $x \in \varphi(t)$, since every neighborhood of x contains a compact neighborhood of x. Conversely, suppose that (a_2) holds. Under the assumptions of (a_1) there are $x_i \in \varphi(t_i) \cap C$, and since C is compact, there exists a subsequence $(x_{i_j})_{j\in\mathbb{N}}$ with $x_{i_j} \to x \in C$ for $j \to \infty$. It follows that $x \in \limsup \varphi(t_i) \subset \varphi(t)$, hence $\varphi(t) \cap C \neq \emptyset$.

The equivalence of (a_1) and (a_2) settles (a). Assertion (b) can be proved similarly. □

Now we can show the semicontinuity of some maps.

Theorem 12.2.6. (a) *The map*

$$\begin{aligned} \mathcal{F} \times \mathcal{F} &\to \quad \mathcal{F} \\ (F, F') &\mapsto F \cap F' \end{aligned}$$

is upper semicontinuous.

(b) *The map*

$$\begin{aligned} \mathcal{F} &\to \mathcal{F} \\ F &\mapsto \operatorname{cl} F^c \end{aligned}$$

is lower semicontinuous.

(c) *If E is locally connected, the map*

$$\begin{aligned} \mathcal{F} &\to \mathcal{F} \\ F &\mapsto \operatorname{bd} F \end{aligned}$$

is lower semicontinuous.

Proof. For the proof of (a), let $(F_i)_{i\in\mathbb{N}}$ and $(F_i')_{i\in\mathbb{N}}$ be convergent sequences in $\mathcal{F}$ with $F_i \to F$ and $F_i' \to F'$. According to Theorem 12.2.5(a), it suffices to show that $\limsup(F_i \cap F_i') \subset F \cap F'$. Let $x \in \limsup(F_i \cap F_i')$. Then there are a subsequence $(F_{i_k} \cap F_{i_k}')_{k\in\mathbb{N}}$ and points $x_{i_k} \in F_{i_k} \cap F_{i_k}'$ with $x_{i_k} \to x$. By Theorem 12.2.2(c), this implies $x \in F$ and $x \in F'$, hence $x \in F \cap F'$.

For the proof of (b), let $c : \mathcal{F} \to \mathcal{F}$ be defined by $c(F) := \operatorname{cl} F^c$. For $G \in \mathcal{G}$ we have

$$\begin{aligned} c^{-1}(\mathcal{F}^G) &= \{F \in \mathcal{F} : G \cap \operatorname{cl} F^c = \emptyset\} \\ &= \{F \in \mathcal{F} : G \cap F^c = \emptyset\} \\ &= \{F \in \mathcal{F} : G \subset F\}. \end{aligned}$$

The set $\{F \in \mathcal{F} : G \subset F\}$ is closed (since $F_i \to F$, $G \subset F_i$, $x \in G$ implies $x_i := x \in F_i$ and $x_i \to x$, hence $x \in F$); therefore, its complement $c^{-1}(\mathcal{F}_G)$ is open. Thus c is lower semicontinuous.

For the proof of (c), let $\partial : \mathcal{F} \to \mathcal{F}$ be defined by $\partial(F) := \operatorname{bd} F$. First let $B \subset E$ be an open connected set. For $F \in \mathcal{F}$ the condition $B \cap \operatorname{bd} F \neq \emptyset$ is equivalent to $B \cap F \neq \emptyset$ and $B \cap F^c \neq \emptyset$. Therefore,

$$\begin{aligned} \partial^{-1}(\mathcal{F}_B) &= \{F \in \mathcal{F} : B \cap \operatorname{bd} F \neq \emptyset\} \\ &= \{F \in \mathcal{F} : B \cap F \neq \emptyset \text{ and } B \cap F^c \neq \emptyset\} \\ &= \mathcal{F}_B \cap \{F \in \mathcal{F} : B \subset F\}^c. \end{aligned}$$

The set $\{F \in \mathcal{F} : B \subset F\}$ is closed, hence $\partial^{-1}(\mathcal{F}_B)$ is open. Since E is locally connected, an arbitrary open set $G \in \mathcal{G}$ can be written as the union $G = \bigcup B_i$ of open connected sets B_i. It follows that $\partial^{-1}(\mathcal{F}_G) = \partial^{-1}(\bigcup \mathcal{F}_{B_i}) = \bigcup \partial^{-1}(\mathcal{F}_{B_i})$ is open. Thus ∂ is lower semicontinuous. □

Recall that $\mathbf{1}_F$ is the indicator function of the set F.

Theorem 12.2.7. *The map*

$$\begin{aligned} \mathcal{F} \times E &\to \mathbb{R} \\ (F, x) &\mapsto \mathbf{1}_F(x) \end{aligned}$$

is upper semicontinuous.

Proof. Suppose that $(F_i, x_i) \to (F, x)$ in $\mathcal{F} \times E$. If $\limsup \mathbf{1}_{F_i}(x_i) = 1$, then there is a subsequence $(F_{i_j}, x_{i_j})_{j\in\mathbb{N}}$ with $x_{i_j} \in F_{i_j}$ for $j \in \mathbb{N}$, and Theorem 12.2.2 implies $x \in F$. Hence,

$$\limsup_{i\to\infty} \mathbf{1}_{F_i}(x_i) \le \mathbf{1}_F(x).$$

If $\limsup \mathbf{1}_{F_i}(x_i) = 0$, this inequality holds trivially. □

Note for Section 12.2

The topology of closed convergence for sequences of sets appears already in the work of Hausdorff [323]. The name 'Fell topology' became common after the generalization established in Fell [230]. For general treatments of topologies on spaces of subsets of a topological space, we refer to Michael [513], Klein and Thompson [419], Beer [87]. Convergence of subsets of $\mathbb{R}^d$ is thoroughly treated in Chapter 4 of Rockafellar and Wets [646]. The use of the Fell topology in stochastic geometry, where it is sometimes called the hit-and-miss topology, was emphasized by Matheron [462].

12.3 Euclidean Spaces and Hausdorff Metric

The special case where the underlying space E is $\mathbb{R}^d$, the d-dimensional Euclidean space, shows additional features, due to the linear structure of $\mathbb{R}^d$ and the properties of its standard metric. In this subsection, $\mathcal{F}, \mathcal{G}, \mathcal{C}$ denote, respectively, the spaces $\mathcal{F}(\mathbb{R}^d)$, $\mathcal{G}(\mathbb{R}^d)$, $\mathcal{C}(\mathbb{R}^d)$.

Theorem 12.3.1. *The maps*

$$\begin{aligned} \mathcal{F} &\to \mathcal{F} \\ F &\mapsto -F \end{aligned} \qquad \text{and} \qquad \begin{aligned} \mathbb{R}^+ \times \mathcal{F} &\to \mathcal{F} \\ (\alpha, F) &\mapsto \alpha F \end{aligned}$$

are continuous. The maps

$$\begin{aligned} \mathcal{F} &\to \mathcal{F} \\ F &\mapsto \mathrm{cl\,conv}\, F \end{aligned} \qquad \text{and} \qquad \begin{aligned} \mathcal{F} \times \mathcal{F} &\to \mathcal{F} \\ (F, F') &\mapsto \mathrm{cl}\,(F + F') \end{aligned}$$

are lower semicontinuous.

Proof. The proofs of the first two assertions are analogous to the proof of Theorem 12.2.3.

For the proof of the third assertion, let $h : \mathcal{F} \to \mathcal{F}$ denote the map defined by $h(F) := \mathrm{cl\,conv}\, F$. Let $(F_i)_{i\in\mathbb{N}}$ be a sequence in $\mathcal{F}$ with $F_i \to F$. We have to show that $h(F) \subset \liminf h(F_i)$. For this, let $x \in \mathrm{conv}\, F$. There is a representation

$$x = \sum_{k=1}^{m} \lambda_k x_k \quad \text{with } x_k \in F,\ \lambda_k \geq 0,\ \sum_{k=1}^{m} \lambda_k = 1,$$

with some $m \in \mathbb{N}$. For each $k \in \{1, \ldots, m\}$ we have, by Theorem 12.2.2,

$$x_k = \lim_{j \to \infty} x_{k,j}$$

with suitable $x_{k,j} \in F_j$ $(j \geq j_0)$. Defining $x_j := \sum_{k=1}^{m} \lambda_k x_{k,j}$, we get $x_j \in \operatorname{conv} F_j \subset h(F_j)$ and $x_j \to x$. By Theorem 12.2.4, this gives $x \in \liminf h(F_j)$. This shows that $\operatorname{conv} F \subset \liminf h(F_j)$. Since the inferior limit of a sequence of sets is always closed, we also have $h(F) = \operatorname{cl}\operatorname{conv} F \subset \liminf h(F_j)$. Hence, the map h is lower semicontinuous.

The remaining assertion is proved in a similar manner. □

Now we take into account the Euclidean metric $d(\cdot,\cdot)$ on $\mathbb{R}^d$, which is defined via the standard scalar product $\langle \cdot,\cdot \rangle$ and the corresponding norm $\|\cdot\|$, by $d(x,y) := \|x-y\|$ for $x, y \in \mathbb{R}^d$. It induces the Hausdorff metric on the set $\mathcal{C}' := \mathcal{C} \setminus \{\emptyset\}$ of nonempty compact subsets of $\mathbb{R}^d$, and we want to compare the corresponding topology on $\mathcal{C}'$ with the one induced by the topology of $\mathcal{F}$. This could be done for more general metric spaces, but it is sufficient for our purposes to consider only $\mathbb{R}^d$.

For $C, C' \in \mathcal{C}'$, the **Hausdorff distance** $\delta(C, C')$ is defined by

$$\delta(C, C') := \max \left\{ \max_{x \in C} \min_{y \in C'} d(x,y),\ \max_{x \in C'} \min_{y \in C} d(x,y) \right\}.$$

Writing

$$d(x, C) := \min_{y \in C} d(x,y)$$

for the distance of x from C and

$$C_\epsilon := \{x \in \mathbb{R}^d : d(x, C) \leq \epsilon\}$$

for $\epsilon \geq 0$, we have

$$\delta(C, C') = \min\{\epsilon \geq 0 : C \subset C'_\epsilon,\ C' \subset C_\epsilon\}.$$

This can also be written as

$$\delta(C, C') = \min\{\epsilon \geq 0 : C \subset C' + \epsilon B^d,\ C' \subset C + \epsilon B^d\},$$

where B^d is the closed unit ball with center at the origin. It is easy to see that δ is a metric on $\mathcal{C}'$. We extend it to a ($\mathbb{R} \cup \{\infty\}$-valued) metric on $\mathcal{C}$ by putting $\delta(C, C') := \infty$ if precisely one of the sets C, C' is the empty set, and $\delta(\emptyset, \emptyset) := 0$.

In the following, we equip $\mathcal{C}$ with the topology induced by the (extended) Hausdorff metric. In this topology, $\emptyset$ is an isolated point of $\mathcal{C}$, but not in the topology induced by $\mathcal{F}$.

Theorem 12.3.2. *The topology of the Hausdorff metric on $\mathcal{C}$ is strictly finer than the trace topology induced by $\mathcal{F}$. On every set*

$$\{C \in \mathcal{C} : C \subset K\}, \qquad K \in \mathcal{C},$$

both topologies are the same.

Proof. Suppose that $C_i \to C$ in the Hausdorff metric, where $C, C_i \in \mathcal{C}'$, w.l.o.g. We have to show that $C_i \to C$ also in the topology of $\mathcal{F}$.

(α) Let $x \in C$. Since $d(x, C_i) \to 0$, there exist $x_i \in C_i$ with $x_i \to x$.

(β) Let $(C_{i_j})_{j\in\mathbb{N}}$ be a subsequence, let $x_{i_j} \in C_{i_j}$ and $x_{i_j} \to x$. Since $\delta(C_i, C) \to 0$, there are $y_{i_j} \in C$ with $d(x_{i_j}, y_{i_j}) \to 0$. It follows that $y_{i_j} \to x$ and hence that $x \in C$.

Now Theorem 12.2.2 shows that $C_i \to C$ in $\mathcal{F}$. Thus the topology of the Hausdorff metric on $\mathcal{C}$ is finer than the topology induced by $\mathcal{F}$. That it is strictly finer follows from the fact that there are sequences $(C_i)_{i\in\mathbb{N}}$ in $\mathcal{C}$ that converge in $\mathcal{F}$, but not in $(\mathcal{C}, \delta)$. An example is given by $(\{x, x_i\})$, where (x_i) is a sequence of points with $\|x_i\| \to \infty$.

Let $K \in \mathcal{C}$, and suppose that $C_i \to C$ in $\mathcal{F}$ with $C, C_i \in \mathcal{C}'$ and $C, C_i \subset K$. Let $0 < \epsilon < 1$ and $\widetilde{C} := \mathrm{cl}\,(K \setminus C_\epsilon)$; then $\widetilde{C} \in \mathcal{C}$. Since $C \cap \widetilde{C} = \emptyset$, Theorem 12.2.2 implies $C_i \cap \widetilde{C} = \emptyset$ for almost all i. For these i we have $C_i \subset C_\epsilon$. Assume that $C \subset (C_i)_\epsilon$ does not hold for almost all i. Then, for infinitely many i there exists $x_i \in C$ with $d(x_i, C_i) \geq \epsilon$. Since C is compact, there is a subsequence $(x_{i_j})_{j\in\mathbb{N}}$ with $x_{i_j} \to x \in C$. By Theorem 12.2.2, there are $y_i \in C_i$ with $y_i \to x$. We conclude that $\epsilon \leq d(x_{i_j}, y_{i_j}) \leq d(x_{i_j}, x) + d(x, y_{i_j}) \to 0$, a contradiction. Thus $C \subset (C_i)_\epsilon$ and, therefore, $\delta(C, C_i) \leq \epsilon$ for almost all i. Since $\epsilon < 1$ was arbitrary, we deduce that $C_i \to C$ in the Hausdorff metric, thus the second assertion holds. □

When working with $\mathcal{C}$ in this book, we often make use of the metric space $(\mathcal{C}, \delta)$. Here the metric δ is induced by the Euclidean metric of $\mathbb{R}^d$. The latter could be replaced by any metric generating the topology of $\mathbb{R}^d$, without changing the topology of $(\mathcal{C}, \delta)$; this follows from Theorem 12.3.2. As an immediate consequence of Theorem 12.3.2, we have:

Theorem 12.3.3. *The convergence of a sequence $(C_i)_{i\in\mathbb{N}}$ in $(\mathcal{C}, \delta)$ is equivalent to* (a) *and* (b)*:*

(a) *$(C_i)_{i\in\mathbb{N}}$ converges in $\mathcal{F}$.*
(b) *$(C_i)_{i\in\mathbb{N}}$ is uniformly bounded, that is, there exists a set $K \in \mathcal{C}$ with $C_i \subset K$ for all i.*

Since $\mathcal{F}$ is compact, Theorem 12.3.3 implies in particular that every uniformly bounded sequence in $\mathcal{C}'$ possesses a convergent subsequence (hence every uniformly bounded set in $\mathcal{C}'$ is relatively compact). This assertion (together with the observation that a limit of compact convex sets is convex) is often called the **Blaschke selection theorem**.

For special sequences in $\mathcal{C}'$, no distinction between convergence in $\mathcal{F}$ and in $(\mathcal{C}, \delta)$ is necessary.

Theorem 12.3.4. *If $C_i \to C$ in $\mathcal{F}$ for $C_i, C \in \mathcal{C}'$ and if each C_i is connected, then $C_i \to C$ in the Hausdorff metric.*

On the set $\mathcal{K}'$ of nonempty compact convex sets, the topology of the Hausdorff metric and the trace topology induced by $\mathcal{F}$ coincide.

Proof. Let B be a ball containing C in its interior. Choose $x \in C$. Since $C_i \to C$ in $\mathcal{F}$, there exist $x_i \in C_i$ with $x_i \to x$, and we have $x_i \in \operatorname{int} B$ for almost all i. Assume that $C_i \not\subset B$ for infinitely many i. Then there is a sequence $(y_{i_j})_{j\in\mathbb{N}}$ such that $y_{i_j} \in C_{i_j} \setminus B$ and $x_{i_j} \in \operatorname{int} B$. Since C_{i_j} is connected, we can choose a point $z_{i_j} \in C_{i_j} \cap \operatorname{bd} B$. The sequence $(z_{i_j})_{j\in\mathbb{N}}$ has a subsequence converging to some $z \in \operatorname{bd} B$. Then $z \in C$ by Theorem 12.2.2, a contradiction. Thus $C_i \subset B$ for almost all i, and Theorem 12.3.3 completes the proof of the first assertion.

The second assertion follows from the first one together with Theorem 12.3.2. □

Next, we show the continuity of some set operations with respect to the Hausdorff metric.

Theorem 12.3.5. *If $\mathcal{C}$ is equipped with the topology of the Hausdorff metric, then the maps*

$$\begin{aligned} \mathcal{C} \times \mathcal{F} &\to \mathcal{F} \\ (C, F) &\mapsto C \cup F \end{aligned} \qquad \begin{aligned} \mathcal{C} \times \mathcal{C} &\to \mathcal{C} \\ (C, C') &\mapsto C \cup C \end{aligned}$$

$$\begin{aligned} \mathcal{C} \times \mathcal{F} &\to \mathcal{F} \\ (C, F) &\mapsto C + F \end{aligned} \qquad \begin{aligned} \mathcal{C}' \times \mathcal{C} &\to \mathcal{C} \\ (C, C') &\mapsto C + C' \end{aligned}$$

$$\begin{aligned} \mathcal{C} &\to \mathcal{C} \\ C &\mapsto -C \end{aligned} \qquad \begin{aligned} \mathbb{R}^+ \times \mathcal{C} &\to \mathcal{C} \\ (\alpha, C) &\mapsto \alpha C \end{aligned}$$

and

$$\begin{aligned} \mathcal{C} &\to \mathcal{C} \\ C &\mapsto \operatorname{conv} C \end{aligned}$$

are continuous.

Proof. We treat only the third and the last map; for the remaining ones, the proofs are similar.

For the proof of the third assertion we first remark that for $C \in \mathcal{C}$ and $F \in \mathcal{F}$ we indeed have $C + F \in \mathcal{F}$. Now suppose that $C_i \to C$ in $\mathcal{C}$ and $F_i \to F$ in $\mathcal{F}$. We use Theorem 12.2.2.

(α) Let $x \in C + F$, thus $x = y + z$ with $y \in C$ and $z \in F$. Then $y = \lim y_i$ holds with suitable $y_i \in C_i$ and $z = \lim z_i$ with $z_i \in F_i$ $(i \geq i_0)$. It follows that $y_i + z_i \in C_i + F_i$ and $y_i + z_i \to x$.

(β) Let $(C_{i_j} + F_{i_j})_{j\in\mathbb{N}}$ be a subsequence, let $x_{i_j} \in C_{i_j} + F_{i_j}$ and $x_{i_j} \to x$. Then $x_{i_j} = y_{i_j} + z_{i_j}$ with $y_{i_j} \in C_{i_j}$ and $z_{i_j} \in F_{i_j}$. By Theorem 12.3.3, there is a set $K \in \mathcal{C}$ with $C_i \subset K$ for all i. Hence, $(y_{i_j})_{j\in\mathbb{N}}$ has a subsequence $(y_{m_j})_{j\in\mathbb{N}}$ converging to some $y \in K$. By $C_{m_j} \to C$ we have $y \in C$, and it follows that $z_{m_j} = x_{m_j} - y_{m_j} \to x - y =: z$ and $z \in F$, hence $x \in C + F$.

We conclude that $C_i + F_i \to C + F$.

For the proof of the last assertion we first remark that $C \in \mathcal{C}$ implies $\operatorname{conv} C \in \mathcal{C}$, by Carathéodory's theorem. Suppose that $C_i \to C$ in $\mathcal{C}$. Then $\delta(C, C_i) \le \epsilon$ implies

$$C_i \subset C + \epsilon B^d \qquad \text{and} \qquad C \subset C_i + \epsilon B^d,$$

hence

$$\operatorname{conv} C_i \subset \operatorname{conv} C + \epsilon B^d, \quad \operatorname{conv} C \subset \operatorname{conv} C_i + \epsilon B^d$$

and thus

$$\delta(\operatorname{conv} C, \operatorname{conv} C_i) \le \epsilon.$$

The assertion follows. □

We note a semicontinuity property of the volume. The **volume functional** V_d on $\mathcal{F}$ is defined by

$$\begin{aligned} V_d : \mathcal{F} &\to \mathbb{R} \cup \{\infty\} \\ F &\mapsto \quad \lambda(F) \end{aligned},$$

where λ is the Lebesgue measure on $\mathbb{R}^d$. On $\mathcal{F}$, the volume functional is neither upper nor lower semicontinuous, but on $(\mathcal{C}, \delta)$ the situation is different.

Theorem 12.3.6. *On $(\mathcal{C}, \delta)$, the volume functional is upper semicontinuous.*

Proof. Let $C_i \to C$ in $(\mathcal{C}, \delta)$, and let $x \in \mathbb{R}^d$. By Theorems 12.2.7 and 12.3.2,

$$\limsup \mathbf{1}_{C_i}(x) \le \mathbf{1}_C(x).$$

This gives

$$\begin{aligned} V_d(C) = \int_{\mathbb{R}^d} \mathbf{1}_C(x)\,\lambda(\mathrm{d}x) &\ge \int_{\mathbb{R}^d} \limsup \mathbf{1}_{C_i}(x)\,\lambda(\mathrm{d}x) \\ &\ge \limsup \int_{\mathbb{R}^d} \mathbf{1}_{C_i}(x)\,\lambda(\mathrm{d}x) = \limsup V_d(C_i), \end{aligned}$$

by Fatou's lemma. Since $C_i \subset C + \epsilon B^d$ for suitable $\epsilon > 0$ and all i, we have $\mathbf{1}_{C_i} \le \mathbf{1}_{C+\epsilon B^d}$ and thus

$$\int_{\mathbb{R}^d} \mathbf{1}_{C+\epsilon B^d}(x)\,\lambda(\mathrm{d}x) < \infty,$$

which justifies the application of Fatou's lemma. □

13

Invariant Measures

Integral geometry, as it is used in this book for the treatment of random geometric structures with stationarity or isotropy properties, is based on the notion of an invariant measure. Here invariance refers to a group operation and thus to a homogeneous space. Invariant measures on topological groups and homogeneous spaces are known as Haar measures. The general theory of such measures can be found, for example, in Hewitt and Ross [342] and Nachbin [571]. However, we do not presuppose here any knowledge of the theory of Haar measure (with the exception of Section 13.3, which is rarely used in this book and could be dispensed with). For the topological groups and homogeneous spaces that are relevant for integral geometry in Euclidean spaces, the existence and uniqueness of invariant measures will be proved in Section 13.2 in a direct and elementary way, starting from Lebesgue measure and assuming only basic facts from measure theory.

13.1 Group Operations and Invariant Measures

A **topological group** is a group G together with a topology on G such that the map from $G \times G$ to G defined by $(x, y) \mapsto xy$ (the product of x and y) and the map from G to G defined by $x \mapsto x^{-1}$ are continuous. The topologies of all topological groups occurring in this book are assumed to be locally compact and second countable.

Let G be a group and E a nonempty set. An **operation** of G on E is a map $\varphi : G \times E \to E$ satisfying

$$\varphi(g, \varphi(g', x)) = \varphi(gg', x), \qquad \varphi(e, x) = x$$

for all $g, g' \in G$, the unit element e of G, and all $x \in E$. One also says that G **operates on** E, by means of φ. For $\varphi(g, x)$ one usually writes gx, provided that the operation is clear from the context. The group G **operates transitively** on E if for any $x, y \in E$ there exists $g \in G$ such that $y = gx$.

If G is a topological group, E is a topological space, and the operation φ is continuous, one says that G **operates continuously** on E.

The following situation often occurs: E is a nonempty set and G is a group of transformations (bijective mappings onto itself) of E, with the composition as group multiplication; the operation of G on E is given by $(g, x) \mapsto gx :=$ image of x under g. When transformation groups occur in the following, multiplication and operation are always understood in this sense.

A basic situation considered in integral geometry is the operation of a transformation group on some space together with the induced operation on a space of geometrically significant subsets. Generally, let the space E be as in Section 12.2, that is, locally compact and second countable, and let $\mathcal{F}$ be the space of closed subsets of E. Let G be a topological group operating continuously on E. For $g \in G$ and $F \subset E$, let

$$gF := \{gx : x \in F\}. \tag{13.1}$$

For each $g \in G$, the bijective map $x \mapsto gx$ is continuous, and so is its inverse, thus it is a homeomorphism. It follows that $gF \in \mathcal{F}$ for $F \in \mathcal{F}$. Hence, (13.1) defines an operation of G on the space $\mathcal{F}$. This operation is continuous.

Theorem 13.1.1. *If the topological group G operates continuously on E, then the map*

$$\begin{aligned} G \times \mathcal{F} &\to \mathcal{F} \\ (g, F) &\mapsto gF \end{aligned}$$

is continuous.

Proof. Suppose that $(g_i, F_i) \to (g, F)$ in $G \times \mathcal{F}$. We have to show that $g_i F_i \to gF$ in $\mathcal{F}$, and for this we use Theorem 12.2.2.

(α) Let $x \in gF$, thus $x = gy$ with $y \in F$. Since $F_i \to F$ in $\mathcal{F}$, there are $y_i \in F_i$ with $y_i \to y$. For $x_i := g_i y_i$ we have $x_i \in g_i F_i$ and $x_i \to x$.

(β) Let $(g_{i_k} F_{i_k})_{k \in \mathbb{N}}$ be a subsequence and let $x_{i_k} = g_{i_k} y_{i_k}$ with $y_{i_k} \in F_{i_k}$ be such that $x_{i_k} \to x$. Then $y_{i_k} \to y := g^{-1}x$, and $y \in F$, hence $x = gy \in gF$. □

Let G be a topological group. A **homogeneous G-space** is a pair (E, φ) with the following properties: E is a topological space, $\varphi : G \times E \to E$ is a transitive continuous operation of G on E, and for (one and hence for all) $p \in E$, the mapping $\varphi(\cdot, p)$ is open. Up to isomorphism, all homogeneous G-spaces are obtained in the following way. Let H be a subgroup of G (with the trace topology) and let G/H be the factor space, that is, the space $\{aH : a \in G\}$ of left cosets of H in G, equipped with the quotient topology. The map $\pi : G \to G/H$ defined by $\pi(a) := aH$ for $a \in G$ is called the **natural projection**. The quotient topology on G/H is characterized by the properties that π is continuous and open. By

$$\zeta(g, aH) := gaH \qquad \text{for } g \in G,\ aH \in G/H,$$

one obtains a transitive continuous operation ζ of G on G/H; it is called the **natural operation** of G on G/H. The pair $(G/H, \zeta)$ is a homogeneous G-space. Conversely, let (E, φ) be a homogeneous G-space. For an arbitrarily chosen point $p \in E$ let S_p be the **stability group** of p, that is, the subgroup $S_p := \{g \in G : gp = p\}$ (with $gp := \varphi(g, p)$). Then the map

$$\begin{aligned} \beta : G/S_p &\to E \\ gS_p &\mapsto gp \end{aligned}$$

is a homeomorphism from G/S_p onto E with the property that $\beta(gaS_p) = g\beta(aS_p)$ for all $g \in G$ and all $aS_p \in G/S_p$. In this sense, the homogeneous G-spaces (E, φ) and $(G/S_p, \zeta)$ are isomorphic. Hence, if a homogeneous G-space is given, one can always assume that it is of the form G/H with a subgroup H and that the operation is the natural one. The subgroup H is closed if and only if G/H is a Hausdorff space.

We turn to invariant measures. Let the topological group G operate continuously on the topological space E. A Borel measure ρ on E is called G-**invariant** (or briefly **invariant**, if G is clear from the context) if

$$\rho(gA) = \rho(A) \qquad \text{for all } A \in \mathcal{B}(E) \text{ and all } g \in G.$$

This definition makes sense: for each $g \in G$, the mapping $x \mapsto gx$ is a homeomorphism, hence $A \in \mathcal{B}(E)$ implies $gA \in \mathcal{B}(E)$. An invariant regular Borel measure on a locally compact homogeneous space which is not identically zero is called a **Haar measure.**

For a measure on a group, several notions of invariance are natural. A topological group G operates on itself by means of the mapping $(g, x) \mapsto gx$ (multiplication in G) for $(g, x) \in G \times G$. The corresponding invariance of a measure on G is called left invariance. More generally, for $g \in G$ and $A \subset G$ we write

$$\begin{aligned} gA &:= \{ga : a \in A\}, \\ Ag &:= \{ag : a \in A\}, \\ A^{-1} &:= \{a^{-1} : a \in A\}. \end{aligned}$$

If $A \in \mathcal{B}(G)$, then also gA, Ag, A^{-1} are Borel sets, because multiplication from the left or the right and taking the inverse are homeomorphisms. Now let ρ be a measure on G. It is called **left invariant** if $\rho(gA) = \rho(A)$, and **right invariant** if $\rho(Ag) = \rho(A)$, for all $A \in \mathcal{B}(G)$ and all $g \in G$. The measure ρ is **inversion invariant** if $\rho(A^{-1}) = \rho(A)$ for all $A \in \mathcal{B}(G)$. If ρ has all three invariance properties, it is called **invariant**. A left invariant (right invariant, invariant) regular Borel measure which is not identically zero is called a **left Haar measure (right Haar measure, Haar measure)**.

Invariance properties of measures are equivalent to invariance properties of integrals. Let ρ be a regular Borel measure on the topological group G. If ρ is left invariant, then every measurable function $f \geq 0$ on G satisfies

$$\int_G f(ag)\,\rho(\mathrm{d}g) = \int_G f(g)\,\rho(\mathrm{d}g) \tag{13.2}$$

for all $a \in G$. This follows immediately from the definition of the integral. Conversely, if (13.2) holds for all continuous functions $f \geq 0$ with compact support, then the left invariance of ρ is obtained from (12.1), (12.2). Similarly, the right invariance of ρ is equivalent to

$$\int_G f(ga)\,\rho(\mathrm{d}g) = \int_G f(g)\,\rho(\mathrm{d}g)$$

for all $a \in G$, and the inversion invariance of ρ is equivalent to

$$\int_G f(g^{-1})\,\rho(\mathrm{d}g) = \int_G f(g)\,\rho(\mathrm{d}g),$$

in each case for all nonnegative functions $f \in \mathbf{C}_c(G)$.

We prove some uniqueness results for invariant measures. They are only needed for the groups and homogeneous spaces of Euclidean geometry, but without additional effort we can prove them in a more general setting.

Theorem 13.1.2. *Every left Haar measure on a compact group G with a countable base is invariant.*

Proof. Let ν be a left Haar measure on a group G satisfying the assumptions. Since ν is finite on compact sets, we may assume $\nu(G) = 1$, without loss of generality. For a continuous function $f \geq 0$ on G and for $x \in G$ we have

$$\int f(y^{-1}x)\,\nu(\mathrm{d}y) = \int f((x^{-1}y)^{-1})\,\nu(\mathrm{d}y) = \int f(y^{-1})\,\nu(\mathrm{d}y). \tag{13.3}$$

Here the integrations extend over all of G; similar conventions will be adopted in the following. Fubini's theorem gives

$$\begin{aligned}\int f(y^{-1})\,\nu(\mathrm{d}y) &= \int\int f(y^{-1}x)\,\nu(\mathrm{d}y)\,\nu(\mathrm{d}x)\\ &= \int\int f(y^{-1}x)\,\nu(\mathrm{d}x)\,\nu(\mathrm{d}y) = \int f(x)\,\nu(\mathrm{d}x).\end{aligned}$$

Hence, the measure ν is inversion invariant. Using this fact and (13.3), we obtain for $x \in G$ that

$$\begin{aligned}\int f(yx)\,\nu(\mathrm{d}y) &= \int f(y^{-1}x)\,\nu(\mathrm{d}y)\\ &= \int f(y^{-1})\,\nu(\mathrm{d}y) = \int f(y)\,\nu(\mathrm{d}y),\end{aligned}$$

which shows that ν is also right invariant. □

Clearly, in Theorem 13.1.2 the assumption 'left invariant' can be replaced by 'right invariant'.

The following uniqueness result for invariant measures makes special assumptions, but in this form it is sufficient for our purposes and is easy to prove.

Theorem 13.1.3. *Let G be a locally compact group with a countable base, let ν be a Haar measure and μ a left Haar measure on G. Then $\mu = c\nu$ with a constant c.*

Proof. For measurable functions $f, g \geq 0$ on G we have

$$\begin{aligned}\int f \, \mathrm{d}\nu \int g \, \mathrm{d}\mu &= \int \int f(xy)g(y) \, \nu(\mathrm{d}x) \, \mu(\mathrm{d}y) \\ &= \int \int f(xy)g(y) \, \mu(\mathrm{d}y) \, \nu(\mathrm{d}x) = \int \int f(y)g(x^{-1}y) \, \mu(\mathrm{d}y) \, \nu(\mathrm{d}x) \\ &= \int f(y) \int g(x^{-1}y) \, \nu(\mathrm{d}x) \, \mu(\mathrm{d}y) = \int g \, \mathrm{d}\nu \int f \, \mathrm{d}\mu.\end{aligned}$$

Here we have used, besides Fubini's theorem, the right and inversion invariance of ν and the left invariance of μ.

Since $\nu \neq 0$, there is a compact set $A_0 \subset G$ with $\nu(A_0) > 0$. For arbitrary $A \in \mathcal{B}(G)$ we put $f := \mathbf{1}_{A_0}$ and $g := \mathbf{1}_A$ and obtain $\nu(A_0)\mu(A) = \nu(A)\mu(A_0)$, hence $\mu = c\nu$ with $c := \mu(A_0)/\nu(A_0)$. □

Next, we prove a formula of integral geometric type, generalizing Theorem 5.2.1, which is useful for obtaining uniqueness results. It is slightly more general than needed.

Theorem 13.1.4. *Suppose that the compact group G operates continuously and transitively on the Hausdorff space E and that G and E have countable bases. Let ν be a Haar measure on G with $\nu(G) = 1$.*

Let $\rho \neq 0$ and α be locally finite Borel measures on E, let ρ be G-invariant. Then

$$\int_G \alpha(A \cap gB) \, \nu(\mathrm{d}g) = \alpha(A)\rho(B)/\rho(E)$$

for all $A, B \in \mathcal{B}(E)$.

Proof. If φ denotes the operation of G on E and if $x \in E$, the mapping $\varphi(\cdot, x) : G \to E$ is continuous and surjective, hence E is compact. Therefore, the measures α and ρ are finite. Let $A, B \in \mathcal{B}(E)$. The mapping $(g, x) \mapsto g^{-1}x$ from $G \times E$ to E is continuous and thus measurable, hence the mapping $(g, x) \mapsto \mathbf{1}_B(g^{-1}x)$ is measurable. From

$$\alpha(A \cap gB) = \int_E \mathbf{1}_{A \cap gB} \, \alpha(\mathrm{d}x) = \int_E \mathbf{1}_A(x)\mathbf{1}_B(g^{-1}x) \, \alpha(\mathrm{d}x)$$

it follows that $g \mapsto \alpha(A \cap gB)$ is a measurable mapping. Fubini's theorem yields

$$\int_G \alpha(A \cap gB)\, \nu(\mathrm{d}g) = \int_E \mathbf{1}_A(x) \int_G \mathbf{1}_B(g^{-1}x)\, \nu(\mathrm{d}g)\, \alpha(\mathrm{d}x). \tag{13.4}$$

The integral $\int_G \mathbf{1}_B(g^{-1}x)\, \nu(\mathrm{d}g)$ does not depend on x, since for $y \in E$ there exists $h \in G$ with $y = hx$ and therefore

$$\int_G \mathbf{1}_B(g^{-1}y)\, \nu(\mathrm{d}g) = \int_G \mathbf{1}_B((h^{-1}g)^{-1}x)\, \nu(\mathrm{d}g) = \int_G \mathbf{1}_B(g^{-1}x)\, \nu(\mathrm{d}g).$$

Hence, we obtain

$$\begin{aligned}\rho(E) \int_G \mathbf{1}_B(g^{-1}x)\, \nu(\mathrm{d}g) &= \int_E \int_G \mathbf{1}_B(g^{-1}x)\, \nu(\mathrm{d}g)\, \rho(\mathrm{d}x) \\ &= \int_G \int_E \mathbf{1}_B(g^{-1}x)\, \rho(\mathrm{d}x)\, \nu(\mathrm{d}g) = \int_G \rho(gB)\, \nu(\mathrm{d}g) = \rho(B).\end{aligned}$$

Inserting this in (13.4), we complete the proof. □

Theorem 13.1.5. *Suppose that the compact group G operates continuously and transitively on the Hausdorff space E and that G and E have countable bases. Let ν be a Haar measure on G with $\nu(G) = 1$.*

Then there exists a unique G-invariant Borel measure ρ on E with $\rho(E) = 1$. It can be defined by

$$\rho(B) = \nu(\{g \in G : gx_0 \in B)\}, \qquad B \in \mathcal{B}(E),$$

with arbitrary $x_0 \in E$.

Proof. Let ρ be a G-invariant Borel measure on E with $\rho(E) = 1$. We choose $x_0 \in E$ and let α be the Dirac measure on E concentrated at x_0. Then Theorem 13.1.4 with $A := \{x_0\}$ gives

$$\rho(B) = \nu(\{g \in G : g^{-1}x_0 \in B\})$$

for $B \in \mathcal{B}(E)$. Thus ρ is unique. Conversely, if ρ is defined in this way, it is clear that it is a G-invariant normalized measure. □

Notes for Section 13.1

1. For an extensive treatment of topological groups and homogeneous spaces, we refer to Hewitt and Ross [342], Nachbin [571], Gaal [242]. Information on invariant measures is also found in Bourbaki [119] and Cohn [177].

2. Results in the spirit of Theorem 13.1.4 (which extends Theorem 5.2.1) go back to Balanzat [55]. More general versions and further references are in Groemer [290] and Schneider [683].

13.2 Homogeneous Spaces of Euclidean Geometry

In this section we introduce the transformation groups and homogeneous spaces that occur in Euclidean integral geometry. Our main aim is to construct their Haar measures in an elementary way, presupposing only the knowledge of Lebesgue measure and its properties.

We consider three groups of bijective affine maps of $\mathbb{R}^d$ onto itself, the **translation group** T_d, the **rotation group** SO_d, and the **rigid motion group** G_d. The **translations** $t \in T_d$ are the maps of the form $t = t_x$ with $x \in \mathbb{R}^d$, where $t_x(y) = y + x$ for $y \in \mathbb{R}^d$. The mapping $\tau : x \mapsto t_x$ is an isomorphism of the additive group $\mathbb{R}^d$ onto T_d. Hence, we can identify T_d with $\mathbb{R}^d$, which we shall often do tacitly. In particular, T_d carries the topology inherited from $\mathbb{R}^d$ via τ. Since $t_x \circ t_y = t_{x+y}$ and $t_x^{-1} = t_{-x}$, composition and inversion are continuous, hence T_d is a topological group. In view of the topological properties of $\mathbb{R}^d$ we can thus state the following.

Theorem 13.2.1. *The translation group T_d is an abelian, locally compact topological group with countable base. The operation of T_d on $\mathbb{R}^d$ is continuous.*

The elements of the rotation group SO_d are the linear mappings $\vartheta : \mathbb{R}^d \to \mathbb{R}^d$ that preserve scalar product and orientation; they are called (**proper**) **rotations**. With respect to the standard (orthonormal) basis of $\mathbb{R}^d$, every rotation ϑ is represented by an orthogonal matrix $M(\vartheta)$ with determinant 1. The mapping $\mu : \vartheta \mapsto M(\vartheta)$ is an isomorphism of the group SO_d onto the group $\mathcal{SO}(d)$ of orthogonal (d,d)-matrices with determinant 1 under matrix multiplication. If we identify a (d,d)-matrix with the d^2-tuple of its entries, we can consider $\mathcal{SO}(d)$ as a subset of $\mathbb{R}^{d^2}$ (this identification defines the topology of $\mathcal{SO}(d)$). This set is bounded, since the rows of an orthogonal matrix are normalized, and it is closed in $\mathbb{R}^{d^2}$, hence compact. The mappings $(M,N) \mapsto MN$ from $\mathcal{SO}(d) \times \mathcal{SO}(d)$ to $\mathcal{SO}(d)$ and $M \mapsto M^{-1}$ from $\mathcal{SO}(d)$ to $\mathcal{SO}(d)$ are continuous, and so is the mapping $(M,x) \mapsto Mx$ (where x is considered as a $(d,1)$-matrix) from $\mathcal{SO}(d) \times \mathbb{R}^d$ into $\mathbb{R}^d$. Using the mapping μ^{-1} to transfer the topology from $\mathcal{SO}(d)$ to SO_d, we thus obtain the following.

Theorem 13.2.2. *The rotation group SO_d is a compact topological group with countable base. The operation of SO_d on $\mathbb{R}^d$ is continuous.*

The elements of the motion group G_d are the affine maps $g : \mathbb{R}^d \to \mathbb{R}^d$ that preserve the distances between points and the orientation; they are called **(rigid) motions**. Every rigid motion $g \in G_d$ can be represented uniquely as the composition of a rotation ϑ and a translation t_x, that is, $g = t_x \circ \vartheta$, or $gy = \vartheta y + x$ for $y \in \mathbb{R}^d$. The mapping

$$\begin{aligned} \gamma : \mathbb{R}^d \times SO_d &\to G_d \\ (x, \vartheta) &\mapsto t_x \circ \vartheta \end{aligned} \tag{13.5}$$

is bijective. We employ it to transfer the topology from $\mathbb{R}^d \times SO_d$ to G_d. Using Theorems 13.2.1 and 13.2.2, it is then easy to show the following.

Theorem 13.2.3. *G_d is a locally compact topological group with countable base. Its operation on $\mathbb{R}^d$ is continuous.*

After these topological groups, we consider the homogeneous spaces that play a role in Euclidean integral geometry.

The unit sphere S^{d-1} is obviously a homogeneous SO_d-space.

For $q \in \{0, \dots, d\}$, let $G(d,q)$ be the set of all q-dimensional linear subspaces of $\mathbb{R}^d$, and let $A(d,q)$ be the set of all q-dimensional affine subspaces of $\mathbb{R}^d$. The natural operation of SO_d on $G(d,q)$ is given by $(\vartheta, L) \mapsto \vartheta L :=$ image of L under ϑ. Similarly, the natural operation of G_d on $A(d,q)$ is given by $(g, E) \mapsto gE :=$ image of E under g. We introduce suitable topologies on $G(d,q)$ and $A(d,q)$. For this, let $L_q \in G(d,q)$ be a fixed subspace and $L_q^\perp$ its orthogonal complement. The mappings

$$\begin{aligned} \beta_q : SO_d &\to G(d,q) \\ \vartheta &\mapsto \vartheta L_q \end{aligned} \tag{13.6}$$

and

$$\begin{aligned} \gamma_q : L_q^\perp \times SO_d &\to A(d,q) \\ (x, \vartheta) &\mapsto \vartheta(L_q + x) \end{aligned} \tag{13.7}$$

are surjective (but not injective). We endow $G(d,q)$ with the finest topology for which β_q is continuous, and $A(d,q)$ with the finest topology for which γ_q is continuous. Thus, a subset $A \subset A(d,q)$, for example, is open if and only if $\gamma_q^{-1}(A)$ is open. It is an elementary task to prove the following.

Theorem 13.2.4. *$G(d,q)$ is compact and has a countable base, the map β_q is open, and the operation of SO_d on $G(d,q)$ is continuous and transitive.*

Theorem 13.2.5. *$A(d,q)$ is locally compact and has a countable base, the map γ_q is open, and the operation of G_d on $A(d,q)$ is continuous and transitive.*

It should be remarked that the topologies on $G(d,q)$ and $A(d,q)$, as well as the invariant measures to be introduced soon, do not depend on the special choice of the subspace L_q. This follows easily from the fact that SO_d operates transitively on $G(d,q)$ and G_d operates transitively on $A(d,q)$.

The topological spaces $G(d,q)$ are called **Grassmann manifolds** or **Grassmannians**, and the spaces $A(d,q)$ are called **affine Grassmannians**.

It was convenient here to introduce the topologies on $G(d,q)$ and $A(d,q)$ as described. Generally in this book, we equip $\mathcal{F}(\mathbb{R}^d)$, the set of closed subsets of $\mathbb{R}^d$, with the topology of closed convergence, as summarized in Section 12.2. The trace of this topology on $G(d,q)$ or $A(d,q)$ coincides with the topology introduced above. To see this, for example for the case of $A(d,q)$, we first note that the mapping $g \mapsto gL_q$ from G_d into $\mathcal{F}$ is continuous, by Theorem 13.1.1. In order to show that the topology of closed convergence on $A(d,q)$ coincides with the one introduced above, it therefore suffices to show the following. If

$E_i, E \in A(d,q)$ and $E_i \to E$ in $\mathcal{F}$, then there exist motions $g_i, g \in G_d$ such that $E_i = g_i L_q$, $E = gL_q$ and $g_i \to g$ (in G_d). This is easy to see with the aid of Theorem 12.2.2.

By Theorem 13.1.1, the induced operation of the motion group G_d on the space $\mathcal{F} = \mathcal{F}(\mathbb{R}^d)$ of closed subsets of $\mathbb{R}^d$ is also continuous. From this fact we draw two conclusions, which are used occasionally. First, applying Theorem 13.1.1 again, but now to the space $E = \mathcal{F}'$ $(= \mathcal{F}(\mathbb{R}^d) \setminus \{\emptyset\})$, we get:

Theorem 13.2.6. *The map*

$$(g, A) \mapsto gA := \{gF : F \in A\}, \qquad g \in G_d,\ A \in \mathcal{F}(\mathcal{F}'),$$

is continuous.

The operation of G_d on $\mathbb{R}^d$ induces also an operation on the space $\mathcal{C}'$ of nonempty compact subsets.

Theorem 13.2.7. *The map*

$$\begin{aligned} G_d \times \mathcal{C}' &\to \mathcal{C}' \\ (g, C) &\mapsto gC \end{aligned}$$

where $\mathcal{C}'$ is equipped with the Hausdorff metric, is continuous.

Proof. Suppose that $(g_i, C_i) \to (g, C)$ in $G_d \times \mathcal{C}'$. Then $C_i \to C$ in $(\mathcal{C}', \delta)$, hence $C_i \to C$ in $\mathcal{F}$ by Theorem 12.3.2. From Theorem 13.1.1 it follows that $g_i C_i \to gC$ in $\mathcal{F}$. Since the sequence $(C_i)_{i\in\mathbb{N}}$ is uniformly bounded and $g_i \to g$, also the sequence $(g_i C_i)_{i\in\mathbb{N}}$ is uniformly bounded. By Theorem 12.3.3, $g_i C_i \to gC$ in $(\mathcal{C}', \delta)$. □

Now we construct invariant measures on the introduced groups and homogeneous spaces. (Since these are locally compact, second countable spaces, all Borel measures on them are regular; see, for example, Cohn [177, Proposition 7.2.3].) We start from Lebesgue measure on $\mathbb{R}^d$ and construct further measures by means of continuous (and hence measurable) mappings. The local finiteness of the image measures has to be checked in every case. We shall, however, not mention this fact explicitly, when it is easy to see.

The measures ρ to be considered below will depend on the dimension d of the space $\mathbb{R}^d$. If different dimensions occur, corresponding measures and other objects will be distinguished by lower indices. Symbols without lower index always refer to the dimension d.

We suppose that the reader is familiar with the construction and the properties of the Lebesgue measure λ on $\mathbb{R}^d$, including the following uniqueness theorem. Recall that $C^d = [0,1]^d$ is the unit cube in $\mathbb{R}^d$.

Theorem 13.2.8. *The Lebesgue measure λ is the only translation invariant measure on $\mathbb{R}^d$ with $\lambda(C^d) = 1$.*

Since the Lebesgue measure λ is rigid motion invariant (as well as invariant under reflections), it is the Haar measure on the homogeneous G_d-space $\mathbb{R}^d$, normalized in a special way. We note that

$$\lambda(B^d) =: \kappa_d = \frac{\pi^{d/2}}{\Gamma\left(1+\frac{d}{2}\right)}.$$

The Haar measure on the homogeneous SO_d-space S^{d-1}, the unit sphere, is easily derived from the Lebesgue measure. For $A \in \mathcal{B}(S^{d-1})$ we define

$$\widehat{A} := \{\alpha x \in \mathbb{R}^d : x \in A,\ 0 \le \alpha \le 1\}.$$

A standard argument shows that $\widehat{A} \in \mathcal{B}(\mathbb{R}^d)$, hence we can define $\sigma(A) := d\lambda(\widehat{A})$. This yields a finite measure σ on S^{d-1} for which

$$\sigma(S^{d-1}) =: \omega_d = d\kappa_d = \frac{2\pi^{\frac{d}{2}}}{\Gamma(\frac{d}{2})}.$$

The rotation invariance of λ implies the rotation invariance of σ. As the Lebesgue measure, σ is invariant under the full group $O(d)$ of orthogonal transformations (proper and improper rotations). We call σ the **spherical Lebesgue measure.** Up to a constant factor, σ is the only rotation invariant Borel measure on S^{d-1}. This follows from Theorem 13.1.5.

The spherical Lebesgue measure can be used to show the existence of the Haar measure for the rotation group.

Theorem 13.2.9. *On the rotation group SO_d, there is a unique Haar measure ν with $\nu(SO_d) = 1$.*

Proof. The uniqueness follows from Theorem 13.1.3. To prove the existence, we denote by LI_d the set of linearly independent d-tuples of vectors from S^{d-1}. We define a map $\psi : LI_d \to SO_d$ in the following way. Let $(x_1, \dots, x_d) \in LI_d$. By Gram–Schmidt orthonormalization we transform $(x_1, \dots, x_d)$ into the d-tuple $(z_1, \dots, z_d)$; then we denote by $(\bar{z}_1, \dots, \bar{z}_d)$ the positively oriented d-tuple for which $\bar{z}_i = z_i$ for $i = 1, \dots, d-1$ and $\bar{z}_d = z_d$ or $-z_d$. If $(e_1, \dots, e_d)$ denotes the standard orthonormal basis of $\mathbb{R}^d$, there is a unique rotation $\vartheta \in SO_d$ satisfying $\vartheta e_i = \bar{z}_i$ for $i = 1, \dots, d$. We define $\psi(x_1, \dots, x_d) := \vartheta$.

Explicitly, we have $z_i = y_i/\|y_i\|$ with $y_1 = x_1$ and

$$y_k = x_k - \sum_{j=1}^{k-1} \langle x_k, y_j \rangle \frac{y_j}{\|y_j\|^2}, \qquad k = 2, \dots, d.$$

From this representation, the following is evident. If $\rho \in SO_d$ is a rotation and if the d-tuple $(x_1, \dots, x_d) \in LI_d$ is transformed into $(z_1, \dots, z_d)$ and then into $(\bar{z}_1, \dots, \bar{z}_d)$, then the d-tuple $(\rho x_1, \dots, \rho x_d)$ is transformed into $(\rho z_1, \dots, \rho z_d)$ and subsequently into $(\rho \bar{z}_1, \dots, \rho \bar{z}_d)$. Thus we have

$$\psi(\rho x_1,\dots,\rho x_d)=\rho\psi(x_1,\dots,x_d).$$

For $(x_1,\dots,x_d)\in(S^{d-1})^d\setminus LI_d$ we define $\psi(x_1,\dots,x_d):=\mathrm{id}$. For the product measure

$$\sigma^d:=\underbrace{\sigma\otimes\ldots\otimes\sigma}_{d},$$

the set $(S^{d-1})^d\setminus LI_d$ has measure zero; hence for any $\rho\in SO_d$ the equality $\psi(\rho x_1,\dots,\rho x_d)=\rho\psi(x_1,\dots,x_d)$ holds σ^d-almost everywhere. The mapping $\psi:(S^{d-1})^d\to SO_d$ is measurable, since LI_d is open and ψ is continuous on LI_d and constant on $(S^{d-1})^d\setminus LI_d$.

Now we define $\bar\nu$ as the image measure of σ^d under ψ, thus $\bar\nu=\psi(\sigma^d)$. Then $\bar\nu$ is a finite measure on SO_d, and for $\rho\in SO_d$ and measurable $f\geq 0$ we obtain

$$\begin{aligned}
&\int_{SO_d} f(\rho\vartheta)\,\bar\nu(\mathrm{d}\vartheta)\\
&=\int_{(S^{d-1})^d} f(\rho\psi(x_1,\dots,x_d))\,\sigma^d(\mathrm{d}(x_1,\dots,x_d))\\
&=\int_{S^{d-1}}\cdots\int_{S^{d-1}} f(\psi(x_1,\dots,x_d))\,\sigma(\mathrm{d}x_1)\cdots\sigma(\mathrm{d}x_d)\\
&=\int_{SO_d} f(\vartheta)\,\bar\nu(\mathrm{d}\vartheta).
\end{aligned}$$

Here we have used the rotation invariance of the spherical Lebesgue measure. We have proved that the measure $\bar\nu$ is left invariant and thus invariant, by Theorem 13.1.2. The measure $\nu:=\bar\nu/\bar\nu(SO_d)$ is invariant and normalized. □

From now on, ν will always denote the normalized invariant measure on SO_d.

The following special result on ν null sets is needed at several instances. Two linear subspaces L,L' of $\mathbb{R}^d$ are in **general position** if

$$\dim(L\cap L')=\max\{0,\dim L+\dim L'-d\},$$

otherwise they are said to be in **special position**. The latter is equivalent to

$$\mathrm{lin}\,(L\cup L')\neq\mathbb{R}^d\qquad\text{and}\qquad\dim(L\cap L')>0.$$

Lemma 13.2.1. *Let L,L' be linear subspaces of $\mathbb{R}^d$, and let $A\subset SO_d$ be the set of all rotations ϑ for which L and $\vartheta L'$ are in special position. Then $\nu(A)=0$.*

Proof. We may assume that $\dim L+\dim L'<d$, since otherwise we can pass to orthogonal complements. Let $v_1,\dots,v_m$ be an orthonormal basis of L' and put $L_i:=\mathrm{lin}\,\{v_1,\dots,v_i\}$ for $i=1,\dots,m$ and $L_0:=\{0\}$. Then

$$\begin{aligned}
&\nu(\{\vartheta \in SO_d : \dim(L \cap \vartheta L') > 0\}) \\
&= \nu\left(\bigcup_{i=1}^{m} \{\vartheta \in SO_d : \dim(L \cap \vartheta L_{i-1}) = 0,\ \dim(L \cap \vartheta L_i) > 0\}\right) \\
&= \sum_{i=1}^{m} \nu(\{\vartheta \in SO_d : \vartheta v_i \in \operatorname{lin}(L \cup \vartheta L_{i-1})\}).
\end{aligned}$$

We shall show that here each summand is zero. Let $i \in \{1, \dots, m\}$ be fixed and write $\operatorname{lin}(L \cup \vartheta L_{i-1}) =: M(\vartheta)$. Put $H := L_{i-1} \cap S^{d-1}$ and $H' := L_{i-1}^{\perp} \cap S^{d-1}$. For $x \in S^{d-1} \setminus (H \cup H')$ there is a unique decomposition $x = tv + \sqrt{1-t^2}\, x'$ with $v \in H$, $x' \in H'$, $t \in (0,1)$. Since $\vartheta v \in M(\vartheta)$ for $v \in H$, we have $\vartheta x \in M(\vartheta)$ if and only if $\vartheta x' \in M(\vartheta)$. Moreover, for $x' \in H'$ there exists a rotation $\rho_x \in SO_d$ with $\rho_x v_i = x'$ and $\rho_x L_{i-1} = L_{i-1}$. We obtain

$$\begin{aligned}
\nu(\{\vartheta \in SO_d : \vartheta x \in M(\vartheta)\}) &= \nu(\{\vartheta \in SO_d : \vartheta x' \in M(\vartheta)\}) \\
&= \nu(\{\vartheta \in SO_d : \vartheta \rho_x v_i \in M(\vartheta\rho_x)\}) \\
&= \nu(\{\vartheta \in SO_d : \vartheta v_i \in M(\vartheta)\})
\end{aligned}$$

by the invariance of ν. Integration with the spherical Lebesgue measure and Fubini's theorem yield

$$\begin{aligned}
&\sigma(S^{d-1})\nu(\{\vartheta \in SO_d : \vartheta v_i \in M(\vartheta)\}) \\
&= \int_{S^{d-1}\setminus(H \cup H')} \nu(\{\vartheta \in SO_d : \vartheta x \in M(\vartheta)\})\, \sigma(\mathrm{d}x) \\
&= \int_{SO_d} \int_{S^{d-1}\setminus(H \cup H')} \mathbf{1}_{M(\vartheta)}(\vartheta x)\, \sigma(\mathrm{d}x)\, \nu(\mathrm{d}\vartheta) = 0,
\end{aligned}$$

since $\dim M(\vartheta) \le \dim L + i - 1 \le d - 1$. □

With the aid of the invariant measures λ and σ, we can construct the Haar measure on the rigid motion group G_d. Since G_d it is not compact, an invariant measure μ on G_d cannot be finite, as is easy to see. In order to normalize μ, we specify the compact set $A_0 := \gamma(C^d \times SO_d)$ and require that $\mu(A_0) = 1$.

Theorem 13.2.10. *On the motion group G_d there is a Haar measure μ with $\mu(A_0) = 1$. Up to a constant factor, it is the only left Haar measure on G_d.*

Proof. The uniqueness assertion is a special case of Theorem 13.1.3. We define μ as the image measure of the product measure $\lambda \otimes \nu$ under the homeomorphism $\gamma : \mathbb{R}^d \times SO_d \to G_d$ defined by (13.5). Then μ is a Borel measure on G_d with $\mu(\gamma(C^d \times SO_d)) = \lambda(C^d)\nu(SO_d) = 1$.

To show the left invariance of μ, let $f \ge 0$ be a continuous function on G_d and let $g' \in G_d$. With $g' = \gamma(t', \vartheta')$ we have

$$\begin{aligned}\int_{G_d} f(g'g)\,\mu(\mathrm{d}g) &= \int_{SO_d}\int_{\mathbb{R}^d} f(\gamma(t',\vartheta')\gamma(t,\vartheta))\,\lambda(\mathrm{d}t)\,\nu(\mathrm{d}\vartheta)\\ &= \int_{SO_d}\int_{\mathbb{R}^d} f(\gamma(t'+\vartheta' t,\vartheta'\vartheta))\,\lambda(\mathrm{d}t)\,\nu(\mathrm{d}\vartheta)\\ &= \int_{SO_d}\int_{\mathbb{R}^d} f(\gamma(t,\vartheta))\,\lambda(\mathrm{d}t)\,\nu(\mathrm{d}\vartheta)\\ &= \int_{G_d} f(g)\,\mu(\mathrm{d}g),\end{aligned}$$

where we have used the motion invariance of λ and the left invariance of ν. Hence, μ is left invariant. Analogously, the right invariance of ν implies via

$$\int_{G_n} f(gg')\,\mu(\mathrm{d}g) = \int_{SO_d}\int_{\mathbb{R}^d} f(\gamma(t+\vartheta t',\vartheta\vartheta'))\,\lambda(\mathrm{d}t)\,\nu(\mathrm{d}\vartheta)$$
$$= \int_{SO_d}\int_{\mathbb{R}^d} f(\gamma(t,\vartheta))\,\lambda(\mathrm{d}t)\,\nu(\mathrm{d}\vartheta) = \int_{G_d} f(g)\,\mu(\mathrm{d}g)$$

the right invariance of μ, and from

$$\int_{G_d} f(g^{-1})\,\mu(\mathrm{d}g) = \int_{SO_d}\int_{\mathbb{R}^d} f(\gamma(-\vartheta^{-1}t,\vartheta^{-1}))\,\lambda(\mathrm{d}t)\,\nu(\mathrm{d}\vartheta)$$
$$= \int_{SO_d}\int_{\mathbb{R}^d} f(\gamma(t,\vartheta))\,\lambda(\mathrm{d}t)\,\nu(\mathrm{d}\vartheta) = \int_{G_d} f(g)\,\mu(\mathrm{d}g),$$

where the inversion invariance of ν was used, we obtain the inversion invariance of μ. ☐

The notation μ for the Haar measure on G_d, normalized as above, will be maintained in the following. The decomposition of the measure μ inherent in its construction is often used in the form

$$\begin{aligned}\int_{G_d} f\,\mathrm{d}\mu &= \int_{SO_d}\int_{\mathbb{R}^d} f(t_x\circ\vartheta)\,\lambda(\mathrm{d}x)\,\nu(\mathrm{d}\vartheta)\\ &= \int_{SO_d}\int_{\mathbb{R}^d} f(\vartheta\circ t_x)\,\lambda(\mathrm{d}x)\,\nu(\mathrm{d}\vartheta) \qquad (13.8)\end{aligned}$$

for μ-integrable functions f on G_d; here t_x is the translation by the vector x. The last equality follows from $\vartheta\circ t_x = t_{\vartheta x}\circ\vartheta$ and the invariance of the Lebesgue measure λ under the rotation ϑ.

Now we turn to invariant measures on the Grassmannian $G(d,q)$ of q-dimensional linear subspaces and on the affine Grassmannian $A(d,q)$ of q-dimensional affine subspaces of $\mathbb{R}^d$. As above, we suppose that $q\in\{0,\dots,d\}$ and $L_q\in G(d,q)$ is a fixed q-dimensional linear subspace, and we use the maps β_q and γ_q defined by (13.6) and (13.7), respectively.

On $G(d,q)$ and $A(d,q)$, some of the transformation groups introduced above operate continuously, for example on $A(d,q)$ the groups T_d, SO_d and G_d. Only the operations of G_d on $A(d,q)$ and of SO_d on $G(d,q)$ are transitive. Therefore, by an **invariant measure** on $A(d,q)$ we understand a rigid motion invariant (G_d-invariant) measure on $A(d,q)$, and an **invariant measure** on $G(d,q)$ is rotation invariant (SO_d-invariant).

Theorem 13.2.11. *On $G(d,q)$ there is a unique Haar measure ν_q, normalized by $\nu_q(G(d,q)) = 1$.*

This is just a special case of Theorem 13.1.5. We also notice that ν_q is the image measure of ν under the mapping β_q.

A corresponding assertion for $A(d,q)$ requires a normalization on a suitable compact subset A_0^q. We choose

$$A_0^q := \{E \in A(d,q) : E \cap B^d \neq \emptyset\}.$$

Theorem 13.2.12. *On $A(d,q)$ there is a unique Haar measure μ_q, normalized by $\mu_q(A_0^q) = \kappa_{d-q}$.*

It satisfies

$$\int_{A(d,q)} f \,\mathrm{d}\mu_q = \int_{G(d,q)} \int_{L^\perp} f(L+y)\,\lambda_{d-q}(\mathrm{d}y)\,\nu_q(\mathrm{d}L) \tag{13.9}$$

for every measurable function $f \geq 0$ on $A(d,q)$.

Proof. We define

$$\mu_q := \gamma_q(\lambda_{d-q} \otimes \nu).$$

If $A \subset A(d,q)$ is compact, the sets

$$\gamma_q(\{x \in L_q^\perp : \|x\| < k\} \times SO_d), \qquad k \in \mathbb{N},$$

constitute an open covering of A, hence A is included in one of these sets. It follows that $\mu_q(A) < \infty$.

Let $g = \gamma(x,\vartheta) \in G_d$ and let $f \geq 0$ be a measurable function on $A(d,q)$. Denoting by Π the orthogonal projection to $L_q^\perp$, we obtain

$$\begin{aligned}
&\int_{A(d,q)} f(gE)\,\mu_q(\mathrm{d}E) \\
&= \int_{SO_d} \int_{L_q^\perp} f(g\rho(L_q + y))\,\lambda_{d-q}(\mathrm{d}y)\,\nu(\mathrm{d}\rho) \\
&= \int_{SO_d} \int_{L_q^\perp} f(\vartheta\rho(L_q + y + \Pi(\rho^{-1}\vartheta^{-1}x)))\,\lambda_{d-q}(\mathrm{d}y)\,\nu(\mathrm{d}\rho) \\
&= \int_{SO_d} \int_{L_q^\perp} f(\vartheta\rho(L_q + y))\,\lambda_{d-q}(\mathrm{d}y)\,\nu(\mathrm{d}\rho)
\end{aligned}$$

$$= \int_{SO_d} \int_{L_q^\perp} f(\rho(L_q + y)) \, \lambda_{d-q}(\mathrm{d}y) \, \nu(\mathrm{d}\rho)$$

$$= \int_{A(d,q)} f(E) \, \mu_q(\mathrm{d}E),$$

where we have used the invariance properties of λ_{d-q} and ν. This shows the invariance of μ_q.

We observe that we may also write

$$\int_{A(d,q)} f \, \mathrm{d}\mu_q = \int_{SO_d} \int_{L_q^\perp} f(\rho(L_q + x)) \, \lambda_{d-q}(\mathrm{d}x) \, \nu(\mathrm{d}\rho)$$
$$= \int_{SO_d} \int_{(\rho L_q)^\perp} f(\rho L_q + y) \, \lambda_{d-q}(\mathrm{d}y) \, \nu(\mathrm{d}\rho).$$

Since ν_q is the image measure of ν under β_q, this can be written in the form (13.9).

From the representation (13.9) we infer that μ_q does not depend on the choice of the subspace L_q. We also deduce that $\mu(A_0^q) = \kappa_{d-q}$.

To prove the uniqueness, we assume that τ is another Haar measure on $A(d,q)$. Let $\widetilde{G}(d,q)$ (respectively $\widetilde{A}(d,q)$) be the open set of all $L \in G(d,q)$ (respectively $E \in A(d,q)$) that intersect $L_q^\perp$ at precisely one point. The mapping

$$\begin{aligned} \delta_q : L_q^\perp \times \widetilde{G}(d,q) &\to \widetilde{A}(d,q) \\ (x, L) \quad &\mapsto L + x \end{aligned}$$

is a homeomorphism. For fixed $B \in \mathcal{B}(\widetilde{G}(d,q))$ and arbitrary $A \in \mathcal{B}(L_q^\perp)$ we define $\eta(A) := \tau(\delta_q(A \times B))$. Then η is a Borel measure on $L_q^\perp$, which is invariant under the translations of $L_q^\perp$ into itself. Theorem 13.2.8 implies that $\eta(A) = \lambda_{d-q}(A)\alpha(B)$ with a constant $\alpha(B) \geq 0$. Hence we have

$$\tau(\delta_q(A \times B)) = \lambda_{d-q}(A)\alpha(B)$$

for arbitrary $A \in \mathcal{B}(L_q^\perp)$ and $B \in \mathcal{B}(\widetilde{G}(d,q))$. Obviously this equality defines a finite measure α on $\mathcal{B}(\widetilde{G}(d,q))$, and $\delta_q^{-1}(\tau) = \lambda_{d-q} \otimes \alpha$. For a measurable function $f \geq 0$ on $\widetilde{A}(d,q)$ we obtain

$$\int_{\widetilde{A}(d,q)} f \, \mathrm{d}\tau = \int_{\widetilde{G}(d,q)} \int_{L_q^\perp} f(L + x) \, \lambda_{d-q}(\mathrm{d}x) \, \alpha(\mathrm{d}L)$$
$$= \int_{\widetilde{G}(d,q)} \int_{L^\perp} f(L + y) \, \lambda_{d-q}(\mathrm{d}y) \, \varphi(\mathrm{d}L) \qquad (13.10)$$

with a new measure φ on $\widetilde{G}(d,q)$, defined by $\mathrm{d}\varphi(L) = D(L_q^\perp, L^\perp)^{-1}\mathrm{d}\alpha(L)$, where $D(L_q^\perp, L^\perp)$ is the absolute determinant of the orthogonal projection from $L_q^\perp$ onto $L^\perp$.

Now let $B \in \mathcal{B}(G(d,q))$ and

$$B' := \{L + y : L \in B,\ \ y \in L^\perp \cap B^d\}.$$

By $\beta(B) := \tau(B')$ we define a rotation invariant finite measure β on $G(d,q)$. According to Theorem 13.2.11 it is a multiple of ν_q. On the other hand, (13.10) gives $\tau(B') = \kappa_{d-q}\varphi(B)$ for $B \subset \widetilde{G}(d,q)$. Hence there is a constant c with $\varphi(B) = c\nu_q(B)$ for all Borel sets $B \subset \widetilde{G}(d,q)$. From (13.10) and (13.9) we deduce that $\tau(A) = c\mu_q(A)$ for all Borel sets $A \subset \widetilde{A}(d,q)$. Since μ_q does not depend on the choice of the subspace $L_q \in G(d,q)$, it is easy to see that $\tau = c\mu_q$. □

From the introduced homogeneous spaces of flats and their invariant measures, we derive other ones, which are used occasionally in integral geometry.

For $p, q \in \{0, \dots, d\}$ and a fixed linear subspace $L \in G(d,p)$, let

$$G(L,q) := \begin{cases} \{L' \in G(d,q) : L' \subset L\} & \text{if } q \le p, \\ \{L' \in G(d,q) : L' \supset L\} & \text{if } q > p. \end{cases}$$

Similarly, for $E \in A(d,p)$ let

$$A(E,q) := \begin{cases} \{E' \in A(d,q) : E' \subset E\} & \text{if } q \le p, \\ \{E' \in A(d,q) : E' \supset E\} & \text{if } q > p. \end{cases}$$

In the case $q \le p$ the spaces $G(L,q)$ and $A(E,q)$ are obviously homeomorphic to $G(p,q)$ and $A(p,q)$, respectively. For $q > p$, the situation is slightly different. Here $G(L,q)$ is homeomorphic to $G(d-p, q-p)$, because each $L' \in G(L,q)$ is of the form $L' = L + L''$ with a unique subspace $L'' \in G(L^\perp, q-p)$, so that $G(L,q)$ is homeomorphic, in a natural way, to $G(L^\perp, q-p)$; the latter space is homeomorphic to $G(d-p, q-p)$, by the preceding remark. The space $A(E,q)$ with $q > p$ is evidently homeomorphic to $G(L,q)$, where L is the translate of E through the origin. Thus $A(E,q)$, too, is homeomorphic to $G(d-p, q-p)$.

On these spaces, we introduce invariant measures in the natural way. For a linear subspace $L \in G(d,q)$ we first put

$$SO(L) := \{\rho \in SO_d : \rho L = L,\ \rho x = x \text{ for } x \in L^\perp\},$$

which is the subgroup of all proper rotations of $\mathbb{R}^d$ mapping L into itself and fixing each point of $L^\perp$. Since $SO(L)$ is isomorphic to SO_p, it carries a unique normalized invariant measure, which we denote by ν_L. As usual, we consider ν_L as a measure defined on the whole group SO_d. We have

$$\nu_{\vartheta L}(\vartheta A \vartheta^{-1}) = \nu_L(A) \tag{13.11}$$

for $A \in \mathcal{B}(SO_d)$ and arbitrary rotations $\vartheta \in SO_d$; this can be deduced, for example, from Theorem 13.1.3.

Let $p, q \in \{0, \dots, d\}$ and $L \in G(d,q)$. We fix a subspace $L_q \in G(L,q)$. By means of the map (13.6), that is, $\beta_q : SO_d \to G(d,q)$, $\vartheta \mapsto \vartheta L_q$, we define

$$\nu_q^L := \beta_q(\nu_L)$$

for $q < p$ and

$$\nu_q^L := \beta_q(\nu_{L^\perp})$$

for $q \geq p$. Then we have

$$\nu_q^L(A) = \nu_L(\{\rho \in SO(L) : \rho L_q \in A\})$$

if $q < p$ and

$$\nu_q^L(A) = \nu_{L^\perp}(\{\rho \in SO(L^\perp) : \rho L_q \in A\})$$

if $q \geq p$, in each case for all $A \in \mathcal{B}(G(d,q))$. Thus ν_q^L is a normalized measure concentrated on $G(L,q)$; it does not depend on the choice of L_q and is invariant under $SO(L)$ and $SO(L^\perp)$. Moreover,

$$\nu_q^{\vartheta L}(\vartheta A) = \nu_q^L(A) \tag{13.12}$$

for $A \in \mathcal{B}(G(d,q))$ and all rotations $\vartheta \in SO_d$, as follows from (13.11).

For a fixed flat $E \in G(d,p)$ we choose $t \in \mathbb{R}^d$ with $E - t =: L \in G(d,p)$ and then a subspace $L_q \in G(L,q)$. If $q < p$, let λ_{p-q} be the Lebesgue measure on $L_q^\perp \cap L$. We define

$$\begin{aligned} \gamma_{q,t} : (L_q^\perp \cap L) \times SO(L) &\to \quad A(d,q) \\ (x, \vartheta) \quad &\mapsto \vartheta(L_q + x) + t \end{aligned}$$

and

$$\mu_q^E := \gamma_{q,t}(\lambda_{p-q} \otimes \nu_L).$$

If $q \geq p$, we define

$$\begin{aligned} \gamma_{q,t} : SO(L^\perp) &\to A(d,q) \\ \vartheta \quad &\mapsto \vartheta L_q + t \end{aligned}$$

and

$$\mu_q^E := \gamma_{q,t}(\nu_{L^\perp}).$$

The measure μ_q^E is independent of the choice of t and L_q; it is concentrated on $A(E,q)$ and is invariant under the rigid motions of $\mathbb{R}^d$ that map E into itself. Moreover,

$$\mu_q^{gE}(gA) = \mu_q^E(A)$$

for $A \in \mathcal{B}(A(d,q))$ and all rigid motions $g \in G_d$.

Let $L \in G(d,q)$. In analogy to (13.9) we see that for given $t \in \mathbb{R}^d$ and for measurable functions $f \geq 0$ on $A(d,q)$ we have for $q < p$ the representation

$$\int_{A(L+t,q)} f \,\mathrm{d}\mu_q^{L+t} = \int_{G(L,q)} \int_{M^\perp \cap L} f(M + x + t)\, \lambda_{p-q}(\mathrm{d}x)\, \nu_q^L(\mathrm{d}M). \tag{13.13}$$

The corresponding equality for $q \geq p$ is

$$\int_{A(L+t,q)} f \, d\mu_q^{L+t} = \int_{G(L,q)} f(M+t) \, \nu_q^L(dM). \tag{13.14}$$

For the following measurability assertion we recall that the measure μ_p^F is concentrated on $A(F,p)$, but defined on all of $A(d,p)$. A similar remark concerns the measure ν_p^L. We write

$$A(d,p,q) := \{(E,F) \in A(d,p) \times A(d,q) : E \subset F\},$$
$$G(d,p,q) := \{(L,M) \in G(d,p) \times G(d,q) : L \subset M\}.$$

Lemma 13.2.2. *Let $0 \leq p < q \leq d$, and let $f : A(d,p,q) \to \mathbb{R}$ be a nonnegative measurable function. Then the maps*

$$F \mapsto \int_{A(F,p)} f(E,F) \, \mu_p^F(dE), \qquad F \in A(d,q), \tag{13.15}$$

and

$$E \mapsto \int_{A(E,q)} f(E,F) \, \mu_q^E(dF), \qquad E \in A(d,p), \tag{13.16}$$

are measurable.

Analogous statements hold for the measures ν_p^M, ν_q^L and nonnegative measurable functions on $G(d,p,q)$.

Proof. First let $f : A(d,p) \to \mathbb{R}$ be a continuous function with compact support. Let $(F_i)_{i\in\mathbb{N}}$ be a sequence in $A(d,q)$ converging to F. Then there is a sequence $(g_i)_{i\in\mathbb{N}}$ in the motion group G_d, converging to the identity id and such that $g_i^{-1}F = F_i$. We have

$$\begin{aligned}\int_{A(d,p)} f(E) \, \mu_p^{F_i}(dE) &= \int_{A(d,p)} f(E) \, \mu_p^{g_i^{-1}F}(dE) \\ &= \int_{A(d,p)} f(E) \, \mu_p^F(dg_i E) \\ &= \int_{A(d,p)} f(g_i^{-1}E) \, \mu_p^F(dE).\end{aligned}$$

The functions $f_i : E \mapsto f(g_i^{-1}E)$ converge to f for $i \to \infty$. If A denotes the compact support of f and $C \subset G_d$ is a compact set with $g_i \in C$ for all i, then CA is compact and $|f(g_i^{-1}E)| \leq \mathbf{1}_{CA}(E) \max |f|$. The dominated convergence theorem yields

$$\int_{A(d,p)} f(E) \, \mu_p^{F_i}(dE) \to \int_{A(d,p)} f(E) \, \mu_p^F(dE)$$

for $i \to \infty$. Thus the map

$$F \mapsto \int_{A(d,p)} f(E)\, \mu_p^F(\mathrm{d}E), \qquad F \in A(d,q),$$

is continuous and hence measurable.

We remark that Lemma 12.1.1 now shows that the function $F \mapsto \mu_p^F(B)$ is measurable for each $B \in \mathcal{B}(A(d,p))$, hence the mapping $(F,B) \mapsto \mu_p^F(B)$, $F \in A(d,q)$, $B \in \mathcal{B}(A(d,p))$, is a kernel.

Lemma 12.1.2 with $E = A(d,p)$ and $T = A(d,q)$ gives the measurability of the map (13.15) for nonnegative measurable functions f on $A(d,p,q)$.

The measurability of the map (13.16) and the proofs of the remaining assertions follow in a completely analogous manner. □

Notes for Section 13.2

1. The proof of Lemma 13.2.1 was communicated to us by Jürgen Kampf. A different proof appears in Goodey and Schneider [274]. The point in both cases was to give an elementary proof, avoiding an explicit representation of the invariant measure ν and using little more than its invariance.

2. Arguments similar to those used in the uniqueness proof of Theorem 13.2.12 appear, for example, also in the book by Ambartzumian [35].

13.3 A General Uniqueness Theorem

It was our aim in the preceding two sections to treat the special Haar measures that are needed for the purposes of Euclidean integral geometry, in a direct and elementary way. We present now a proof of a general uniqueness theorem for relatively invariant measures. This proof, in which the existence of general Haar measures is taken for granted, is adapted from Nachbin [571, pp. 138 ff].

In the following, we assume that G is a locally compact topological group and H is a closed subgroup; then the factor space G/H is locally compact. The following result is fundamental. *On every locally compact group there is a left Haar measure, and it is unique up to a positive factor.* The uniqueness admits a fairly quick proof (see, for example, Cohn [177], also for a lucid existence proof). Not every locally compact homogeneous space carries a Haar measure. This is shown, for example, by the standard operation of the affine group G_{aff} on $\mathbb{R}^d$. Any affine-invariant Borel measure on $\mathbb{R}^d$ is, in particular, translation invariant and thus, if it is locally finite, a multiple of the Lebesgue measure λ, but this is not affine-invariant. The Lebesgue measure satisfies

$$\lambda(gA) = |\det g|\lambda(A)$$

for $g \in G_{\mathrm{aff}}$ and $A \in \mathcal{B}(\mathbb{R}^d)$, and here the factor is independent of A. This motivates the following definition.

For the rest of this section, a measure on G or G/H is always a locally finite Borel measure. The measure ρ on the homogeneous space G/H is called **relatively invariant** if it is regular and not identically zero and if there exists a function $\chi : G \to \mathbb{R}$ such that

$$\rho(gA) = \chi(g)\rho(A) \qquad \text{for } g \in G \text{ and } A \in \mathcal{B}(G/H).$$

In that case, the function χ is called the **multiplier** of ρ. Obviously, χ is a homomorphism from G into the multiplicative group of positive real numbers. It can be shown that χ is continuous (cf. Hewitt and Ross [342, p. 204] or Gaal [242, p. 265]).

The study of relatively invariant measures is equivalent to the study of relatively invariant integrals. Here an **integral** on the locally compact space E is a positive linear functional on the space $\mathbf{C}_c(E)$ that is not identically zero. For $f \in \mathbf{C}_c(G)$ and $a \in G$ one writes $(a.f)(x) := f(a^{-1}x)$ for $x \in G$; then $a.f \in \mathbf{C}_c(G)$. The integral I on G is **left invariant** if $I(a.f) = I(f)$ for all $f \in \mathbf{C}_c(G)$ and all $a \in G$. For $f \in \mathbf{C}_c(G/H)$ and $a \in G$ one defines $(a.f)(xH) := f(a^{-1}xH)$, then $a.f \in \mathbf{C}_c(G/H)$. The integral I on G/H is called **relatively invariant with multiplier** χ if $I(a.f) = \chi(a)I(f)$ for all $f \in \mathbf{C}_c(G/H)$ and all $a \in G$. Every measure $\rho \neq 0$ on G/H induces an integral I via $I(f) = \int_{G/H} f \, \mathrm{d}\rho$ for $f \in \mathbf{C}_c(G/H)$. Conversely, the Riesz representation theorem implies that each integral I on G/H is generated in this way, by a uniquely determined regular Borel measure ρ. An integral is relatively invariant with multiplier χ if and only if the same holds for the corresponding measure.

We want to show that on a locally compact homogeneous space G/H there is, up to a constant factor, at most one relatively invariant measure with a given multiplier. For this, we first establish a relation between the spaces $\mathbf{C}_c(G)$ and $\mathbf{C}_c(G/H)$. For a function $f \in \mathbf{C}_c(G)$ we define

$$f'(x) := \int_H f(xy)\, \eta(\mathrm{d}y) \qquad \text{for } x \in G,$$

where η is a left Haar measure on H (so here we make use of its existence). The function f' is constant on the left cosets of H, since for $x \in zH$, which means $x = zh$ with $h \in H$, we have $f'(x) = \int f(zhy)\, \eta(\mathrm{d}y) = \int f(zy)\, \eta(\mathrm{d}y) = f'(z)$. Hence, there is a unique function $f^+ : G/H \to \mathbb{R}$ satisfying $f'(x) = f^+(xH)$ and thus

$$f^+(\pi(x)) = \int_H f(xy)\, \eta(\mathrm{d}y).$$

In this way, a linear map $f \mapsto f^+$ from $\mathbf{C}_c(G)$ into the vector space of real functions on G/H has been defined.

Lemma 13.3.1. *The correspondence $f \mapsto f^+$ maps $\mathbf{C}_c(G)$ onto $\mathbf{C}_c(G/H)$.*

Proof. Let $f \in \mathbf{C}_c(G)$. The function f' is continuous, because f is uniformly continuous. Since π is an open map, f^+ is continuous. If $f^+(xH) \neq 0$, there is an element $y \in H$ with $f(xy) \neq 0$ and thus $xy \in \operatorname{supp} f$ (where supp denotes the support), hence $xH \in \pi(\operatorname{supp} f)$. We conclude that $\operatorname{supp} f^+ \subset \pi(\operatorname{supp} f)$ and therefore $f^+ \in \mathbf{C}_c(G/H)$.

To prove that the mapping $f \mapsto f^+$ is surjective, let $h \in \mathbf{C}_c(G/H)$ and $K := \operatorname{supp} h$. Let V be a compact neighborhood of the unit element of G. For $g \in G$ the set $\pi(Vg)$ is a neighborhood of $\pi(g)$. Since K is compact, there are finitely many elements $g_1, \ldots, g_k \in G$ with $K \subset \bigcup_{j=1}^k \pi(Vg_j)$. Then the set $A := (Vg_1 \cup \ldots \cup Vg_k) \cap \pi^{-1}(K)$ is a compact subset of G with the property that $\pi(A) = K$. We can choose a function $u \in \mathbf{C}_c(G)$ with $u(A) = \{1\}$ and $0 \leq u \leq 1$. For $z \in G/H$ we define

$$\psi(z) := \begin{cases} \dfrac{h(z)}{u^+(z)} & \text{if } \ u^+(z) \neq 0, \\ 0 & \text{if } \ u^+(z) = 0. \end{cases}$$

Then $\psi u^+ = h$. Since ψ vanishes outside K and K is contained in the open set $\{z \in G/H : u^+(z) = 0\}$, the function ψ is continuous. Let $f := (\psi \circ \pi)u$; then $f \in \mathbf{C}_c(G)$ and

$$\begin{aligned} f^+(xH) &= \int_H f(xy)\, \eta(\mathrm{d}y) = \int_H \psi(xyH)u(xy)\, \eta(\mathrm{d}y) \\ &= \psi(xH) \int_H u(xy)\, \eta(\mathrm{d}y) \\ &= h(xH). \end{aligned}$$

Thus $f^+ = h$, which completes the proof. □

Theorem 13.3.1. *On a locally compact homogeneous space G/H there is, up to a constant factor, at most one relatively invariant measure with a given multiplier.*

Proof. Let ρ be a relatively invariant measure on G/H with multiplier χ. Let $a \in G$. The relative invariance of ρ implies that

$$\int_{G/H} a.h \, \mathrm{d}\rho = \chi(a) \int_{G/H} h \, \mathrm{d}\rho$$

for $h \in \mathbf{C}_c(G/H)$. For $f \in \mathbf{C}_c(G)$ we have $(a.f)^+ = a.f^+$, as follows immediately from the definitions. Since χ is a homeomorphism, we have $\chi = \chi(a)a.\chi$ and hence

$$\left(\frac{a.f}{\chi}\right)^+ = \frac{1}{\chi(a)} \left(\frac{a.f}{a.\chi}\right)^+ = \chi(a^{-1})a. \left(\frac{f}{\chi}\right)^+.$$

Now we define

$$I(f) := \int_{G/H} \left(\frac{f}{\chi}\right)^+ \mathrm{d}\rho \qquad \text{for } f \in \mathbf{C}_c(G).$$

Then I is a positive linear functional on $\mathbf{C}_c(G)$. For $a \in G$ we get

$$I(a.f) = \int_{G/H} \left(\frac{a.f}{\chi}\right)^+ \mathrm{d}\rho = \chi(a^{-1}) \int_{G/H} a.\left(\frac{f}{\chi}\right)^+ \mathrm{d}\rho$$
$$= \chi(a^{-1})\chi(a) \int_{G/H} \left(\frac{f}{\chi}\right)^+ \mathrm{d}\rho = I(f).$$

Thus I is a left invariant integral on $\mathbf{C}_c(G)$ and is, therefore, unique up to a constant factor; as mentioned, this uniqueness is equivalent to that of the left Haar measure. If now $\bar{\rho}$ is another relatively invariant measure on G/H with multiplier χ, then

$$\int_{G/H} \left(\frac{f}{\chi}\right)^+ \mathrm{d}\rho = c \int_{G/H} \left(\frac{f}{\chi}\right)^+ \mathrm{d}\bar{\rho}$$

for all $f \in \mathbf{C}_c(G)$ with some constant c. Since by Lemma 13.3.1 the function $(f/\chi)^+$ can be any element of $\mathbf{C}_c(G/H)$, we conclude that $\rho = c\bar{\rho}$. This completes the proof of the theorem. □

14

Facts from Convex Geometry

In this book, the more concrete examples of random sets are generated as unions of random systems of convex bodies. The quantitative description of such random sets is based on functionals of convex bodies which are particularly adapted to taking unions: they are additive. In Section 14.2 we collect the basic facts about the most important of these functionals, the rigid motion invariant intrinsic volumes, and their local counterparts, the curvature measures. In Section 14.4 we provide general information about additive functionals, as far as needed.

Aside from their use in the generation of random sets, convex bodies play an important role in this book, for example as associated convex bodies, and results from convex geometry find various applications. Section 14.3 lists the needed notions and results, for the purpose of convenient references.

Another particular class of sets, besides convex bodies, that are used to construct special particle processes or random sets, are k-dimensional surfaces, or curves if $k = 1$. In Section 14.5 we collect a few notions and results from geometric measure theory, leading to general notions of k-surfaces.

In Section 14.1 we introduce a metric invariant of pairs or more general tuples of linear subspaces, which appears in several formulas based on integral-geometric transformations.

14.1 The Subspace Determinant

We introduce a function of linear subspaces that occurs frequently in integral geometry. We consider k-tuples of linear subspaces $L_1, \dots, L_k$ of $\mathbb{R}^d$ satisfying either

$$\sum_{i=1}^{k} \dim L_i =: m \leq d \tag{14.1}$$

or

$$\sum_{i=1}^{k} \dim L_i \geq (k-1)d. \tag{14.2}$$

Linear subspaces $L_1, \ldots, L_k$ of $\mathbb{R}^d$ are said to be in **general position**, in case (14.1) if

$$\dim \operatorname{lin}(L_1 \cup \ldots \cup L_k) = \dim L_1 + \ldots + \dim L_k,$$

and in case (14.2) if

$$\dim(L_1 \cap \ldots \cap L_k) = \dim L_1 + \ldots + \dim L_k - (k-1)d.$$

Thus, $L_1, \ldots, L_k$ are in general position if and only if $L_1^\perp, \ldots, L_k^\perp$ are in general position.

In particular, L_1 and L_2 are in general position if and only if

$$L_1 \cap L_2 = \{0\} \qquad \text{or} \qquad \operatorname{lin}(L_1 \cup L_2) = \mathbb{R}^d.$$

We define a function $[L_1, \ldots, L_k]$, the **subspace determinant**, in the following way. Suppose, first, that (14.1) holds. We choose an orthonormal basis in each L_i (the empty set if $\dim L_i = 0$) and let $[L_1, \ldots, L_k]$ be the m-dimensional volume of the parallelepiped spanned by the union of these bases (1, by definition, if $\dim L_i = 0$ for $i = 1, \ldots, k$). If (14.2) holds, we define

$$[L_1, \ldots, L_k] := [L_1^\perp, \ldots, L_k^\perp].$$

Note that this is consistent if $k = 2$ and $\dim L_1 + \dim L_2 = d$. Obviously, any d-dimensional argument of $[L_1, \ldots, L_k]$ can be deleted without changing the value. We also note that $[L] = 1$ and that $[L_1, \ldots, L_k] = 0$ if and only if $L_1, \ldots, L_k$ are not in general position.

Let $k = 2$ and $\dim L_1 + \dim L_2 = d$. Then $[L_1, L_2]$ is equal to the factor by which the $(\dim L_1)$-dimensional volume is multiplied under the orthogonal projection from L_1 to $L_2^\perp$, and this is equal to the factor by which the $(\dim L_2)$-dimensional volume is multiplied under the orthogonal projection from L_2 to $L_1^\perp$. If $\dim L_1 + \dim L_2 \geq d$, we may choose an orthonormal basis of $L_1 \cap L_2$ and extend it to an orthonormal basis of L_1 and also to an orthonormal basis of L_2; then $[L_1, L_2]$ is the d-dimensional volume of the parallelepiped spanned by the obtained vectors.

The function $[\cdot, \cdot]$ is sometimes called the 'generalized sine function'. Other notations to be found in the literature are

$$[L_1, L_2] = |\langle L_1, L_2^\perp \rangle| = |\cos(L_1, L_2^\perp)|.$$

The following recursion formula for the subspace determinant is needed in Section 6.4.

Lemma 14.1.1. *If $k \in \mathbb{N}$ and if $L_1, \ldots, L_k$ are subspaces of $\mathbb{R}^d$ satisfying*

$$\sum_{i=1}^{k} \dim L_i \geq (k-1)d, \tag{14.3}$$

then

$$\begin{aligned}[L_1, \ldots, L_k] &= [L_1, L_2][L_1 \cap L_2, L_3, \ldots, L_k] \\ &= \ldots = [L_1, \ldots, L_{k-1}][L_1 \cap \ldots \cap L_{k-1}, L_k]. \end{aligned} \tag{14.4}$$

Proof. We first notice that, if $L_1, \ldots, L_k$ satisfy (14.3), then

$$\dim L_1 + \dim L_2 \geq (k-1)d - \sum_{i=3}^{k} \dim L_i \geq d$$

and

$$\dim(L_1 \cap L_2) + \sum_{i=3}^{k} \dim L_i \geq \sum_{i=1}^{k} \dim L_i - d \geq (k-2)d,$$

hence the invariants $[L_1, L_2]$ and $[L_1 \cap L_2, L_3, \ldots, L_k]$ are defined, and the first equation in (14.4) is equivalent to

$$[L_1^\perp, \ldots, L_k^\perp] = [L_1^\perp, L_2^\perp][L_1^\perp \vee L_2^\perp, L_3^\perp \ldots, L_k^\perp],$$

where $L_1^\perp \vee L_2^\perp$ denotes the linear hull of $L_1^\perp$ and $L_2^\perp$. The latter equation is now a familiar property of determinants.

Iterating the first equation, we get, in the same manner, the other results in (14.4). □

For tuples of linear subspaces of dimensions 1 or $d-1$, the subspace determinant is more conveniently expressed as a function of vectors.

For $k \in \{1, \ldots, d\}$ and $x_1, \ldots, x_k \in \mathbb{R}^d$ we denote by $\nabla_k(x_1, \ldots, x_k)$ the k-dimensional volume of the parallelepiped spanned by the vectors $x_1, \ldots, x_k$. For use in affine formulas, it is also convenient to define, for $x_0, x_1, \ldots, x_k \in \mathbb{R}^d$, the number

$$\Delta_k(x_0, \ldots, x_k) := \frac{1}{k!} \nabla_k(x_1 - x_0, \ldots, x_k - x_0).$$

Thus, $\Delta_k(x_0, \ldots, x_k)$ is the k-dimensional volume of the convex hull of $\{x_0, \ldots, x_k\}$.

14.2 Intrinsic Volumes and Curvature Measures

By a **convex body** in $\mathbb{R}^d$ we understand here a compact, convex subset of $\mathbb{R}^d$. The set of all convex bodies in $\mathbb{R}^d$ is denoted by $\mathcal{K}$, and $\mathcal{K}'$ is the subset of nonempty convex bodies. This set is equipped with the Hausdorff metric

δ (see Section 12.3), and topological statements about $\mathcal{K}'$ always refer to the topology induced by this metric.

The interior of a nonempty convex set $K \subset \mathbb{R}^d$ with respect to its affine hull, aff K, is not empty; it is called the **relative interior** of K and denoted by relint K. The boundary of K with respect to aff K is the **relative boundary** of K and is denoted by relbd K.

Associated with a convex body $K \in \mathcal{K}'$ are two useful functions. The **support function** of K is defined by

$$h(K,u) := \max\{\langle x,u\rangle : x \in K\} \qquad \text{for } u \in \mathbb{R}^d.$$

The function $h(K,\cdot)$ is convex and positively homogeneous, and every function on $\mathbb{R}^d$ with these properties is the support function of a uniquely determined convex body.

For $x \in \mathbb{R}^d$, the point $p(K,x)$ is defined as the unique point in K nearest to x, thus $d(x,K) = \|x - p(K,x)\|$ is the distance of x from K. The mapping $p(K,\cdot) : \mathbb{R}^d \to K$ is the **nearest-point map** or **metric projection** onto K; it is non-expansive.

We introduce a series of basic functionals on convex bodies. For $K \in \mathcal{K}'$ and $\epsilon > 0$, the set

$$K_\epsilon := K + \epsilon B^d = \{x \in \mathbb{R}^d : d(x,K) \le \epsilon\}$$

is the **parallel body** of K at distance ϵ. Its volume is a polynomial in ϵ of degree at most d. This result, known as the **Steiner formula**, is commonly written in either of the forms

$$V_d(K + \epsilon B^d) = \sum_{j=0}^{d} \epsilon^{d-j} \kappa_{d-j} V_j(K) \tag{14.5}$$

$$= \sum_{i=0}^{d} \epsilon^i \binom{d}{i} W_i(K). \tag{14.6}$$

This defines the **intrinsic volumes** $V_0, \dots, V_{d-1}$ and the **quermassintegrals** $W_0\ (= V_d), W_1, \dots, W_d$, also often called the **Minkowski functionals**. Thus, these two sets of functionals differ only by their normalizations. In this book, we prefer the first one. One advantage of that normalization is the fact that $V_j(K)$ depends only on K and not of the dimension of its surrounding space, that is, if $\dim K < d$, then the computation of $V_j(K)$ in $\mathbb{R}^d$ or alternatively in the affine hull of K (regarded as a Euclidean space) leads to the same result. We extend the definition by $V_j(\emptyset) := 0$. The intrinsic volumes inherit from the volume the important property of being additive. A real function φ on $\mathcal{K}$ is called **additive** if

$$\varphi(K \cup L) + \varphi(K \cap L) = \varphi(K) + \varphi(L)$$

for all $K, L \in \mathcal{K}$ with $K \cup L \in \mathcal{K}$, and $\varphi(\emptyset) = 0$. Besides additivity, the functional $V_j : \mathcal{K}' \to \mathbb{R}$ has the following properties: it is invariant under rigid motions, continuous, nonnegative, monotone under set inclusion, and locally bounded. The latter means that V_j is bounded on the set of convex bodies contained in a given convex body (and thus on every compact subset of $\mathcal{K}'$). Hadwiger's characterization theorem (Theorem 14.4.6) says that any additive, motion invariant, continuous real function on $\mathcal{K}'$ is a linear combination of the intrinsic volumes.

Additive functions are treated in Section 14.4, but we note here already some relevant properties of the intrinsic volumes. We denote by $\mathcal{R}$ the lattice generated by $\mathcal{K}$, thus $\mathcal{R}$ is the system of all unions of finitely many convex bodies in $\mathbb{R}^d$; this system is usually called the **convex ring**, and its elements are called **polyconvex** sets. The **extended convex ring** $\mathcal{S}$ consists of all sets $M \subset \mathbb{R}^d$ with $M \cap K \in \mathcal{R}$ for all $K \in \mathcal{K}$; such sets are called **locally polyconvex**. The functional V_j has an additive extension to the convex ring $\mathcal{R}$. This extension, also denoted by V_j, is motion invariant and measurable; however, the continuity is lost, and $V_0, \ldots, V_{d-2}$ attain also negative values on $\mathcal{R}$. Concerning the geometric meaning of the extensions, the following can be said. The extended function V_d is still the volume on $\mathcal{R}$, that is, the d-dimensional Lebesgue measure. If the set $K \in \mathcal{R}$ is the closure of its interior, then

$$V_{d-1}(K) = \frac{1}{2}\mathcal{H}^{d-1}(\operatorname{bd} K),$$

where $\mathcal{H}^{d-1}$ is the $(d-1)$-dimensional Hausdorff measure. Thus $2V_{d-1}$ is the **surface area**. If $K \in \mathcal{R}$ is the union of finitely many j-dimensional convex bodies, then $V_j(K)$ is the ordinary j-dimensional volume (j-dimensional Hausdorff measure) of K. On convex bodies, the functional V_1 is proportional to the **mean width** b; for $K \in \mathcal{K}'$,

$$\frac{d\kappa_d}{2} b(K) = \kappa_{d-1} V_1(K) = \int_{S^{d-1}} h(K, u)\, \sigma(\mathrm{d}u). \tag{14.7}$$

This connection is no longer valid on the convex ring. The functional V_0, which on $\mathcal{K}'$ is identically equal to one, is called the **Euler characteristic**; on $\mathcal{R}$ it coincides with the equally coined topological invariant and is, therefore, often denoted by χ. If all connected components of $K \in \mathcal{R}$ have characteristic one (for instance, if they are convex), then $V_0(K)$ equals the number of connected components. In the plane, $V_0(K)$ can be described as the number of connected components minus the number of 'holes' (bounded connected components of the complement) of $K \in \mathcal{R}$.

A frequently occurring constant is

$$V_k(B^d) = \binom{d}{k} \frac{\kappa_d}{\kappa_{d-k}}. \tag{14.8}$$

We repeat the geometric meaning of the intrinsic volumes on the convex ring for dimensions 2 and 3 and add the special symbols which are often used in the applied literature:

$d = 2$	V_2 area A
	$2V_1$ boundary length (perimeter) L
	V_0 Euler characteristic χ
$d = 3$	V_3 volume V
	$2V_2$ surface area S
	$\frac{1}{2}V_1$ mean width b (only for convex bodies)
	V_0 Euler characteristic χ

The following lemma computes the intrinsic volumes of a direct orthogonal sum. We give a proof, since this does not appear in standard monographs on convex bodies.

Lemma 14.2.1. *If $L \in G(d,q)$ with $q \in \{1,\dots,d-1\}$ and if $K \subset L$, $M \subset L^\perp$ are convex bodies, then*

$$V_j(K+M) = \sum_{k=0}^{j} V_k(K)V_{j-k}(M)$$

for $j \in \{0,\dots,d\}$.

Proof. For a polytope $P \in \mathcal{P}'$, we put

$$W(P) := \int_{\mathbb{R}^d} \mathrm{e}^{-\pi d(x,P)^2}\,\lambda(\mathrm{d}x).$$

It is not difficult to see (introducing, for each face F of P, generalized cylinder coordinates on the pre-image $p(P,\cdot)^{-1}(F)$; compare also the proof of Lemma 8.5.1) that

$$W(P) = V_d(P) + 2\int_0^\infty V_{d-1}(P + rB^d)\mathrm{e}^{-\pi r^2}\,\mathrm{d}r.$$

With (14.5), this gives

$$W(P) = V_d(P) + \sum_{j=0}^{d-1}(d-j)\kappa_{d-j}V_j(P)\int_0^\infty r^{d-1-j}\mathrm{e}^{-\pi r^2}\,\mathrm{d}r = \sum_{j=0}^{d} V_j(P).$$

Now let $K \subset L$ and $M \subset L^\perp$ be polytopes, and let $P = K + M$. If $x \in \mathbb{R}^d$ and $x = y + z$ with $y \in L$ and $z \in L^\perp$, then

$$d(P,x)^2 = d(K,y)^2 + d(M,z)^2.$$

It follows that

$$W(K+M) = \int_{L^\perp}\int_L \mathrm{e}^{-\pi d(y,K)^2}\mathrm{e}^{-\pi d(z,M)^2}\,\lambda_q(\mathrm{d}y)\,\lambda_{d-q}(\mathrm{d}z) = W(K)W(M).$$

Applying this to αK, αM with $\alpha > 0$, we get

$$\begin{aligned}\sum_{j=0}^d \alpha^j V_j(K+M) &= \left(\sum_{r=0}^d \alpha^r V_r(K)\right)\left(\sum_{s=0}^d \alpha^s V_s(M)\right)\\ &= \sum_{j=0}^d \alpha^j \sum_{r=0}^j V_r(K)V_{j-r}(M).\end{aligned}$$

Comparing the coefficients, we obtain the assertion for polytopes, and by approximation it is obtained for general convex bodies. □

The Steiner formula, leading to the intrinsic volumes, can be generalized in different ways. One possibility, replacing the Minkowski sum $K+\epsilon B^d$ by a Minkowski combination $\lambda_1 K_1 + \ldots + \lambda_m K_m$, leads to the mixed volumes, which are briefly mentioned in Section 14.3. Another possibility, a local version of the Steiner formula, leads to the curvature measures and will now be explained.

The parallel body K_ϵ at distance ϵ of a convex body K consists of all points x with $d(x,K) \le \epsilon$. Instead, we now take only those points of K_ϵ into account for which the nearest point $p(K,x)$ in K belongs to some specified Borel set and, if $x \notin K$, the unit vector pointing from $p(K,x)$ to x belongs to some specified Borel set of directions. Again, the volume of the local parallel set obtained in this way has a polynomial expansion in ϵ. The coefficients define the support measures or generalized curvature measures. By specialization, one obtains curvature measures and area measures.

Let $K \subset \mathbb{R}^d$ be a nonempty closed convex set. A **support element** of K is a pair (x,u) where x is a boundary point of K and u is an outer unit normal vector of K at x. The set of all support elements of K is denoted by $\operatorname{Nor} K$ and is called the **generalized normal bundle** of K. It gets its topology as a subset of the product space $\Sigma := \mathbb{R}^d \times S^{d-1}$. For $x \in K$, the set

$$N(K,x) := p(K-x,\cdot)^{-1}(0)$$

is the **normal cone** of K at x. It consists of all outer normal vectors of K at x, together with the zero vector. For $x \notin K$ we have $d(x,K) > 0$, and the unit vector

$$u(K,x) := \frac{x - p(K,x)}{d(x,K)}$$

is an outer normal vector to K at $p(K,x)$, thus $(p(K,x), u(K,x)) \in \operatorname{Nor} K$.

Definition 14.2.1. *For a nonempty closed convex set $K \subset \mathbb{R}^d$, for $\epsilon > 0$ and a Borel set $A \subset \Sigma$, the set*

$$M_\epsilon(K, A) := \{x \in K_\epsilon \setminus K : (p(K, x), u(K, x)) \in A\}$$

is the **local parallel set** *of K at distance ϵ determined by A.*

The following result extends the Steiner formula.

Theorem 14.2.1 (Local Steiner formula). *For $K \in \mathcal{K}'$, there are finite measures $\Xi_0(K, \cdot), \ldots, \Xi_{d-1}(K, \cdot)$ on Σ such that, for $\epsilon \geq 0$ and every $A \in \mathcal{B}(\Sigma)$,*

$$\lambda(M_\epsilon(K, A)) = \sum_{m=0}^{d-1} \epsilon^{d-m} \kappa_{d-m} \Xi_m(K, A). \tag{14.9}$$

The measures $\Xi_0(K, \cdot), \ldots, \Xi_{d-1}(K, \cdot)$ are concentrated on $\operatorname{Nor} K$.

The measures $\Xi_0(K, \cdot), \ldots, \Xi_{d-1}(K, \cdot)$ are called the **support measures** or **generalized curvature measures** of K.

Results on support measures are often proved for polytopes first, hence we introduce some relevant notation. For a polyhedral set $P \subset \mathbb{R}^d$ (a nonempty intersection of finitely many closed halfspaces), we denote by $\mathcal{F}_m(P)$ the set of m-dimensional faces of P, $m = 0, \ldots, \dim P$. The set P is considered as a face of itself. Each face of P is a polyhedral set, and if C is a polyhedral cone, then each face of C is a polyhedral cone. We write

$$\mathcal{F}_\bullet(P) := \bigcup_{m=0}^{\dim P} \mathcal{F}_m(P)$$

for the set of all faces of P. If $F \in \mathcal{F}_\bullet(P)$ and $x \in \operatorname{relint} F$, then the normal cone $N(P, x)$ does not depend on x; it is denoted by $N(P, F)$ and is called the **normal cone** of P at F. For $F \in \mathcal{F}_m(P)$, the number

$$\gamma(F, P) := \frac{\lambda_{d-m}(N(P, F) \cap B^d)}{\kappa_{d-m}} = \frac{\sigma_{d-m-1}(N(P, F) \cap S^{d-1})}{(d-m)\kappa_{d-m}} \tag{14.10}$$

is the **external angle** of P at F. By σ_j we have denoted the j-dimensional spherical Lebesgue measure. We also put $\gamma(P, P) = 1$ and $\gamma(F, P) = 0$ if either $F = \emptyset$ or F is not a face of P.

Now we sketch the proof of Theorem 14.2.1, to make the reader familiar with the main arguments, but for the details we refer to Schneider [695, sects. 4.1 and 4.2]. We define

$$\mu_\epsilon(K, A) := \lambda(M_\epsilon(K, A)) \qquad \text{for } A \in \mathcal{B}(\Sigma),\ \epsilon \geq 0.$$

Thus, $\mu_\epsilon(K, \cdot)$ is the image measure of the Lebesgue measure, restricted to $K_\epsilon \setminus K$, under the map $x \mapsto (p(K, x), u(K, x))$ from $K_\epsilon \setminus K$ to Σ. This map is continuous and hence measurable. In particular, $\mu_\epsilon(K, \cdot)$ is a finite measure on $\mathcal{B}(\Sigma)$. We call it the **local parallel volume** of K at distance ϵ. The measure

$\mu_\epsilon(K,\cdot)$ is concentrated on $\operatorname{Nor} K$, that is, $\mu_\epsilon(K,A) = \mu_\epsilon(K, A \cap \operatorname{Nor} K)$ for $A \in \mathcal{B}(\Sigma)$.

We first quote some fundamental properties of the mapping $\mu_\epsilon : \mathcal{K} \times \mathcal{B}(\Sigma) \to \mathbb{R}$. In the following, $\xrightarrow{w}$ denotes weak convergence of finite measures.

Lemma 14.2.2. *If $(K_j)_{j\in\mathbb{N}}$ is a sequence in $\mathcal{K}'$ with $K_j \to K \in \mathcal{K}'$ for $j \to \infty$, then*

$$\mu_\epsilon(K_j,\cdot) \xrightarrow{w} \mu_\epsilon(K,\cdot), \qquad j \to \infty,$$

for every $\epsilon > 0$.

Lemma 14.2.3. *For any Borel set $A \in \mathcal{B}(\Sigma)$ and for any $\epsilon > 0$, the function $\mu_\epsilon(\cdot, A) : \mathcal{K}' \to \mathbb{R}$ is measurable.*

Lemma 14.2.4. *For any $A \in \mathcal{B}(\Sigma)$ and for any $\epsilon > 0$, the function $\mu_\epsilon(\cdot, A)$ is additive, that is*

$$\mu_\epsilon(K_1 \cup K_2, A) + \mu_\epsilon(K_1 \cap K_2, A) = \mu_\epsilon(K_1, A) + \mu_\epsilon(K_2, A)$$

whenever $K_1, K_2, K_1 \cup K_2 \in \mathcal{K}'$.

Now we compute the local parallel volume in the case of a polytope. Let $P \in \mathcal{K}'$ be a polytope. For $x \in \mathbb{R}^d \setminus P$, the nearest point $p(P,x)$ belongs to the relative interior $\operatorname{relint} F$ of a unique face F of P. Therefore,

$$M_\epsilon(P,A) = \bigcup_{m=0}^{d-1} \bigcup_{F\in\mathcal{F}_m(P)} \left[M_\epsilon(P,A) \cap p(P,\cdot)^{-1}(\operatorname{relint} F)\right]$$

is a disjoint decomposition of the local parallel set $M_\epsilon(P,A)$. For $m \in \{0,\dots,d-1\}$ and $F \in \mathcal{F}_m(P)$ it follows from the properties of the nearest-point map that

$$M_\epsilon(P,A) \cap p(P,\cdot)^{-1}(\operatorname{relint} F) = \operatorname{relint} F \oplus ([N(P,F)\setminus\{0\}] \cap \epsilon B^d),$$

where $\oplus$ denotes a direct orthogonal sum. An application of Fubini's theorem gives

$$\lambda(M_\epsilon(P,A) \cap p(P,\cdot)^{-1}(\operatorname{relint} F)) = \epsilon^{d-m} \int_F \lambda_{d-m}(N(P,F) \cap A_x)\, \lambda_m(\mathrm{d}x)$$

with

$$A_x := \{ru : (x,u) \in A,\ 0 < r \le 1\}.$$

Hence, if we define

$$\Xi_m(P,A) := \frac{1}{\kappa_{d-m}} \sum_{F\in\mathcal{F}_m(P)} \int_F \lambda_{d-m}(N(P,F) \cap A_x)\, \lambda_m(\mathrm{d}x), \tag{14.11}$$

then we obtain

$$\mu_\epsilon(P, A) = \sum_{m=0}^{d-1} \epsilon^{d-m} \kappa_{d-m} \Xi_m(P, A).$$

This is already a polynomial expansion. Together with approximation by polytopes and Lemma 14.2.2 it yields the proof of Theorem 14.2.1.

In the following theorem, we collect some properties of the support measures. These properties follow from the corresponding properties of $\mu_\epsilon(K, \cdot)$ and from obvious properties of $\Xi_m(P, \cdot)$ for polytopes P, together with the previous lemmas.

Theorem 14.2.2. *For $m = 0, \ldots, d-1$, the map $\Xi_m : \mathcal{K}' \times \mathcal{B}(\Sigma) \to \mathbb{R}$ has the following properties:*

(a) **Motion covariance:** $\Xi_m(gK, g.A) = \Xi_m(K, A)$ *for* $g \in G_d$, *where* $g.A := \{(gx, g_0 u) : (x, u) \in A\}$, g_0 *denoting the rotation part of* g,
(b) **Homogeneity:** $\Xi_m(\alpha K, \alpha \cdot A) = \Xi_m(K, A)$ *for* $\alpha \geq 0$, *where* $\alpha \cdot A := \{(\alpha x, u) : (x, u) \in A\}$,
(c) **Weak continuity:** $K_j \to K$ *implies* $\Xi_m(K_j, \cdot) \xrightarrow{w} \Xi_m(K, \cdot)$,
(d) $\Xi_m(\cdot, A)$ *is* **additive**, *for each fixed* $A \in \mathcal{B}(\Sigma)$,
(e) $\Xi_m(\cdot, A)$ *is* **measurable**, *for each fixed* $A \in \mathcal{B}(\Sigma)$.

It is compatible with the additivity to extend the definition by

$$\Xi_m(\emptyset, \cdot) := 0, \qquad m = 0, \ldots, d-1.$$

In applications, often only the marginal measures of the support measures are required. Let $K \in \mathcal{K}'$. By

$$\Phi_m(K, A) := \Xi_m(K, A \times S^{d-1}), \qquad A \in \mathcal{B}(\mathbb{R}^d),$$

one defines the mth **curvature measure** $\Phi_m(K, \cdot)$ of K, for $m = 0, \ldots, d-1$. Thus, $\Phi_m(K, \cdot)$ is the image measure of the support measure $\Xi_m(K, \cdot)$ under the projection $(x, u) \mapsto x$. One extends the definition by

$$\Phi_d(K, \cdot) := \lambda(K \cap \cdot),$$

and then has the local Steiner formula

$$\lambda(\{x \in K_\epsilon : p(K, x) \in A\}) = \sum_{m=0}^{d} \epsilon^{d-m} \kappa_{d-m} \Phi_m(K, A) \tag{14.12}$$

for $A \subset \mathcal{B}(\mathbb{R}^d)$. Further, by

$$\Psi_m(K, B) := \Xi_m(K, \mathbb{R}^d \times B), \qquad B \in \mathcal{B}(S^{d-1}),$$

one defines the mth **area measure** $\Psi_m(K, \cdot)$ of K, for $m = 0, \ldots, d-1$. Thus, $\Psi_m(K, \cdot)$ is the image measure of $\Xi_m(K, \cdot)$ under the projection $(x, u) \mapsto u$.

Clearly, for $m = 0, \ldots, d-1$ and $K \in \mathcal{K}'$,

$$\Xi_m(K, \Sigma) = \Phi_m(K, \mathbb{R}^d) = \Psi_m(K, S^{d-1}) = V_m(K).$$

In the literature, the names curvature measure and area measure are also used for differently normalized versions of the measures defined here. We introduce only the two frequently used notations

$$C_{d-1}(K, \cdot) := 2\Phi_{d-1}(K, \cdot), \qquad S_{d-1}(K, \cdot) := 2\Psi_{d-1}(K, \cdot).$$

These measures have simple intuitive interpretations. Let $\mathcal{H}^{d-1}$ denote the $(d-1)$-dimensional Hausdorff measure. If $\dim K \neq d-1$, then, for $A \in \mathcal{B}(\mathbb{R}^d)$,

$$C_{d-1}(K, A) = 2\Phi_{d-1}(K, A) = \mathcal{H}^{d-1}(A \cap \operatorname{bd} K).$$

Therefore, the measure $C_{d-1}(K, \cdot)$ is called the **boundary measure** of K. For $\dim K \leq d-1$, one trivially has $\Phi_{d-1}(K, A) = \mathcal{H}^{d-1}(A \cap \operatorname{bd} K)$. For $B \in \mathcal{B}(S^{d-1})$, let $\tau(K, B)$ be the set of boundary points of K at which there exists an outer normal vector belonging to B. For a convex body K of dimension $\dim K \neq d-1$,

$$S_{d-1}(K, B) = 2\Psi_{d-1}(K, B) = \mathcal{H}^{d-1}(\tau(K, B)).$$

The measure $S_{d-1}(K, \cdot)$ is called the **surface area measure** of K. If $\dim K = d-1$, then $S_{d-1}(K, B) = 2\mathcal{H}^{d-1}(K)$, $\mathcal{H}^{d-1}(K)$, or 0, according to whether both, one, or none of the unit normal vectors of the affine hull of K belong to B.

Under special assumptions on the convex bodies, also the curvature measures and area measures of orders less than $d-1$ have intuitive interpretations. If the boundary $\operatorname{bd} K$ of the convex body K is a regular hypersurface of class C^2, then the local parallel volume can be computed by differential-geometric means. Let $A \in \mathcal{B}(\mathbb{R}^d)$. For $m = 0, \ldots, d-1$ one obtains for the mth curvature measure the representation

$$\Phi_m(K, A) = \frac{\binom{d}{m}}{d\kappa_{d-m}} \int_{A \cap \operatorname{bd} K} H_{d-1-m} \,\mathrm{d}S,$$

where H_k denotes the kth normalized elementary symmetric function of the principal curvatures of $\operatorname{bd} K$ and where $\mathrm{d}S$ is the volume form on $\operatorname{bd} K$. This, of course, explains the name 'curvature measure'. If P is a polytope, the explicit representation (14.11) specializes to

$$\Phi_m(P, A) = \sum_{F \in \mathcal{F}_m(P)} \gamma(F, P) \lambda_m(F \cap A), \tag{14.13}$$

where $\gamma(F, P)$ is the external angle of P at its face F.

An important special case of (14.13) is the representation

$$V_m(P) = \sum_{F \in \mathcal{F}_m(P)} \gamma(F, P) \lambda_m(F) \tag{14.14}$$

for the mth intrinsic volume of a polytope P.

For arbitrary $K \in \mathcal{K}'$, it is clear from (14.12) that the curvature measures $\Phi_j(K, \cdot)$ are concentrated on the boundary of K, for $j = 0, \ldots, d-1$.

Let $\sigma(K, A) \subset S^{d-1}$ denote the set of all outer unit normal vectors to K at points of $A \cap \operatorname{bd} K$; then

$$\Phi_0(K, A) = \frac{1}{d\kappa_d} \mathcal{H}^{d-1}(\sigma(K, A)).$$

Thus, the measure Φ_0 is the normalized area of the spherical image and is, therefore, also known as the **Gaussian curvature measure**.

For the area measures at a Borel set $B \in \mathcal{B}(S^{d-1})$, the following representations in special cases are obtained. For a convex body K with a regular C^2 boundary of positive Gauss–Kronecker curvature, the mth area measure is given by

$$\Psi_m(K, B) = \frac{\binom{d}{m}}{d\kappa_{d-m}} \int_B s_m \, d\sigma.$$

Here s_m is the mth normalized elementary symmetric function of the principal radii of curvature of $\operatorname{bd} K$, as a function of the outer unit normal vector (recall that $\sigma := \sigma_{d-1}$, by convention). For a polytope P, we get

$$\Psi_m(P, B) = \sum_{F \in \mathcal{F}_m(P)} \frac{\sigma_{d-1-m}(N(P, F) \cap B)}{(d-m)\kappa_{d-m}} \lambda_m(F).$$

For general convex bodies K, the measure $\Psi_0(K, \cdot)$ is independent of K and is equal to $\sigma/\sigma(S^{d-1})$.

It is important to notice that the support measure Ξ_m has an additive extension to the convex ring $\mathcal{R}$ (denoted by the same symbol). Thus, there is a mapping $\Xi : \mathcal{R} \times \mathcal{B}(\Sigma) \to \mathbb{R}$ extending the mth support measure, which is additive in the first argument. The existence of the extension follows either from Theorem 14.4.2, or by an extension of the local Steiner formula (see Schneider [695, sect. 4.4], and also Note 3 of Section 14.4). If $K = K_1 \cup \ldots \cup K_k$ with $K_i \in \mathcal{K}'$, $i = 1, \ldots, k$, then the inclusion–exclusion principle gives

$$\Xi_m(K, \cdot) = \sum_{r=1}^{k} (-1)^{r-1} \sum_{1 \le i_1 < \ldots < i_r \le k} \Xi_m(K_{i_1} \cap \ldots \cap K_{i_r}, \cdot).$$

This shows that the extension $\Xi_m(K, \cdot)$ is a signed measure. In general, it is not a positive measure, with the exception that always

$$\Xi_{d-1}(K, \cdot) \ge 0 \qquad \text{for } K \in \mathcal{R}.$$

As a consequence, also Φ_m and Ψ_m, and in particular C_{d-1} and S_{d-1}, have extensions to the convex ring that are additive in the first argument and signed measures in the second, and

$$C_{d-1}(K,\cdot) \geq 0, \quad S_{d-1}(K,\cdot) \geq 0 \qquad \text{for } K \in \mathcal{R}. \tag{14.15}$$

If $K \in \mathcal{R}$ is the closure of its interior, then $C_{d-1}(K,A) = \mathcal{H}^{d-1}(A \cap \operatorname{bd} K)$, as for convex bodies. This was proved in [695, Th. 4.4.1], and by extending the argument, Weil [797] showed (14.15).

The following property is not surprising, but occasionally useful.

Theorem 14.2.3. *The curvature measures and the area measures are* **locally determined**, *in the following sense. Let $K, M \in \mathcal{K}'$.*

(a) *If $A \subset \mathbb{R}^d$ is open and $K \cap A = M \cap A$, then $\Phi_m(K,B) = \Phi_m(M,B)$ for every Borel set $B \subset A$, $m = 0,\dots,d$.*
(b) *If $B \in \mathcal{B}(S^{d-1})$ is a Borel set with $\tau(K,B) = \tau(M,B)$, then $\Psi(K,B) = \Psi_m(M,B)$, $m = 0,\dots,d-1$.*

Since the mth curvature measure is locally determined, the value $\Phi_m(M,B)$ is well defined for a locally polyconvex set $M \in \mathcal{S}$ and a bounded Borel set B. Indeed, we may choose a convex body K with $B \subset \operatorname{int} K$, then $\Phi_m(M,B) := \Phi_m(M \cap K, B)$ does not depend on the choice of K. We call this extension of Φ_m to the extended convex ring a **signed Radon measure**, since

$$f \mapsto \int_{\mathbb{R}^d} f(x)\, \Phi_m(M, \mathrm{d}x), \qquad f \in \mathbf{C}_c(\mathbb{R}^d),$$

defines a linear functional on the vector space $\mathbf{C}_c(\mathbb{R}^d)$ of continuous functions on $\mathbb{R}^d$ with compact support, and this linear functional is bounded on $\mathbf{C}(C)$ for every compact set $C \subset \mathbb{R}^d$. If $M \in \mathcal{S}$ is convex, hence a closed convex set, the curvature measures $\Phi_m(M,\cdot)$ are nonnegative, they can therefore be extended to all Borel sets B, as locally finite measures.

Since the support measures appear in the polynomial expansion (14.9), they satisfy themselves Steiner type formulas. For $\epsilon \geq 0$, we define a map $T_\epsilon : \Sigma \to \Sigma$ by $T_\epsilon(x,u) := (x + \epsilon u, u)$.

Theorem 14.2.4. *If $K \in \mathcal{K}'$, $A \subset \mathcal{B}(\Sigma)$, $\epsilon \geq 0$ and $k \in \{0,\dots,d-1\}$, then*

$$\Xi_k(K + \epsilon B^d, T_\epsilon A) = \sum_{m=0}^{k} \epsilon^{k-m} \frac{1}{(k-m)!} c_{d-k}^{d-m}\, \Xi_m(K,A).$$

Special cases are obtained for curvature measures and area measures. If $A \subset K$ is a Borel set, then

$$\Phi_k(K + \epsilon B^d, A + \epsilon B^d) = \sum_{m=0}^{k} \epsilon^{k-m} \frac{1}{(k-m)!} c_{d-k}^{d-m}\, \Phi_m(K,A). \tag{14.16}$$

If $A \subset S^{d-1}$ is a Borel set, then

$$\Psi_k(K + \epsilon B^d, A) = \sum_{m=0}^{k} \epsilon^{k-m} \frac{1}{(k-m)!} c_{d-k}^{d-m}\, \Psi_m(K,A).$$

Notes for Section 14.2

1. Detailed proofs of the statements of this section are found in Schneider [695]. An exception is Lemma 14.2.1, for which we gave a proof using an idea of Hadwiger [308]; see also Federer [229, p. 272].

2. For the early history of the Steiner formula, and for information about the introduction of area measures and curvature measures, we refer to the Notes in Schneider [695, pp. 211 ff.]. The support measures or generalized curvature measures were first introduced in Schneider [678] and further studied in Schneider [679].

14.3 Mixed Volumes and Inequalities

The Steiner formula is just a special case of a general polynomial expansion. For convex bodies $K_1, \dots, K_m \in \mathcal{K}'$ and numbers $\lambda_1, \dots, \lambda_m \geq 0$, there is a representation

$$V_d(\lambda_1 K_1 + \ldots + \lambda_m K_m) = \sum_{i_1,\dots,i_d=1}^{m} \lambda_{i_1} \cdots \lambda_{i_d} V(K_{i_1}, \dots, K_{i_d}) \qquad (14.17)$$

with uniquely determined symmetric coefficients $V(K_{i_1}, \dots, K_{i_d})$. The function $(K_1, \dots, K_d) \mapsto V(K_1, \dots, K_d)$ is called the **mixed volume**. We use the notation

$$V(K[k], M[d-k]) := V(\underbrace{K, \dots, K}_{k}, \underbrace{M, \dots, M}_{d-k}).$$

In particular, the intrinsic volumes are given by

$$V_j(K) = \frac{\binom{d}{j}}{\kappa_{d-j}} V(K[j], B^d[d-j]). \qquad (14.18)$$

Also the volume of the orthogonal projection $K|L$ of a convex body K to a subspace $L \in G(d,j)$, $j \in \{1, \dots, d-1\}$, can be expressed as a mixed volume. If $B_{L^\perp} \subset L^\perp$ is a ball of $(d-j)$-dimensional volume 1, then

$$V_j(K|L) = \binom{d}{j} V(K[j], B_{L^\perp}[d-j]). \qquad (14.19)$$

A special case of the expansion (14.17), slightly more general than the Steiner formula, is the expression

$$V_d(K + \epsilon M) = \sum_{j=0}^{d} \epsilon^{d-j} \binom{d}{j} V(K[j], M[d-j]) \qquad (14.20)$$

for $K, M \in \mathcal{K}'$.

The mixed volume has an important integral representation, given by

$$V(M,K_1,\dots,K_{d-1}) = \frac{1}{d}\int_{S^{d-1}} h(M,u)\, S(K_1,\dots,K_{d-1},\mathrm{d}u). \tag{14.21}$$

Here, $S(K_1,\dots,K_{d-1},\cdot)$ is a uniquely determined finite measure on S^{d-1}, the **mixed area measure** of $K_1,\dots,K_{d-1}$. By specializing, one obtains the surface area measure of K,

$$S_{d-1}(K,\cdot) = S(K,\dots,K,\cdot), \tag{14.22}$$

and the integral representation

$$V(M,K,\dots,K) = \frac{1}{d}\int_{S^{d-1}} h(M,u)\, S_{d-1}(K,\mathrm{d}u). \tag{14.23}$$

It is important to know which measures can occur as surface area measures of convex bodies. The answer is given by the following existence and uniqueness theorem, which goes back to Hermann Minkowski.

Theorem 14.3.1 (Minkowski). *Let φ be a finite measure on the sphere S^{d-1} with the properties that*

$$\int_{S^{d-1}} u\,\varphi(\mathrm{d}u) = 0$$

and $\varphi(S) < \varphi(S^{d-1})$ for every great subsphere $S \subset S^{d-1}$. Then there exists a convex body $K \in \mathcal{K}'$ with interior points such that $S_{d-1}(K,\cdot) = \varphi$. It is uniquely determined up to translations.

For some applications, we need a generalization of the support measures, where the Euclidean distance used in the definition of local parallel sets is replaced by a (not necessarily symmetric) distance notion defined by some **gauge body** B (also called **structuring element** in certain applications). By this, we understand here a convex body containing the origin 0. The B**-distance** of a point x from a nonempty closed set F is defined by

$$\begin{aligned} d_B(x,F) &:= \min\{r \geq 0 : (x + rB) \cap F \neq \emptyset\} \\ &= \min\{r \geq 0 : x \in F - rB\} \end{aligned}$$

(with $\min\emptyset := \infty$). For $K \in \mathcal{K}'$ we say that K and B have **independent support sets** if

$$\dim F(K - B, u) = \dim F(K,u) + \dim F(-B,u), \qquad u \in S^{d-1},$$

where $F(K,u)$ is the intersection of K with its support hyperplane with outer unit normal vector u. Suppose that this holds. Then, for each $x \in \mathbb{R}^d$ there is a unique point $y \in K$ for which $(x + d_B(x,K)B) \cap K = \{y\}$. This point is denoted by $p_B(K,x)$, and the vector $u_B(K,x)$ is defined by

$$u_B(K,x) := \frac{x - p_B(K,x)}{d_B(x,K)}.$$

Note that $u_B(K,x) \in \mathrm{bd}\,(-B)$. The **Minkowski normal bundle** of K is the set

$$N_B(K) := \{(p_B(K,x), u_B(K,x)) : x \in \mathrm{bd}\,(K - rB)\},$$

where $r > 0$ is arbitrary.

For $A \in \mathcal{B}(\mathbb{R}^d \times \mathbb{R}^d)$ and $\epsilon \geq 0$ we define a relative local parallel set by

$$M_\epsilon^B(K,A) := \{x \in \mathbb{R}^d : 0 < d_B(x,K) \leq \epsilon,\ (p_B(K,x), u_B(K,x)) \in A\}.$$

Its Lebesgue measure has a polynomial expansion in ϵ,

$$\lambda(M_\epsilon^B(K,A)) = \sum_{m=0}^{d-1} \epsilon^{d-m} \kappa_{d-m} \Xi_m(K;B;A). \tag{14.24}$$

This generalizes the local Steiner formula (14.9) and defines the **relative support measures** $\Xi_m(K;B;\cdot)$, $m = 0, \ldots, d-1$, of K with respect to the gauge body B. The measure $\Xi_m(K;B;\cdot)$ is concentrated on the Minkowski normal bundle $N_B(K)$. Notice that

$$\Xi_m(K;B;\mathbb{R}^d \times \mathbb{R}^d) = \frac{\binom{d}{m}}{\kappa_{d-m}} V(K[m], -B[d-m]). \tag{14.25}$$

The polynomial expansion (14.24) can be generalized as follows.

Theorem 14.3.2. *Let $K, B \in \mathcal{K}'$ be convex bodies with $0 \in B$ and such that K, B have independent support sets. If f is a nonnegative measurable function on $\mathbb{R}^d$, then*

$$\int_{\mathbb{R}^d} \mathbf{1}\{0 < d_B(x,K) < \infty\} f(x)\, \lambda(\mathrm{d}x) \tag{14.26}$$

$$= \sum_{m=0}^{d-1} (d-m)\kappa_{d-m} \int_0^\infty \int_{\mathbb{R}^d \times \mathbb{R}^d} t^{d-1-m} f(y+tb)\, \Xi_m(K;B;\mathrm{d}(y,b))\, \mathrm{d}t.$$

The following consequence is formulated in a way that is convenient for applications.

Theorem 14.3.3. *Let $K, B \in \mathcal{K}'$ be convex bodies with $0 \in B$ and such that K, B have independent support sets. If g is a nonnegative measurable function on $[0,\infty] \times \mathbb{R}^d \times \mathbb{R}^d$ and f is a nonnegative measurable function on $\mathbb{R}^d$, then*

$$\int_{\mathbb{R}^d} \mathbf{1}\{0 < d_B(x,K) < \infty\} g(d_B(x,K), u_B(K,x), p_B(K,x)) f(x)\, \lambda(\mathrm{d}x)$$

$$= \sum_{m=0}^{d-1} (d-m)\kappa_{d-m}$$

$$\times \int_0^\infty \int_{\mathbb{R}^d \times \mathbb{R}^d} t^{d-1-m} g(t,b,y) f(y+tb)\, \Xi_m(K;B;\mathrm{d}(y,b))\, \mathrm{d}t. \tag{14.27}$$

In deriving (14.27) from (14.26), one has to observe that for (y,b) in the support of $\Xi_m(K;B;\cdot)$ one has $d_B(y+tb,K) = t$, $p_B(K,y+tb) = y$ and $u_B(K,y+tb) = b$.

The intrinsic volumes and also the mixed volumes have vector-valued counterparts. Of these, we need only the **Steiner point** s. For $K \in \mathcal{K}'$, it is defined by

$$s(K) := \frac{1}{\kappa_d} \int_{S^{d-1}} h(K,u)u\,\sigma(\mathrm{d}u). \tag{14.28}$$

Because of $h(K+t,u) = h(K,u) + \langle t,u \rangle$, the Steiner point has the property

$$s(K+t) = s(K) + t, \qquad t \in \mathbb{R}^d.$$

The Steiner point is the centroid of the Gaussian curvature measure, thus

$$s(K) = \int_{\mathbb{R}^d} x\,\Phi_0(K,\mathrm{d}x)$$

for $K \in \mathcal{K}'$. For a polytope P, this reads

$$s(K) = \sum_{e \in \mathrm{vert}\,(P)} \gamma(\{e\},P)e, \tag{14.29}$$

where vert (P) is the set of vertices of P.

The mixed volumes satisfy a set of inequalities, with many applications. For arbitrary convex bodies $K, M, K_3, \dots, K_d$, the **Aleksandrov–Fenchel inequality** states that

$$V(K,M,K_3,\dots,K_d)^2 \geq V(K,K,K_3,\dots,K_d)V(M,M,K_3,\dots,K_d).$$

Further inequalities can be deduced from this, such as **Minkowski's inequality**

$$V(M,K,\dots,K)^d \geq V_d(M)V_d(K)^{d-1} \tag{14.30}$$

and the inequalities

$$\left(\frac{\kappa_{d-j}}{\binom{d}{j}} V_j(K)\right)^k \geq \kappa_d^{k-j} \left(\frac{\kappa_{d-k}}{\binom{d}{k}} V_k(K)\right)^j \tag{14.31}$$

for $0 < j < k \leq d$ and, as a special case of the **general Brunn–Minkowski theorem**,

$$V_j(K+M)^{1/j} \geq V_j(K)^{1/j} + V_j(M)^{1/j} \tag{14.32}$$

for $K, M \in \mathcal{K}'$ and $j = 1, \dots, d$. Equality in (14.30) for $V_d(K) > 0$ and $\dim M > 0$ holds if and only if K and M are homothetic. Equality in (14.31) for $V_j(K) > 0$ characterizes balls. For $j = d$ and d-dimensional bodies K, M, equality in (14.32) holds if and only if K and M are homothetic.

In this book, the preceding inequalities are applied to special convex bodies, the zonoids. The convex body $Z \in \mathcal{K}'$ is called a (centered) **zonoid** if its support function has a representation of the form

$$h(Z,u) = \int_{S^{d-1}} |\langle u,v\rangle|\,\rho(\mathrm{d}v), \qquad u \in \mathbb{R}^d, \tag{14.33}$$

with an even finite measure ρ on the sphere S^{d-1}. Here the measure ρ is called **even** if $\rho(A) = \rho(-A)$ for all $A \in \mathcal{B}(S^{d-1})$. Every translate of a centered zonoid is called a zonoid. If the measure ρ in (14.33) has finite support, then Z is a Minkowski sum of finitely many segments. Such a convex body is a special polytope, called a **zonotope**. Zonoids are precisely the convex bodies that can be approximated by zonotopes. When (14.33) holds and ρ is even, then ρ is called the **generating measure** of Z. It is uniquely determined, due to the following theorem.

Theorem 14.3.4. *If ρ is an even finite signed measure on S^{d-1} with*

$$\int_{S^{d-1}} |\langle u,v\rangle|\,\rho(\mathrm{d}v) = 0$$

for all $v \in S^{d-1}$, then $\rho = 0$.

For mixed volumes of zonoids there are special integral representations in terms of the generating measures. Let Z_i be a zonoid with generating measure ρ_i $(i = 1,\dots,d)$. For vectors $u_1,\dots,u_j$, let $\nabla_j(u_1,\dots,u_j)$ denote the j-dimensional volume of the parallelepiped spanned by these vectors. Then the mixed volume of $Z_1,\dots,Z_d$ is given by

$$V(Z_1,\dots,Z_d) = \frac{2^d}{d!}\int_{S^{d-1}}\cdots\int_{S^{d-1}} \nabla_d(u_1,\dots,u_d)\,\rho_1(\mathrm{d}u_1)\cdots\rho_d(\mathrm{d}u_d).$$

If some of the zonoids are equal to the unit ball, this formula simplifies. For $j \in \{1,\dots,d\}$, we have

$$\begin{aligned} &V(Z_1,\dots,Z_j,B^d,\dots,B^d) \\ &= \frac{2^j(d-j)!\kappa_{d-j}}{d!}\int_{S^{d-1}}\cdots\int_{S^{d-1}} \nabla_j(u_1,\dots,u_j)\,\rho_1(\mathrm{d}u_1)\cdots\rho_j(\mathrm{d}u_j). \end{aligned} \tag{14.34}$$

(The proof is the obvious modification of the one given in [695, p. 300] for the subsequent formula (14.35).) In particular, the intrinsic volumes of the zonoid Z with generating measure ρ have the representation

$$V_j(Z) = \frac{2^j}{j!}\int_{S^{d-1}}\cdots\int_{S^{d-1}} \nabla_j(u_1,\dots,u_j)\,\rho(\mathrm{d}u_1)\cdots\rho(\mathrm{d}u_j) \tag{14.35}$$

for $j = 1,\dots,d-1$.

It is convenient here to associate with the generating measure ρ a measure $\rho_{(j)}$ on the Grassmannian $G(d,j)$ of j-dimensional linear subspaces. For this, let LI_j be the set of linearly independent j-tuples of unit vectors in S^{d-1}, and define the map $\Lambda_j : LI_j \to G(d,j)$ by

$$\Lambda_j(u_1,\dots,u_j) := \mathrm{lin}\{u_1,\dots,u_j\}.$$

We define $\rho_{(j)}$ as the image measure under Λ_j, of the measure appearing in (14.35), thus

$$\rho_{(j)} := \Lambda_j\left(\frac{2^j}{j!\kappa_j}\int_{(\cdot)} \nabla_j(u_1,\dots,u_j)\,\rho^j(\mathrm{d}(u_1,\dots,u_j))\right). \tag{14.36}$$

Since the function $(u_1,\dots,u_j) \mapsto \nabla_j(u_1,\dots,u_j)$ vanishes on $(S^{d-1})^j \setminus LI_j$, we can write

$$V_j(Z) = \kappa_j \rho_{(j)}(G(d,j)). \tag{14.37}$$

The measure $\rho_{(j)}$ on $G(d,j)$ is called the jth **projection generating measure** of Z, since

$$V_j(Z|E) = \kappa_j \int_{G(d,j)} |\langle E,L\rangle|\,\rho_{(j)}(\mathrm{d}L) \quad \text{for } E \in G(d,j). \tag{14.38}$$

Recall that $Z|E$ denotes the image of Z under the orthogonal projection to E, and $|\langle E,L\rangle| = [E,L^\perp]$ is the absolute j-dimensional determinant of the orthogonal projection from E to L (or L to E). More general versions of equation (14.38) are found in Weil [781].

For the preceding results, it is not necessary that ρ is a positive measure. If the support function of a convex body Z with center 0 has a representation (14.33) with a finite signed measure ρ, then Z is called a **generalized zonoid**. If the signed measure $\rho_{(j)}$ is defined by (14.36), then (14.38) still holds (the proof given by Weil [781] works also for signed measures).

For the spherical Lebesgue measure σ, the measure $\sigma_{(j)}$ is rotation invariant and hence a multiple of the invariant measure ν_j on $G(d,j)$. In fact, with the notation (5.5),

$$\sigma_{(j)} = (2^j\kappa_{d-1})^j c^{d,0}_{j,d-j}\,\nu_j. \tag{14.39}$$

Zonoids with nonempty interior can be interpreted as projection bodies. The **projection body** of the convex body $K \in \mathcal{K}'$ is defined as the convex body Π_K with support function given by

$$h(\Pi_K,u) = V_{d-1}(K|u^\perp), \qquad u \in S^{d-1}. \tag{14.40}$$

Explicitly,

$$V_{d-1}(K|u^\perp) = \frac{1}{2}\int_{S^{d-1}} |\langle u,v\rangle|\,S_{d-1}(K,\mathrm{d}v), \qquad u \in S^{d-1}, \tag{14.41}$$

so that the projection body Π_K of K is the zonoid having the even part of $S_{d-1}(K, \cdot)/2$ as its generating measure. It follows from Theorem 14.3.1 that every d-dimensional centered zonoid is the projection body of a unique convex body with 0 as center of symmetry.

Besides the inequalities from the theory of mixed volumes, we employ volume estimates involving polar bodies. Let $K \in \mathcal{K}'$ be a convex body having 0 as an interior point. The **polar body** K^o of K is defined by

$$K^o := \{x \in \mathbb{R}^d : \langle x, y \rangle \leq 1 \text{ for all } y \in K\}.$$

It is again a convex body with 0 as an interior point. We have

$$\rho(K^o, u) = \frac{1}{h(K, u)}, \qquad u \in S^{d-1}, \tag{14.42}$$

where ρ is the **radius function.** It is defined by

$$\rho(M, u) := \sup\{\lambda \geq 0 : \lambda u \in M\} \qquad \text{for } u \in S^{d-1};$$

here M can be any set star-shaped with respect to 0, that is, satisfying $[0, x] \subset M$ for all $x \in M$.

According to (14.42), the volume of the polar body K^o of the convex body K can be expressed by

$$V_d(K^o) = \frac{1}{d} \int_{S^{d-1}} h(K, u)^{-d} \, \sigma(\mathrm{d}u).$$

Applying Jensen's inequality and observing (14.7), we obtain the inequality

$$V_d(K^o) \geq \kappa_d \left(\frac{\kappa_{d-1}}{d\kappa_d} V_1(K) \right)^{-d}. \tag{14.43}$$

Here equality holds if and only if K is a ball with center 0.

The polar projection body (that is, the polar body of the projection body) of a convex body $K \in \mathcal{K}'$ with nonempty interior satisfies **Petty's projection inequality**

$$V_d((\Pi_K)^o) V_d(K)^{d-1} \leq \left(\frac{\kappa_d}{\kappa_{d-1}} \right)^d, \tag{14.44}$$

with equality if and only if K is an ellipsoid.

If Z is a d-dimensional zonoid with center 0, then

$$\frac{4^d}{d!} \leq V_d(Z) V_d(Z^o) \leq \kappa_d^2. \tag{14.45}$$

In the right inequality, which is known as the **Blaschke–Santaló inequality** and holds for general convex bodies with 0 as center of gravity, the equality sign holds if and only if Z is an ellipsoid. In the left inequality, equality holds if and only if Z is a parallelepiped (for a short proof, see Gordon, Meyer and Reisner [282]).

Notes for Section 14.3

1. All results stated here without references, with the exception of those to which Note 2 refers, can be found in Schneider [695].

2. The relative support measures were introduced independently by Kiderlen and Weil [409] and (for strictly convex gauge bodies) by Hug and Last [357]. There one also finds versions of Theorems 14.3.2 and 14.3.3. In the form we presented them here, both theorems follow from more general results proved in Hug, Last and Weil [358]. It should be observed that the papers on non-symmetric distances and relative support measures do not always use the same terminology. In the notation of Kiderlen and Weil [409], $d_B(x, K) = d(K, -B, x)$ and

$$\Xi_j(K; B; \cdot) = \frac{1}{d\kappa_{d-j}} \binom{d}{j} \Theta_{j;d-j}(K; -B; \cdot).$$

Our distance $d_B(x, K)$ is denoted by $d_B(K, x)$ in [357], by $d_{-B}(K, x)$ in Hug [356], and by $d_B(\{x\}, K)$ in [358]. Hug and Last [357] denote our $\Xi_j(K; B; \cdot)$ by $C_j^B(K, \cdot)$.

14.4 Additive Functionals

A function φ on an intersectional family $\mathcal{V}$ of sets with values in an abelian group is called **additive** or a **valuation** if it satisfies

$$\varphi(K \cup L) + \varphi(K \cap L) = \varphi(K) + \varphi(L), \tag{14.46}$$

whenever $K, L, K \cup L \in \mathcal{V}$. If φ is an additive function on $\mathcal{V}$, we may assume, without loss of generality, that $\emptyset \in \mathcal{V}$ and $\varphi(\emptyset) = 0$. Therefore, if an additive function on an intersectional family $\mathcal{V}$ with $\emptyset \notin \mathcal{V}$ is given, we always extend the definition to $\mathcal{V} \cup \{\emptyset\}$, putting $\varphi(\emptyset) := 0$. The extension is then additive on $\mathcal{V} \cup \{\emptyset\}$.

The function φ on $\mathcal{V}$ is called **generally additive** if it satisfies the general version of (14.46), namely

$$\varphi(K_1 \cup \ldots \cup K_m) = \sum_{r=1}^{m} (-1)^{r-1} \sum_{1 \le i_1 < \ldots < i_r \le m} \varphi(K_{i_1} \cap \ldots \cap K_{i_r}), \tag{14.47}$$

whenever $K_1, \ldots, K_m \in \mathcal{V}$ and $K_1 \cup \ldots \cup K_m \in \mathcal{V}$. If φ is generally additive, it is also said to satisfy the **inclusion–exclusion principle**, (14.47), on $\mathcal{V}$.

We often write (14.47) in the more concise form

$$\varphi(K_1 \cup \ldots \cup K_m) = \sum_{v \in S(m)} (-1)^{|v|-1} \varphi(K_v), \tag{14.48}$$

where we use the following notation: $S(m)$ is the set of nonempty subsets of $\{1,\dots,m\}$, $|v|$ is the number of elements in $v \in S(m)$, and for $v \in \{i_1,\dots,i_k\}$ we write

$$K_v := K_{i_1} \cap \ldots \cap K_{i_k}.$$

It is a natural question to ask whether a valuation φ on $\mathcal{V}$ can be extended to an additive function on the lattice $U(\mathcal{V})$ consisting of all finite unions of elements of $\mathcal{V}$. If such an extension (also denoted by φ) to $U(\mathcal{V})$ exists, it satisfies the inclusion–exclusion principle on $U(\mathcal{V})$, as follows from (14.46) by induction. This implies, in particular, that an additive extension is unique. However, one cannot simply use (14.47) to define an extension of φ from $\mathcal{V}$ to $U(\mathcal{V})$, since the representation of an element $K \in U(\mathcal{V})$ in the form $K = K_1 \cup \ldots \cup K_m$ with $K_1,\dots,K_m \in \mathcal{V}$ is in general not unique. It is one of the main purposes of this section to prove some extension theorems for additive functions. The next two theorems are essentially due to H. Groemer.

If $\mathcal{V}$ is an intersectional family, we write $E(\mathcal{V})$ for the union of the sets in $\mathcal{V}$, and for $K \subset E(\mathcal{V})$ we denote (only in this section) by K^* the indicator function of K on $E(\mathcal{V})$, that is,

$$K^*(x) := \begin{cases} 1 \text{ if } x \in K, \\ 0 \text{ if } x \in E(\mathcal{V}) \setminus K. \end{cases}$$

The indicator function satisfies

$$(K_1 \cup \ldots \cup K_m)^* = \sum_{r=1}^{m} (-1)^{r-1} \sum_{1 \le i_1 < \ldots < i_r \le m} (K_{i_1} \cap \ldots \cap K_{i_r})^* \qquad (14.49)$$

for arbitrary sets $K_1,\dots,K_m \subset E(\mathcal{V})$, as follows immediately from the binomial theorem. In particular, the map $K \mapsto K^*$ is generally additive on $\mathcal{V}$.

We denote by $M(\mathcal{V})$ the $\mathbb{Z}$-module spanned by the indicator functions of the elements of $\mathcal{V}$, and $\overline{U}(\mathcal{V})$ is defined as the family of all subsets of $E(\mathcal{V})$ whose indicator function belongs to $M(\mathcal{V})$. Then $\overline{U}(\mathcal{V})$ is intersectional and closed under finite unions. Clearly $U(\mathcal{V}) \subset \overline{U}(\mathcal{V})$, as follows from (14.49).

Theorem 14.4.1 (Groemer). *Let φ be a function from an intersectional class $\mathcal{V}$ (with $\emptyset \in \mathcal{V}$) into an abelian group, satisfying $\varphi(\emptyset) = 0$. The following statements are equivalent:*

(a) *φ has an additive extension to $\overline{U}(\mathcal{V})$.*
(b) *φ is generally additive on $\mathcal{V}$.*
(c) *$a_1K_1^* + \ldots + a_mK_m^* = 0$ with $K_i \in \mathcal{V}$ and $a_i \in \mathbb{Z}$ $(i = 1,\dots,m)$ implies $a_1\varphi(K_1) + \ldots + a_m\varphi(K_m) = 0$.*

Proof. The implication (a) $\Rightarrow$ (b) is clear from the previous remarks.

To prove the implication (b) $\Rightarrow$ (c), we assume that (b) holds, but that (c) is false. Then there are sets $K_1,\dots,K_m \in \mathcal{V}$ $(m \ge 1)$ and integers $a_1,\dots,a_m$ so that

$$a_1K_1^* + \ldots + a_mK_m^* = 0, \tag{14.50}$$

but

$$a_1\varphi(K_1) + \ldots + a_m\varphi(K_m) \neq 0. \tag{14.51}$$

Let $D_i := K_i$ for $i = 1, \ldots, m$, let $D_{m+1}, \ldots, D_{m_1}$ be the intersections $K_i \cap K_j$ with $i < j$ (in some order), let $D_{m_1+1}, \ldots, D_{m_2}$ be the intersections $K_i \cap K_j \cap K_k$ with $i < j < k$, and so on, until $D_p = K_1 \cap \ldots \cap K_m$. There exist integers c_i so that

$$\sum_{i=r}^{p} c_i D_i^* = 0, \qquad c_r \neq 0, \tag{14.52}$$

$$\sum_{i=r}^{p} c_i \varphi(D_i) \neq 0, \tag{14.53}$$

for example by (14.50), (14.51). We choose the coefficients in such a way that r is maximal. Then $r < p$, since otherwise $c_pD_p^* = 0$ with $c_p \neq 0$, hence $D_p = \emptyset$, but $c_p\varphi(D_p) \neq 0$, a contradiction. By (14.52), any point of D_r is contained in some D_i with $i > r$, hence

$$D_r = (D_r \cap D_{r+1}) \cup (D_r \cap D_{r+2}) \cup \ldots \cup (D_r \cap D_p).$$

Since φ is generally additive, this gives

$$\varphi(D_r) = \sum_{r<i\leq p} \varphi(D_r \cap D_i) - \sum_{r<i<j\leq p} \varphi(D_r \cap D_i \cap D_j) + - \ldots$$

By (14.49),

$$D_r^* = \sum_{r<i\leq p} (D_r \cap D_i)^* - \sum_{r<i<j\leq p} (D_r \cap D_i \cap D_j)^* + - \ldots$$

Each intersection appearing here is some set D_s with $s > r$. Hence, there are integer coefficients d_s such that

$$D_r^* = \sum_{s=r+1}^{p} d_s D_s^*$$

and

$$\varphi(D_r) = \sum_{s=r+1}^{p} d_s \varphi(D_s).$$

If these representations are inserted in (14.52) and (14.53), we obtain expressions of the same type as (14.52), (14.53), again with integer coefficients, but with r replaced by a larger number. This contradicts the choice of r and thus proves (c).

Suppose that (c) holds. For given $f \in M(\mathcal{V})$ we choose a representation

$$f = \sum_{i=1}^{m} a_i K_i^*$$

with $m \in \mathbb{N}$, $K_i \in \mathcal{V}$ and $a_i \in \mathbb{Z}$, and we define

$$\widetilde{\varphi}(f) := \sum_{i=1}^{m} a_i \varphi(K_i).$$

Since (c) holds, this definition is possible and does not depend on the special representation of f. In this way, we have defined a map $\widetilde{\varphi}$ on $M(\mathcal{V})$ satisfying $\widetilde{\varphi}(K^*) = \varphi(K)$ for $K \in \mathcal{V}$ and $\widetilde{\varphi}(f+g) = \widetilde{\varphi}(f) + \widetilde{\varphi}(g)$ for $f, g \in M(\mathcal{V})$. Now we extend φ from $\mathcal{V}$ to $\overline{U}(\mathcal{V})$ by setting

$$\varphi(K) := \widetilde{\varphi}(K^*) \qquad \text{for } K \in \overline{U}(\mathcal{V}).$$

For $K, L \in \overline{U}(\mathcal{V})$ we then have

$$\begin{aligned}
&\varphi(K \cup L) + \varphi(K \cap L) = \widetilde{\varphi}((K \cup L)^*) + \widetilde{\varphi}((K \cap L)^*) \\
&= \widetilde{\varphi}((K \cup L)^* + (K \cap L)^*) = \widetilde{\varphi}(K^* + L^*) \\
&= \widetilde{\varphi}(K^*) + \widetilde{\varphi}(L^*) = \varphi(K) + \varphi(L).
\end{aligned}$$

Thus, the extension is additive on $\overline{U}(\mathcal{V})$. We have proved that (a) holds. □

From Theorem 14.4.1 we shall draw two important conclusions. Recall that the convex ring $\mathcal{R}$ is the system $U(\mathcal{K})$ of finite unions of convex bodies in $\mathbb{R}^d$.

Theorem 14.4.2 (Groemer's extension theorem). *Every continuous valuation $\varphi : \mathcal{K} \to X$ with values in a topological vector space X has an additive extension to the convex ring $\mathcal{R}$.*

Proof. By Theorem 14.4.1 it suffices to show that

$$\sum_{i=1}^{m} a_i K_i^* = 0 \tag{14.54}$$

with $m \in \mathbb{N}$, $a_i \in \mathbb{Z}$, $K_i \in \mathcal{K}$ implies

$$\sum_{i=1}^{m} a_i \varphi(K_i) = 0.$$

Assume that this were false. Then there exists a minimal counterexample, that is, a smallest number $m \geq 2$ for which there are integers $a_1, \dots, a_m$ and convex bodies $K_1, \dots, K_m \in \mathcal{K}'$ such that (14.54) holds together with

$$\sum_{i=1}^{m} a_i \varphi(K_i) =: c \neq 0. \tag{14.55}$$

Let $H \subset \mathbb{R}^d$ be a hyperplane with $K_1 \subset \operatorname{int} H^+$, where H^+, H^- are the two closed halfspaces bounded by H. Since (14.54) holds at each point, we have

$$\sum_{i=1}^m a_i (K_i \cap H^-)^* = 0, \quad \sum_{i=1}^m a_i (K_i \cap H)^* = 0.$$

By the choice of H, each of these sums has at most $m-1$ non-zero summands. Since m was minimal and $\varphi(\emptyset) = 0$, we conclude that

$$\sum_{i=1}^m a_i \varphi(K_i \cap H^-) = 0, \quad \sum_{i=1}^m a_i \varphi(K_i \cap H) = 0.$$

Since $K_i = (K_i \cap H^+) \cup (K_i \cap H^-)$ and $(K_i \cap H^+) \cap (K_i \cap H^-) = K_i \cap H$, the additivity of φ on $\mathcal{K}$ yields

$$\sum_{i=1}^m a_i \varphi(K_i \cap H^+) = c. \tag{14.56}$$

We can choose a sequence $(H_j)_{j \in \mathbb{N}}$ of hyperplanes with $K_1 \subset \operatorname{int} H_j^+$ for $j \in \mathbb{N}$ and

$$K_1 = \bigcap_{j=1}^\infty H_j^+.$$

Repetition of the argument leading from (14.54) and (14.55) to (14.56) shows that

$$\sum_{i=1}^m a_i \varphi\left(K_i \cap \bigcap_{j=1}^k H_j^+ \right) = c$$

for $k = 1, 2, \dots$ We have

$$\lim_{k \to \infty} K_i \cap \bigcap_{j=1}^k H_j^+ = K_i \cap K_1$$

if $K_i \cap K_1 \neq \emptyset$, and if $K_i \cap K_1 = \emptyset$, then

$$K_i \cap \bigcap_{j=1}^k H_j^+ = \emptyset$$

for all sufficiently large k. From the continuity of φ and from $\varphi(\emptyset) = 0$ we conclude that

$$\sum_{i=1}^m a_i \varphi(K_i \cap K_1) = c. \tag{14.57}$$

By (14.54) we still have

$$\sum_{i=1}^{m} a_i (K_i \cap K_1)^* = 0. \tag{14.58}$$

We repeat the procedure leading from (14.54) and (14.55) to (14.58) and (14.57), replacing the bodies K_i and K_1 first by $K_i \cap K_1$ and K_2, then by $K_i \cap K_1 \cap K_2$ and K_3, and so on. Finally, this results in the relations

$$\sum_{i=1}^{m} a_i (K_1 \cap \ldots \cap K_m)^* = 0$$

and

$$\sum_{i=1}^{m} a_i \varphi(K_1 \cap \ldots \cap K_m) = c.$$

Now $c \neq 0$ implies $\sum_{i=1}^{m} a_i \neq 0$, hence $(K_1 \cap \ldots \cap K_m)^* = 0$ and thus $\varphi(K_1 \cap \ldots \cap K_m) = 0$, a contradiction. This completes the proof. □

A second extension theorem concerns valuations on polytopes. In this case, no continuity assumption is required.

We denote by $\mathcal{P}$ the set of (compact, convex) polytopes in $\mathbb{R}^d$ (including $\emptyset$) and by $\mathcal{P}_{ro}$ the set of relatively open polytopes, **ro-polytopes** for short. By definition, every $Q \in \mathcal{P}_{ro}$ is the relative interior of a convex polytope. The system $U(\mathcal{P}_{ro})$ consists of all finite unions of ro-polytopes. The elements of $U(\mathcal{P}_{ro})$ are called **ro-polyhedra**. Every polytope $P \in \mathcal{P}$ is a ro-polyhedron; in fact, P is the disjoint union of the relative interiors of its faces. We write $\mathcal{P}'$ for the set of nonempty polytopes and $\mathcal{P}'_{ro}$ for the set of nonempty ro-polytopes.

Theorem 14.4.3. *Every valuation on $\mathcal{P}$ has an additive extension to a valuation on the class $U(\mathcal{P}_{ro})$ of ro-polyhedra.*

Proof. Let φ be a valuation on $\mathcal{P}$. We shall show that it is generally additive on $\mathcal{P}$. If that is proved, Theorem 14.4.1 shows that φ has an additive extension to $\overline{U}(\mathcal{P})$. It follows easily by induction that the indicator function of a ro-polytope is a linear combination, with integer coefficients, of indicator functions of polytopes, thus $\mathcal{P}_{ro} \subset \overline{U}(\mathcal{P})$. Since every ro-polyhedron can be represented as a finite disjoint union of ro-polytopes, it follows that $U(\mathcal{P}_{ro}) \subset \overline{U}(\mathcal{P})$. Therefore, the additive extension of φ to $\overline{U}(\mathcal{P})$ induces an additive extension to $U(\mathcal{P}_{ro})$.

The proof that φ is generally additive on $\mathcal{P}$ follows Volland [771]. If $\tau = (P_1, \ldots, P_m)$ is a tuple of polytopes with $P_1 \cup \ldots \cup P_m = P \in \mathcal{P}'$, we write

$$\varphi(P, \tau) := \sum_{r=1}^{m} (-1)^{r-1} \sum_{1 \le i_1 < \ldots < i_r \le m} \varphi(P_{i_1} \cap \ldots \cap P_{i_r}).$$

We have to show that

$$\varphi(P) = \varphi(P, \tau). \tag{14.59}$$

The proof is by induction with respect to $\dim P$. For $\dim P = 0$, there is nothing to prove. Suppose that $P = P_1 \cup \ldots \cup P_m$ with $P, P_i \in \mathcal{P}'$ and $\dim P \geq 1$ and that the assertion is true for polytopes of dimension less than $\dim P$. Without loss of generality, we can assume $\dim P = d$. We prove (14.59) by induction with respect to the number m. For $m = 1$, there is nothing to prove. Suppose that $m > 1$ and that the assertion is true for all representations of d-dimensional polytopes as unions of less than m polytopes. First we consider two special cases.

Case 1. At least one polytope of τ, say P_m, is of dimension less than d.

Case 2. At least one polytope of τ, say P_1, is equal to P.

Putting $Q := \bigcup_{i=1}^{m-1} P_i$, in either case we have $P = Q \cup P_m$ and $Q \in \mathcal{P}'$. The additivity of φ gives

$$\varphi(P) = \varphi(P_m) + \varphi(Q) - \varphi(Q \cap P_m).$$

Inserting $Q = \bigcup_{i=1}^{m-1} P_i$ and $Q \cap P_m = \bigcup_{i=1}^{m-1} (P_i \cap P_m)$ and using the (second) induction hypothesis, we immediately obtain (14.59).

Now we assume that none of the two cases occurs. Then $\dim P_1 = d$, and there is a hyperplane H such that $P_1 \cap H$ is a facet of P_1 and $H \cap \operatorname{int} P \neq \emptyset$. Let H^+, H^- be the two closed halfspaces bounded by H, where $P_1 \subset H^+$. By additivity,

$$\varphi(P) = \varphi(P \cap H^+) + \varphi\left(\bigcup_{i=1}^{m} (P_i \cap H^-)\right) - \varphi\left(\bigcup_{i=1}^{m} (P_i \cap H)\right).$$

Relation (14.59) can be applied to the second term on the right, since $\dim (P_1 \cap H^-) < d$ (Case 1). It can also be applied to the third term, since it holds in lower dimensions, by the first induction hypothesis. Carrying this out, using that

$$\varphi(Q \cap H^-) - \varphi(Q \cap H) = \varphi(Q) - \varphi(Q \cap H^+)$$

holds for any $Q \in \mathcal{P}'$, and rearranging the terms, we obtain the relation

$$\varphi(P) - \varphi(P, \tau) = \varphi(P \cap H^+) - \varphi(P \cap H^+, \tau^+), \tag{14.60}$$

where $\tau^+ := (P_1 \cap H^+, \ldots, P_m \cap H^+)$. By construction, $P_1 \subset P \cap H^+$. If $P_1 \neq P \cap H^+$, we can continue the procedure with a different facet of P_1. After finitely many steps, the remaining polytope coincides with P_1. This situation is covered by Case 2, hence in the final relation corresponding to (14.60), the right side vanishes. This completes the proof. □

The additive extension of a continuous valuation on $\mathcal{K}$ to the convex ring $\mathcal{R}$, which exists by Groemer's extension theorem, will in general no longer be continuous (with respect to the topology induced by the Hausdorff metric on $\mathcal{C}$). This is already shown by the example of the volume, together with the fact that a convex body of positive volume can be approximated, in the Hausdorff metric, by finite sets, which belong to the convex ring and have volume zero. More importantly, measurability does extend. This is shown by the following theorem.

Theorem 14.4.4. *Let $\varphi : \mathcal{R} \to \mathbb{R}$ be an additive functional. If the restriction of φ to $\mathcal{K}$ is measurable, then φ is measurable.*

Proof. Every element $K \in \mathcal{R}$ can be represented in the form $K = \bigcup_{i=1}^m K_i$ with $m \in \mathbb{N}$ and $K_i \in \mathcal{K}$ for $i = 1, \ldots, m$. For $k \in \mathbb{N}$ we denote by $\mathcal{R}_k$ the set of all elements $K \in \mathcal{R}'$ having such a representation with $m \leq k$ (and hence also with $m = k$, admitting $K_i = K_j$ for $i \neq j$, if necessary). We show first that $\mathcal{R}_k$ is closed. Let $K \in \mathcal{C}$ be in the closure of $\mathcal{R}_k$. Then there is a sequence $(K_j)_{j\in\mathbb{N}}$ in $\mathcal{R}_k$ that converges to K. For each j, there is a representation $K_j = \bigcup_{i=1}^k K_{ji}$ with suitable convex bodies $K_{ji} \in \mathcal{K}'$. Since $K_j \to K$ for $j \to \infty$, the bodies K_{ji} are uniformly bounded. By the Blaschke selection theorem (mentioned before Theorem 12.3.4) there are a subsequence $(j_r)_{r\in\mathbb{N}}$ of $\mathbb{N}$ and convex bodies $\overline{K}_i \in \mathcal{K}$ such that $K_{j_r i} \to \overline{K}_i$ for $r \to \infty$ $(i = 1, \ldots, k)$. Using Theorem 12.3.5, we infer that

$$K_{j_r} = \bigcup_{i=1}^k K_{j_r i} \to \bigcup_{i=1}^k \overline{K}_i$$

and hence $K = \bigcup_{i=1}^k \overline{K}_i \in \mathcal{R}_k$. Thus $\mathcal{R}_k$ is closed; in particular, it is a Borel set.

Now $\mathcal{R} = \{\emptyset\} \cup \bigcup_{k\in\mathbb{N}} \mathcal{R}_k$, hence it suffices to prove, for given $k \in \mathbb{N}$, the measurability of the restriction of φ to $\mathcal{R}_k$. Define $\gamma_k : \mathcal{K}^k \to \mathcal{R}_k$ by

$$\gamma_k(K_1, \ldots, K_k) := \bigcup_{i=1}^k K_i \qquad \text{for } (K_1, \ldots, K_k) \in \mathcal{K}^k.$$

The mapping γ_k is continuous and hence measurable. Let

$$\Gamma_k(K) := \gamma_k^{-1}(K) \qquad \text{for } K \in \mathcal{R}_k.$$

Then $\Gamma_k(K)$ is a compact subset of $\mathcal{K}^k$. Let $\mathcal{C}(\mathcal{K}^k)$ be the space of nonempty compact subsets of $\mathcal{K}^k$, equipped with the metric induced from the Hausdorff metric and the corresponding σ-algebra of Borel sets. We assert that the map $\Gamma_k : \mathcal{R}_k \to \mathcal{C}(\mathcal{K}^k)$ just defined is measurable. For the proof, let $\mathcal{A} \subset \mathcal{K}^k$ be a closed set. For $m \in \mathbb{N}$ we put

$$\mathcal{R}_{km} := \{K \in \mathcal{R}_k : K \subset mB^d\},$$
$$\mathcal{A}_m := \{(K_1, \ldots, K_k) \in \mathcal{A} : K_1, \ldots, K_k \subset mB^d\};$$

then

$$\{K \in \mathcal{R}_{km} : \Gamma_k(K) \cap \mathcal{A} \neq \emptyset\} = \mathcal{R}_k \cap \gamma_k(\mathcal{A}_m).$$

Since $\mathcal{A}_m$ is compact and γ_k is continuous, the intersection $\mathcal{R}_k \cap \gamma_k(\mathcal{A}_m)$ is closed; therefore the set

$$\{K \in \mathcal{R}_k : \Gamma_k(K) \cap \mathcal{A} \neq \emptyset\} = \bigcup_{m \in \mathbb{N}} \{K \in \mathcal{R}_{km} : \Gamma_k(K) \cap \mathcal{A} \neq \emptyset\}$$

is measurable. Since this holds for all closed sets $\mathcal{A} \subset \mathcal{K}^k$, the mapping Γ_k is measurable (see Castaing and Valadier [165, Theorem III. 2]).

From [165, Theorem III. 6] we now deduce the existence of a measurable mapping $\xi_k : \mathcal{R}_k \to \mathcal{K}^k$ with $\xi_k(K) \in \Gamma_k(K)$, hence $\gamma_k(\xi_k(K)) = K$, for all $K \in \mathcal{R}_k$.

Concerning the additive functional φ, we know from (14.48) that

$$\varphi(K) = \sum_{v \in S(k)} (-1)^{|v|-1} \varphi(K_v) \tag{14.61}$$

whenever $K = \bigcup_{i=1}^k K_i$ and $K_i \in \mathcal{K}$. For each $v = \{i_1, \dots, i_j\} \subset \{1, \dots, k\}$, the map $f_v : \mathcal{K}^k \to \mathcal{K}$ defined by

$$f_v(K_1, \dots, K_k) := K_v = K_{i_1} \cap \dots \cap K_{i_j}$$

is measurable, as follows from Theorems 12.2.6 and 2.4.1. For $K \in \mathcal{R}_k$, (14.61) yields

$$\varphi(K) = \sum_{v \in S(k)} (-1)^{|v|-1} \varphi(f_v(\xi_k(K))).$$

Since φ is measurable on $\mathcal{K}$, the measurability of φ on $\mathcal{R}_k$ is proved. □

As a first application of Groemer's extension theorem, we can extend the trivial valuation χ on $\mathcal{K}$, defined by $\chi(K) = 1$ for $K \in \mathcal{K}'$ and $\chi(\emptyset) = 0$, to a valuation, also denoted by χ, on the convex ring $\mathcal{R}$. This valuation is the **Euler characteristic**, known from topology, but obtained here, for polyconvex sets, in an elementary way. The extension theorem 14.4.3 shows that the Euler characteristic has also an additive extension to the class of ro-polyhedra. For both extension results, there are shorter proofs. For the second one, we shall give such a direct proof, which is independent of Theorems 14.4.1 and 14.4.3.

Theorem 14.4.5. *There is a unique additive function χ on $U(\mathcal{P}_{ro})$ with*

$$\chi(P) = 1 \qquad \text{for } P \in \mathcal{P}'.$$

It satisfies

$$\chi(Q) = (-1)^{\dim Q} \qquad \text{for } Q \in \mathcal{P}'_{ro}.$$

Proof. The uniqueness is clear. We prove the existence by induction with respect to the dimension. In the zero-dimensional space $\{0\}$, the set $\{0\}$ is the only nonempty polytope and ro-polytope, hence there is nothing to prove. Let $d \geq 1$ and assume that the existence of χ has been proved for real affine spaces of dimension $d-1$. We choose $u \in \mathbb{R}^d \setminus \{0\}$ and use the hyperplanes $H_\lambda := \{x \in \mathbb{R}^d : \langle x, u \rangle = \lambda\}$, $\lambda \in \mathbb{R}$. If P is a polytope (a ro-polytope), then $P \cap H_\lambda$ is either empty or a polytope (a ro-polytope). For a ro-polyhedron $Q \in U(\mathcal{P}_{ro})$ we define

$$\chi(Q) := \sum_{\lambda \in \mathbb{R}} \left[\chi(Q \cap H_\lambda) - \lim_{u \downarrow \lambda} \chi(Q \cap H_\mu) \right], \tag{14.62}$$

where χ on the right side is the function which exists by the induction hypothesis. For $Q \neq \emptyset$, there is a representation $Q = Q_1 \cup \ldots \cup Q_k$ with ro-polytopes $Q_1, \ldots, Q_k$; these generate only finitely many intersections. Hence, there are numbers $\lambda_1 < \lambda_2 < \ldots < \lambda_m$ such that each function $\lambda \mapsto \chi(Q_i \cap H_\lambda)$ is constant on each open interval $(\lambda_r, \lambda_{r+1})$, as follows from the induction hypothesis and the inclusion–exclusion principle. Together with the induction hypothesis, this shows that the limits in (14.62) exist and that the sum is finite, so that the definition is possible. The induction hypothesis also implies that χ is additive on $U(\mathcal{P}_{ro})$. If $Q \in \mathcal{P}'$ is a polytope, the right side of (14.62) gives $(1-1)+(1-0) = 1$ if Q is not contained in some H_λ, and $1 - 0 = 1$ otherwise. If $Q \in \mathcal{P}'_{ro}$ is a ro-polytope, the right side of (14.62) gives either $0 - (-1)^{(\dim Q)-1} = (-1)^{\dim Q}$ or $(-1)^{\dim Q} - 0 = (-1)^{\dim Q}$. This completes the induction and thus proves the assertion. □

An immediate consequence is the **Euler relation**

$$\sum_{i=0}^{n} (-1)^i f_i(P) = 1, \tag{14.63}$$

where $P \in \mathcal{P}'$ is a polytope and $f_i(P)$ denotes the number of its i-dimensional faces. In fact, since $P = \bigcup_{F \in \mathcal{F}_\bullet(P)} \operatorname{relint} F$ is a disjoint union, the additivity of χ on $U(\mathcal{P}_{ro})$ gives

$$\chi(P) = \sum_{F \in \mathcal{F}_\bullet(P)} \chi(\operatorname{relint} F),$$

where $\mathcal{F}_\bullet(P)$ denotes the set of all faces of P (including P). This yields (14.63).

In the following, the notation $F \leq P$ for a polytope P means that F is a face of P (and $F < P$ means that F is a face of P different from P). Sums of the form

$$\sum_{F \leq P} \quad \text{and} \quad \sum_{G \leq F \leq P}$$

extend over all faces F of P (including P), respectively over all faces F of P containing the given face G (including P and G). A similar notation is used for faces of a mosaic m.

The Euler relation can be extended. If $P \in \mathcal{P}'$ and $G < P$ is a face of P (possibly the empty face, of dimension -1), then

$$\sum_{G \leq F \leq P} (-1)^{\dim F} = 0. \tag{14.64}$$

To prove this, we can assume that $\dim P = d$ and $0 \in \operatorname{int} P$ and use the dual polytope P^o. The faces of P and P^o are in a bijective correspondence $F \leftrightarrow \widehat{F}$, where $F \leq G \Leftrightarrow \widehat{G} \leq \widehat{F}$ and $\dim \widehat{F} = d - 1 - \dim F$. Therefore,

$$\sum_{G \leq F \leq P} (-1)^{\dim F} = (-1)^d + \sum_{G \leq F < P} (-1)^{\dim F} = (-1)^d + \sum_{\widehat{F} \leq \widehat{G}} (-1)^{d-1-\dim \widehat{F}}$$

$$= (-1)^{d-1} \left[-1 + \sum_{\widehat{F} \leq \widehat{G}} (-1)^{\dim \widehat{F}} \right].$$

The last sum is equal to 1, by the Euler relation for the polytope $\widehat{G}$. This proves (14.64).

A similar relation holds for face-to-face mosaics (as defined in Section 10.1). Let m be such a mosaic in $\mathbb{R}^d$. Let S be a j-dimensional face of m, $j \in \{0, \ldots, d-1\}$, and let $f_i(\mathsf{m}, S)$ be the number of i-dimensional faces of m that contain S. Then

$$\sum_{i=j}^{d} (-1)^{d-i} f_i(\mathsf{m}, S) = 1. \tag{14.65}$$

For the proof, we choose a d-dimensional ro-polytope Q with $Q \cap S \neq \emptyset$ and $Q \cap F = \emptyset$ for every face F of m with $S \not\subset F$. Then

$$Q = \bigcup_{F \leq \mathsf{m}} (Q \cap \operatorname{relint} F)$$

is a disjoint decomposition of Q. Using the Euler characteristic χ on $U(\mathcal{P}_{ro})$, we get

$$(-1)^{\dim Q} = \chi(Q) = \sum_{F \leq \mathsf{m}} \chi(Q \cap \operatorname{relint} F) = \sum_{S \leq F \leq \mathsf{m}} (-1)^{\dim F},$$

which gives (14.65).

We add another combinatorial relation for face-to-face mosaics m. Let F be a face of such a mosaic, and let $S_1, \ldots, S_m$ be the cells of m containing F. For $r \in \mathbb{N}$, let $\nu(F, r)$ denote the number of r-tuples $(S_{i_1}, \ldots, S_{i_r})$, $1 \leq i_1 < \ldots < i_r \leq m$, for which $S_{i_1} \cap \ldots \cap S_{i_r} = F$. Then

$$\sum_{r=1}^{m}(-1)^{r-1}\nu(F,r) = (-1)^{d-\dim F}. \tag{14.66}$$

For the proof, we consider first the case where $\dim F = 0$, say $F = \{x\}$. We choose a d-polytope P with $x \in \operatorname{int} P$ and such that P does not meet a cell of m different from $S_1, \dots, S_m$. We make use of two facts about the Euler characteristic on finite unions of polytopes. First, $\chi(\operatorname{bd} P) = 1-(-1)^d$. Second, if C is a convex polyhedral cone with a vertex at x, but $C \neq \{x\}$, then $\chi(C \cap \operatorname{bd} P) = 1$. Both facts are easy to prove, using the additivity of χ. Now

$$\chi(S_{i_1} \cap \ldots \cap S_{i_r} \cap \operatorname{bd} P) = \begin{cases} 0, \text{ if } S_{i_1} \cap \ldots \cap S_{i_r} = \{x\}, \\ 1 \ \text{ otherwise,} \end{cases}$$

and hence

$$\begin{aligned}\sum_{r=1}^{m}(-1)^{r-1}\nu(\{x\},r) &= \sum_{r=1}^{m}(-1)^{r-1} \sum_{i_1<\ldots<i_r} [1-\chi(S_{i_1} \cap \ldots \cap S_{i_r} \cap \operatorname{bd} P)] \\ &= 1-\chi(\operatorname{bd} P) = 1-(1-(-1)^d) = (-1)^d.\end{aligned}$$

If now $\dim F > 0$, we apply the foregoing in a suitable $(d-\dim F)$-dimensional plane intersecting the relative interior of F and thus obtain (14.66).

Talking of combinatorial formulas, we note a result which is applied in Section 10.3. It refers to the decomposition of a d-dimensional convex body K by a finite system $\mathcal{H}$ of hyperplanes. Let ν_k denote the number of k-dimensional faces of the cell decomposition of $\mathbb{R}^d$ induced by $\mathcal{H}$ that meet the interior of K. For $j \in \{1, \dots, d\}$, let α_j be the number of j-tuples from $\mathcal{H}$ whose intersection meets the interior of K; put $\alpha_0 := 1$. If the system $\mathcal{H}$ is in general position, then

$$\nu_k = \sum_{j=d-k}^{d} \binom{j}{d-k} \alpha_j \tag{14.67}$$

for $k \in \{0, \dots, d\}$; see Miles [532].

In Section 14.2 we have already mentioned that the intrinsic volumes can be characterized by simple geometric properties. We now give a proof for this celebrated theorem due to Hugo Hadwiger.

Theorem 14.4.6 (Hadwiger's characterization theorem). *If the function $\psi : \mathcal{K}' \to \mathbb{R}$ is additive, continuous, and invariant under rigid motions, then*

$$\psi = \sum_{i=0}^{d} c_i V_i$$

with constants $c_0, \dots, c_d$.

Proof. The core of the proof consists in establishing the following proposition.

Proposition 14.4.1. *If the function* $\psi : \mathcal{K}' \to \mathbb{R}$ *is additive, continuous, motion invariant and satisfies* $\psi(K) = 0$ *whenever either* $\dim K < d$ *or* K *is a unit cube, then* $\psi = 0$.

Let us first assume that this has been proved. Then Hadwiger's theorem is proved by induction with respect to the dimension. For $d = 0$, the assertion is trivial. Suppose that $d > 0$ and the assertion has been proved in dimensions less than d. Let $H \subset \mathbb{R}^d$ be a hyperplane. The restriction of ψ to the convex bodies lying in H is additive, continuous and invariant under motions of H into itself. By the induction hypothesis, there are constants $c_0, \dots, c_{d-1}$ so that $\psi(K) = \sum_{i=0}^{d-1} c_i V_i(K)$ holds for convex bodies $K \subset H$. The intrinsic volume $V_i(K)$ is the same whether computed in H or in $\mathbb{R}^d$. By the motion invariance of ψ and V_i, the relation $\psi(K) = \sum_{i=0}^{d-1} c_i V_i(K)$ holds for all $K \in \mathcal{K}'$ of dimension less than d. It follows that the function ψ' defined by

$$\psi'(K) := \psi(K) - \sum_{i=0}^{d} c_i V_i(K)$$

for $K \in \mathcal{K}'$, where c_d is chosen so that ψ' vanishes at a fixed unit cube, satisfies the assumptions of Proposition 14.4.1. Hence $\psi' = 0$, which completes the proof of Hadwiger's theorem.

Now we turn to the proof of Proposition 14.4.1. Again we use induction with respect to the dimension. For $d = 0$, there is nothing to prove. If $d = 1$, the additive function ψ vanishes on one-pointed sets and on segments (that is, closed line segments) of unit length, hence on segments of length $1/k$ for $k \in \mathbb{N}$ and therefore on segments of rational length. By continuity, ψ vanishes on all segments and thus on $\mathcal{K}'$.

Now let $d > 1$ and suppose that the assertion has been proved in dimensions less than d. Let $H \subset \mathbb{R}^d$ be a hyperplane and I a segment of length 1, orthogonal to H. For convex bodies $K \subset H$, we define $\varphi(K) := \psi(K + I)$. It is easy to see that φ has, relative to H, the properties of ψ in the proposition, hence the induction hypothesis yields $\varphi = 0$. For fixed $K \subset H$, we thus have $\psi(K + I) = 0$, and an argument similar to that used above for $d = 1$ shows that $\psi(K+S) = 0$ for any closed segment S orthogonal to H. Thus ψ vanishes on right convex cylinders.

Let $K \subset H$ be a convex body again, and let $S = \operatorname{conv}\{0, s\}$ be a segment not parallel to H. If $m \in \mathbb{N}$ is sufficiently large, the cylinder $Z := K + mS$ can be cut by a hyperplane H' orthogonal to S so that the two closed halfspaces H^-, H^+ bounded by H' satisfy $K \subset H^-$ and $K + ms \subset H^+$. Then $\overline{Z} := [(Z \cap H^-) + ms] \cup (Z \cap H^+)$ is a right cylinder, and we deduce that $m\mu(K + S) = \mu(Z) = \mu(\overline{Z}) = 0$. Thus ψ vanishes on arbitrary convex cylinders.

By Groemer's extension theorem 14.4.2, the continuous additive function ψ has an additive extension to the convex ring, hence

$$\psi\left(\bigcup_{i=1}^{k} K_i\right) = \sum_{i=1}^{k} \psi(K_i)$$

whenever $K_1, \dots, K_k$ are convex bodies such that $\dim(K_i \cap K_j) < d$ for $i \neq j$.

Let P be a polytope and S a segment. The Minkowski sum $P + S$ has a decomposition

$$P + S = \bigcup_{i=1}^{k} P_i$$

with $P_1 = P$, where the polytope P_i is a convex cylinder for $i > 1$ and where $\dim(P_i \cap P_j) < d$ for $i \neq j$. It follows that $\psi(P + S) = \psi(P)$. By induction, we deduce that $\psi(P + Z) = \psi(P)$ if Z is a finite sum of segments. Since the function ψ is continuous, it follows that $\psi(K + Z) = \psi(K)$ for arbitrary convex bodies K and zonoids Z.

Let K be a centrally symmetric convex body with a support function of class C^∞ (on $\mathbb{R}^d \setminus \{0\}$). Then there exist zonoids Z_1, Z_2 so that $K + Z_1 = Z_2$ (this can be seen from Section 3.5 in Schneider [695], especially Theorem 3.5.3). We conclude that $\psi(K) = \psi(K + Z_1) = \psi(Z_2) = 0$. Since every centrally symmetric convex body K can be approximated by bodies which are centrally symmetric and have support functions of class C^∞ (for example, [695, Theorem 3.3.1]), it follows from the continuity of ψ that $\psi(K) = 0$ for all centrally symmetric convex bodies.

Now let Δ be a simplex, say $\Delta = \operatorname{conv}\{0, v_1, \dots, v_d\}$, without loss of generality. Let $v := v_1 + \dots + v_d$ and $\Delta' := \operatorname{conv}\{v, v - v_1, \dots, v - v_d\}$, then $\Delta' = -\Delta + v$. The vectors $v_1, \dots, v_d$ span a parallelotope P. It is the union of Δ, Δ' and the part of P, denoted by Q, that lies between the hyperplanes spanned by $v_1, \dots, v_d$ and $v - v_1, \dots, v - v_d$, respectively. The polytope Q is centrally symmetric, and $\Delta \cap Q$, $\Delta' \cap Q$ are of dimension $d - 1$. We deduce that $0 = \psi(P) = \psi(\Delta) + \psi(Q) + \psi(\Delta')$, thus $\psi(-\Delta) = -\psi(\Delta)$. If the dimension d is even, then $-\Delta$ is obtained from Δ by a proper rigid motion, and the motion invariance of ψ yields $\psi(\Delta) = 0$. If the dimension $d > 1$ is odd, we decompose Δ as follows. Let z be the center of the inscribed ball of Δ, and let p_i be the point where this ball touches the facet F_i of Δ ($i = 1, \dots, d+1$). For $i \neq j$, let Q_{ij} be the convex hull of the face $F_i \cap F_j$ and the points z, p_i, p_j. The polytope Q_{ij} is invariant under reflection in the hyperplane spanned by $F_i \cap F_j$ and z. If $Q_1, \dots, Q_m$ are the polytopes Q_{ij} for $1 \leq i < j \leq d + 1$ in any order, then $\Delta = \bigcup_{r=1}^{m} Q_r$ and $\dim(Q_r \cap Q_s) < d$ for $r \neq s$. Since $-Q_r$ is the image of Q_r under a proper rigid motion (a reflection in a hyperplane followed by a reflection in a point), we have $\psi(-\Delta) = \sum \psi(-Q_r) = \sum \psi(Q_r) = \psi(\Delta)$. Thus $\psi(\Delta) = 0$ for every simplex Δ.

Decomposing a polytope P into simplices, we obtain $\psi(P) = 0$. The continuity of ψ now implies $\psi(K) = 0$ for all convex bodies K. This finishes the induction and hence the proof of Proposition 14.4.1. □

Finally, we prove a characterization of the spherical volume of polytopes in the sphere S^{d-1}, by additivity and invariance properties. A subset $P \subset S^{d-1}$ is a **spherical polytope** (a polytope for short) if it is the nonempty intersection of finitely many closed hemispheres. We denote by $\mathcal{P}_s$ the set of all spherical polytopes. The **dimension** of $P \in \mathcal{P}_s$ is defined by $\dim P := \dim \operatorname{lin} P - 1$ where the linear hull refers to $\mathbb{R}^d$.

Theorem 14.4.7. *Let $\varphi : \mathcal{P}_s \to \mathbb{R}$ be a valuation which is simple (that is, vanishes on spherical polytopes of dimension less than $d-1$), rotation invariant and nonnegative. Then there is a constant $c \geq 0$ such that $\varphi(P) = c\sigma(P)$ for all $P \in \mathcal{P}_s$.*

Proof. Let $\mathcal{Q}_s$ be the system of all finite unions of elements of $\mathcal{P}_s$. The simple valuation φ has an additive extension to $\mathcal{Q}_s$. This can be deduced by proving the spherical analog of Theorem 14.4.3. But as φ is simple, there is also an easier way, namely arguing similarly to Hadwiger [307, p. 81]. The extension, also denoted by φ, is still rotation invariant and nonnegative. It is increasing under set inclusion, since to $A, B \in \mathcal{Q}_s$ with $A \subset B$ there exists $A' \in \mathcal{Q}_s$ such that $A \cup A' = B$ and $A \cap A'$ is a finite union of polytopes of dimensions less than $d-1$.

We take the following fact for granted. If f is a Riemann integrable real function on S^{d-1}, then for every $\epsilon > 0$ there exist a number $k \in \mathbb{N}$ and k rotations $\vartheta_i \in SO_d$ such that

$$\left| \frac{1}{k} \sum_{i=1}^{k} f(\vartheta_i x) - \frac{1}{\omega_d} \int_{S^{d-1}} f \, d\sigma \right| < \epsilon \qquad \text{for each } x \in S^{d-1}$$

(this follows from Pontrjagin [608, §29], where it is obtained for continuous functions, and approximation, or from Hadwiger [303, §3]).

Let $P \in \mathcal{Q}_s$, and let $\epsilon > 0$ be given. Applying the preceding assertion to the indicator function of P, we obtain

$$\left| \frac{1}{k} \nu(x) - \frac{1}{\omega_d} \sigma(P) \right| < \epsilon \qquad \text{for each } x \in S^{d-1}, \tag{14.68}$$

where $\nu(x)$ denotes the number of sets $\vartheta_i P$ containing x.

For $j \in \{1, \dots, k\}$, let $U_j := \{x \in S^{d-1} : \nu(x) \geq j\}$, then

$$U_j = \bigcup_{1 \leq i_1 < \dots < i_j \leq k} (\vartheta_{i_1} P \cap \dots \cap \vartheta_{i_j} P) \in \mathcal{Q}_s.$$

Since

$$\sum_{i=1}^{k} \mathbf{1}_{\vartheta_i P}(x) = \sum_{j=1}^{k} \mathbf{1}_{U_j}(x) \qquad \text{for each } x \in S^{d-1},$$

Theorem 14.4.1 implies the right equality in

$$k\varphi(P) = \sum_{i=1}^{k} \varphi(\vartheta_i P) = \sum_{j=1}^{k} \varphi(U_j). \tag{14.69}$$

The left equality follows from the rotation invariance of φ.

Let $y, z \in S^{d-1}$ be points at which the function ν attains its minimum and maximum, respectively. Then $U_j = S^{d-1}$ for $j = 1, \dots, \nu(y)$, hence $\sum_{j=1}^k \varphi(U_j) \geq \nu(y)\varphi(S^{d-1})$. Further, $U_j = \emptyset$ for $j > \nu(z)$, hence $\sum_{j=1}^k \varphi(U_j) \leq \nu(z)\varphi(S^{d-1})$. Together with (14.68) and (14.69) this gives

$$\left(\frac{\sigma(P)}{\omega_d} - \epsilon\right)\varphi(S^{d-1}) \leq \varphi(P) \leq \left(\frac{\sigma(P)}{\omega_d} + \epsilon\right)\varphi(S^{d-1}).$$

Since $\epsilon > 0$ was arbitrary, we conclude that $\varphi(P) = (\varphi(S^{d-1})/\omega_d)\sigma(P)$. □

Notes for Section 14.4

1. The important role that valuations play in the geometry of convex bodies can be seen from the survey articles by McMullen and Schneider [474] and by McMullen [471].

2. Theorems 14.4.1 and 14.4.2 and their proofs are due to Groemer [289]. Volland [771] has proved that every valuation on the class $\mathcal{P}$ of polytopes has an additive extension to the class $U(\mathcal{P})$ of polyhedra; his proof is reproduced in our proof of Theorem 14.4.3, together with the slight (but useful) extension to the class $U(\mathcal{P}_{ro})$ of ro-polyhedra, which was mentioned in McMullen and Schneider [474, p. 192]. For this result, it is sufficient to assume that the function φ on $\mathcal{P}$ is **weakly additive**, that is, it satisfies $\varphi(P) = \varphi(P \cap H^+) + \varphi(P \cap H^-) - \varphi(P \cap H)$ for every polytope $P \in \mathcal{P}$ and every hyperplane H, where H^+, H^- are the two closed halfspaces bounded by H; see Schneider [690].

2. We wish to point out that the proof given here for the Euler (or Euler–Schläfli) relation (14.63) is particularly short. Relation (14.62) was used by Hadwiger for an elementary existence proof of the Euler characteristic on the convex ring. That the extension of the Euler characteristic to relatively open polytopes allows a simple approach to the Euler relation, was pointed out by Nef [580].

3. A local Steiner formula for the convex ring. As a particular case of Groemer's extension theorem, the curvature measure mapping $K \mapsto \Phi_j(K, \cdot)$ has an additive extension to the convex ring $\mathcal{R}$; see the end of Section 5.2. This extension can also be achieved in a more concrete and intuitive way. For a Borel set $A \in \mathcal{B}(\mathbb{R}^d)$ and for $\epsilon \geq 0$, we define $U_\epsilon(K, A) := M_\epsilon(K, A \times S^{d-1})$ and put $\rho_\epsilon(K, A) := \lambda(U_\epsilon(K, A))$. Then a special case of (14.9) says that

$$\rho_\epsilon(K, A) = \sum_{j=0}^{d} \epsilon^{d-j}\kappa_{d-j}\Phi_j(K, A). \tag{14.70}$$

This local Steiner formula is now extended to the convex ring $\mathcal{R}$, by introducing local parallel sets with multiplicity. Let $B(z, \alpha)$ denote the closed ball with center z and radius α. For $K \in \mathcal{R}$ and points $q, x \in \mathbb{R}^d$ with $q \neq x$, we define the **index of** K **at** q **with respect to** x by

$$j(K, q, x) := \begin{cases} \lim_{\delta \to 0+} \lim_{\epsilon \to 0+} \chi(K \cap B(x, \|x - q\| - \epsilon) \cap B(q, \delta)), \text{ if } q \in K, \\ 0, \text{ if } q \notin K. \end{cases}$$

Here we use that the Euler characteristic χ has an additive extension to the convex ring. The existence of the limits in the definition of the index is easy to see. For $A \in \mathcal{B}$ and $\epsilon > 0$ we define

$$c_\epsilon(K, A, x) := \sum_{q \in A \setminus \{x\}} j(K \cap B(x, \epsilon), q, x),$$

here only finitely many summands are different from zero. If K is convex, then

$$c_\epsilon(K, A, x) = \mathbf{1}_{U_\epsilon(K,A)}(x), \qquad x \in \mathbb{R}^d,$$

and hence

$$\rho_\epsilon(K, A) = \int_{\mathbb{R}^d} c_\epsilon(K, A, x) \, \lambda(\mathrm{d}x).$$

We use this equation to define $\rho_\epsilon(K, A)$ for $K \in \mathcal{R}$. We can interpret this as a local parallel volume with multiplicity. It follows from the additivity of the Euler characteristic on the convex ring that the function $j(\cdot, q, x)$ is additive on $\mathcal{R}$, hence the same is true for the functions $c_\epsilon(\cdot, A, x)$ and $\rho_\epsilon(\cdot, A)$. Now it follows that formula (14.70) holds for all sets $K \in \mathcal{R}$, with signed measures $\Phi_j(K, \cdot)$ which can be obtained from the curvature measures on $\mathcal{K}$ by means of the inclusion–exclusion formula. Thus, formula (14.70) yields an interpretation of the additive extensions of the curvature measures as coefficients in a generalized local Steiner formula. This approach comes from Schneider [679].

4. Theorem 14.4.4 and its proof are taken from Weil and Wieacker [804].

5. Hadwiger proved his characterization theorem for dimension three in [304] and for general dimensions in [305]; the proof is reproduced in [307]. The simpler proof of Theorem 14.4.6 as presented here is due to Klain [414]; see also [416].

For analogs of Hadwiger's characterization theorem, where the rotation group is replaced by a compact subgroup that is still transitive on the sphere, see Note 4 of Section 5.1 and the references given there.

6. Theorem 14.4.7 was proved by Schneider [676, Th. (6.2)]. A general version for invariant measures on compact homogeneous spaces appears in Schneider [682].

It would be interesting to know whether a counterpart to Theorem 14.4.7 is true where the assumption of nonnegativity of φ is replaced by continuity (with respect to the Hausdorff metric on the sphere).

7. An extension theorem for valuations on the (non-intersectional) class of lattice polytopes was proved by McMullen [473].

14.5 Hausdorff Measures and Rectifiable Sets

Occasionally in this book, we refer to general notions of a k-dimensional surface. For example, the boundary of a d-dimensional convex body is a $(d-1)$-dimensional surface which is, in general, not smooth. The theory of fiber and surface processes makes use of general lower-dimensional surfaces. We collect here briefly some definitions and results from geometric measure theory which are suitable for the introduction of quite general classes of k-dimensional surfaces. For more information and proofs, we refer to books on geometric measure theory (Federer [229], Mattila [466]).

Let $k \in \{0, \ldots, d\}$. The k-dimensional **Hausdorff measure** of a set $A \subset \mathbb{R}^d$ is defined by

$$\mathcal{H}^k(A) := \frac{\kappa_k}{2^k} \liminf_{\delta \to 0+} \left\{ \sum_{j=1}^{\infty} (\operatorname{diam} M_j)^k : A \subset \bigcup_{j=1}^{\infty} M_j, \ \operatorname{diam} M_j \leq \delta \right\},$$

where the M_j can be arbitrary subsets of $\mathbb{R}^d$, and diam denotes the diameter. This defines an outer measure $\mathcal{H}^k$ on $\mathbb{R}^d$. Borel sets are $\mathcal{H}^k$-measurable, hence the restriction of $\mathcal{H}^k$ to $\mathcal{B}(\mathbb{R}^d)$ is a measure. The k-dimensional Hausdorff measure of a k-dimensional C^1 submanifold coincides with its usual differential-geometric k-dimensional volume measure. In particular, $\mathcal{H}^0$ is the counting measure, and $\mathcal{H}^d$ coincides on Lebesgue measurable sets with the Lebesgue measure λ_d.

A map f from a metric space (E_1, δ_1) to a metric space (E_2, δ_2) is a **Lipschitz map** if there exists a constant L such that $\delta_2(f(x), f(y)) \leq L\delta_1(x, y)$ for all $x, y \in E_1$. For a positive integer k, a subset $M \subset \mathbb{R}^d$ is called k-**rectifiable** if it is the image of some bounded subset of $\mathbb{R}^k$ under some Lipschitz map. A set M is 0-rectifiable if it is finite.

A set $M \subset \mathbb{R}^d$ is called $(\mathcal{H}^k, k)$**-rectifiable** if it is $\mathcal{H}^k$-measurable, satisfies $\mathcal{H}^k(M) < \infty$, and there are countably many k-rectifiable sets $M_1, M_2, \ldots$ such that

$$\mathcal{H}^k\left(M \setminus \bigcup_{i \in \mathbb{N}} M_i \right) = 0.$$

The set M is $\mathcal{H}^k$**-rectifiable** if $M \cap C$ is $(\mathcal{H}^k, k)$-rectifiable for every compact set C.

Let $S \subset \mathbb{R}^d$ and $a \in S$. The **tangent cone** of S at a, denoted by $\operatorname{Tan}(S, a)$, is the closed convex cone consisting of all vectors $v \in \mathbb{R}^d$ with the property that, for each $\epsilon > 0$, there exist $x \in S$ and $\alpha > 0$ with $\|x - a\| < \epsilon$ and $\|\alpha(x - a) - v\| < \epsilon$.

Let M be $(\mathcal{H}^k, k)$-rectifiable. The cone of **approximate tangent vectors** of M at a is defined by

$$\operatorname{Tan}^k(\mathcal{H}^k \llcorner M, a) := \bigcap \left\{ \operatorname{Tan}(S, a) : S \subset \mathbb{R}^d, \ \Theta^{*k}(M \setminus S, a) = 0 \right\},$$

where

$$\Theta^{*k}(A, a) := \limsup_{r \to 0+} \frac{1}{\kappa_k r^k} \mathcal{H}^k(A \cap B(a, r))$$

and $B(a, r)$ denotes the closed ball with center a and radius r. Let $0 < k < d$. The $(\mathcal{H}^k, k)$-rectifiable set M has the property that for $\mathcal{H}^k$-almost all $a \in M$ the cone $\operatorname{Tan}^k(\mathcal{H}^k \llcorner M, a)$ is a k-dimensional vector subspace of $\mathbb{R}^d$.

If M is $(\mathcal{H}^k, k)$-rectifiable, there exist C^1 submanifolds $S_1, S_2, \ldots$ of $\mathbb{R}^d$ such that

$$\mathcal{H}^k\left(M \setminus \bigcup_{i\in\mathbb{N}} S_i\right) = 0.$$

Let $i \in \mathbb{N}$. For $\mathcal{H}^k$-almost all $a \in S_i \cap M$, the cone $\mathrm{Tan}^k(\mathcal{H}^k \llcorner M, a)$ coincides with the tangent space $T_a S_i$ of S_i at a, in the usual sense of differential geometry. It follows that there is a subset $M' \subset M$ with $\mathcal{H}^k(M \setminus M') = 0$ such that the mapping $a \mapsto \mathrm{Tan}^k(\mathcal{H}^k \llcorner M, a)$ is an $\mathcal{H}^k$-measurable mapping from M' into the Grassmannian $G(d, k)$.

For many purposes, $(\mathcal{H}^k, k)$-rectifiable Borel sets provide an appropriate general notion of k-dimensional surfaces.

Note for Section 14.5

Besides the books of Federer [229] and Mattila [466], we recommend the articles by Zähle [822] and Wieacker [816], for typical applications of the above notions.

References

1. Adler, R.J.: The Geometry of Random Fields. Wiley, New York (1981)
2. Adler, R.J., Taylor, J.E.: Random Fields and Geometry. Springer, New York (2007)
3. Affentranger, F.: On polytopes defined by random points in a convex body and a generalization of Sylvester's problem. Doctoral Thesis, Buenos Aires (1987)
4. Affentranger, F.: The expected volume of a random polytope in a ball. J. Microsc., **151**, 277–287 (1988)
5. Affentranger, F.: Generalization of a formula of C. Buchta about the convex hull of random points. Elem. Math., **43**, 39–45 (1988)
6. Affentranger, F.: Remarks on the note "Generalization of a formula of C. Buchta about the convex hull of random points". Elem. Math., **43**, 151–152 (1988)
7. Affentranger, F.: Pairs of non-intersecting random flats. Probab. Theory Related Fields, **79**, 47–50 (1988)
8. Affentranger, F.: Random circles in the d-dimensional unit ball. J. Appl. Prob., **26**, 408–412 (1989)
9. Affentranger, F.: Random spheres in a convex body. Arch. Math., **55**, 74–81 (1990)
10. Affentranger, F.: The convex hull of random points with spherically symmetric distributions. Rend. Sem. Mat. Univers. Politecn. Torino, **49**, 359–383 (1991)
11. Affentranger, F., Schneider, R.: Random projections of regular simplices. Discrete Comput. Geom., **7**, 219–226 (1992)
12. Affentranger, F., Wieacker, J.A.: On the convex hull of uniform random points in a simple d-polytope. Discrete Comput. Geom., **6**, 291–305 (1991)
13. Alagar, V.S.: On the distribution of a random triangle. J. Appl. Prob., **14**, 284–297 (1977)
14. Aldous, D.J., Fristedt, J., Griffin, P.S., Pruitt, W.E.: The number of extreme points in the convex hull of a random sample. J. Appl. Prob., **28**, 287–304 (1991)
15. Aleman, A., Stoka, M., Zamfirescu, T.: Convex bodies instead of needles in Buffon's experiment. Geom. Dedicata, **67**, 301–308 (1997)
16. Alesker, S.: Continuous rotation invariant valuations on convex sets. Ann. of Math., **149**, 977–1005 (1999)

17. Alesker, S.: Description of continuous isometry covariant valuations on convex sets. Geom. Dedicata, **74**, 41–248 (1999)
18. Alesker, S.: On P. McMullen's conjecture on translation invariant valuations. Adv. Math., **155**, 239–263 (2000)
19. Alesker, S.: Hard Lefschetz theorem for valuations, complex integral geometry, and unitarily invariant valuations. J. Differential Geom., **63**, 63–95 (2003)
20. Alesker, S.: The multiplicative structure on continuous polynomial valuations. Geom. Funct. Anal., **14**, 1–26 (2004)
21. Alesker, S.: $SU(2)$-invariant valuations. In: Milman, V.D., Schechtman, G. (eds.) Geometric Aspects of Functional Analysis, Israel Seminar 2002–2003, Lecture Notes in Math., **1850**, pp. 21–29, Springer, Berlin (2004)
22. Alesker, S.: Theory of valuations on manifolds: a survey. Geom. Funct. Anal. (to appear)
23. Allendoerfer, C.B.: Steiner's formula on a general S^{n+1}. Bull. Amer. Math. Soc., **54**, 128–135 (1948)
24. Allendoerfer, C.B., Weil, A.: The Gauss–Bonnet theorem for Riemannian polyhedra. Trans. Amer. Math. Soc., **53**, 101–129 (1943)
25. Álvarez Paiva, J.C., Berck, G.: What is wrong with the Hausdorff measure in Finsler spaces. Adv. Math., **204**, 647–663 (2006)
26. Álvarez Paiva, J.C., Fernandes, E.: Crofton formulas in projective Finsler spaces. Electronic Res. Announcements AMS, **4**, 91–100 (1998)
27. Álvarez Paiva, J.C., Fernandes, E.: Fourier transforms and the Holmes–Thompson volume of Finsler manifolds. Int. Math. Res. Notices **19**, 1032–1042 (1999)
28. Álvarez Paiva, J.C., Fernandes, E.: Gelfand transforms and Crofton formulas. Selecta Math. (N.S.), **13**, 369–390 (2007)
29. Álvarez Paiva, J.C., Fernandes, E.: What is a Crofton formula? Math. Notae, **42**, 95–108 (2003/04)
30. Álvarez Paiva, J.C., Gelfand, I.M., Smirnov, M.M.: Crofton densities, symplectic geometry, and Hilbert's fourth problem. In: Arnold, V.I. *et al.* (eds) Arnold–Gelfand Mathematical Seminars, Geometry and Singularity Theory. Birkhäuser, Boston (1997)
31. Álvarez Paiva, J.C., Thompson, A.C.: Volumes on normed and Finsler spaces. In: Bao, D., Bryant, R.L., Chern, S.–S., Shen, Z. (eds) A Sampler of Riemann–Finsler Geometry, MSRI Publ., vol. 50. Cambridge University Press (2004)
32. Ambartzumian, R.V.: On random plane mosaics. Soviet Math. Dokl., **12**, 1349–1353 (1971)
33. Ambartzumian, R.V.: Convex polygons and random tessellations. In: Harding, E.F., Kendall, D.G. (eds) Stochastic Geometry, pp. 176–191. Wiley, New York (1974)
34. Ambartzumian, R.V.: Combinatorial Integral Geometry. With Applications to Mathematical Stereology. Wiley, Chichester (1982)
35. Ambartzumian, R.V.: Factorization Calculus and Geometric Probability. Cambridge Univ. Press, Cambridge (1990)
36. Ambartzumian, R.V., Mecke, J., Stoyan, D.: Geometrische Wahrscheinlichkeiten und Stochastische Geometrie. Akademie-Verlag, Berlin (1993)
37. Araujo, A., Giné, E.: The Central Limit Theorem for Real and Banach Valued Random variables. Wiley, Chichester (1980)
38. Arbeiter, E., Zähle, M.: Kinematic relations for Hausdorff moment measures in spherical spaces. Math. Nachr., **153**, 333–348 (1991)

39. Arbeiter, E., Zähle, M.: Geometric measures for random mosaics in spherical spaces. Stoch. Stoch. Rep., **46**, 63–77 (1994)
40. Artstein, Z., Vitale, R.A.: A strong law of large numbers for random compact sets. Ann. Probab., **5**, 879–882 (1979)
41. Averkov, G., Bianchi, G.: Confirmation of Matheron's conjecture on the covariogram of a planar convex body. arXiv:0711.0572v2 [math.MG]
42. Avram, F., Bertsimas, D.: On central limit theorems in geometric probability. Ann. Appl. Probab., **3**, 1033–1046 (1993)
43. Baddeley, A.J.: A fourth note on recent research in geometrical probability. Adv. Appl. Prob., **9**, 824–860 (1977)
44. Baddeley, A.J.: Stochastic geometry: an introduction and reading list. Int. Stat. Rev., **50**, 179–193 (1982)
45. Baddeley, A.J.: Stochastic geometry and image analysis. CWI Newslett., **4**, 2–20 (1984)
46. Baddeley, A.J.: Vertical sections. In: Ambartzumian, R.V., Weil, W. (eds) Proc. Conf. Stochastic Geometry, Geometric Statistics, Stereology, Oberwolfach 1983, pp. 43–52. Teubner, Leipzig (1984)
47. Baddeley, A.J.: An anisotropic sampling design. In: Nagel, W. (ed) Geobild '85, pp. 92–97. Friedrich-Schiller-Universität, Jena (1985)
48. Baddeley, A.J.: Stereology and image analysis for anisotropic plane sections. In: Proc. of the 13th International Biometric Conference, Seattle, pp. 1–9 (1987)
49. Baddeley, A.J.: A crash course in stochastic geometry. In [80], pp. 1–35 (1999)
50. Baddeley, A.J., Bárány, I., Schneider, R., Weil, W.: Stochastic Geometry. C.I.M.E. Summer School, Martina Franca, Italy, 2004 (W. Weil, ed.), Lecture Notes in Math., **1892**, Springer, Berlin (2007)
51. Baddeley, A., Gregori, P., Mateu, J., Stoica, R., Stoyan, D. (eds.): Case Studies in Spatial Point Process Models. Lecture Notes in Statist., **185**, Springer, New York (2006)
52. Baddeley, A.J., Gundersen, H.J.G., Cruz–Orive, L.M.: Estimation of surface area from vertical sections. J. Microsc., **142**, 259–276 (1986)
53. Baddeley, A.J., Jensen, E.B.V.: Stereology for Statisticians. Chapman & Hall/CRC, Boca Raton (2005)
54. Badertscher, E.: An explicit formula about the convex hull of random points. Elem. Math., **44**, 104–106 (1989)
55. Balanzat, M.: Sur quelques formules de la géométrie intégrale des ensembles dans un espace à n dimensions. Portugaliae Math., **3**, 87–94 (1942)
56. Bárány, I.: Intrinsic volumes and f-vectors of random polytopes. Math. Ann., **285**, 671–699 (1989)
57. Bárány, I.: Random polytopes in smooth convex bodies. Mathematika, **39**, 81–92 (1992). Corrigendum: Mathematika, **51**, 31 (2005)
58. Bárány, I.: Approximation by random polytopes is almost optimal. Rend. Circ. Mat. Palermo, Ser. II, Suppl., **50**, 43–50 (1997)
59. Bárány, I.: Affine perimeter and limit shape. J. reine angew. Math., **484**, 71–84 (1997)
60. Bárány, I.: Sylvester's question: the probability that n points are in convex position. Ann. Probab., **27**, 2020–2034 (1999)
61. Bárány, I.: The technique of M-regions and cap-coverings: a survey. Rend. Circ. Mat. Palermo, Ser. II, Suppl., **65**, 21–38 (1999)
62. Bárány, I.: A note on Sylvester's four-point problem. Studia Sci. Math. Hungar., **38**, 733–77 (2001)

63. Bárány, I.: Approximation by random polytopes has low complexity. Rend. Circ. Mat. Palermo, Ser. II, Suppl, **70**, 53–56 (2002)
64. Bárány, I.: Random polytopes, convex bodies, and approximation. In [50], pp. 77–118 (2007)
65. Bárány, I.: The chance that a convex body is lattice-point free: a relative of Buffon's needle problem. Random Structures Algorithms, **30**, 414–426 (2007)
66. Bárány, I.: Random points and lattice points in convex bodies. Bull. Amer. Math. Soc., **45**, 339–365 (2008)
67. Bárány, I., Buchta, C.: On the convex hull of uniform random points in an arbitrary d-polytope. Anz. Österreich. Akad. Wiss. Math.-Natur. Kl., **127**, 25–27 (1990)
68. Bárány, I., Buchta, C.: Random polytopes in a convex polytope, independence of shape, and concentration of vertices. Math. Ann. **297**, 467–497 (1993)
69. Bárány, I., Dalla, L.: Few points to generate a random polytope. Mathematika, **44**, 325–331 (1997)
70. Bárány, I., Füredi, Z.: On the shape of the convex hull of random points. Probab. Theory Related Fields, **77**, 231–240 (1988)
71. Bárány, I., Larman, D.G.: Convex bodies, economic cap coverings, random polytopes. Mathematika, **35**, 274–291 (1988)
72. Bárány, I., Pór, A.: $0-1$ polytopes with many facets. Adv. Math., **161**, 209–228 (2001)
73. Bárány, I., Reitzner, M.: Random polytopes. Preprint (2006)
74. Bárány, I., Rote, G., Steiger, W., Zhang, C.–H.: A central limit theorem for convex chains in the square. Discrete Comput. Geom., **23**, 35–50 (2000)
75. Bárány, I., Vitale, R.A.: Random convex hulls: Floating bodies and expectations. J. Approx. Theory, **75**, 130–135 (1993)
76. Bárány, I., Vu, V.: Central limit theorems for Gaussian polytopes. Ann. Probab., **35**, 1593–1621 (2007)
77. Barbier, E.: Note sur le problème de l'aiguille et le jeu du joint couvert. J. math. pures appl., **5**, 273–286 (1860)
78. Barbour, A.D., Schmidt, V.: On Laslett's transform for the Boolean model. Adv. Appl. Prob. (SGSA), **33**, 1–5 (2001)
79. Barndorff–Nielsen, O.E., Blæsild, P., Eriksen, P.S.: Decomposition and Invariance of Measures, and Statistical Transformation Models. Lecture Notes in Stat., **58**, Springer, New York (1989)
80. Barndorff–Nielsen, O.E., Kendall, W.S., van Lieshout, M.N.M. (eds.): Stochastic Geometry: Likelihood and Computation. Monographs on Statistics and Applied Probability 80, Chapman & Hall/CRC, Boca Raton (1999)
81. Baryshnikov, Y.M.: Gaussian samples, regular simplices, and exchangeability. Discrete Comput. Geom., **17**, 257–261 (1997)
82. Baryshnikov, Y.M., Vitale, R.A.: Regular simplices and Gaussian samples. Discrete Comput. Geom., **11**, 141–147 (1994)
83. Bauer, C., Schneider, R.: Extremal problems for geometric inequalities involving convex bodies. Adv. Appl. Prob. (SGSA), **27**, 20–34 (1995)
84. Baumstark, V.: Stoppmengen und Palmsche Maße für Poissonsche Modelle der Stochastischen Geometrie. Doctoral Thesis, Universität Karlsruhe (2007)
85. Baumstark, V., Last, G.: Some distributional results for Poisson–Voronoi tessellations. Adv. Appl. Prob. (SGSA), **39**, 16–40 (2007)
86. Baumstark, V., Last, G.: Gamma distributions for stationary Poisson flat processes. Adv. Appl. Prob. (SGSA) (to appear)

87. Beer, G.: Topologies on Closed and Closed Convex Sets. Kluwer, Dordrecht (1993)
88. Beisbart, C., Barbosa, M.S., Wagner, H., da F. Costa, L.: Extended morphometric analysis of neuronal cells with Minkowski valuations. Eur. Phys. J. B, **52**, 531–546 (2006)
89. Beisbart, C., Dahlke, R., Mecke, K., Wagner, H.: Vector- and tensor-valued descriptors for spatial patterns. In: Mecke, K., Stoyan, D. (eds) Morphology of Condensed Matter, Lecture Notes in Physics, **600**, pp. 238–260. Springer, Berlin (2002)
90. Beneš, V., Rataj, J.: Stochastic Geometry. Kluwer, Dordrecht (2004)
91. Berg, C., Christensen, J.P.R., Ressel, P.: Harmonic Analysis on Semigroups. Springer, New York (1984)
92. Bernig, A.: A Hadwiger-type theorem for the special unitary group. Geom. Funct. Anal. (to appear)
93. Bernig, A., Bröcker, L.: Lipschitz–Killing invariants. Math. Nachr., **245**, 5–25 (2002)
94. Bernig, A., Bröcker, L.: Courbures intrinsèques dans les catégories analytico-géométriques. Ann. Inst. Fourier, **53**, 1897–1924 (2003)
95. Bernig, A., Fu, J.H.G.: Convolution of convex valuations. Geom. Dedicata, **123**, 153–169 (2006)
96. Bernig, A., Fu, J.H.G.: Hermitian integral geometry. Preprint (2008)
97. Bertrand, J.: Calcul des probabilités. Gauthier–Villars, Paris (1888)
98. Berwald, L., Varga, O.: Integralgeometrie 24. Über die Schiebungen im Raum. Math. Z., **42**, 710–736 (1937)
99. Betke, U., Henk, M.: Intrinsic volumes and lattice points of crosspolytopes. Monatsh. Math., **115**, 27–33 (1993)
100. Bianchi, G.: Matheron's conjecture for the covariogram problem. J. London Math. Soc., **71**, 203–220 (2005)
101. Bianchi, G.: The covariogram determines three-dimensional convex polytopes. arXiv.0805.1605v1
102. Blaschke, W.: Über affine Geometrie XI: Lösung des "Vierpunktproblems" von Sylvester aus der Theorie der geometrischen Wahrscheinlichkeiten. Ber. Verh. Sächs. Akad. Wiss., Math.-Phys. Kl., **69**, 436–453 (1917). Also in: Wilhelm Blaschke, Gesammelte Werke (Burau, W. *et al.*, eds.), vol. 3, pp. 284–301, Thales, Essen (1985)
103. Blaschke, W.: Eine isoperimetrische Eigenschaft des Kreises. Math. Z., **1**, 52–57 (1918)
104. Blaschke, W.: Vorlesungen über Differentialgeometrie II, Affine Differentialgeometrie. Springer, Berlin (1923)
105. Blaschke, W.: Integralgeometrie 2: Zu Ergebnissen von M.W. Crofton. Bull. Math. Soc. Roum. Sci., **37**, 3–11 (1935)
106. Blaschke, W.: Integralgeometrie 21. Über Schiebungen. Math. Z., **42**, 399–410 (1937)
107. Blaschke, W.: Vorlesungen über Integralgeometrie. 3. Aufl., VEB Deutsch. Verl. d. Wiss., Berlin, (1955) (first ed.: Part I, 1935; Part II, 1937)
108. Blaschke, W.: Gesammelte Werke (Burau, W. *et al.*, eds.), vol. 2: Kinematik und Integralgeometrie. Thales, Essen (1985)
109. Blaschke, W., Varga, O: Integralgeometrie 9. Über Mittelwerte an Eikörpern. Mathematica (Cluj), **12**, 65–80 (1936)

110. Böhm, J., Hertel, E.: Polyedergeometrie in n-dimensionalen Räumen konstanter Krümmung. VEB Deutsch. Verl. d. Wiss., Berlin (1980)
111. Bokowski, J., Hadwiger, H., Wills, J.M.: Eine Erweiterung der Croftonschen Formeln für konvexe Körper. Mathematika, **23**, 212–219 (1976)
112. Bol, G.: Zur Theorie der Eikörper. Jahresber. Deutsche Math. Ver., **52**, 250–266 (1942)
113. Böröczky, K.J., Fodor, F., Reitzner, M., Vígh, V.: Mean width of inscribed random polytopes in a reasonably smooth convex body. Preprint (2007)
114. Böröczky Jr., K., Henk, M.: Random projections of regular polytopes. Arch. Math., **73**, 465–473 (1999)
115. Böröczky, K.J., Hoffmann, L.M., Hug, D.: Expectation of mean projections of random polytopes. Preprint (2008)
116. Böröczky Jr., K., Reitzner, M.: Approximation of smooth convex bodies by random circumscribed polytopes. Ann. Appl. Probab., **14**, 239–273 (2004)
117. Böröczky K.J., Schneider, R.: The mean width of circumscribed random polytopes. Preprint (2007)
118. Bosetto, E.: Systems of stochastically independent and normally distributed random points in the Euclidean space E_3. Beiträge Algebra Geom., **40**, 291–301 (1999)
119. Bourbaki, N.: Éléments de Mathématique. Livre VI, Intégration; Chapitre 7, Mesure de Haar. Hermann, Paris (1963)
120. Bräker, H., Hsing, T.: On the area and perimeter of a random convex hull in a bounded convex set. Probab. Theory Related Fields, **111**, 517–550 (1998)
121. Bräker, H., Hsing, T., Bingham, N.H.: On the Hausdorff distance between a convex set and an interior random convex hull. Adv. Appl. Prob. (SGSA), **30**, 295–316 (1998)
122. Bröcker, L., Kuppe, M.: Integral geometry of tame sets. Geom. Dedicata, **82**, 285–328 (200)
123. Buchta, C.: Über die konvexe Hülle von Zufallspunkten in Eibereichen. Elem. Math., **38**, 153–156 (1983)
124. Buchta, C.: Stochastische Approximation konvexer Polygone. Z. Wahrscheinlichkeitsth. verw. Geb., **67**, 283–304 (1984)
125. Buchta, C.: Zufallspolygone in konvexen Vielecken. J. reine angew. Math., **347**, 212–220 (1984)
126. Buchta, C.: Zufällige Polyeder – Eine Übersicht. In: Hlawka, E. (ed) Zahlentheoretische Analysis, pp. 1–13, Lecture Notes in Math., **1114**. Springer, Berlin (1985)
127. Buchta, C.: On a conjecture of R.E. Miles about the convex hull of random points. Monatsh. Math., **102**, 91–102 (1986)
128. Buchta, C.: On the number of vertices of random polyhedra with a given number of facets. SIAM J. Alg. Discr. Meth., **8**, 85–92 (1987)
129. Buchta, C.: Distribution-independent properties of the convex hull of random points. J. Theoret. Probab., **3**, 387–393 (1990)
130. Buchta, C.: On the probability that two sets of points have disjoint convex hulls. Rend. Circ. Mat. Palermo, Ser. II, Suppl., **35**, 67–74 (1994)
131. Buchta, C.: An identity relating moments of functionals of convex hulls. Discrete Comput. Geom., **33**, 125–142 (2005)
132. Buchta, C.: The exact distribution of the number of vertices of a random convex chain. Mathematika, **53**, 247–254 (2006)

133. Buchta, C., Müller, J.: Random polytopes in a ball. J. Appl. Prob., **21**, 753–762 (1984)
134. Buchta, C., Müller, J., Tichy, R.F.: Stochastical approximation of convex bodies. Math. Ann., **271**, 225–235 (1985)
135. Buchta, C., Reitzner, M.: What is the expected volume of a tetrahedron whose vertices are chosen at random from a given tetrahedron? Anz. Österreich. Akad. Wiss. Math.-Natur. Kl., **129**, 63–68 (1992)
136. Buchta, C., Reitzner, M.: Equiaffine inner parallel curves of a plane convex body and the convex hulls of randomly chosen points. Probab. Theory Related Fields, **108**, 385–415 (1997)
137. Buchta, C., Reitzner, M.: On a theorem of G. Herglotz about random polygons. Rend. Circ. Mat. Palermo, Ser. II, Suppl., **50**, 89–102 (1997)
138. Buchta, C., Reitzner, M.: The convex hull of random points in a tetrahedron: Solution of Blaschke's problem and more general results. J. reine angew. Math., **536**, 1–29 (2001)
139. Buffon, G.–L.L. Comte de: Editor's note concerning a lecture given 1733 by Mr. Leclerc Buffon to the Royal Academy of Sciences in Paris. Histoire de l'Acad. Roy. des Sci., 43–45 (1733)
140. Buffon, G.–L.L. Comte de: Essai d'Arithmétique Morale. In: Histoire naturelle, générale et particulière, Supplément 4, pp. 46–123. Imprimerie Royale, Paris (1777)
141. Busemann, H.: Volume in terms of concurrent cross-sections. Pacific J. Math., **3**, 1–12 (1953)
142. Busemann, H.: The Geometry of Geodesics. Academic Press, New York (1955)
143. Busemann, H.: Areas in affine spaces III. The integral geometry of affine area. Rend. Circ. Mat. Palermo **9**, 226–242 (1960)
144. Busemann, H., Straus, E.G.: Area and normality. Pacific J. Math., **10**, 35–72 (1960)
145. Cabo, A., Baddeley, A.: Line transects, covariance functions and set convergence. Adv. Appl. Prob. (SGSA), **27**, 585–605 (1995)
146. Cabo, A., Baddeley, A.: Estimation of mean particle volume using the set covariance function. Adv. Appl. Prob. (SGSA), **35**, 27–46 (2003)
147. Cabo, A.J., Groeneboom, P.: Limit theorems for functionals of convex hulls. Probab. Theory Related Fields, **100**, 31–55 (1994)
148. Calka, P.: Mosaïques poissoniennes de l'espace euclidien. Une extension d'un résultat de R.E. Miles. C. R. Acad. Sci., Paris, Sér. I, Math., **332**, 557–562 (2001)
149. Calka, P.: Poissonian tessellations of the Euclidean space. An extension of a result of R.E. Miles. In: De nouveaux résultats sur la géométrie des mosaïques de Poisson–Voronoi et des mosaïques poissoniennes d'hyperplans. Doctoral Thesis, Université Lyon 1 (2002).
150. Calka, P.: The distributions of the smallest disks containing the Poisson–Voronoi typical cell and the Crofton cell in the plane. Adv. Appl. Prob. (SGSA), **34**, 702–717 (2002)
151. Calka, P.: An explicit expression for the distribution of the number of sides of the typical Poisson–Voronoi cell. Adv. Appl. Prob. (SGSA), **35**, 863–970 (2003)
152. Calka, P.: Precise formulae for the distributions of the principal geometric characteristics of the typical cells of a two-dimensional Poisson–Voronoi tessellation and a Poisson line process. Adv. Appl. Prob. (SGSA), **35**, 551–562 (2003)

153. Calka, P., Schreiber, T.: Limit theorems for the typical Poisson–Voronoi cell and the Crofton cell with a large inradius. Ann. Probab., **33**, 1625–1642 (2005)
154. Calka, P., Schreiber, T.: Large deviation probabilities for the number of vertices of random polytopes in the ball. Adv. Appl. Prob. (SGSA), **38**, 47–58 (2006)
155. Campi, S., Colesanti, A., Gronchi, P.: A note on Sylvester's problem for random polytopes in a convex body. Rend. Istit. Mat. Univ. Trieste, **31**, 79–94 (1999)
156. Campi, S., Gronchi, P.: Extremal convex sets for Sylvester–Busemann type functionals. Applicable Analysis, **85**, 129–141 (2006)
157. Capasso, V., DeGiosa, M., Micheletti, A., Mininni, R.: Stochastic modelling and statistics of polymer crystallization processes. Surveys Math. Ind., **6**, 109–132 (1996)
158. Capasso, V., Villa, E.: Survival functions and contact distribution functions for inhomogeneous, stochastic geometric marked point processes. Stoch. Anal. Appl., **23**, 79–96 (2005)
159. Capasso, V., Villa, E.: On the evolution equations of mean geometric densities for a class of space and time inhomogeneous stochastic birth-and-growth processes. In [50], pp. 185–245 (2007)
160. Carleman, T.: Über eine isoperimetrische Aufgabe und ihre physikalischen Anwendungen. Math. Z., **3**, 1–7 (1919)
161. Carlsson, S., Grenander, U.: Statistical approximation of plane convex sets. Skand. Aktuarietidsskr., **3/4**, 113–127 (1967)
162. Carnal, H.: Die konvexe Hülle von n rotationssymmetrisch verteilten Punkten. Z. Wahrscheinlichkeitsth. verw. Geb., **15**, 168–176 (1970)
163. Carnal, H., Hüsler, J.: On the convex hull of n random points on a circle. J. Appl. Prob., **28**, 231–237 (1991)
164. Cartan, É.: Le principe de dualité et certaines intégrales multiples de l'espace tangentiel et de l'espace réglé. Bull. Soc. Math. France, **24**, 140–177 (1896); Œvres complètes, Partie II, vol. 1, 265–302, Gauthier–Villars, Paris (1953)
165. Castaing, C., Valadier, M.: Convex Analysis and Measurable Multifunctions. Lecture Notes in Math., **580**, Springer, Berlin (1977)
166. Cauchy, A.: Note sur divers théorèmes relatifs à la rectification des courbes, et à la quadrature des surfaces. C. R. Acad. Sci., Paris, **13**, 1060–1065 (1841); Œvres complètes (1), vol. 6, pp. 369–375. Gauthier–Villars, Paris (1888)
167. Cauchy, A.: Mémoire sur la rectification des courbes et la quadrature des surfaces courbes. Mém. Acad. Sci. Paris, **22**, 3–13 (1850); Œvres complètes (1), vol. 2, pp. 167–177. Gauthier–Villars, Paris (1908)
168. Černý, R.: Laslett's transform for the Boolean model in $\mathbb{R}^d$. Kybernetika (Prague), **42**, 569–594 (2006)
169. Chern, S.–S.: On the kinematic formula in the euclidean space of n dimensions. Amer. J. Math., **74**, 227–236 (1952)
170. Chern, S.–S.: On the kinematic formula in integral geometry. J. Math. Mech., **16**, 101–118 (1966)
171. Chern, S.–S., Yien, C.T.: Sulla formula principale cinematica dello spazio ad n dimensioni. Boll. Un. Mat. Ital., (2) **2**, 434–437 (1940)
172. Chiu, S.N., van de Weygaert, R., Stoyan, D.: The sectional Poisson Voronoi tessellation is not a Voronoi tessellation. Adv. Appl. Prob. (SGSA), **28**, 356–376 (1996)
173. Choquet, G.: Theory of capacities. Ann. Inst. Fourier V (1953–54) , 131–295 (1955)

174. Choquet, G.: Lectures on Analysis, vol. 1. W.A. Benjamin, Reading, Mass. (1969)
175. Chow, Y.S., Teicher, H.: Probability Theory. 3rd ed., Springer, New York (1997)
176. Chu, D.P.T.: Random r-content of an r-simplex from beta-type-2 random points. Canad. J. Statist., **21**, 285–293 (1993)
177. Cohn, D.L.: Measure Theory. Birkhäuser, Boston (1980)
178. Coleman, R.: Random paths through convex bodies. J. Appl. Prob., **6**, 430–441 (1969)
179. Cover, T.M., Efron, B.: Geometrical probability and random points on a hypersphere. Ann. of Math. Stat., **38**, 213–220 (1971)
180. Cowan, R.: The use of the ergodic theorems in random geometry. Adv. Appl. Prob., Suppl., **10**, 47–57 (1978)
181. Cowan, R.: Properties of ergodic random mosaic processes. Math. Nachr., **97**, 89–102 (1980)
182. Cowan, R.: A more comprehensive complementary theorem for the analysis of Poisson point processes. Adv. Appl. Prob. (SGSA), **38**, 581–601 (2006)
183. Cowan, R.: Identities linking volumes of convex hulls. Adv. Appl. Prob. (SGSA), **39**, 1–15 (2007)
184. Cowan, R., Tsang, A.K.L.: The falling leaf mosaic and its equilibrium properties. Adv. Appl. Prob., **26**, 54–62 (1994)
185. Cressie, N.: Statistics for Spatial Data. Wiley, New York (1993)
186. Crofton, M.W.: On the theory of local probability, applied to straight lines drawn at random in a plane; the methods used being also extended to the proof of certain new theorems in the integral calculus. Phil. Trans. Roy. Soc. London, **158**, 181–199 (1868)
187. Crofton, M.W.: Sur quelques théorèmes de calcul intégral. C. R. Acad. Sci. Paris, **68**, 1469–1470 (1869)
188. Crofton, M.W.: Probability. In: Enyclopedia Britannica, 9th ed., **19**, pp. 768–788. Cambridge University Press, Cambridge (1885)
189. Cruz–Orive, L.M.: Stereology: meeting point of integral geometry, probability, and statistics. Math. Notae, **41**, 49–98 (2002)
190. Cruz–Orive, L.M.: A new stereological principle for test lines in three-dimensional space. J. Microsc., **219**, 18–28 (2005)
191. Czuber, E.: Geometrische Wahrscheinlichkeiten und Mittelwerte. Teubner, Leipzig (1884)
192. Dafnis, N., Giannopoulos, A., Guédon, O.: On the isotropic constant of random polytopes. Preprint (2008)
193. Daley, D., Last, G.: Descending chains, the lilypond model, and mutual nearest neighbour matching. Adv. Appl. Prob. (SGSA), **37**, 604–628 (2004)
194. Daley, D., Vere–Jones, D.: An Introduction to the Theory of Point Processes. Springer, New York (1988)
195. Daley, D., Vere–Jones, D.: An Introduction to the Theory of Point Processes, Vol. 1: Elementary Theory and Methods. 2nd ed. 2003, Corr. 2nd printing, Springer, New York (2005)
196. Daley, D., Vere–Jones, D.: An Introduction to the Theory of Point Processes, Vol. 2, Springer, New York (2008)
197. Dalla, L., Larman, D.G.: Volumes of a random polytope in a convex set. In: Gritzmann, P., Sturmfels, B. (eds) Applied Geometry and Discrete Mathematics: The Victor Klee Festschrift, pp. 175–180. DIMACS, Ser. Discrete Math. Theor. Comput. Sci., **4** (1991)

198. Davy, P.: Projected thick sections through multi-dimensional particle aggregates. J. Appl. Prob., **13**, 714–722 (1976); Correction: J. Appl. Prob., **15**, 456 (1978)
199. Davy, P.: Stereology – A Statistical Viewpoint. Ph.D. Thesis, Australian National University, Canberra (1978)
200. Davy, P.: Inequalities for moments of secant length. Z. Wahrscheinlichkeitsth. verw. Geb., **68**, 243–246 (1984)
201. Davy, P., Miles, R.E.: Sampling theory for opaque spatial specimens. J. R. Statist. Soc. B **39**, 56–65 (1977)
202. Deltheil, R.: Probabilités Géométriques. Gauthier–Villars, Paris (1926)
203. Devroye, L.: On the oscillation of the expected number of extreme points of a random set. Statistics & Prob. Letters, **11**, 281–286 (1991)
204. Diggle, P.J.: Statistical Analysis of Spatial Point Patterns. Academic Press, London (1983)
205. Döge, G., Mecke, K., Møller, J., Stoyan, D., Waagepetersen, R.: Grand canonical simulations of hard-disk systems by simulated tempering. Int. J. Modern Physics C, **15**, 129–147 (2004).
206. Donoho, D.L.: Neighborly polytopes and sparse solutions of underdetermined linear equations. Technical Report, Department of Statistics, Stanford University (2004)
207. Donoho, D.L.: High-dimensional centrally symmetric polytopes with neighborliness proportional to dimension. Discrete Comput. Geom., **35**, 617–652 (2006)
208. Donoho, D.L., Tanner, J.: Sparse nonnegative solutions of underdetermined linear equations by linear programming. Proc. Natl. Acad. Sci. USA, **102**, 9446–9451 (2005)
209. Donoho, D.L., Tanner, J.: Neighborliness of randomly-projected simplices in high dimensions. Proc. Natl. Acad. Sci. USA, **102**, 9452–9457 (2005)
210. Donoho, D.L., Tanner, J.: Counting faces of randomly-projected polytopes when the projection radically lowers dimension. Technical Report No. 2006-11, Dept. of Statistics, Stanford University (2006)
211. Dümbgen, L., Walther, G.: Rates of convergence for random approximations of convex sets. Adv. Appl. Prob. (SGSA), **28**, 384–393 (1996)
212. Duma, A., Stoka, I.: Geometrical probabilities for convex test bodies. Beiträge Algebra Geom., **40**, 15-25 (1999)
213. Dwyer, R.A.: Convex hulls of samples from spherically symmetric distributions. Discrete Appl. Math., **31**, 113–132 (1991)
214. Dyer, M.E., Füredi, Z., McDiarmid, C.: Volume spanned by random points in the hypercube. Random Structures Algorithms, **3**, 91–106 (1992)
215. Efron, B.: The convex hull of a random set of points. Biometrika, **52**, 331–343 (1965)
216. El–Ekhtiar, A.: Integral geometry in Minkowski space. Ph.D. Thesis, University of California, Davis, CA (1992)
217. Enns, E.G., Ehlers, P.F.: Random paths through a convex region. J. Appl. Prob., **15**, 144–152 (1978)
218. Enns, E.G., Ehlers, P.F.: Random paths originating within a convex region and terminating on its surface. Austral. J. Statist., **22**, 60–68 (1980)
219. Enns, E.G., Ehlers, P.F.: Random secants of a convex body generated by surface randomness. J. Appl. Prob., **18**, 157–166 (1981)
220. Enns, E.G., Ehlers, P.F.: Chords through a convex body generated from within an embedded body. J. Appl. Prob., **25**, 700–707 (1988)

221. Enns, E.G., Ehlers, P.F.: Notes on random chords in convex bodies. J. Appl. Prob., **30**, 889–897 (1993)
222. Fallert, H.: Intensitätsmaße und Quermaßdichten für (nichtstationäre) zufällige Mengen und geometrische Punktprozesse. Dissertation, Universität Karlsruhe (1992)
223. Fallert, H.: Quermaßdichten für Punktprozesse konvexer Körper und Boolesche Modelle. Math. Nachr., **181**, 165–184 (1996)
224. Favis, W.: Extremaleigenschaften und Momente für stationäre Poissonsche Hyperebenenmosaike. Doctoral Thesis, Friedrich-Schiller-Universität, Jena (1995)
225. Favis, W.: Inequalities for stationary Poisson cuboid processes. Math. Nachr., **178**, 117–127 (1996)
226. Favis, W., Weiss, V.: Mean values of weighted cells of stationary Poisson hyperplane tessellations of R^d. Math. Nachr., **193**, 37–48 (1998)
227. Federer, H.: Some integral geometric theorems. Trans. Amer. Math. Soc., **77**, 238–261 (1954)
228. Federer, H.: Curvature measures. Trans. Amer. Math. Soc., **93**, 418–491 (1959)
229. Federer, H.: Geometric Measure Theory. Springer, Berlin (1969)
230. Fell, J.M.G.: A Hausdorff topology for the closed subsets of a locally compact space. Proc. Amer. Math. Soc., **13**, 472–476 (1962)
231. Fernandes, E.: Double fibrations: a modern approach to integral geometry and Crofton formulas in projective Finsler spaces. Doctoral Thesis, Université de Louvain (2002)
232. Finch, S., Hueter, I.: Random convex hulls: a variance revisited. Adv. Appl. Prob. (SGSA), **36**, 981–986 (2004)
233. Firey, W.J.: An integral-geometric meaning for lower order area functions of convex bodies. Mathematika, **19**, 205–212 (1972)
234. Firey, W.J.: Kinematic measures for sets of support figures. Mathematika, **21**, 270–281 (1974)
235. Firey, W.J.: Inner contact measures. Mathematika, **26**, 106–112 (1979)
236. Fu, J.H.G.: Kinematic formulas in integral geometry. Indiana Univ. Math. J., **39**, 1115–1154 (1990)
237. Fu, J.H.G.: Curvature of singular spaces via the normal cycle. In: Greene, R. *et al.* (eds) Differential geometry. Part 2: Geometry in mathematical physics and related topics (Los Angeles, CA, 1990), Proc. Symp. Pure Math. **54**, Part 2, pp. 211–221. Amer. Math. Soc., Providence, RI (1993)
238. Fu, J.H.G.: Curvature measures of subanalytic sets. Amer. J. Math., **116**, 819–880 (1994)
239. Fu, J.H.G.: Structure of the unitary valuation algebra. J. Differential Geom., **72**, 509–533 (2006)
240. Füredi, Z.: Random polytopes in the d-dimensional cube. Discrete Comput. Geom., **1**, 315–319 (1986)
241. Furstenberg, H., Tzkoni, I.: Spherical functions and integral geometry. Israel J. Math., **10**, 327–338 (1971)
242. Gaal, S.A.: Linear Analysis and Representation Theory. Springer, Berlin (1973)
243. Gao, F., Hug, D., Schneider, R.: Intrinsic volumes and polar sets in spherical space. Math. Notae, **41**, 159–176 (2002)
244. Gardner, R.J.: Geometric Tomography. 2nd ed., Cambridge University Press, Cambridge (2006)
245. Gardner, R.J.: The dual Brunn–Minkowski theory for bounded Borel sets: Dual affine quermassintegrals and inequalities. Adv. Math., **216**, 358–386 (2007)

246. Gardner, R.J., Pfeffer, W.F.: Borel measures. In: Kunen, K., Vaughan, J.E. (eds) Handbook of Set-theoretic Topology, pp. 961–1043. North-Holland, Amsterdam (1984)
247. Gates, J.: Recognition of triangles and quadrilaterals by chord length distributions. J. Appl. Prob., **19**, 873–879 (1982)
248. Gates, J.: Bounds for the probability of complete intersection of random chords in a circle. J. Appl. Prob., **21**, 419–424 (1984)
249. Gates, J.: Some properties of chord length distributions. J. Appl. Prob., **24**, 863–874 (1987)
250. Gatzouras, D., Giannopoulos, A.: A large deviation approach to the geometry of random polytopes. Mathematika, **53**, 173–211 (2006)
251. Gatzouras, D., Giannopoulos, A.: Threshold for the volume spanned by random points with independent coordinates. Israel J. Math. (to appear)
252. Gatzouras, D., Giannopoulos, A., Markoulakis, N.: Lower bound for the maximal number of facets of a 0/1 polytope. Discrete Comput. Geom., **34**, 331–349 (2005)
253. Gatzouras, D., Giannopoulos, A., Markoulakis, N.: On the maximal number of facets of 0/1 polytopes. Geometric Aspects of Functional Analysis (Israel Seminar 2004–2005), Lecture Notes in Math., **1910**, 117–125, Springer, Berlin (2007)
254. Geffroy, J.: Localisation asymptotique du polyèdre d'appui d'un échantillon laplacien à k dimensions. Publ. Inst. Statist. Univ. Paris X, 3, 213–228 (1961)
255. Gelfand, I.M., Smirnov, M.M.: Lagrangians satisfying Crofton formulas, Radon transforms, and nonlocal differentials. Adv. Math., **109**, 188–227 (1994)
256. Geman, D., Horowitz, J.: Random shifts which preserve measure. Proc. Amer. Math. Soc., **49**, 143–150 (1975)
257. Giannopoulos, A.: On the mean value of the area of a random polygon in a plane convex body. Mathematika, **39**, 279–290 (1992)
258. Giannopoulos, A., Hartzoulaki, M.: Random spaces generated by vertices of the cube. Discrete Comput. Geom., **28**, 255–273 (2002)
259. Giannopoulos, A., Tsolomitis, A.: Volume radius of a random polytope in a convex body. Math. Proc. Cambridge Philos. Soc., **134**, 13–21 (2003)
260. Giger, H., Hadwiger, H.: Über Treffzahlwahrscheinlichkeiten im Eikörperfeld. Z. Wahrscheinlichkeitsth. verw. Geb., **10**, 329–334 (1968)
261. Gilbert, E.N.: Random subdivisions of space into crystals. Ann. of Math. Statist., **33**, 958–972 (1962)
262. Giné, E., Hahn, M.G.: The Lévy-Hinčin representation for random compact convex subsets which are infinitely divisible under Minkowski addition. Z. Wahrscheinlichkeitsth. verw. Geb., **70**, 271–287 (1985)
263. Giné, E., Hahn, M.G, Zinn, J.: Limit theorems for random sets: an application of probability in Banach space results. (Probability in Banach spaces, IV, Oberwolfach 1982) Lecture Notes in Math., **990**, 112–135, Springer, New York (1983)
264. Glasauer, S.: Integralgeometrie konvexer Körper im sphärischen Raum. Doctoral Thesis, Albert-Ludwigs-Universität, Freiburg i. Br. (1995)
265. Glasauer, S.: Integral geometry of spherically convex bodies. Diss. Summ. Math., **1**, 219–226 (1996)
266. Glasauer, S.: A generalization of intersection formulae of integral geometry. Geom. Dedicata, **68**, 101–121 (1997)

267. Glasauer, S.: Translative and kinematic integral formulae concerning the convex hull operation. Math. Z., **229**, 493–518 (1998)
268. Glasauer, S.: Kinematic formulae for support measures of convex bodies. Beiträge Algebra Geom., **40**, 113–124 (1999)
269. Glasauer, S., Schneider, R.: Asymptotic approximation of smooth convex bodies by polytopes. Forum Math., **8**, 363–377 (1996)
270. Goldman, A.: Sur une conjecture de D.G. Kendall concernant la cellule de Crofton du plan et sur sa contrepartie brownienne. Ann. Probab., **26**, 1727–1750 (1998)
271. Goodey, P., Howard, R.: Processes of flats induced by higher-dimensional processes. Adv. Math., **80**, 92–109 (1990)
272. Goodey, P., Howard, R.: Processes of flats induced by higher-dimensional processes II. Contemp. Math., **113**, 111–119 (1990)
273. Goodey, P., Howard, R., Reeder, M.: Processes of flats induced by higher-dimensional processes III. Geom. Dedicata, **61**, 257–269 (1996)
274. Goodey, P.R., Schneider, R.: On the intermediate area functions of convex bodies. Math. Z., **173**, 185–194 (1980)
275. Goodey, P.R., Schneider, R., Weil, W.: Projection functions on higher rank Grassmannians. In: Lindenstrauss, J., Milman, V. (eds) Geometric Aspects of Functional Analysis (Israel Seminar 1992–1994), Operator Theory: Advances and Applications, pp. 75–90. Birkhäuser, Basel (1995)
276. Goodey, P.R., Schneider, R., Weil, W.: On the determination of convex bodies by projection functions. Bull. London Math. Soc., **29**, 82–88 (1997)
277. Goodey, P., Weil, W.: Translative integral formulae for convex bodies. Aequationes Math., **34**, 64–77 (1987)
278. Goodey, P., Weil, W.: Integral geometric formulae for projection functions. Geom. Dedicata, **41**, 117–126 (1992)
279. Goodey, P., Weil, W.: The determination of convex bodies from the mean of random sections. Math. Proc. Camb. Phil. Soc., **112**, 419–430 (1992)
280. Goodey, P., Weil, W.: Representations of mixed measures with applications to Boolean models. Rend. Circ. Mat. Palermo, Ser. II, Suppl., **70**, 325–346 (2002)
281. Goodey, P., Weil, W.: Translative and kinematic integral formulae for support functions II. Geom. Dedicata, **99**, 103–125 (2003)
282. Gordon, Y., Meyer, M., Reisner, S.: Zonoids with minimal volume product – a new proof. Proc. Amer. Math. Soc., **104**, 273–276 (1988)
283. Goudsmit, S.: Random distributions of lines in a plane. Rev. Modern Phys., **17**, 321–322 (1945)
284. Goutsias, J., Mahler, R.P.S., Nguyen, H.T.: Random Sets: Theory and Applications. Springer, New York (1997)
285. Gridgeman, N.T.: Geometric probability and the number π. Scripta Mathematica, **25**, 183–195 (1960)
286. Grinberg, E.L.: Isoperimetric inequalities and identities for k-dimensional cross-sections of convex bodies. Math. Ann., **291**, 75–87 (1991)
287. Groemer, H.: On some mean values associated with a randomly selected simplex in a convex set. Pacific J. Math., **45**, 525–533 (1973)
288. Groemer, H.: On the mean value of the volume of a random polytope in a convex set. Arch. Math., **25**, 86–90 (1974)
289. Groemer, H.: On the extension of additive functionals on classes of convex sets. Pacific J. Math., **75**, 397–410 (1978)

290. Groemer, H.: The average measure of the intersection of two sets. Z. Wahrscheinlichkeitsth. verw. Geb., **54**, 15–20 (1980)
291. Groemer, H.: The average distance between two convex sets. J. Appl. Prob., **17**, 415–422 (1980)
292. Groemer, H.: On the average size of polytopes in a convex set. Geom. Dedicata, **13**, 47–62 (1982)
293. Groeneboom, P.: Limit theorems for convex hulls. Probab. Theory Related Fields, **79**, 327–368 (1988)
294. Groeneboom, P.: Convex hulls of uniform samples from a convex polygon. Preprint (2006)
295. Gruber, P.M.: In most cases approximation is irregular. Rend. Sem. Mat. Univers. Politecn. Torino, **41**, 19–33 (1983)
296. Gruber, P.M.: Expectation of random polytopes. Manuscr. math., **91**, 393–419 (1996)
297. Gruber, P.M.: Comparisons of best and random approximations of convex bodies by polytopes. Rend. Circ. Mat. Palermo, Ser. II, Suppl., **50**, 189–216 (1997)
298. Gruber, P., Schneider, R.: Problems in geometric convexity. In: Tölke, J., Wills, J.M. (eds) Contributions to Geometry, pp. 255–278. Birkhäuser, Basel (1979)
299. Grünbaum, B.: Convex Polytopes. Interscience, London (1967)
300. Grünbaum, B.: Grassmann angles of convex polytopes. Acta Math., **121**, 293–302 (1968)
301. Guderlei, R., Klenk, S., Mayer, J., Schmidt, V., Spodarev, E.: Algorithms for the computation of the Minkowski functionals of deterministic and random polyconvex sets. Image and Vision Computing, **25**, 464–474 (2007)
302. Gutkowski, P., Jensen, E.B.V., Kiderlen, M.: Directional analysis of digitized three-dimensional images by configuration counts. J. Microsc., **216**, 175–185 (2004)
303. Hadwiger, H.: Über eine Mittelwertformel für Richtungsfunktionale im Vektorraum und einige Anwendungen. J. reine angew. Math., **185**, 241–252 (1943)
304. Hadwiger, H: Beweis eines Funktionalsatzes für konvexe Körper. Abh. Math. Sem. Univ. Hamburg, **17**, 69–76 (1951)
305. Hadwiger, H: Additive Funktionale k-dimensionaler Eikörper, I. Arch. Math., **3**, 470–478 (1952)
306. Hadwiger, H: Integralsätze im Konvexring. Abh. Math. Sem. Univ. Hamburg, **20**, 136–154 (1956)
307. Hadwiger, H.: Vorlesungen über Inhalt, Oberfläche und Isoperimetrie. Springer, Berlin (1957)
308. Hadwiger, H.: Das Wills'sche Funktional. Monatsh. Math., **79**, 213–221 (1975)
309. Hadwiger, H.: Eine Erweiterung der kinematischen Hauptformel der Integralgeometrie. Abh. Math. Sem. Univ. Hamburg, **44**, 84–90 (1975)
310. Hadwiger, H.: Eikörperrichtungsfunktionale und kinematische Integralformeln. Studienvorlesung (manuscript), Universität Bern (1975)
311. Hadwiger, H.: Gitterpunktzahl im Simplex und Wills'sche Vermutung. Math. Ann., **239**, 271–288 (1979)
312. Hadwiger, H., Schneider, R.: Vektorielle Integralgeometrie. Elem. Math., **26**, 49–57 (1971)
313. Hadwiger, H., Streit, F.: Über Wahrscheinlichkeiten räumlicher Bündelungserscheinungen. Monatsh. Math., **74**, 30–40 (1970)
314. Hahn, U., Micheletti, A., Pohlink, R., Stoyan, D., Wendrock, H.: Stereological analysis and modelling of gradient structures. J. Microsc., **195**, 113–124 (1999)

315. Hall, G.R.: Acute triangles in the n-ball. J. Appl. Prob., **19**, 712–715 (1982)
316. Hall, P.: Counting methods for inference in binary mosaics. Biometrics, **41**, 1049–1052 (1985)
317. Hall, P.: Introduction to the Theory of Coverage Processes. Wiley, New York (1988)
318. Halmos, P.R.: Measure Theory. Van Nostrand, Princeton, N.J. (1950)
319. Hanisch, K.–H.: On classes of random sets and point processes. Serdica, **7**, 160–166 (1981)
320. Hanisch, K.–H.: On inversion formulae for n-fold Palm distributions of point processes in LCS-spaces. Math. Nachr., **106**, 171–179 (1982)
321. Harding, E.F., Kendall, D.G.: Stochastic Geometry. Wiley, London (1974)
322. Hartzoulaki, M., Paouris, G.: Quermassintegrals of a random polytope in a convex body. Arch. Math., **80**, 430–438 (2003)
323. Hausdorff, F.: Grundzüge der Mengenlehre. Verlag von Veit, Leipzig (1914)
324. Heinrich, L.: On existence and mixing properties of germ-grain models. Statistics, **23**, 271–286 (1992)
325. Heinrich, L.: Contact and chord length distribution of a stationary Voronoi tessellation. Adv. Appl. Prob. (SGSA), **30**, 603–618 (1998)
326. Heinrich, L.: Large deviations of the empirical volume fraction for stationary Poisson grain models. Ann. Appl. Probab., **15**, 392–420 (2005)
327. Heinrich, L.: Limit distributions of some stereological estimators in Wicksell's corpuscle problem. Image Anal. Stereol., **26**, 63–71 (2007)
328. Heinrich, L.: Central limit theorems for motion-invariant Poisson hyperplanes in expanding convex bodies. Rend. Circ. Mat. Palermo (to appear)
329. Heinrich, L., Molchanov, I.S.: Central limit theorems for a class of random measures associated with germ-grain models. Adv. Appl. Prob. (SGSA), **31**, 283–314 (1999)
330. Heinrich, L., Muche, L.: Second-order properties of the point process of nodes in a stationary Voronoi tessellation. Math. Nachr., **281**, 350–375 (2008)
331. Heinrich, L., Schmidt, H., Schmidt, V.: Limit theorems for stationary tessellations with random inner cell structures. Adv. Appl. Prob. (SGSA), **37**, 25–47 (2005)
332. Heinrich, L., Schmidt, H., Schmidt, V.: Central limit theorems for Poisson hyperplane tessellations. Ann. Appl. Prob. (SGSA), **16**, 919–950 (2006)
333. Heinrich, L., Schmidt, H., Schmidt, V.: Limit theorems for functionals on the facets of stationary random tessellations. Bernoulli, **13**, 868–891 (2007)
334. Henze, N.: Random triangles in convex regions. J. Appl. Prob., **20**, 111–125 (1983)
335. Herglotz, G.: Geometrische Wahrscheinlichkeiten. Vorlesungsausarbeitung. Göttingen, 156 S. (1933)
336. Herglotz, G.: Über die Steinersche Formel für Parallelflächen. Abh. Math. Sem. Univ. Hamburg, **15**, 165–177 (1943)
337. Heveling, M.: Bijective point maps, point-stationarity and characterization of Palm measures. Doctoral Thesis, Universität Karlsruhe 2005. Universitätsverlag, Karlsruhe (2006)
338. Heveling, M., Hug, D., Last, G.: Does polynomial parallel volume imply convexity? Math. Ann., **328**, 469–479 (2004)
339. Heveling, M., Last, G.: Characterization of Palm measures via bijective point-shifts. Ann. Probab., **33**, 1698–1715 (2005)

340. Heveling, M., Last, G.: Existence, uniqueness and algorithmic computation of general lilypond systems. Random Structures Algorithms, **29**, 338–350 (2006)
341. Heveling, M., Last, G.: Point shift characterization of Palm measures on abelian groups. Electron. J. Probab., **12**, 122–137 (2007)
342. Hewitt, E., Ross, K.A.: Abstract Harmonic Analysis. Springer, Berlin (1963)
343. Hilhorst, H.J.: The perimeter of large planar Voronoi cells: a double-stranded random walk. J. Stat. Mech. L02003 (2005)
344. Hilhorst, H.J.: Asymptotic statistics of the n-sided planar Poisson–Voronoi cell. I. Exact results. J. Stat. Mech. P09005 (2005)
345. Hoffmann, L.M.: Schnittdichten inhomogener Poissonprozesse. Doctoral Thesis, Unversität Karlsruhe (2006)
346. Hoffmann, L.M.: Intersection densities of nonstationary Poisson processes of hypersurfaces. Adv. Appl. Prob. (SGSA), **39**, 307–317 (2007)
347. Hoffmann, L.M.: On weak stationarity and weak isotropy of processes of convex bodies and cylinders. Adv. Appl. Prob. (SGSA), **39**, 864–882 (2007)
348. Hoffmann, L.M.: Mixed measures of convex cylinders and quermass densities of Boolean models. Preprint (2007)
349. Holroyd, A., Peres, Y.: Extra heads and invariant allocations. Ann. Probab., **33**, 31–52 (2005)
350. Hostinský, B.: Sur les probabilités géométriques. Publ. Fac. Sci. Univ. Masaryk (Brno), **50**, 1–26 (1925)
351. Hsing, T.: On the asymptotic distribution of the area outside a random convex hull in a disk. Ann. Appl. Probab., **4**, 478–493 (1994)
352. Hueter, I.: The convex hull of a normal sample. Adv. Appl. Prob., **26**, 855–875 (1994)
353. Hueter, I.: Limit theorems for the convex hull of random points in higher dimensions. Trans. Amer. Math. Soc., **351**, 4337–4363 (1999)
354. Hueter, I.: Random convex hulls and the Stein method. Manuscript (2004)
355. Hug, D.: Measures, curvatures and currents in convex geometry. Habilitation Thesis, Universität Freiburg i. Br. (1999)
356. Hug, D.: Contact distributions of Boolean models. Rend. Circ. Mat. Palermo, Ser. II, Suppl., **65**, 137–181 (2000)
357. Hug, D., Last, G.: On support measures in Minkowski spaces and contact distributions in stochastic geometry. Ann. Probab., **28**, 796–850 (2000)
358. Hug, D., Last, G., Weil, W.: Generalized contact distributions of inhomogeneous Boolean models. Adv. Appl. Prob. (SGSA), **34**, 21–47 (2002)
359. Hug, D., Last, G., Weil, W.: A survey of contact distributions. In: Mecke, K., Stoyan, D. (eds) Morphology of Condensed Matter, pp. 317–357, Lecture Notes in Physics, **600**. Springer, Berlin (2002)
360. Hug, D., Last, G., Weil, W.: Distance measurements on processes of flats. Adv. Appl. Prob. (SGSA), **35**, 70–95 (2003)
361. Hug, D., Last, G., Weil, W.: A local Steiner-type formula for general closed sets and applications. Math. Z., **246**, 237–272 (2004)
362. Hug, D., Last, G., Weil, W.: Polynomial parallel volume, convexity and contact distributions of random sets. Probab. Theory Related Fields, **135**, 169–200 (2006)
363. Hug, D., Mani–Levitska, P., Schätzle, R.: Almost transversal intersections of convex surfaces and translative integral formulae. Math. Nachr., **246–247**, 121–155 (2002)

364. Hug, D., Munsonius, G.O., Reitzner, M.: Asymptotic mean values of Gaussian polytopes. Beiträge Algebra Geom., **45**, 531–548 (2004)
365. Hug, D., Reitzner, M.: Gaussian polytopes: variance and limit theorems. Adv. Appl. Prob (SGSA), **37**, 297–320 (2005)
366. Hug, D., Reitzner, M., Schneider, R.: The limit shape of the zero cell in a stationary Poisson hyperplane tessellation. Ann. Probab., **32**, 1140–1167 (2004)
367. Hug, D., Reitzner, M., Schneider, R.: Large Poisson–Voronoi cells and Crofton cells. Adv. Appl. Prob. (SGSA), **36**, 667–690 (2004)
368. Hug, D., Schätzle, R.: Intersections and translative integral formulas for boundaries of convex bodies. Math. Nachr., **226**, 99–128 (2001)
369. Hug, D., Schneider, R.: Kinematic and Crofton formulae of integral geometry: recent variants and extensions. In: Barceló i Vidal, C. (ed) Homenatge al professor Lluís Santaló i Sors. Universitat de Girona, Girona (2002)
370. Hug, D., Schneider, R.: Large cells in Poisson–Delaunay tessellations. Discrete Comput. Geom., **31**, 503–514(2004)
371. Hug, D., Schneider, R.: Large typical cells in Poisson–Delaunay mosaics. Rev. Roumaine Math. Pures Appl., **50**, 657–670 (2005)
372. Hug, D., Schneider, R.: Asymptotic shapes of large cells in random tessellations. Geom. Funct. Anal., **17**, 156–191 (2007)
373. Hug, D., Schneider, R.: Typical cells in Poisson hyperplane tessellations. Discrete Comput. Geom., **38**, 305–319 (2007)
374. Hug, D., Schneider, R., Schuster, R.: The space of isometry covariant tensor valuations. St. Petersburg Math. J., **19**, 137–158 (2008)
375. Hug, D., Schneider, R., Schuster, R.: Integral geometry of tensor valuations. Adv. in Appl. Math. (to appear)
376. Illian, J., Penttinen, A., Stoyan, H., Stoyan, D.: Statistical Analysis and Modelling of Spatial Point Patterns. Wiley, Chichester (2008)
377. Isokawa, Y.: Poisson–Voronoi tessellations in three-dimensional hyperbolic space. Adv. Appl. Prob. (SGSA), **32**, 648–662 (2000)
378. Janson, S., Kallenberg, O.: Maximizing the intersection density of fibre processes. J. Appl. Prob., **18**, 820–828 (1981)
379. Jensen, E.B.V.: Local Stereology. World Scientific, Singapore (1998)
380. Jensen, E.B., Gundersen, H.J.G.: Fundamental stereological formulae based on isotropically oriented probes through fixed points with applications to particle analysis. J. Microsc., **153**, 249–267 (1989)
381. Jensen, E.B.V., Kiderlen, M.: Directional analysis of digitized planar sets by configuration counts. J. Microsc., **212**, 158–168 (2003)
382. Jensen, E.B., Kiêu, K.: A new integral geometric formula of the Blaschke–Petkantschin type. Research Reports, **180**, Dept. of Theor. Statistics, Univ. of Aarhus (1991)
383. Jensen, E.B., Kiêu, K., Gundersen, H.J.G.: On the stereological estimation of reduced moment measures. Ann. Inst. Statist. Math., **42**, 445–461 (1990)
384. Jeulin, D. (ed.): Advances in Theory and Applications of Random Sets. World Scientific, Singapore (1997)
385. Kallenberg, O.: Random Measures. 4th ed., Akademie Verlag, Berlin and Academic Press, London (1986)
386. Kallenberg, O.: Foundations of Modern Probability. 2nd ed., Springer, New York (2002)
387. Kallenberg, O.: Invariant measures and disintegrations with applications to Palm and related kernels. Probab. Theory Related Fields, **139**, 285–310 (2007)

388. Kaltenbach, F.J.: Asymptotisches Verhalten zufälliger konvexer Polyeder. Doctoral Thesis, Albert-Ludwigs-Universität, Freiburg i. Br. (1990)
389. Karr, A.F.: Point Processes and their Statistical Inference. 2nd ed., Marcel Dekker, New York (1991)
390. Kellerer, A.M.: On the number of clumps resulting from the overlap of randomly placed figures in the plane. J. Appl. Prob., **20**, 126–135 (1983)
391. Kellerer, A.M.: Counting figures in planar random configurations. J. Appl. Prob., **22**, 68–81 (1985)
392. Kellerer, H.G.: Minkowski functionals of Poisson processes. Z. Wahrscheinlichkeitsth. verw. Geb., **67**, 63–84 (1984)
393. Kelley, J.L.: General Topology. Van Nostrand, Princeton (1955)
394. Kelly, D.G., Tolle, J.W.: Expected number of vertices of a random convex polyhedron. SIAM J. Alg. Discr. Meth., **2**, 441–451 (1981)
395. Kendall, D.G.: Foundations of a theory of random sets. In: Harding, E.F., Kendall, D.G. (eds) Stochastic Geometry, pp. 322–376. Wiley, New York (1974)
396. Kendall, D.G., Barden, D., Carne, T.K., Le, H.: Shape and Shape Theory. Wiley, Chichester (1999)
397. Kendall, M.G., Moran, P.A.P.: Geometrical Probability. Charles Griffin, London (1963)
398. Kendall, W.S., van Lieshout, M.N.M., Baddeley, A.J.: Quermass-interaction processes: conditions for stability. Adv. Appl. Prob. (SGSA), **31**, 315–342 (1999)
399. Kendall, W.S., Mecke, J.: The range of the mean-value quantities of planar tessellations. J. Appl. Prob., **24**, 411–421 (1987)
400. Kerstan, J., Matthes, K., Mecke, J.: Unbegrenzt teilbare Punktprozesse. Akademie-Verlag, Berlin (1974)
401. Keutel, J.: Ein Extremalproblem für zufällige Ebenen und für Ebenenprozesse in höherdimensionalen Räumen. Doctoral Thesis, Friedrich-Schiller-Universität, Jena (1991)
402. Khmaladze, E., Toronjadze, N.: On the almost sure coverage property of Voronoi tessellation: the $\mathbb{R}^1$ case. Adv. Appl. Prob. (SGSA), **33**, 756–764 (2001)
403. Kiderlen, M.: Non-parametric estimation of the directional distribution of stationary line and fibre processes. Adv. Appl. Prob. (SGSA), **33**, 6–24 (2001)
404. Kiderlen, M.: Determination of the mean normal measure from isotropic means of flat sections. Adv. Appl. Prob. (SGSA), **34**, 505–519 (2002)
405. Kiderlen, M.: Estimating the Euler characteristic of a planar set from a digital image. J. Vis. Commun. Image R., **17**, 1237–1255 (2006)
406. Kiderlen, M.: Estimation of the mean normal measure from flat sections. Adv. Appl. Prob. (SGSA), **40**, 31–48 (2008)
407. Kiderlen, M., Jensen, E.B.V.: Estimation of the directional measure of planar random sets by digitization. Adv. Appl. Prob. (SGSA), **35**, 583–602 (2003)
408. Kiderlen, M., Pfrang, A.: Algorithms to estimate the rose of directions of a spatial fiber system. J. Microsc., **219**, 50–60 (2005)
409. Kiderlen, M., Weil, W.: Measure-valued valuations and mixed curvature measures of convex bodies. Geom. Dedicata, **76**, 291–329 (1999)
410. Kiêu, K.: A coarea formula for multiple geometric integrals. Math. Nachr., **156**, 75–86 (1992)
411. Kingman, J.F.C.: Mean free paths in a convex reflecting region. J. Appl. Prob., **2**, 162–168 (1965)

412. Kingman, J.F.C.: Random secants of a convex body. J. Appl. Prob., **6**, 660–672 (1969)
413. Kingman, J.F.C.: Poisson Processes. Clarendon Press, Oxford (1993)
414. Klain D.: A short proof of Hadwiger's characterization theorem. Mathematika, **42**, 329–339 (1995)
415. Klain D.: Kinematic formulas for finite lattices. Ann. Combinatorics, **1**, 353–366 (1997)
416. Klain D., Rota, G.–C.: Introduction to Geometric Probability. Cambridge University Press, Cambridge (1997)
417. Klartag, B., Kozma, G.: On the hyperplane conjecture for random convex sets. arXiv:math.MG/0612517v1 (2006) Israel J. Math. (to appear)
418. Klee, V.: What is the expected volume of a simplex whose vertices are chosen at random from a given convex body? (Research problem) Amer. Math. Monthly, **76**, 286–288 (1969)
419. Klein, E., Thompson, A.C.: Theory of Correspondences. Wiley, New York (1984)
420. Klenk, S., Schmidt, V., Spodarev, E.: A new algorithmic approach to the computation of Minkowski functionals of germ-grain models. Comput. Geom., **34**, 127-148 (2006)
421. Knothe, H.: Über Ungleichungen bei Sehnenpotenzintegralen. Deutsche Math., **2**, 544–551 (1937)
422. Kohlmann, P.: Curvature measures and Steiner formulae in space forms. Geom. Dedicata, **40**, 191–211 (1991)
423. König, D., Schmidt, V.: Zufällige Punktprozesse. Teubner, Stuttgart (1992)
424. Kötzer, S., Jensen, E.B.V., Baddeley, A.: Geometric identities in stereological particle analysis. Thiele Centre Res. Rep. 2005–10 (2005)
425. Kovalenko, I.N.: A proof of a conjecture of David Kendall on the shape of random polygons of large area. (Russian) Kibernet. Sistem. Anal. 1997, 3–10, 187; Engl. transl.: Cybernet. Systems Anal., **33**, 461–467 (1997)
426. Kovalenko, I.N.: An extension of a conjecture of D.G. Kendall concerning shapes of random polygons to Poisson Voronoï cells. In: Engel, P. *et al.* (eds) Voronoï's impact on modern science, Book I. Transl. from the Ukrainian, Kyiv: Institute of Mathematics. Proc. Inst. Math. Natl. Acad. Sci. Ukr., Math. Appl., **212**, 266–274 (1998)
427. Kovalenko, I.N.: A simplified proof of a conjecture of D.G. Kendall concerning shapes of random polygons. J. Appl. Math. Stochastic Anal., **12**, 301–310 (1999)
428. Krengel, U.: Ergodic Theorems. W. de Gruyter, Berlin (1985)
429. Küfer, K.–H.: On the approximation of a ball by random polytopes. Adv. Appl. Prob., **26**, 876–892 (1994)
430. Kurtz, T.G.: Point processes and completely monotone set functions. Z. Wahrscheinlichkeitsth. verw. Geb., **31**, 57–67 (1974)
431. Lang, C., Ohser, J., Hilfer, R.: On the analysis of spatial binary images. J. Microsc., **203**, 303–313 (2001)
432. Laplace, P.S.: Théorie analytique des probabilités. Courcier, Paris (1812); Œvres complètes, tome 7, Gauthier–Villars, Paris
433. Last, G.: Stationary partitions and Palm probabilities. Adv. Appl. Prob. (SGSA), **38**, 602–620 (2006)
434. Last, G., Brandt, A.: Marked Point Processes on the Real Line. The Dynamic Approach. Springer, New York (1995)

435. Last, G., Thorisson, H.: Invariant transports of stationary random measures and mass-stationarity. Ann. Probab. (to appear)
436. Lautensack, C.: Random Laguerre tessellations. Doctoral Thesis, Universität Karlsruhe (2007)
437. Lebesgue, H.: Exposition d'un mémoire de M.W. Crofton. Nouv. Ann. of Math., (4) **12**, 481–502 (1912)
438. Leistritz, L., Zähle, M.: Topological mean value relations for random cell complexes. Math. Nachr., **155**, 57–72 (1992)
439. van Lieshout, M.N.M.: Markov Point Processes and their Applications. Imperial College Press, London (2000)
440. Little, D.V.: A third note on recent research in geometrical probability. Adv. Appl. Prob., **6**, 103–130 (1974)
441. Litvak, A.E., Pajor, A., Rudelson, M., Tomczak–Jaegermann, N.: Smallest singular value of random matrices and geometry of random polytopes. Adv. Math., **195**, 491–523 (2005)
442. Lonke, Y.: On random sections of the cube. Discrete Comput. Geom., **23**, 157–169 (2000)
443. Lutwak, E.: Inequalities for Hadwiger's harmonic quermass integrals. Math. Ann., **280**, 165–175 (1988)
444. Lyashenko, N.N.: Statistics of random compact sets in Euclidean space. J. Soviet. Math., **21**, 76–92 (1983)
445. Mackisack, M.S., Miles, R.E.: Homogeneous rectangular tessellations. Adv. Appl. Prob. (SGSA), **28**, 993–1013 (1996)
446. Maier, R., Schmidt, V.: Stationary iterated tessellations. Adv. Appl. Prob. (SGSA), **35**, 337–353 (2003)
447. Maier, R., Mayer, J., Schmidt, V.: Distributional properties of the typical cell of stationary iterated tessellations. Math. Methods Oper. Res., **59**, 287–302 (2004)
448. Mallows, C.L., Clark, J.M.C.: Linear-intercept distributions do not characterize plane sets. J. Appl. Prob., **7**, 240–244 (1970)
449. Mani–Levitska, P.: Random Steiner symmetrizations. Studia Sci. Math. Hungar., **21**, 373–378 (1986)
450. Mannion, D.: The volume of a tetrahedron whose vertices are chosen at random in the interior of a parent tetrahedron. Adv. Appl. Prob., **26**, 577–596 (1994)
451. Mase, S.: Random compact convex sets which are infinitely divisible with respect to Minkowski addition. Adv. Appl. Prob., **11**, 834–850 (1979)
452. Massé, B.: On the variance of the number of extreme points of a random convex hull. Statist. Probab. Lett., **44**, 123–130 (1999)
453. Massé, B.: On the LLN for the number of vertices of a random convex hull. Adv. Appl. Prob. (SGSA), **32**, 675–681 (2000)
454. Matérn, B.: Spatial Variation. Lecture Notes in Statist., **36**, Springer, Berlin (1986)
455. Mathai, A.M.: On a conjecture in geometric probability regarding asymptotic normality of a random simplex. Ann. Probab., **10**, 247–251 (1982)
456. Mathai, A.M.: An Introduction to Geometrical Probability. Distributional Aspects with Applications. Gordon and Breach, Amsterdam (1999)
457. Mathai, A.M., Tracy, D.S.: On a random convex hull in an n-ball. Commun. Statist. Theory Methods, **12**, 1727–1736 (1983)
458. Matheron, G.: Les variables régionalisées et leur estimation. Masson, Paris (1965)

459. Matheron, G.: Théorie des ensembles aléatoires. In: Cahiers du Centre de Morph. Math., Fasc., **4**, Fontainebleau (1969)
460. Matheron, G.: Ensembles aléatoires, ensembles semi-markoviens et polyèdres poissoniens. Adv. Appl. Prob., **4**, 508–541 (1972)
461. Matheron, G.: Hyperplans Poissoniens et compact de Steiner. Adv. Appl. Prob., **6**, 563–579 (1974)
462. Matheron, G.: Random Sets and Integral Geometry. Wiley, New York (1975)
463. Matheron, G.: Le covariogramme géométrique des compacts convexes de $\mathbb{R}^2$. Technical Report N-2/86/G, Centre de Géostatistique, École Nationale Supérieure des Mines de Paris. Available at http://cg.ensmp.fr/bibliotheque/1986/MATHERON/Rapport/DOC_00258/MATHERON_Rapport_00258.pdf (1986)
464. Matschinski, M.: Considérations statistiques sur les polygones et les polyèdres. Publ. Inst. Stat. Univ. Paris, **3**, 179–201 (1954)
465. Matthes, K., Kerstan, J., Mecke, J.: Infinitely Divisible Point Processes. Wiley, Chichester (1978)
466. Mattila, P.: Geometry of Sets and Measures in Euclidean Spaces. Cambridge University Press, Cambridge (1995)
467. Mayer, M., Molchanov, I.: Limit theorems for the diameter of a random sample in the unit ball. Extremes, **10**, 139–150 (2007)
468. McMullen, P.: A dice probability problem. Mathematika, **21**, 193–198 (1974)
469. McMullen, P.: Non-linear angle sum relations for polyhedral cones and polytopes. Math. Proc. Camb. Phil. Soc., **78**, 247–261 (1975)
470. McMullen, P.: On the axiomatic characterization of spherical volume. Private communication to Rolf Schneider (1981)
471. McMullen, P.: Valuations and dissections. In: Gruber, P.M., Wills, J.M. (eds) Handbook of Convex Geometry, vol. B, pp. 933–988. North-Holland, Amsterdam (1993)
472. McMullen, P.: Isometry covariant valuations on convex bodies. Rend. Circ. Mat. Palermo, Ser. II, Suppl., **50**, 259–271 (1997)
473. McMullen, P.: Valuations on lattice polytopes. Preprint (2008)
474. McMullen, P., Schneider, R.: Valuations on convex bodies. In: Gruber, P.M., Wills, J.M. (eds) Convexity and Its Applications, pp. 170–247. Birkhäuser, Basel (1983)
475. McMullen, P., Shephard, G.C.: Convex Polytopes and the Upper Bound Conjecture. London Math. Soc. Lecture Note Ser. 3, Cambridge University Press (1971)
476. Mecke, J.: Stationäre zufällige Maße auf lokalkompakten Abelschen Gruppen. Z. Wahrscheinlichkeitsth. verw. Geb., **9**, 36–58 (1967)
477. Mecke, J.: Invarianzeigenschaften allgemeiner Palmscher Maße. Math. Nachr., **65**, 335–344 (1975)
478. Mecke, J.: Palm methods for stationary random mosaics. In: Ambartzumian, R.V. (ed) Combinatorial Principles in Stochastic Geometry, pp. 124–132. Armenian Acad. Sci. Publ., Erevan (1980)
479. Mecke, J.: Stereological formulas for manifold processes. Prob. and Math. Statist., **2**, 31–35 (1981)
480. Mecke, J.: Inequalities for intersection densities of superpositions of stationary Poisson hyperplane processes. In: Jensen, E.B., Gundersen, H.J.G. (eds) Proc. Second Int. Workshop Stereology, Stochastic Geometry, pp. 115–124. Aarhus (1983)

481. Mecke, J.: Parametric representation of mean values for stationary random mosaics. Math. Operationsforsch. Statist., Ser. Statist., **15**, 437–442 (1984)
482. Mecke, J.: Isoperimetric properties of stationary random mosaics. Math. Nachr., **117**, 75–82 (1984)
483. Mecke, J.: Random tessellations generated by hyperplanes. In: Ambartzumian, R.V., Weil, W. (eds) Stochastic Geometry, Geometric Statistics, Stereology (Proc. Conf. Oberwolfach, 1983), pp. 104–109. Teubner, Leipzig (1984)
484. Mecke, J.: On some inequalities for Poisson networks. Math. Nachr., **128**, 81–86 (1986)
485. Mecke, J.: Extremal properties of some geometric processes. Acta Appl. Math., **9**, 61–69 (1987)
486. Mecke, J.: An extremal property of random flats. J. Microsc., **151**, 205–209 (1988)
487. Mecke, J.: Random r-flats meeting a ball. Arch. Math., **51**, 378–384 (1988)
488. Mecke, J.: An isoperimetric inequality for random quadrangle tessellations. Statistics, **19**, 57–65 (1988)
489. Mecke, J.: On the intersection density of flat processes. Math. Nachr., **151**, 69–74 (1991)
490. Mecke, J.: Inequalities for the anisotropic Poisson polytope. Adv. Appl. Prob. (SGSA), **27**, 56–62 (1995)
491. Mecke, J.: Inequalities for mixed stationary Poisson hyperplane tessellations. Adv. Appl. Prob. (SGSA), **30**, 921–928 (1998)
492. Mecke, J.: On the relationship between the 0-cell and the typical cell of a stationary random tessellation. Pattern Recognition, **32**, 1645–1648 (1999)
493. Mecke, J.: Networks generated by random tessellations. Lecture at Oberwolfach (2002)
494. Mecke, J., Muche, L.: The Poisson Voronoi tessellation I. A basic identity. Math. Nachr., **176**, 199–208 (1995)
495. Mecke, J., Nagel, W.: Stationäre räumliche Faserprozesse und ihre Schnittzahlrosen. Elektron. Informationsverarb. Kyb., **16**, 475–483 (1980)
496. Mecke, J., Nagel, W., Weiss, V.: Length distributions of edges in planar stationary and isotropic STIT tessellations. J. Contemp. Math. Anal., **42**, 28–43 (2007)
497. Mecke, J., Nagel, W., Weiss, V.: A global construction of homogeneous planar tessellations that are stable under iterations. Stochastics Internat. J. Prob. Stoch. Process., **80**, 51–67 (2008)
498. Mecke, J., Nagel, W., Weiss, V.: The iteration of random tessellations and a construction of a homogeneous process of cell divisions. Adv. Appl. Prob. (SGSA), **40**, 49–59 (2008)
499. Mecke, J., Osburg, I.: On the shape of large Crofton parallelotopes. Math. Notae, **41** (2001/02), 149–157 (2003)
500. Mecke, J., Schneider, R., Stoyan, D., Weil, W.: Stochastische Geometrie. DMV-Seminar **16**, Birkhäuser, Basel (1990)
501. Mecke, J., Stoyan, D.: Formulas for stationary planar fibre processes I – General theory. Math. Operationsforsch. Statist., Ser. Statistics, **11**, 267–279 (1980)
502. Mecke, J., Stoyan, D.: Stereological problems for spherical particles. Math. Nachr., **96**, 311–317 (1980)
503. Mecke, J., Stoyan, D.: The specific connectivity number of random networks. Adv. Appl. Prob. (SGSA), **23**, 576–583 (2001)

504. Mecke, J., Thomas, C.: On an extreme value problem for flat processes. Commun. Stat., Stochastic Models (2) **2**, 273–280 (1986)
505. Mecke, K.: Integralgeometrie in der Statistischen Physik. Verlag Harri Deutsch, Thun (1994)
506. Mecke, K.: Integral geometry in statistical physics. Int. J. Mod. Phys. B, **12**, 861–899 (1998)
507. Meckes, M.W.: Volumes of symmetric random polytopes. Arch. Math., **82**, 85–96 (2004)
508. Meckes, M.W.: Sylvester's problem for symmetric convex bodies and related problems. Monatsh. Math., **145**, 307–319 (2005)
509. Meesters, R., Roy, R.: Continuum Percolation. Cambridge University Press, New York (1996)
510. Meijering, J.L.: Interface area, edge length, and number of vertices in crystal aggregates with random nucleation. Philips Res. Rep., **8**, 270–290 (1953)
511. Meilijson, I.: The expected value of some functions of the convex hull of a random set of points sampled in R^d. Israel J. Math., **72**, 341–352 (1990)
512. Mendelson, S., Pajor, A., Rudelson, M.: The geometry of random $\{-1,1\}$-polytopes. Discrete Comput. Geom., **10**, 365–379 (2005)
513. Michael, E.: Topologies on spaces of subsets. Trans. Amer. Math. Soc., **71**, 152–182 (1951)
514. Michel, J., Paroux, K.: Local convergence of the Boolean shell model towards the thick Poisson hyperplane process in the Euclidean space. Adv. Appl. Prob. (SGSA), **35**, 354–361 (2003)
515. Micheletti, A.: Mathematical modelling and simulation of polymer crystallization processes. Nonlinear Anal., **47**, 1761–1772 (2001)
516. Micheletti, A., Stoyan, D.: Volume fraction and surface density for inhomogeneous random sets. Quaderno N. 17, Dipartimento di Matematica 'F. Enriques', Università degli Studi di Milano (1998)
517. Miles, R.E.: Random polytopes: the generalisation to n dimensions of the intervals of a Poisson process. Ph.D. Thesis, Cambridge University (1961)
518. Miles, R.E.: Random polygons determined by random lines in a plane. Proc. Natl. Acad. Sci., **52**, 902–907 (1964)
519. Miles, R.E.: Random polygons determined by random lines in a plane II. Proc. Natl. Acad. Sci., **52**, 1157–1160 (1964)
520. Miles, R.E.: Poisson flats in Euclidean spaces. I: A finite number of random uniform flats. Adv. Appl. Prob., **1**, 211–237 (1969)
521. Miles, R.E.: A synopsis of 'Poisson flats in Euclidean spaces'. Izv. Akad. Nauk Arm. SSR, Mat., **5**, 263–285 (1970) (Reprinted in: Harding, E.F., Kendall, D.G. (eds) Stochastic Geometry, pp. 202–227. Wiley, New York 1974)
522. Miles, R.E.: On the homogeneous planar Poisson point process. Math. Biosci., **6**, 85–127 (1970)
523. Miles, R.E.: Poisson flats in Euclidean spaces. II: Homogeneous Poisson flats and the complementary theorem. Adv. Appl. Prob., **3**, 1–43 (1971)
524. Miles, R.E.: Random points, sets and tessellations on the surface of a sphere. Sankhyā, Ser. A, **33**, 145–174 (1971)
525. Miles, R.E.: Isotropic random simplices. Adv. Appl. Prob., **3**, 353–382 (1971)
526. Miles, R.E.: The random division of space. Adv. Appl. Prob., Suppl., 243–266 (1972)
527. Miles, R.E.: The various aggregates of random polygons determined by random lines in a plane. Adv. Math., **10**, 256–290 (1973)

528. Miles, R.E.: A simple derivation of a formula of Furstenberg and Tzkoni. Israel J. Math., **14**, 278–280 (1973)
529. Miles, R.E.: The fundamental formula of Blaschke in integral geometry and geometric probability, and its iteration, for domains with fixed orientations. Austral. J. Statist., **16**, 111–118 (1974)
530. Miles, R.E.: Estimating aggregate and overall characteristics from thick sections by transmission microscopy. J. Microsc., **107**, 227–233 (1976)
531. Miles, R.E.: Some new integral geometric formulae, with stochastic applications. J. Appl. Prob., **16**, 592–606 (1979)
532. Miles, R.E.: A generalization of a formula of Steiner. Z. Wahrscheinlichkeitsth. verw. Geb., **61**, 375–378 (1982)
533. Miles, R.E.: Sectional Voronoi tessellations. Rev. Unión Mat. Argentina, **29**, 310–327 (1984)
534. Miles, R.E.: Random tessellations. In: Kotz, S., Johnson, N.L. (eds) Encyclopedia of Statistical Sciences, vol. 7, pp. 567–572. Wiley, New York (1986)
535. Miles, R.E.: A heuristic proof of a long-standing conjecture of D.G. Kendall concerning the shapes of certain large random polygons. Adv. Appl. Prob. (SGSA), **27**, 397–417 (1995)
536. Miles, R.E., Davy, P.: Precise and general conditions for the validity of a comprehensive set of stereological fundamental formulae. J. Microsc., **107**, 211–226 (1976)
537. Miles, R.E., Mackisack, M.S.: A large class of random tessellations with the classic Poisson polygon distributions. Forma, **17**, 1–17 (2002)
538. Miles, R.E., Maillardet, R.J.: The basic structures of Voronoi and generalized Voronoi polygons. J. Appl. Prob., **19 A**, 97–112 (1982)
539. Miles, R.E., Serra, J. (eds.): Geometrical Probability and Biological Structures: Buffon's 200th Anniversary (Proceedings, Paris 1977), Lect. Notes in Biomath., **23**, Springer, Berlin (1978)
540. Milman, E.: Generalized intersection bodies. J. Funct. Anal., **240**, 530–567 (2006)
541. Milman, V., Pajor, A.: Isotropic position and inertia ellipsoids and zonoids of the unit ball of a normed n-dimensional space. Lecture Notes in Math., **1376**, 64–104, Springer, Berlin (1989)
542. Molchanov, I.S.: Random sets. A survey of results and applications. (Russian) Ukr. Mat. Zh., **43**, 1587–1599 (1991) Engl. transl.: Ukr. Math. J., **43**, 1477–1487 (1992)
543. Molchanov, I.S.: Limit Theorems for Unions of Random Closed Sets. Lecture Notes in Math., **1561**, Springer, Berlin (1993)
544. Molchanov, I.S.: Statistics of the Boolean model: from the estimation of means to the estimation of distributions. Adv. Appl. Prob. (SGSA), **27**, 63–86 (1995)
545. Molchanov, I.S.: A limit theorem for scaled vacancies of the Boolean model. Stoch. Stoch. Rep., **58**, 45–65 (1996)
546. Molchanov, I.S.: Statistics of the Boolean Model for Practitioners and Mathematicians. Wiley, Chichester (1997)
547. Molchanov, I.S.: Random closed sets: results and problems. In: Barndorff–Nielsen, O.E., Kendall, W.S., van Lieshout, M.N.M. (eds) Stochastic Geometry. Likelihood and Computation. Chapman & Hall, Boca Raton (1999)
548. Molchanov, I.S.: Theory of Random Sets. Springer, London (2005)
549. Molchanov, I., Stoyan, D.: Asymptotic properties of estimators for parameters of the Boolean model. Adv. Appl. Prob., **26**, 301–323 (1994)

550. Møller, J.: A simple derivation of a formula of Blaschke and Petkantschin. Research Reports, **138**, Dept. of Theor. Statistics, Univ. of Aarhus (1985)
551. Møller, J.: Random tessellations in $\mathbb{R}^d$. Adv. Appl. Prob., **21**, 37–73 (1989)
552. Møller, J.: Random Johnson–Mehl tessellations. Adv. Appl. Prob., **24**, 814–844 (1992)
553. Møller, J.: Lectures on Random Voronoi Tessellations. Lecture Notes in Statist., **87**, Springer, New York (1994)
554. Møller, J.: A review on probabilistic models and results for Voronoi tessellations. In: Engel. P., Syta, H. (eds) Voronoi's Impact on Modern Science, vol. I, pp. 254–265. Inst. Math. Nat. Acad. Sci. Ukraine, Kyev (1998)
555. Møller, J., Zuyev, S.: Gamma-type results and other related properties of Poisson processes. Adv. Appl. Prob. (SGSA), **28**, 662–673 (1996)
556. Møller, J., Waagepetersen, R.P.: Statistical Inference and Simulation for Spatial Point Processes. Chapman & Hall/CRC, Boca Raton (2004)
557. Molter, U.: Tangential measure on the set of convex infinite cylinders. J. Appl. Prob., **23**, 961–972 (1986)
558. Moran, P.A.P.: A note on recent research in geometrical probability. J. Appl. Prob., **3**, 453–463 (1966)
559. Moran, P.A.P.: A second note on recent research in geometrical probability. Adv. Appl. Prob., **1**, 73–89 (1969)
560. Moran, P.A.P.: Another quasi-Poisson plane point process. Z. Wahrscheinlichkeitsth. verw. Geb., **33**, 269–272 (1975)
561. Muche, L.: Untersuchung von Verteilungseigenschaften des Poisson–Voronoi-Mosaiks. Doctoral Thesis, Bergakademie Freiberg (1993)
562. Muche, L.: Distributional properties of the Poisson–Voronoi tessellation. Acta Stereol., **12**, 125–130 (1993)
563. Muche, L.: The Poisson Voronoi tessellation II. Edge length distribution functions. Math. Nachr., **178**, 271–283 (1996)
564. Muche, L.: The Poisson Voronoi tessellation III. Miles' formula. Math. Nachr., **191**, 247–267 (1998)
565. Muche, L.: The Poisson–Voronoi tessellation: relationships for edges. Adv. Appl. Prob. (SGSA), **37**, 279–296 (2005)
566. Muche, L., Stoyan, D.: Contact and chord length distributions of the Poisson Voronoi tessellation. J. Appl. Prob., **29**, 467–471 (1992)
567. Müller, H.R: Über Momente ersten und zweiten Grades in der Integralgeometrie. Rend. Circ. Mat. Palermo, Ser. II, **2**, 1–21 (1953)
568. Müller, J.S.: On the mean width of random polytopes. Probab. Theory Related Fields, **82**, 33–37 (1989)
569. Müller, J.S.: Approximation of a ball by random polytopes. J. Approx. Theory, **63**, 198–209 (1990)
570. Mycielski, J.: On random convex hulls. Colloq. Math., **51**, 263–265 (1987)
571. Nachbin, L.: The Haar Integral. Van Nostrand, Princeton (1965)
572. Nagel, W.: Orientation-dependent chord length distributions characterize convex polygons. J. Appl. Prob., **30**, 730–736 (1993)
573. Nagel, W., Ohser, J., Pischang, K.: An integral-geometric approach for the Euler–Poincaré characteristic of spatial images. J. Microsc., **189**, 54–62 (2000)
574. Nagel, W., Weiss, V.: Interdependences of directional quantities of planar tessellations. Adv. Appl. Prob. (SGSA), **31**, 664–678 (1999)
575. Nagel, W., Weiss, V.: Limits of sequences of stationary planar tessellations. Adv. Appl. Prob. (SGSA), **35**, 123–138 (2003)

576. Nagel, W., Weiss, V.: Crack STIT tessellations – existence and uniqueness of tessellations that are stable with respect to iterations. Izvest. Nac. Akad. Nauk. Armenii (Mat.), **39**, 84–114 (2004)
577. Nagel, W., Weiss, V.: Crack STIT tessellations: characterization of stationary random tessellations stable with respect to iteration. Adv. Appl. Prob. (SGSA), **37**, 859–883 (2005)
578. Nagel, W., Weiss, V.: STIT tessellations in the plane. Rend. Circ. Mat. Palermo, Ser. II, Suppl., **77**, 441–458 (2006)
579. Nagel, W., Weiss, V.: Mean values for homogeneous STIT tessellations in 3D. Image Anal. Stereol., **27**, 29–37 (2008)
580. Nef, W.: Zur Einführung der Eulerschen Charakteristik. Monatsh. Math., **92**, 41–46 (1981)
581. Neveu, J.: Mathematical Foundations of the Calculus of Probability. Holden-Day, San Francisco (1965)
582. Neveu, J.: Processus ponctuels. In: Hennequin, P.–L. (ed) Ecole d'Eté de Probabilités de Saint-Flour VI-1976, pp. 249–447, Lecture Notes in Math., **598**. Springer, Berlin (1977)
583. Nguyen, H.T.: An Introduction to Random Sets. Chapman & Hall/CRC, Boca Raton (2006)
584. Nguyen, X.X., Zessin, H.: Ergodic theorems for spatial processes. Z. Wahrscheinlichkeitsth. verw. Geb., **48**, 133–158 (1979)
585. Nijenhuis, A.: On Chern's kinematic formula in integral geometry. J. Differential Geom., **9**, 475–482 (1974)
586. Norberg, T.: Existence theorems for measures on continuous posets, with applications to random set theory. Math. Scand., **64**, 15–51 (1989)
587. Ohser, J., Mücklich, F.: Statistical Analysis of Microstructures in Materials Science. Wiley, Chichester (2000)
588. Ohser, J., Nagel. W.: The estimation of the Euler–Poincaré characteristic from observations on parallel sections. J. Microsc., **184**, 117–126 (1996)
589. Ohser, J., Nagel, W., Schladitz, K.: The Euler number of discretized sets – on the choice of adjacency in homogeneous lattices. In: Mecke, K., Stoyan, D. (eds) Morphology of Condensed Matter, pp. 275–298, Lecture Notes in Physics, **600**. Springer, Berlin (2002)
590. Ohser, J., Nagel, W., Schladitz, K.: The Euler number of discretised sets – surprising results in three dimensions. Image Anal. Stereol., **22**, 11–19 (2003)
591. Okabe, A., Boots, B., Sugihara, K., Chiu, S.N.: Spatial Tessellations; Concepts and Applications of Voronoi Diagrams. 2nd ed., Wiley, Chichester (2000)
592. Olsbo, V.: On the correlation between the volumes of the typical Poisson–Voronoi cell and the typical Stienen sphere. Adv. Appl. Prob. (SGSA), **39**, 883–892 (2007)
593. Palm, C.: Intensitätsschwankungen im Fernsprechverkehr. Ericsson Technics **44**, 189 pp. (1943)
594. Pantle, U., Schmidt, V., Spodarev, E.: A central limit theorem for functionals of germ-grain models. Adv. Appl. Prob. (SGSA), **38**, 76–94 (2006)
595. Papaderou–Vogiatzaki, I., Schneider, R.: A collision probability problem. J. Appl. Prob., **25**, 617–623 (1988)
596. Paroux, K.: Quelques théorèmes centraux limites pour les processus Poissoniens de droites dans le plan. Adv. Appl. Prob. (SGSA), **30**, 640–656 (1998)
597. Pederzoli, G.: Series forms of exact densities associated with uniformly distributed random points in an n-ball. S. Afr. Statist. J., **20**, 19–28 (1986)

598. Penrose, M.: Random Geometric Graphs. Oxford University Press, Oxford (2003)
599. Penrose, M.: Laws of large numbers in stochastic geometry with statistical applications. Bernoulli, **13**, 1124–1150 (2007)
600. Penrose, M.D., Yukich, J.E.: Central limit theorems for some graphs in computational geometry. Ann. Appl. Probab., **11**, 1005–1041 (2001)
601. Petkantschin, B.: Integralgeometrie 6. Zusammenhänge zwischen den Dichten der linearen Unterräume im n-dimensionalen Raum. Abh. Math. Sem. Univ. Hamburg, **11**, 249–310 (1936)
602. Pfiefer, R.E.: The historical development of J.J. Sylvester's four point problem. Math. Mag., **62**, 309–317 (1989)
603. Pfiefer, R.E.: Maximum and minimum sets for some geometric mean values. J. Theoret. Probab., **3**, 169–179 (1990)
604. Piefke, F.: Beziehungen zwischen der Sehnenlängenverteilung und der Verteilung des Abstandes zweier zufälliger Punkte im Eikörper. Z. Wahrscheinlichkeitsth. verw. Geb., **43**, 129–134 (1978)
605. Piefke, F.: Zwei integralgeometrische Formeln für Paare konvexer Körper. Z. angew. Math. Phys., **29**, 664–669 (1978)
606. Pohlmann, S., Mecke, J., Stoyan, D.: Stereological formulas for stationary surface processes. Math. Operationsforsch. Statist., **12**, 429–440 (1981).
607. Poincaré, H.: Calcul des probabilités. Georges Carré, Paris (1896)
608. Pontrjagin, L.S.: Topologische Gruppen, Teil 1. Teubner, Leipzig (1957)
609. Quintanilla, J., Torquato, S.: Microstructure functions for a model of statistically inhomogeneous random media. Phys. Rev. E, **55**, 1558–1565 (1997)
610. Quintanilla, J., Torquato, S.: Clustering in a continuum percolation model. Adv. Appl. Prob. (SGSA), **29**, 327–336 (1997)
611. Radecke, W.: Some mean value relations on stationary random mosaics in the space. Math. Nachr., **97**, 203–210 (1980)
612. Rataj, J.: Estimation of oriented direction distribution of a planar body. Adv. Appl. Prob. (SGSA), **28**, 394–404 (1996)
613. Rataj, J.: The iterated version of a translative integral formula for sets of positive reach. Rend. Circ. Mat. Palermo, Ser. II, Suppl., **46**, 129–138 (1997)
614. Rataj, J.: Remarks on a translative formula for sets of positive reach. Geom. Dedicata, **65**, 59–62 (1997)
615. Rataj, J.: Translative and kinematic formulae for curvature measures of flat sections. Math. Nachr., **197**, 89–101 (1999)
616. Rataj, J.: On estimation of the Euler number by projections of thin slabs. Adv. Appl. Prob. (SGSA), **36**, 715–724 (2004)
617. Rataj, J., Zähle, M.: Mixed curvature measures for sets of positive reach and a translative integral formula. Geom. Dedicata, **57**, 259–283 (1995)
618. Rataj, J., Zähle, M.: Curvatures and currents for unions of sets with positive reach, II. Ann. Global Anal. Geom., **20**, 1–21 (2001)
619. Rataj, J., Zähle, M.: A remark on mixed curvature measures of sets with positive reach. Beiträge Algebra Geom., **43**, 171–179 (2002)
620. Rataj, J., Zähle, M.: Normal cycles of Lipschitz manifolds by approximation with parallel sets. Diff. Geom. Appl., **19**, 113–126 (2003)
621. Rataj, J., Zähle, M.: General normal cycles and Lipschitz manifolds of bounded curvature. Ann. Global Anal. Geom., **27**, 135–156 (2005)
622. Rathie, P.N.: On the volume distribution of the typical Poisson–Delaunay cell. J. Appl. Prob., **29**, 740–744 (1992)

623. Raynaud, H.: Sur l'enveloppe convexe des nuages de points aléatoires dans $\mathbb{R}^n$. J. Appl. Prob., **7**, 35–48 (1970)
624. Reed, W.J.: Random points in a simplex. Pacific J. Math., **54**, 183–198 (1974)
625. Reisner, S.: Zonoids with minimal volume product. Math. Z., **192**, 339–346 (1986)
626. Reiss, R.–D.: A Course on Point Processes. Springer, New York (1993)
627. Reitzner, M.: The floating body and the equiaffine inner parallel curve of a plane convex body. Geom. Dedicata, **84**, 151–167 (2001)
628. Reitzner, M.: Random points on the boundary of smooth convex bodies. Trans. Amer. Math. Soc., **354**, 2243–2278 (2002)
629. Reitzner, M.: Random polytopes are nearly best approximating. Rend. Circ. Mat. Palermo, Ser. II, Suppl., **70**, 263–278 (2002)
630. Reitzner, M.: Random polytopes and the Efron–Stein jackknife inequality. Ann. Probab., **31**, 2136–2166 (2003)
631. Reitzner, M.: Stochastical approximation of smooth convex bodies. Mathematika, **51**, 11–29 (2004)
632. Reitzner, M.: The combinatorial structure of random polytopes. Adv. Math., **191**, 178–208 (2005)
633. Reitzner, M.: Central limit theorems for random polytopes. Probab. Theory Related Fields, **133**, 483–507 (2005)
634. Reitzner, M., Heveling, M.: Poisson–Voronoi approximation. Preprint (2007)
635. Ren, D.: Topics in Integral Geometry. World Scientific, Singapore (1994)
636. Ren D., Zhang G.: Random convex sets in a lattice of parallelograms. Acta Math. Sci., (English Ed.) **11**, 317–326 (1991)
637. Rényi, A.: On a new axiomatic theory of probability. Acta Math. Acad. Sci. Hung., **6**, 285–335 (1955)
638. Rényi, A.: Remarks on the Poisson process. Studia Sci. Math. Hungar., **2**, 119–123 (1967)
639. Rényi, A., Sulanke, R.: Über die konvexe Hülle von n zufällig gewählten Punkten. Z. Wahrscheinlichkeitsth. verw. Geb., **2**, 75–84 (1963)
640. Rényi, A., Sulanke, R.: Über die konvexe Hülle von n zufällig gewählten Punkten. II. Z. Wahrscheinlichkeitsth. verw. Geb., **3**, 138–147 (1964)
641. Rényi, A., Sulanke, R.: Zufällige konvexe Polygone in einem Ringgebiet. Z. Wahrscheinlichkeitsth. verw. Geb., **9**, 146–157 (1968)
642. Richards, P.I.: Averages for polygons formed by random lines. Proc. Natl. Acad. Sci., **52**, 1160–1164 (1964)
643. Richardson, R.M., Vu, V.H., Wu, L.: An inscribing model for random polytopes. Discrete Comput. Geom., **39**, 469–499 (2008)
644. Ripley, B.D.: Spatial Statistics. Wiley, New York (1981)
645. Ripley, B.D.: Statistical Inference for Spatial Processes. Cambridge University Press, Cambridge (1988)
646. Rockafellar, R.T., Wets, R.J.–B.: Variational Analysis. Springer, Berlin (1998)
647. Rogers, L.C.G.: The probability that two samples in the plane will have disjoint convex hulls. J. Appl. Prob., **15**, 790–802 (1978)
648. Ross, D.: Random sets without separability. Ann. Probab., **14**, 1064–1069 (1986)
649. Rother, W., Zähle, M.: A short proof of a principal kinematic formula and extensions. Trans. Amer. Math. Soc., **321**, 547–558 (1990)

650. Ruben, H.: On the geometrical moments of skew-regular simplices in hyperspherical space, with some applications in geometry and mathematical statistics. Acta Math., **193**, 1–23 (1960)
651. Ruben, H.: The volume of a random simplex in an n-ball is asymptotically normal. J. Appl. Prob., **14**, 647–653 (1977)
652. Ruben, H., Miles, R.E.: A canonical decompositon of the probability measure of sets of isotropic random points in $\mathbb{R}^n$. J. Multivariate Anal., **10**, 1–18 (1980)
653. Rubin, B.: Inversion formulas for the spherical Radon transform and the generalized cosine transform. Adv. Appl. Math., **29**, 471–497 (2002)
654. Rudin, W.: Real and Complex Analysis. 2nd ed., McGrawHill, New York (1974)
655. Salinetti, G., Wets, R.J.–B., On the convergence in distribution of measurable multifunctions (random sets), normal integrands, stochastic processes and stochastic infima. Math. Operations Res., **11**, 385–419 (1986)
656. Santaló, L.A.: Über das kinematische Maß im Raum. Actualités Sci. Indust., **357**, Hermann, Paris (1936)
657. Santaló, L.A.: On parallel hypersurfaces in the elliptic and hyperbolic n-dimensional space. Proc. Amer. Math. Soc., **1**, 325–330 (1950)
658. Santaló, L.A.: Geometria integral en espacios de curvatura constante. Publ. Com. Nac. Energia Atomica, Ser. Mat., **1**, fasc. 1 (1952)
659. Santaló, L.A.: Introduction to Integral Geometry. Hermann, Paris (1953)
660. Santaló, L.A.: Cuestiones de geometria differencial e integral en espacios de curvatura constante. Rend. Sem. Mat. Torino, **14**, 277–295 (1954–55)
661. Santaló, L.A.: Sobre la formula de Gauss–Bonnet para poliedros en espacios de curvatura constante. Revista Un. Mat. Argentina, **20**, 79–91 (1962)
662. Santaló, L.A.: Integral Geometry and Geometric Probability. Addison–Wesley, Reading, Mass. (1976)
663. Santaló, L.A.: Mixed random mosaics. Math. Nachr., **117**, 129–133 (1984)
664. Santaló, L.A.: On the measure of line segments entirely contained in a convex body. In: Barroso, J.A. (ed) Aspects of Math. and Its Appl., pp. 677–687. North-Holland, Amsterdam (1986)
665. Scheike, T.H.: Anisotropic growth of Voronoi cells. Adv. Appl. Prob., **26**, 43–53 (1994)
666. Schladitz, K.: Estimation of the intensity of stationary flat processes. Adv. Appl. Prob. (SGSA), **32**, 114–139 (2000)
667. Schläfli, L.: Gesammelte Mathematische Abhandlungen, I. Birkhäuser, Basel (1950)
668. Schlather, M.: A formula for the edge length distribution function of the Poisson Voronoi tessellation. Math. Nachr., **214**, 113–119 (1999)
669. Schmidt, V., Spodarev, E.: Joint estimators for the specific intrinsic volumes of stationary random sets. Stochastic Process. Appl., **115**, 959–981 (2005)
670. Schmitt, M.: Estimation of the density in a stationary Boolean model. J. Appl. Prob., **28**, 702–708 (1991)
671. Schmitt, M.: Estimation of intensity and shape in a non-stationary Boolean model. In: Jeulin, D. (ed) Advances in Theory and Applications of Random Sets (Proc. Int. Symp. Fontainebleau, 1996), pp. 251–267. World Scientific, Singapore (1997)
672. Schneider, R.: Kinematische Berührmaße für konvexe Körper. Abh. Math. Sem. Univ. Hamburg, **44**, 12–23 (1975)
673. Schneider, R.: Kinematische Berührmaße für konvexe Körper und Integralrelationen für Oberflächenmaße. Math. Ann., **218**, 253–267 (1975)

674. Schneider, R.: Bestimmung eines konvexen Körpers durch gewisse Berührmaße. Arch. Math., **27**, 99–105 (1976)
675. Schneider, R.: Eine kinematische Integralformel für konvexe Körper. Arch. Math., **28**, 217–220 (1977)
676. Schneider, R.: Curvature measures of convex bodies. Ann. Mat. Pura Appl., **116**, 101–134 (1978)
677. Schneider, R.: Kinematic measures for sets of colliding convex bodies. Mathematika, **25**, 1–12 (1978)
678. Schneider, R.: Bestimmung konvexer Körper durch Krümmungsmaße. Comment. Math. Helvet., **54**, 42–60 (1979)
679. Schneider, R.: Parallelmengen mit Vielfachheit und Steiner-Formeln. Geom. Dedicata, **9**, 111–127 (1980)
680. Schneider, R.: Curvature measures and integral geometry of convex bodies. Rend. Sem. Mat. Univers. Politecn. Torino, **38**, 79–98 (1980)
681. Schneider, R.: Crofton's formula generalized to projected thick sections. Rend. Circ. Mat. Palermo, **30**, 157–160 (1981)
682. Schneider, R.: A uniqueness theorem for finitely additive invariant measures on a compact homogeneous space. Rend. Circ. Mat. Palermo, **30**, 341–344 (1981)
683. Schneider, R.: A local formula of translative integral geometry. Arch. Math., **36**, 466–469 (1981)
684. Schneider, R.: Zur optimalen Approximation konvexer Hyperflächen durch Polyeder. Math. Ann., **256**, 289–301 (1981)
685. Schneider, R.: Random polytopes generated by anisotropic hyperplanes. Bull. London Math. Soc., **14**, 549–553 (1982)
686. Schneider, R.: Random hyperplanes meeting a convex body. Z. Wahrscheinlichkeitstheorie verw. Geb., **61**, 379–387 (1982)
687. Schneider, R.: Inequalities for random flats meeting a convex body. J. Appl. Prob., **22**, 710–716 (1985)
688. Schneider, R.: Curvature measures and integral geometry of convex bodies, II. Rend. Sem. Mat. Univers. Politecn. Torino, **44**, 263–275 (1986)
689. Schneider, R.: Geometric inequalities for Poisson processes of convex bodies and cylinders. Results Math., **11**, 165–185 (1987)
690. Schneider, R.: Equidecomposable polyhedra. In: Böröczky, K., Fejes Tóth, G. (eds) Intuitive Geometry (Siófok 1985). Colloq. Math. Soc. János Bolyai, **48**, pp. 481–501. North-Holland, Amsterdam (1987)
691. Schneider, R.: Tessellations generated by hyperplanes. Discrete Comput. Geom., **2**, 223–232 (1987)
692. Schneider, R.: Approximation of convex bodies by random polytopes. Aequationes Math., **32**, 304–310 (1987)
693. Schneider, R.: Curvature measures and integral geometry of convex bodies, III. Rend. Sem. Mat. Univers. Politecn. Torino, **46**, 111–123 (1988)
694. Schneider, R.: Random approximation of convex sets. J. Microsc., **151**, 211–227 (1988)
695. Schneider, R.: Convex Bodies: the Brunn-Minkowski Theory. Cambridge University Press, Cambridge (1993)
696. Schneider, R.: An extension of the principal kinematic formula of integral geometry. Rend. Circ. Mat. Palermo, Ser. II, Suppl. **35**, 275–290 (1994)
697. Schneider, R.: Isoperimetric inequalities for infinite hyperplane systems. Discrete Comput. Geom., **13** , 609–627 (1995)

698. Schneider, R.: On areas and integral geometry in Minkowski spaces. Beiträge Algebra Geom., **38**, 73–86 (1997)
699. Schneider, R.: A duality for Poisson flats. Adv. Appl. Prob. (SGSA), **31**, 63–68 (1999)
700. Schneider, R.: Convex bodies in exceptional relative positions. J. London Math. Soc., **60**, 617–629 (1999)
701. Schneider, R.: Tensor valuations on convex bodies and integral geometry. Rend. Circ. Mat. Palermo, Ser. II, Suppl., **65**, 295–316 (2000)
702. Schneider, R.: Mixed functionals of convex bodies. Discrete Comput. Geom., **24**, 527–538 (2000)
703. Schneider, R.: On the Busemann area in Minkowski spaces. Beiträge Algebra Geom., **42**, 263–273 (2001)
704. Schneider, R.: On the mean normal measures of a particle process. Adv. Appl. Prob. (SGSA), **33**, 25–38 (2001)
705. Schneider, R.: Crofton measures in hypermetric projective Finsler spaces. Arch. Math., **77**, 85–97 (2001)
706. Schneider, R.: On integral geometry in projective Finsler spaces. J. Contemp. Math. Anal., **37**, 30–46 (2002)
707. Schneider, R.: Nonstationary Poisson hyperplanes and their induced tessellations. Adv. Appl. Prob. (SGSA), **35**, 139–158 (2003)
708. Schneider, R.: An integral geometric theorem for simple valuations. Beiträge Algebra Geom., **44**, 487–492 (2003)
709. Schneider, R.: Mixed polytopes. Discrete Comput. Geom., **29**, 575–593 (2003)
710. Schneider, R.: Crofton measures in projective Finsler spaces. In: Grinberg, E.L., Li, S., Zhang, G., Zhou, J. (eds) Integral Geometry and Convexity, pp. 67–98. World Scientific, New Jersey (2006)
711. Schneider, R.: Crofton measures in polytopal Hilbert geometries. Beiträge Algebra Geom., **47**, 479–488 (2006)
712. Schneider, R.: Integral geometric tools for stochastic geometry. In: [50], pp. 119–184 (2007)
713. Schneider, R., Schuster, R.: Tensor valuations on convex bodies and integral geometry, II. Rend. Circ. Mat. Palermo, Ser. II, Suppl., vol. II, **70**, 295–314 (2002)
714. Schneider, R., Schuster, R.: Particle orientation from section stereology. Rend. Circ. Mat. Palermo, Ser. II, Suppl., **77**, 623–633 (2006)
715. Schneider, R., Weil, W.: Translative and kinematic integral formulae for curvature measures. Math. Nachr., **129**, 67–80 (1986)
716. Schneider, R., Weil, W.: Integralgeometrie. Teubner, Stuttgart (1992)
717. Schneider, R., Weil, W.: Stochastische Geometrie. Teubner, Stuttgart (2000)
718. Schneider, R., Wieacker, J.A.: Random polytopes in a convex body. Z. Wahrscheinlichkeitstheorie. verw. Geb., **52**, 69–73 (1980)
719. Schneider, R., Wieacker, J.A.: Random touching of convex bodies. In: Ambartzumian, R.V., Weil, W. (eds) Proc. Conf. Stochastic Geometry, Geometric Statistics, Stereology (Oberwolfach 1983), pp. 154–169. Teubner-Verlag, Leipzig (1984)
720. Schneider, R., Wieacker, J.A.: Integral geometry. In: Gruber, P.M., Wills, J.M. (eds) Handbook of Convex Geometry, vol. B., pp. 1349–1390. North-Holland, Amsterdam (1993)
721. Schneider, R., Wieacker, J.A.: Integral geometry in Minkowski spaces. Adv. Math., **129**, 222–260 (1997)

722. Schöpf, P.: Gewichtete Volumsmittelwerte von Simplices, welche zufällig in einem konvexen Körper des R^n gewählt werden. Monatsh. Math., **83**, 331–337 (1977)
723. Schreiber, T.: Limit theorems for certain functionals of unions of random closed sets. Teor. Veroyatnost. i Primenen., **47**, 130–142 (2002)
724. Schütt, C.: Random polytopes and affine surface area. Math. Nachr., **170**, 227–249 (1994)
725. Schütt, C., Best and random approximation of convex bodies by polytopes. Rend. Circ. Mat. Palermo, Ser. II, Suppl., **70**, 315–334 (2002)
726. Schütt, C., Werner, E.: Random polytopes with vertices on the boundary of a convex body. C. R. Acad. Sci. Paris, **331**, 1–5 (2000)
727. Schütt, C., Werner, E.: Polytopes with vertices chosen randomly from the boundary of a convex body. In: Milman, V., Schechtman, G. (eds) Israel Seminar 2001–2002, pp. 241–422, Lecture Notes in Math., **1807**. Springer, New York (2003)
728. Seneta, E., Parshall, K., Jongmans, F.: Nineteenth-century developments in geometric probability: J.J. Sylvester, M.W. Crofton, J.–E. Barbier, and J. Bertrand. Arch. Hist. Exact Sci., **55**, 501-524 (2001)
729. Serra, J.A.: Image Analysis and Mathematical Morphology. Academic Press, London (1982)
730. Small, C.G.: Reconstructing convex bodies from random projected images. Canad. J. Statist., **19**, 341–347 (1991)
731. Solomon, H.: Geometric Probability. Soc. Industr. Appl. Math., Philadelphia (1978)
732. Spiess, M., Spodarev, E.: Anisotropic dilated Poisson k-flat processes. Preprint (2007)
733. Spodarev, E.: Selected topics in the theory of spatial stationary flat processes. Doctoral Thesis, Friedrich-Schiller-Universität, Jena (2001)
734. Spodarev, E.: On the rose of intersections of stationary flat processes. Adv. Appl. Prob. (SGSA), **33**, 584–599 (2001)
735. Spodarev, E.: Isoperimetric problems and roses of neighborhood for stationary flat processes. Math. Nachr., **251**, 88–100 (2003)
736. Steiner, J.: Einige Gesetze über die Theilung der Ebene und des Raumes. J. reine angew. Math., **1**, 349–364 (1826)
737. Stewart, G.W.: The efficient generation of random orthogonal matrices with an application to condition estimators. SIAM J. Numer. Anal., **17**, 403–409 (1980)
738. Stoka, M.: Géométrie intégrale. Gauthier–Villars, Paris (1968)
739. Stoyan, D.: Proofs of some fundamental formulas of stereology for non-Poisson grain models. Math. Operationsforsch. Statist., Ser. Optimization, **10**, 575–583 (1979)
740. Stoyan, D.: Stereological formulae for size distributions via marked point processes. Probab. Math. Stat., **2**, 161–166 (1982)
741. Stoyan, D.: Stereology and stochastic geometry. Int. Stat. Rev., **58**, 227–242 (1990)
742. Stoyan, D.: Random sets: Models and statistics. Int. Stat. Rev., **66**, 1–27 (1998)
743. Stoyan, D., Kendall, W.S., Mecke, J.: Stochastic Geometry and Its Applications. 2nd ed., Wiley, Chichester (1995)
744. Stoyan, D., Molchanov, I.S.: Set-valued means of random particles. J. Math. Imaging Vision, **7**, 111–121 (1997)

745. Stoyan, D., Schlather, M.: Random sequential adsorption: Relationship to dead leaves and characterization of variability. J. Stat. Physics, **100**, 969–979 (2000)
746. Stoyan, D., Stoyan, H.: Fractals, Random Shapes and Point Fields. Wiley, Chichester (1994)
747. Streit, F.: On multiple integral geometric integrals and their applications to probability theory. Canad. J. Math., **22**, 151–163 (1970)
748. Sulanke, R.: Schnittpunkte zufälliger Geraden. Arch. Math., **16**, 320–324 (1965)
749. Sulanke, R., Wintgen, P.: Differentialgeometrie und Faserbündel. Birkhäuser, Basel (1972)
750. Sylvester, J.J.: Question 1491. Educational Times, London, April (1864)
751. Sylvester, J.J.: On a funicular solution of Buffon's "problem of the needle" in its most general form. Acta Math., **14**, 185–205 (1891)
752. Tanner, J.C.: The proportion of quadrilaterals formed by random lines in a plane. J. Appl. Prob., **20**, 400–404 (1983)
753. Tanner, J.C.: Polygons formed by random lines in a plane: Some further results. J. Appl. Prob., **20**, 778–787 (1983)
754. Taylor, J.E.: A Gaussian kinematic formula. Ann. Probab., **34**, 122–158 (2006)
755. Tempel'man, A.A.: Ergodic theorems for general dynamical systems (in Russian). Trudy Moskov. Mat. Obsc., **26**, (1972). Engl. transl.: Trans. Moscow Math. Soc., **26**, 94–132.
756. Thomas, C.: Extremum properties of the intersection densities of stationary Poisson hyperplane processes. Math. Operationsforsch. Statist., Ser. Statist., **15**, 443–449 (1984)
757. Thorisson, H.: Point-stationarity in d dimensions and Palm theory. Bernoulli **5**, 797–831 (1999)
758. Thorisson, H.: Coupling, stationarity, and regeneration. Springer, New York (2000)
759. Torquato, S.: Random Heterogeneous Materials. Microstructure and Macroscopic Properties. Springer, New York (2002)
760. Valtr, P.: Probability that n random points are in convex position. In: Bárány, I., Pach, J. (eds) The László Fejes Tóth Festschrift. Discrete Comput. Geom., **13**, 637–643 (1995)
761. Valtr, P.: The probability that n random points in a triangle are in convex position. Combinatorica, **16**, 567–573 (1996)
762. Varga, O.: Integralgeometrie 3. Croftons Formeln für den Raum. Math. Z., **40**, 387–405 (1936)
763. Vershik, A.M.: Topology of the convex polytopes' manifold, the manifold of the projective configurations of a given combinatorial type and representation of lattices. In: Viro, O. Ya. (ed) Topology and Geometry – Rohlin Seminar. Lecture Notes in Math., **1346**, pp. 557–581. Springer, Berlin (1988)
764. Vershik, A.M., Chernyakov, A.G.: Critical points of fields of convex polytopes and the Pareto–Smale optimum with respect to a convex cone. Soviet Math. Dokl., **26**, 353–356 (1982)
765. Vershik, A.M., Sporyshev, P.V.: An estimate of the average number of steps in the simplex method, and problems in asymptotic integral geometry. Soviet Math. Dokl., **28**, 195–199 (1983)
766. Vershik, A.M., Sporyshev, P.V.: An asymptotic estimate of the average number of steps of the parametric simplex method. U.S.S.R. Comput. Math. and Math. Phys., **26**, 104–113 (1986)

767. Vershik, A.M., Sporyshev, P.V.: Asymptotic behavior of the number of faces of random polyhedra and the neighborliness problem. Selecta Math. Soviet., **11**, vol. 2, 181–201 (1992)
768. Vitale, R.A.: On Gaussian random sets. In: Ambartzumian, R.V., Weil, W. (eds) Proc. Conf. Stochastic Geometry, Geometric Statistics, Stereology (Oberwolfach 1983), pp. 222–224. Teubner, Leipzig (1984)
769. Vitale, R.: Stochastic smoothing of convex bodies: two examples. Rend. Circ. Mat. Palermo, Ser. II, Suppl., **35**, 315–322 (1994)
770. Vitale, R.A.: The Brunn–Minkowski inequality for random sets. J. Multivariate Anal., **33**, 286–293 (1990)
771. Volland, W.: Ein Fortsetzungssatz für additive Eipolyederfunktionale im euklidischen Raum. Arch. Math., **8**, 144–149 (1957)
772. Voss, K.: Integralgeometrie für Stereologie und Bildrekonstruktion. Springer, Berlin (2007)
773. Vu, V.: Sharp concentration of random polytopes. Geom. Funct. Anal., **15**, 1284–1318 (2005)
774. Vu, V.: Central limit theorems for random polytopes in a smooth convex set. Adv. Math., **207**, 221–243 (2006)
775. Wagner, U., Welzl, E.: A continuous analogue of the Upper Bound Theorem. Discrete Comput. Geom., **26**, 205–219 (2001)
776. Waksman, P.: Plane polygons and a conjecture of Blaschke's. Adv. Appl. Prob., **17**, 774–793 (1985)
777. Waymire, E.: Buffon noodles. Amer. Math. Monthly, **101**, 550–559 (1994)
778. Weibel, E.R.: Stereological Methods, I, II. Academic Press, London (1980)
779. Weil, W.: Berührwahrscheinlichkeiten für konvexe Körper. Z. Wahrscheinlichkeitsth. verw. Geb., **48**, 327–338 (1979)
780. Weil, W.: Kinematic integral formulas for convex bodies. In: Tölke, J., Wills, J.M. (eds) Contributions to Geometry (Proc. Geometry Symp. Siegen 1978), pp. 60–76. Birkhäuser, Basel (1979)
781. Weil, W.: Centrally symmetric convex bodies and distributions. II. Israel J. Math., **32**, 173–182 (1979)
782. Weil, W.: Zufällige Berührung konvexer Körper durch q-dimensionale Ebenen. Resultate Math., **4**, 84–101 (1981)
783. Weil, W.: Inner contact probabilities for convex bodies. Adv. Appl. Prob., **14**, 582–599 (1982)
784. Weil, W.: An application of the central limit theorem for Banach-space-valued random variables to the theory of random sets. Z. Wahrscheinlichkeitstheorie verw. Geb., **60**, 203–208 (1982)
785. Weil, W.: Stereology: A survey for geometers. In: Gruber, P.M., Wills, J.M. (eds) Convexity and Its Applications, pp. 360–412. Birkhäuser, Basel (1983)
786. Weil, W.: Stereological results for curvature measures. Bull. Int. Statist. Inst., **50**, 872–883 (1983)
787. Weil, W.: Densities of quermassintegrals for stationary random sets. In: Ambartzumian, R.V., Weil, W. (eds) Stochastic Geometry, Geometric Statistics, Stereology (Proc. Conf. Oberwolfach, 1983), pp. 233–247. Teubner, Leipzig (1984)
788. Weil, W.: Point processes of cylinders, particles and flats. Acta Appl. Math., **9**, 103–136 (1987)
789. Weil, W.: Collision probabilities for convex sets. J. Appl. Prob., **26**, 649–654 (1989)

790. Weil, W.: Translative integral geometry. In: Hübler, A. *et al.* (eds) Geobild 89, pp. 75–86, Math. Research, vol. 51. Akademie-Verlag, Berlin (1989)
791. Weil, W.: Lectures on translative integral geometry and stochastic geometry of anisotropic random geometric structures. In: Atti del Primo Convegno Italiano di Geometria Integrale, Rend. Sem. Mat. Messina (2) **13**, 79–97 (1990)
792. Weil, W.: Iterations of translative integral formulae and non-isotropic Poisson processes of particles. Math. Z., **205**, 531–549 (1990)
793. Weil, W.: Support functions in the plane and densities of random sets and point processes. Rend. Circ. Mat. Palermo, Ser. II, Suppl., **35**, 323–344 (1994)
794. Weil, W.: The estimation of mean shape and mean particle number in overlapping particle systems in the plane. Adv. Appl. Prob. (SGSA), **27**, 102–119 (1995)
795. Weil, W.: Translative and kinematic integral formulae for support functions. Geom. Dedicata, **57**, 91–103 (1995)
796. Weil, W.: On the mean shape of particle processes. Adv. Appl. Prob. (SGSA), **29**, 890–908 (1997)
797. Weil, W.: Mean bodies associated with random closed sets. Rend. Circ. Mat. Palermo, Ser. II, Suppl., **50**, 387–412 (1997)
798. Weil, W.: Intensity analysis of Boolean models. Pattern Recognition, **32**, 1675–1684 (1999)
799. Weil, W.: A uniqueness problem for non-stationary Boolean models. Rend. Circ. Mat. Palermo, Ser. II, Suppl., **65**, 329–344 (2000)
800. Weil, W.: Mixed measures and functionals of translative integral geometry. Math. Nachr., **223**, 161–184 (2001)
801. Weil, W.: Densities of mixed volumes for Boolean models. Adv. Appl. Prob. (SGSA), **33**, 39–60 (2001)
802. Weil, W.: Contact distributions of Boolean models. In: Lecherová, R., Saxl, J., Beneš, V. (eds) Proc. S^4G, International Conf. on Stereology, Spatial Statistics and Stochastic Geometry (Prague 2006), pp. 41–51. Union Czech Math. Phys. (2006)
803. Weil, W.: Random sets (in particular Boolean models). In: [50], pp. 185–245 (2007)
804. Weil, W., Wieacker, J.A.: Densities for stationary random sets and point processes. Adv. Appl. Prob., **16**, 324–346 (1984)
805. Weil, W., Wieacker, J.A.: A representation theorem for random sets. Probab. Math. Stat., **9**, 147–151 (1988)
806. Weil, W., Wieacker, J.A.: Stochastic geometry. In: Gruber, P.M, Wills, J.M. (eds) Handbook of Convex Geometry, pp. 1391–1438. North-Holland, Amsterdam (1993)
807. Weiss, V.: Relations between mean values for stationary random hyperplane mosaics of R^d. Forschungsergeb. Friedrich-Schiller-Univ. Jena, N/86/33, 12p. (1986)
808. Weiss, V.: Second-order quantities for random tessellations of $\mathbb{R}^d$. Stochastics Stochastics Reports, **55**, 195–205 (1995)
809. Weiss, V., Zähle, M.: Geometric measures for random curved mosaics of R^d. Math. Nachr., **138**, 313–326 (1988)
810. Wendel, J.G.: A problem in geometric probability. Math. Scand., **11**, 109–111 (1962)
811. Wheeden, R.L., Zygmund, A.: Measure and Integral. Marcel Dekker, New York (1977)

812. Whittaker, E.T., Watson, G.N.: A Course of Modern Analysis. 4th. ed., Cambridge University Press, Cambridge (1952)
813. Wicksell, S.D.: The corpuscle problem. A mathematical study of a biometric problem. Biometrika, **17**, 84–89 (1925)
814. Wieacker, J.A.: Einige Probleme der polyedrischen Approximation. Diploma Thesis, Albert-Ludwigs-Universität, Freiburg i. Br. (1978)
815. Wieacker, J.A.: Translative stochastische Geometrie der konvexen Körper. Doctoral Thesis, Albert-Ludwigs-Universität, Freiburg i. Br. (1982)
816. Wieacker, J.A.: Translative Poincaré formulae for Hausdorff rectifiable sets. Geom. Dedicata, **16**, 231–248 (1984)
817. Wieacker, J.A.: Intersections of random hypersurfaces and visibility. Probab. Theory Related Fields, **71**, 405–433 (1986)
818. Wieacker, J.A.: Geometric inequalities for random surfaces. Math. Nachr., **142**, 73–106 (1989)
819. Winkler, G.: Image Analysis, Random Fields and Dynamic Monte Carlo Methods: a Mathematical Introduction. Springer, Berlin (1995)
820. Wu, L.C.: Pairs of non-intersecting random flats meeting two convex bodies. In: Gu, C.H. (ed) Differential Geometry, Proc. Symp. in honour of Prof. Su Buchin, pp. 254–261. World Scientific, Singapore (1993)
821. Xie, P., Jiang, J.: Double chord-power integrals of a convex body and their applications. In: Grinberg, E.L., Li, S., Zhang, G., Zhou, J. (eds) Integral Geometry and Convexity, pp. 177–188. World Scientific, New Jersey (2006)
822. Zähle, M.: Random processes of Hausdorff rectifiable closed sets. Math. Nachr., **108**, 49–72 (1982)
823. Zähle, M.: Ergodic properties of random fields and images with point imbedded processes. Theory Probab. Appl., **27**, 536–550 (1982)
824. Zähle, M.: Curvature measures and random sets, I. Math. Nachr., **119**, 327–339 (1984)
825. Zähle, M.: Integral and current representation of Federer's curvature measures. Arch. Math., **46**, 557–567 (1986)
826. Zähle, M.: Curvature measures and random sets, II. Probab. Theory Related Fields, **71**, 37–58 (1986)
827. Zähle, M.: Curvatures and currents for unions of sets with positive reach. Geom. Dedicata, **23**, 155–171 (1987)
828. Zähle, M.: Random cell complexes and generalised sets. Ann. Probab., **16**, 1742–1766 (1988)
829. Zähle, M.: A kinematic formula and moment measures of random sets. Math. Nachr., **149**, 325–340 (1990)
830. Zähle, M.: Wicksell's corpuscle problem in spherical space. J. Appl. Prob., **27**, 701–706 (1990)
831. Zähle, M.: Nonosculating sets of positive reach. Geom. Dedicata, **76**, 183–187 (1999)
832. Zhang, G.: Geometric inequalities and inclusion measures of convex bodies. Mathematika, **41**, 95–116 (1994)
833. Zhang, G.: Dual kinematic formulas. Trans. Amer. Math. Soc., **351**, 985–995 (1999)
834. Zhang, G., Zhou, J.: Containment measures in integral geometry. In: Grinberg, E.L., Li, S., Zhang, G., Zhou, J. (eds) Integral Geometry and Convexity, pp. 153–168. World Scientific, New Jersey (2006)

835. Ziegel, J., Kiderlen, M.: Estimation of surface area and surface area measure of three-dimensional sets from digitizations. Thiele Centre Res. Rep., 01, January (2008)
836. Ziezold, H.: Über die Eckenanzahl zufälliger Polygone. Izv. Akad. Nauk Armjansk. SSR, **5**, 296–312 (1970)
837. Ziezold, H.: The mean breadth of a random polytope in a convex body. Z. Wahrscheinlichkeitsth. verw. Geb., **68**, 121–125 (1984)
838. Zuyev, S.A.: Estimates for distributions of the Voronoi polygon's geometric characteristics. Random Structures Algorithms, **3**, 149–162 (1992)

Author Index

Subject Index

Notation Index

MIX
Papier aus verantwortungsvollen Quellen
Paper from responsible sources
FSC® C105338

If you have any concerns about our products,
you can contact us on
ProductSafety@springernature.com

In case Publisher is established outside the EU,
the EU authorized representative is:
Springer Nature Customer Service Center GmbH
Europaplatz 3, 69115 Heidelberg, Germany

Printed by Libri Plureos GmbH
in Hamburg, Germany